Generalized Geologic and Tectonic Map of North America

SEDIMENTARY UNITS

Thick deposits in structurally negative areas

Paleozoic and Mesozoic active margin deposits

Synorogenic and postorogenic deposits

Paleozoic and Mesozoic passive margin deposits

Late Precambrian deposits
Of Middle and Upper Proterozoic ages

VOLCANIC AND PLUTONIC UNITS

Postorogenic volcanic cover

Ultramafic rocks

Granitic plutons
Ages are generally within the span of the tectonic cycle of the foldbelt in which they lie

SPECIAL UNITS

Former subduction complex rocks of the Pacific border

Exposed parts of Ouachita foldbelt

Probable western extention of Innuitian foldbelt
In cores of northern Alaska Ranges

PLATFORM AREAS

Ice cap of Quaternary age
On Precambrian and Paleozoic basement

Platform deposits on Precambrian basement
In central craton

Platform deposits on Paleozoic basement
In Atlantic and Gulf Coastal Plains

Platform deposits within the Precambrian
Mainly in the Canadian Shield

PRECAMBRIAN

Basement igneous and metamorphic complexes mainly of Precambrian age

Grenville foldbelt
Deformed 880–1,000 m.y. ago

Hudsonian foldbelts
Deformed 1,640–1,820 m.y. ago

Kenoran foldbelts
Deformed 2,390–2,600 m.y. ago

Anorthosite bodies
Plutons composed almost entirely of plagioclase

STRUCTURAL SYMBOLS

Normal fault
Hachures on downthrown side

Subsea fault

Strike-slip fault
Arrows show relative lateral movement

Salt domes and salt diapirs
In Gulf Coastal Plain and Gulf of Mexico

▲
Volcano

Thrust fault
Barbs on upthrown side

———1000———
———0———
———+1000———

Contours on basement surfaces beneath platform areas
All contours are below sea level except where marked with plus symbols. Interval is 1,000 meters

Axes of sea-floor spreading

Modified from the Generalized Tectonic Map of North America by P.B. King and Gertrude J. Edmonston, U.S. Geological Survey Map I-688

Meet the Authors

David McGeary, Diane Carlson, and Charles Plummer at an outcrop of a Sierra Nevadan intrusive body.

Charles Plummer

Professor Charles "Carlos" Plummer grew up in the shadows of volcanoes in Mexico City. There, he developed a love for mountains and mountaineering that eventually led him into geology. He received his B.A. degree from Dartmouth College. After graduation, he served in the U.S. Army as an artillery officer. He resumed his geological education at the University of Washington where he received his M.S. and PhD degrees. His geologic work has been in mountainous and polar regions, notably Antarctica (where a glacier is named in his honor). He taught at Olympic Community College in Washington before joining the faculty at California State University, Sacramento. At CSUS, he teaches introductory geology courses as well as optical mineralogy, metamorphic petrology, and field courses. *(plummercc@csus.edu)*

David McGeary

Professor David McGeary was born and raised in central Pennsylvania, learning to love the Nittany Lions as well as shoofly pie for breakfast. He earned his B.S. degree at Williams College, M.S. at University of Illinois, and PhD at Scripps Institution of Oceanography. He helped pay for grad school by teaching scuba diving. David has been married to his wife for more than 30 years, has two grown sons, and a handmade house he built with his brother on 10 country acres. He taught geology at California State University for more than 23 years, specializing in physical geology, sedimentary petrology, oceanography, and field methods. He retired in 1995, turning his portion of the text over to the excellent hands of fellow professor Diane Carlson. He is now a theater actor, appearing in 5 productions in 1997. His favorite role was as Colonel Jessep in *A Few Good Men*.

Diane Carlson

Professor Diane Carlson grew up on the glaciated Precambrian shield of northern Wisconsin and received an A.A. degree at Nicolet College in Rhinelander and her B.S. in geology at the University of Wisconsin at Eau Claire. She continued her studies at the University of Minnesota–Duluth where she studied the structural complexities of high-grade metamorphic rocks along the margin of the Idaho batholith for her master's thesis. The lure of the West and an opportunity to work with the U.S. Geological Survey to map the Colville batholith in northeastern Washington, led her to Washington State University for her PhD. Dr. Carlson accepted a position at California State University, Sacramento after her PhD and teaches physical geology, structural geology, environmental and applied geology, and field geology. Professor Carlson is a recent recipient of the Outstanding Teacher Award from the CSUS School of Arts and Sciences. She is currently Chair of the Geology Department, and is actively engaged in researching the structural and tectonic evolution of part of the Foothill Fault System in the northern Sierra Nevada of California. *(carlsondh@csus.edu)*

Eighth Edition

PHYSICAL
GEOLOGY

Featuring The Smithsonian Institution's *Journey Through Geology*
CD-ROM

Charles C. Plummer

California State University at Sacramento

David McGeary

Emeritus of California State University at Sacramento

Diane H. Carlson

California State University at Sacramento

WCB
McGraw-Hill

Boston   Burr Ridge, IL   Dubuque, IA   Madison, WI   New York   San Francisco   St. Louis
Bangkok   Bogotá   Caracas   Lisbon   London   Madrid
Mexico City   Milan   New Delhi   Seoul   Singapore   Sydney   Taipei   Toronto

WCB/McGraw-Hill

*A Division of The **McGraw·Hill** Companies*

PHYSICAL GEOLOGY, EIGHTH EDITION

  This book is printed on recycled paper.

1 2 3 4 5 6 7 8 9 0 QPD/QPD 9 3 2 1 0 9 8

ISBN 0-697-37404-1

Vice president and editorial director: *Kevin T. Kane*
Publisher: *Edward E. Bartell*
Sponsoring editor: *Robert Smith*
Senior developmental editor: *Lynne M. Meyers*
Developmental editor: *Lu Ann Weiss*
Marketing manager: *Lisa L. Gottschalk*
Project manager: *Donna Nemmers*
Production supervisor: *Deborah Donner*
Designer: *K. Wayne Harms*
Senior photo research coordinator: *Carrie K. Burger*
Art editor: *Jodi K. Banowetz*
Compositor: *Precision Graphics*
Typeface: *11/12 AGaramond*
Printer: *Quebecor Printing Book Group/Dubuque, IA*

Cover image: *Galen Rowell/Mountain Light*
Illustrator: *Precision Graphics*

Library of Congress Cataloging In Publication Data

Plummer, Charles C., 1937–
    Physical geology / Charles C. Plummer, David McGeary, Diane H. Carlson. —8th ed.
        p. cm.
    Includes index.
    ISBN 0-697-37404-1
    1. Physical geology. I. McGeary, David. II. Carlson, Diane H. III. Title.
    QE28.2.P58 1999
    550—dc21   97-51330
                                                          CIP

www.mhhe.com

Contents

About the Authors ii
Preface x
The Learning System xii

CHAPTER 1
Introduction to Physical Geology 2
Who Needs Geology? 4
  Avoiding Geologic Hazards 6
  Supplying Things We Need 7
  Protecting the Environment 8
  Understanding Our Surroundings 11
An Overview of Physical Geology—Important
    Concepts 11
  Internal Processes: How the Earth's Internal Heat
    Engine Works 14
  The Earth's Interior 15
  The Theory of Plate Tectonics 16
  Surficial Processes: The Earth's External Heat
    Engine 18
Geologic Time 21
Summary 22
Terms to Remember* 22
Testing Your Knowledge* 23
Expanding your Knowledge* 24
Exploring Resources* 24
Interacting with Journey Through Geology
    CD-ROM* 25

CHAPTER 2
Atoms, Elements, and Minerals 26
Atoms and Elements 28
  Chemical Activity 30
  Ions 30
Chemical Composition of the Earth's Crust 31
Crystallinity 32
  The Silicon-Oxygen Tetrahedron 33
Minerals 37
  Crystalline Solids 37
  Natural and Inorganic Substances 37
  Definite Chemical Composition 37
  The Important Minerals 38
The Physical Properties of Minerals 40
  Color 40
  Streak 40
  Luster 40
  Hardness 40

*These end-of-chapter sections appear in every chapter.

  External Crystal Form 41
  Cleavage 43
  Fracture 45
  Specific Gravity 45
  Other Properties 46
  Simple Chemical Tests 46
The Rock Cycle 47
Summary 49

CHAPTER 3
**Igneous Rocks, Intrusive Activity, and the
Origin of Igneous Rocks 52**
Igneous Rocks 55
  Identification of Igneous Rocks 55
  Varieties of Granite 58
  Chemistry of Igneous Rocks 58
Intrusive Bodies 59
  Shallow Intrusive Structures 60
  Intrusives That Crystallize at Depth 61
Abundance and Distribution of Plutonic Rocks 63
How Magma Forms 63
  Heat for Melting Rock 64
  Factors That Control Melting Temperatures 64
How Magmas of Different Compositions Evolve 66
  Differentiation and Bowen's Reaction Theory 66
  Partial Melting 68
  Assimilation 69
  Mixing of Magmas 69
Explaining Igneous Activity by Plate Tectonics 70
  Igneous Processes at Divergent Boundaries 70
  Intraplate Igneous Activity 70
  Igneous Processes at Convergent Boundaries 71
Summary 73

CHAPTER 4
Volcanism and Extrusive Rocks 76

Effects on Humans 78
    The Growth of Hawaii 78
    Geothermal Energy 78
    Effect on Climate 78
    Volcanic Catastrophes 81
    Eruptive Violence and Physical Characteristics of
        Lava 82
Extrusive Rocks and Gases 83
    Scientific Investigation of Volcanism 83
    Gases 84
Extrusive Rocks 84
Types of Volcanoes 87
    Shield Volcanoes 88
    Cinder Cones 88
    Composite Volcanoes 90
    Volcanic Domes 93
Lava Floods 95
Submarine Eruptions 96
    Pillow Basalts 97
Summary 99

CHAPTER 5
Weathering and Soil 102

Weathering, Erosion, and Transportation 104
How Weathering Alters Rocks 104
    Effects of Weathering 104
Mechanical Weathering 104
    Frost Action 104
    Abrasion 106
    Pressure Release 107
Chemical Weathering 108
    Role of Oxygen 108
    Role of Acid 109
    Solution Weathering 110
    Chemical Weathering of Feldspar 111
    Chemical Weathering of Other Minerals 112
    Weathering and Climate 113
    Weathering Products 113
Soil 113
    Soil Horizons 114
    Residual and Transported Soils 114
    Soils, Parent Rock, Time, and Slope 115
    Organic Activity 116
    Soils and Climate 116
    Buried Soils 117
Summary 118

CHAPTER 6
Sediments and Sedimentary Rocks 120

Sediment 122
    Transportation 122
    Deposition 123
    Preservation 124
    Lithification 125

Types of Sedimentary Rocks 126
Clastic Rocks 126
    Breccia and Conglomerate 126
    Sandstone 127
    The Fine-Grained Rocks 128
Chemical Sedimentary Rocks 130
    Carbonate Rocks 130
    Chert 133
    Evaporites 134
Organic Sedimentary Rocks 135
    Coal 135
The Origin of Oil and Gas 135
Sedimentary Structures 135
Formations 139
Interpretation of Sedimentary Rocks 140
    Source Area 140
    Environment of Deposition 142
    Plate Tectonics and Sedimentary Rocks 145
Summary 146

CHAPTER 7
Metamorphism, Metamorphic Rocks, and Hydrothermal Rocks 150

Factors Controlling the Characteristics of Metamorphic
    Rocks 152
    Composition of the Parent Rock 152
    Temperature 153
    Pressure 153
    Fluids 155
    Time 156
Classification of Metamorphic Rocks 156
Types of Metamorphism 156
    Contact Metamorphism 156
    Regional Metamorphism 158
Plate Tectonics and Metamorphism 162
Hydrothermal Processes 163
    Hydrothermal Activity at Diverging Plate
        Boundaries 163
    Metasomatism 165
    Hydrothermal Rocks and Minerals 166
    Sources of Water 167
Summary 169

CHAPTER 8
Time and Geology 172

The Key to the Past 174
Relative Time 174
    Principles Used to Determine Relative Age 174
    Correlation 179
    The Standard Geologic Time Scale 182
Unconformities 183
    Disconformities 183
    Angular Unconformities 185
    Nonconformities 186
Absolute Age 186
    Isotopic Dating 186
    Uses of Isotopic Dating 189

Combining Relative and Absolute Ages 190
Age of the Earth 191
    Comprehending Geologic Time 191
Summary 195

CHAPTER 9
Mass Wasting 198

Classification of Mass Wasting 200
    Rate of Movement 200
    Type of Material 200
    Type of Movement 202
Controlling Factors in Mass Wasting 203
    Gravity 204
    Water 204
Common Types of Mass Wasting 205
    Creep 205
    Debris Flow 206
    Rockfalls and Rockslides 212
Preventing Landslides 215
    Preventing Mass Wasting of Debris 215
    Preventing Rockfalls and Rockslides on
        Highways 216
Summary 218

CHAPTER 10
Streams and Floods 220

The Hydrologic Cycle 222
Channel Flow and Sheet Flow 222
Drainage Basins 224
Drainage Patterns 224
Factors Affecting Stream Erosion and Deposition 225
    Velocity 225
    Gradient 226
    Channel Shape and Roughness 226
    Discharge 227
Stream Erosion 228
Stream Transportation of Sediment 229
Stream Deposition 230
    Bars 230
    Braided Streams 231
    Meandering Streams and Point Bars 235
    Flood Plains 237
    Deltas 239
    Alluvial Fans 241
Flooding 242
    Urban Flooding 242
    Flash Floods 245
    Controlling Floods 246
    The Great Flood of 1993 247
Stream Valley Development 248
    Downcutting and Base Level 248
    The Concept of a Graded Stream 250
    Lateral Erosion 251
    Headward Erosion and Stream Piracy 251
Stream Terraces 252

Incised Meanders 253
Superposed Streams 255
Summary 255

CHAPTER 11
Ground Water 258

Porosity and Permeability 260
The Water Table 260
The Movement of Ground Water 262
Aquifers 263
Wells 264
Springs and Streams 266
Pollution of Ground Water 266
Balancing Withdrawal and Recharge 272
Effects of Ground-Water Action 274
    Caves, Sinkholes, and Karst Topography 274
    Other Effects 276
Hot Water Underground 277
    Geothermal Energy 278
Summary 278

CHAPTER 12
Glaciers and Glaciation 282

The Theory of Glacial Ages 284
Glaciers—Where They Are, How They Form and
    Move 284
    Distribution of Glaciers 284
    Types of Glaciers 285
    Formation and Growth of Glaciers 285
    Movement of Valley Glaciers 288
    Movement of Ice Sheets 293
Glacial Erosion 294
    Erosional Landscapes Associated with Alpine
        Glaciation 295
    Erosional Landscapes Associated with Continental
        Glaciation 298
Glacial Deposition 299
    Moraines 300
    Outwash 302
    Glacial Lakes and Varves 303
Effects of Past Glaciation 303
    The Glacial Ages 303
    Direct Effects of Past Glaciation in North
        America 305
    Indirect Effects of Past Glaciation 310
    Evidence for Older Glaciation 310
Summary 311

CHAPTER 13
Deserts and Wind Action 314

Distribution of Deserts 316
Some Characteristics of Deserts 316
Desert Features in the Southwestern United States 319

Wind Action 322
    Wind Erosion and Transportation 324
    Wind Deposition 327
Summary 332

CHAPTER 14
Waves, Beaches, and Coasts 336

Water Waves 338
    Surf 339
Nearshore Circulation 340
    Wave Refraction 340
    Longshore Currents 340
    Rip Currents 341
Beaches 341
Longshore Drift of Sediment 342
    Human Interference with Sand Drift 344
    Sources of Sand on Beaches 344
Coasts and Coastal Features 344
    Erosional Coasts 345
    Depositional Coasts 348
    Drowned Coasts 349
    Uplifted Coasts 352
    Coasts Shaped by Organisms 352
Summary 353

CHAPTER 15
Geologic Structures 356

Tectonic Forces at Work 358
    Stress and Strain in the Earth's Crust 358
    Behavior of Rocks to Stress and Strain 359
    Present Deformation of the Crust 360
Structures as a Record of the Geologic Past 361
    Geologic Maps and Field Methods 361
Folds 363
    Geometry of Folds 363
    Interpreting Folds 366
Fractures in Rock 367
    Joints 368
    Faults 370
Summary 379

CHAPTER 16
Earthquakes 382

Causes of Earthquakes 384
Seismic Waves 386
    Body Waves 386
    Surface Waves 386
Locating and Measuring Earthquakes 386
    Determining the Location of
        an Earthquake 389
    Measuring the Size of an Earthquake 390
    Location and Size of Earthquakes in the United
        States 393

Effects of Earthquakes 395
    Tsunamis 399
World Distribution of Earthquakes 401
First-Motion Studies
    of Earthquakes 404
Earthquakes and Plate Tectonics 404
    Earthquakes at Plate Boundaries 405
    Subduction Angle 408
Earthquake Prediction 408
Earthquake Control 409
Summary 411

CHAPTER 17
The Earth's Interior 414

Evidence from Seismic Waves 416
The Earth's Internal Structure 417
    The Earth's Crust 418
    The Mantle 418
    The Core 420
Isostasy 423
Gravity Measurements 425
The Earth's Magnetic Field 427
    Magnetic Reversals 428
    Magnetic Anomalies 429
Heat Within the Earth 432
    Geothermal Gradient 432
    Heat Flow 433
Summary 433

CHAPTER 18
The Sea Floor 436

Origin of the Ocean 438
Methods of Studying the Sea Floor 439
Features of the Sea Floor 440
Continental Shelves and Continental Slopes 440
Submarine Canyons 442
    Turbidity Currents 443
Passive Continental Margins 443
    The Continental Rise 443
    Abyssal Plains 444
Active Continental Margins 444
    Oceanic Trenches 445
The Mid-Oceanic Ridge 445
    Geologic Activity on the Ridge 446
    Biologic Activity on the Ridge 446
Fracture Zones 449
Seamounts, Guyots,
    and Aseismic Ridges 449
Reefs 450
Sediments of the Sea Floor 452
Oceanic Crust and Ophiolites 452
The Age of the Sea Floor 454
The Sea Floor and Plate Tectonics 455
Summary 455

Contents

CHAPTER 19
Plate Tectonics 458

The Early Case for Continental Drift 460
  Skepticism about Continental Drift 462
Paleomagnetism and the Revival
    of Continental Drift 463
  Recent Evidence for Continental Drift 464
  History of Continental Positions 465
Sea-Floor Spreading 465
  Hess's Driving Force 466
  Explanations 466
Plates and Plate Motion 467
How Do We Know
    That Plates Move? 468
  Marine Magnetic Anomalies 468
  Another Test: Fracture Zones and Transform
      Faults 471
  Measuring Plate Motion Directly 472
Divergent Plate Boundaries 472
Transform Boundaries 475
Convergent Plate Boundaries 476
  Ocean-Ocean Convergence 477
  Ocean-Continent Convergence 478
  Continent-Continent Convergence 480
Backarc Spreading 481
The Motion of Plate Boundaries 482
Plate Size 482
The Attractiveness of
    Plate Tectonics 482
What Causes Plate Motions? 483
  Mantle Plumes and Hot Spots 486
The Relationship Between Plate Tectonics and Ore
    Deposits 488
A Final Note 492
Summary 492

CHAPTER 20
**Mountain Belts and the
Continental Crust 496**

Characteristics of
    Major Mountain Belts 499
  Size and Alignment 499
  Ages of Mountain Belts and Continents 499
  Thickness and Characteristics
      of Rock Layers 501
  Patterns of Folding and Faulting 502
  Metamorphism and Plutonism 504
  Normal Faulting 504
  Thickness and Density of Rocks 504
  Features of Active Mountain Ranges 505
The Evolution of a Mountain Belt 505
  The Accumulation Stage 505
  The Orogenic Stage 506
  The Uplift and Block-faulting Stage 509

The Growth of Continents 514
  Suspect and Exotic Terranes 515
Summary 518

CHAPTER 21
Geologic Resources 520

Types of Resources 522
  Resources and Reserves 523
Energy Use 523
Oil and Natural Gas 523
  The Occurrence of Oil and Gas 524
  Recovering the Oil 526
  How Much Oil Do We Have Left? 527
Heavy Crude and Oil Sands 528
Oil Shale 528
Coal 529
  Varieties of Coal 529
  Occurrence of Coal 530
  Environmental Effects 532
  Reserves and Resources 532
Uranium 532
Alternative Sources of Energy 533
Metals and Ores 533
Origin of Metallic Ore Deposits 533
  Ores Associated with Igneous Rocks 534
  Ores Formed by Surface Processes 535
Metal Ores and Plate Tectonics 536
Mining 536
  Environmental Effects 537
Some Important Metals 538
  Iron 538
  Copper 538
  Aluminum 538
  Lead 539
  Zinc 539
  Silver 539
  Gold 539
  Other Metals 540
Nonmetallic Resources 540
  Construction Materials 540
  Fertilizers and Evaporites 541
  Other Nonmetallics 542
Some Future Trends 542
Summary 543

Appendixes A–G 546

Glossary 556

Index 569

Preface

Publication of the eighth edition of Physical Geology is a milestone as it celebrates the book's twentieth anniversary. Another milestone is the addition of a third author, Diane Carlson, to the team as Dave McGeary assumes a less active role.

Physical Geology is a straightforward, easy-to-read introduction to geology both for nonscience majors and for students contemplating majoring in geology. The organization of the book is traditional and matches the organization of most lab manuals. In this edition chapters 3 and 4 have been reversed. Chapter 3 now covers igneous rocks, intrusive activity and the origin of igneous rocks, whereas Chapter 4 is about volcanic activity. We have taken pains to be sure either chapter could be covered before the other. Other chapters or groups of chapters have been written so as to be as self-contained as possible, allowing the instructor to reorganize the chapter sequence, if desired.

We are excited about the new CD-ROM that accompanies this book. *Journey Through Geology CD-ROM* was produced in partnership with the Smithsonian Institution and should set a new standard for CD-ROMs that accompany textbooks.

The end-of-chapter material has been expanded to include selected websites, CD-ROMs, and videotapes that are available. The website maintained for this book should provide students and instructors with up-to-date additional resources and help maintain enthusiasm for geology.

The eighth edition is new and improved in a number of ways. Some of the more significant changes are: We have taken a number of descriptions and examples of geologic resources from the book's final chapter and integrated them into appropriate chapters elsewhere in the book; plate tectonics has also been more thoroughly integrated throughout the book. The rock cycle (in chapter 2) has been expanded to include a plate tectonic example. The discussion of the origin of magmas at convergent boundaries places more emphasis on the current view by researchers that mafic magmas are generated in the asthenosphere above the subducted oceanic crust. All diagrams showing magma generation at convergent boundaries have been redone. In the chapter on geologic time we have introduced the term "actualism" and discussed why it might be preferable to "uniformitarianism". Lateral continuity and inclusions have been added to the principles used for determining relative time relationships. In the metamorphic chapter we have related foliation to the modern concept of gravitational collapse and spreading. We have also moved unconformities from the structural geology chapter to the geological time chapter. A section on changing concepts of the age of the earth has been added. The relationship between isotopic dating and the geologic time scale has been expanded. The 1996 rockfall at Yosemite is used as an example of mass wasting. In the glaciation chapter we clarify and expand on the conversion of snow to glacier ice. The stream chapter has undergone a major revision and now includes an expanded discussion of flooding with examples from the 1997 floods in the upper Midwest and California and also a comparison of the historic Big Thompson Creek, Colorado flash flood with the more recent Fort Collins flood. In the structure chapter, the section on stress and strain and the behavior of rocks has been rewritten and new examples and figures are included to clarify these difficult concepts; the fault and fold sections have also been expanded and include new figures and photographs. Love and Rayleigh waves are now discussed and illustrated in the earthquake chapter. We also introduce the idea that some mountains may not have thick roots but are instead buoyed by warm, less dense mantle. In the chapter on mountains and the continental crust we added a section on the disparity of the height of the Rocky Mountains and the thickness of the crust and describe recent work that indicates that the Basin and Range mountains were three kilometers higher than at present. An appendix listing commonly used prefixes, suffixes and root words has been added. The geologic map of North America has been moved from the appendix to the inside front cover.

The text has new boxes on water and ice—molecules and crystals, flight hazards associated with volcanoes, the eruptions on Montserrat compared to the disastrous eruption that destroyed St. Pierre on Martinique in 1902, the Bingham Canyon copper mine, highlights of biological evolution through time, the meteorite from Mars with possible signs of former life. A box on water beneath glaciers describes the recently discovered lake beneath the East Antarctic Ice Sheet, surging glaciers and subglacial volcanism and flooding in Iceland. The stream chapter has been revised and reorganized and includes boxes on the planned flood in the Grand Canyon and how the recurrence interval of large floods is calculated; the astrogeology box includes a photo of Mars and the Sojourner Rover. The structure chapter includes a box on how to find oil and the salt dome box has been expanded. The interior of the earth chapter now includes a box on the spinning inner core. In the mountains chapter we have added a box on a systems approach to understanding mountains and expanded a former box, retitling it "Dance on the Continents (with SWEAT)."

The eighth edition has a new look with many of the diagrams redone or replaced. New photos have also been added and include the volcanic eruption at Soufrière Hills, Montserrat, the 1997 Yosemite rockfall, rock slabs bolted to prevent landsliding, glacially carved features in the Teton range, giant stream ripples, the 1997 Fort Collins flash flood and the flooding along the Red River. Photos of many rocks and minerals and geologic structures have been replaced.

Supplements to Accompany *Physical Geology, Eighth Edition:*

- *Journey Through Geology two CD-ROM set*
- *Instructor's Manual*
- *computerized testing software*
- *224 transparencies and 350 slides*
- *Visual Resource Library CD-ROM*
- *Student Study Guide*
- *Physical Geology* and *Journey Through Geology Web sites*

Additional classroom tools include:

- *The American Geological Institute's Videodisc*
- *JLM Visuals Physical Geology Photo CD*
- *Interactive Plate Tectonics CD-ROM, Annual Editions: Geology 98/99*
- *Student Atlas of Environmental Issues*
- *McGraw-Hill Learning Architecture*

For additional information on *Physical Geology, Eighth Edition* or *Journey Through Geology CD-ROM,* please visit our Web sites at **www.mhhe.com/ plummer** or **www.mhhe.com/jtg.**

We have tried to write a book that will be useful to both students and instructors. We would be grateful for any comments by users, especially regarding mistakes within the text or sources of good geological photographs.

We would like to thank Susan Slaymaker for writing the original boxed material on planetary geology.

We are also very grateful to the following reviewers of the seventh and eighth editions of this text for their careful evaluation and useful suggestions for improvement.

John M. Alderson *Marymount College*

N. L. Archbold *Western Illinois University*

Victor R. Baker *The University of Texas*

Joan Baldwin *El Camino College*

Alexander R. Ball *Los Angeles Valley College*

Paul G. Bauer *Cuesta College*

Kenneth A. Beem *Montgomery College*

Robert E. Behling *Western Virginia University*

Terrill R. Berkland *Central Missouri State University*

David J. Berner *Normandale Community College*

Peter E. Borella *Riverside City College*

David P. Bucke *University of Vermont*

Gary Carlson *Midland Lutheran College*

Greg S. Conrad *Sam Houston State University*

Peter Copeland *University of Houston*

Paul A. Dike *Glassboro State College*

Steven F. Dodin *Community College of Allegheny County*

Robert J. Elias *University of Manitoba*

Stanley C. Fagerlin *Southwest Missouri State University*

Peter Fisher *California State University*

Ronald C. Flemal *Northern Illinois University*

Richard A. Flory *California State University at Chico*

Robert D. Forester *Collin County Community College*

M. G. Frey *University of New Orleans*

John S. Galehouse *San Francisco State University*

Heather L. Gallacher *Cleveland State University*

Lloyd Glawe *Northeast Louisiana University*

Andrew J. Hajash *Texas A&M*

Frank M. Hanna *California State University— Northridge*

Stephen B. Harper *East Carolina University*

Stephen L. Harris *California State University at Sacramento*

Barry Haskell *Los Angeles Pierce College*

Miles O. Hayes *University of South Carolina*

Richard A. Heimlich *Kent State University*

Mary S. Hubbard *Kansas State University*

Roy L. Ingram *University of North Carolina*

Clark M. Johnson *University of Wisconsin— Madison*

Norris W. Jones *University of Wisconsin— Oshkosh*

Lawrence W. Knight *William Rainey Harper College*

Ronald H. Konig *University of Arkansas*

Albert M. Kudo *The University of New Mexico*

Howard Level *Ventura College*

David N. Lumsden *Memphis State University*

Harmon D. Maher, Jr. *University of Nebraska at Omaha*

Donald Marchand, Jr. *Old Dominion University*

Kathleen Marsaglia *University of Texas at El Paso*

James McLelland *Colgate University*

William S. McLoda *Mountain View College*

Margaret E. McMillan *University of Arkansas at Little Rock*

C. Daniel Miller *U.S. Geological Survey*

William D. Orndorff *Concord College*

Bruce C. Panuska *Mississippi State University*

Jacqueline Patterson *California State University, Fullerton*

David R. Privette *Central Piedmont Community College*

Frederick J. Rich *Georgia Southern University*

Gary D. Rosenberg *Indiana University— Purdue University at Indianapolis*

Robert A. Schiffman *Bakersfield College*

Vernon P. Scott *Oklahoma State University*

Charles R. Singer *Youngstown State University*

Kenneth G. Smith *Dallas Baptist University*

William A. Smith *Grand Valley State University*

Richard Smosna *University of West Virginia*

Steven Stearns *College of Charleston*

Don W. Steeples *University of Kansas*

Dion C. Stewart *Adams State College*

M. Ali Tabidian *California State University, Northridge*

Norman W. Ten Brink *Grand Valley State College*

J. Robert Thompson *Glendale Community College*

Daniel R. Tucker *University of Southwestern Louisiana*

Sherwood D. Tuttle *University of Iowa*

W. R. Van Schmus *University of Kansas*

Kenneth J. Van Dellen *Macomb Community College*

Stephen Wareham *California State University at Fullerton*

Stephen H. Watts *Sir Sanford Fleming College*

William J. Wayne *University of Nebraska*

Thomas H. Wolosz *SUNY College of Plattsburgh*

A special thank you to the '97 GSA Focus Group:

Scott Babcock *Western Washington University*

Drew Coleman *Boston University*

Ralph Davis *University of Arkansas*

Megan Jones *University of Minnesota and Inver Hills Community College*

Michael Katuna *College of Charleston*

Peter Kresan *University of Arizona*

James Miller *Southwest Missouri State University*

The proven *Physical Geology* Learning System.

This anniversary edition of *Physical Geology* has been thoroughly updated to bring you the most current information on how physical geology is working in your world today—information you're more than likely watching on the 6 o'clock news.

10.2

IN GREATER DEPTH

Estimating the Size and Frequency of Floods

Because people have encroached on the flood plains of many rivers, flooding is one of the most universally experienced geologic hazards. To minimize flood damage and loss of life, it is useful to know the potential size of large floods and how often they might occur. This is often a difficult task because of the lack of long-term records for most rivers. The U.S. Geological Survey monitors the stage (water elevation) and discharge of rivers and streams throughout the U.S. in order to collect data that can be used to attempt to predict the size and frequency of flooding and to make estimates of water supply.

Hydrologists designate floods based on their *recurrence interval,* or *return period.* For example, a 100-year flood is the largest flood expected to occur within a period of 100 years. This does not mean that a 100-year flood occurs once every century, but that there is 1-in-100 chance, or a one percent probability, each year that a flood of this size will occur. Usually flood control systems are built to accommodate a 100-year flood because that is the minimum margin of safety required by the federal government if an individual wants to obtain flood insurance subsidized by the Federal Emergency Management Agency (FEMA).

To calculate the recurrence interval of flooding for a river, the annual peak discharges (largest discharge of the year) are collected and ranked according to size (box figure 1 and table 1). The largest annual peak discharge is assigned a rank (m) of one, the second a two, and so on until all of the discharges are assigned a rank number. The *recurrence interval* (R) of each annual peak discharge is then calculated by adding one to the *number of years of record* (n) and dividing by its *rank* (m).

$$R = \frac{n+1}{m}$$

For example, the Cosumnes River in California has 90 years of record (n = 90), and in 1907 the second largest peak discharge (m = 2) of 71,000 cfs occurred. The recurrence interval (R), or expected frequency of occurrence, for a discharge this large is 45.5 years:

$$R = \frac{90+1}{2} = 45.5 \text{ years}$$

That is, there is a 1-in-45.5, or 2 percent, chance each year of a peak discharge of 71,000 cfs or greater occurring on the Cosumnes River.

The flood of record (largest recorded discharge) occurred on January 2, 1997 when heavy, unseasonably warm rains rapidly melted snow in the Sierra Nevada and caused flooding in much of northern California. A peak discharge of 93,000 cfs in the Cosumnes River resulted in levee breaks and widespread flooding of homes and agricultural areas (box figure 2). The recurrence interval for the 1997 flood (93,000 cfs) is 91 years:

$$R = \frac{90+1}{1} = 91 \text{ years}$$

Box 10.2 Figure 1
Annual peak discharge for the Cosumnes River.
After U.S. Geological Survey Water-Data Report, CA-97-3.

Streams and Floods

243

In Greater Depth Box 10.2
Estimating the Size and Frequency of Floods

10.1 Astrogeology

Stream Features on the Planet Mars

There is no liquid water on the surface of Mars today. With the present surface temperatures, atmospheric pressures, and water content in the Martian atmosphere, any liquid water would immediately evaporate. There are some indications, however, that conditions may have been different in the past and that liquid water existed on Mars, at least temporarily. Certain features on Mars, called *channels*, closely resemble certain types of stream channels on Earth. They have tributary systems and meanders and are sometimes braided. The channels trend downslope and tend to get wider toward their mouths. These Martian channels are restricted to certain areas and appear to have been formed by intermittent episodes of erosion.

One type of stream channel on Mars appears to have formed by large flooding events and is similar in appearance to those observed in the Channeled Scablands of Washington state. The Channeled Scablands were formed by extensive flooding during the Pleistocene glacial ages when a naturally formed ice dam broke and released water from a large lake. The mouth of Ares Vallis, an ancient Martian flood channel similar to those observed in the Channeled Scablands, was selected for the July 4, 1997 landing of the Pathfinder spacecraft and Sojourner Rover. It was postulated that a variety of rock types should be present in the mouth of what appeared to be an ancient flood channel. The first photos from the Mars Pathfinder Lander and Sojourner Rover, a "robotic field geologist", revealed a variety of rock types (box figure 1) and also an imbricate or overlapping arrangement of the rocks similar to that seen in stream beds here on Earth.

A second kind of Martian channel (box figure 2) occurs mostly on the sides of some volcanoes and some meteorite impact craters, and is more typical on Earth. Channels of this type occur on the older surfaces of Mars (more than 3.5 billion years old) and may indicate that early in the history of Mars temperature and atmospheric conditions were such that rainfall could have occurred and long-lived river systems could have existed. Possibly as the climate of Mars cooled, the water froze and formed the extensive deposit of underground ice known to exist in the surface layers of Mars today. (See Astrogeology Box 9.1 for a look at underground ice.)

Astrogeology Box 10.1 Figure 2
Viking Orbiter view of a portion of channel Nirgal Vallis and its tributary system.
Photo by Dr. Michael H. Carr/National Space Science Data Center

Astrogeology Box 10.1 Figure 1
View from the Mars Pathfinder Lander showing the Sojourner Rover and a variety of rocks from Ares Vallis.
Photo Courtesy of Jet Propulsion Laboratory

Chapter 10

254

10.1 Environmental Geology

A Controlled Flood in the Grand Canyon: A Bold Experiment to Restore Sediment Movement in the Colorado River

On March 26, 1996, one of the largest experiments ever conducted on a river took place along the Colorado River below the Glen Canyon Dam (box figure 1). For six days the discharge from the Glen Canyon Dam was increased from 8,000 cfs to 45,000 cfs (a spike flow) to emulate the effects of a flood on the Colorado River (box figure 2). One of the main goals of this controlled flooding experiment was to determine whether the higher flows would result in bed scour and redeposition of sand bars and beaches along the sides of the channel. Another goal was to measure and observe how rocks move along the bed of the river bed with increasing discharge and velocity of floodwaters.

The Colorado River had not experienced its usual summertime floods since the Glen Canyon Dam was completed in 1963. The construction of the dam controlled peak discharges or flows on the Colorado River, which resulted in sand being deposited mainly along the bed or bottom of the river and erosion of beaches along the banks of the river. The Glen Canyon Dam cuts off a significant percentage of the sand and erosion of lower Colorado River such that most of the downstream sand is supplied by two tributary streams, the Paria and Little Colorado Rivers. In August 1992 the Paria River flooded and deposited 330,000 tons of sand into the Colorado River, and in January 1993 a flood on the Little Colorado River, and in 10 million tons of sediment below its confluence with the Colorado River. The influx of sediment, coupled with the relatively low discharges from the dam (8–20,000 cfs), resulted in sand being concentrated along the bed of the Colorado River.

One of the main predictions of the experiment proved true. That is, sand caught in deep pools in the bottom of the main channel was scoured and carried in suspension downstream where it was redeposited along the river banks as beaches (box figure 3). Deposition of sand and along the banks occurs due to back eddies that upwell and move upstream along the river banks and decrease the velocity of the downstream flow so that deposition can occur. Most of the scouring and deposition occurred in the first three days of the experiment; however, when flows were reduced

Box 10.1 Figure 1
Location map of the Grand Canyon controlled-flood experiment.
U.S. Geological Survey

Streams and Floods

233

High-tech, interactive learning from the best.

The Smithsonian Institution's *Journey Through Geology* CD-ROM.

Packaged free with this text, this brand new CD-ROM was developed from a series of interactive kiosks at the Smithsonian Institution's new Geology, Gems, and Minerals Hall Exhibit. State-of-the-art animations and video footage of actual events are complemented by navigational information, an overview of each segment, interactive quizzing, and Web site links to related geologic resources.

Multimedia in your textbook.

CD-ROM icons are selectively placed throughout the text to identify areas that correlate directly with the new **Smithsonian Institution's *Journey Through Geology* CD-ROM** packaged with this text. Find out more about this CD-ROM. Visit the Journey Through Geology Web site:

www.mhhe.com/jtg

Related Web resources are also noted in boxes and at chapter end where appropriate.

Additional support helps you make the grade.

Use these helpful end-of-chapter learning aids to prepare for tests and quizzes.

Summary—overviews of chapter content.

Terms to Remember—important terms to review and remember.

Testing Your Knowledge— realistic sample tests you can use to prepare for exams and improve your grades.

Expanding Your Knowledge—questions that help you develop critical thinking skills.

Exploring Resources—Supplemental references in a number of different media.

 Textbook reference

CD-ROM

Videotape

 World Wide Web addresses

*Interacting with **Journey Through Geology** CD-ROM*—questions that help tie the concepts of the chapter to modules of **The Smithsonian Institution's** *Journey Through Geology* two CD-ROM set.

Learn more about this text.

Visit the *Physical Geology* Web site:

www.mhhe.com/plummer

Natural rock sculpture in Paria Plateau, Arizona. Sandstone formed from ancient sand dunes. Running water has eroded the rock into the present distinctive shapes.

Photo © Kerrick James

CHAPTER 1

Introduction to Physical Geology

Geology uses the scientific method to explain natural aspects of the earth—for example, how mountains form or why oil resources are concentrated in some rocks and not in others. This chapter briefly explains how and why the earth's surface, and its interior, are constantly changing. It relates this constant change to the major geological topics of interaction of the atmosphere, water and rock, the modern theory of plate tectonics and geologic time. These concepts form a framework for the rest of the book. Understanding the "big picture" presented here will aid you in comprehending the chapters that follow.

- As authors, we try to be thorough in our coverage of topics so the textbook can serve you as a resource. However, your instructor may choose to concentrate only on certain topics for *your* course. Find out which topics and chapters you should focus upon in your studying and concentrate your energies there.
- Your instructor may present additional material that is not in the textbook. Take good notes in class.
- Do not get overwhelmed by terms. (Every discipline has its own language.) Don't just memorize each term and its definition. If you associate a term with a concept or mental picture, remembering the term comes naturally when you understand the concept. (You remember names of people you know because you associate personality and physical characteristics with a name.) You may find it helpful to learn the meanings of frequently used prefixes and suffixes for geologic terms. These can be found in Appendix G.
- **Boldfaced** terms are ones you are likely to need to understand because they are important to the entire course.
- *Italicized* terms are not as important, but may be necessary to understand the material in a particular chapter.
- Pay particular attention to illustrations. Geology is a visually oriented science and the photos and artwork are at least as important as the text. You should be able to sketch important concepts from memory.
- Find out to what extent your instructor expects you to learn the material in the boxes. They offer an interesting perspective on geology and how it is used, but much of the material might well be considered optional for an introductory course and not vital to your understanding of major topics. Many of the "In Depth" boxes are meant to be challenging—do not be discouraged if you

need your instructor's help in understanding them.
- Read through the appropriate chapter before going to class. Reread it after class, concentrating on the topics covered in the lecture or discussion. Especially concentrate on concepts that you do not fully understand. Return to previously covered chapters to refresh your memory on necessary background material.
- Use the end of chapter material for review. The Summary is just that, a summary. Don't expect to get through an exam by only reading the summary but not the rest of the chapter. Use the Terms to Remember to see if you can visually or verbally associate the appropriate concept with each term. Answer the Testing Your Knowledge questions in writing. Be honest with yourself. If you are fuzzy on an answer, return to that portion of the chapter and reread it. Remember that these are just a sampling of the kind of questions that might be on an exam.
- Geology, like most science, builds on previously acquired knowledge. You must retain what you learn from chapter to chapter. If you forget or did not learn significant concepts covered early in your course, you will find it frustrating later in the course. (To verify this, turn to chapter 20 and you will probably find it intimidating; but if you build on your knowledge as you progress through your course, the chapter material will fall nicely into place.)
- The Smithsonian Institution's *Journey Through Geology* CD-ROM accompanying this textbook is an exciting innovation. Try it out. The animation and film clips in it should provide a real boost to visualizing difficult concepts.
- Be curious. Geologists are motivated by a sense of discovery. We hope you will be too.

Who Needs Geology?

Imagine yourself as a student at California State University, Northridge (CSUN) in greater Los Angeles. At 4:30 A.M. on January 17, 1994, you are jolted awake when your apartment begins to sway violently. Dishes, bookcases, and ceiling plaster crash to the floor. The noise and shaking are terrifying as you struggle to stand up. In less than a minute the shaking stops and silence momentarily returns. You realize you survived an earthquake, but you are still scared and disoriented.

This seismic event took place because of a sudden shift of bedrock eighteen kilometers beneath Northridge. Shock waves spread in all directions with damaging effects in many parts of Los Angeles. Northridge and vicinity suffered the heaviest damage. You feel fortunate, you are safe and your apartment building was not destroyed. Others nearby were not so fortunate. Two University students were among the sixteen killed in

one of the forty apartment buildings that collapsed (figure 1.1). (Emergency crews drove right past one apartment building because it appeared undamaged; but it had collapsed straight downward, completely crushing the first floor but leaving the upper floors standing.)

After the earthquake you and your neighbors do not have electricity (power was lost, temporarily, as far away as Edmonton, Canada). Water is in short supply because of broken water mains. Leaving the area is not a good option. Eleven major highway structures have been destroyed, including a segment of the Santa Monica Freeway, the nation's busiest highway. Despite the deprivations and the jitters caused by frequent aftershocks, you are comforted by how people pitch in to help each other. A tremendous sense of community grows from the shared adversity.

A

B

Figure 1.1

Damage from the Northridge earthquake. (*A*) The Northridge Meadows apartments where 16 people or roughly one-third of all those killed in the quake died when the first floor was crushed by the weight of the top two floors. (*B*) The parking garage at California State University, Northridge that was destroyed by the earthquake.

Photo A © Michael Edwards/LA Times; Photo B by Frank M. Hanna

Weeks later, the spring semester begins at CSUN. You are among the 27,000 students that must cope with the fact that all the buildings on campus were damaged and some were destroyed. But at least the most severely mangled building was a large, prefabricated parking structure (figure 1.1*B*). Repairing your campus will take time and an estimated $350 million. Meanwhile, some of your classes are held outdoors, others in hastily erected temporary buildings. As the weeks pass, you share the pride of the university community and are determined that the quality of your education shall not be undermined by the earth-

quake and its damage. But your earthquake jitters continue even though the aftershocks now are infrequent and barely perceptible.

Sixty people were killed in the Northridge earthquake. The monetary cost was estimated at between $20 and $40 billion—financially the most costly natural disaster in North America's history. The loss in lives could have been far worse. For instance, 8,000 people died in the 1985 Mexico City earthquake and 5,000 were killed in 1995 at Kobe, Japan. The Northridge death toll was low due mainly to good planning.

Preparing for major earthquakes has been an ongoing process in California for decades. The quake was expected because of what geologists have learned about how the earth works. Skyscrapers in the Los Angeles area survived because they were built to meet high standards for seismic-resistance—the fruits of many decades of engineering studies and design. The apartment buildings that collapsed in Northridge probably would have survived if, during their construction, building codes had been enforced. It is likely that thousands of lives were saved because of adherence to building codes. Pillars for freeway bridges and underpasses were known to be vulnerable. The Santa Monica and other freeways were scheduled to have their pillars reinforced later in 1994, but the quake came first.

Luck also played a role in keeping the casualties down. Los Angeles' normally densely packed freeways were almost empty, because of the early hour and it being a holiday, Martin Luther King's birthday. (The San Francisco Bay area was equally lucky when, four years earlier, the Loma Prieta earthquake took place as the area's two baseball teams were about to begin a World Series game; the normally heavy rush hour traffic was the lightest in anyone's memory when a freeway and part of a bridge collapsed.)

The Northridge earthquake was a reminder that our solid earth does not stand still.

According to the theory of plate tectonics, the earth's rigid outer shell is broken into a series of *plates.* Adjoining plates may slide past, move away from, or collide with one another. Plates generally move from 1 to 18 centimeters a year. But the motion may not be smooth and continuous. Plates may be "locked" against one another for many years and move suddenly.

Sudden motion along a fault caused the Northridge earthquake (figure 1.2). The mountain ranges that rise above Los Angeles are another product of relentless plate motion. It's as if a giant vise is slowly closing, forcing bedrock upward into mountains. The Santa Susana Mountains that border Northridge grew higher by 38 centimeters during the earthquake.

The awesome energy released by an earthquake is a product of forces within the earth that move firm rock. Earthquakes are only one consequence of the ongoing changing of the earth. Ocean basins open and close. Mountain ranges rise and are worn down to plains through slow, but very effective, processes. Studying how the earth works can be as exciting as watching a great theatrical performance. Understanding the changes that take place in and on the earth, and the reasons for

Figure 1.2

Compression (indicated by large arrows) in the Los Angeles area causes movement along a fault and the January 17, 1994 earthquake.

those changes, is the challenging objective of **geology,** the scientific study of the earth.

Physical geology is the division of geology concerned with earth materials, changes in the surface and interior of the earth, and the dynamic forces that cause those changes. Put another way, physical geology is about how the earth works.

Earthquakes and other aspects of geology are interesting, but how does geology benefit you, as an inhabitant of this planet? Some of the ways are discussed next.

Avoiding Geologic Hazards

Geology can have a direct application in ensuring people's safety and well-being. For example, if you were building a house in an earthquake-prone area, you would want to know how to minimize danger to yourself and your home. You would want to build the house on a type of ground not likely to be shaken apart by an earthquake. You would want the house designed and built to absorb the kind of vibrations given off by earthquakes.

People instinctively regard volcanoes as dangerous. They are, but the hazards are not immediately apparent to the nongeologist. (One is not likely to get killed by a lava flow or by a boulder ejected from a volcano.) The 1985 eruption of Nevado del Ruiz in Colombia is a tragic example. No one died from the eruption itself, but 23,000 people were killed from the indirect effects. The hot rocks blasted out of the volcano caused part of the ice and snow capping the peak to melt. The

Figure 1.3

Armero, Colombia, after the 1985 mudflow. The buildings are the small portion of the town that survived the mudflow.

Photo by U.S. Geological Survey

water mixed with loose rock on the flank of the mountain and flowed down stream channels as a *mudflow.* At the base of the volcano, the mudflow overwhelmed the town of Armero, killing most of its inhabitants (figure 1.3). Colombian geolo-

1.1
ENVIRONMENTAL GEOLOGY

The 1991 Eruption of Mount Pinatubo—Geologists Save Thousands of Lives

In contrast to the disaster in Colombia, geologists were much more successful at saving lives when Mount Pinatubo erupted in the Philippines in 1991. When minor steam eruptions began in April, 1991, Pinatubo was a vegetation-covered mountain that had last erupted 400 years earlier. As the eruptions intensified, Filipino geologists thought a major eruption might be developing. Geologic fieldwork completed in earlier years indicated that prehistoric eruptions of the volcano tended to be violent. Under a previous arrangement for cooperation, American geologists joined their Philippine colleagues and deployed portable seismographs to detect and locate small earthquakes within the volcano and tiltmeters for measuring the bulging of the volcano. Data were analyzed by state-of-the-art computer programs.

Fortunately, it took two months for the volcano to reach its climactic eruption, allowing time for the scientists to work with local officials and develop emergency evacuation plans. Geologists had to educate the officials about the principal hazards—mudflows and pyroclastic flows. *Pyroclastic flows* (as explained in the chapter on volcanoes) are hot, turbulent mixtures of exploding gases and volcanic ash that flow rapidly down the flanks of a volcano. Pyroclastic flows can reach speeds of over 100 kilometers per hour and are extremely destructive.

In June, explosions, ash eruptions and minor pyroclastic flows indicated that magma (molten rock) was not far underground and a major eruption was imminent. Some 80,000 people were evacuated from the vicinity of the volcano. The U.S. military evacuated and later abandoned Clark Air Base, which was buried by ash. The climactic eruption occurred on June 15 when huge explosions blasted the top off the volcano and resulted in large pyroclastic flows. Volcanic debris was propelled high into the atmosphere. A typhoon 50 km away brought heavy rains, which mixed with the ash and resulted in numerous, large mudflows.

The estimated volume of magma that erupted from the climactic eruption is 5 km^3, making it the world's largest eruption since 1917. Its effects extended beyond the Philippines. Fine volcanic dust and gas blasted into the high atmosphere were carried around the world and would take years to settle out. For a while, we got more colorful sunsets worldwide. Because of the filtering effect for solar radiation, worldwide average temperature was estimated to drop by 0.5°C for two years, more than countering the long-term warming trend of the earth's climate.

The death toll from the eruption was 374. Of these, 83 were killed in mudflows. Most of the rest died because roofs collapsed from the weight of ash. In addition, 358 people died from illness related to the eruptions. More than 108,000 homes were partly or totally destroyed. The death toll probably would have been in the tens of thousands had the prediction and warning system not been so successful.

Related Web Resources:
Volcano World http://volcano.und.nodak.edu/
The site contains a wealth of information on volcanoes, including Pinatubo.

---

gists had previously predicted such a mudflow could occur and published maps showing the location and extent of expected mudflows. The actual mudflow that wiped out the town matched that shown on the geologists' map almost exactly. Unfortunately, government officials had ignored the map and the geologists' report; otherwise the tragedy could have been averted. A case where geologists were successful at saving lives is described in Box 1.1.

Geologic hazards, other than earthquakes and volcanoes, that geologists investigate include floods, wave erosion at coastlines, collapsing ground surfaces, and landslides. (In the United States and Canada, far more property and lives have been lost due to landslides and floods than to earthquakes and volcanoes.)

Supplying Things We Need

We depend upon the earth for energy resources and the raw materials we need for survival, comfort, and pleasure. The earth, at work for billions of years, has distributed material into localized concentrations that humans can mine or extract. By learning how the earth works and how different kinds of substances are distributed and why, we can intelligently search for metals, sources of energy, and gems. Even maintaining a supply of sand and gravel for construction purposes depends upon geology.

The economic systems of Western civilization currently depend on abundant and cheap energy sources. Nearly all our vehicles and machinery are powered by petroleum, coal, or nuclear power and depend on energy sources concentrated unevenly in

Box 1.1 Figure 1

The major eruption of Mount Pinatubo on June 15, 1991 as seen from Clark Air Force Base, Philippines.

Photo by Robert Lapointe, U.S. Air Force

the earth. The United States economy in particular is geared to petroleum as a cheap source of energy. In a few decades Americans have used up most of the country's known petroleum reserves, which took nature hundreds of millions of years to store in the earth. Americans are now heavily dependent on imported oil. (The Gulf War of 1991 was at least partially fought because of the industrialized nations' petroleum requirements.) To find more of this diminishing resource will require more money and increasingly sophisticated knowledge of geology. Although many people are not aware of it, we face similar problems with diminishing resources of other materials, notably metals such as iron, aluminum, copper, and tin, each of which has been concentrated in a particular environment by the action of the earth's geologic forces.

Protecting the Environment

Our demands for more energy and metals have, in the past, led us to extract them with little regard for effects on the balance of nature within the earth, and therefore on us, earth's residents. Mining of coal, if done carelessly, for example, can release acids into water supplies. Understanding geology can help us lessen or prevent damage to the environment—just as it can be used to find the resources in the first place.

The environment is further threatened because these are nonrenewable resources. Petroleum and metal deposits do not grow back after being harvested. As demands for these commodities increase, so does the pressure to disregard the ecological damage caused by the extraction of the remaining deposits.

Problems involving petroleum illustrate this. Oil companies employ geologists to discover new oil fields, while the public and government depend on other geologists to assess the potential environmental impact of petroleum's removal from the ground, the transportation of petroleum (see Box 1.2), and disposal of any toxic wastes from petroleum products.

1.2
ENVIRONMENTAL GEOLOGY

Delivering Alaskan Oil— The Environment vs. the Economy

When the tanker *Exxon Valdez* ran aground in 1989, over 240,000 barrels of crude oil were spilled into the waters of Alaska's Prince William Sound. It was the worst ever oil spill in U.S. waters. The spill, with its devastating effects on wildlife and the fishing industry, dramatically highlighted the conflicts between maintaining the energy demands of the American economy and conservation of the environment. The *Exxon Valdez* was only one of many tankers that carry oil from the southern end of the Alaskan pipeline to refineries on the west coast of the United States. Oil spills, such as this one, were predicted by the 1972 environmental impact statement for the Alaska pipeline prepared by the U.S. Geological Survey.

In the late 1970s the United States was importing almost half its petroleum, at a loss of billions of dollars per year to the national economy. (By 1997, the United States was importing more than half of the petroleum it uses.) This drain on the country's economy and the increasing cost of energy can be major causes of inflation, lower industrial productivity, unemployment, and the erosion of standards of living.

In the 1960s, geologists discovered oil beneath the shores of the Arctic Ocean on Alaska's North Slope. It is now the United States' largest oil field. Thanks to the Alaska pipeline, completed in 1977, Alaska has supplied as much as 20% of the United States' domestic oil.

Despite its important role in the American economy, some people consider the Alaska pipeline and the use of tankers as unacceptable threats to the area's ecology. The

Box 1.2 Figure 1

Road bulldozed during the winter in Arctic Alaska became a useless quagmire after thawing.

Photo by U.S. Geological Survey

1989 oil spill demonstrated the hazards of the marine portion of the oil transportation system. Oil spilled from a ruptured pipe could have a devastating effect on the fragile Arctic plant and animal life.

Geologists with the U.S. Geological Survey conducted the official environmental impact investigation of the proposed pipeline route. After an exhaustive study, they recommended against its construction, partially because of the hazards to oil tankers and partially because of the geologic hazards of the pipeline route. Their report was overruled. The Congress and the President of the United States exempted the pipeline from laws that require a favorable environmental impact statement before a major project can begin. The 1.5 to 2 million barrels of oil a day that flow from the Arctic oil fields mean that over $10 billion a year that would have been lost through the purchase of foreign oil instead remains in the American economy.

The 1,250-kilometer-long pipeline, through which 88,000 barrels of oil an hour flow, crosses regions of ice-saturated, frozen ground and major earthquake-prone

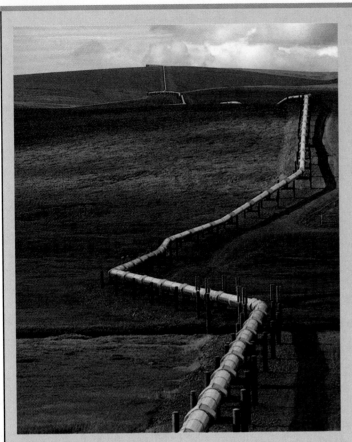

Box 1.2 Figure 2

The Alaska pipeline.

Photo © Steve McCutcheon/Alaska Pictorial Service

mountain ranges that geologists regard as serious hazards to the structure. Building anything on frozen ground creates problems. For example, the road in box figure 1 was built during the exploration of Alaska's North Slope oil fields. When the road was being built, the protective vegetation was scraped off. During the summer thaw the road became a quagmire. As thawing continued, the flooded tracks shown in the photo grew into ponds so that the road will never be passable.

Building the pipeline over such terrain also presented enormous engineering problems. If the pipeline were placed on the ground, the hot oil flowing through it could melt the frozen ground. On a slope, mud could easily slide and rupture the pipeline. Careful (and costly) engineering minimized these hazards. Much of the pipeline is elevated above the ground (box figure 2). Radiators conduct heat out of the structure. In some places refrigeration equipment in the ground protects against melting.

Records indicate that a strong earthquake can be expected every few years in the earthquake belts crossed by the pipeline. An earthquake could rupture a pipeline—especially a conventional pipe as in the original design. However, when the Alaska pipeline was built, in several places sections were specially jointed to allow the pipe to shift as much as 6 meters without rupturing.

The original estimated cost of the pipeline was $900 million, but the final cost was $7.7 billion, making it the costliest privately financed construction project in history. The redesigning and construction that minimized the potential for an environmental disaster were among the reasons for the increased cost. There have been some minor spills from the pipeline. For instance, in January 1981, 5,000 barrels of oil were lost when a valve ruptured.

Thus far, the pipeline company's claim that there is virtually no chance of a major spill from a pipeline rupture seems vindicated. However, the risks of marine transportation remain. In hindsight, a pipeline through Canada to the American Midwest might have been better in the long run (as advocated by geologists who prepared the environmental impact statement). Although longer, it would traverse less hazardous terrain, would avoid the long marine leg, and would bring the oil to refineries in the central part of the United States. A joint venture with Canada would have benefited both countries, particularly since Canada has discovered major petroleum resources in its Arctic territory.

Related Web Resources
http://www.alyeska-pipe.com/

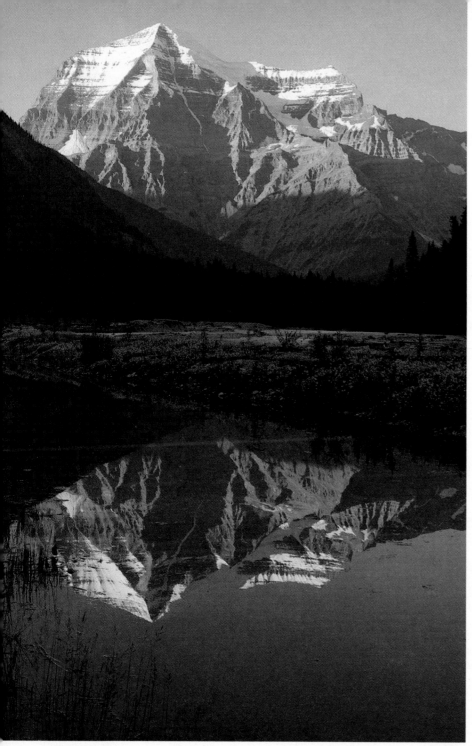

Understanding Our Surroundings

Understanding geology leads to a greater appreciation of your surroundings. If, for instance, you were traveling through the Canadian Rockies, you might see the scene in figure 1.4 and wonder how the landscape came to be.

You might wonder: (1) why there are layers in the rock exposed in the cliffs; (2) why the peaks are so jagged; (3) why there is a glacier in a valley carved into the mountain; (4) why this is part of a mountain belt that extends northward and southward for thousands of kilometers; (5) why there are mountain ranges here and not in the central part of the continent. After completing a course in physical geology, you should be able to answer these questions as well as understand how other kinds of landscapes formed.

People who have completed a physical geology course find that their understanding of geological processes makes travel much more interesting. Figure 1.5 is a map showing the landforms of the contiguous United States and southernmost Canada. Major features discussed or depicted in this book are named with black lettering. (In addition, on the inside, front cover is a generalized geologic map of North America that will be a useful reference as you progress through this book.) In your future air or surface travel, you may want to take along your book with these maps to enhance your appreciation of the scenery.

An Overview of Physical Geology—Important Concepts

The remainder of this chapter is an overview of physical geology that should provide a framework for most of the material in this book. Although the concepts probably are totally new to you, it is important that you comprehend what follows. You may want to reread portions of this chapter while studying later chapters when you need to expand or reinforce your comprehension of this basic material.

The earth can be visualized as a giant machine driven by two engines, one internal and the other external. Both are *heat engines,* devices that convert heat energy into mechanical

Figure 1.4

Mount Robson, the highest peak in the Canadian Rocky Mountains.

Photo © Superstock

Figure 1.5

Raisz physiographic map of the 48 contiguous
states and southernmost Canada.

© Erwin Raisz; Courtesy of Raisz Landform Maps, PO
Box 2254, Jamaica Plain, MA 02130. (Large copies may
be purchased from them.)

CANADIAN SHIELD

Laurentian upland of low hills and many lakes

Ontario

Quebec

New Brunswick

Newfoundland

Maine

ADIRONDACK MOUNTAINS

New Hampshire

Minnesota

Wisconsin

Michigan

New York

Massachusetts

Rhode Island

Connecticut

Valley & Ridge Province

Iowa

Pennsylvania

New Jersey

Ohio

Maryland

Delaware

Illinois

Indiana

West Virginia

Missouri

Kentucky

Virginia

MISSISSIPPI RIVER

APPALACHIAN MOUNTAINS

Tennessee

North Carolina

Arkansas

OUACHITA MOUNTAINS

South Carolina

Mississippi

Alabama

Georgia

Louisiana

Mississippi Delta

Florida

LANDFORMS OF THE UNITED STATES

by ERWIN RAISZ   *Sixth, revised edition, 1957*

Scale   50   100   150   200 Miles
         0   100   200 Kilometers

Miami

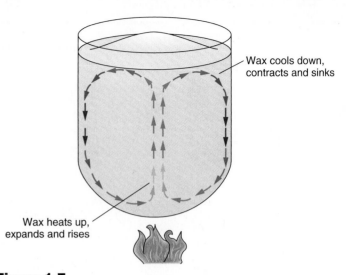

Figure 1.7

Movement of wax due to density differences caused by heating and cooling.

Figure 1.6

Two examples of simple heat engines. (*A*) A "lava lamp." Blobs are heated from below and rise. Blobs cool off at the top of the lamp and sink. (*B*) A pinwheel held over steam. Heat energy is converted to mechanical energy.

energy. Two simple heat engines are shown in figure 1.6. An automobile is powered by a heat engine. When gasoline is ignited in the cylinders, the resulting hot gases expand, driving pistons to the far end of cylinders. In this way, the heat energy of the expanding gas has been converted to the mechanical energy of the moving pistons, then transferred to the wheels, where the energy is put to work moving the car.

The *internal* heat engine of the earth is powered by heat moving from the hot interior of the earth toward the cooler exterior. Moving plates and earthquakes are products of this heat engine.

The earth's *external* heat engine is driven by solar power and gravity. Heat from the sun provides the energy for circulating the atmosphere and oceans. Water, especially from the

oceans, is evaporated due to solar heating. When moist air cools, rain or snow falls.

Over long periods of time, moisture at the earth's surface helps rock disintegrate. Water washing down hillsides and flowing in streams loosens and carries away the rock particles. In this way mountains originally raised by the earth's internal forces are worn away by processes driven by the external heat engine.

We will look at how the earth's heat engines work and show how some of the major topics of physical geology are related to the *surficial* (on the earth's surface) and *internal* processes powered by the heat engines.

Internal Processes: How the Earth's Internal Heat Engine Works

The earth's internal heat engine works because hot, buoyant material deep within the earth moves slowly upward toward the cool surface and cold, denser material moves downward. Visualize a vat of hot wax, heated from below (figure 1.7). As the wax immediately above the fire gets hotter, it expands, becomes less dense (that is, a given volume of the material will weigh less) and will rise. Wax at the top of the vat loses heat to the air, cools, contracts, becomes denser, and will sink. A similar process takes place in the earth's interior. Rock that is deep within the earth and is very hot rises slowly toward the surface while rock that has cooled near the surface is denser and sinks downward. Instinctively, we don't want to believe that rock can flow like hot wax. However, experiments have shown that, under the right conditions, rocks behave *plastically*, meaning they are capable of being molded (like wax or putty). The hotter a rock and the greater the pressure on the rock, the more likely a rock is to behave plastically. Pressure and temperature, even a few kilometers beneath the earth's surface, can be high enough so that rock may behave plastically and can flow very slowly.

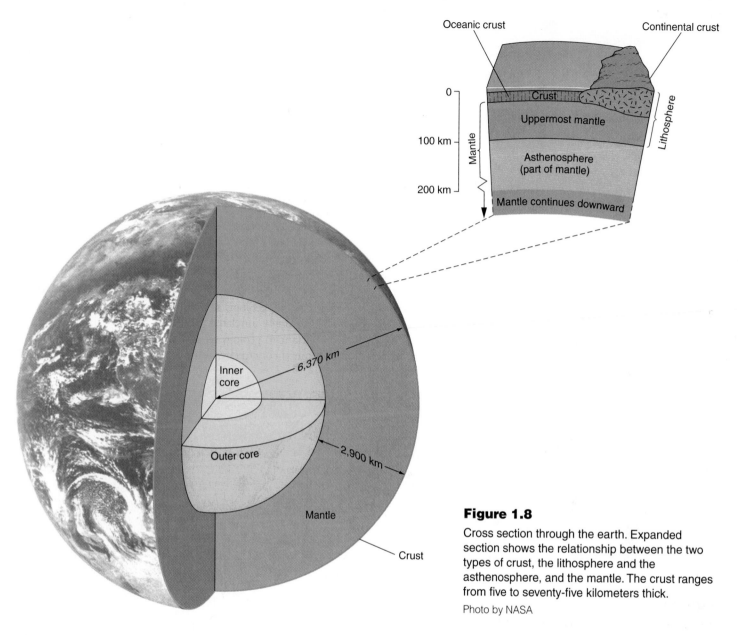

Oceanic crust   Continental crust

0

Crust

Uppermost mantle   Lithosphere

100 km

Mantle

Asthenosphere
(part of mantle)

200 km

Mantle continues downward

6,370 km

Inner
core

Outer core

2,900 km

Mantle

Crust

Figure 1.8

Cross section through the earth. Expanded
section shows the relationship between the two
types of crust, the lithosphere and the
asthenosphere, and the mantle. The crust ranges
from five to seventy-five kilometers thick.

Photo by NASA

The Earth's Interior

Rock flowage is believed to take place in part of the interior of
the earth in the zone known as the **mantle,** the largest, by volume, of the earth's three major concentric zones (see figure
1.8). The mantle is solid (except in a few spots) and probably
composed of rock not very different from some kinds of rock
found at the earth's surface.

The other two zones are the **crust** and the **core.** The crust
of the earth is analogous to the skin on an apple. The thickness
of the crust is insignificant compared to the whole earth. We
have direct access to only the crust, and not much of the crust
at that. We are like microbes crawling on an apple, without the
ability to penetrate its skin. Because it is our home and we
depend on it for resources, we are concerned more with the
crust than with the inaccessible mantle and core. The crust
varies in thickness. Two major types of crust are *oceanic crust*
and *continental crust.* The crust under the oceans is much thin-

ner. It is made of rock that is somewhat denser than the rock
that underlies the continents.

The lower parts of the crust and the entire mantle are inaccessible to direct observation. No mine or oil well has penetrated through the crust, so our concept of the earth's interior
is based on indirect evidence (the topic of chapter 17).

The crust and the uppermost part of the mantle are relatively rigid. Collectively they make up the **lithosphere.** (To
help you remember terms, the meanings of commonly used
prefixes and suffixes are given in Appendix G. For example,
lith means "rock" in Greek. You will find *lith* to be part of
many geologic terms.) The upper mantle underlying the
lithosphere, called the **asthenosphere,** is soft and therefore
flows. It provides a "lubricating" layer over which the lithosphere moves (*asthenos* means "weak" in Greek). Where hot
mantle material wells upward, it will uplift the lithosphere.
Where the lithosphere is coldest and densest, it will sink
down through the asthenosphere and into the deeper man

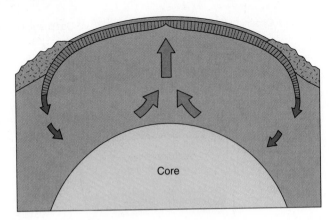

Figure 1.9

Hot mantle travels upward. Cold crust and mantle sink.

(figure 1.9). The effect of this internal heat engine on the crust is of great significance to geology. The forces generated inside the earth, called **tectonic forces,** cause deformation of rock as well as vertical and horizontal movement of portions of the earth's crust. Mountain ranges are evidence of tectonic

forces strong enough to outdo gravitational forces. (Mount Everest, the world's highest peak, is made of rock that formed beneath an ancient sea.) Mountain ranges are built over extended periods, as portions of the earth's crust are squeezed, stretched and raised.

Most tectonic forces are mechanical forces. Some of the energy from these forces is put to work deforming rock, bending and breaking it, and raising mountain ranges. The mechanical energy may be stored (an earthquake is a sudden release of stored mechanical energy) or converted to heat energy (rock may melt, resulting in volcanic eruptions). The working of the machinery of the earth is elegantly demonstrated by plate tectonics.

The Theory of Plate Tectonics

From time to time a theory emerges within a science that revolutionizes that field. The plate tectonic theory, currently accepted by virtually all geologists, is a unifying theory that accounts for many seemingly unrelated geological phenomena. The theory is as important to geology as the theory of relativity is to physics, the atomic theory to chemistry, or evolution is to biology.

Figure 1.10

(A) Plates of the world. Large arrows indicate direction of plate motion. (B) Plate motion away from a divergent boundary toward a convergent boundary.

After W. Hamilton, U.S. Geological Survey.

A

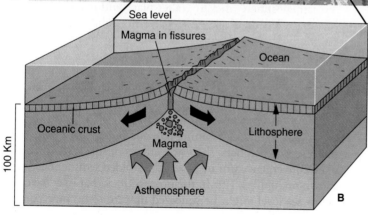

B

Figure 1.11

(A) The sea floors of the world. (B) A diverging boundary at a mid-oceanic ridge. Hot asthenosphere wells upward beneath the ridge crest. Magma forms and squirts into fissures. Solid material that does not melt remains as mantle in lower part of lithosphere. As lithosphere moves away from spreading axis, it cools, becomes denser, and sinks to a lower level.

From *World Ocean Floor* by Bruce C. Heezen and Marie Tharp, 1977. © by Marie Tharp 1977. Reproduced by permission of Marie Tharp, 1 Washington Ave., South Nyack, NY 10960.

Plate tectonics was seriously proposed as a hypothesis in the early 1960s, though the idea was based on earlier work—notably, the hypothesis of *continental drift*. In the chapters on igneous, sedimentary, and metamorphic rocks, as in the chapter on earthquakes, we will expand upon what you learn about the theory here to explain the origin of some rocks and why volcanoes and earthquakes occur. Chapter 19 is devoted to plate tectonics and will show that what you learned in many previous chapters is interrelated and explained by plate tectonic theory.

Plate tectonics regards the lithosphere as broken into *plates* that are in motion (figure 1.10). The plates, which are much like segments of the cracked shell on a boiled egg, move relative to one another, sliding on the underlying asthenosphere. Much of what we observe in the rock record can be explained by what takes place along *plate boundaries,* where

two plates are pulling away from each other, sliding past each other, or moving toward each other.

According to plate tectonics, **divergent boundaries** exist where plates are moving apart. Most divergent boundaries coincide with the crests of submarine mountain ranges, called **mid-oceanic ridges** (figures 1.10*B* and 1.11).

A mid-oceanic ridge is higher than deep ocean floor (figure 1.11*A*) because the rocks, being hotter there, are less dense. Tensional cracks develop along the ridge crest (figure 1.11*B*). These cracks tap localized **magma** (molten rock) chambers in the underlying asthenosphere, and the magma squeezes into fissures (cracks through the lithosphere). Some magma erupts along the ridge crest, and the rest solidifies in the fissure. Continued pulling apart of the ridge crest develops new cracks, and the process of filling and cracking continues indefinitely. Thus,

Figure 1.12

A convergent boundary.

new oceanic crust is continuously created at a diverging boundary. Not all of the mantle material melts; a solid residue remains under the newly created crust. New crust and underlying solid mantle make up the lithosphere that moves away from the ridge crest, traveling like the top of a conveyor belt. The rate of motion is generally 1 to 18 centimeters per year—slow in human terms but quite fast by geologic standards.

As the lithosphere moves away from the divergent boundary, the material slowly cools. As it cools it contracts and becomes denser. The contraction of the lithosphere and its slow sinking (because of the increased density) cause the floors beneath oceans to deepen away from ridge crests.

The top of a plate may be composed exclusively of oceanic crust or include a continent or part of a continent. For example, if you live on the North American plate, you are riding westward relative to Europe because the plate's divergent boundary is along the mid-oceanic ridge in the North Atlantic Ocean (figures 1.10 and 1.11). The western half of the North Atlantic sea floor and North America are moving together in a westerly direction away from the mid-Atlantic ridge plate boundary.

A second type of boundary, a **transform boundary,** occurs where two plates slide past each other. The San Andreas fault in California is an example of this type of boundary, and the earthquakes along the fault are a result of plate motion.

The third type of boundary, one resulting in a wide range of geologic activities, is a **convergent boundary,** where plates move toward each other (figure 1.12). If one plate is capped by oceanic crust and the other by continental crust, the less dense, more buoyant continental plate will override the denser, oceanic plate. The oceanic plate sinks along what is known as a **subduction zone,** a zone where an oceanic plate descends into the mantle beneath an overriding plate. In the

region where the top of the subducting plate slides beneath the asthenosphere, melting takes place and magma is created. Magma is less dense than the overlying solid rock. Therefore, the magma created along the subduction zone works its way upward and either erupts on the earth's surface to solidify as *extrusive* **igneous rock** or solidifies within the crust to become *intrusive* igneous rock. Hot rock, under high pressure, near the subduction zone that does not melt may change in the solid state to a new rock—**metamorphic rock.**

In addition to containing igneous and metamorphic rocks, major mountain belts show the effects of squeezing caused by plate convergence (for instance, the "folded sedimentary rocks" shown on figure 1.12). In the process, rock that may have been below sea level might be squeezed upward to become part of a mountain range.

Box 1.3 describes how plate tectonic theory was developed through the *scientific method.* If you do not have a thorough comprehension of how the scientific method works, be sure to study the box.

Surficial Processes: The Earth's External Heat Engine

When tectonic forces shove a portion of the earth's crust above sea level, rocks are exposed to the atmosphere. The earth's external heat engine, driven by solar power and gravity, then comes into play. Our weather patterns are largely a product of this heat engine. For instance, hot air rises near the equator and sinks in cooler zones to the north and south. Solar heating of air creates wind; ocean waves are, in turn, produced by wind. When moist air cools, it rains or snows. Rainfall on hillsides flows down slopes and into streams. Streams flow to lakes

Although the hypothesis was proposed only a few decades ago, plate tectonics has been so widely accepted and disseminated that most people have at least a rough idea of what it is about. Most nonscientists can understand the television and newspaper reports (and occasional comic strip, such as that in box figure 1) that include plate tectonics in reports on earthquakes and volcanoes. Our description of plate tectonics implies little doubt about the existence of the process. The theory of plate tectonics has been accepted by nearly all geologists (this does not mean it is "true"). Plate tectonic theory, like all knowledge gained by science, has evolved through the processes of the **scientific method.** We will illustrate the scientific method by showing how plate tectonics has evolved from a vague idea into a plausible theory.

The basis for the scientific method is the belief that the universe is orderly and that by *objectively* analyzing phenomena, we can discover its workings. The technique is best illustrated as a series of steps, although a scientist does not ordinarily go through a formal checklist.

1. A question is raised or a problem is presented.
2. Available information pertinent to the question or problem is analyzed. Facts, which scientists call **data,** are gathered.
3. After the data have been analyzed, tentative explanations or solutions, called **hypotheses,** are proposed.

4. One predicts what would occur in given situations if a hypothesis were correct.
5. Predictions are tested. Incorrect hypotheses are discarded.
6. A hypothesis that passes the testing becomes a **theory,** which is regarded as having an excellent chance of being true. In science, however, nothing is considered proven absolutely. All theories remain open to scrutiny, further testing, and refinement.

Like any human endeavor, the scientific method is not infallible. Objectivity is needed throughout. Someone can easily become attached to the hypothesis he or she has created and so tend subconsciously to find only supporting evidence. As in a court of law, every effort is made to have observers objectively examine the logic of both procedures and conclusions. Courts sometimes make wrong decisions; science, likewise, is not immune to error.

How the concept of plate tectonics evolved into a theory is outlined below.

Step 1: A question asked or problem raised.
Actually, a number of questions were being asked about seemingly unrelated geological phenomena.

What caused the submarine ridge that extends through most of the oceans of the world? Why are rocks in mountain belts intensely deformed? What sets off earthquakes? What causes rock to melt underground and erupt as volcanoes?

Step 2: Gathering of data. Early in the century, the amount of data was limited. But through the decades, the information gathered increased enormously. New data, most notably information gained from exploration of the sea floor in the mid-1900s, forced scientists to discard old hypotheses and come up with new ideas.

Step 3: Hypotheses proposed. Most of the questions being asked were treated as separate problems wanting separate hypotheses. Some appeared

FRANK & ERNEST® by Bob Thaves

Box 1.3 Figure 1

Plate tectonics sometimes show up in comic strips.

Reprinted by permission of Bob Thaves.

Box 1.3 Figure 2

Ages of rocks from holes drilled into the oceanic crust. (Vertical scale of diagram is exaggerated.)

interrelated. One hypothesis, **continental drift,** did address several questions. It was advocated by Alfred Wegener, a German scientist, in a book published in the early 1900s.

Wegener postulated that the continents were all once part of a single supercontinent called Pangaea. The hypothesis explained why the coastlines of Africa and South America look like separated parts of a jigsaw puzzle. Some 200 million years ago this supercontinent broke up, and the various continents slowly drifted into their present positions. The hypothesis suggested that the rock within mountain belts becomes deformed as the leading edge of a continental crust moves against and over the stationary oceanic crust. Earthquakes were presumably caused by continuing movement of the continents.

It was not until the 1960s, after new data on the nature of the sea floor became available, that the idea of continental drift was incorporated into the concept of plate tectonics. What was added in the plate tectonic hypothesis was the idea that oceanic crust, as well as continental crust, was shifting.

Step 4: Prediction. An obvious prediction, if plate tectonics is correct, is that if Europe and North America are moving away from each other, the distance measured between the two continents is greater from one year to the next. But, we can't stretch a tape measure across oceans and, until recently, we have not had the technology to accurately measure distances between continents. So, in the 1960s, other testable predictions had to be made. One prediction was that the rocks of the oceanic crust will be progressively older the farther they are from the crest of a mid-oceanic ridge.

Step 5: Predictions are tested. Experiments were conducted in which holes were drilled in the deep sea floor from a specially designed ship. Rocks and

sediment were collected from these holes, and the ages of these materials were determined. As the hypothesis predicted, the youngest sea floor (generally less than a million years old) is near the mid-oceanic ridges, whereas the oldest sea floor (up to about 200 million years old) is farthest from the ridges (box figure 2).

This test was only one of a series. Various other tests, described in some detail later in this book, tended to confirm the hypothesis of plate tectonics. Some tests did not work out exactly as predicted. Because of this, and more detailed study of data, the original concept was, and continues to be, modified. The basic premise, however, is generally regarded as valid.

Step 6: The hypothesis becomes a theory. Most geologists in the world considered the results of this and other tests as positive, indicating that the concept is probably true. It can now be called the plate tectonic theory.

During the last few years, plate tectonic theory has been further confirmed by the results of very accurate satellite surveys that determine where points on separate continents are relative to one another. The results indicate that the continents are indeed moving relative to one another. Europe and North America *are* moving farther apart.

Although it is unlikely that plate tectonic theory will be replaced by something we haven't thought of yet, aspects that fall under plate tectonics' umbrella continue to be analyzed and revised as new data become available. For instance, the question of how magma is generated deep down in a subduction zone has been addressed by several hypotheses (and is discussed in chapter 4). Does melting take place in the descending plate or in the overlying mantle? Is water carried down by the descending plate an aid to melting? Similar questions are raised in many other or seas. Glaciers grow where there is abundant snowfall at colder, high elevations and flow downhill due to gravity.

Where moving water, ice, or wind loosens and removes material, **erosion** is taking place. Streams flowing toward oceans remove some of the land over which they run. Crashing waves carve back a coastline. Glaciers grind and carry away underlying rock as they move. In each case, rock originally raised by the earth's internal processes is worn down by surficial processes (figure 1.13).

Rocks formed at high temperature and under high pressure deep within the earth and pushed upward by tectonic forces are unstable in their new environment. Air and water tend to cause the once deep-seated rocks to break down and form new materials. The new materials, stable under conditions at the earth's surface, are said to be in **equilibrium,** that is, adjusted to the physical and chemical conditions of their environment so that they do not change or alter with time. For example, much of an igneous rock (such as granite) that formed at a high temperature tends to break down chemically to clay. Clay is in equilibrium, that is to say it is stable, at the earth's surface.

The product of the breakdown of rock is **sediment,** loose material. Sediment may be transported by an agent of erosion, such as running water in a stream. Sediment is deposited when the transporting agent loses its carrying power. For example, when a river slows down as it meets the sea, the sand being transported by the stream is deposited as a layer of sediment.

In time a layer of sediment deposited on the sea floor becomes buried under another layer. This process may continue burying our original layer progressively deeper. The pressure from overlying layers compresses the sediment, helping to consolidate the loose material. With the cementation of the loose particles, the sediment becomes *lithified* (cemented or otherwise consolidated) into a **sedimentary rock.**

Geologic Time

We have mentioned the great amount of time required for geological processes. As humans, we think in units of time related to personal experience—seconds, hours, years, a human lifetime. It stretches our imagination to contemplate

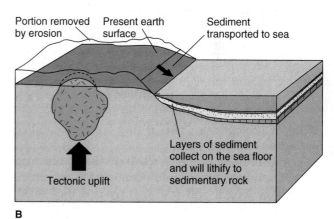

Figure 1.13

Uplift, erosion, and deposition. (A) Magma has solidified underground to become igneous rock. (B) Land is uplifted. Upper portion is eroded. Sediment is transported to the sea to become sedimentary rock.

ancient history that involves 1,000 or 2,000 years. Geology involves vastly greater amounts of time, often referred to as *deep time.*

To be sure, some geological processes occur quickly, such as a great landslide or a volcanic eruption. These events occur when stored energy (like the energy stored in a car battery) is suddenly released. Most geological processes, how-

Table 1.1   Some Important Ages in the Development of Life on Earth

Millions of Years Before Present	Animal Life on Earth	Eras	Periods
3	First humans	Cenozoic	{ Quaternary / Tertiary
65	First important mammals / Extinction of dinosaurs		
	First dinosaurs	Mesozoic	{ Cretaceous / Jurassic / Triassic
245			
300	First reptiles	Paleozoic	{ Permian / Pennsylvanian / Mississippian / Devonian / Silurian / Ordovician / Cambrian
400	Fish becomes abundant		
	First abundant fossils		
545			
3,500	Earliest single-celled fossils	Precambrian	(The Precambrian accounts for the vast majority of geologic time.)
4,500	Origin of the earth		

ever, are slow but relentless, reflecting the pace at which the heat engines work. It is unlikely that a hill will visibly change in shape or height during your lifetime (unless through human activity). However, in a geologic time frame, the hill probably is eroding away quite rapidly. "Rapidly" to a geologist may mean that within a few million years the hill will be reduced nearly to a plain. Similarly, in the geologically "recent" past of several million years ago, a sea may have existed where the hill is now. Some processes are regarded by geologists as "fast" if they are begun and completed within a million years.

The rate of plate motion is relatively fast. If new magma erupts and solidifies along a mid-oceanic ridge, we can easily calculate how long it will take that igneous rock to move 1,000 kilometers away from the spreading center. At the rate of 1 centimeter per year, it will take 100 million years for the presently forming part of the crust to travel the 1,000 kilometers.

Although we will discuss geologic time in detail in chapter 8, table 1.1 shows some reference points to keep in mind. The earth is estimated to be about 4.5 billion years old (4,500,000,000 years). Fossils in rocks indicate that complex forms of animal life have existed in abundance on the earth for about the past 545 million years. Reptiles became abundant about 230 million years ago. Dinosaurs evolved from reptiles and became extinct about 65 million years ago. Humans have been here only about the last 3 million years. The eras and periods shown in table 1.1 comprise a kind of calendar for geologists into which geologic events are placed (as explained in the chapter on geologic time).

Not only are the immense spans of geologic time difficult to comprehend, but very slow processes are impossible to duplicate. A geologist who wants to study a certain process cannot repeat in a few hours a chemical reaction that takes a million years to occur in nature. As Mark Twain wrote in *Life on the Mississippi*, "Nothing hurries geology."

Summary

Geology is the scientific study of the earth. Geological investigations indicate that the earth is changing because of internal and surficial processes. Internal processes are driven mostly by temperature differences within the earth's mantle. Surficial processes are driven by solar energy and gravity. Internal forces cause the crust of the earth to move. Plate tectonic theory visualizes the lithosphere (the crust and uppermost mantle) as broken into plates that move relative to each other over the asthenosphere. The plates are moving *away* from divergent boundaries usually located at the crests of mid-oceanic ridges where new crust is being created. Plates move *toward* convergent boundaries. Convergence results in subduction of one plate. Plates slide past one another at transform boundaries.

Erosion takes place at the earth's surface where rocks are exposed to air and water. Rocks that formed under high pressure and temperature inside the earth are out of equilibrium at the earth's surface and tend to alter to substances that are stable at the earth's surface. Sediment is transported to a lower elevation where it is deposited (commonly on a sea floor in layers). When sediment is cemented it becomes sedimentary rock.

Although the earth is changing constantly, the rates of change are generally extremely slow by human standards.

Terms to Remember

asthenosphere 15
continental drift 20
convergent boundary 18
core 15
crust 15
data 19
divergent boundary 17
equilibrium 21
erosion 21

geology 6
hypothesis 19
igneous rock 18
lithosphere 15
magma 17
mantle 15
metamorphic rock 18
mid-oceanic ridge 17
physical geology 6

plate tectonics 17
scientific method 19
sediment 21
sedimentary rock 21
subduction zone 18
tectonic forces 16
theory 19
transform boundary 18

Testing Your Knowledge

Use the questions below to prepare for exams based on this chapter.

1. Plate tectonics is a result of the earth's internal heat engine, powered by (choose all that apply) (a) the sun (b) gravity (c) heat flowing from the earth's interior outward

2. What tectonic plate are you presently on? Where is the nearest plate boundary and what kind of boundary is it?

3. What is the most likely geologic hazard in your part of the country?

4. What are the three major types of rocks?

5. What are the relationships among the mantle, the crust, the asthenosphere, and the lithosphere?

6. What would the surface of the earth be like if there were no tectonic activity?

7. Explain why cavemen never saw a dinosaur.

8. What is meant by equilibrium? What happens when rocks are forced out of equilibrium?

9. A typical rate of plate motion is (a) 3–4 meters per year (b) 1 kilometer per year (c) 1–10 centimeters per year (d) 1000 kilometers per year

10. Earthquakes may be caused by (a) movement of plates (b) motion along faults (c) shifting of bedrock (d) all of the above

11. The division of geology concerned with earth materials, changes in the surface and interior of the earth, and the dynamic forces that cause those changes is (a) physical geology (b) historical geology (c) geophysics (d) paleontology

12. Which is a geologic hazard? (a) earthquake (b) volcano (c) mudflows (d) floods (e) wave erosion at coastlines (f) landslides (g) all of the above

13. The largest zone of the earth's interior by volume is the (a) crust (b) mantle (c) outer core (d) inner core

14. Oceanic and continental crust differ in (a) composition (b) density (c) thickness (d) all of the above

15. The forces generated inside the earth that cause deformation of rock as well as vertical and horizontal movement of portions of the earth's crust are called (a) erosional forces (b) gravitational forces (c) tectonic forces (d) all of the above

16. Plate tectonics is a (a) conjecture (b) opinion (c) hypothesis (d) theory

17. Which is a type of a plate boundary? (a) divergent (b) transform (c) convergent (d) all of the above

18. The lithosphere is (a) the same as the crust (b) the layer beneath the crust (c) the crust and upper mantle (d) part of the mantle

19. Erosion is a result of the earth's external heat engine, powered by (choose all that apply) (a) the sun (b) gravity (c) heat flowing from the earth's interior outward

Expanding Your Knowledge

1. Why are some parts of the lower mantle hotter than other parts?
2. According to plate tectonic theory, where are crustal rocks created? Why doesn't the earth keep getting larger if rock is continually created?
3. What percentage of geologic time is accounted for by the last century?
4. What would the earth be like without solar heating?
5. What are some of the technical difficulties you would expect to encounter if you tried to drill a hole to the center of the earth?

Exploring Resources

 Allègre, C. 1988. *The behavior of the Earth.* Boston: Harvard University Press.

Drake, E. T., and W. M. Jordan, eds. 1985. *Geologists and ideas: A history of North American geology.* Boulder, Colo.: Geological Society of America.

McPhee, J. 1981. *Basin and range.* New York: Farrar, Straus & Giroux.

Moores, E. M., ed. 1990. *Shaping the earth: Tectonics of continents and oceans.* New York: W. H. Freeman and Company.

Nuhfer, E. B., R. J. Proctor, and P. H. Moser. 1993. *The citizens guide to geologic hazards.* Arvada, Colorado: American Institute of Professional Geologists.

Officer, C. B. and J. Page. 1993. *Tales of the earth: Paroxysms and perturbations of the blue planet.* New York: Oxford University Press.

Pirsig, R. M. 1974. *Zen and the art of motorcycle maintenance.* New York: Bantam Books (paperback). This book contains an exceptionally good exposition of the scientific method as well as considerable insight into the philosophy of science.

Rhodes, F. H. T., and R. O. Stone, eds. 1981. *Language of the earth.* New York: Pergamon Press.

 Earth Revealed. Southern California Consortium. A series of videotapes produced for a televised physical geology course.

Interactive Plate Tectonics by Plummer/McGeary. 1996. WCB/McGraw-Hill.

http://www.mhhe.com/sciencemath/earthscience/jtg/
Physical Geology by Plummer/McGeary/Carlson.

http://www.usgs.gov
United States Geological Survey includes a wide range of geologic information.

http://www.nrcan.gc.ca/gsc/
The Geological Survey of Canada

http://atlas.es.mq.edu.au/users/pingram/v_earth.html
The Virtual Earth—Tour of the Web for Earth Scientists (from Australia).

Expand your knowledge of the concepts presented in this chapter by using the CD-ROM to answer the following questions.

1. The earth has changed greatly over time through the processes of plate tectonics. Watch the animations on the Changing Earth module to find examples of each of these plate processes:

   - continents splitting apart at divergent boundaries
   - oceans shrinking as the ocean floor is consumed at subduction zones
   - continents colliding at convergent boundaries

2. Transform Faults, Earthquake Footage: What were the effects of the Loma Prieta earthquakes? How was this earthquake related to plate processes?

Large cubic pyrite crystals and small black sphalerite crystals.

CHAPTER
2

Atoms, Elements, and Minerals

This chapter is the first of six on the material of which the earth is made. The following chapters are mostly about rocks. Nearly all rocks are made of minerals. Therefore, to be ready to learn about rocks, you must first understand what minerals are as well as the characteristics of some of the most common minerals.

In this chapter, you are introduced to some basic principles of chemistry (this is for those of you who have not had a chemistry course). This will help you understand material covered in the chapters on rocks, weathering and the composition of the earth's crust and its interior. You will discover that each mineral is composed of specific chemical elements, the atoms of which are in a remarkably orderly arrangement. A mineral's chemistry and the architecture of its internal structure determine the physical properties used to distinguish it from other minerals. You should learn how to readily determine physical properties and use them to identify common minerals. (Appendix A is a further guide to identifying minerals.)

To launch you into the chapters on rocks that follow, we close this chapter by introducing you to a conceptual device called the rock cycle. We show one way in which plate tectonics relates to the rock cycle.

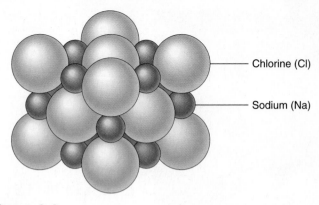

Figure 2.2

Model of the crystal structure of halite (or table salt).

5 mm

Feldspar
Quartz
Biotite

0.0000001 mm

Silicon and oxygen atoms in crystalline structure

=

Diagrammatic representation of crystalline structure

Figure 2.1

Specimen of granite showing the relationship among rock, minerals, and crystal structures of atoms. The diagrammatic representation as tetrahedrons is explained later in the chapter.

Rock is naturally formed, consolidated material composed of grains of one or more minerals (this definition has a few exceptions).

Figure 2.1 shows a specimen of the common rock *granite.* It forms from magma solidifying within the earth's crust. Granite is made up mostly of the minerals *feldspar* and *quartz.* What, then, is a mineral? A **mineral** is defined as a naturally occurring, inorganic, crystalline substance having a definite chemical composition. We will return to the first three parts of the definition later in this chapter after discussing the chemical composition of minerals. Quartz is made up exclusively of oxygen and silicon atoms. More precisely, quartz contains twice as many oxygen atoms as silicon atoms. Therefore, the chemical formula for quartz is SiO_2.

Rock salt is another rock, quite different from granite. It forms when salt water is evaporated. Unlike granite, it is made of grains of only one mineral. Rock salt is a consolidated aggre-

gate of grains of the mineral *halite*—you know it as table salt. Halite's definite chemical composition is indicated by its formula, NaCl. This means that it is composed of equal numbers of sodium (Na) and chlorine (Cl) atoms. These atoms are arranged in an orderly 3-dimensional pattern (meaning halite is crystalline). Each sodium atom is surrounded by six chlorine atoms and each chlorine atom is surrounded by six sodium atoms. Billions of each type of atom are necessary to form a salt crystal the size of a pinhead. Halite tends to form as cubic crystals because of the particular orderly way in which chlorine and sodium atoms are packed together (figure 2.2). This and other physical and chemical properties of halite are caused by (1) the pattern of repeating atoms, (2) the way the atoms are bonded to neighboring atoms, and (3) the characteristics of the elements chlorine and sodium.

Atoms and Elements

Halite (or table salt) crystals can be separated chemically into sodium and chlorine, both of which are elements. An **element** is a substance that cannot be broken down into other substances by ordinary chemical methods.

An **atom** is the smallest possible particle of an element that retains the properties of that element. All atoms of the element chlorine are essentially identical to all other chlorine atoms; the same is true for sodium. Water (H_2O) can be chemically broken down to the elements oxygen and hydrogen (two hydrogen atoms for every oxygen atom). When two hydrogen atoms and one oxygen atom are bonded, as in liquid water (see Box 2.1), they are part of a *molecule,* the smallest possible unit of a compound that has the properties of that substance.

Atoms are far too small to be seen with even the most powerful optical microscope. (Recently developed technology has produced electronic images, such as the one shown in Box 2.5.) Our concepts of atoms are really models. A *model* in science is an image—graphic, mathematical, or verbal—that is consistent with the known data.

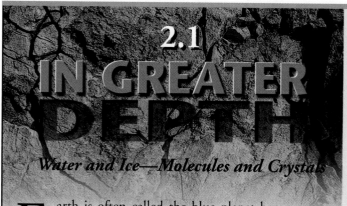

Water and Ice—Molecules and Crystals

Earth is often called the blue planet because oceans cover most of its surface. Ice dominates our planet's polar regions. It is fortunate that water is so abundant because life would be impossible without it. In fact, we humans are made up mostly of water. The nature and behavior of water molecules helps explain why water is vital to life on earth.

In a water molecule the two hydrogen atoms are tightly bonded to the oxygen atom. However, the shape of the molecule is asymmetrical, with the two hydrogen atoms on the same side of the atom (box figure 1). This means the molecule is polarized, with a slight excessive positive charge at the hydrogen side of the molecule and a slight excess negative charge at the opposite side. Because of the slight electrical attraction of water molecules, other substances are readily attracted to the molecules and dissolved or carried

Box 2.1 Figure 1

Water molecule.

away by water. Water has been called the universal solvent. Dirt washes out of clothing; water, in blood, carries nutrients to our muscles and transports waste to our kidneys and out of our bodies.

When water is in its liquid state, the molecules are moving about. However, because of the polarity, molecules are slightly attracted to one another. For this reason, water molecules are closer together than they are in most other liquids. However, in ice the water molecules are not as tightly packed together as in liquid water.

When water freezes, the bonding hydrogen atoms are shared between adjacent water molecules resulting in an orderly three-dimensional pattern that is hexagonal, like in a honeycomb. (This explains the hexagonal shape of snowflakes.) The openness of the honeycomb-like crystalline structure of ice contrasts with the more closely packed molecules in liquid water. For this reason, ice is less dense than liquid water. This is an unusual solid-liquid relationship. For most substances, the solid is denser than its liquid phase.

The fact that ice is less dense than liquid water has profound implications. Ice floats rather than sinks in liquid water. Icebergs float in the ocean. Lakes freeze from the top down. Ice on a lake surface acts as an insulating layer that retards the freezing of underlying water. If ice sank, lakes would freeze much more readily and thaw much more slowly. Our climate would be very different if ice sank. The Arctic Ocean surface freezes during the winter but only at its surface. If the ice were to sink, more ocean water would be exposed to the cold atmosphere and would freeze and sink. Eventually, the entire Arctic Ocean would freeze and would not thaw during the summer. If this were the case, life, as we know it, probably would not exist.

When water freezes, it expands. A bottled beverage placed in a freezer breaks its container upon freezing. When water trapped in cracks in rock freezes it will expand and will help break up the rock (as explained in the chapter on weathering).

Models of atoms constructed by chemists and physicists show three types of subatomic particles—protons, neutrons, and electrons. A **proton** is a subatomic particle that contributes mass and a single positive electrical charge to an atom. A **neutron** is a subatomic particle that contributes mass to an atom but is electrically neutral. An **electron** is a single negative electric charge that contributes a tiny percentage of mass to an atom. Electrons are regarded as moving very rapidly within specific energy levels, which are depicted as shells (figure 2.3A).

Protons and neutrons form the **nucleus** of an atom. Although the nucleus occupies an extremely tiny fraction of the volume of the entire atom, practically all the mass of the atom is concentrated in the nucleus. The **atomic mass number** is the total number of neutrons and protons in an atom. The atomic mass number of the oxygen atom in figure 2.3B is 16 (8 protons plus 8 neutrons). Heavier elements have more neutrons and protons than do lighter ones. For example, the heavy element gold has an atomic mass number of 197, whereas helium has 4.

The number of protons controls the "character" of an element more than does the number of other subatomic particles. The **atomic number** of an element is the number of protons

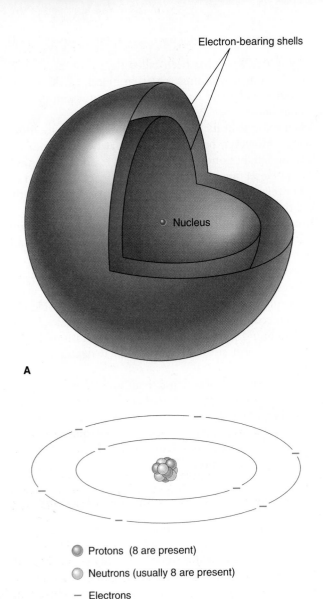

Nucleus

A

Protons (8 are present)

Neutrons (usually 8 are present)

— Electrons

B

Figure 2.3

(*A*) Model of an oxygen atom. The nucleus, composed of neutrons and protons, is actually much smaller than indicated relative to the volume of the atom. The hollow spheres represent the two electron-bearing shells. (*B*) Schematic representation of the oxygen atom. The two circles containing electrons represent the electron-bearing shells.

in each atom. We can refine our earlier definition of an element by adding that each atom of an element has the *same number of protons*. Gold has an atomic number of 79, or 79 protons per atom; oxygen always has 8 protons; hydrogen always has 1 proton; chlorine has 17; and sodium has 11. (Other atomic numbers are listed in Appendix C as well as in Appendix D, the periodic table of elements.)

The number of neutrons (and therefore the mass of an element) can vary within limits. **Isotopes** of an element are atoms *containing different numbers of neutrons* but the *same number of protons*. For example, the most common isotope of oxygen has

8 neutrons; far less abundant is the oxygen isotope with 10 neutrons. In geology, radioactive isotopes are important for determining how old rocks are (chapter 8).

An element's atomic weight is closely related to the mass number. **Atomic weight** is the weight of an *average* atom of an element, given in atomic mass units. Because sodium has only one naturally occurring isotope, its atomic mass number and its atomic weight are the same—23. On the other hand, chlorine has two common isotopes, with mass numbers of 35 and 37. The atomic weight of chlorine, which takes into account the abundance of each isotope, is 35.5 (the lighter isotope is more common than the heavier one).

Electrons, each having 1/1846 the mass of a proton, are visualized as occupying shells around the nucleus. Virtually the entire volume of an atom is taken up by the space in which these tiny, electrically negative charges move. The number of electrons in an atom is related to the number of protons in the nucleus. Each electron is opposite in charge but equal in strength to a proton.

If the number of electrons in orbit around a nucleus is equal to the number of protons in the nucleus, the positive and negative charges balance each other and the atom is electrically neutral. Helium has 2 electrons, exactly balancing its 2 protons. Most elements, however, are not able to maintain an electrical balance between protons and electrons within a single atom.

Chemical Activity

Many geological processes can be explained as chemical reactions. Some rocks form as a result of chemical reactions between substances. Chemical reactions also play a part in weathering (see chapter 5). Understanding a few basic concepts of chemistry will clarify why and under what conditions chemical reactions occur.

Atoms that are not electrically neutral tend to react (or combine) with other atoms to neutralize the electrical imbalance. Each atom not only seeks electrical neutrality but wants each of its shells to be full of electrons. The innermost shell is full when it possesses 2 electrons. Outer shells each require 8 electrons for an atom to be nonreactive (this is true for the second and third shells; elements having additional shells are more complicated).

Helium, for example, is a *stable* element because 2 protons are balanced by 2 electrons, and the 2 electrons exactly fill one shell. Neon is also stable; its 10 protons are balanced by 2 electrons in the inner shell and 8 electrons in the next shell (see figure 2.4). Normally, neither of these elements reacts with other elements.

Ions

Chlorine and sodium are more typical elements because if an electron shell is complete, the atom is electrically out of balance. Note that sodium in figure 2.5 has a complete inner shell with 2 electrons and a second shell, also filled, with 8 electrons. One more electron would neutralize all 11 protons in the

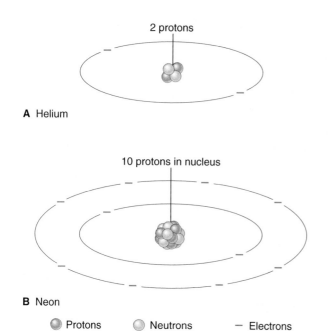

A Helium

10 protons in nucleus

B Neon

| ● Protons | ○ Neutrons | − Electrons |

Figure 2.4

(*A*) Helium atom and (*B*) neon atom.

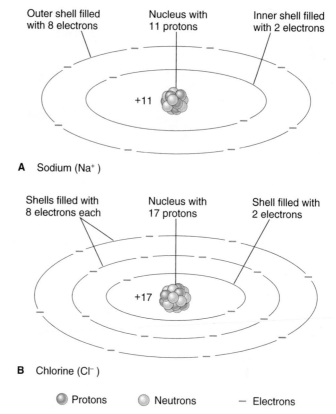

A Sodium (Na⁺)

B Chlorine (Cl⁻)

| ● Protons | ○ Neutrons | − Electrons |

Figure 2.5

(*A*) Sodium (Na^+) ion. Ten electrons fill two shells. The nucleus contains eleven protons. (*B*) Chlorine (Cl^-) ion. Electron shown in red completes the outer shell of the chlorine atom, making it an ion.

nucleus, but an eleventh electron alone in a shell is extremely unstable, so the sodium atom normally does without it. In each sodium atom, then, the 11 protons (11+) and 10 electrons (10−) add up to a single excess positive charge (+1). Such an atom is an **ion,** an electrically charged atom or group of atoms. The sodium ion can be abbreviated as Na^+.

Chlorine, with an atomic number of 17, has a complete inner shell with 2 electrons and a complete second shell of 8 electrons around this. A neutral chlorine atom would have only 7 electrons in the third shell, but this shell requires 8 electrons so an extra electron is captured and incorporated in it. The chlorine ion then contains 18 electrons and 17 protons, and so has a single excess negative charge (Cl^-).

Positive and negative ions are attracted to each other. In a crystalline structure the mutual attraction is one way atoms are held in place or bonded to one another. **Bonding** is the attachment of an atom to one or more adjacent atoms (see Box 2.2).

Chemical Composition of the Earth's Crust

Estimates of the chemical composition of the earth's crust are based on many chemical analyses of the rocks exposed on the earth's surface. (Models for the composition of the interior of the earth—the core and the mantle—are based on more indirect evidence.) Table 2.1 lists the generally accepted estimates of the abundance of elements in the earth's crust. At first glance, the chemical composition of the crust (and, therefore, the average rock) seems quite surprising.

We think of oxygen as the O_2 molecules in the air we breathe. Yet most rocks are composed largely of oxygen, as it is the most abundant element in the earth's crust. Unlike the oxy-

gen gas in air, oxygen in minerals is strongly bonded to other elements. By weight, oxygen accounts for almost half the crust, but it takes up 93% of the volume of an average rock. This is because oxygen's electron shells take up a large amount of space relative to its weight. It is not an exaggeration to regard the crust as a mass of oxygen with other elements occupying positions in crystalline structures between oxygen atoms.

Silica is a term for *oxygen combined with silicon.* Because silicon is the second most abundant element in the crust, most minerals contain silica. The common mineral quartz (SiO_2) is pure silica that has crystallized. Quartz is one of many minerals that are **silicates,** substances that contain silica (as indicated by their chemical formulas). Most silicate minerals also contain one or more other elements.

Note that the third most abundant element is aluminum, which is more common in rocks than iron. Knowing this, one might assume that aluminum would be less expensive than iron, but of course this is not the case. Common rocks are not mined for aluminum because it is so strongly bonded to oxygen and other elements. The amount of energy required to break these bonds and separate the aluminum makes the process too costly for commercial production. Aluminum is mined from the uncommon deposits where aluminum-bearing rocks have been weathered, producing compounds in which the crystalline bonds are not so strong.

IN GREATER DEPTH

Bonding

Ions may be regarded as tiny spheres that behave much like magnets. Positively charged ions attract negatively charged ions so that their electrical charges can be neutralized. In salt water, equal numbers of sodium ions (Na$^+$) and chlorine ions (Cl$^-$) move about freely. The electrical neutrality of the water is maintained because positive sodium ions exactly balance negative chlorine ions. If the water evaporates, the sodium and chlorine are electrically attracted to each other and crystallize into halite. The crystalline structure is the most orderly way for chlorine and sodium ions to pack themselves together and neutralize their collective charges.

A chlorine ion and a sodium ion are fixed in place by their electrical attraction to each other. This is called **ionic bonding** because it is brought about by an attraction between positively and negatively charged ions (box figure 1).

Ionic bonding is the most common type of bonding in minerals. However, in most minerals the bonds between atoms are not purely ionic. Atoms are also commonly bonded together by **covalent bonding,** or bonding in which adjacent atoms *share* electrons. Diamond is composed exclusively of covalently bonded carbon atoms (box figure 2). Carbon has an atomic number of 6, which means that the innermost shell is full with 2 electrons. Four more electrons are required to maintain electrical neutrality. In a diamond, each carbon atom has 4 electrons in the outer shell to maintain neutrality, while the need for 8 electrons in that shell is satisfied by electrons that are shared with adjacent carbon atoms. Neighboring carbon atoms are so close together that each of the outer-shell electrons spends half its time orbiting one atom and half orbiting an adjacent atom. Electrical neutrality is maintained, and each atom, in a sense, has 8 electrons in the outer shell (even though they are not all there at the same time). Covalent bonds in the diamond are extremely strong, and diamond is the hardest natural substance on earth. However, covalent bonds are not necessarily stronger than ionic bonds.

A third type of bonding, *metallic bonding,* is not as important to geology. In metals, such as iron or gold, the atoms are closely packed and the electrons move freely throughout the crystal so as to hold the atoms together. The ease with which electrons move accounts for the high electrical conductivity of metals.

Finally, residual forces (called van der Waal's forces) remaining after atoms have bonded together may result in very weak bonds, such as those that hold adjacent sheets of graphite or mica together.

Box 2.2 Figure 1

Ionic bonding between sodium (Na$^+$) and chlorine (Cl$^-$).

Box 2.2 Figure 2

Carbon atoms covalently bonded, as in diamond.

Collectively, the eight elements listed in table 2.1 account for more than 98% of the weight of the crust. All the other elements total only about 1.5%. Absent from the top eight elements are such vital elements as hydrogen (tenth by weight) and carbon (seventeenth by weight).

The element copper is only twenty-seventh in abundance, but our industrialized society is highly dependent on this metal. Most of the wiring in electronic equipment is copper, as are many of the telephone and power cables that crisscross the continent. However, the earth's crust is not homogeneous, and geological processes have created concentrations of elements such as copper in a few places. Exploration geologists are employed by mining companies to discover where (as well as why) ore deposits of copper and other metals occur (see chapter 21).

Crystallinity

By definition, a mineral must be crystalline. In fact, most solids are crystalline. A **crystalline** substance is one in which the atoms are arranged in a three-dimensional, regularly repeating, orderly pattern. The print by M.C. Escher (figure 2.6) vividly expresses the principle of crystallinity. You can visualize what crystallinity is

Table 2.1 — Crustal Abundance of Elements

Element	Symbol	Percentage by Weight	Percentage by Volume	Percentage of Atoms
Oxygen	O	46.6	93.8	60.5
Silicon	Si	27.7	0.9	20.5
Aluminum	Al	8.1	0.8	6.2
Iron	Fe	5.0	0.5	1.9
Calcium	Ca	3.6	1.0	1.9
Sodium	Na	2.8	1.2	2.5
Potassium	K	2.6	1.5	1.8
Magnesium	Mg	2.1	0.3	1.4
All other elements		1.5	—	3.3

in nature by mentally substituting identical clusters of atoms for each fish and imagining the clusters packed together.

As you can tell from figure 2.2, halite (or table salt) is crystalline. What controls the type (or "architecture") of crystal structure of a substance is principally the relative size of adjacent atoms. Thus, if all the sodium atoms in figure 2.2 were about the same size as the chlorine atoms, the particular crystal structure would be different (bearing in mind that neighboring atoms must "touch" one another). Liquids are not crystalline because their atoms or molecules are free to move about; nor is glass, because the atoms in glass are as randomly arranged as those in a liquid, only "frozen" into place. The structure of glass is comparable to what would happen if fish like those in the Escher drawing (figure 2.6) were swimming freely and randomly distributed when the water suddenly froze.

Of particular importance are the crystal structures derived from the two most common elements in the earth's crust—oxygen and silicon.

The Silicon-Oxygen Tetrahedron

The two most abundant elements of the crust, silicon and oxygen, combine to form the basic building block for most common minerals. In each "building block," four oxygen atoms are packed together around a single, much smaller, silicon atom, as shown in figure 2.7A. The four-sided, pyramidal, geometric shape called a *tetrahedron* is used to represent the four oxygen atoms surrounding a silicon atom. Each *corner* of the tetrahedron represents the *center* of an oxygen atom (figure 2.7B). This basic building block of a crystal is called a **silicon-oxygen tetrahedron** (also known as the silica tetrahedron).

The atoms of the tetrahedron are strongly bonded together. Within a silicon-oxygen tetrahedron the negative charges exceed the positive charges (see figures 2.7A and 2.8A). A single silicon-oxygen tetrahedron is a complex ion with a formula of SiO_4^{-4} because silicon has a charge of +4 and the four oxygen ions have 8 negative charges (–2 for each oxygen atom).

Figure 2.6

Depth, print by M. C. Escher.

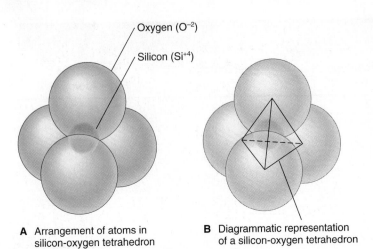

A Arrangement of atoms in silicon-oxygen tetrahedron

B Diagrammatic representation of a silicon-oxygen tetrahedron

Figure 2.7

(*A*) The silicon-oxygen tetrahedron. (*B*) The silicon-oxygen tetrahedron showing the corners of the tetrahedron coinciding with the centers of oxygen ions.

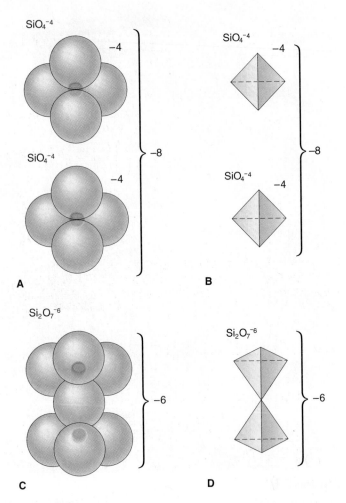

Figure 2.8

Two single tetrahedrons (*A* and *B*) require more positively charged ions to maintain electrical neutrality than two tetrahedrons sharing an oxygen atom (*C* and *D*). *B* and *D* are the schematic representations of *A* and *C* respectively.

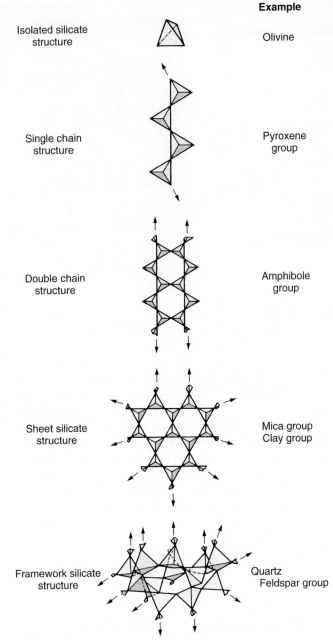

Figure 2.9

Common silicate structures. Arrows indicate directions in which structure repeats indefinitely.

For the silicon-oxygen tetrahedron to be stable within a crystal structure, it must either (1) be balanced by enough positively charged ions or (2) share oxygen atoms with adjacent tetrahedrons (as shown in figure 2.8*C* and *D*) and therefore reduce the need for extra, positively charged ions. The structures of silicate minerals range from an *isolated silicate structure*, which depends entirely on positively charged ions to hold the tetrahedrons together, to *framework silicates* (quartz, for example), in which all oxygen atoms are shared by adjacent tetrahedrons. The various types of silicate structures are shown diagrammatically in figure 2.9 and are discussed next.

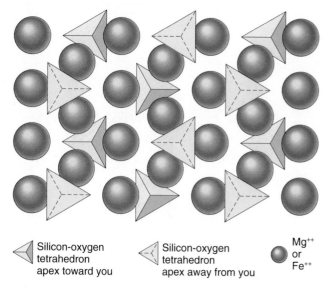

Figure 2.10

Diagram of the crystal structure of olivine, as seen from one side of the crystal.

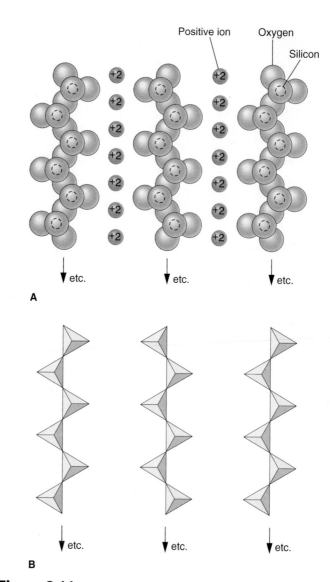

Figure 2.11

Single-chain silicate structure. (*A*) Model of a single-chain silicate mineral. (*B*) The same chain silicate shown diagrammatically as linked tetrahedrons; positive ions between the chains are not shown.

Isolated Silicate Structure

Silicate minerals that are structured so that none of the oxygen atoms are shared by tetrahedrons have an **isolated silicate structure.** The individual silicon-oxygen tetrahedrons are bonded together by positively charged ions (figure 2.10). The common mineral **olivine,** for example, contains two ions of either magnesium (Mg^{+2}) or iron (Fe^{+2}) for each silicon-oxygen tetrahedron. The formula for olivine is $(Fe, Mg)_2SiO_4$.

Chain Silicates

If two oxygen atoms of each tetrahedron are shared with adjacent tetrahedrons, the result may be a chain of tetrahedrons, a **chain silicate structure.** Each chain, which extends indefinitely, has a net excess of negative charges. The ratio of silicon to oxygen (as figure 2.11 shows) is 1:3; therefore, each mineral in this group (the *pyroxene* group) incorporates SiO_3^{-2} in its formula, and it must be electrically balanced by the positive ions that hold the parallel chains together.

One pyroxene mineral, for example, has a formula of $MgSiO_3$. This pyroxene may form in a cooling magma when earlier formed crystals of olivine, Mg_2SiO_4, react with silica (SiO_2) in the remaining melt (as discussed in the chapter on the origin of igneous rocks). To accommodate the additional silica, olivine's isolated silicate structure is rearranged into the single-chain silicate structure of pyroxene. In this case, Mg^{+2} ions occupy the positive ion positions between chains as shown in figure 2.11*A*.

The *amphibole* group is characterized by *two* parallel chains (double-chain silicate structure) in which every other

tetrahedron along a chain shares an oxygen ion with the adjacent chain (see figure 2.9). In even a small amphibole crystal, millions of parallel double chains are bonded together by positively charged ions.

Minerals with chain silicate structure tend to be shaped like columns, needles, or even fibers. The long structure of the external form corresponds to the linear dimension of the chain structure. Fibrous aggregates of minerals are called *asbestos* (see Box 2.3).

Sheet Silicates

When each tetrahedron shares three oxygen ions, the result is a **sheet silicate structure,** characteristic of the *mica* group and

2.3

ENVIRONMENTAL GEOLOGY

Asbestos—How Hazardous?

A *sbestos* is a generic name for fibrous aggregates of minerals (box figure 1). Because it does not ignite or melt in fire, asbestos has a number of valuable industrial applications. Woven into cloth, it may be used to make suits for firefighters. It can also be used as a fireproof insulation for homes and other buildings and has commonly been used in plaster for ceilings. Five of the six commercial varieties of asbestos are amphiboles, known commercially as "brown" and "blue" asbestos. The sixth variety is *chrysotile,* which is not a chain silicate and belongs to the *serpentine* family of minerals, and is more commonly known as "white asbestos." White asbestos is, by far, the most commonly used in North America (about 95% of that used in the United States).

Public fear of asbestos in the United States has resulted in its being virtually outlawed by the federal government. Tens of billions of dollars have been spent (probably unnecessarily) to remove or seal off asbestos from schools and other public buildings.

Asbestos' bad reputation comes from the high death rate among asbestos workers exposed, without protective attire, to extremely high levels of asbestos dust. Some of these workers, whose bodies were covered with fibers, were called "snowmen." In Manville, New Jersey, children would catch the "snow" (white asbestos dust released from the asbestos factory there) in their mouths. The high death rates among asbestos workers are attributed to *asbestosis* and lung cancer. Asbestosis is similar to silicosis contracted by miners; essentially, the lungs become clogged with asbestos dust after prolonged heavy exposure. The incidence of cancer has been especially high among asbestos workers who were also smokers. It's not clear that heavy exposure to white asbestos caused cancer among nonsmoking asbestos workers. However, brown and blue (amphibole) varieties, which are not mined for asbestos in North America, have been linked to cancer for heavy exposure (even if for a short term).

What are the hazards of asbestos to an individual in a building where walls or ceilings contain asbestos? Recent studies from a wide range of scientific disciplines indicate that the risks are minimal to nonexistent, at least for exposure to white asbestos. The largest asbestos mines in the

Box 2.3 Figure 1

Chrysotile asbestos.

Photo by C. C. Plummer

world are at Thetford Mines, Quebec. A study of longtime Thetford Mines residents, whose houses border the waste piles from the asbestos mines, indicated that their incidence of cancer was no higher than that of Canadians overall. Nor have studies in the United States been able to link nonoccupational exposure to asbestos and cancer. One estimate of the risk of death from cancer due to exposure to asbestos dust is one per 100,000 lifetimes. (Compare this to the risk of death from lightning of 4 per 100,000 lifetimes or automobile travel—1,600 deaths per 100,000 lifetimes.)

In California, a closed-down white asbestos mining site designated for Environmental Protection Agency Superfund cleanup is a short distance from where asbestos is being mined cleanly and efficiently. The asbestos is packaged and shipped to Japan. It cannot be used in the United States, because the U.S. is the only industrialized nation whose laws do not distinguish between asbestos types and permit the use of chrysotile.

A reason why chrysotile is less hazardous than amphibole asbestos is that chrysotile fibers will dissolve in lungs and amphibole will not. Recent experiments by scientists at Virginia Polytechnic Institute indicate that it takes about a year for chrysotile fibers to dissolve in lung fluids, whereas, glass fibers of the same size will dissolve only after several hundred years. Yet fiberglass is being used increasingly as a substitute for asbestos.

Further Reading

Alleman, J. E., and B. T. Mossman. 1997. Asbestos Revisited. *Scientific American,* v. 277, n. 1 (July 1997), pp. 70–75.

the *clay* group of minerals (figure 2.9). The positive ions that hold the sheets together in mica are "sandwiched" between the silicate sheets.

Framework Silicates

When all four oxygen ions are shared by adjacent tetrahedrons, a **framework silicate structure** is formed. Quartz is a framework silicate mineral. A feldspar is a framework silicate as well. However, its structure is slightly more complex because aluminum substitutes for some of the silicon atoms in some of the tetrahedrons. The same kind of substitution also takes place in amphiboles and micas, which helps account for the wide variety of silicate minerals.

Minerals

According to our earlier definition, four conditions must be satisfied for a substance to be a mineral:

1. It must be a *crystalline solid.*
2. It must *occur naturally.*
3. It must be *inorganic.*
4. It must have a *definite chemical composition.*

There are other meanings for the word *mineral* that conflict with the geologist's definition. The "minerals" listed on cereal boxes, for instance, have nothing to do with what geologists or chemists mean when they talk about minerals. Nor, for that matter, does the geologist's definition agree with the miner's definition of a mineral. To a miner, a "mineral" is anything of commercial value that is extracted from the ground.

In this book, the term *mineral* is used strictly in the geologic sense.

Crystalline Solids

We have already discussed crystallinity. In particular, we have described the various orderly arrangements of atoms for the silicate minerals. Nature is not always accommodating to definitions, however, and some substances exist that are not crystalline but otherwise meet the criteria for a mineral. These not-quite-minerals are called *mineraloids.* An example of a mineraloid is *opal,* which does not have a very orderly arrangement of its atoms.

Natural and Inorganic Substances

The second and third criteria need little explanation. Human-made crystalline compounds are not regarded as minerals. Nor are organic substances, which are compounds of carbon, oxygen, and hydrogen made by living organisms. Some organisms build shells or other hard parts that are considered minerals. Shells composed of the mineral *calcite* are common in sedimentary rocks.

Definite Chemical Composition

The fourth criterion, definite chemical composition, is to be expected because of the inherent orderliness of crystalline substances. Essentially, this means that chemical analysis of any sample of a given mineral will always produce the same ratios of elements (in quartz, for example, two atoms of oxygen for every atom of silicon). In other words, the composition of any mineral can be expressed as a chemical formula. Quartz and halite have simple formulas, SiO_2 and $NaCl$, respectively. Some formulas are more complex because atoms of one element may substitute for those of another element in a particular mineral. This is known as *solid solution.* For instance, we gave the formula for the mineral olivine as $(Fe, Mg)_2SiO_4$. Because magnesium and iron ions are about the same size, they can substitute freely for each other without distorting the crystal structure and significantly altering the properties of the mineral. Some chemical formulas appear much more complex than the crystal structures they represent because several such substitutions can occur.

Orderly variations in chemical compositions can exist within single crystals. This property is known as *zoning.* Zoning is very common in plagioclase feldspar (the crust's most abundant mineral, described later in this chapter). In plagioclase, calcium (Ca) and sodium (Na) can substitute for one another within the crystal structure (figure 2.12). Commonly, plagioclase in igneous rocks is zoned with the center of the crystal being high in Ca; toward the edges, the crystal is progressively richer in Na (and poorer in Ca). This is interpreted as indicating that the crystal started growing when crystallization of magma began at a high temperature. (High tempera-

Figure 2.12

Zoning in plagioclase feldspar, as seen under a polarizing microscope. The concentric color bands each indicate different amounts of Ca and Na in the crystal structure.

tures favor Ca and low temperatures favor Na in plagioclase.) As the magma cooled, the plagioclase crystal continued to grow, but the new layers of plagioclase incorporated progressively more Na and less Ca.

Chemical analysis can aid the identification of minerals. But, for a variety of reasons, including the cost and difficulty of chemical analysis, physical properties (which reflect chemical composition, as well as bonding and crystalline structure) are more generally used in identifying minerals. These properties are described later in this chapter.

The Important Minerals

It is useful to be able to associate the names of important minerals with the physical properties that identify them. Of course, what constitutes an "important" mineral depends on your perspective. To a miner or prospector, an important mineral is one that is commercially valuable (and is, by implication, relatively uncommon). A "rock hound" is interested in collecting any mineral that is pretty or unusual. A gemologist specializes in those varieties of minerals that are of gem quality (diamonds, emeralds, etc.). A mineralogist is a scientist who studies the chemistry and crystallographic structure of minerals. In this book, the minerals we regard as important are those that help us understand the nature of the earth. We are particularly interested in the *rock-forming minerals* because they make up most of the rocks of the earth's crust.

Of the several thousand identifiable minerals on earth, most are rare and not important to geology (many occur at only a single site on the globe). Only a few hundred are classified as rock-forming minerals. Even most of these are relatively uncommon in comparison with the few minerals that make up the vast bulk of the earth's crust. The five mineral groups listed in the upper third of table 2.2 account for well over 90% of the earth's crust. These are the minerals whose names recur most often in this book. All are silicates.

Quartz may be the only familiar name among the most common minerals, unless you have already had some exposure to geology. However, like people, each mineral has its own character or physical properties. As you become more familiar with them, they will become more than just strange names.

As shown in table 2.2, minerals with similar crystal structures and compositions are grouped under a common name. The most abundant group of minerals in the earth's crust is the **feldspar group.** Like quartz, the feldspars are framework silicates. However, aluminum has substituted for some of the silicon in the linked tetrahedrons (aluminum ions and silicon ions are close to the same size). In addition to silicon, aluminum, and oxygen, feldspars contain sodium, calcium, or potassium. Those that contain potassium are called **potassium feldspar.** (Two potassium feldspars that have subtle crystallographic differences are *microcline* and *orthoclase*.) If sodium or calcium (or both) are incorporated into the

Table 2.2	Minerals of the Earth's Crust	
Name	Chemical Composition	Type of Silicate Structure or Chemical Group
The most common rock-forming minerals. *(These make up more than 90% of the earth's crust.)*		
Feldspar group		
Plagioclase	Ca and Na Al silicate	Framework silicate
Potassium feldspar (orthoclase, microcline)	K Al silicate	Framework silicate
Pyroxene group (augite most common)	Fe, Mg silicate (some with Al, Na, Ca)	Single-chain silicate
Amphibole group (hornblende most common)	Complex Fe, Mg, Al silicate hydroxide	Double-chain silicate
Quartz	Silica	Framework silicate
Mica group		
Muscovite	K Al silicate hydroxide	Sheet silicate
Biotite	K Fe, Mg Al silicate hydroxide	Sheet silicate
Other common rock-forming minerals.		
Silicates		
Olivine (especially common in the mantle)	Mg, Fe silicate	Isolated silicate
Garnet group	Complex silicates	Isolated silicate
Clay minerals group (especially common at the earth's surface)	Complex Al silicate hydroxides	Sheet silicate
Nonsilicates		
Calcite	$CaCO_3$	Carbonate
Dolomite	$CaMg (CO_3)_2$	Carbonate
Gypsum	$CaSO_4 \cdot 2H_2O$	Sulfate
Much less common minerals of commercial value.		
Halite	NaCl	Chloride
Diamond	C	Native element
Gold	Au (gold)	Native element
Hematite	Iron oxide (Fe_2O_3)	Oxide
Magnetite	Iron oxide (Fe_3O_4)	Oxide
Chalcopyrite	Cu, Fe sulfide	Sulfide
Sphalerite	Zn sulfide	Sulfide
Galena	Pb sulfide	Sulfide

Clay Minerals That Swell

Clay minerals are very common at the earth's surface; they are a major component of soil. There are a great number of different clay minerals. What they all have in common is that they are sheet silicates. They differ by which ions hold sheets together and by the number of sheets "sandwiched" together. Surprisingly, some clay minerals are edible; some are used in the manufacturing of pills. *Kaolinite,* a clay mineral, is the main ingredient in Kaopectate, a remedy for upset stomachs. Popular fast-food chains use clay minerals as a thickener for shakes (you can tell which ones, because the chains do not call them "milk shakes"—they do not use milk).

Montmorillonite is one of the more interesting clay minerals. It is better known as *expansive clay* or *swelling clay.* If water is added to the montmorillonite, the water molecules are adsorbed into the spaces between silicate layers (box figure 1). This results in a large increase in volume, sometimes up to several hundred percent. The pressure generated can be up to 50,000 kilograms per square meter. This is sufficient to lift a good-sized building.

If a building is erected on expansive clay that subsequently gets wet, a portion of the building will be shoved upward. In all likelihood the building will break. Some people think that expansive soils have caused more damage than earthquakes and landslides combined.

On the other hand, swelling clays can be put to use. Montmorillonite, mixed with water, can be pumped into fractured rock or concrete. When the water is adsorbed, swelling clay expands to fill and seal the crack. The technique is particularly useful where dams have been built against fractured bedrock. Sealing the cracks with expansive clays ensures that water will stay in the reservoir behind the dam.

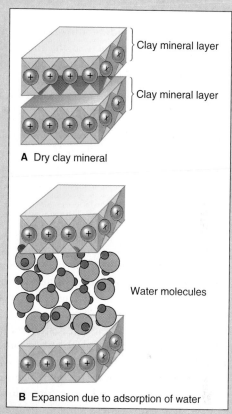

Box 2.4 Figure 1
Expansive clays.

---

feldspar crystal structure, then the mineral is a **plagioclase feldspar.**

The **pyroxene group** and the **amphibole group,** which are single- and double-chain silicates, respectively, each contain a number of minerals. *Augite* is the most common pyroxene, and *hornblende* is the most common amphibole.

The **mica group** is characterized by minerals with a sheet silicate structure. The two most common micas are biotite and muscovite. **Biotite** is a dark-colored, iron/magnesium-bearing mica. **Muscovite** mica lacks iron and magnesium and is transparent or white.

The **clay mineral group** is another group of sheet silicates (see Box 2.4). Clays are abundant on the earth's surface and in sedimentary rocks but make up only a minor percentage of the crust as a whole.

Nonsilicate minerals include *native elements,* which are minerals composed of only one element. Gold is a native element, as are diamond and graphite, both of which are composed solely of carbon. Other nonsilicates are classified according to the predominant negatively charged ions in their crystal structures. For instance, halite is a chloride because the negatively charged ions in the crystal are Cl^-. If the mineral contains CO_3^{-2} ions, it is a *carbonate. Sulfides* have S^{-2} ions, *sulfates* SO_4^{-2}, and *oxides* O^{-2} (but without Si, S, or C bonded to the oxygen atoms).

Nonsilicate minerals are also more abundant on the earth's surface than in the crust as a whole. **Calcite** (calcium carbonate, or $CaCO_3$) is the most common nonsilicate mineral and is usually found at or near the earth's surface. Limestone and marble are rocks composed mainly of calcite.

Ore minerals, or economic minerals, are minerals of commercial value; most are not silicates. Among the ore minerals are iron oxides (the minerals magnetite and hematite) mined for iron and a copper-iron sulfide (the mineral chalcopyrite) that is the main source of copper. Lead comes from galena (lead sulfide) and zinc is derived from sphalerite (a zinc sulfide).

The Physical Properties of Minerals

The best approach to understanding physical properties of minerals is to obtain a sample of each of the most common rock-forming minerals named in table 2.2. The properties described can then be identified in these samples.

To identify an unknown mineral, you should first determine its physical properties, then match the properties with the appropriate mineral, using a mineral identification key or chart such as the ones included in Appendix A of this book. With a bit of experience, you may get to know the few diagnostic tests for each common mineral and no longer need to refer to an identification table.

Color

The first thing most people notice about a mineral is its color. For some minerals, color is a useful property. Muscovite mica is white or colorless. Most of the **ferromagnesian minerals** (iron/magnesium-bearing), such as augite, hornblende, olivine, and biotite, are either green or black.

Because color is so obvious, beginning students tend to rely too heavily on it as a key to mineral identification. Unfortunately, color is also apt to be the most ambiguous of physical properties. If you look at a number of quartz crystals, for instance, you may find specimens that are white, pink, black, yellow, or purple. Color is extremely variable in quartz and many other minerals because even minute chemical impurities can strongly influence it. Obviously, it is poor procedure to attempt to identify quartz strictly on the basis of color.

Streak

A pulverized mineral gives a color, called a **streak,** that usually is more reliable than the color of the specimen itself. Scraping the edge of a mineral sample across an unglazed porcelain plate leaves a streak that may be diagnostic of the mineral. For instance, hematite always leaves a reddish brown streak though the sample may be brown or red or silver.

Unfortunately, few of the silicate minerals—the most common minerals—leave an identifying streak because most are harder than the porcelain streak plate.

Luster

The quality and intensity of *light* that is reflected from the surface of a mineral is termed **luster.** (A photograph cannot show this quality.) The luster of a mineral is described by comparing it to familiar substances.

Luster is either *metallic* or *nonmetallic*. A **metallic luster** gives a substance the appearance of being made of metal. Metallic luster may be very shiny, like a chrome car part, or less shiny, like the surface of a broken piece of iron.

Nonmetallic luster is more common. The most important type is **glassy** (also called **vitreous**) **luster,** which gives a substance a glazed appearance, like glass or porcelain. Most silicate minerals have this characteristic. The feldspars, quartz, the micas, and the pyroxenes and amphiboles all have a glassy luster.

Less common is an **earthy luster.** This resembles the surface of unglazed pottery and is characteristic of the various clay minerals. Some uncommon lusters include *resinous* luster (appearance of resin), *silky* luster, and *pearly* luster.

Hardness

The property of "scratchability," or **hardness,** can be tested fairly reliably. For a true test of hardness, the harder mineral or substance must be able to make a groove or scratch on a smooth, fresh surface of the softer mineral. For example, quartz can always scratch calcite or feldspar. Substances can be compared to **Mohs' hardness scale** (table 2.3), on which ten minerals are designated as standards of hardness. The softest mineral, talc (used for talcum powder because of its softness), is designated as 1. Diamond, the hardest natural substance on earth, is 10 on the scale.

Rather than carry samples of the ten standard minerals, a geologist doing fieldwork usually relies on common objects to test for hardness (table 2.3). A fingernail usually has a hardness of about $2\frac{1}{2}$. If you can scratch the smooth surface of a mineral with your fingernail, the hardness of the mineral must be less than $2\frac{1}{2}$ (figure 2.13). A copper coin or a penny has a hardness between 3 and 4; however, the brown oxidized surface of most pennies is much softer, so check for a groove into the coin. A knife blade or a steel nail generally has a hardness slightly greater than 5, but it depends on the particular steel alloy used. A geologist uses a knife blade to distinguish between softer minerals, such as calcite, and similarly appearing harder minerals, such as quartz. Ordinary window glass, usually slightly harder than a knife blade (although some glass, such as that containing lead, is much softer), can be used in the same way as a knife blade for hardness tests. A file (one made of tempered steel for filing metal, not a fingernail file) can be

Table 2.3	Mohs' Hardness Scale
1. Talc	6. Feldspar
	File
2. Gypsum	
Fingernail	7. Quartz
3. Calcite	
Copper coin	8. Topaz
4. Fluorite	9. Corundum
5. Apatite	10. Diamond
Knife blade	
Glass	

A

B

Figure 2.13

Fingernail (hardness of $2^1/_2$) easily scratches gypsum (hardness of 2).

used for a hardness of between 6 and 7. A porcelain streak plate also has a hardness of around $6^1/_2$.

External Crystal Form

The **crystal form** of a mineral is a set of faces that have a definite geometric relationship to one another. What most people call a "crystal" is a mineral with well-developed crystal faces. To geologists a *crystal* usually means any homogenous solid that is crystalline (with or without crystal faces).

If minerals were always able to develop their characteristic crystal forms, mineral identification would be a much simpler task. In rocks, however, most minerals grow while competing for space with other minerals. In fact, the orderly sets of faces that make up a crystal form can develop only under rather specialized conditions. Specifically, most minerals are able to develop their characteristic crystal faces only if they are surrounded by a fluid that can be easily displaced as the crystal grows. On the other hand, a few minerals, notably garnet, are able to overpower and displace surrounding solid material during growth, so that they almost always develop their characteristic crystal faces. Some common minerals with well-developed crystal forms are shown in figure 2.14.

Minerals displaying well-developed crystal faces have played an important role in the development of chemistry and physics. Steno, a Danish naturalist of the seventeenth century, first noted that the angle between two adjacent faces of quartz is always exactly the same, no matter what part of the world the quartz sample comes from or the color or size of the quartz. As shown in figure 2.15, the angle between any two adjacent sides of the six-sided "pillar" (which is called a prism by mineralogists) is always exactly 120°, while between a face of the "pillar" and one of the "pyramid" faces (actually part of a rhombohedron) the angle is always exactly 141°45′.

The discovery of such regularity in nature usually has profound implications. When minerals other than quartz were stud-

C

Figure 2.14

Crystal forms. (*A*) Quartz. (*B*) Feldspar. (*C*) Garnet in white quartz.

A   **B**

Figure 2.15

Quartz crystals showing how two interfacial angles remain the same in perfectly proportioned crystals (*A*) and misshapen (*B*).

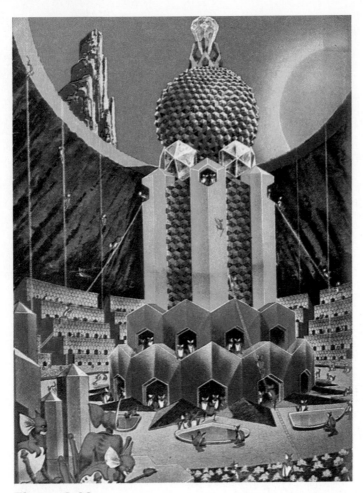

Figure 2.16

Quartz City on Mercury by science fiction artist Frank R. Paul. Crystals have been a recurring theme in science fiction. According to the story, this city on Mercury was built entirely of quartz crystals by insect men.

Amazing Stories, September 1941; reprinted by permission of Ultimate Publishing Company.

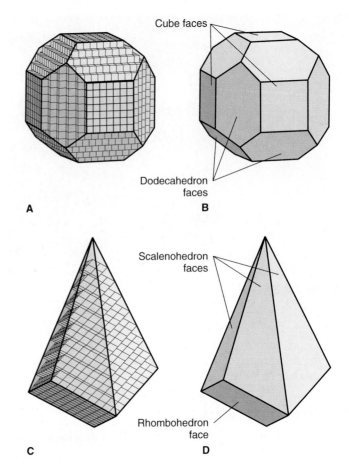

Figure 2.17

Geometric forms built by stacking cubes (*A* and *B*) and rhombohedrons (*C* and *D*). *A* and *C* are from a diagram published in 1801 by Haüy, a French mathematician. *A* and *B* show how cubes can be stacked for cubic and dodecahedral (12-sided) crystal forms. *C* and *D* show the relationship of stacked rhombohedrons to a "dog-tooth" (scalenohedron) form and a rhombohedral face.

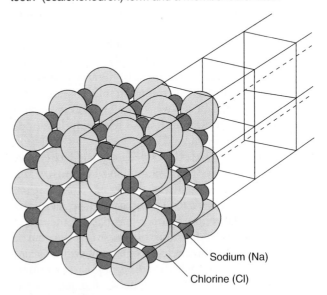

Figure 2.18

Relationship of cubes to part of a halite crystal. The crystal structure of halite can be represented as cubes stacked together in three dimensions.

ied, they too were found to have sets of angles for adjacent faces that never varied from sample to sample. This observation became formalized as the *law of constancy of interfacial angles.* Later the discovery of X-ray beams and their behavior in crystals confirmed Steno's theory about the structure of crystals.

Steno suspected that each type of mineral was composed of many tiny, identical building blocks, with the geometric shape of the crystal being a function of how these building blocks are put together. If you are stacking cubes, you can build a structure having only a limited variety of planar forms. Likewise, stacking rhombohedrons in three dimensions limits you to other geometric forms (figure 2.17).

Steno's law was really a precursor of atomic theory, developed centuries later. Our present concept of crystallinity is that atoms are clustered into geometric forms—cubes, bricks, hexagons, and so on—and that a crystal is essentially an orderly three-dimensional stacking of these tiny geometric forms (see Box 2.5). Halite, for example, may be regarded as a series of cubes stacked in three dimensions (figure 2.18). Because of the

Unit Cells and Crystal Systems

The cubes in figure 2.18 and rhombohedrons of figure 2.17 are known to crystallographers as *unit cells,* small volumes of identical shape, size, and orientation that collectively form the crystal. Each unit cell has a chemical composition identical to its neighboring unit cells.

Because a crystal is a three-dimensionally repeating pattern of atoms, each unit cell must share its faces and corners with adjoining unit cells (as shown in figure 2.17). Therefore, unit cells must be six-sided polyhedrons (such as cubes or rhombohedrons). The faces of a unit cell may be square, rectangular, or rhomboidal (see box figure 1). These restrictions result in only six basic shapes of unit cells being possible. The shape of a mineral's unit cell determines its *crystal system.* Halite, with its cubic unit cell, belongs in one system, whereas calcite, with its rhombohedral unit cells belongs in another. For minerals, the most common of the six crystal systems has a unit cell with two opposite sides being rhombs while the remaining sides are rectangles.

Unit cells cannot, of course, be seen. Their shape must be inferred by the symmetry of patterns that result from X rays beamed through the crystal (described later in this chapter). Each of the crystal systems will have a particular symmetry to distinguish it from the other systems. When a perfect crystal (such as those in figure 2.14) is found, a trained person can infer its crystal system by visually determining its symmetry.

Box 2.5 Figure 1

The surface atoms of silicon metal are shown magnified about ten million times by a scanning tunneling microscope. One side of a unit cell is outlined.

Courtesy IBM Corporation.

cubic "building block," its usual crystal form is a cube with crystal faces at 90° angles to each other.

Cleavage

The internal order of a crystal may be expressed externally by crystal faces, or it may be indicated by the mineral's tendency to split apart along certain preferred directions. **Cleavage** is the ability of a mineral to break, when struck, along preferred directions.

A mineral tends to break along certain planes because the bonding between atoms is weaker there. In quartz, the bonds are equally strong in all directions; therefore quartz has no cleavage. The micas, however, are easily split apart into sheets (figure 2.19). If we could look at the arrangement of atoms in the crystalline structure of micas, we would see that the individual silica tetrahedrons are strongly bonded to one another within each of the silicate sheets. The bonding *between* adjacent sheets, however, is very weak; therefore it is easy to pull the mineral apart parallel to the plane of the sheets.

Cleavage is one of the most useful diagnostic tools because it is identical for a given mineral from one sample to another. Cleavage is especially useful for identifying minerals when they are small grains in rocks.

The wide variety of combinations of cleavage and *quality* of cleavage also increases the diagnostic value of this property. Mica has a single direction of cleavage, and its quality is perfect. Other minerals are characterized by one, two, or more cleavage directions; the quality can range from perfect to poor (poor cleavage is very hard for anyone but a well-trained mineralogist to detect).

Three of the most common mineral groups—the feldspars, the amphiboles, and the pyroxenes—have two directions of cleavage (figure 2.20*B* and *C*). In feldspars, the two directions are at angles of about 90° to each other, and both directions are of very good quality. In pyroxenes, the two directions are also at about right angles, but the quality is only fair. In amphiboles (figure 2.21), the quality of the cleavage is very good and the two directions are at an angle of 56° (or 124° for the obtuse angle).

A

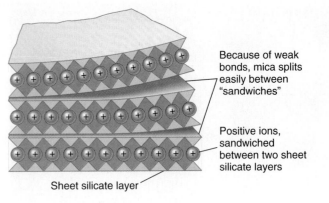

B

Because of weak bonds, mica splits easily between "sandwiches"

Positive ions, sandwiched between two sheet silicate layers

Sheet silicate layer

Figure 2.19

(*A*) Mica pulled apart along cleavage planes. (*B*) Relationship of mica to cleavage. Mica crystal structure is simplified in this diagram.

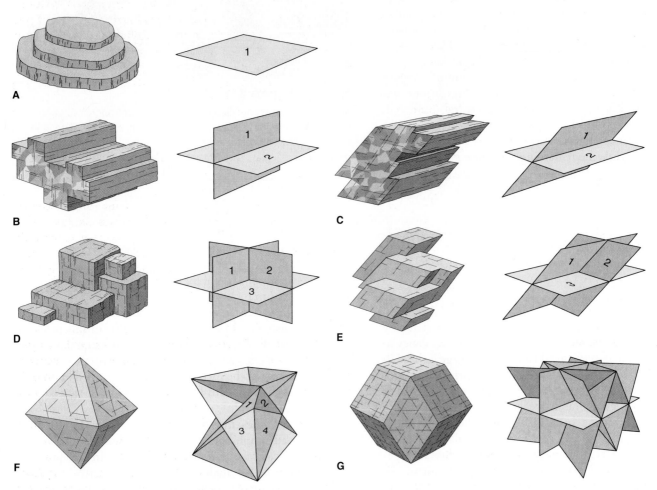

A

B

C

D

E

F

G

Figure 2.20

Possible types of mineral cleavage. (*A*) One direction of cleavage. (*B*) Two directions of cleavage that intersect at 90° angles. Feldspar is an example. (*C*) Two directions of cleavage that do not intersect at 90° angles. Amphibole is an example. (*D*) Three directions of cleavage that intersect at 90° angles. Halite is an example. (*E*) Three directions of cleavage that do not intersect at 90° angles. Calcite is an example. (*F*) Four directions of cleavage. Diamond is an example. (*G*) Six directions of cleavage. Sphalerite is an example.

Reprinted by permission from R. D. Dallmeyer, *Physical Geology Laboratory Manual,* Dubuque, Iowa: Kendall-Hunt Publishing Company, 1978.

Figure 2.21

Amphibole cleavage as seen under a microscope.

Figure 2.23

Conchoidal fracture in glass.

Figure 2.22

Cleavage fragments of calcite.

Halite is an example of a mineral with three excellent cleavage directions, all at 90° to each other. This is called *cubic cleavage* (figure 2.20*D*). Halite's cleavage tells us that the bonds are weak in the planes parallel to the cube faces shown in figure 2.18.

Calcite also has three cleavage directions, each excellent. But the angles between them are clearly not right angles. Calcite's cleavage is known as *rhombohedral* cleavage (figure2.20*E* and 2.22).

Some minerals have more than three directions of cleavage (figure 2.20*F* and *G*). Diamond has very good cleavage in four directions (ironically, the hardest natural substance on earth can be easily shattered into small cleavage fragments). Sphalerite, the principal ore of zinc, has six directions.

Recognizing cleavage and determining angular relationships between cleavage directions takes some practice. Students new to mineral identification tend to ignore cleavage because it is not as immediately apparent to the eye as color. But determining cleavage is frequently the key to identifying a mineral, so the small amount of practice needed to develop this skill is worthwhile.

Fracture

Fracture is the way a substance breaks where not controlled by cleavage. Minerals that have no cleavage commonly have an *irregular fracture*.

Some minerals break along curved fracture surfaces known as *conchoidal fractures* (figure 2.23). These look like the inside of a clam or conch shell. This type of fracture is commonly observed in quartz and garnet (but these minerals also show irregular fractures). Conchoidal fracture is particularly common in glass, including obsidian (volcanic glass).

Specific Gravity

It is easy to tell that a brick is heavier than a loaf of bread just by hefting each of them. The brick has a higher **density,** weight per given volume, than the bread. Density is commonly expressed as **specific gravity,** the ratio of a mass of a substance to the mass of an equal volume of water.

Liquid water has a specific gravity of 1. (Ice, being lighter, has a specific gravity of about 0.9.) Most of the common silicate minerals weigh about two and a half times as much as equal volumes of water: quartz has a specific gravity of 2.65; the feldspars range from 2.56 to 2.76. Special scales are needed to determine specific gravity precisely. However, a person can easily distinguish by hand very dense minerals such as galena (a lead sulfide with a specific gravity of 7.5) from the much less dense silicate minerals.

Gold, with a specific gravity of 19.3, is much heavier than galena. Because of its high density, gold can be collected by

Figure 2.24
Plagioclase striations.

Figure 2.25
Double refraction in calcite. Two images of the letters are seen through the transparent calcite crystal.

"panning." While the lighter clay and silt particles in the pan are sloshed out with the water, the gold dust lags behind in the bottom of the pan.

Other Properties

Properties that are useful in only a few instances include taste and smell. Halite obviously tastes salty; few other minerals have any taste at all. An "earthy" smell is characteristic of some clay minerals when they are moistened.

Plagioclase feldspar commonly exhibits **striations**—straight, parallel lines on the flat surfaces of one of the two cleavage directions (figure 2.24). The lines appear to be etched by a delicate scriber. In plagioclase, they are caused by a systematic change within the pattern of crystalline structure.

The mineral **magnetite** (an iron oxide) owes its name to its characteristic physical property of being attracted to a magnet. Where large bodies of magnetite are found in the earth's crust, compass needles point toward the magnetite body rather than to magnetic north. Airplanes navigating by compass have become lost because of the influence of large magnetite bodies. Some other minerals are weakly magnetic; their magnetism can only be detected by specialized magnetometers, similar to metal detectors in airports.

A clear crystal of calcite exhibits an unusual property. If you place transparent calcite over an image on paper, you will see two images (figure 2.25). This phenomenon is known as *double refraction* and is caused by light splitting into two components when it enters some crystalline materials.

Highly specialized equipment is needed to detect some properties. Perhaps most important are the characteristic

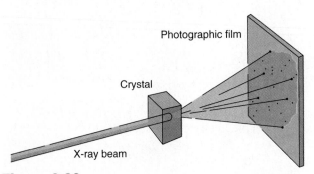

Figure 2.26
An X-ray beam passes through a crystal and is deflected by the atoms into a pattern of beams. The dots exposed on the film reveal the orderly pattern for the particular mineral.

effects of minerals on X rays, which we can explain only briefly here. X rays entering a crystalline substance are deflected by planes of atoms within the crystal. The X rays leave the crystal at precise and measurable angles controlled by the orientation of the planes of atoms that make up the internal crystalline structure (figure 2.26). The pattern of X rays exiting can be recorded on photographic film or by various recording instruments. Each mineral has its own pattern of reflected X rays, which serves as an identifying "fingerprint."

Simple Chemical Tests

One chemical reaction is routinely used for identifying minerals. The mineral calcite, as well as some other carbonate minerals

Chances are your watch says "quartz" on it. It may have been quite inexpensive, yet it keeps time much more accurately than even the most expensive spring-driven watch of a decade or more ago. What is a quartz watch, and why is it so much more accurate than the older kinds of watches?

Each quartz watch contains a thin wafer of quartz sliced along a particular crystallographic orientation. The quartz acts as a "pacemaker" to keep the hands or the electronic display precisely on time. The property of quartz that makes this work is called *piezoelectricity*, which means that pressure applied to a crystal creates an electric current (box figure 1*A*). Quartz is used to make gauges that measure high pressures (for instance, the water pressure on a deep-ocean submersible vessel). The greater the pressure, the stronger the electric current. With tension (outward pulling), the electric current flows in the opposite direction (box figure 1*B*). Electrons move one way in an electric circuit if the quartz crystal is under pressure and the other way if the quartz is under tension.

If you reverse the process and apply an electric current to a quartz crystal, the crystal changes shape slightly. More precisely, it expands and compresses (box figure 1*C*); and it will do so extremely rapidly and with remarkable regularity. The expansions and compressions are tiny vibrations that occur at the rate of about 100,000 vibrations per second. Thus for each minute that your watch runs, the quartz wafer in it shrinks and expands 6 million times (the electricity for this is supplied by the watch's battery). These vibrating crystals are so accurate that they are off by no more than one vibration out of 10 billion. Precision-manufactured quartz clocks used in observatories lose or gain no more than one second every ten years. Your watch is probably not that accurate

A Compression

B Tension—electrons flow in opposite direction

C Electric current applied to quartz crystal causes it to vibrate rapidly

Box 2.6 Figure 1
Piezoelectricity.

because of imperfections in the mechanical parts or the electronic circuitry. But even if your watch is a few seconds off each month, it is a vast improvement over the windup watches, even the finest of which are now obsolete.

(those containing CO_3^{-2}), reacts with a weak acid to produce carbon dioxide gas. In this test, a drop of dilute hydrochloric acid applied to the sample of calcite bubbles vigorously, indicating that CO_2 gas is being formed. Normally this is the only chemical test that geologists do during field research.

The Rock Cycle

With few exceptions, rocks are made of minerals. You will see how some minerals break down chemically and form new minerals when a rock finds itself in a new physical setting. For

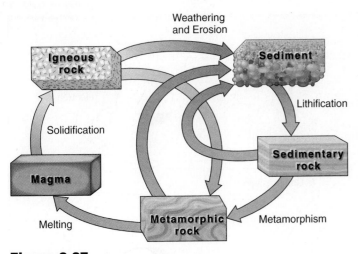

Figure 2.27

The rock cycle.

instance, feldspars that may have formed at high temperatures deep within the earth can react with surface waters to become clay minerals at the earth's surface.

As mentioned in chapter 1, the earth changes because of its internal and external heat engines. If the earth's internal engine had died (and tectonic forces had therefore stopped operating), the external engine plus gravity would long ago have leveled the continents, and the resulting sediment would have been deposited on the sea floor. Everything would be at rest. Nothing would be changing. That is to say, everything would be in *equilibrium* (and geology would be a dull subject). But this is not the case. The internal and external forces continue to interact, forcing substances out of equilibrium. Therefore, the earth has a highly varied and ever-changing surface. Minerals and rocks change as well.

A useful aid in visualizing these relationships is the **rock cycle** shown in figure 2.27. The three major rock types—igneous, metamorphic, and sedimentary—are shown. As you see, each may form at the expense of another if it is forced out of equilibrium with its physical or climatic environment by either internal or surficial forces.

As described in chapter 1, *magma* is molten rock. *Igneous rocks* form when magma solidifies. If the magma is brought to the surface by a volcanic eruption, it may solidify into an *extrusive* igneous rock. Magma may also solidify very slowly beneath the surface. The resulting *intrusive* igneous rock may be exposed later after uplift and erosion remove the overlying rock (as shown in figure 1.13). The igneous rock, being out of equilibrium, may then undergo *weathering* and *erosion,* and the debris produced is transported and ultimately deposited (usu-

ally on a sea floor) as *sediment.* If the unconsolidated sediment becomes *lithified* (cemented or otherwise consolidated into a rock), it becomes a *sedimentary rock.* As the rock is buried by additional layers of sediment and sedimentary rock, heat and pressure increase. Tectonic forces may also increase the temperature and pressure. If the temperature and pressure become high enough, usually at depths greater than several kilometers below the surface, the original sedimentary rock is no longer in equilibrium and recrystallizes. The new rock that forms is called a *metamorphic rock.* If the temperature gets very high, the rock melts and becomes magma again, completing the cycle.

The cycle can be repeated, as implied by the arrows in figure 2.27. However, there is no reason to expect all rocks to go through each step in the cycle. For instance, sedimentary rocks might be uplifted and exposed to weathering, creating new sediment.

A Plate Tectonic Example

One way of relating the rock cycle to plate tectonics is illustrated by an example from what happens at a convergent plate boundary (figure 2.28). *Magma* is created in the zone of melting above the subduction zone. The magma, being less dense than adjacent rock, works its way upward. A volcanic eruption takes place if magma reaches the surface. The magma solidifies into *igneous rock.* The igneous rock is exposed to the atmosphere and subjected to *weathering* and *erosion.* The resulting *sediment* is transported and then deposited in low-lying areas. In time, the buried layers of sediment solidify into *sedimentary rock.* The sedimentary rock becomes increasingly more deeply buried as more sediment accumulates. After the sedimentary rock is buried to depths of several kilometers, the heat and pressure become too great and the rock recrystallizes into a *metamorphic rock.* As the depth of burial becomes even greater (several tens of kilometers), the metamorphic rock may find itself in a zone of melting. Here, temperatures are high enough to partially melt the metamorphic rock and magma is created, thus completing the cycle.

We should emphasize that the rock cycle is a conceptual device to help students place the common rocks and how they form in perspective. As such it is a simplification and does not encompass all geologic processes. For instance (as you will learn from the next two chapters) most magma comes from the mantle, rather than from melting of metamorphic rocks of the earth's crust that had gone through the cycle.

The rock cycle diagram will reappear on the opening pages of chapters 3 through 7. The highlighted portion of the diagram will indicate where the material covered in each chapter fits into the rock cycle.

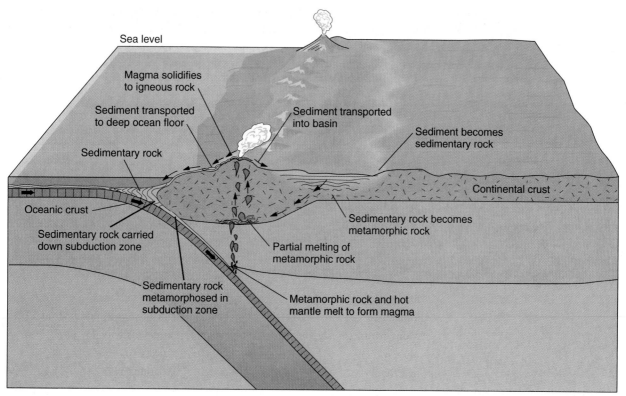

Figure 2.28

The rock cycle with respect to a convergent plate boundary. Magma solidifies as igneous rock at the volcano. Sediment from the eroded volcano collects in the basin to the right of the diagram. Sediment converts to sedimentary rock as it is buried by more sediment. Deeply buried sedimentary rocks are metamorphosed. The most deeply buried metamorphic rocks partially melt and the magma moves upward. An alternate way the rock cycle works is shown on the left of the diagram. Sediment from the continent (and volcano) becomes sedimentary rock, some of which is carried down the subduction zone. It is metamorphosed as it descends. Eventually, it contributes to the magma that forms at the top of the subduction zone.

Summary

Atoms are composed of *protons* (+), *neutrons,* and *electrons* (–). A given element always has the same number of protons. An atom in which the positive and negative electric charges do not balance is an *ion.*

Ions or atoms bond together in very orderly, three-dimensional structures that are *crystalline.*

A crystalline substance is considered a *mineral* (in geologic terms) if it is naturally occurring and inorganic and has a definite chemical composition.

The three most abundant elements in the earth's crust are oxygen, silicon, and alu-minum. Most minerals are silicates, with the silicon-oxygen tetrahedron as the basic building block.

Feldspars are the most common miner-als in the earth's crust. The next most abun-dant minerals are quartz, the pyroxenes, the amphiboles, and the micas. All are silicates.

Minerals are usually identified by their physical properties. Cleavage is perhaps the most useful physical property for identifica-tion purposes. Other important physical prop-erties are external crystal form, fracture, hardness, luster, color, streak, and specific gravity.

The interaction between the internal and external forces of the earth is illustrated by the rock cycle, a conceptual device relat-ing igneous, sedimentary, and metamorphic rocks to each other, to surficial processes such as weathering and erosion, and to inter-nal processes such as tectonic forces. Changes take place when one or more processes force earth's material out of equi-librium.

Terms to Remember

amphibole group 39
atom 28
atomic mass number 29
atomic number 29
atomic weight 30
biotite 39
bonding 31
calcite 39
chain silicate structure 35
clay mineral group 39
cleavage 43
covalent bonding 32
crystal form 41
crystalline 32
density 45
earthy luster 40
electron 29
element 28

feldspar group 38
ferromagnesian mineral 40
fracture 45
framework silicate structure 37
glassy (vitreous) luster 40
hardness 40
ion 31
ionic bonding 32
isolated silicate structure 35
isotope 30
luster 40
magnetite 46
metallic luster 40
mica group 39
mineral 28
Mohs' hardness scale 40
muscovite 39
neutron 29

nonmetallic luster 40
nucleus 29
olivine 35
ore mineral 40
plagioclase feldspar 39
potassium (orthoclase) feldspar 38
proton 29
pyroxene group 39
quartz 38
rock 28
rock cycle 48
sheet silicate structure 35
silica 31
silicates 31
silicon-oxygen tetrahedron 33
specific gravity 45
streak 40
striations 46

Testing Your Knowledge

Use the questions below to prepare for exams based on this chapter.

1. Compare feldspar and quartz.
   How do they differ chemically?
   What type of silicate structure does each have?
   How would you distinguish between them on the basis of cleavage?

2. How do the crystal structures of pyroxenes and amphiboles differ from one another?

3. How do the various feldspars differ from one another chemically?

4. Distinguish the following terms:
   silica   silicate
   silicon   silicon-oxygen tetrahedron

5. What is the distinction between cleavage and external crystal form?

6. How would you distinguish the following on the basis of physical properties? (You might refer to Appendix A.)
   feldspar/quartz
   muscovite/feldspar
   calcite/feldspar
   pyroxene/feldspar

7. Using triangles to represent tetrahedrons, start with a single triangle (to represent isolated silicate structure) and, by drawing more triangles, build on the triangle to show a single-chain silicate structure. By adding more triangles, convert that to a double-chain structure. Turn your double-chain structure into a sheet silicate structure.

8. What major factor controls chemical activity between atoms?

9. What are the three most common elements (by number and approximate percentage) in the earth's crust?

10. What are the next five most common elements?

11. A substance that cannot be broken down into other substances by ordinary chemical methods is a(n) (a) atom (b) element (c) molecule (d) compound

12. The subatomic particle that contributes mass and a single positive electrical charge is the (a) proton (b) neutron (c) electron

13. Atoms containing different numbers of neutrons but the same number of protons are called (a) compounds (b) ions (c) elements (d) isotopes

14. Atoms with either a positive or negative charge are called (a) compounds (b) ions (c) elements (d) isotopes

15. The most common type of chemical bonding in minerals is (a) ionic (b) covalent (c) metallic (d) van der Waal's forces

16. Which is not true of a single silicon-oxygen tetrahedron? (a) the atoms of the tetrahedron are strongly bonded together (b) it has a net negative charge (c) the formula is SiO_4 (d) it has four silicon atoms

17. Which is not a type of silicate structure? (a) isolated (b) single chain (c) double chain (d) sheet (e) framework (f) pentagonal

18. The most common mineral in the earth's crust is (a) quartz (b) feldspar (c) pyroxene (d) amphibole (e) biotite

19. On Mohs' hardness scale ordinary window glass has a hardness of about (a) 2–3 (b) 3–4 (c) 5–6 (d) 7–8

20. The ability of a mineral to break along preferred directions is called (a) fracture (b) crystal form (c) hardness (d) cleavage

21. Which is not a major rock type? (a) magmatic (b) igneous (c) metamorphic (d) sedimentary

22. Glass is (a) atoms randomly arranged (b) crystalline (c) ionically bonded (d) covalently bonded

23. Crystalline substances are always (a) ionically bonded (b) minerals (c) made of repeating patterns of atoms (d) made of glass

Expanding Your Knowledge

1. Why are nonsilicate minerals more common on the surface of the earth than within the crust?

2. How does oxygen in the atmosphere differ from oxygen in rocks and minerals?

3. Using Appendix A, can you make any generalization relating physical properties to chemical composition? (For instance, carbonate minerals tend to be softer than most silicate minerals.)

4. What happens to the atoms in water when it freezes? Is ice a mineral? Is a glacier a rock?

5. How would you expect the appearance of a rock high in iron and magnesium to differ from a rock with very little iron and magnesium?

Exploring Resources

Blackburn, W. H., and W. H. Dennen. 1993. *Principles of mineralogy.* 2d ed. Dubuque, Iowa: Wm. C. Brown Publishers.

Chesterman, C. W. 1978. *The Audubon Society field guide to North American rocks and minerals.* New York: Alfred A. Knopf.

Gaines, R. V., H. C. W. Skinner, E. E. Foord, B. Mason, and A. Rosenzweig. 1997. *Dana's new mineralogy,* 8th ed. Somerset, New Jersey: John Wiley & Sons.

Klein, C., and C. S. Hurlbut. 1993. *Manual of mineralogy (after J. D. Dana).* 21st ed. New York: John Wiley & Sons.

Prinz, M., G. Harlow, and J. Peters. 1978. *Simon & Schuster's guide to rocks and minerals.* New York: Simon & Schuster.

Rocks & Minerals. A bimonthly magazine mainly for professional and amateur mineral and rock collectors. Washington, D.C.: Heldref Publications.

Schumann, W. 1993. *Handbook of rocks, minerals and gemstones.* New York: Houghton Mifflin Co.

http://www.rockhounds.com/ Bob's Rock Shop. Contains a great amount of information for mineral collectors. Click on crystallography and mineral crystal systems for a more in-depth study of crystallography than presented in this book.

http://www.theimage.com/ Minerals: The Image. Click on Mineral Gallery for photos of minerals.

http://galaxy.einet.net/images/gems/ gems-icons.html Website for the Smithsonian Museum mineral and gem collection.

 The study of minerals. TASA Graphic Arts. For preview, visit this website: http://www.swcp.com/~tasa

Interacting with Journey Through Geology CD-ROM

Expand your knowledge of the concepts presented in this chapter by using the CD-ROM to answer the following questions.

1. Using the Pegmatite Pocket module of the CD-ROM, answer this question:

Pegmatites are composed of very large, often dramatic crystals. Why do these crystals grow so large?

2. Spreading Ridge module, Iceland, Eruption Footage: Think about the rock cycle as you watch Surtsey form. What portions of the rock cycle are captured in this video?

A geologist investigating intrusive rocks in northern Victoria Land, Antarctica.
Photo by C.C. Plummer

Igneous Rocks, Intrusive Activity, and the Origin of Igneous Rocks

*C*hapters 3 and 4 are about igneous rocks and igneous processes. (Either chapter may be read first.) Chapter 4 focuses on volcanoes and igneous activity that takes place at the earth's surface. Chapter 3 describes igneous processes that take place underground. However, you will learn early in this chapter how volcanic as well as intrusive rocks are classified based on their grain size and mineral content. The chapter continues with an explanation of the structural relationships between bodies of intrusive rock and other rocks in the earth's crust. This is followed by a discussion of how magmas form and are altered. We conclude by discussing various hypotheses that relate igneous activity to plate tectonic theory.

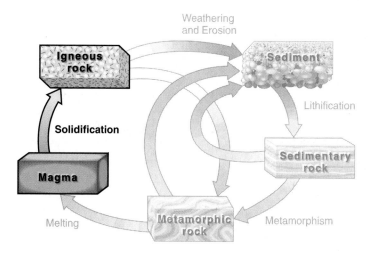

If you go to the island of Hawaii, you might observe red hot lava flowing over the land, and, as it cools, solidifying into the fine-grained, black rock we call basalt. Basalt is an **igneous rock,** rock that has solidified from magma. **Magma** is molten rock, usually rich in silica and containing dissolved gasses. (**Lava** is magma on the earth's surface.) Igneous rocks may be either **extrusive** if they form at the earth's surface (e.g., basalt) or **intrusive** if magma solidifies underground. **Granite,** a coarse-grained rock composed predominantly of feldspar and quartz, is an intrusive rock. In fact, granite is the most abundant intrusive rock found in the continents.

Unlike the volcanic rock in Hawaii, nobody has ever seen magma solidify into intrusive rock. So what evidence suggests that bodies of granite (and other intrusive rocks) solidified underground from magma? (1) Mineralogically and chemically, intrusive rocks are essentially identical to volcanic rocks. (2) Volcanic rocks are fine-grained (or glass) due to their rapid solidification; intrusive rocks are generally coarse-grained, which is inferred to mean that the magna crystallized slowly underground. (3) Experiments have confirmed that most of the minerals in these rocks can form only at high temperatures. Other experiments indicate that some of the minerals could have formed only under high pressures, implying they were deeply buried. More evidence comes from examining *intrusive contacts,* such as shown in figures 3.1 and 3.2. (A **contact** is a surface separating different rock types. Other types of contacts are described elsewhere in this book.) (4) Preexisting solid rock, *country rock,* appears to have been forcibly broken by an intruding liquid, with the magma flowing into the fractures that developed. **Country rock,** inciden-

Figure 3.2

Granite (light-colored rock) solidified from magma that intruded dark-colored country rock. Torres del Paine, Chile (see also this book's cover photo).

Photo by Kay Kepler

tally, is an accepted term for any older rock into which an igneous body intruded. (5) Close examination of the country rock immediately adjacent to the intrusive rock usually indicates that it appears "baked" (*metamorphosed* is the correct term) close to the contact with the intrusive rock. (6) Rock types of the country rock often match **xenoliths,** fragments of rock that are distinct from the body of igneous rocks in which they are enclosed. (7) In the intrusive rock adjacent to contacts with country rock are **chill zones,** finer-grained rocks that indicate magma solidified more quickly here because of the rapid loss of heat to cooler rock.

Although laboratory experiments have greatly increased our understanding of how igneous rocks form, geologists have not been able to create in the laboratory an artificial rock identical to granite. Only very fine-grained rocks containing the minerals of granite have been made from artificial magmas, or "melts." The temperature and pressure at which granite apparently forms can be duplicated in the laboratory—but not the time element. According to calculations, a large body of magma requires over a million years to solidify completely. This very gradual cooling causes the coarse-grained texture of most intrusive rocks. Chemical processes involving silicates are known to be exceedingly slow. Yet another problem in trying to apply experimental procedures to real rocks is determining the role of water vapor and other gases in the crystallization of rocks such as granite. Only a small amount of gases are retained in rock crystallized underground from a magma, but large amounts of gas (especially water vapor) are released during volcanic eruptions. No one has seen an intrusive rock forming; hence we can only speculate about the role these gases might have played before they escaped. One example shows why gases are important. Laboratory studies have shown that granite can melt at temperatures as low as 650°C if water

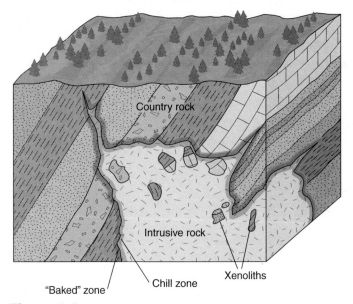

Figure 3.1

Igneous rock apparently intruded preexisting rock (country rock) as a liquid.

1mm.

A

B

Figure 3.3

(*A*) Coarse-grained texture characteristic of plutonic rock. Plagioclase is white and quartz is gray; potassium feldspar has been stained yellow. Small gradations on ruler are millimeters. (*B*) A similar rock seen through a polarizing microscope. Note the interlocking crystal grains of individual minerals.

Photos by C. C. Plummer

is present and under high pressure. Without the water, the melting temperature is several hundred degrees higher. Not knowing how much water was present during crystallization makes accurate determination of temperatures difficult and speculative.

Igneous Rocks

Extrusive rocks are **fine-grained rocks,** in which most of the grains are smaller than 1 millimeter. This is because magma cools rapidly at the earth's surface and large crystals do not have time to form. Some intrusive rocks are also fine-grained; these occur as smaller bodies that apparently solidified near the surface upon intrusion into relatively cold country rock (probably within a couple kilometers of the earth's surface). *Basalt, andesite,* and *rhyolite* are the common fine-grained igneous rocks. Igneous rocks that formed at considerable depth—usually more than several kilometers—are called **plutonic rocks** (after Pluto, the Roman god of the underworld). Characteristically, these rocks are coarse-grained, reflecting the slow cooling and solidification of magma. For our purposes, **coarse-grained rocks** are defined as those in which most of the grains are larger than 1 millimeter. The crystalline grains of plutonic rocks are commonly interlocked in a mosaic pattern (figure 3.3).

Some rocks are *porphyritic* and have larger crystals enclosed in a much finer-grained matrix (as described more fully in chapter 4). If the matrix is fine-grained, extrusive rock names are used. For instance, figure 3.5E shows a *porphyritic*

andesite. Porphyritic extrusive rocks are usually interpreted as having begun crystallizing slowly underground followed by eruption and rapid solidification of the remaining magma at the earth's surface.

Identification of Igneous Rocks

Igneous rock names are based on texture (notably grain size) and mineralogical composition (which reflects chemical composition). Mineralogically (and chemically) equivalent rocks are *granite-rhyolite, diorite-andesite,* and *gabbro-basalt.* The relationships between igneous rocks are shown in figure 3.4.

Because of their larger mineral grains, plutonic rocks are easier to identify than extrusive rocks. The physical properties of each mineral in a plutonic rock can be determined more readily. And, of course, knowing what minerals are present makes rock identification a simpler task. For instance, **gabbro** is formed of coarse-grained ferromagnesian minerals and gray, plagioclase feldspar. One can positively identify the feldspar on the basis of cleavage and, with practice, verify that no quartz is present. Gabbro's fine-grained counterpart is **basalt,** which is also composed of ferromagnesian minerals and plagioclase. However, the individual minerals cannot be identified by the naked eye and one must use the less reliable attribute of color—basalt is usually dark gray to black.

As you can see from figure 3.4, granite and **rhyolite** are composed predominantly of feldspars (usually white or pink) and quartz. Granite, being coarse-grained, can be positively identified by verifying that quartz is present. Rhyolite

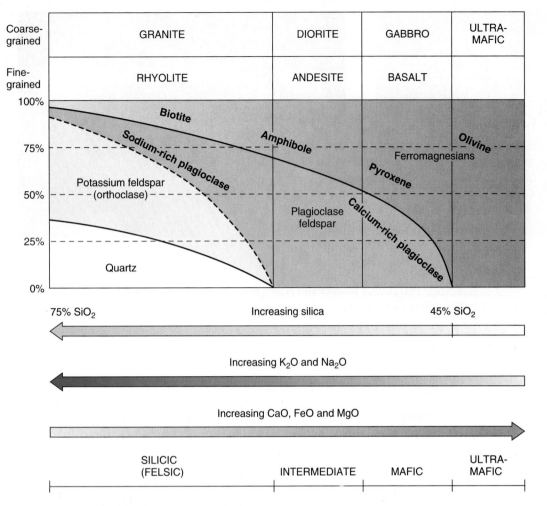

Figure 3.4

Classification chart for the most common igneous rocks. Rock names based on special textures are not shown. Sodium-rich plagioclase is associated with silicic rocks whereas calcium-rich plagioclase is associated with mafic rocks. The names of the particular ferromagnesian minerals indicate the approximate composition of the rocks they are most likely to be found in.

Table 3.1	Identification of Igneous Rocks			
Coarse-Grained	Granite	Diorite	Gabbro	Ultramafic Rocks
Fine-Grained	Rhyolite	Andesite	Basalt	—
Mineral Content	Quartz, feldspars (white, light gray, or pink). Minor ferromagnesian minerals.	Feldspars (white or gray) and about 35–50% ferromagnesian minerals. No quartz.	Predominance of ferromagnesian minerals. Rest of rock is plagioclase feldspar (medium to dark gray).	Entirely ferromagnesian minerals (usually olivine and pyroxene).
Color of Rock (most commonly)	Light-colored	Medium-gray or medium-green	Dark gray to black	Green to black

is usually cream-colored, tan, or pink. Its light color indicates that ferromagnesian minerals are not abundant. **Diorite** and **andesite** are composed of feldspars and significant amounts of ferromagnesian minerals (30–50%). The minerals can be identified and their percentages estimated to indicate diorite. Andesite, being fine-grained, can usually be identified by its medium-gray or medium-green color. Its appearance is intermediate between light-colored rhyolite and dark basalt.

Use the chart in figure 3.4 along with table 3.1 to identify common igneous rocks. You may also find it helpful to turn to Appendix B, which includes a key for identifying common igneous rocks. (Photos of typical igneous rocks are shown in figure 3.5.)

A

B

C

D

E

F

Figure 3.5

Igneous rocks. (*A*) Granite. (*B*) Diorite. (*C*) Gabbro. (*D*) Rhyolite. (*E*) Andesite. (*F*) Basalt.

3.1

IN GREATER DEPTH

Pegmatite—A Rock Made of Giant Crystals

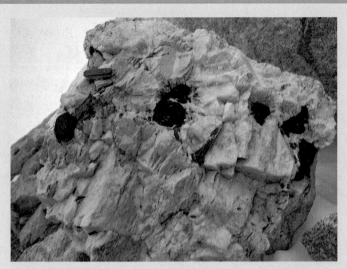

Box 3.1 Figure 1

Pegmatite in northern Victoria Land, Antarctica. The knife is 8 centimeters long. The black crystals are tourmaline. Quartz and feldspar are light colored.

Photo by C. C. Plummer

Pegmatites are extremely coarse-grained igneous rocks. In some pegmatites, crystals are as large as 10 meters across. Strictly speaking, a pegmatite can be of diorite, gabbro, or granite. However, the vast majority of pegmatites are felsic, with very large crystals of potassium feldspar, sodium-rich plagioclase feldspars, and quartz. Hence the term *pegmatite* generally refers to a rock of granitic composition (if otherwise, a term like *gabbroic pegmatite* is used). Pegmatites are interesting as geological phenomena and important as minable resources.

The extremely coarse texture of pegmatites is attributed to both slow cooling and the low viscosity of the fluid from which they form. Lava solidifying to rhyolite is very viscous. Magma solidifying to granite, being chemically similar, should be viscous.

Pegmatites, however, probably crystallize from a fluid composed largely of water under high pressure. Water molecules and ions from the parent, granitic magma make up a residual magma. Geologists believe the following sequence of events accounts for most pegmatites.

As a granite pluton cools, increasingly more of the magma solidifies into the minerals of a granite. By the time the pluton is well over 90% solid, the residual magma contains a very high amount of silica and ions of

Varieties of Granite

Granite and rhyolite occupy a larger area in the classification chart than do the other rocks. This reflects their greater variation in composition. For instance, a granite whose composition corresponds to the right side of the field in figure 3.4 (that is, nearer to diorite/andesite) contains much more plagioclase than potassium feldspar and somewhat more ferromagnesian minerals than does a rock whose composition plots in the left side of the granite field. Geologists have arbitrarily subdivided the field of granite and named each of the varieties; a rock in the right portion of the field, for example, is called granodiorite.

Any classification system is, of course, a human device, and for this reason, classification systems differ somewhat among groups of geologists. We define the boundary between granite and diorite by the presence or absence of quartz; but we could just as easily have placed the boundary slightly to the left, so that a rock with 10% or less quartz would be diorite.

Chemistry of Igneous Rocks

The chemical composition of the magma determines which minerals and how much of each will crystallize when an igneous rock forms. For instance, the presence of quartz in a rock indicates that the magma was enriched in silica (SiO_2).

The lower part of figure 3.4 shows the relationship of chemical composition to rock type. Chemical analyses of rocks are reported as weight percentages of oxides (e.g., SiO_2, MgO, Na_2O, etc.) rather than as separate elements (e.g., Si, O, Mg, Na). For virtually all igneous rocks, SiO_2 (silica) is the most abundant component. The amount of SiO_2 varies from about 45% to 75% of the total weight of common volcanic rocks. The variations between these extremes account for striking differences in the appearance and mineral content of the rocks.

Mafic Rocks

Rocks with a silica content close to 50% (by weight) are considered *silica-deficient,* even though SiO_2 is, by far, the most abundant constituent. Chemical analyses show that the remainder is composed mostly of the oxides of aluminum (Al_2O_3), calcium (CaO), magnesium (MgO), and iron (FeO and Fe_2O_3). (These oxides generally combine with SiO_2 to form the silicate minerals as described in chapter 2.) Rocks in this group are called **mafic**—silica-deficient igneous rocks with a relatively high content of magnesium, iron, and calcium. (The term *mafic* comes from magnesium and ferric.) Basalt and gabbro are, of course, mafic rocks.

elements that will crystallize into potassium and sodium feldspars. Also present are elements that could not be accommodated into the crystal structures of the common minerals that formed during the normal solidification phase of the pluton. Fluids, notably water, that were in the original magma are left over as well. If no fracture above the pluton permits the fluids to escape, they are sealed in, as in a pressure cooker. The watery residual magma has a low viscosity which allows appropriate atoms to migrate easily toward growing crystals. The crystals add more and more atoms and grow very large.

Pegmatite bodies are generally quite small. Many are podlike structures, located either within the upper portion of a granite pluton or within the overlying country rock near the contact with granite, the fluid body evidently having squeezed into the country rock before solidifying. Pegmatite dikes are fairly common, especially within granite plutons, where they apparently filled cracks that developed in the already solid granite. Some pegmatites form small dikes along contacts between granite and country rock, filling cracks that developed as the cooling granite pluton contracted.

Most pegmatites contain only quartz, feldspar, and perhaps mica. Minerals of considerable commercial value are found in a few pegmatites. Large crystals of muscovite mica are mined from pegmatites. These crystals are called "books" because the cleavage flakes (tens of centimeters across) look like pages. Because muscovite is an excellent insulator, the cleavage sheets are used in electrical devices, such as toasters, to separate uninsulated electrical wires. Even the large feldspar crystals in pegmatites are mined for various industrial uses, notably the manufacture of ceramics.

Many rare elements are mined from pegmatites. These elements were not absorbed by the minerals of the main pluton and so were concentrated in the residual pegmatitic magma, where they crystallized as constituents of unusual minerals. Minerals containing the element lithium are mined from pegmatites. Lithium becomes part of a sheet silicate structure to form a pink or purple variety of mica (called lepidolite). Uranium ores, similarly concentrated in the residual melt of magmas, are also extracted from pegmatites.

Some pegmatites are mined for gemstones. Emerald and aquamarine, varieties of the mineral beryl, occur in pegmatites that crystallized from a solution containing the element beryllium. A large number of the world's very rare minerals are found only in pegmatites, many of these in only one known pegmatite body. These rare minerals are mainly of interest to collectors and museums.

Hydrothermal veins (described in chapter 7) are closely related to pegmatites. Veins of quartz are common in country rock near granite. Many of these are believed to be caused by water that escapes from the magma. Silica dissolved in the very hot water cakes on the walls of cracks as the water cools while traveling surfaceward. Sometimes valuable metals such as gold, silver, lead, zinc, and copper are deposited with the quartz in veins.

Silicic (Felsic) Rocks

At the other extreme, the *silica-rich* (65% or more of SiO_2) rocks tend to have only very small amounts of the oxides of calcium, magnesium, and iron. The remaining 25% to 35% of these rocks is mostly aluminum oxide (Al_2O_3) and oxides of sodium (Na_2O) and potassium (K_2O). These are called **silicic** or **felsic rocks**—silica-rich igneous rocks with a relatively high content of potassium and sodium (the *fel* part of the name comes from *feldspar,* which crystallizes from the potassium, sodium, aluminum, and silicon oxides; *si* in *felsic* is for silica). The silicic rocks rhyolite and granite are light-colored because of the low amount of ferromagnesian minerals.

Intermediate Rocks

Rocks with a chemical content between that of felsic and mafic are classified as **intermediate rocks.** *Andesite,* which is usually green or medium gray, is the most common intermediate volcanic rock.

Ultramafic Rocks

An **ultramafic rock** is composed entirely or almost entirely of ferromagnesian minerals. No feldspars are present and, of course, no quartz. Most ultramafics are composed of coarse-grained pyroxene and/or olivine. Chemically, these rocks contain less than 45% silica.

Note that the chart (figure 3.4) does not include a fine-grained counterpart. This is because ultramafic extrusive rocks are restricted to the very early history of the earth and are quite rare. For our purposes they can be ignored.

Some ultramafic rocks form from differentiation (explained later in this chapter) of a basaltic magma at very high temperatures. Most ultramafic rocks come from the mantle, rather than from the earth's crust (see Box 17.2).

Where we find large bodies of ultramafic rocks, the usual interpretation is that a part of the mantle has traveled upward as solid rock.

Intrusive Bodies

Intrusions, or **intrusive structures,** are bodies of intrusive rock whose names are based on their size and shape, as well as their relationship to surrounding rocks. They are important aspects of the architecture, or *structure,* of the earth's crust. The various intrusions are named and classified on the basis of the

A

following considerations: (1) Is the body large or small? (2) Does it have a particular geometric shape? (3) Did the rock form at a considerable depth or was it a shallow intrusion? (4) Does it follow layering in the country rock or not?

Shallow Intrusive Structures

Some igneous bodies apparently solidified near the surface of the earth (at depths of probably less than 2 kilometers). These bodies appear to have solidified in the subsurface "plumbing systems" of volcanoes or lava flows. Shallow intrusive structures tend to be relatively small compared with those that formed at considerable depth. Because the country rock near the earth's surface generally is cool, intruded magma tends to chill and solidify relatively rapidly. Also, smaller magma bodies will cool faster than larger bodies, regardless of depth. For both these reasons, shallow intrusive bodies are likely to be fine-grained.

A **volcanic neck** is an intrusive structure apparently formed from magma that solidified within the throat of a volcano. One of the best examples is Ship Rock in New Mexico (figure 3.6). Here is how geologists interpret the history of this feature. A volcano formed above what is now Ship Rock. The magma for the volcano moved upward through a more or less cylindrical

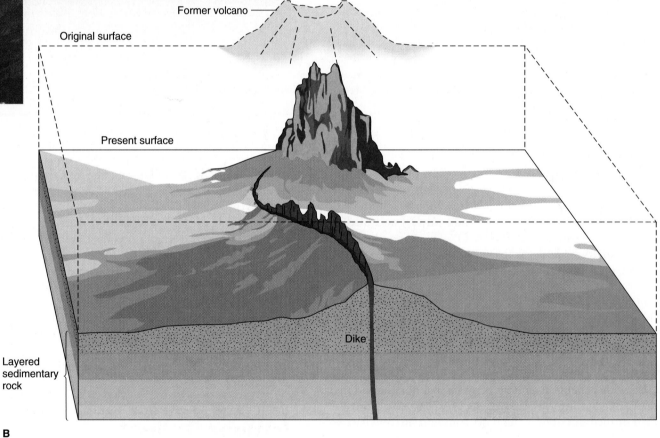

B

Figure 3.6

(*A*) Ship Rock in New Mexico, which rises 420 meters (1400 feet) above the desert floor. (*B*) Relationship to the former volcano.

Photo by Frank M. Hanna

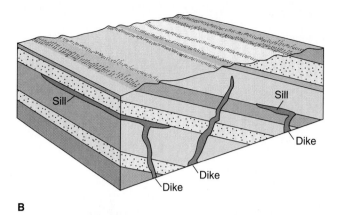

A   B

Figure 3.7

(*A*) Cracks or planes of weakness before intrusion of magma. (*B*) The concordant intrusion where magma has intruded between sedimentary layers is a sill; the discordant intrusions are dikes.

"plumbing system." Eruptions ceased and the magma underground solidified into what is now Ship Rock. In time, the volcano and its underlying rock—the country rock around Ship Rock—eroded away. The more resistant igneous body eroded more slowly into its present shape. Weathering and erosion are continuing (falling rock has been a serious hazard to rock climbers).

Dikes and Sills

Another, and far more common, intrusive structure can also be seen at Ship Rock. The low, wall-like ridge extending outward from Ship Rock is an eroded dike. A **dike** is a tabular (shaped like a tabletop), discordant, intrusive structure (figure 3.7). **Discordant** means that the body is not parallel to any layering in the country rock. (Think of a dike as cutting across layers of country rock.) Dikes may form at shallow depths and be fine-grained, such as those at Ship Rock, or form at greater depths and be coarser-grained. Dikes need not appear as walls protruding from the ground (figure 3.8). The ones at Ship Rock do so only because they are more resistant to weathering and erosion than the country rock.

Figure 3.8

Dikes (light-colored rocks) in northern Victoria Land, Antarctica.
Photo by C. C. Plummer

A **sill** is also a tabular intrusive structure, but it is **concordant**. That is, sills, unlike dikes, are parallel to any planes or layering in the country rock (figures 3.7 and 3.9). Typically, the country rock bounding a sill is layered sedimentary rock. As magma squeezes into a crack between two layers, it solidifies into a sill.

If the country rock is not layered, a tabular intrusion is regarded as a dike.

Intrusives That Crystallize at Depth

A **pluton** is a body of magma or igneous rock that crystallized at considerable depth within the crust. Most plutons have no particular shape, unlike dikes and sills. Where plutons are exposed at the earth's surface, they are arbitrarily distinguished by size. A **stock** is a small discordant pluton with an outcrop area (i.e., the area over which it is exposed to the atmosphere) of less than 100 square kilometers. If the outcrop area is greater than 100 square kilometers, the body is called a **batholith** (figure 3.10), a large discordant pluton.

Most batholiths crop out over areas vastly greater than the minimum 100 square kilometers.

Although batholiths often contain mafic and intermediate rocks, they almost always are predominantly composed of granite. Detailed studies of batholiths indicate that they are formed

Figure 3.9

Sill (dark rock) in sedimentary layered rock, Grand Canyon, Arizona.

of numerous, coalesced plutons. Apparently, large blobs of magma worked their way upward through the lower crust and collected 5 to 30 kilometers below the surface where they solidified (figure 3.11). These blobs of magma, known as **diapirs,** are less dense than the surrounding rock that is shouldered aside as the magma rises. Batholiths occupy large portions of North America, particularly in the west. Over half of California's Sierra Nevada mountains (figure 3.12 and map, figure 1.5) is a batholith whose individual plutons were emplaced during a period of over 100 million years. An even larger batholith extends almost the entire length of the mountain ranges of Canada's west coast and southeastern Alaska—a distance of 1,800 kilometers. Smaller batholiths are also found in the Appalachian Mountains in eastern North America. (The extent and location of North American batholiths are shown on the geologic map on the inside cover.)

Figure 3.10

(A) The first of numerous magma diapirs has worked its way upward and is emplaced in the country rock. (B) Other magma diapirs have intruded, coalesced, and solidified into a solid mass of plutonic rock. (C) After erosion, surface exposures of plutonic rock are a batholith and a stock.

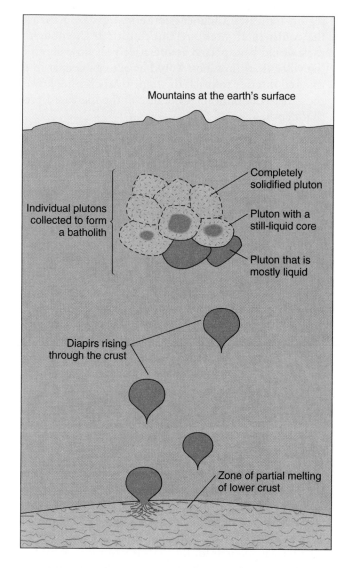

Mountains at the earth's surface

Individual plutons collected to form a batholith

Completely solidified pluton

Pluton with a still-liquid core

Pluton that is mostly liquid

Diapirs rising through the crust

Zone of partial melting of lower crust

Figure 3.11

Diapirs of magma travel upward from the lower crust and solidify in the upper crust.

Granite is considerably more common than rhyolite, its volcanic counterpart. Why is this? Silicic magma is much more *viscous* (that is, more resistant to flow) than mafic magma. Therefore, a silicic magma body will travel upward through the crust more slowly and with more difficulty than mafic or intermediate magma. Unless it is exceptionally hot a silicic magma will not be able to work its way through the relatively cool and rigid rocks of the upper few kilometers of crust. Instead, it is much more likely to solidify slowly into a pluton.

Abundance and Distribution of Plutonic Rocks

Granite is the most abundant igneous rock in mountain ranges. It is also the most commonly found igneous rock in the interior lowlands of continents. Throughout the lowlands of much of Canada, very old plutons have intruded even older metamorphic rock. As explained in chapter 20, very old mountain ranges have, over time, eroded and become the stable interior of a continent. Metamorphic and plutonic rocks similar in age and complexity to those in Canada are found in the Great Plains of the United States. Here, however, they are mostly covered by a veneer (a kilometer or so) of younger, sedimentary rock. These "basement" rocks are exposed to us in only a few places. In Grand Canyon, Arizona, the Colorado River has eroded through the layers of sedimentary rock to expose the ancient plutonic and metamorphic basement. In the Black Hills of South Dakota, local uplift and subsequent erosion has exposed similar rocks.

Granite, then, is the predominant igneous rock of the continents. As described in chapter 4, basalt and gabbro are the predominant rocks underlying the oceans. Andesite (usually along continental margins) is the building material of most young volcanic mountains. Underneath the crust, ultramafic rocks make up the upper mantle.

How Magma Forms

If a rock is heated sufficiently, it begins melting to form magma. Under ideal conditions, rock can melt and yield a granitic magma at temperatures as low as 625°C. Temperatures

Figure 3.12

Part of the Sierra Nevada batholith. All light-colored rock shown here (including that under the distant snow-covered mountains) is granite.

Photo by C. C. Plummer

Figure 3.13

Geothermal gradients at two parts of the earth's crust.

over 1,000°C are required to create basaltic magma. However, there are several factors that control the melting temperature of rock. Pressure, amount of gas (particularly water) present, and the particular mix of minerals all influence when melting takes place. These factors are discussed later in this chapter.

Heat for Melting Rock

Most of the heat that contributes to the generation of magma comes from the very hot earth's core (where temperatures are estimated to be greater than 5,000°C). Heat is conducted toward the earth's surface through the mantle and crust. This is comparable to the way heat is conducted through the wall from a hot room into a cooler room or through the metal of a frying pan. Heat is also brought from the lower mantle when part of the mantle flows upward, either through convection (described in chapter 1) or by hot mantle plumes. The geothermal gradient, described below, is a manifestation of heat transfer in the mantle.

Geothermal Gradient

A miner descending a mine shaft notices a rise in temperature. This is due to the **geothermal gradient,** the rate at which temperature increases with increasing depth beneath the surface. Data show the geothermal gradient, on the average, to be about 3°C for each 100 meters (30°C/km) of depth in the upper part of the crust. The geothermal gradient is not the same everywhere. Figure 3.13 shows geothermal gradients for two regions. The curve for the volcanic

region indicates a higher geothermal gradient than that for the continental interior. Temperatures high enough to melt rock would be expected at a relatively shallow depth beneath the volcanic region. You would have to go deeper in the continental interior to reach the same temperature; however, the rock there does not melt because of the increased pressure at that depth.

One reason for a higher geothermal gradient is that deeper, and therefore hotter, mantle rock has worked its way upward closer to the earth's surface due either to mantle convection or mantle plumes. "Hot spots" in the crust (where the geothermal gradient is locally very high) are believed to be caused by hot **mantle plumes,** which are narrow upwellings of hot material within the mantle. Hot mantle plumes help account for some igneous activity, such as the oceanic eruptions that built up the Hawaiian Islands. Volcanism in the middle of continents may also be attributable to mantle plumes. Yellowstone National Park is a product of silicic eruptions. The eruptions were much larger and more violent than any of historical time. Geologists attribute these eruptions to a hot mantle plume that caused melting of the crust beneath this area.

Factors That Control Melting Temperatures

Pressure

The melting point of a mineral generally *increases* with increasing pressure. Pressure increases with depth in the earth's crust, just as temperature does. So a rock that melts at a given temperature at the surface of the earth requires a higher temperature to melt deep underground. Rock will not melt where the geothermal gradient for the plate interior is applicable, because the melting temperature is always going to be higher than the temperature of rock at any given depth.

Hawaiian volcanic activity attributable to the underlying mantle plume illustrates how *reduced* pressure contributes to the creation of magma. Solid rock that was once very deep in the mantle (and, therefore, very hot) has worked its way upward. Most of its heat has been retained during the upward journey. However, the pressure decreases as the rock body travels upward. As it approaches 50 kilometers or so from the earth's surface, pressure is sufficiently reduced so that melting takes place.

Water Under Pressure

If enough gas, especially water vapor, is present and under high pressure, a dramatic change occurs in the melting process. Water vapor sealed in under high pressure by overlying rocks helps break down crystal structures. High water pressure can significantly lower the melting points of minerals. (Figure 3.14 shows the relationship between water pressure and the melting temperature of a plagioclase.)

3.2 ENVIRONMENTAL GEOLOGY

Harnessing Magmatic Energy

Buried magma chambers indirectly contribute the heat for today's geothermal electric generating plants. As explained in chapter 11 (Ground Water) water becomes heated in hot rocks. The heat source is usually presumed to be an underlying magma chamber. The rocks containing the hot water are penetrated by drilling. Steam exiting the hole is used to generate electricity.

Why not drill into and tap magma itself for energy? The amount of energy stored in a body of magma is enormous. The U.S. Geological Survey estimates that magma chambers in the United States within 10 kilometers of the earth's surface contain about 5,000 times as much energy as the country consumes each year. Our energy problems could largely be solved if significant amounts of this energy were harnessed.

There are some formidable technical difficulties in drilling into a magma chamber and converting the heat into useful energy. Despite these difficulties, the United States has considered developing magmatic energy. Experimental drilling has been carried out in Hawaii through the basalt crust of a lava lake that formed in 1960.

As drill bits approach a magma chamber, they must penetrate increasingly hotter rock. The drill bit must be made of special alloys to prevent it from becoming too soft to cut rock. The rock immediately adjacent to a basaltic magma chamber is around 1000°C, even though that rock is solid. Drilling into the magma would require a special technique. One device currently being experimented with is a jet-augmented drill. As the drill enters the magma chamber, it simultaneously cools and solidifies the magma in front of the drill bit. Thus the drill bit creates a column of rock that extends downward into the magma chamber and simultaneously bores a hole down the center of this column. Once the hollow column is deep enough within the magma chamber, a boiler is placed in the hole. The boiler is protected from the magma by the jacket of the column of rock. Water would be pumped down the hole and turned to water vapor in the boiler by heat from the magma. Steam emerging from the hole would be used to generate electricity.

In principle, the idea is fairly simple, but there are serious technical problems. For one thing, high pressures would have to be maintained on the drill bit during drilling and while the boiler system was being installed; otherwise, gases within the magma might blast the magma out of the drill hole and create a man-made volcano. (The closest thing to a man-made volcano occurred in Iceland when a small amount of magma broke into a geothermal steam well and erupted briefly at the well head, showering the area with a few tons of volcanic debris.)

Experiments have shown that, under moderately high pressure, water mixed with granite lowers the melting point of granite from over 900°C (when dry) to as low as 650°C when saturated with water under pressure equivalent to that of 10,000 atmospheres or *bars*. (Pressure at depth is usually expressed in *kilobars;* one kilobar is equal to 1,000 bars.)

Effect of Mixed Minerals

Two metals—as in solder—can be mixed in a ratio that lowers their melting temperature far below that of the melting points of

Figure 3.14

Melting temperature of a mineral relative to water pressure. If no water is present temperatures to the right of the red line are required to melt plagioclase at corresponding pressures. Note that if water pressure is zero, a temperature of 1,100°C is required for the mineral to melt, whereas with 2 kilobars of water pressure, the mineral will melt at temperatures just over 800°C.

After Tuttle and Brown, Geologic Society of America, 1958.

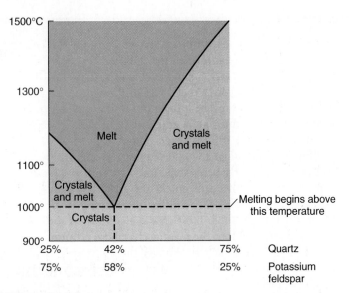

Figure 3.15

Melting temperatures for mixtures of quartz and potassium feldspar at atmospheric pressure.

Modified from Schairer and Bowen, 1956. V 254, p. 16. *American Journal of Science.*

the pure metals. Minerals behave similarly. Experiments have shown that in some cases mixed fragments of two minerals melt at a lower temperature than either mineral alone. Figure 3.15 shows the melting temperatures for quartz and potassium feldspar mixed in various proportions. If the mixture is 42% quartz (58% potassium feldspar), melting takes place at just above 1,000°C. On the other hand, 1,500°C is needed to liquify a mixture of 75% quartz and 25% potassium feldspar (corresponding to the right edge of the diagram). Pure quartz requires even higher temperatures to melt.

How Magmas of Different Compositions Evolve

A major topic of investigation for geologists is why igneous rocks are so varied in composition. On a global scale magma composition is clearly controlled by geologic setting. But why? Why are basaltic magmas associated with oceanic crust whereas granitic magmas are common in the continental crust? On a local scale igneous bodies often show considerable variation in rock type. For instance, individual plutons typically display a considerable range of compositions, mostly varieties of granite, but many also will contain minor

amounts of gabbro or diorite. In this section, we describe processes that result in differences in composition of magmas. The following section relates these processes to plate tectonics for the larger view of igneous activity.

Differentiation and Bowen's Reaction Theory

Differentiation is the process by which different ingredients separate from an originally homogenous mixture. An example is the separation of whole milk into cream and nonfat milk. In the early part of the twentieth century, N. L. Bowen conducted a series of laboratory experiments demonstrating that differentiation is a plausible way for silicic and mafic rocks to form from a single parent magma.

Bowen's reaction series, shown in figure 3.16, is the sequence in which minerals crystallize from a cooling magma, as demonstrated by Bowen's laboratory experiments. In simplest terms, Bowen's reaction series shows that those minerals with the highest melting temperatures crystallize from the cooling magma before those with lower melting points. However, the concept is a bit more complicated than that.

Crystallization begins along two branches, the *discontinuous branch* and the *continuous branch*. In the discontinuous branch, one mineral changes to another at discrete temperatures during cooling and solidification of the magma. Changes in the continuous branch occur gradationally through a range in temperatures and affect only the one

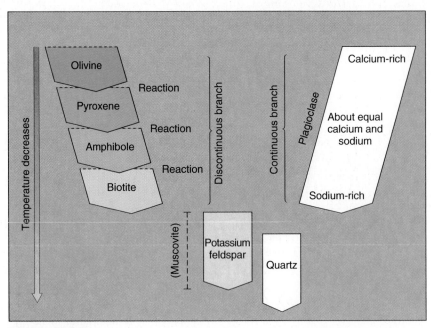

Figure 3.16

Bowen's reaction series.

mineral, plagioclase. Although crystallization takes place simultaneously along both branches, we must explain each separately.

Discontinuous Branch

All the minerals in the discontinuous branch are ferromagnesian. In this branch, as the completely liquid magma slowly cools, it reaches the temperature at which *olivine* begins to crystallize from the magma. Olivine is a mineral with an exceptionally high proportion (2:1) of iron and magnesium to silicon—its formula is $(Fe, Mg)_2SiO_4$. The liquid left after olivine has crystallized is relatively depleted in iron and magnesium and relatively enriched in silicon (because only one part of Si is used for two parts of Fe and Mg).

As the melt cools further, the melting temperature for the next mineral of the series is reached, and *pyroxene* begins to crystallize at olivine's expense. The previously formed olivine now *reacts* with the remaining melt, and the original crystal structure of olivine rearranges into that of pyroxene (from isolated silicon-oxygen tetrahedrons to single chains of tetrahedrons). The crystal structure of pyroxene, with a formula of $(Mg, Fe)SiO_3$, accommodates a higher amount of silicon relative to the iron and magnesium—a ratio of 1 to 1. After all of the olivine has reacted with the melt to form pyroxene, the temperature of the magma can decrease and pyroxene will crystallize directly from the melt.

However, if the original melt was basaltic, all of the liquid would likely be used up before all of the olivine has reacted with the melt. In this case the rock formed would have only olivine and pyroxene as its ferromagnesian minerals, which (along with the plagioclase that crystallized simultaneously in the continuous branch) would be a basalt. If, on the other hand, the original melt were more silicic or if early formed ferromagnesian minerals were removed from the melt, there would still be melt left after pyroxene crystallized and the next mineral (*amphibole*) could crystallize.

Assuming there is melt remaining when the crystallization temperature for amphibole is reached, pyroxene reacts with that melt. Its crystal structure rearranges into amphibole's double chains of silicon-oxygen tetrahedrons. More of the silicon leaves the melt (along with aluminum, calcium and minor amounts of sodium) and is incorporated into the newly developing amphibole crystals.

If melt is left after amphibole has formed, on further cooling amphibole reacts with the melt to produce *biotite* (which is a sheet silicate). Biotite is the last of the ferromagnesian minerals to crystallize. Any magma remaining after biotite has finished crystallizing contains very little iron or magnesium.

Continuous Branch

Plagioclase feldspar is the only mineral in the continuous branch. Silicon and aluminum, which are part of all feldspars, combine with calcium and sodium to form plagioclase. Calcium-rich plagioclase will crystallize first and, upon slow cooling, increasingly more sodic plagioclase will crystallize. If a basaltic melt, which is enriched in calcium relative to sodium, is cooled slowly, a very calcium-rich plagioclase will crystallize first. With progressive cooling, the plagioclase crystals react with the melt and grow larger. The growing plagioclase crystals will have an increasingly higher amount of sodium relative to calcium. Crystallization will stop when the plagioclase crystals have the same calcium-to-sodium ratio as did the original magma. In the case of a basaltic magma this will be at a fairly high temperature (approximately the temperature at which pyroxene crystallizes in the discontinuous branch). If there is a lower ratio of calcium to sodium (or if calcium-rich plagioclase is removed from the melt), plagioclase will continue to crystallize through lower temperatures.

Any magma left after the crystallization is completed along the two branches is richer in silicon than the original magma and also contains abundant potassium and aluminum. The potassium and aluminum combine with silicon to form *potassium feldspar*. (If the water pressure is high, *muscovite* may also form at this stage.) Excess SiO_2 crystallizes as *quartz*.

Normally a newly erupted cooling basalt lava progresses only a short distance down the reaction series before all the magma is consumed by growing crystals. Olivine develops, but only part of it reacts with the melt to form pyroxene before all the magma is solidified. Simultaneously, calcium-rich plagioclase grows and becomes increasingly sodic; but its growth ceases when all liquid is consumed. The rock becomes a completely solid aggregate of calcium-rich plagioclase, pyroxene, and olivine—in other words, what one expects to find in a basalt.

However, Bowen used this experimentally determined reaction series to support the hypothesis that all magmas (mafic, intermediate, and silicic) derive from a single parent (mafic) magma by differentiation. The early-developing minerals are separated from the remaining magma. These minerals collectively result in a rock that is more mafic than the original magma. The remaining magma is deficient in iron, magnesium, and calcium; therefore, upon cooling, it solidifies into a silicic or intermediate rock.

Crystal Settling

Only if the original basaltic magma cools slowly, and the earliest-formed minerals physically separate from the magma, can the minerals on the lower part of the reaction series crystallize. **Crystal settling** is the downward movement of minerals that are denser (heavier) than the magma from which they crystallized. What is pictured happening is that as the olivine crystallizes from the magma, the crystals settle to the

A

Chill zones

B

Accumulation of mafic minerals

C

Figure 3.17

Cross section showing the process of differentiation by crystal setting in a sill and dike. (*A*) Recently intruded magma is completely liquid. (*B*) Upon slow cooling, minerals such as olivine crystallize first. (*C*) The heavier crystals that formed early sink, leaving the remaining magma depleted in mafic constituents.

bottom of the magma chamber (figure 3.17). Calcium-rich plagioclase also separates as it forms. The remaining magma is, therefore, depleted of calcium, iron, and magnesium. Because these minerals were economical in using the relatively abundant silica, the remaining magma becomes richer in silica as well as in sodium and potassium. If enough mafic ingredients are removed in this manner, the remaining residue of magma eventually solidifies into a granite.

Undoubtedly this method of differentiation does take place in nature, though probably not to the extent that Bowen envisioned. The lowermost portions of some large sills are composed predominantly of olivine, whereas upper levels are considerably less mafic. Even in large sills, however, differentiation has rarely progressed far enough to produce any granite within the sill.

If we assume that the mafic minerals settle ever deeper in large magma bodies, there is still a problem in trying to explain the origin of granite by Bowen's theory. Calculations show that to produce a given volume of granite, about ten times as much mafic rock first has to form and settle out. If this is true, we would expect to find far more mafic plutonic rock than granite in the continental crust.

This is not to say that Bowen's work is discredited. Quite the opposite. His work has led to other theories on the behavior of magmas. Moreover, differentiation does occur and can explain relatively minor compositional variations within intrusive bodies, even if it does not satisfactorily explain the origin of large granite bodies.

Ore Deposits Due to Crystal Settling

Crystal settling accounts for important ore deposits that are mined for chromium and platinum. Most of the world's chromium and platinum come from a huge sill in South Africa. The sill, the famous Bushveldt Complex, is 8 kilometers thick and 500 kilometers long. Layers of chromite (a chromium-bearing mineral) up to 2 meters thick are found, and mined, at the base of the sill. Layers containing platinum overlie the chromite-rich layers.

Partial Melting

A granitic magma could be created by partial melting of rock—visualize a progression *upward* through part of Bowen's reaction series (going from cool to hot). As might be expected, the first portion of a rock to melt as temperatures rise forms a liquid with the chemical composition of quartz and potassium feldspar. The oxides of silicon plus potassium and aluminum "sweated out" of the solid rock could accumulate into a pocket of felsic magma. If higher temperatures prevailed, more mafic magmas would be created. Small pockets of magma could merge and form a large enough mass to rise as a diapir. The lower part of the continental crust is a very plausible source for felsic magma generated by partial melting.

Geologists generally regard basaltic magma (Hawaiian lava, for example) as the product of partial melting of ultramafic rock in the mantle, at temperatures hotter than those in the crust. The solid residue left behind in the mantle when the basaltic magma is removed is an even more silica-deficient ultramafic rock.

A

B

C

Figure 3.18

Assimilation. Magma formed is intermediate in composition between the original magma and the absorbed country rock. (A) Ascending magma breaks off blocks of country rock (the process is called *stoping*). (B) Xenoliths of country rock with melting temperatures lower than the magma melt. (C) The molten country rock blends with the original magma, leaving unmelted portions as inclusions.

Assimilation

A very hot magma may melt some of the country rock and *assimilate* the newly molten material into the magma (figure 3.18). This is like putting a few ice cubes into a cup of hot coffee. The ice melts and the coffee cools as it becomes diluted. Similarly, if a hot basaltic magma, perhaps generated from the mantle, melts portions of the continental crust, the magma

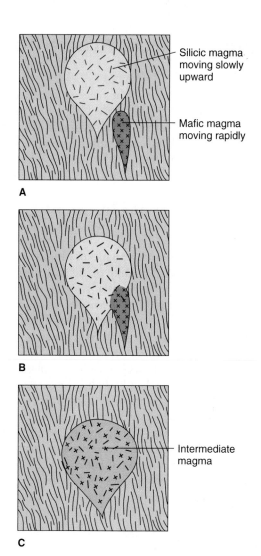

Figure 3.19

Mixing of magmas. (A) Two bodies of magma moving surfaceward. (B) The mafic magma catches up with the silicic magma. (C) The two magmas combine and become an intermediate magma.

simultaneously becomes richer in silica and cooler. Possibly intermediate magmas such as are associated with circum-Pacific andesite volcanoes may derive from assimilation of some crustal rocks by a basaltic magma.

Mixing of Magmas

The idea that some of our igneous rocks may be "cocktails" of different magmas is currently receiving more attention by geologists than it did in the past. The concept is quite simple. If two magmas meet and merge within the crust, the combined magma will be compositionally intermediate (figure 3.19). If you had approximately equal amounts of a granitic magma mixing with a basaltic magma, the resulting magma should crystallize underground as diorite or erupt on the surface to solidify as andesite.

Explaining Igneous Activity by Plate Tectonics

One of the appealing aspects of the theory of plate tectonics is that it accounts reasonably well for the variety of igneous rocks and their distribution patterns. Divergent boundaries are associated with creation of basalt and gabbro of the oceanic crust. Andesite and granite are associated with convergent boundaries.

Igneous Processes at Divergent Boundaries

Geologists agree that basaltic magma produced at divergent boundaries is due to partial melting of the asthenosphere. The *asthenosphere*, as described in chapter 1, is the plastic zone of the mantle beneath the rigid *lithosphere* (the upper mantle and crust that make up a plate). Along divergent boundaries, the asthenosphere is relatively close (5 to 10 kilometers) to the surface (figure 3.20).

The probable reason the asthenosphere is plastic or "soft" is that temperatures there are only slightly lower than the temperatures required for partial melting of mantle rock.

If extra heat is added, or pressure is reduced, partial melting should take place. The asthenosphere beneath divergent boundaries probably represents mantle material that has welled upward from deeper levels of the mantle. As the hot asthenosphere gets close to the surface, pressure is reduced sufficiently for partial melting. The resulting magma is mafic and will solidify as basalt or gabbro. The portion that did not melt remains behind as a silica-depleted, iron and magnesium enriched ultramafic rock.

Some of the basaltic magma erupts along a submarine ridge to form pillow basalts (described in chapter 4), while some fills near-surface fissures to create dikes. Deeper down, magma solidifies more slowly into gabbro. The newly solidified rock is pulled apart by spreading plates; more magma fills the new fracture and erupts on the sea floor. The process is repeated, resulting in a continuous production of mafic crust.

The basalt magma that builds the oceanic crust is removed from the underlying mantle, depleting the mantle beneath the ridge of much of its calcium, aluminum, and silicon oxides. The unmelted residue (olivine and pyroxene) becomes depleted mantle, but it is still a variety of ultramafic rock. The rigid ultramafic rock, the overlying gabbro and basalt, and any sediment that may have deposited on the basalt collectively are the lithosphere of an oceanic plate, which moves away from a spreading center over the asthenosphere. (The nature of the oceanic crust is described in more detail in chapter 18.)

Intraplate Igneous Activity

Hawaiian volcanism is unusual because Hawaii is not at a plate boundary. Most geologists think that basaltic magma there is created because the Pacific plate is overriding a hot mantle plume in a process described and illustrated in chapter 19.

Figure 3.20

Schematic representation of how basaltic oceanic crust and the underlying ultramafic mantle rock form at a diverging boundary. The process is more continuous than the two-step diagram implies. (*A*) Partial melting of asthenosphere takes place beneath a mid-oceanic ridge. (*B*) The magma squeezes into the fissure system. Solid mafic minerals are left behind as ultramafic rock.

The huge volume of mafic magma that erupted to form the Columbia plateau basalts (described in chapter 4) is attributed to a past hot mantle plume, according to a recent hypothesis (figure 3.21). In this case, the large volume of basalt is due to the arrival beneath the lithosphere of a mantle plume with a large head on it (sort of a mega-diapir).

The very explosive rhyolitic eruptions at Yellowstone National Park are also credited to a hot mantle plume that caused extensive partial melting of the continental crust. The resulting silicic magma bodies reached the surface (unlike most granitic magmas that solidify as plutons).

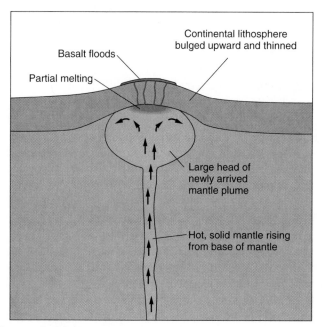

Figure 3.21

A hot mantle plume with a large head rises from the lower mantle. When it reaches the base of the lithosphere it uplifts and stretches the overlying lithosphere. The reduced pressure results in partial melting, producing basaltic magma. Large volumes of magma travel through fissures and flood the earth's surface.

Igneous Processes at Convergent Boundaries

Intermediate and silicic magmas are clearly related to the convergence of two plates and subduction. However, exactly what takes place is debated by geologists. Compared to divergent boundaries, there is much less agreement about how magmas are generated at converging boundaries. The scenarios that follow are currently believed by geologists to be the best explanations of the data.

The Origin of Andesite

Magma for most of our andesitic composite volcanoes (such as are found along the west coast of the Americas) seems to originate from a depth of about 100 kilometers. This coincides with the depth at which we would expect the subducted oceanic plate to slide under the asthenosphere (figure 3.22). Partial melting of the asthenosphere takes place, resulting in a mafic magma. A plausible reason for the melting to occur is that the subducted oceanic crust releases water into the asthenosphere. The water would have collected in the oceanic crust when it was beneath the ocean and is driven out because of heating of the descending plate. The water lowers the melting temperature of the ultramafic rocks in this part of the mantle. Partial melting produces a mafic magma. On its slow journey through the crust, the mafic magma evolves into an intermediate magma by differentiation, assimilation of silicic crustal rocks, and by magma mixing (see box 3.3).

The Origin of Granite

To explain the great volumes of granitic plutonic rocks, most geologists think that partial melting of the lower continental crust must take place. The continental crust contains the high amount of silica needed for a silicic magma. As the silicic rocks of the continental crust have relatively low melting temperatures (especially if water is present), partial melting of the lower continental crust is likely. Currently, geologists think that *magmatic underplating* by andesitic or basaltic magma

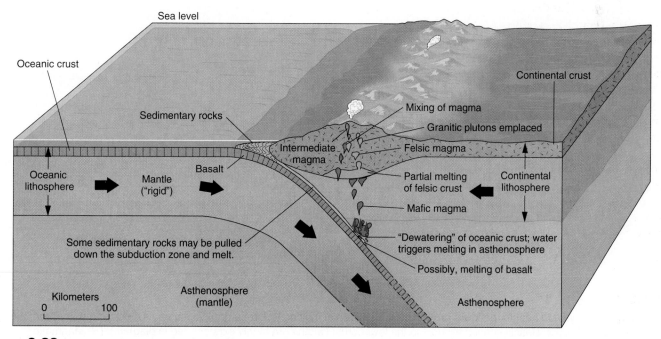

Figure 3.22

Some processes that may contribute to magma generation at a convergent boundary.

Geologists do not all agree on how intermediate and silicic magmas are generated at convergent boundaries. Opinions change as scientists analyze and weigh a wide range of complex data. Below are some alternatives that have had varying degrees of support from geologists.

Partial Melting of Basalt

Some geologists think that partial melting of basalt (as opposed to mantle) takes place. As the slab of oceanic lithosphere descends, the rock gets hotter (figure 3.22). Partial melting of basalt produces a magma that is less mafic than basalt. The resulting magma is likely to be intermediate. The magma would travel through the lithosphere to form composite volcanoes. However, because this process, at best, could create only a minor amount of granitic melt,

it is not regarded as the source of magma for the great granite batholiths.

Assimilation of Crustal Rocks

Rocks of the overlying continental crust are more silicic than those of the descending oceanic plate. A mafic magma is generated, either in the asthenosphere or from oceanic crust. While moving upward through continental crust, the more mafic magma absorbs some of the more silica-rich rocks, becoming intermediate in composition.

Mixing of Magmas

Intermediate magmas that form composite volcanoes could be products of magma mixing. Mafic magma working its way from the asthenosphere through the crust could mix with silicic magma generated at the base of the crust. The resulting intermediate magma would reach the surface to erupt as andesite, forming our large, composite volcanoes.

Melting of Sedimentary Rock

The sedimentary rock that rests on the basalt of the descending oceanic plate is much richer in silica than the basalt itself. The sedimentary rock therefore melts at a lower temperature than the basalt and could form a silicic magma. (This alone could probably not account for the huge volume of granitic batholiths in the continental crust.) It could also mix with mafic magma generated from the asthenosphere or from the basalt to form an intermediate magma.

A

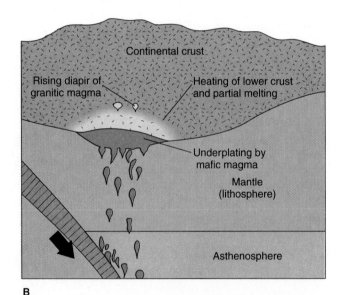

B

Figure 3.23

How mafic magma could add heat to the lower crust and result in partial melting to form a granitic magma. (*A*) Mafic magma from the asthenosphere rises through closely spaced fissures in the lower crust (widths are highly exaggerated in diagram). (*B*) Magmatic underplating of the continental crust.

plays an important role providing the extra heat source needed to generate granitic magmas in the lower continental crust. Initially, some of the mafic magma coming from the asthenosphere works its way through fissures systems in the lower crust (figure 3.23A). As the lower crust gets hotter, the rock becomes more plastic and melting begins. Fissures are sealed. The denser, mafic magma then pools under (underplates) the lighter, partially molten, lowest crust (figure 3.23B). Heat from the cooling and crystallizing mafic magma is conducted upward to create larger volumes of silicic magma by partially melting more of the continental crust. The silicic magma, in turn, separates from its solid residue and works its way upward in diapirs to a higher level of the crust where it slowly solidifies to a pluton, usually as part of a batholith.

We should emphasize that the picture we have presented is not an observation but a reasoned interpretation of available data. There are a number of variations of this picture based on different interpretations of the data (box 3.3). Plate tectonics has not provided final answers to the problems raised by analyzing igneous phenomena. But it has provided a broad framework in which to work toward solutions to the problems.

Summary

Igneous rocks form from solidification of magma. If the rock forms at the earth's surface it is extrusive. Intrusive rocks are igneous rocks that formed underground. Some intrusive rocks have solidified near the surface as a direct result of volcanic activity. Volcanic necks solidified within volcanoes. Fine-grained dikes and sills may also have formed in cracks during local extrusive activity. A sill is concordant—parallel to the planes within the country rock. A dike is discordant. Both are tabular bodies. Coarser grains in either a dike or a sill indicate that it probably formed at considerable depth.

Most intrusive rock is plutonic—that is, coarse-grained rock that solidified slowly at considerable depth. Most plutonic rock exposed at the earth's surface is in batholiths—large plutonic bodies with no particular shape. A smaller, irregular body is called a stock.

Silicic (or felsic) rocks are rich in silica, whereas mafic rocks are silica deficient. Most igneous rocks are named on the basis of their mineral content, which in turn reflects the chemical composition of the magmas from which they formed, and on grain sizes. Granite, diorite, and gabbro are the coarse-grained equivalents of rhyolite, andesite, and basalt, respectively. Ultramafic rocks are made entirely of ferromagnesian minerals and are mostly associated with the mantle.

Basalt and gabbro are strongly predominant in the oceanic crust. Granite strongly predominates in the continental crust. Younger granite batholiths occur mostly within younger mountain belts. Andesite is largely restricted to narrow zones along convergent plate boundaries.

The geothermal gradient is the increase in temperature with increase in depth. Hot mantle plumes, from the lower mantle, and magma at shallow depths in volcanic regions locally raise the geothermal gradient.

No single process can satisfactorily account for all igneous rocks. Several hypotheses adequately explain some igneous rocks. In the process of differentiation, based on Bowen's reaction series, a residual magma more silicic than the original mafic magma is created when the early-forming minerals separate out of the magma. In assimilation, a hot, original magma is contaminated by picking up and absorbing rock of a different composition. Magma mixing produces a magma whose composition is intermediate, between that of the two types of magma that were mixed.

Partial melting of the mantle usually produces basaltic magma whereas granitic magma is most likely produced by partial melting of the lower crust.

The theory of plate tectonics incorporates various parts of previous theories. Basalt is generated where hot mantle rock partially melts, most notably along divergent boundaries. The fluid magma rises easily through fissures, if present. The ferromagnesian portion that stays solid remains in the mantle as ultramafic rock. Granite and andesite are associated with subduction. Differentiation, assimilation, partial melting, and mixing of magmas may each play a part in creating the appropriate rocks.

Terms to Remember

andesite 56
basalt 55
batholith 61
Bowen's reaction series 66
chill zone 54
coarse-grained rock 55
concordant 61
contact 54
country rock 54

crystal settling 67
diapir 62
differentiation 66
dike 61
diorite 56
discordant 61
extrusive rock 54
fine-grained rock 55
gabbro 55

geothermal gradient 64
granite 54
igneous rock 54
intermediate 59
intrusion (intrusive structure) 59
intrusive rock 54
lava 54
mafic 58
magma 54

mantle plume 64
pluton 61
plutonic rock 55
rhyolite 55

silicic (felsic) 59
sill 61
stock 61
ultramafic rock 59

volcanic neck 60
xenolith 54

Testing Your Knowledge

Use the questions below to prepare for exams based on this chapter.

1. Why do mafic magmas tend to reach the surface much more often than silicic magmas?

2. What role does the asthenosphere possibly play in generating magma at (a) a converging boundary; (b) a diverging boundary?

3. How do batholiths form?

4. How would you distinguish, on the basis of minerals present, among granite, gabbro, and diorite?

5. How would you distinguish andesite from a diorite?

6. What rock would probably form if magma that was feeding composite volcanoes solidified at considerable depth?

7. Why is a higher temperature required to form magma at the oceanic ridges than in the continental crust?

8. What is the difference between feldspar found in gabbro and feldspar found in granite?

9. What is the difference between a dike and a sill?

10. Describe the differences between the continuous and the discontinuous branches of Bowen's reaction series.

11. A surface separating different rock types is called a (a) xenolith (b) contact (c) chill zone (d) none of the above

12. The major difference between intrusive igneous rocks and extrusive igneous rocks is (a) where they solidify (b) chemical composition (c) type of minerals (d) all of the above

13. Which is not an intrusive igneous rock? (a) gabbro (b) diorite (c) granite (d) andesite

14. By definition, stocks differ from batholiths in (a) size (b) shape (c) chemical composition (d) all of the above

15. Which is not a source of heat for melting rock? (a) geothermal gradient (b) the hotter mantle (c) mantle plumes (d) water under pressure

16. The geothermal gradient is, on the average, about (a) 1°C/km (b) 10°C/km (c) 30°C/km (d) 50°C/km

17. The continuous branch of Bowen's reaction series contains the mineral (a) pyroxene (b) plagioclase (c) amphibole (d) biotite

18. The discontinuous branch of Bowen's reaction series contains the mineral (a) pyroxene (b) amphibole (c) biotite (d) all of the above

19. The most common igneous rock of the continents is (a) basalt (b) granite (c) rhyolite (d) ultramafic

20. Granite and rhyolite are different in (a) texture (b) chemistry (c) mineralogy (d) the kind of magma that each crystallized from

21. The difference in texture between intrusive and extrusive rocks is primarily due to (a) different mineralogy (b) different rates of cooling and crystallization (c) different amounts of water in the magma

22. How can geologists recognize shallow intrusives from deeper intrusives (choose all that apply) (a) mineralogy (b) size (c) rock texture (d) chemistry

23. A change in magma composition due to melting of surrounding country rock is called (a) magma mixing (b) assimilation (c) crystal setting (d) partial melting

Expanding Your Knowledge

1. In parts of major mountain belts there are sequences of rocks that geologists interpret as slices of ancient oceanic lithosphere. Assuming that such a sequence formed at a divergent boundary and was moved toward a convergent boundary by plate motion, what rock types would you expect to make up this sequence, going from the top downward?

2. What would happen, according to Bowen's reaction series, under the following circumstances: olivine crystals form and only the surface of each crystal reacts with the melt to form a coating of pyroxene that prevents the interior of olivine from reacting with the melt?

Exploring Resources

 Barker, D. S. 1983. *Igneous rocks.* Englewood Cliffs, N.J.: Prentice-Hall.

Blatt, H. and R. C. Tracy. 1996. *Petrology: Igneous, Sedimentary, and Metamorphic.* 2d ed. New York: W. H. Freeman.

Hibbard, M. J. 1995. *Petrography to petrogenesis.* Englewood Cliffs, N.J.: Prentice-Hall.

MacKenzie, W. S., C. H. Donaldson, and C. Guilford. 1982. *Atlas of igneous rocks and their textures.* New York: Halsted Press.

 http://uts.cc.utexas.edu/~rmr/ Rob's Granite Page.

This site has a lot of information related to granite and igneous activity. Many photos as well as information useful to people new to geology and for professionals. There are many links to other sites.

Interacting with Journey Through Geology CD-ROM

Expand your knowledge of the concepts presented in this chapter by using the CD-ROM to answer the following questions.

1. Pegmatite Pocket module: What conditions promote the formation of large crystals in pegmatites? Give 3 examples of gemstones that are mined from pegmatites.

2. Hot Spots module, Yellowstone, Watch Hot Spot Animation: Yellowstone volcanic rocks are primarily rhyolites, yet the Yellowstone volcanic system is fueled by a hot plume from the mantle. Explain how an ultramafic magma could produce rhyolite eruptions.

3. Rocks Melt module: Experiment to find the crystallization (not melting) temperatures of plagioclase, orthoclase, and quartz at 1 atmosphere and at 5000 atmospheres. On Venus, which has no water, would rock melt at deeper, shallower, or the same depths as on the Earth?

Eruption of La Fournaise Volcano, Réunion Island, Indian Ocean.
© Adam G. Sylvester/ Photo Researchers, Inc.

CHAPTER

4

Volcanism and Extrusive Rocks

Chapters 3 and 4 cover igneous activity. Either may be read before the other. Chapter 3 emphasizes intrusive activity, but it also covers igneous rock classification and the origin of magmas, which are applicable both to volcanic and intrusive phenomena. Chapter 4 concentrates on volcanoes and related extrusive activity.

Volcanic eruptions, while awesome natural spectacles, also provide important information on the workings of the earth's interior. Volcanic eruptions vary in nature and in degree of explosive violence. A strong correlation exists between the chemical composition of magma (or lava) and the violence of an eruption. The size and shape of volcanoes and lava flows and their pattern of distribution on the earth's surface also correspond to the composition of the lavas.

Understanding volcanism provides a background for theories relating to mountain building, the development and evolution of continental and oceanic crust, and how the crust is deformed. Our observations of volcanic activity fit nicely into plate-tectonic theory as described in chapter 3.

The May 18, 1980 eruption of Mount St. Helens (Box 4.1) was a spectacular release of energy from the earth's interior. The plate-tectonic explanation is that North America, moving westward, is overriding a portion of the Pacific Ocean floor. Melting of previously solid rock takes place at depth, just above the subduction zone. (This was described briefly in chapter 1 and more thoroughly in chapter 3.) At least some of the **magma** (molten rock or liquid that is mostly silica) works its way upward to the earth's surface to erupt. Magma does not always reach the earth's surface before solidifying, but when it does it is called **lava.**

At Mount St. Helens the lava solidified quickly as it was blasted explosively by gases into the air, producing rock fragments known as **pyroclasts** (from the Greek *pyro,* "fire," and *clast,* "broken"). Pyroclastic debris is also known as *tephra.* Elsewhere in the world (notably Hawaii) lava extrudes out of fissures in the ground as **lava flows.** Pyroclastic debris and rock formed by solidification of lava are collectively regarded as **extrusive rock,** surface rock resulting from volcanic activity.

The most obvious landform created by **volcanism** is a **volcano,** a hill or mountain formed by the extrusion of lava or ejection of rock fragments from a vent. However, volcanoes are not the only volcanic landforms. Very fluid lava may flow out of the earth and flood an area, solidifying into a nearly horizontal layer of extrusive rock. Successive layers of lava flows may accumulate into a lava plateau.

Volcanic activity is important to geology for several reasons. Landforms are created and portions of the earth's surface built up. Less commonly, as at Mount St. Helens, landforms are destroyed by violent eruptions. Volcanoes are important to the science of geology because they provide clues about the nature of the earth's inaccessible interior and help us understand how the earth's internal processes work. By studying the magma, gases, and rocks from eruptions, we can infer the chemical conditions as well as the temperatures and pressures within the earth's crust or underlying mantle.

Effects on Humans

Volcanism is also significant in human affairs. Its effects can be catastrophic or, surprisingly, beneficial.

The Growth of Hawaii

The overall effects of volcanism have been favorable to humans in Hawaii, although occasionally a field or village is overrun by outpourings of lava. Kilauea volcano is especially active. During the 1980s and 1990s, over 1.5 billion cubic meters of lava erupted—enough to build a highway that circles the world four times. There were 181 houses destroyed in the 1980s and 1990s, but no one was killed or injured. Nevertheless, the weathered volcanic ash and lava produce excellent fertile soil.

Figure 4.1

Volcanic eruptions in Hawaii, 1969. A lava fountain is supplying the lava cascading over the cliff.

Photo by D. A. Swanson, U.S. Geological Survey

Moreover, Hawaii's periodically erupting volcanoes (which are relatively safe to watch) are great spectacles that attract both tourists and scientists, benefiting the island's economy (figure 4.1).

Were it not for volcanic activity, Hawaii would not exist. The islands are the crests of a series of volcanoes that have been built up from the bottom of the Pacific Ocean over millions of years (the vertical distance from the summit of Mauna Loa volcano to the ocean floor greatly exceeds the height above sea level of Mount Everest). When lava flows into the sea and solidifies, more land is added to the islands. Hawaii is, quite literally, growing.

Geothermal Energy

In some other areas of geologically recent volcanic activity, underground heat generated by igneous activity is harnessed for human needs. In Italy, Mexico, New Zealand, Argentina, Japan, and California, geothermal installations produce electric power. Steam or superheated water trapped in layers of hot volcanic rock is tapped by drilling and then piped out of the ground to power turbines that generate electricity. Naturally heated geothermal fluids can also be tapped for space or domestic water heating or industrial use, as in paper manufacturing.

Effect on Climate

Occasionally, a volcano will spew large amounts of fine, volcanic dust and gas into the high atmosphere. Winds can keep fine particles suspended over the earth for years. The 1991 eruption of Mount Pinatubo in the Philippines produced

ENVIRONMENTAL GEOLOGY

Mount St. Helens, 1980—North America's Eruption of the Century

O ver a decade has passed since Mount St. Helens blew up. Before 1980 not many people living in Washington or Oregon regarded the volcanic cones of the Cascade Range as hazardous. Mount Rainier, Mount Hood, and Mount Baker (and about ten other peaks) were fixtures of the scenery. These picturesque snow-capped cones towering over evergreen forests seemed unlikely to change. Geologists knew otherwise, that it was exceedingly unlikely that volcanoes that had shown intermittent activity over hundreds of thousands of years were dead. In fact, California's Lassen Peak at the southern extreme of the Cascade Range had erupted from 1914 to 1917.

Mount St. Helens (box figure 1), in southern Washington, had last erupted in 1857. Geologists were not surprised when it erupted in 1980. Its inactivity had lasted little more than a century—a trivial amount of time in geology. Between March and May of 1980 Mount St. Helens was transformed from a serene, snow-covered cone to a darkened, gouged-out stump of a peak; from a thing of beauty to a killer.

The 1980 eruptions were preceded by thousands of small earthquakes. On March 27 ash and steam eruptions began and continued for the next six weeks. The steam explosions and the pattern of earthquakes indicated that magma was working its way upward beneath the volcano.

During this early stage of steam and ash eruptions, St. Helens provided a fine show. "Volcano fever" spread, particularly in nearby Portland, Oregon. Ice-cream dishes and cocktails resembling volcanoes and T-shirts and commemorative trinkets sold briskly.

A few years earlier a team from the U.S. Geological Survey completed a study of the potential hazards of renewed eruption, based on the peak's geologic record for the past 38,000 years. On the advice of the Survey geologists, U.S. Forest Service and state officials blocked off the potentially most hazardous areas. Geologists were particularly worried about large mudflows. Small mudflows occurred after the eruptions began. The mudflows were slurries of loose ash and water (from rainfall or melting glacier ice or snow) that flowed down the volcano. A large mudflow might follow a stream valley beyond the mountain and overwhelm a populated area. For this reason, water levels were lowered in reservoirs in order to impound potential mudflows.

Box 4.1 Figure 1

Mount St. Helens before the 1980 eruption, as seen from the north.

Photo by Jim Nieland, U.S. Forest Service

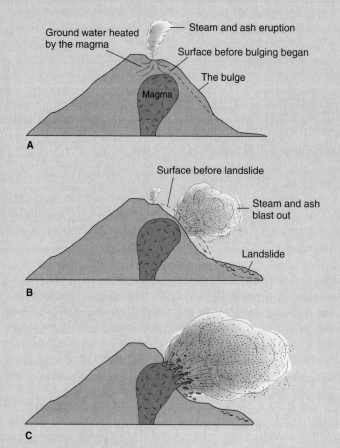

Box 4.1 Figure 2

Sequence of events at Mount St. Helens, May 18, 1980. (*A*) Just before the May 18, 1980 eruption. (*B*) The landslide relieves the pressure on the underlying magma. (*C*) Magma blasts outward.

Box 4.1 Figure 3

May 18, 1980. The side of the volcano has been blasted away as magmatic explosions continue.

© 1980 Keith Ronnholm

As the steam-driven eruptions continued, the peak began swelling—like a balloon being inflated—indicating that the magma was moving upward. The northern flank of the mountain bulged outward at a rate of 1.5 meters per day. Bulging continued until the surface of the northern slope was displaced outward over a hundred meters from its original position (see box figure 2A). The bulge was gravitationally unstable and the Geological Survey warned of another hazard—a mammoth landslide.

The huge blast (box figure 3) that marked the arrival of the magma at the surface on May 18 destroyed the summit and north flank of St. Helens almost instantly. Seconds after the eruption began, an area extending outward 10 kilometers was stripped of all vegetation and soil. Downslope forests were leveled; from the air the leafless, scorched trunks looked like thousands of matchsticks strewn on the ground.

Although the sequence of events was exceedingly rapid, it is now clear what happened (box figure 2). A fairly strong earthquake shook the bulging north slope loose. The resulting landslide stripped away the rock that sealed in the magma. With the protective lid removed, gases that had been dissolved in the magma were suddenly released. On a grand scale, it was comparable to shaking a bottle of warm beer or pop and removing the top. The magma exploded into a froth while blasting out of the north flank of the volcano. The mass of gases and volcanic rock fragments roared down the side of the volcano and continued beyond its base, destroying trees and all organic matter close to the volcano and knocking over forests farther downslope.

For the next thirty hours, exploding gases continued to propel frothing magma and volcanic ash vertically into the atmosphere. The dense, mushroom-shaped cloud of ash was blown northeastward by the winds. Fallout of ash went on for days, causing serious damage as far away as Montana. In some towns, there was nightlike darkness at midday. Crops were destroyed. Breathing became difficult. Automobile engines were destroyed by fine volcanic particles that were sucked into cylinders.

Damage was in the hundreds of millions of dollars, and 63 people were killed. The death toll might have been much worse. For comparison, 29,000 people were killed during an eruption of Mount Pelée (described later in this chapter), and 23,000 lives were lost in a 1985 volcanically triggered mudflow in Colombia.

Many lives were saved, no doubt, by the U.S. Geological Survey advice to public officials. Just as important, the officials heeded the advice (perhaps it helped that Washington's governor was a scientist) and resisted pressure from the public, notably the people that wanted access to the closed lands. Only two days before the catastrophic eruption, a group whose homes were in the area north of the peak staged a protest demanding access. They owe their lives to scientists whose knowledge kept them out of an area that would be utterly destroyed.

After the eruption on May 18, 1980, there were occasional, less intense eruptions. A volcanic dome grew as magma periodically welled upward into the floor of the new depression (see figure 4.24).

More than a decade has gone by since the big eruption and the mountain is quiet. We do not know when eruptions will resume—perhaps in a few years, perhaps in a century. Nor do we know which of the other Cascade volcanoes (see figure 4.5) will erupt next. Seattle, with Mount Rainier nearby, and Portland, with Mount Hood almost in its suburbs, could be endangered. Vancouver might suffer if either Mount Garibaldi in British Columbia or Mount Baker in Washington erupt.

Related Web Resources

http://vulcan.wr.usgs.gov/volcanoes/MSH/framework.html

---

noticeably more colorful sunsets worldwide (see box in chapter 1). More significantly, it reduced solar radiation that penetrates the atmosphere. Measurements indicated a worldwide average temperature drop of approximately half a degree Celsius for a couple years. While this may not seem like much, it is enough to temporarily offset the global warming trend of the past 100 years.

The 1815 eruption of Tambora in Indonesia was the largest, single eruption in a millenium—40 cubic kilometers of material was blasted out of a volcanic island, leaving a six-kilometer-wide depression. The following year, 1816, became known as "the year without summer." In New England, snow in June was widespread and frosts throughout the summer ruined crops. Parts of Europe suffered famine because of the cold weather effects on agriculture.

Volcanic Catastrophes

While the eruption of Mount St. Helens in 1980 was indeed awesome, its effects were not nearly as disastrous as a number of historical eruptions elsewhere in the world. For instance, the Roman city of Pompeii and at least four other towns near Naples in Italy were destroyed in A.D. 79 when Mount Vesuvius erupted (figure 4.2). Before that time vineyards on the flanks of the apparently "dead" volcano extended to the summit. Pompeii was buried under 5 to 8 meters of hot ash from the surprise eruption of Vesuvius. Seventeen centuries later the town was rediscovered. Excavation revealed molds of people suffocated by the ashfall, many with facial expressions of terror. This eruption was not the end of Vesuvius's activity. The volcano was active almost continually from 1631 to 1944, with major twentieth-century eruptions in 1906, 1929, and 1944.

The island of Krakatoa in the western Pacific, composed of three apparently inactive volcanoes, erupted in 1883 with the force of several hydrogen bombs. This Indonesian island, which formerly rose 800 meters above sea level, was blown apart. Only one-third of the island remained after the eruption. An esti-

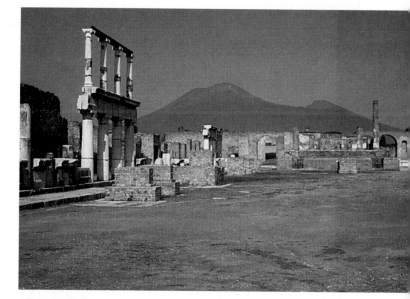

Figure 4.2

Pompeii with Mount Vesuvius in the background.
Photo by R. W. Decker

mated 13 cubic kilometers of rock collapsed into the subsurface magma chamber that had been emptied by the eruption, leaving an underwater depression 300 meters deep where the major part of the island had been. The explosion was heard 5,000 kilometers away. Over 34,000 people died as a result of the giant sea waves (tsunamis) generated by the explosion.

A similar series of explosions in prehistoric time (about 6,600 years ago) was at least partially responsible for creating Crater Lake in Oregon. Volcanic debris covering more than a million square kilometers in Oregon and neighboring states has been traced to those eruptions. The original volcano, named Mount Mazama by geologists, is estimated to have been about 2,000 meters higher than the present rim of Crater

Figure 4.3

Crater Lake, Oregon.

Photo by © Greg Vaughn/Tom Stack & Associates

Lake. Collapse of the volcano, as well as explosions, accounts for the depression the present-day lake occupies (figures 4.3 and 4.4).

The southern Cascade Mountains, where Crater Lake is located, have been built up by eruptions over the past 30 to 40 million years (figure 4.5; see also the geologic map, inside cover). Only the youngest peaks (those built within the past 2 million years), such as Mount St. Helens, Mount Rainier, Mount Shasta, and Mount Hood, still stand out as cones. As we know from Mount St. Helens, any of these could again become active.

Eruptive Violence and Physical Characteristics of Lava

What determines the degree of violence associated with volcanic activity? Why can we state confidently that active volcanism in Hawaii poses only slight danger to humans, but expect violent explo-

A

B

C

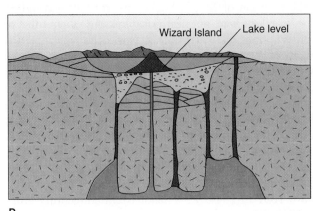

D

Figure 4.4

The development of Crater Lake. (*A*) Cluster of overlapping volcanoes form. (*B*) Collapse into the partially emptied magma chamber is accompanied by violent eruptions. (*C*) Volcanic activity ceases, but steam explosions take place in the caldera. (*D*) Water fills the caldera to become Crater Lake, and minor renewed volcanism builds a cinder cone (Wizard Island).

After C. Bacon, U.S. Geological Survey

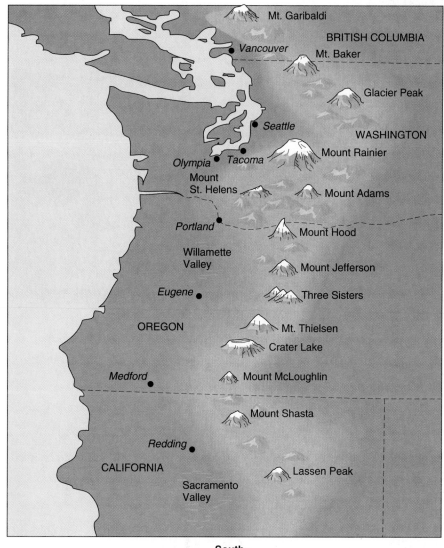

North

Mt. Garibaldi

BRITISH COLUMBIA

Vancouver

Mt. Baker

Glacier Peak

Seattle

WASHINGTON

Mount Rainier

Olympia Tacoma

Mount
St. Helens

Mount Adams

Portland

Mount Hood

Willamette
Valley

Mount Jefferson

Eugene

Three Sisters

OREGON

Mt. Thielsen

Crater Lake

Medford

Mount McLoughlin

Mount Shasta

Redding

CALIFORNIA

Lassen Peak

Sacramento
Valley

South

Figure 4.5

The Cascade volcanoes. See also the map in chapter 1 (figure 1.5).

sions to occur in the Cascade Mountains? Whether eruptions are very explosive or relatively "quiet" is largely determined by two factors: (1) the amount of gas in the lava or magma and (2) the ease or difficulty with which the gas can escape to the atmosphere. The **viscosity,** or resistance to flow, of a lava determines how easily the gas escapes. The more viscous the lava and the greater the volume of gas trying to escape, the more violent the eruption. Later we will show how these factors not only determine the degree of violence of an eruption but also influence the shape and height of a volcano.

The two most important factors that influence viscosity are (1) the temperature of the lava relative to the cooler temperature at which it solidifies and (2) the silica (SiO_2) content of the lava. (A third factor is gas dissolved in magma—the greater the dissolved gas content, the more fluid the lava.) If the lava being extruded is considerably hotter than its solidification temperature, the lava is less viscous (more fluid) than when its temperature is near its solidification point. Temperatures at which lavas solidify range from about 700°C for silicic rocks to 1200°C for mafic rocks.

Volcanic rocks, and the magma from which they formed, have a silica content that ranges from 45 to 75 percent by weight. **Silicic** (or felsic) rocks are silica-rich (65% or more SiO_2) rocks. *Rhyolite* is the most abundant silicic volcanic rock. **Mafic rocks** are *silica-deficient* rocks. Their silica content is close to 50%. *Basalt* is the most common mafic rock. **Intermediate rocks** have a chemical content between that of felsic and mafic rocks. The most common intermediate rock is *andesite.* A more complete description of the chemistry of igneous rocks and their relationship to the mineral content of rocks is given in chapter 3.

Mafic lavas, which are relatively low in SiO_2, tend to flow easily. Conversely, felsic lavas are very viscous and flow sluggishly. Lavas rich in silica are more viscous because even before they have cooled enough to allow crystallization of minerals, silicon-oxygen tetrahedrons have begun to form small frameworks in the lava. Although too few atoms are involved for the structures to be considered crystals, the total effect of these silicate structures is to make the liquid lava more viscous, much the way that flour or cornstarch thickens gravy.

Because silicic magmas are the most viscous, they are associated with the most violent eruptions. Mafic magmas are the least viscous and commonly erupt as lava flows (such as in Hawaii). Eruption associated with intermediate magma can be violent or can produce lava flows. (The Cascade volcanoes are predominantly composed of intermediate rock.)

Extrusive Rocks and Gases

Scientific Investigation of Volcanism

Volcanoes and lava flows, unlike many other geologic phenomena, can be observed directly, and samples can be collected without great difficulty (at least for the quiet, Hawaiian-type of eruption). We can measure the temperature of lava flows, collect samples of gases being given off, observe the lava solidifying into rock, and take newly-formed rock samples into the laboratory for analysis and study. By comparing rocks observed

Figure 4.6

Pyroclastic flow descending Mayon volcano, Philippines, in 1984.

Photo by Chris Newhall, U.S. Geological Survey

Figure 4.7

The ruins of St. Pierre in 1902. Mount Pelée is in the clouds.

Photo by Underwood & Underwood, courtesy Library of Congress

solidifying from lava with similar ones from other areas of the world (and even with samples from the moon) where volcanism is no longer active, we can determine the nature of volcanic activity that took place in the geologic past.

Gases

From active volcanoes we have learned that most of the gas released during eruptions is water vapor, which condenses as steam. Other gases, such as carbon dioxide, sulfur dioxide, hydrogen sulfide (which smells like rotten eggs), and hydrochloric acid, are given off in lesser amounts with the steam.

As exemplified by the devastation of the island of Krakatoa (described earlier), surface water introduced into a volcanic system can greatly increase the explosivity of an eruption.

Gases and Pyroclastics

During an eruption, expanding, hot gases may propel pyroclastics high into the atmosphere as a column rising from a volcano. At high altitudes, the pyroclastics often spread out into a dark, mushroom cloud. The fine particles are transported downwind by high atmosphere winds. Eventually debris settles back to earth under gravity's influence as *ashfall* (or sometimes *pumice fall*) deposits.

A **pyroclastic flow** is a mixture of gas and pyroclastic debris that is so dense that it hugs the ground as it flows rapidly into low areas (figure 4.6). These turbulent masses can travel over 100 kilometers per hour and are extremely dangerous. In 1991, a pyroclastic flow at Japan's Mount Unzen killed 31 people, including 3 geologists. Far worse was the destruction of St. Pierre (figure 4.7) on the Caribbean island of Martinique where about 28,000 people were killed by a pyroclastic flow in 1902 (see box 4.3).

Extrusive Rocks

Most extrusive rocks are named and identified on the basis of their composition and texture. However, some are named exclusively on texture (e.g., pumice).

Composition

The amount of silica in a lava largely controls not only the viscosity of lava and the violence of eruptions but also which particular rock is formed. Chapter 3 describes how igneous rocks are identified based on the minerals present and their relative abundance in the rock. Because extrusive igneous rocks are generally fine-grained, a specialized microscope is usually needed for precise identification of the component minerals. In most cases, however, we can guess the probable mineral content by noting how dark or light in color an extrusive rock is. Most silicic rocks are light-colored because they contain abundant feldspar and quartz (both of which are silica-rich) and few dark minerals (which contain iron and magnesium and are silica-deficient). Mafic rocks, on the other hand, tend to be dark because of the abundance of ferromagnesian minerals.

Rhyolite, a silicic rock, is usually cream-colored, tan, or pink; it is made up mostly of feldspar but always includes some quartz. Note that the rhyolite (and granite) portion of figure 3.4 is larger than the areas shown for andesite and basalt. Geologists commonly subdivide this portion of the classification system. For example, *dacite*, the rock associated with the 1980 St. Helens eruptions, contains more ferromagnesian minerals and plagioclase but less potassium feldspar and quartz than the average rhyolite. In our classification system, dacite corresponds to the right portion of the area in figure 3.4 assigned to rhyolite.

A **basalt** has a relatively low amount (about 50% by weight) of SiO_2. Much of that silica is bonded to iron and magnesium to form ferromagnesian minerals, such as *olivine* or *pyroxene,* which are green or black. The remaining silica plus aluminum is bonded predominantly with calcium to form calcium-rich *plagioclase feldspar* (which tends to be darker gray than the white or pink potassium or sodium feldspars associated with felsic rocks). Basalt does not contain quartz because no silica is left over after the other minerals have formed. Because of the preponderance of dark minerals in basalt, this rock is usually dark gray to black.

Andesite, which crystallizes from an intermediate lava, can be recognized by its moderately gray or green color. It is this color because a little over half the rock is light- to medium-gray plagioclase feldspar, while the rest is ferromagnesian minerals.

Textures

Texture refers to a rock's appearance with respect to the size, shape, and arrangement of its grains or other constituents.

Some extrusive rocks (such as obsidian and pumice) are classified solely on the basis of their textures, but most are classified by composition *and* texture. *Grain size* is a rock's most important textural characteristic. For the most part, extrusive rocks are fine-grained or else made of glass.

A **fine-grained rock** is one in which most of the mineral grains are smaller than 1 millimeter. In some, the individual minerals are distinguishable only with a microscope. **Obsidian** (figure 4.8), which is volcanic glass, is one of the few rocks that is not composed of minerals. A fine-grained or glassy texture distinguishes extrusive rocks from most intrusive rocks.

Two critical factors determine grain size during the solidification of igneous rocks: rate of cooling and viscosity. If lava cools rapidly, the atoms have time to move only a short distance; they bond with nearby atoms, forming only small crystals. With extremely rapid or almost instantaneous cooling, individual atoms in the lava are "frozen" in place, forming glass rather than crystals.

Grain size is controlled to a lesser extent by the viscosity of the lava. The atoms in a very viscous lava cannot move as freely

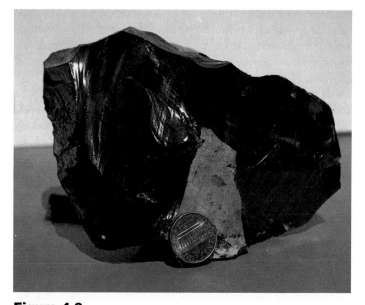

Figure 4.8

Obsidian.

Photo by C. C. Plummer

A

B

1 mm

Figure 4.9

Porphyritic andesite. A few large crystals (phenocrysts) are surrounded by a great number of fine grains. (*A*) Hand specimen. (*B*) Photomicrograph (using polarized light) of the same rock. The black and white striped phenocrysts are plagioclase and the green ones are ferromagnesian minerals.

Photos by C. C. Plummer

as those in a very fluid lava. Hence, a rock formed from viscous lava is more likely to be obsidian or of finer grains than one formed from more fluid lava. Most obsidian, when chemically analyzed, has a very high silica content and is silicic, and the chemical equivalent of rhyolite.

Porphyritic Textures

Extrusive rock that does not have a uniformly fine-grained texture throughout is described as porphyritic. A **porphyritic rock** is one in which larger crystals are enclosed in a *matrix* (or *ground mass*) of much finer-grained minerals or obsidian. The larger crystals are termed **phenocrysts.** A porphyritic rock looks rather like raisin bread; the matrix or ground mass is the bread, the phenocrysts are the raisins. In the porphyritic andesite shown in figure 4.9, phenocrysts of feldspar and ferromagnesian minerals are enclosed in a matrix of crystals too fine-grained to distinguish with the naked eye but visible under a microscope.

Porphyritic texture usually indicates two stages of solidification. Slow cooling takes place while the magma is underground. Minerals that form at higher temperatures crystallize and grow to form phenocrysts in the still partly-fluid magma. If the entire mass is then erupted, the remaining liquid portion cools rapidly at the earth's surface and forms the fine-grained matrix.

Textures Due to Trapped Gas

A magma deep underground is under high pressure, generally high enough to keep all its gases in a dissolved state. On eruption, the pressure is suddenly released and the gases come out of solution. This is analogous to what happens when a bottle of beer or soda is opened. Because the drink was bottled under pressure, the gas (carbon dioxide) is in solution. Uncapping the drink relieves the pressure, and the carbon dioxide separates from the liquid as gas bubbles. If you freeze the newly opened drink very quickly,

Figure 4.10

Vesicular basalt.

Photo by C. C. Plummer

you have a piece of ice with small, bubble-shaped holes. Similarly, when a lava solidifies while gas is bubbling through it, holes are trapped in the rock, creating a distinctive *vesicular* texture. **Vesicles** are cavities in extrusive rock resulting from gas bubbles that were in lava. A vesicular rock has the appearance of Swiss cheese (whose texture is caused by trapped carbon dioxide gas). *Vesicular basalt* is quite common (figure 4.10). *Scoria,* a highly vesicular basalt, actually contains more gas space than rock.

Figure 4.11

Pumice.

Photo by C. C. Plummer

Figure 4.12

Volcanic bombs.

Photo by C. C. Plummer

Figure 4.13

Photomicrograph of a tuff. Fragments of different rocks and minerals are angular and variously colored.

Photo by C. C. Plummer

In more viscous lavas, where the gas cannot escape as easily, the lava is churned into a froth (like the head of a glass of beer). When cooled quickly, it forms **pumice** (figure 4.11), a frothy glass with so much void space that it floats in water. Powdered pumice is used as an abrasive because it can scratch metal or glass.

Fragmental Textures

Pyroclasts, the fragments formed by volcanic explosion, can be almost any size. *Dust* and *ash* are the finest particles; *cinders* are about the size of sand grains; *bombs* and *blocks* are large pyroclasts. When solid rock has been blasted apart by a volcanic explosion, the pyroclastic fragments are *angular,* with no rounded edges or corners and are called **blocks.** If lava is ejected into the air, a molten blob becomes streamlined during flight, solidifies, and falls to the ground as a **bomb,** a spindle- or lens-shaped pyroclast (figure 4.12).

When loose pyroclastic material (ash, bombs, etc.) is cemented or otherwise consolidated, the new rock is called *tuff* or *volcanic breccia,* depending on the size of the fragments. A **tuff** (figure 4.13) is a rock composed of fine-grained pyroclastic particles. A **volcanic breccia** is a rock that includes larger pieces of volcanic rock (blocks, bombs).

Types of Volcanoes

Volcanic material that is ejected from and deposited around a central vent produces the conical shape typical of volcanoes. The **vent** is the opening through which an eruption takes place. The **crater** of a volcano is a basinlike depression over a

vent at the summit of the cone (figure 4.14). Material is not always ejected from the central vent. In a **flank eruption,** lava pours from a vent on the side of a volcano.

A **caldera** is a volcanic depression much larger than the original crater, having a diameter of at least one kilometer. (The most famous caldera in the United States is misnamed "Crater Lake.") A caldera can be created when a volcano's summit is blown off by exploding gases, as occurred at Mount St. Helens in May 1980, or, as in the case of Crater Lake, when a volcano (or several volcanoes) collapses into a vacated magma chamber (see figure 4.4).

The three major types of volcanoes (shield, cinder cone, and composite) discussed below are markedly distinct from one another in size, shape, and, usually, composition. Although volcanic domes are not cones, they are associated with volcanoes and are also examined in this section.

Figure 4.14

Crater and caldera in Kamchatka, Russia. In the foreground is the crater on Karymsky volcano. In the background is a lake-filled caldera.

Photo by C. Dan Miller, U.S. Geological Survey

Shield Volcanoes

Shield volcanoes are broad, gently sloping cones constructed of solidified lava flows. During eruptions, the lava spreads widely and thinly due to its low viscosity. Because the lava flows from a central vent, without building up much near the vent, the slopes are usually between 2° and 10° from the horizontal, producing a volcano in the shape of a flattened dome or "shield" (figure 4.15).

The islands of Hawaii are essentially a series of shield volcanoes built upward from the ocean floor by intermittent eruptions over millions of years (figure 4.15B). Although spectacular to observe, the eruptions are relatively nonviolent because the lavas are fairly fluid (less viscous). By implication, then, the shield volcanoes of the Hawaiian Islands are composed of a series of layers of basalt.

Hawaiian names have been given to two distinctive surfaces of basalt flows. *Pahoehoe* (pronounced *pah-hoy-hoy*) is characterized by a ropy or billowy surface (figure 4.16). The surface is formed by the quick cooling and solidification from the surface downward of a lava flow or pool of lava that was fully liquid. By contrast, basalt that is cool enough to have partially solidified moves as a slow, pasty mass. Its largely solidified front is shoved forward as a pile of rubble. A flow such as this is called *aa* (pronounced *ah-ah*) and has a spiny, rubbly surface (figure 4.17).

A minor feature called a **spatter cone,** a small, steep-sided cone built from lava sputtering out of a vent (figure 4.18), will occasionally develop on a solidifying lava flow. When a small concentration of gas is trapped in a cooling lava flow, lava is belched out of a vent through the solidified surface of the flow. Falling lava plasters itself onto the developing cone and solidifies. The sides of a spatter cone can be very steep, but they are rarely over 10 meters high.

Cinder Cones

A **cinder cone** (less commonly called a *pyroclastic cone*) is a volcano constructed of pyroclasts ejected from a central vent (figure 4.19). In contrast to the gentle slopes of shield volcanoes, cinder cones commonly have slopes of about 30°. Most of the ejected material lands near the vent during an eruption, building up the cone to a peak. The steepness of slopes of accumulating loose material is limited by gravity to about 33°. Cinder cones tend to be very much smaller than shield volcanoes. In fact, cinder cones are commonly found on the flanks and in the calderas of Hawaii's shield volcanoes. Few cinder cones exceed a height of 500 meters.

Cinder cones are a product of concentration of gas in magma rather than of a particular chemical composition of

New lava

Feeders

Layers of basalt

A

B

Figure 4.15

(*A*) Cutaway view of a shield volcano. (*B*) The top of Mauna Loa, a shield volcano in Hawaii, and its summit caldera. The smaller depressions are pit craters. In the distance is Mauna Kea, another shield volcano that last erupted about 3,000 years ago.

Photo by D. W. Peterson, U.S. Geological Survey

Figure 4.16

Pahoehoe from a 1972 eruption in Hawaii.

Photo by D. W. Peterson, U.S. Geological Survey

Figure 4.17

An *aa* flow in Hawaii, 1983.

Photo by J. D. Griggs, U.S. Geological Survey

Figure 4.18

A spatter cone (approximately 1 meter high) erupting in Hawaii.

Photo by J. B. Judd, U.S. Geological Survey

A

B

Figure 4.19

Cerro Negro, a cinder cone in Nicaragua. (*A*) View from the air; (*B*) nighttime eruption of pyroclastics at the summit.

Photo A by Mark Hurd Aerial Surveys Corp. courtesy California Division of Mines and Gregory; Photo B by R. W. Decker magma. Most cinder cones are associated with mafic or intermediate lava. Silicic cinder cones, which are made of fragments of pumice, are also known as pumice cones.

The life span of an active cinder cone tends to be short. The local concentration of gas is depleted rather quickly during the eruptive periods. Moreover, as landforms, cinder cones are temporary features in terms of geologic time. The unconsolidated pyroclasts are eroded relatively easily.

Composite Volcanoes

A **composite volcano** (also called **stratovolcano**) is one constructed of alternating layers of pyroclastics and rock solidified from lava flows (figure 4.20A). The slopes are intermediate in steepness compared with cinder cones and shield volcanoes. Pyroclastic layers build steep slopes as debris collects near the vent, just as in cinder cones. However, subsequent lava flows partially flatten the profile of the cone as the downward flow builds up the height of the flanks more than the summit area. The solidified lava acts as a protective cover over the loose pyroclastic layers, making composite volcanoes less vulnerable to erosion than cinder cones.

Composite volcanoes are built over long spans of time. Eruption is intermittent, with hundreds or thousands of years of inactivity separating a few years of intense activity. During the quiet intervals between eruptions, composite volcanoes may be eroded by running water, landslides, or glaciers. These surficial processes tend to alter the surface, shape, and form of the cone. But because of their long lives and relative resistance to erosion, composite cones can become very large. Aconcagua, a composite volcano in the Andes, is 6,960 meters (22,835 feet) above sea level and the highest peak in the western hemisphere.

The extrusive material that builds composite cones is predominantly of intermediate composition, although there may be local minor silicic and mafic eruptions. Therefore, *andesite* is the rock most associated with composite volcanoes. If the temperature of an andesite lava is considerably above the temperature at which it would be completely solidified, the relatively nonviscous fluid flows easily from the crater down the slopes. On the other hand, if enough gas pressure exists, an explosion may litter the slopes with pyroclastic andesite, particularly if the lava has fully or partially solidified and clogged the volcano's vent.

The composition as well as eruptive history of individual volcanoes can vary considerably. For instance, Mount Rainier is composed of 90% lava flows and only 10% pyroclastic layers. On the other hand, Mount St. Helens was built mostly

Crater

New lava flow

Older lava flow

A

B

Figure 4.20

(*A*) Cutaway view of a composite volcano. Light-colored layers are pyroclastics. (*B*) Mount Shasta, a composite volcano in California. Shastina on Mount Shasta's flanks is a subsidiary cone, largely made of pyroclastics. Note the lava flow that originated on Shasta and extends beyond the volcano's base.

Photo by B. Amundson

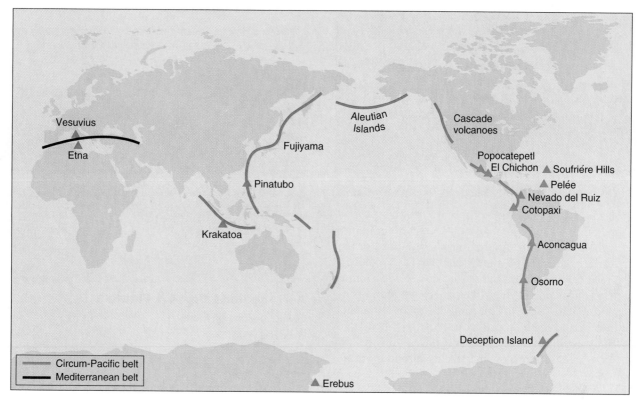

Figure 4.21

Map of the world showing the major volcanic belts.

from pyroclastic eruptions—reflecting a more violent history. As would be expected, the composition of the rocks formed during the 1980 eruptions of Mount St. Helens is somewhat higher in silica than average for Cascade volcanoes.

Distribution of Composite Volcanoes

Nearly all the larger and better known volcanoes of the world are composite volcanoes. They tend to align along two major belts on the earth (figure 4.21). The **circum-Pacific belt,** or "Ring of Fire," is the larger. The Cascade Range volcanoes described earlier make up a small segment of the circum-Pacific belt.

Several composite volcanoes in Mexico rise higher than 5,000 meters, including Orizaba (third highest peak in North America) and Popocatepetl.

The circum-Pacific belt includes many volcanoes in Central America, western South America (including Nevado del Ruiz in Colombia), and Antarctica. Mount Erebus, in Antarctica, is the southernmost active volcano in the world (figure 4.22).

The western portion of the Pacific belt includes volcanoes in New Zealand, Indonesia, the Philippines (with Pinatubo, Mayon, whose 1993 eruptions killed over 30 people, and many other volcanoes), and Japan. The beautifully symmetri-

Figure 4.22

Mount Erebus, Antarctica, the southernmost active volcano in the world.

Photo by Philip R. Kyle

4.1
Astrogeology

Extraterrestrial Volcanic Activity

Volcanic activity has been a common geologic process operating on the Moon and on several other bodies in the solar system. Approximately one-third of the Moon's surface consists of nearly circular, dark-colored, smooth, relatively flat lava plains. The lava plains, found mostly on the near side of the Moon, are called *maria* (singular, *mare;* literally, "seas"). This kind of terrain represents a significant period in the Moon's early history when large lava flows flooded parts of the Moon's surface, filling in low places. There are also a few extinct shield volcanoes on the Moon.

Elongate trenches or cracklike valleys called *rilles* are found mainly in the smoother portions of the lunar maria. They range in length from a few kilometers to hundreds of kilometers. Some are arc-shaped or crooked and may be collapsed lava tubes, channels eroded by pyroclastic flows, or fractures along which gas has escaped.

Mercury, the innermost planet, also has extensive areas covered by maria.

Radar images of Venus show a surface that is young and volcanically active. More than three-fourths of that surface is covered by continuous plains formed by enormous floods

Astrogeology Box 4.1 Figure 1
Olympus Mons on Mars.
Photo by NASA

of lava. Close examination of these plains reveals extensive networks of lava channels and individual lava flows as much as 300 kilometers long.

Large shield volcanoes, some in chains along a great fault, have been identified on Venus, and molten lava lakes may exist. In other places, thick lavas have oozed out to form kilometer-high pancake-shaped domes. Radar studies have shown that some of these domes are composed of a glassy substance mixed with bubbles of trapped gas. Fan-shaped deposits adjacent to some volcanoes may be pyroclastic debris.

---

cal Fujiyama, in Japan, is probably the most frequently painted volcano in the world (figure 4.23). The northernmost part of the circum-Pacific belt includes active volcanoes in Russia (see figure 4.14) and on Alaska's Aleutian Islands.

The second major volcanic belt is the **Mediterranean belt,** which includes Mount Vesuvius. An exceptionally violent eruption of Mount Thera, an island in the Mediterranean, may have destroyed an important site of early Greek civilization. (Some archaeologists consider Thera the original "lost continent" of Atlantis.) Mount Etna, on the island of Sicily, has been called the "lighthouse of the Mediterranean," because of its frequent eruptions throughout the centuries. The largest eruption in 300 years began in 1991 and lasted for 473 days. Some 250 million cubic meters of lava covered 7 km² of land. A town was saved from the lava by heroic efforts that included building a dam to retain the lava (the lava quickly overtopped it), plugging some natural channels, and diverting the lava into other, newly constructed channels.

Figure 4.23
Mount Fuji, woodblock print by Japanese artist Hiroshige (1797–1858).

Astrogeology Box 4.1 Figure 2

A volcanic eruption on Jupiter's moon Io.

Photo by NASA

to the lunar maria and a number of volcanoes, some with associated lava flows.

Mars has at least nineteen large shield volcanoes, probably composed of basalt. The largest one, Olympus Mons (box figure 1), is three times the height of Mount Everest and wider than the state of California. Its caldera is more than 90 kilometers across.

Jupiter's moon Io is the only object in the solar system besides the Earth that is certain to have currently active volcanoes (box figure 2). Ten have been observed by spacecraft and at least seven of those have erupted for periods of at least four months. Material rich in sulfur compounds is thrown at least 500 kilometers into space at speeds of up to 3,200 kilometers per hour. This material often forms umbrella-shaped clouds as it spreads out and falls back to the surface. Lakes of molten sulfur and huge multicolored (black, yellow, red, orange, and brown) lava flows of sulfur or a sulfur-silicate mixture are common. More than 100 calderas larger than 25 kilometers across have been observed, including one that vents sulfur gases. Clouds of sulfur and sulfur dioxide often form and precipitate reddish sulfur dioxide "snow." The energy source for Io's volcanoes may be the gravitational pulls of Jupiter and two of its other larger satellites, causing Io to heat up much as a piece of wire will do if it is flexed continuously. Neptune's moon Triton may have active volcanoes that erupt nitrogen.

Several of Venus' volcanoes may be active, emitting large amounts of sulfur gases and causing the almost continuous lightning that has been observed by spacecraft.

Nearly half the planet Mars may be covered with volcanic material. There are areas of extensive lava flows similar

Volcanic Domes

Volcanic domes are steep-sided, dome- or spine-shaped masses of volcanic rock formed from viscous lava that solidifies in or immediately above a volcanic vent. A volcanic dome grew within the caldera of Mount St. Helens after the climactic eruption of May 1980 (figure 4.24). This was expected because of the exceptionally high viscosity of the lava from the St. Helens eruptions. In 1983 alone, the dome increased its elevation by 200 meters. Most of the viscous lavas that form volcanic domes are high in silica. They solidify as rhyolite or, less commonly, andesite if minerals crystallize, or as obsidian if no minerals crystallize.

Figure 4.24

The dome in Mount St. Helens crater as seen from a biplane flying inside the crater in 1993.

Photo by C. C. Plummer

4.3 ENVIRONMENTAL GEOLOGY

A Tale of Two Volcanoes—Lives Lost and Lives Saved in the Caribbean

Montserrat and Martinique are two of the tropical islands that are part of a volcanic island arc (box figure 1). Both islands had major eruptions during the twentieth century that destroyed towns. Violent and deadly pyroclastic flows associated with growth of volcanic domes caused most of the destruction. For one island the death toll was huge and for the other it was minimal.

In 1902, the port city of St. Pierre on the island of Martinique was destroyed after a period of dome growth on Mount Pelée. A series of pyroclastic flows broke out of the dome and flowed down the sides of the volcano. Searingly hot pyroclastic flows can travel at over 100 km per hour and will destroy any living things in their paths. After the pyroclastic flows began, the residents of St. Pierre became fearful and many wanted to leave the island. The authorities claimed there was no danger and prevented evacuation. The climax came on the morning of May 8, when great exploding clouds descended like an avalanche down the mountainside, raced down a stream valley, through the port city and onto the harbor. St. Pierre and the ships anchored in the harbor were incinerated (see fig. 4.7, pg. 84). Temperatures within the pyroclastic flow were estimated at 700°C. About 28,000 people were burned to death or suffocated (of the four survivors, one was a condemned prisoner in a poorly ventilated dungeon).

Almost a century later, in July 1995, small steam-ash eruptions began at Soufrière Hills volcano on the neighboring island of Montserrat. As a major eruption looked increasingly likely, teams of volcanologists from France, the U.K., the U.S. (including members of the U.S. Geological Survey's Volcano Disaster Assistance Team that had successfully predicted the eruption of Mt. Pinatubo in the Philippines, as described in chapter 1), and elsewhere flew in to study the volcano and help assess the hazards. An unprecedented array of modern instruments (including seismographs, tiltmeters, gas analyzers) were

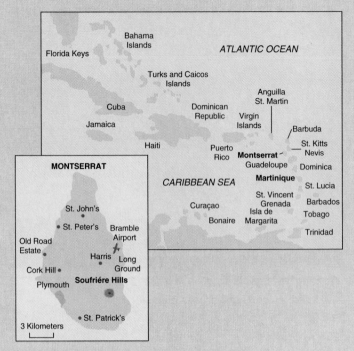

Box 4.3 Figure 1

Map of the West Indies showing location of Montserrat, Martinique, and Soufriére Hills volcano.

deployed around the volcano. In November 1995 viscous, andesitic lava built a dome over the vent. Pyroclastic flows began when the dome collapsed in March 1996. Pyroclastic flows continued with more dome building and collapsing. The populace of Montserrat became, volcanologically, the world's most highly educated citizenry. By 1997, nearly all of the people in the southern part of the island were evacuated, following advice from the scientific teams. In June 1997 large eruptions took place and pyroclastic flows destroyed the capital city of Plymouth. In contrast to the tragedy of St. Pierre only 19 people were killed.

In August, 1997, major eruptions resumed. This time the northern part of the island, previously considered safe, was faced with pyroclastic flows (box figure 2) and more people were evacuated from the island.

Montserrat Volcano Observatory home page
http://www.geo.mtu.edu/volcanoes/west.indies/soufriere/govt

Box 4.3 Figure 2

Eruption of Soufriére Hills volcano on Montserrat, August 4, 1997. An ash cloud billows upward above a ground-hugging pyroclastic flow.

Photo by AP/Kevin West

Because the thick, pasty lava that squeezes from a vent is too viscous to flow, it builds up a steep-sided dome or spine (figure 4.25). Some volcanic domes act like champagne corks, keeping gases from escaping. If the plug is removed or broken, the gas escapes suddenly and violently. Some of the most destructive volcanic explosions known have been associated with volcanic domes.

Lava Floods

Not all extrusive rocks are associated with volcanoes. Lava that is very nonviscous and flows almost as easily as water, does not build a cone around its vents. Such lava is, of course, mafic (low in silica). It is extruded at a temperature well above the melting point of basalt to which it solidifies.

Plateau basalts were produced during the geologic past by vast outpourings of lava. The Columbia plateau area of

Viscous lava welling up into a crater

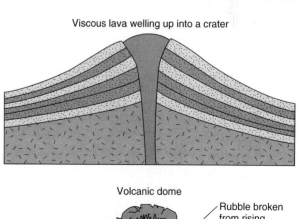

Volcanic dome

Rubble broken from rising dome

Figure 4.25

A volcanic dome forming in the crater of a cinder cone.

Volcanism and Extrusive Rocks   **95**

Figure 4.26

Basalt layers in the Columbia plateau, Washington.

Photo by P. Weis, U.S. Geological Survey

Washington, Idaho, and Oregon (see figure 1.5), for example, is constructed of layer upon layer of basalt (figure 4.26), in places as thick as 3,000 meters. Each individual flood of lava added a layer usually between 15 and 100 meters thick and sometimes thousands of square kilometers in extent.

Basalt layers give the landscape a striking appearance in most places where they are exposed. Instead of stacked-up slabs or tablets of solid, unbroken rock, the individual layers appear to be formed of parallel, vertical columns, mostly six-sided. This characteristic of basalt is called **columnar structure** or **columnar jointing** (figure 4.27). The columns can be explained by the way in which basalt contracts as it cools *after* solidifying. Basalt solidifies completely at temperatures below about 1200°C. The hot layer of rock then continues to cool to temperatures normal for the earth's surface. Like most solids,

basalt contracts as it cools. The layer of basalt is easily able to accommodate the shrinkage in the narrow vertical dimension; but the cooling rock cannot "pull in" its edges, which may be many kilometers away. The tension fractures the rock into an orderly hexagonal pattern.

Submarine Eruptions

Submarine eruptions, notably those occurring along mid-oceanic ridges, almost always consist of mafic lavas that create basalt. As described in chapter 3, basaltic rock, thought to have been formed from lava erupting along mid-oceanic ridges, or solidifying underground beneath the ridges, makes up virtually the entire crust underlying the oceans. In a few places—Iceland, for example—volcanic islands rise above the otherwise submerged system (see Box 4.4).

A

B

Figure 4.27

Columnar jointing at Devil's Postpile, California. (*A*) The columns as seen from above. (Scratches were caused by glacial erosion as described in chapter 12.) (*B*) Side view.

Photos by C. C. Plummer

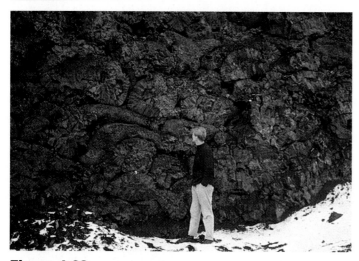

Figure 4.28

Pillow basalt in Iceland.

Photo by R. W. Decker

Figure 4.29

Pillow basalt on a mid-oceanic ridge. Photo taken from a submersible vessel.

Courtesy of Woods Hole Oceanographic Institution

Pillow Basalts

Figure 4.28 shows **pillow structure**—rocks, generally basalt, occurring as pillow-shaped rounded masses closely fitted together. From observations of submarine eruptions by divers, we know how the pillow structure is produced. Elongate blobs of lava break out of a thin skin of solid basalt over the top of a flow that is submerged in water. Each blob is squeezed out like toothpaste, and its surface is chilled to rock within seconds. A new blob forms as more lava inside breaks out. Each new pillow settles down on the pile, with little space left in between. Newly formed pillow basalt (figure 4.29) along the crests of mid-oceanic ridges (regarded as diverging boundaries) is important evidence supporting plate-tectonic theory. When pillow basalts are found exposed in mountain ranges in various parts of the world, they are usually indications of submarine eruption in the geologic past followed, much later, by uplift.

4.4
ENVIRONMENTAL GEOLOGY

An Icelandic Community Battles a Volcano—and Wins

The inhabitants of Iceland are in daily contact with volcanism. Volcanic heat mitigates the harsh cold of these sub-Arctic islands through the geothermal water that provides warmth for homes.

In 1973, eruptions began on one of the tiny Vestmann Islands south of Iceland's main island. Heimaey, one of Iceland's most important and prosperous fishing ports, was nearly destroyed by the eruptions. First ash and then lava flows erupted from a new fissure on the edge of the town. The ashfall quickly buried nearby parts of the town and accumulated on the streets and houses. Flaming spatter ignited roofs, burning some houses. Within a week a cinder cone 150 meters high stood on the edge of the town. Most residents were evacuated, but the townspeople who remained worked hard to save the rest of the town. Ash was shoveled off roofs before a heavy accumulation could cause them to collapse. Windows were sealed off with sheet metal to prevent drifting ash from breaking them. Despite the constant volcanic activity, water and electrical services were maintained. The greatest fear was that the harbor would be blocked by a 40-meter-high tongue of rubbly *aa* lava that began moving slowly away from the main flow and toward the harbor, engulfing twelve houses in one day. The town's livelihood depended on the survival of the harbor.

To halt the lava's advance, fire hoses and large pumps were put to work to cool down the lava with seawater. This unprecedented step apparently worked; the lava front hardened enough so that new lava was deflected elsewhere. The harbor was saved, but the island was covered with black ash and lava.

Five months after they began, the eruptions stopped. Nearly all the evacuated residents returned to resume their lives on the volcano-built island. Ash cleared from the streets and buildings was used to make an extension of the airport runway.

The town benefited by having a unique heating system engineered and built for it. By 1982, nearly every home in the town was being heated by the lava flows. Water is pumped to the lava flow where it percolates through the pyroclastic layers overlying the flows. The water turns to steam, which is collected and goes to a central station. From there, hot water condensed from the steam is circulated through individual homes. The cooled down water is recycled through the lava flows.

A

B

Box 4.4 Figure 1

(*A*) Lava fountaining behind the town on Heimaey. The glow behind the town in the left part of the photo is the lava flow advancing to the harbor. (*B*) Spraying the advancing flow with water.

Photo (A) © Solarfilma; Photo (B) U.S. Geological Survey

Further Reading

McPhee, J. 1989. *The control of nature.* New York: Noonday Press, pp. 95–179.

Summary

Lava is molten rock that reaches the earth's surface, having been formed as *magma* from rock within the earth's crust or from the uppermost part of the mantle.

Lava contains 45% to 75% *silica* (SiO_2). The more silica, the more viscous the lava. Viscosity is also determined by the temperature of the lava. Viscous lavas are associated with more violent eruptions than are fluid lavas. *Volcanic domes* form from the extrusion of very viscous lavas.

A *mafic* lava, relatively low in silica, crystallizes into *basalt,* the most abundant extrusive igneous rock. Basalt, which is dark in color, is composed of minerals that are relatively high in iron, magnesium, and calcium.

Rhyolite, a light-colored rock, forms from *silicic* lavas that are high in silica but contain little iron, magnesium, or calcium. Because potassium and sodium are important elements in rhyolite, its constituent minerals are mostly potassium- and sodium-rich feldspars and quartz.

A lava with a composition between mafic and silicic crystallizes to *andesite,* a moderately dark rock. Andesite contains about equal amounts of ferromagnesian minerals and sodium- and calcium-rich feldspars.

Extrusive rocks are characteristically fine-grained. *Porphyritic* rock contains some larger crystals in an otherwise fine-grained rock. Rocks that solidified too rapidly for crystals to develop form a natural glass called *obsidian.* Gas trapped in rock forms *vesicles.*

Pyroclasts are the result of volcanic explosions. *Tuff* is volcanic ash that has consolidated into a rock. If large pyroclastic fragments have reconsolidated, the rock is a *volcanic breccia.*

A *cinder cone* is composed of loose pyroclastic material that forms steep slopes as it falls back around the crater. Cinder cones are not as large as the other two major types of cones.

A *shield volcano* is built up by successive eruptions of mafic lava. Its slopes are gentle but its volume generally large.

Composite cones are made of alternating layers of pyroclastic material and solidified lava flows. They are not as steep as cinder cones but steeper than shield volcanoes. Young composite volcanoes, predominantly composed of andesite, are aligned along the circum-Pacific belt and, less extensively, in the Mediterranean belt.

Plateau basalts are thick sequences of lava floods. *Columnar jointing* develops in solidified basalt flows. Basalt that erupts underwater forms a *pillow structure.* Pillow basalts are common along the crests of mid-oceanic ridges.

Terms to Remember

andesite 85
basalt 85
block 87
bomb 87
caldera 87
cinder cone 88
circum-Pacific belt 91
columnar structure (columnar jointing) 96
composite volcano (stratovolcano) 90
crater 87
extrusive rock 78
fine-grained rock 85
flank eruption 87
intermediate rock 83

lava 78
lava flows 78
mafic rock 83
magma 78
Mediterranean belt 92
obsidian 85
phenocryst 86
pillow structure (pillow basalts) 97
plateau basalts 95
porphyritic rock 86
pumice 87
pyroclastic flow 84
pyroclasts 78
rhyolite 84

shield volcano 88
silicic (felsic) 83
spatter cone 88
texture 85
tuff 87
vent 87
vesicle 86
viscosity 83
volcanic breccia 87
volcanic dome 93
volcanism 78
volcano 78

Testing Your Knowledge

Use the questions below to prepare for exams based on this chapter.

1. Compare the hazards of lava flows to pyroclastic flows.

2. What roles do gases play in volcanism?

3. What do pillow structures indicate about the environment of volcanism?

4. Name the minerals, and the approximate percentage of each, that you would expect to be present in each of the following rocks: andesite, rhyolite, basalt.

5. What property (or characteristic) of obsidian makes it an exception to the usual geologic definition of *rock?*

6. What determines the viscosity of a lava?

7. What determines whether a series of volcanic eruptions builds a shield volcano, a composite volcano, or a cinder cone? Describe each type of volcanic cone.

8. Explain how a vesicular porphyritic andesite might have formed.

9. Why are extrusive igneous rocks fine-grained?

10. Why don't flood basalts build volcanic cones?

11. Mount St. Helens (a) last erupted violently in 1980 (b) is part of the Cascade Range (c) is located in southern Washington (d) all of the above

12. Volcanic eruptions can affect the climate because (a) they heat the atmosphere (b) volcanic dust and gas can reduce the amount of solar radiation that penetrates the atmosphere (c) they change the elevation of the land (d) all of the above

13. Whether volcanic eruptions are very explosive or relatively quiet is largely determined by (a) the amount of gas in the lava or magma (b) the ease or difficulty with which the gas escapes to the atmosphere (c) the viscosity of a lava (d) all of the above

14. Temperatures at which lavas solidify range from about ___°C for silicic rocks to ___°C for mafic rocks. (a) 100, 200 (b) 300, 1,000 (c) 700, 1,200 (d) 1,000, 2,000

15. One gas typically not released during a volcanic eruption is (a) water vapor (b) carbon dioxide (c) sulfur dioxide (d) hydrogen sulfide (e) oxygen

16. Mafic rocks contain about ___% silica. (a) 10 (b) 25 (c) 50 (d) 65 (e) 80

17. Silicic rocks contain about ___% silica. (a) 10 (b) 25 (c) 50 (d) 70 (e) 80

18. Which is not an extrusive igneous rock? (a) granite (b) rhyolite (c) basalt (d) andesite

19. Which is not a major type of volcano? (a) shield (b) cinder cone (c) composite (d) stratovolcano (e) spatter cone

20. A typical example of a shield volcano is (a) Mount St. Helens (b) Kilauea in Hawaii (c) El Chichón (d) Mount Vesuvius

21. An example of a composite volcano is (a) Mount St. Helens (b) El Chichón (c) Mount Vesuvius (d) all of the above

22. Which volcano is not usually made of basalt? (a) shield (b) composite cone (c) spatter cone (d) cinder cone

23. An igneous rock made of pyroclasts has a texture called (a) fragmental (b) vesicular (c) porphyritic (d) fine-grained

Expanding Your Knowledge

1. What might explain the remarkable alignment of the Cascade volcanoes?

2. What would the present-day environmental effects be for an eruption such as that which created Crater Lake?

3. Why are there no active volcanoes in the eastern parts of the United States and Canada?

4. Why are continental igneous rocks richer in silica than oceanic igneous rocks?

Exploring Resources

Bullard, F. M. 1984. *Volcanoes of the earth.* 2d ed. Austin: University of Texas Press.

Decker, R. W., and B. B. Decker. 1989. *Volcanoes, with CD-ROM.* 3d ed. New York: W. H. Freeman.

Decker, R. W., and B. B. Decker. 1991. *Mountains of fire: The nature of volcanoes.* New York: Cambridge University Press.

Fisher, R. V., G. Heiken, and J. B. Hulen. 1997. *Volcanoes: Crucibles of change.* Princeton, N.J.: Princeton University Press.

Francis, P. 1993. *Volcanoes: A planetary perspective.* New York: Oxford University Press.

Harris, S. L. 1989. *Fire mountains of the west: The Cascade and Mono Lake volcanoes.* Missoula, Montana: Mountain Press.

Krafft, M. 1993. *Volcanoes: Fire from the earth.* New York: Harry N. Abrams. (Maurice Krafft and his wife Katia were the world's foremost photographers of volcanoes before they were killed during the June 1991 eruption of Unzen Volcano, Japan.)

Simpkin, T., and L. Siebert. 1994. *Volcanoes of the world.* 2d ed. Tucson, Arizona: Geoscience Press.

Wright, T. L., and T. C. Pierson. 1992. *Living with volcanoes.* U.S. Geological Survey Circular 1073.

Eruption at sea. VHS. Ka'io Productions; P.O. Box 476, Volcano, Hawaii 96785.

Eruptive phenomena of Kilauea's East Rift Zone. VHS. Ka'io Productions; P.O. Box 476, Volcano, Hawaii 96785.

In the path of a killer volcano. VHS. Films for the Humanities & Sciences, Princeton, NJ. First-rate video on Mt. Pinatubo produced for the NOVA television series.

Rivers of fire. VHS. Hawaii Natural History Association; P.O. Box 74, Hawaii Volcanoes National Park, Hawaii 96718.

http://volcano.und.edu/Volcano World.

This site, supported by NASA, primarily offers educational material, including opportunities to learn how volcanoes work and submit questions to volcanologists.

http://www.dartmouth.edu/~volcano/ The Electronic Volcano.

Emphasis is on active volcano resources in libraries and on other information servers.

http://www.geo.mtu.edu/volcanoes/ Michigan Tech Volcanoes Page.

Focus for this site is to provide scientific and educational information that can lead to volcanic hazard mitigation.

Images of Kilauea East Rift Zone eruption, 1983–93. U.S. Geological Survey. Includes 475 photographs.

Explore Kilauea Volcano. Fire Work Studios, Ho-Ho-Kus, NJ. For more information: www.fireworkstudios.com

Volcanoes: Life on the Edge. Corbis, P.O. Box 2284, South Burlington, VT 05407. Phone 800-246-2065.

Interacting with Journey Through Geology CD-ROM

Expand your knowledge of the concepts presented in this chapter by using the CD-ROM to answer the following questions.

1. Using both this book and the CD-ROM that accompanies it, complete the following chart. The answers can be found in the volcanoes module of the CD-ROM, in the Watch a Volcano Grow activity, and on pages 87–96 of your book. To fill in the chart, answer these questions for each kind of volcano:

   a.) What is the viscosity of the magma that builds this type of volcano?

   b.) Of what composition of magma is this kind of volcano usually composed (basalt, andesite, rhyolite)?

   c.) What materials are produced in the eruption of this type of volcano (lava flows, pyroclasts, volcanic ash, pyroclastic flows)?

   d.) Are eruptions usually explosive or nonexplosive?

   e.) How long does this kind of volcano live (continue to have eruptions) and how frequent are eruptions?

Volcano	Magma viscosity	Magma composition	Products of eruption	Explosivity	Life & frequency
Shield					
Composite cone					
Cinder cone					
Lava dome					
Flood basalt					

2. Try designing a volcano (Volcanoes module, Design a Volcano activity) with these parameters: Low viscosity, high volatility, any volume. What happens? Why do you think these conditions virtually never occur in nature?

3. Convergent Margins module, Mt. St. Helens Erupts: What effects did the 1980 Mt. St. Helens eruption have? How was this eruption related to the motion of tectonic plates?

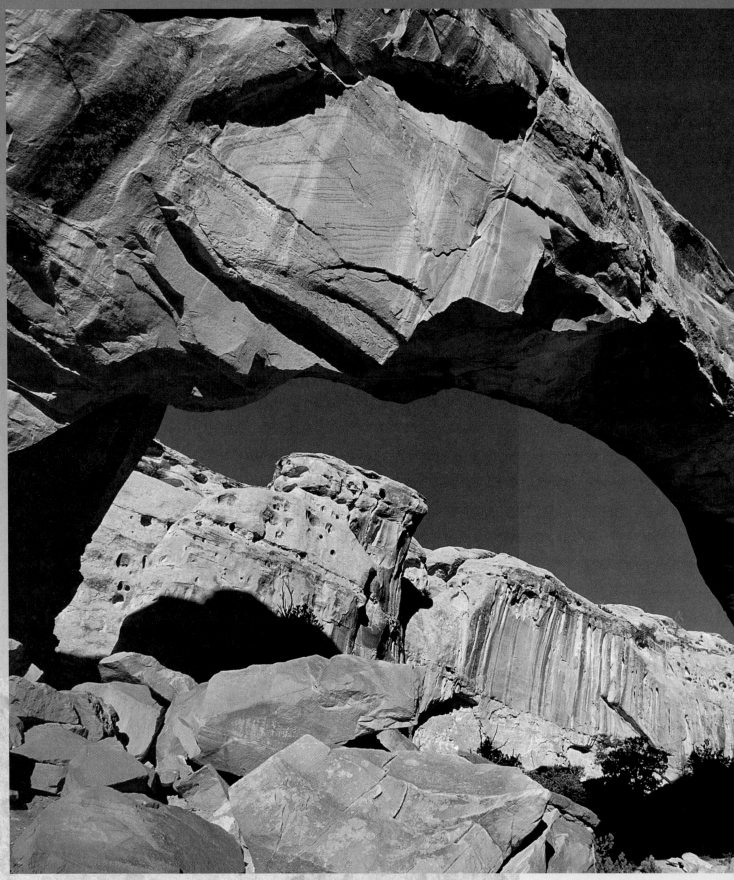

Rock arches formed by weathering of sandstone. Capitol Reef National Park,
Utah.

CHAPTER
5

Weathering and Soil

*I*n this chapter, you will study several visible signs of weathering in the world around you, such as the cliffs and slopes of the Grand Canyon and the rounded edges of boulders. As you study these features, keep in mind that weathering processes made the planet suitable for human use. From the weathering of rock eventually came the development of soil, on which the world's food supply depends.

How and why does rock weather? You learned in chapters 3 and 4 that the minerals making up igneous rocks crystallize at relatively high temperatures and sometimes at high pressures as magma and lava cool. Although these minerals are stable when they form, most of them are not stable during prolonged exposure at the earth's surface. In this chapter you see how minerals and rocks change when they are subjected to the physical and chemical conditions existing at the earth's surface. Rocks undergo mechanical weathering (physical disintegration) and chemical weathering (decomposition) as they are attacked by air and water. Your knowledge of the chemical composition and atomic structure of minerals will help you understand the reactions that occur during chemical weathering.

Weathering processes create sediments (primarily mud and sand) and soil. Sedimentary rocks, which form from sediments, are discussed in chapter 6. In a general sense, weathering prepares rocks for erosion and is a fundamental part of the rock cycle, transforming rocks into the raw material that eventually becomes sedimentary rocks.

Weathering, Erosion, and Transportation

Rocks exposed at the earth's surface are constantly being altered by water, air, changing temperature, and other environmental factors. The term **weathering** refers to the group of destructive processes that change the physical and chemical character of rock at or near the earth's surface. The tightly bound crystals of an igneous rock can be loosened and altered to new minerals by weathering.

It is important to distinguish between weathering and *erosion,* and between erosion and *transportation.* Weathering breaks down rocks that are either stationary or moving. **Erosion** is the picking up or *physical removal* of rock particles by an agent such as streams or glaciers. Weathering helps break down a solid rock into loose particles that are easily eroded. Most eroded rock particles are at least partially weathered, but rock can be eroded before it has weathered at all. A stream can erode weathered or unweathered rock fragments.

After a rock fragment is picked up (eroded), it is transported. **Transportation** is the movement of eroded particles by agents such as rivers, waves, glaciers, or wind. Weathering processes continue during transportation. A boulder being transported by a stream can be physically worn down and chemically altered as it is carried along by the water.

How Weathering Alters Rocks

Rocks undergo both mechanical weathering and chemical weathering. **Mechanical weathering** (or physical disintegration) includes several processes that break rock into smaller pieces. The change in the rock is physical; there is little or no chemical change. For example, water freezing and expanding in cracks can cause rocks to disintegrate physically. **Chemical weathering** is the decomposition of rock from exposure to water and atmospheric gases (principally carbon dioxide, oxygen, and water vapor). As rock is decomposed by these agents, new chemical compounds form.

Mechanical weathering breaks up rock but does not change the composition. A large mass of granite may be broken into smaller pieces by frost action, but its original crystals of quartz, feldspar, and ferromagnesian minerals are unchanged. On the other hand, if the granite is being chemically weathered, some of the original minerals are chemically changed into different minerals. Feldspar, for example, will change into a clay mineral (with a crystal structure similar to mica). In nature, mechanical and chemical weathering usually occur together, and the effects are interrelated.

Weathering is a relatively long, slow process. Typically, cracks in rock are enlarged gradually by frost action or plant growth (as roots pry into rock crevices), and as a result, more surfaces are exposed to attack by chemical agents. Chemical weathering initially works along contacts between mineral grains. Tightly bound crystals are loosened as weathering products form at their contacts. Mechanical and chemical weathering then proceed together, until a once tough rock slowly crumbles into individual grains.

Solid minerals are not the only products of chemical weathering. Some minerals—calcite, for example—dissolve when chemically weathered. We can expect limestone and marble, rocks consisting mainly of calcite, to weather chemically in quite a different way than granite.

Effects of Weathering

The results of chemical weathering are easy to find. Look along the edges or corners of old stone structures for evidence. The inscriptions on statues and tombstones that have stood for several decades may no longer be sharp (figure 5.1). Building blocks of limestone or marble exposed to rain and atmospheric gases may show solution effects of chemical weathering in a surprisingly short time. Granite buildings may also show the effects of weathering, although the effects may not become apparent for centuries. Mineral grains in granite may be loosened, cracks enlarged, and the surface discolored and dulled by the products of weathering. Surface discoloration is also common on rock *outcrops,* where rock is exposed to view, with no plant or soil cover. That is why field geologists carry rock hammers—to break rocks to examine unweathered surfaces.

We tend to think of weathering as destructive because it mars statues and building fronts. As rock is destroyed, however, valuable products can be created. Soil is produced by rock weathering, so most plants depend on weathering for the soil they need in order to grow. In a sense, then, all agriculture depends upon weathering. Weathering products dissolved in the sea make seawater salty and serve as nutrients for many marine organisms. Some metallic ores, such as those of copper and aluminum, are concentrated into economic deposits by chemical weathering.

Many weathered rocks display interesting shapes. A **spheroidally weathered boulder** is a rock that has been rounded by weathering from an initial blocky shape. It is rounded because chemical weathering acts more rapidly or intensely on the corners and edges of a rock than on the smooth rock faces (figures 5.2 and 5.3).

Differential weathering is the term for varying rates of weathering in an area where some rocks are more resistant to weathering than others. Resistant rocks weather slowly, and may protrude above softer rocks that weather rapidly. Figures 5.4 and 5.5 show some striking landforms produced by erosion of rocks that weather at different rates.

Mechanical Weathering

Of the many processes that cause rocks to disintegrate, the most effective are frost action, abrasion, and pressure release.

Frost Action

Frost action—the mechanical effect of freezing water on rocks—commonly occurs as frost wedging or frost heaving.

A

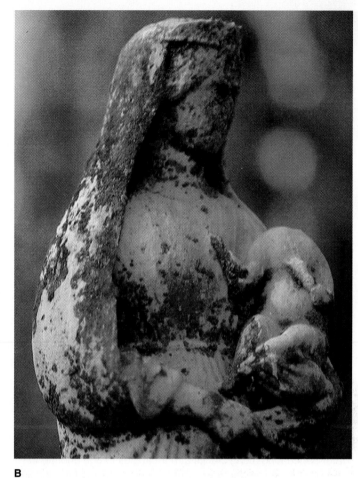

B

Figure 5.1

(*A*) Chemical weathering in a humid climate has pitted and discolored this marble tombstone since it was erected in 1882. (*B*) This marble statue has lost most of the fine detail on the face by chemical weathering.

A

B

Figure 5.2

Spheroidal weathering of granite. (*A*) The edges and corners of an angular rock are attacked by weathering from more than one side and retreat faster than flat rock faces. (*B*) In this way, angular rocks become rounded, often shedding partially weathered minerals in onionlike layers.

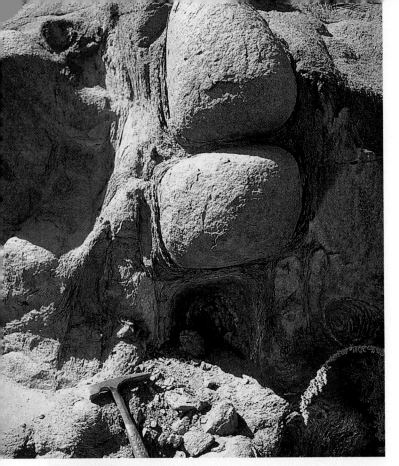

Figure 5.3

Water penetrating along cracks at right angles to one another in this igneous rock produces spheroidal weathering of once-angular blocks underground. Salt River Canyon, Arizona.

Figure 5.4

Pedestal rock near Lees Ferry, Arizona. Resistant sandstone cap protects weak shale pedestal from weathering and erosion. Hammer for scale is barely visible at base of pedestal.

In **frost wedging** the expansion of freezing water pries rock apart. Most rock contains a system of cracks called *joints,* caused by the slow flexing of brittle rock by deep-seated earth forces (see chapter 15). Water that has trickled into a joint in

Figure 5.5

Sedimentary rocks in the Grand Canyon, Arizona. In the foreground layers of sandstone resist weathering and form steep cliffs. Less resistant layers of shale weather to form gentler slopes between cliffs.

a rock can freeze and expand by as much as 9 percent when the temperature drops below 0°C (32°F). The expanding ice wedges the rock apart, extending the joint or even breaking the rock into pieces (figures 5.6 and 5.7). Frost wedging is most effective in regions with many days of freezing and thawing (mountaintops and midlatitude regions with pronounced seasons). Partial thawing during the day adds new water to the ice in the crack; refreezing at night adds new ice to the old ice.

Frost heaving lifts rock and soil vertically. Solid rock conducts heat better than soil, so on a cold winter day the bottom of a partially buried rock will be much colder than soil at the same depth. As the ground freezes in winter, ice forms first under large rock fragments in the soil. The expanding ice layers push boulders out of the ground, a process well known to New England farmers and other residents of rocky soils. Frost heaving bulges the ground surface upward in winter, breaking up roads and leaving lawns spongy and misshapen after the spring thaw.

Abrasion

Another process that can mechanically weather rock is **abrasion,** the grinding away of rock by friction and impact during trans-

A

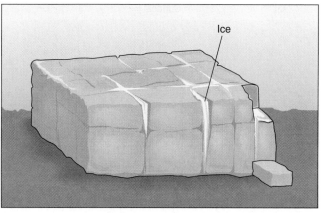

B

Figure 5.6

Frost wedging occurs when (A) water fills joints (cracks) in a rock and then freezes. (B) Expanding ice wedges the rock apart.

portation. As loose fragments of rock are picked up and moved by a stream, they tumble against one another and against the rocky stream bottom (figure 5.8). The fragments gradually grind themselves into smaller and smaller pieces and also grind away the rock of the stream bottom. Glaciers, waves, and even wind are other agents that carry and abrade rock fragments.

Pressure Release

The reduction of pressure on a body of rock can cause it to crack as it expands; **pressure release** is a significant type of mechanical weathering. A large mass of rock, such as a batholith, may originally form under great pressure from the weight of several kilometers of rock above it. This batholith is gradually exposed by tectonic uplift of the region followed by erosion of the overlying rock (figure 5.9). The removal of the great weight of rock above the batholith, usually termed *unloading,* allows the granite to expand upward. Cracks called **sheet joints** develop parallel to the outer surface of the rock as the outer part of the rock expands more than the inner part (figures 5.9 and 5.10). On slopes, gravity may cause the rock between such joints to break loose in concentric slabs from the underlying granite mass. This process of spalling off of rock layers is called **exfoliation;** it is somewhat similar to peeling layers from an onion. **Exfoliation domes** (figure 5.11) are large, rounded landforms developed in massive rock, such as granite, by exfoliation. Some famous examples of exfoliation domes include Stone Mountain in Georgia and Half Dome in Yosemite.

Several other processes mechanically weather rock but in most environments are less effective than frost action, abrasion, and pressure release. *Plant growth,* particularly roots growing in cracks (figure 5.12A), can break up rocks, as can *burrowing animals.* Such activities help to speed up chemical weathering by enlarging passageways for water and air. The *pressure of salt crystals* formed as water evaporates inside small spaces in rock also helps

Figure 5.7

Frost wedging has broken this granite into large, angular fragments as ice expanded in rock joints. Sierra Nevada, California.

Photo by B. Amundson

Figure 5.8

During periods of high flow these granite boulders are reduced in size and corners are rounded by the impact of sand and gravel moving along the bottom or bed of the stream.

Weathering and Soil   **107**

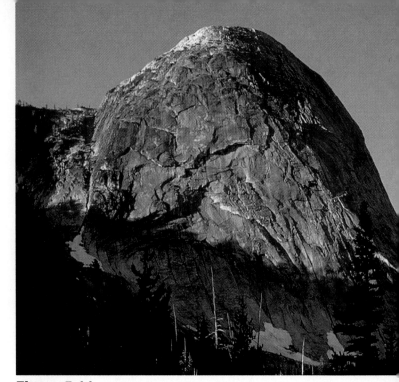

Figure 5.11

Exfoliation dome, Yosemite National Park, California. Onionlike layers of rock are peeling off the dome.

Figure 5.9

Sheet joints caused by pressure release. A granite batholith is exposed by regional uplift followed by the erosion of the overlying rock. Unloading reduces pressure on the granite and causes outward expansion. Sheet joints are closely spaced at the surface where expansion is greatest. Exfoliation of rock layers produces rounded exfoliation domes.

Figure 5.10

Sheet joints in a granite outcrop near the top of the Sierra Nevada, California. The granite formed several kilometers below the surface, and expanded outward when it was exposed by uplift and erosion. Note that the sheet joints are closer together near the top of the outcrop, where the pressure release is the greatest.

to disintegrate desert rocks (figure 5.12*B*). *Extreme changes in temperature,* as in a forest fire, can cause a rock to expand or contract until it cracks. Whatever processes of mechanical weathering are at work, as rocks disintegrate into smaller fragments the total surface area increases (figure 5.13), allowing more extensive chemical weathering by water and air.

Chemical Weathering

The processes of chemical weathering, or *rock decomposition,* transform rocks and minerals exposed to water and air into new chemical products. A mineral that crystallized deep underground from a water-deficient magma may eventually be exposed at the earth's surface, where it can react with the abundant water there to form a new, different mineral. A mineral containing very little oxygen may react with oxygen in the air, extracting oxygen atoms from the atmosphere and incorporating them into its own crystal structure, thus forming a different mineral. These new minerals are weathering products. They have adjusted to physical and chemical conditions at (or near) the earth's surface. Minerals change gradually at the surface until they come into *equilibrium,* or balance, with the surrounding conditions.

Role of Oxygen

Oxygen is abundant in the atmosphere and quite active chemically, so it often combines with minerals or with elements within minerals that are exposed at the earth's surface.

The rusting of an iron nail exposed to air is a simple example of chemical weathering. Oxygen from the atmosphere

A

B

Figure 5.12

(*A*) Tree roots are prying this rock apart as they grow within the rock joints, San Gabriel Mountains, California. (*B*) This rock is being broken by the growth of salt crystals, which precipitate as water evaporates within the cracks in the rock. Death Valley, California.

Photo *A* by © John Shelton
Photo *B* by © by Frank M. Hanna

Figure 5.13

Mechanical weathering can increase the surface area of a rock, increasing the rate of chemical weathering. As a cube breaks up into smaller pieces, its volume remains the same but its surface area increases.

combines with the iron to form iron oxide, the reaction being expressed as follows:

$$4Fe \quad + \quad 3O_2 \quad \rightarrow \quad 2Fe_2O_3$$
$$iron \quad + \quad oxygen \quad \rightarrow \quad iron \; oxide$$

Iron oxide formed in this way is a weathering product of numerous minerals containing iron, such as the ferromagnesian group (pyroxenes, amphiboles, biotite, and olivine). The iron in the ferromagnesian silicate minerals must first be separated from the silica in the crystal structure before it can oxidize. The iron oxide (Fe_2O_3) formed is the mineral **hematite,** which has a brick-red color when powdered. If water is present, as it usually is at the earth's surface, the iron oxide combines with water to form **limonite,** which is the name for a group of mostly amorphous, hydrated iron oxides (often including the mineral *goethite*), which are yellowish-brown when powdered. The general formula for this group is $Fe_2O_3 \cdot nH_2O$ (the *n* represents a small, whole number such as 1, 2, or 3 to show a variable amount of water). The brown, yellow, or red color of soil and many kinds of sedimentary rock is commonly the result of small amounts of hematite and limonite released by the weathering of iron-containing minerals (figure 5.14).

Role of Acid

The most effective agent of chemical weathering is acid. Acids are chemical compounds that give off hydrogen ions (H^+) when they dissociate, or break down, in water. Strong acids

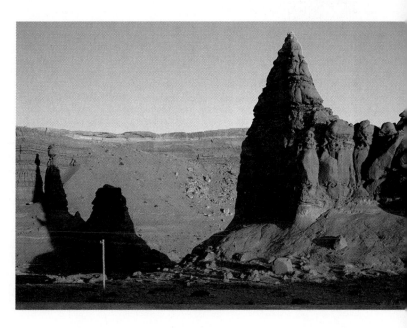

Figure 5.14

This weathered exposure of sandstone has been colored red by hematite, released by the chemical weathering of ferromagnesian minerals, southern Utah.

Weathering and Soil **109**

Figure 5.15

A mudpot of boiling mud is created by intense chemical weathering of the surrounding rock by the acid gases dissolved in a hot spring. Yellowstone National Park, Wyoming.

produce a great number of hydrogen ions when they dissociate, and weak acids produce relatively few such ions.

The hydrogen ions given off by natural acids disrupt the orderly arrangement of atoms within most minerals. Because a hydrogen ion has a positive electrical charge and a very small size, it can substitute for other positive ions (such as Ca^{++}, Na^+ or K^+) within minerals. This substitution changes the chemical composition of the mineral and disrupts its atomic structure. The mineral decomposes, often into a different mineral, when it is exposed to acid.

Some strong acids occur naturally on the earth's surface, but they are relatively rare. Sulfuric acid and hydrofluoric acid are strong acids emitted during many volcanic eruptions. They can kill trees and cause intense chemical weathering of rocks near volcanic vents. The bubbling mud of Yellowstone National Park's mudpots (figure 5.15) is produced by rapid weathering caused by acidic sulfur gases that are given off by some hot springs. Strong acids also drain from some mines as sulfur-containing minerals such as pyrite oxidize and form acids at the surface (figure 5.16). Uncontrolled mine drainage can kill fish and plants downstream and accelerate rock weathering.

The most important natural source of acid for rock weathering at the earth's surface is dissolved carbon dioxide (CO_2) in water. Water and carbon dioxide form *carbonic acid* (H_2CO_3), a weak acid that dissociates into the hydrogen ion and the bicarbonate ion (see equation A in table 5.1). Even though carbonic acid is a weak acid, it is so abundant at the earth's surface that it is the single most effective agent of chemical weathering.

The earth's atmosphere (mostly oxygen and nitrogen) contains 0.03% carbon dioxide. Some of this carbon dioxide dissolves in rain as it falls, so most rain is slightly acidic when it hits the ground. Large amounts of carbon dioxide also dissolve in water that percolates through soil. The openings in soil are filled with a gas mixture that differs from air. Soil gas has a much higher content of carbon dioxide than does air (up to 10%), because carbon dioxide is produced by the decay of organic matter and the respiration of soil organisms, such as worms. Rainwater that has trickled through soil is therefore usually acidic and readily attacks minerals in the unweathered rock below the soil (figure 5.17).

Solution Weathering

Some minerals are completely dissolved by chemical weathering. *Calcite,* for instance, goes into solution when exposed to carbon dioxide and water, as shown in equation B in table 5.1. The carbon dioxide and water combine to form carbonic acid, which dissociates into the hydrogen ion and the bicarbonate ion, as you have seen, so the equation for the solution of calcite can also be written as equation C in table 5.1.

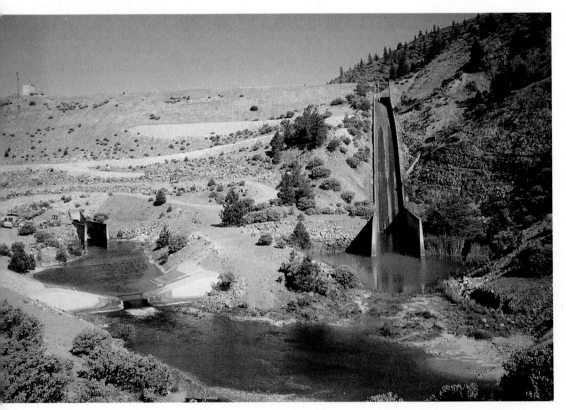

Figure 5.16

Spring Creek debris dam collects acid mine drainage from the Iron Mountain Mines superfund site in northern California

Photo by Charles Alpers, U.S. Geological Survey

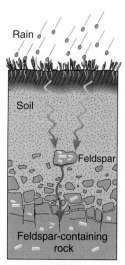

Rain

Rain picks up CO_2 from the atmosphere and becomes acidic

Soil

Water percolating through the ground picks up more CO_2 from the upper part of the soil, becoming more acidic

Feldspar

A rock containing a feldspar crystal, loosened from the rock below, slowly alters to a clay mineral as it reacts with the acidic water

The water carries away soluble ions and SiO_2 to the ground-water supply or to a stream

Feldspar-containing rock

Figure 5.17

Chemical weathering of a feldspar. Water percolating through the soil alters the feldspar to a clay mineral and carries away soluble ions and silica.

There are no solid products in the last part of the equation, indicating that complete solution of the calcite has occurred. Caves can form underground when flowing ground water dissolves the sedimentary rock limestone, which is mostly calcite. Rain can discolor and dissolve statues and tombstones carved from the metamorphic rock marble, which is also mostly calcite (see figure 5.1).

Chemical Weathering of Feldspar

The weathering of feldspar is an example of the alteration of an original mineral to an entirely different type of mineral as the weathered product. When feldspar is attacked by the hydrogen ion of carbonic acid (from carbon dioxide and water), it forms clay minerals. In general, a **clay mineral** is a hydrous aluminum silicate with a sheet-silicate structure like that of mica. Therefore, the entire silicate structure of the feldspar crystal is altered by weathering: feldspar is a framework silicate, but the clay mineral product is a sheet silicate, differing both chemically and physically from feldspar. Partly because of the complexity of the reaction, the chemical weathering of feldspar proceeds at a much slower rate than the solution weathering illustrated by calcite.

Let us look in more detail at the weathering of feldspar (equation D in table 5.1). Rainwater percolates down through soil, picking up carbon dioxide from the atmosphere and the upper part of the soil. The water, now slightly acidic, comes in contact with feldspar in the lower part of the soil (figure 5.17), as shown in the first part of the equation. The acidic water reacts with the feldspar and alters it to a clay mineral.

The hydrogen ion (H^+) attacks the feldspar structure, becoming incorporated into the clay mineral product. When the hydrogen moves into the crystal structure, it releases potassium (K) from the feldspar. The potassium is carried away in solution as a dissolved ion (K^+). The bicarbonate ion from the original carbonic acid does not enter into the reaction; it reappears on the right side of the equation. The soluble potassium and bicarbonate ions are carried away by water (ground water or streams).

All the silicon from the feldspar cannot fit into the clay mineral, so some is left over and is carried away as silica (SiO_2) by the moving water. This excess silica may be carried in solution or as extremely small solid particles.

The weathering process is the same regardless of the type of feldspar: K-feldspar forms potassium ions; Na-feldspar and Ca-feldspar (plagioclase) form sodium ions and calcium ions, respectively. The ions that result from the weathering of Ca-feldspar are calcium ions (Ca^{++}) and bicarbonate ions (HCO_3^-), both of which are very common in rivers and in underground water, particularly in humid regions.

Table 5.1	Chemical Equations Important to Weathering

A. Solution of Carbon Dioxide in Water to Form Acid

$$CO_2 + H_2O \rightleftarrows H_2CO_3 \rightleftarrows H^+ + HCO_3^-$$

carbon dioxide + water ⇄ carbonic acid ⇄ hydrogen ion + bicarbonate ion

B. Solution of Calcite

$$CaCO_3 + CO_2 + H_2O \rightleftarrows Ca^{++} + 2HCO_3^-$$

calcite + carbon dioxide + water ⇄ calcium ion + bicarbonate ion

C. Solution of Calcite

$$CaCO_3 + H^+ + HCO_3^- \rightleftarrows Ca^{++} + 2HCO_3^-$$

D. Chemical Weathering of Feldspar to Form a Clay Mineral

$$2 KAlSi_3O_8 + 2 H^+ + 2 HCO_3^- + H_2O \rightarrow Al_2Si_2O_5(OH)_4 + 2 K^+ + 2 HCO_3^- + 4 SiO_2$$

potassium feldspar + (from CO_2 and H_2O) → clay mineral + (soluble ions) + silica in solution or as fine solid particles

Acid Rain

The burning of coal, oil, and natural gas (the *fossil fuels*) adds a great deal of carbon dioxide to the atmosphere (box figure 1). As you have seen in table 5.1, this carbon dioxide combines with water to form carbonic acid in rain. Coal and oil can also contain nitrogen and sulfur, which are given off as gases (NO_2 and SO_2) when these fuels are burned, forming nitric acid and sulfuric acid in rain. These two acids are much stronger than carbonic acid.

The strength of an acidic solution is measured on the pH scale from 0 to 14 (box figure 2). A solution of pH 7 is chemically neutral, neither acidic nor alkaline. Values below 7 are acidic; the lower the number the more acidic the solution. Values above 7 are alkaline or basic. The pH scale is logarithmic, so a change of 1 on the scale means a change of 10 in the concentration of H^+ ions that make a solution acidic.

Ordinary rain has a pH of about 5.5 to 6 from the small amount of carbon dioxide given off during respiration (every time we exhale, we add a little CO_2 to the atmosphere), and from natural sources of acidic sulfur gases such as volcanoes and coastal marshes. Ordinary rain is about as acidic as milk—hardly a strong acid.

In cities and downwind of industrial smokestacks (often for hundreds of miles) the increased amount of acid gases can reduce the pH of rain to 4, 3, or even 2 (the pH of lemon juice or vinegar). This is the environmental problem termed "acid rain" (although all rain is really acid). This rain in turn can lower the pH of streams and lakes. Such low pH values are hard on organisms; fish may die in streams and lakes polluted by acid rain, and forests suffer under acid

Box 5.1 Figure 1

The burning of fossil fuels releases carbon dioxide and other acid gases, such as sulfur dioxide and nitrous oxide, that combine with water to produce acid rain.

Box 5.1 Figure 2

The pH scale.

rain. This is particularly a problem where rocks and soil do not buffer the acid. For example, areas with exposed limestone are least affected.

Chemical weathering is accelerated by acid rain. Statues and stone buildings in cities weather many times faster than stone structures in rural areas free of acid rain. See figure 5.1.

Chemical Weathering of Other Minerals

The weathering of ferromagnesian or dark minerals is much the same as that of feldspars. Two additional products are found on the right side of the equation—magnesium ions and iron oxides (hematite and limonite).

The susceptibility of the rock-forming minerals to chemical weathering is dependent upon the strength of the mineral's chemical bonding within the crystal framework. Because of the strength of the silicon-oxygen bond, it is quite resistant to chemical weathering. Thus, quartz (SiO_2) is the rock-forming mineral least susceptible to chemical attack at the earth's surface. Ferromagnesian minerals such as olivine, pyroxene, and

amphibole include other positively charged ions such as Al, Fe, Mg, and Ca. The presence of these positively charged ions in the crystal framework makes these minerals vulnerable to chemical attack due to the weaker chemical bonding between these ions and oxygen, as compared to the much stronger silicon-oxygen bonds. For example olivine ($(Fe, Mg)_2SiO_4$) weathers rapidly because its isolated silicon-oxygen tetrahedra are held together by relatively weak ionic bonds between oxygen and iron and magnesium. These ions are replaced by H^+ ions during chemical weathering similar to that described for the feldspars.

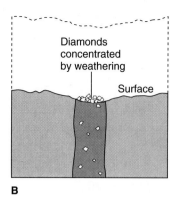

Figure 5.18

Residual concentration by weathering. (*A*) Cross-sectional view of diamonds widely scattered within diamond pipe. (*B*) Diamonds concentrated on surface by removal of rock by weathering and erosion.

Weathering and Diamond Concentration

Diamond is the hardest mineral known and is also extremely resistant to weathering. This is due to the very strong covalent bonding of carbon, as described in chapter 2. However, diamonds are often concentrated by weathering as illustrated in figure 5.18. Diamonds are brought to the surface of the earth in *diamond pipes,* columns of brecciated or broken ultramafic rock that have risen from the upper mantle. Diamonds are widely scattered in diamond pipes when they form. At the earth's surface the ultramafic rock in the pipe is preferentially weathered and eroded away. The diamonds, being more resistant to weathering, are left behind, concentrated in rich deposits on top of the pipes. Rivers may redistribute and reconcentrate the diamonds, as in South Africa. In Canada, diamond pipes have been eroded by glaciers and diamonds may be found widely scattered in glacial deposits.

Weathering and Climate

Chemical weathering is most intense where there is abundant liquid water. If water is scarce, as in deserts, or is frozen into solid ice, chemical weathering is slow or even absent. As a result, the resistance of different rock types to chemical weathering varies from climate to climate. In the wet climate of central Pennsylvania, for example, limestone dissolves readily to form valleys, shale forms low hills, and the ridges of the Appalachian Mountains are mostly formed of resistant, quartz-rich sandstone. In the dry climate of Arizona, however, limestone is as resistant as quartz sandstone, so both limestone and sandstone form the resistant cliffs within the Grand Canyon, while less resistant shale forms the gentle slopes between cliffs (see figure 5.5).

Weathering Products

Table 5.2 summarizes weathering products for the common minerals. Note that quartz and clay minerals commonly are left after complete chemical weathering of a rock. Sometimes other solid products, such as iron oxides, also are left after weathering.

The solution of calcite supplies substantial amounts of calcium ions (Ca^{++}) and bicarbonate ions (HCO_3^-) to underground water. The weathering of Ca-feldspars (plagioclase) into clay minerals can also supply Ca^{++} and HCO_3^- ions, as well as silica (SiO_2), to water. Under ordinary chemical circumstances, the dissolved Ca^{++} and HCO_3^- can combine to form solid $CaCO_3$ (calcium carbonate), the mineral calcite. Dissolved silica can also precipitate as a solid from underground water. This is significant because calcite and silica are the most common materials precipitated as *cement,* which binds loose particles of sand, silt, and clay into solid sedimentary rock (see chapter 6). The weathering of calcite, feldspars, and other minerals is a likely source for such cement.

If the soluble ions and silica are not precipitated as solids, they remain in solution and may eventually find their way into a stream and then into the ocean. Enormous quantities of dissolved material are carried by rivers into the sea (one estimate is 4 billion tons per year). This is the main reason seawater is salty.

Soil

In engineering and construction, soil is the usual name for any kind of loose, unconsolidated earth material; but most geologists commonly use the term **soil** for a layer of *weathered,* unconsolidated material on top of bedrock (a general term for rock beneath soil). Soil scientists further restrict the term *soil* to layers of weathered, unconsolidated material that contains organic matter and is capable of supporting plant growth. (If this definition is used, then the term *regolith* can be applied to any loose surface sediment; soil would be the upper part of the regolith.) A mature, fertile soil is the product of centuries of

Table 5.2	Weathering Products of Common Rock-Forming Minerals		
Original Mineral	**Under Influence of CO_2 and H_2O**	**Main Solid Product**	**Other Products (Mostly Soluble)**
Feldspar	$\longrightarrow$	Clay mineral +	Ions (Na^+, Ca^{++}, K^+), SiO_2
Ferromagnesian minerals (including biotite mica)	$\longrightarrow$	Clay mineral +	Ions (Na^+, Ca^{++}, K^+, Mg^{++}), SiO_2, Fe oxides
Muscovite mica	$\longrightarrow$	Clay mineral +	Ions (K^+), SiO_2
Quartz	$\longrightarrow$	Quartz grains (sand)	
Calcite	$\longrightarrow$	—	Ions (Ca^{++}, HCO_3^-)

Figure 5.19

Negative charges on a clay mineral attract positive ends of water molecules.

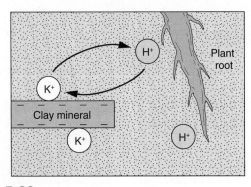

Figure 5.20

Ion exchange between plant root and clay mineral.

mechanical and chemical weathering of rock, combined with the addition and decay of plants and other organic matter.

The term **loam** refers to a soil of approximately equal amounts of sand, silt, and clay. (*Clay-sized* particles usually consist of *clay minerals*). Loamy soils are often well-drained, may contain organic matter, and are often very fertile. *Topsoil* is the upper part of the soil and is more fertile than the underlying *subsoil,* which is often stony and lacks organic matter.

Clay minerals and quartz, the two minerals usually remaining after complete weathering of rock (table 5.2), have important roles in soil development and plant growth. Quartz crystals form sand grains that help keep soil loose and aerated, allowing good water drainage. (Partially weathered crystals of feldspar and other minerals can also form sand-sized grains.)

Clay minerals help to hold water and plant nutrients in a soil. Clay minerals occur as microscopic plates. Because of ion substitution within their sheet silicate structure, most clay minerals have a negative electrical charge on the flat faces of the plates. This negative charge attracts water and nutrient ions to the clay mineral.

The water molecule, made up of two hydrogen atoms and one oxygen atom, is neutral in charge but has a positive end and a negative end. The negative charge on the flat faces of the clay mineral attracts the positive ends of the water molecules to the clay flake (figure 5.19). The clay holds the water loosely enough that most of it is available for uptake by plant roots.

Plant nutrients, such as Ca^{++} and K^+, commonly supplied by the weathering of minerals such as feldspar, are also held loosely on the surface of clay minerals. A plant root is able to release H^+ from organic acids and exchange it for the Ca^{++} and K^+ that the plant needs for healthy growth (figure 5.20).

Soil Horizons

Most soils take a long time to form. The rate of soil formation is controlled by rainfall, temperature, and to some extent the type of bedrock that weathers to form soil. High temperature and abundant rainfall speed up soil formation, but in most places a fully developed soil that can support plant growth takes hundreds or thousands of years to form.

As soils mature, distinct layers appear in them (figure 5.21). Soil layers are called **soil horizons,** and can be distinguished from one another by appearance and chemical composition. Boundaries between soil horizons are usually transitional rather than sharp.

The **O horizon** is the dark-colored soil layer that is rich in organic material and forms just below surface vegetation. This horizon contains decomposed plant material, or *humus,* and contributes to the formation of organic acids that accelerate leaching in the underlying A horizon.

The **A horizon,** or **zone of leaching,** is characterized by the downward movement of water. Part of the rain falling on the ground percolates downward through the soil. This tends to leach, or carry dissolved chemicals downward to lower levels in the soil profile. In a humid (wet) climate, iron oxides and dissolved calcite are most typically leached downward; clays are also transported downward. Leaching may make the A horizon pale and sandy, but the uppermost part is often darkened by humus.

The **B horizon,** or **zone of accumulation,** is a soil layer characterized by the accumulation of material leached downward from the A horizon above. This layer is often quite clayey and stained red or brown by hematite and limonite. Calcite may also build up in B horizons.

The **C horizon** is incompletely weathered parent material, lying below the B horizon. The parent material is commonly the underlying bedrock, which is subjected to mechanical and chemical weathering from frost action, roots, plant acids, and other agents. In such a case, the C horizon is transitional between unweathered bedrock below and developing soil above.

Residual and Transported Soils

A **residual soil** is one that develops from weathering of the rock directly beneath it. Figure 5.21*A* is a diagram of a residual soil developing in a humid climate from a bedrock source. Although this is a typical situation, a number of important agricultural regions in the United States and elsewhere have developed on **transported soils,** which did not form from the local rock but from parent material brought in from some other region. Transported soils usually form on sediment deposited by running water, wind, or glacial ice. For example, mud deposited by a river during times of flooding can form an excellent agricultural soil next to the river after floodwaters recede. The soil-forming mud was not weathered from the rock beneath its present location but was carried downstream from regions perhaps hundreds of miles away. Transported wind deposits called *loess* (see chapter 13) are

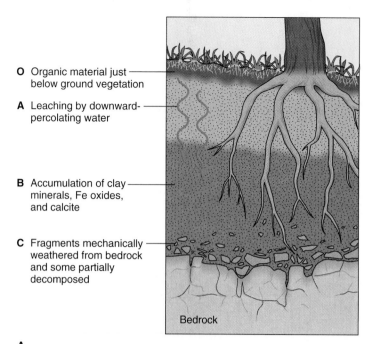

O Organic material just below ground vegetation

A Leaching by downward-percolating water

B Accumulation of clay minerals, Fe oxides, and calcite

C Fragments mechanically weathered from bedrock and some partially decomposed

Bedrock

A

Figure 5.21

(*A*) Soil horizons (*O, A, B,* and *C*) that form in a humid climate. (*B*) A pedalfer soil in Illinois. The *O* horizon is stained dark by humus. Horizon *A* is lighter in color, sandy, and crumbly. The hammer rests on the clayey horizon *B*, which is stained red by hematite, leached downward from horizon *A*.

the parent material for some of the most valuable food-producing soils in the Midwest and the Pacific Northwest.

Soils, Parent Rock, Time, and Slope

The character of a soil depends partly on the parent rock it develops from. A soil developing on weathering granite will be sandy, as sand-sized particles of quartz and partially weathered feldspar are released from the granite. As time passes, the partially weathered feldspar grains weather completely, forming fine-grained clay minerals. The quartz does not weather, so the resulting soil has both sand and clay (and perhaps silt) in it.

A soil forming on basalt may never be sandy, even in its early stages of development (this depends on the relative rates of chemical weathering versus mechanical weathering). The fine-grained feldspars in the basalt weather to fine-grained clay minerals. Since the parent rock had no coarse-grained minerals and no quartz to start with, the resulting soil may lack sand. Such a soil may not drain well, although it can be quite fertile.

Note that the character of a soil changes with time. A soil developing from granite begins as a sandy soil and becomes more clayey with time. Over very long periods, the type of parent rock becomes less and less important. Given enough time, soils forming from many different kinds of igneous, metamorphic, and sedimentary rocks can become quite similar (in the same climate). The presence or absence of coarse grains of quartz in the parent rock becomes the only characteristic of the parent rock to have long-term significance.

B

A   **B**

Figure 5.22

Soil thickness. (*A*) Soil thickens with time. Basalt flow 1 was exposed to soil-forming processes for a longer time than flow 2. The soil developed below flow 2 is an example of a buried soil. (*B*) Steep slopes have thin soil.

With time, soils tend to become thicker (figure 5.22*A*); most modern soils have taken centuries to form. A new deposit of volcanic ash, which is very fine-grained and rich in plant nutrients, may be covered with grass and other low plants in just a few years, but a new lava flow, which weathers much more slowly than ash, may not have enough soil to support grass for many decades. It may take centuries for the lava-flow soil to thicken enough to support shrubs, and thousands of years to support trees. The fertile agricultural soils of the Canadian plains and the northern United States took more than 10,000 years to develop on glacial deposits after the thick continental ice sheets melted (see figure 12.33).

Another factor controlling soil thickness is the slope of the land surface (figure 5.22*B*). Soils tend to be thick on flat land where erosion is slow and water can collect, and thin on steep slopes where gravity pulls water and soil particles downhill.

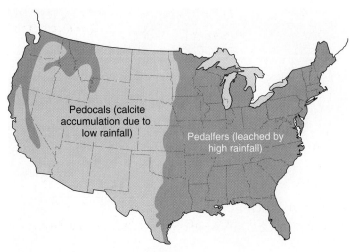

Figure 5.23

Distribution of pedalfer (colored area) and pedocal soils in the United States.

U.S. Department of Agriculture

Accumulation of a calcite layer and nodules as upward-moving ground water evaporates

A
B

Figure 5.24

Pedocal soils are marked by upward-moving ground water that evaporates underground in a dry climate, precipitating calcium carbonate within the soil, sometimes forming a light-colored layer.

Photo by D. Yost, USDA–Soil Conservation Service

Organic Activity

Organisms contribute to soil development. Plant roots break up rocks and burrowing organisms such as ants, worms, and rodents bring soil particles to the surface and create passageways for water and air to get underground, thus speeding up chemical weathering. Respiration of soil organisms and decay of plant and animal material adds carbon dioxide gas to soil, creating carbonic acid. Plants and humus also release organic acids that increase chemical weathering. Once soil begins to develop on a newly exposed rock, it attracts plants and soil organisms that increase chemical weathering, accelerating the rate of soil development. Partially decayed organic matter provides plant nutrients, increasing soil fertility.

Soils and Climate

Climate affects soil thickness and character. Soils in wet climates, as in Europe, most of Canada, and the eastern United States, tend to be thick and are generally characterized by downward movement of water through the earth materials (figure 5.21 shows such a soil). A general term for such soils is **pedalfer** for the high content of aluminum and iron (*Al* and *Fe*) oxides they contain. Pedalfers are marked by effective downward leaching due to high rainfall and to the acids produced by the decay of abundant humus.

In arid (dry) climates, as in many parts of the western United States, soils called **pedocals** form (figures 5.23 and 5.24). These soils tend to be thin and are characterized by little leaching, scant humus, and the *upward* movement of soil water beneath the land surface. The water is drawn up by subsurface evaporation and capillary action.

The evaporation of water beneath the land surface can cause the precipitation of salts within the soil. These salts are usually calcium salts such as calcite (which is the origin of the term pedo*cal*). An extreme example of salt buildup can be found in desert *alkali soils,* in which heavy concentrations of toxic sodium salts may prevent plant growth.

Hardpan

"Hardpan" is a general term for a hard layer of earth material that is difficult to dig or drill. Geologists usually restrict the term to a hard, often clayey, layer of cemented soil particles. Such a layer may be too hard for even backhoes to dig through; planting a tree in a lawn with a hardpan layer may require dynamite. Hardpan layers in wet climates are usually formed of clay minerals, silica, and iron compounds that have accumulated in the B horizon. In arid climates a different type of hardpan, called *caliche* (or sometimes *calcrete*), forms from the cementing of soil by calcium carbonate and other salts that precipitate in the soil as water evaporates. Both types of hardpan are really layers of rock within loose soil. A hardpan layer can break plows, prevent water drainage through the soil, and act as a barrier to plant roots. Tree roots may grow laterally along rather than down through hardpan; such shallow-rooted trees are easily uprooted by wind.

Laterites

In tropical regions where temperatures are high and rainfall is abundant, highly leached soils called **laterites** form. Under such conditions weathering is deep and intense. Laterites are usually red and are composed almost entirely of iron and aluminum oxides, generally the least soluble products of rock weathering in tropical climates (figure 5.25). If the soil is rich in hematite, it can be mined as iron ore, but tropical rainfall usually hydrates the hematite to limonite, which is seldom rich

Figure 5.25

Laterite soil develops in very wet climates, where intense downward leaching carries away all but iron and aluminum oxides. Many laterites are deep red.

Photo by D. Yost, USDA–Soil Conservation Service

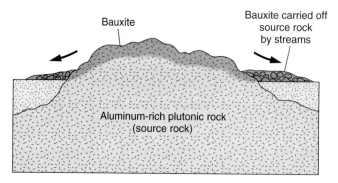

Figure 5.26

Bauxite forms by intense tropical weathering of an aluminum-rich source rock.

Figure 5.27

Soil erosion caused by clear cutting of rain forest, north of Kuantan, Malaysia.

Photo © George Loun/Visuals Unlimited

enough to mine. However, aluminum is sometimes found in nearly pure layers of *bauxite* ($Al_2O_3 \cdot nH_2O$, the principal ore of aluminum), particularly in laterites formed by the weathering of aluminum-rich igneous rocks.

Under tropical conditions of high rainfall and high temperature, most weathering products are soluble—even silica. The least soluble product is aluminum oxide, which remains on top of the weathering rocks, forming bauxite in a soil very rich in aluminum. Like the diamonds, the aluminum has been concentrated residually by the removal of everything else. The aluminum ores may be redistributed slightly by running water (figure 5.26).

Because bauxite forms under conditions of tropical weathering, the United States has very little aluminum ore and depends almost entirely on tropical countries for its aluminum supply. A small percentage of the U.S. aluminum supply has come from bauxite deposits in Arkansas that formed on igneous rock with a high aluminum content approximately 50 million years ago when the region had a tropical climate.

Laterites are relatively nonproductive soils. This may seem strange when you think of the lush jungle growth that often exists on tropical lateritic soils. Jungle vegetation, though, is nourished largely by a layer of humus on top of the soil. If the jungle and the humus layer are cleared away or burned—an increasingly common practice in tropical regions—the laterite quickly becomes incapable of sustaining plant growth, making tropical agriculture very difficult (figure 5.27). Laterite exposed to the sun is apt to bake into a permanent, bricklike layer that makes digging nearly impossible. This hard layer can be quarried, however, and makes a durable building material.

Buried Soils

A soil may become buried by volcanic ash, windblown dust, other sediment, or lava (see figure 5.22). A buried soil is called a *paleosol* (paleo = ancient). Such soils may be distinctive and traceable over wide regions, making them useful for dating rocks and sediments, and for interpreting past climates and topography.

Summary

When rocks that formed deep in the earth become exposed at the earth's surface, they are altered by *mechanical* and *chemical weathering*.

Weathering processes form *spheroidally weathered boulders, differentially weathered* landforms, *sheet joints,* and *exfoliation domes*.

Mechanical weathering, largely caused by *frost action, abrasion* during transportation, and *pressure release* after unloading, disintegrates (breaks) rocks into smaller pieces.

By increasing the exposed surface area of rocks, mechanical weathering helps speed chemical weathering.

Chemical weathering results when a mineral is unstable in the presence of water and atmospheric gases. As chemical weathering proceeds, the mineral's components recombine into new minerals that are more in equilibrium.

Weak acid, primarily from the solution of carbon dioxide in water, is the most effective agent of chemical weathering.

Calcite dissolves when it is chemically weathered. Most of the silicate minerals form *clay minerals* when they chemically weather. Quartz is very resistant to chemical weathering.

Soil develops by chemical and mechanical weathering of a parent material. Some definitions of soil require that it contain organic matter and be able to support plant growth.

Soils, which can be *residual* or *transported,* usually have distinguishable layers, or *horizons,* caused in part by water movement within the soil.

Climate is the most important factor determining soil type. Other factors in soil development are parent rock, time, slope, and organic activity.

Pedalfers are soils characterized by downward leaching. *Pedocals* are soils marked by salt precipitation caused by evaporation of upward-moving soil water.

Laterites form under conditions of intense tropical weathering; they are usually red from concentrated iron oxides. Bauxite, the ore of aluminum, may be found in laterites.

Terms to Remember

abrasion 106
A horizon (zone of leaching) 114
B horizon (zone of accumulation) 114
chemical weathering 104
C horizon 114
clay mineral 111
differential weathering 104
erosion 104
exfoliation 107
exfoliation dome 107

frost action 104
frost heaving 106
frost wedging 106
hematite 109
laterite 116
limonite 109
mechanical weathering 104
O horizon 114
pedalfer 116
pedocal 116

pressure release 107
residual soil 114
sheet joints 107
soil 113
soil horizon 114
spheroidally weathered boulder 104
transportation 104
transported soil 114
weathering 104

Testing Your Knowledge

Use the questions below to prepare for exams based on this chapter.

1. Why are some minerals stable several kilometers underground but unstable at the earth's surface?

2. Describe what happens to each mineral within granite during the complete chemical weathering of granite in a humid climate. List the final products for each mineral.

3. Explain what happens chemically when calcite dissolves. Show the reaction in a chemical equation.

4. Why do stone buildings tend to weather more rapidly in cities than in rural areas?

5. Describe at least three processes that mechanically weather rock.

6. How can mechanical weathering speed up chemical weathering?

7. Name at least three natural sources of acid in solution. Which one is most important for chemical weathering?

8. What is the difference between a residual soil and a transported soil?

9. What is a laterite and how does it form?

10. How do pedocals differ from pedalfers?

11. What are the soil horizons? Why do they form?

12. Physical disintegration of rock into smaller pieces is called (a) chemical weathering (b) transportation (c) deposition (d) mechanical weathering

13. The decomposition of rock from exposure to water and atmospheric gases is called (a) chemical weathering (b) transportation (c) deposition (d) mechanical weathering

14. Which is not a type of mechanical weathering? (a) frost wedging (b) frost heaving (c) abrasion (d) pressure release (e) oxidation

15. The single most effective agent of chemical weathering at the earth's surface is (a) carbonic acid H_2CO_3 (b) water H_2O (c) carbon dioxide CO_2 (d) hydrochloric acid HCl

16. The most common end product of the chemical weathering of feldspar is (a) clay minerals (b) pyroxene (c) amphibole (d) calcite

17. The most common end product of the chemical weathering of quartz is (a) clay minerals (b) pyroxene (c) amphibole (d) calcite (e) quartz does not weather chemically

18. Soil with approximately equal amounts of sand, silt, and clay along with a generous amount of organic matter is called (a) loam (b) inorganic (c) humus (d) caliche

19. Which is characteristic of soil horizons? (a) they can be distinguished from one another by appearance and chemical composition (b) boundaries between soil horizons are usually transitional rather than sharp (c) they are classified by letters (d) all of the above

20. Pedocal soils (a) form in arid (dry) climates (b) tend to be thin (c) are characterized by little leaching, scant humus (d) all of the above

21. The soil horizon rich in organic material is the (a) A horizon (b) B horizon (c) C horizon (d) O horizon

22. Hardpan forms in the (a) A horizon (b) B horizon (c) C horizon (d) O horizon

23. Tropical soils are typically (a) rich in organic material (b) very fertile (c) deeply leached (d) easily replenished

Expanding Your Knowledge

1. Which mineral weathers faster—hornblende or muscovite? Why? Consider Bowen's Reaction Series in your answer.

2. Compare and contrast the weathering rate and weathering products for Ca-rich plagioclase in the following localities:
   a. central Pennsylvania with 40 inches of rain per year;
   b. Death Valley with 2 inches of rain per year;
   c. an Alaskan mountaintop where water is frozen year-round.

3. The amount of carbon dioxide gas has been increasing in the atmosphere for the past 40 years as a result of the burning of fossil fuels. What effect will the increase in CO_2 have on the rate of chemical weathering? The increase in CO_2 may cause global warming in the future. What effect would a warmer climate have on the rate of chemical weathering? Give the reasons for your answers.

4. In a humid climate, is a soil formed from granite the same as one formed from gabbro? Discuss the similarities and possible differences with particular regard to mineral content and soil color.

Exploring Resources

Birkeland, P. W. 1984. *Soils and geomorphology.* New York: Oxford University Press.

Carroll, D. 1970. *Rock weathering.* New York: Plenum Press.

Colman, S. M., and D. P. Dethier. 1986. *Rates of chemical weathering of rocks and minerals.* New York: Academic Press.

Gauri, K. L. 1978. The preservation of stone. *Scientific American* 238(6): 126–36.

Hunt, C. B. 1972. *The geology of soils.* New York: W.H. Freeman.

Keller, W. D. 1957. *The principles of chemical weathering.* Columbia, Mo.: Lucas Brothers.

Keller, W. D. 1969. *Chemistry in introductory geology.* Columbia, Mo.: Lucas Brothers.

Krauskopf, K. K., and D. K. Bird. 1995. *Introduction to geochemistry.* 3d ed. New York: McGraw-Hill.

Loughnan, F. C. 1969. *Chemical weathering of the silicate minerals.* New York: Elsevier.

McNeil, M. 1964. Lateritic soils. *Scientific American* 211(5): 96–102.

Ollier, C. D. 1986. *Weathering.* 2d ed. New York: John Wiley.

Singer, M. J., and D. N. Munns. 1986. *Soils.* New York: Macmillan.

Soil Survey Staff, 1975. *Soil taxonomy: a basic system of soil classification for making and interpreting soil surveys.* U.S. Department of Agriculture Handbook 436.

Soil Survey Staff, 1987. *Keys to soil taxonomy.* 3d printing. Soil Management Support Services Technical Monograph 6. U.S. Department of Agriculture.

Layers of sedimentary rocks exposed at Capitol Reef National Park, Utah.

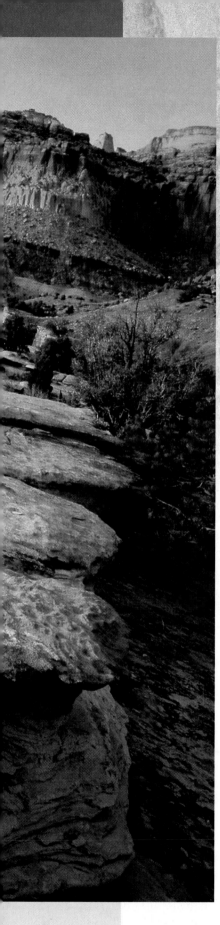

CHAPTER

6

Sediments and Sedimentary Rocks

The rock cycle (chapter 2) is a theoretical model of the constant recycling of rocks as they form, are destroyed, and then reform. We began our discussion of the rock cycle with igneous rock (chapters 3 and 4), and we now discuss sedimentary rocks. Metamorphic rocks, the third major rock type, are the subject of the next chapter.

You saw in chapter 5 how weathering produces sediment. In this chapter we explain more about sediment origin, as well as the erosion, transportation, sorting, deposition, and eventual lithification of sediments to form sedimentary rock. Because they have such diverse origins, sedimentary rocks are difficult to classify. We divide them into clastic, chemical, and organic sedimentary rocks, but this classification is not entirely satisfactory. Furthermore, despite their great variety, only three sedimentary rocks are very common—shale, sandstone, and limestone.

Sedimentary rocks contain numerous clues to their origin and the environment of deposition in which they were deposited. Geologists find out this information from the shape and sequence of rock layers and from the sediment grains and the sedimentary structures such as fossils, cross-beds, ripple marks, and mud cracks that are contained in the rock.

Sedimentary rocks are important because they are widespread and because many of them, such as coal and limestone, are economically important. About three-fourths of the surface of continents is blanketed with a thin skin of sedimentary rocks. Concentrated in sedimentary rocks are important resources such as crude oil, natural gas, ground water, salt, gypsum, uranium, and iron ore.

Sediment

Most sedimentary rocks form from loose grains of sediment. Sediment includes such particles as sand on beaches, mud on a lake bottom, boulders frozen into glaciers, pebbles in streams, and dust particles settling out of the air. An accumulation of clam shells on the sea bottom offshore is sediment, as are coral fragments broken from a reef by large storm waves.

Sediment is the collective name for loose, solid particles that originate from:

1. Weathering and erosion of preexisting rocks.
2. Chemical precipitation from solution, including secretion by organisms in water.

These particles usually collect in layers on the earth's surface. An important part of the definition is that the particles are loose. Sediments are said to be *unconsolidated,* which means that the grains are separate, or unattached to one another.

Sediment particles are classified and defined according to the size of individual fragments. Table 6.1 shows the precise definitions of particles by size.

Gravel includes all rounded particles coarser than 2 mm in diameter, the thickness of a U.S. nickel. (Angular fragments of this size are called *rubble.*) *Pebbles* range from 2 to 64 mm (about 2.5 inches, the size of a tennis ball). *Cobbles* range from 64 to 256 mm (10 inches, about the size of a basketball), and *boulders* are coarser than 256 mm.

Sand grains are from 1/16 mm (about the thickness of a human hair) to 2 mm in diameter. Grains of this size are visible and feel gritty between the fingers. **Silt** grains are from 1/256 to 1/16 mm. They are too small to see without a magnifying device, such as a geologist's hand lens. Silt does not feel gritty between the fingers, but it does feel gritty between the teeth (geologists often bite sediments to test their grain size). **Clay** is the finest sediment, at less than 1/256 mm, too fine to feel gritty to fingers or teeth. *Mud* is a term loosely used for wet silt and clay.

Note that we have two different uses of the word *clay*—a *clay-sized particle* (table 6.1) and a *clay mineral*. A clay-sized particle can be composed of any mineral at all provided its diameter is less than 1/256 mm. A clay mineral, on the other hand, is one of a small group of silicate minerals with a sheet-silicate structure. Clay minerals usually fall into the clay-size range.

Quite often the composition of sediment in the clay-size range turns out to be mostly clay minerals, but this is not always the case. Because of its resistance to chemical weathering, quartz may show up in this fine-size grade. (Most silt is quartz.) Intense mechanical weathering can break down a wide variety of minerals to clay size, and these extremely fine particles may retain their mineral identity for a long time if chemical weathering is slow. The great weight of glaciers is particularly effective at grinding minerals down to the clay-size range, producing "rock flour," which gives a milky appearance to glacial meltwater streams (see chapter 12).

Weathering, erosion, and transportation are some of the processes that affect the character of sediment. Both weathered and chemically unweathered rock and sediment can be eroded, and weathering does not stop after erosion has taken place.

Table 6.1	Sediment Particles and Clastic Sedimentary Rocks		
Diameter (mm)	Sediment		Sedimentary Rock
256	Boulder	Gravel	Breccia (angular particles) or Conglomerate (rounded particles)
64	Cobble		
2	Pebble		
1/16	Sand		**Sandstone**
1/256	Silt	"Mud"	Siltstone (mostly silt)
	Clay		**Shale** or mudstone (mostly clay)

Sandstone and shale are quite common; the others are relatively rare.

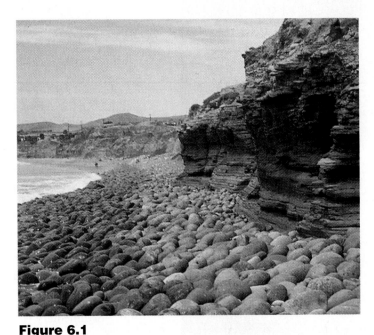

Figure 6.1

These boulders have been rounded by abrasion as wave action rolled them against one another on this beach.

Sand being transported by a river also can be actively weathering, as can mud on a lake bottom. The character of sediment can also be altered by *rounding* and *sorting* during transportation, and by eventual *deposition.*

Transportation

Rounding is the grinding away of sharp edges and corners of rock fragments during transportation. Rounding occurs in sand and gravel as rivers, glaciers, or waves cause particles to hit and scrape against one another (figure 6.1) or against a rock surface, such as a rocky stream bed. Boulders in a stream may show substantial rounding in less than one mile of travel.

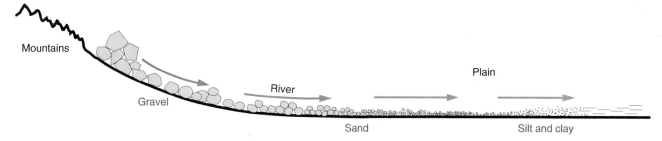

Figure 6.2

Sorting of sediment by a river. The coarse sediment is deposited first, and the finest sediment is carried the farthest.

Because rounding during transportation is so rapid, it is a much more important process than spheroidal weathering (see chapter 5), which also tends to round sharp edges.

Sorting is the process by which sediment grains are selected and separated according to grain size (or grain shape or specific gravity) by the agents of transportation, especially by running water. Because of their high viscosity and manner of flow, glaciers are poor sorting agents. Glaciers deposit all sediment sizes in the same place, so glacial sediment usually consists of a mixture of clay, silt, sand, and gravel. Such glacial sediment is considered *poorly sorted*. Sediment is considered *well-sorted* when the grains are nearly all the same size. A river, for example, is a good sorting agent, separating sand from gravel, and silt and clay from sand. Sorting takes place because of the greater weight of larger particles. Boulders weigh more than pebbles and are more difficult for the river to transport, so a river must flow more rapidly to move boulders than to move pebbles. Similarly, pebbles are harder to move than sand, and sand is harder to move than silt and clay.

Figure 6.2 shows the sorting of sediment by a river as it flows out of steep mountains onto a gentle plain, where the water loses energy and slows down. As the river loses energy, the heaviest particles of sediment are deposited. The boulders come to rest first (figure 6.3). As the river continues to slow down, cobbles and then pebbles are deposited. Sand comes to rest as the river loses still more energy (figure 6.4). Finally, the river is carrying only the finest sediment—silt and clay (figure 6.5). The river has sorted the original sediment mix by grain size.

Deposition

When transported material settles or comes to rest, **deposition** occurs. Sediment is deposited when running water, glacial ice, waves, or wind loses energy and can no longer transport its load.

Deposition also refers to the accumulation of chemical or organic sediment, such as clam shells on the sea floor, or plant material on the floor of a swamp. Such sediments may form as organisms die and their remains accumulate, perhaps with no transportation at all. Deposition of salt crystals can take place as seawater evaporates. A change in the temperature, pressure, or chemistry of a solution may also cause precipitation—hot springs may deposit calcite or silica as the warm water cools.

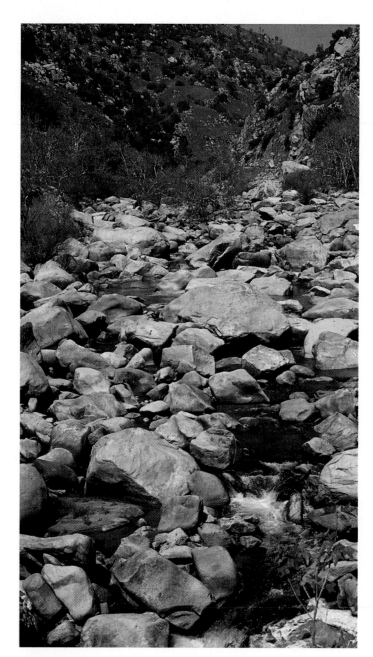

Figure 6.3

Coarse gravel (boulder size) is deposited first along a river's course as the river sorts out the various sediment sizes. River gravel is usually deposited in or near steep mountains.

Sediments and Sedimentary Rocks **123**

Figure 6.4

A river carries sand farther along its course than it carries gravel.

Figure 6.5

The river on the right is carrying only silt and clay as it enters the clear river on the left. This fine sediment may come to rest at the mouth of a river where it enters a lake or the sea.

Photo by C. W. Montgomery

The **environment of deposition** is the location in which deposition occurs. A few examples of environments of deposition are the deep sea floor, a desert valley, a river channel, a coral reef, a lake bottom, a beach, and a sand dune. Each environment is marked by characteristic physical, chemical, and biological conditions. You might expect mud on the sea floor to differ from mud on a lake bottom. Sand on a beach may differ from sand in a river channel. Some differences are due to varying sediment sources and transporting agents, but some are the result of conditions in the environments of deposition themselves.

One of the most important jobs of geologists studying sedimentary rocks is to try to determine the ancient environment of deposition of the sediment that formed the rock. Factors that can help in determining this are a detailed knowledge of modern environments, the shape and vertical sequences of rock layers in the field, the features (including fossils) found within the rock, the mineral composition of the rock, and the size, shape, and surface texture of the individual sediment grains. Later in the chapter we give a few examples of interpreting sedimentary rocks.

Preservation

Not all sediments are preserved as sedimentary rock. Gravel in a river may be deposited when a river is low, but then may be reeroded and retransported by the next flood on the river. Many sediments on land, particularly those well above sea level, are easily reeroded and carried away, so they are not commonly preserved. Sediments on the sea floor are easier to preserve. In general, continental and marine sediments are

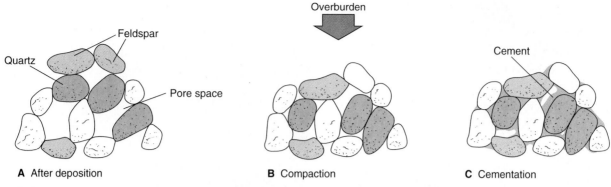

A After deposition **B** Compaction **C** Cementation

Figure 6.6

Lithification of sand grains to become sandstone. (*A*) Loose sand grains are deposited with open pore space between the grains. (*B*) The weight of overburden compacts the sand into a tighter arrangement, reducing pore space. (*C*) Precipitation of cement in the pores by ground water binds the sand into the rock sandstone, which has a clastic texture.

most likely to be preserved if they are deposited in a *subsiding* (sinking) *basin* and if they are *buried by later sediments.*

Lithification

Lithification is the general term for a group of processes that convert loose sediment into sedimentary rock. Most sedimentary rocks are lithified by a combination of *compaction,* which packs loose sediment grains tightly together, and *cementation,* in which the precipitation of cement around sediment grains binds them into a firm, coherent rock. *Crystallization* of minerals from solution, without passing through the loose-sediment stage, is another way that rocks may be lithified.

As sediment grains settle slowly in a quiet environment such as a lake bottom, they form an arrangement with a great deal of open space between the grains (figure 6.6*A*). The open spaces between grains are called *pores,* and in a quiet environment, a deposit of sand may have 40% to 50% of its volume as open **pore space.** (If the grains were traveling rapidly and impacting one another just before deposition, the percentage of pore space will be less.) As more and more sediment grains are deposited on top of the original grains, the increasing weight of this *overburden* packs the original grains together, reducing the amount of pore space. This shift to a tighter packing, with a resulting decrease in pore space, is called **compaction** (figure 6.6*B*). As pore space decreases, some of the interstitial water that usually fills sediment pores is driven out of the sediment.

As underground water moves through the remaining pore space, solid material called **cement** can precipitate in the pore space and bind the loose sediment grains together to form a solid rock. The cement attaches very tightly to the grains, holding them in a rigid framework. As cement partially or completely fills the pores, the total amount of pore space is further reduced (figure 6.6*C*), and the loose sand forms a hard, coherent sandstone by **cementation.**

Sedimentary rock cement is often composed of the mineral calcite or of other carbonate minerals. Dissolved calcium and bicarbonate ions are common in surface and underground waters, as you saw in chapter 5. If the chemical conditions are

Figure 6.7

Crystalline texture. The rock is held together by interlocking crystals, which grew as they precipitated from solution. Such a rock has no cement or pore space.

right, these ions may recombine to form solid calcite, as shown in the following equation.

$$\underbrace{Ca^{++} + 2HCO_3^-}_{\substack{\text{dissolved} \\ \text{ions}}} \rightarrow \underbrace{CaCO_3}_{\text{calcite}} + H_2O + CO_2$$

Silica is another common cement. Iron oxides and clay minerals can also act as cement but are less common than calcite and silica. The dissolved ions that precipitate as cement originate from the chemical weathering of minerals such as feldspar and calcite. This weathering may be local, within the sediments being cemented, or very distant, with the ions being transported tens or even hundreds of miles by water before precipitating as solid cement.

A sedimentary rock that consists of sediment grains bound by cement into a rigid framework is said to have a **clastic texture.** Usually such a rock still has some pore space; cement rarely fills the pores completely (figure 6.6*C*).

Some sedimentary rocks form by **crystallization,** the development and growth of crystals by precipitation from solution at or near the earth's surface (the term is also used for igneous rocks that crystallize as magma cools). These rocks have a **crystalline texture,** an arrangement of interlocking crystals that develops as crystals grow and interfere with each other (figure 6.7). Crystalline rocks lack cement. They are held together by the interlocking of crystals. Such rocks have no pore space because the crystals have grown until they fill all available space. Some sedimentary rocks with a crystalline tex-

Figure 6.8

Breccia is characterized by coarse, angular fragments. The cement in this rock is colored by hematite. The wide green and white bars on the scale are one centimeter long, the small divisions are one millimeter. Note that most grains exceed 2 mm (table 6.1).

Figure 6.9

An outcrop of conglomerate. Note the rounding of cobbles, which vary in composition. Long scale bar 10 cm, short bars 1 cm.

ture are the result of *recrystallization,* the growth of new crystals that form from and then destroy the original clastic grains of a rock that has been buried.

Types of Sedimentary Rocks

Sedimentary rock is rock that has formed from (1) lithification of sediment, (2) precipitation from solution, or (3) consolidation of the remains of plants or animals. These different types of sedimentary rocks are called, respectively, *clastic, chemical,* and *organic* rocks.

Most sedimentary rocks are **clastic sedimentary rocks,** formed from cemented sediment grains that are fragments of preexisting rocks. The rock fragments can be either identifiable pieces of rock, such as pebbles of granite or shale, or individual mineral grains, such as sand-sized quartz and feldspar crystals loosened from rocks by weathering and erosion. Clay minerals formed by chemical weathering are also considered fragments of preexisting rocks. In most cases the sediment has been eroded and transported before being deposited. During transportation the grains may have been rounded and sorted. Table 6.1 shows the clastic rocks, such as conglomerate, sandstone, and shale, and shows how these rocks vary in grain size.

Chemical sedimentary rocks are rocks deposited by precipitation of minerals from solution. An example of inorganic precipitation is the formation of *rock salt* as seawater evaporates. Chemical precipitation can also be induced by organisms. The sedimentary rock *limestone,* for instance, can form by the precipitation of calcite within a coral reef by corals and algae. Such a rock may be classified as a *biochemical* limestone.

Chemical rocks may or may not have once been sediment. Rock salt may form from sediment; individual salt crystals forming in evaporating water act as sediment until they grow large enough to interlock into a solid rock. Minerals that crystal-

lize from solution on the sides of a rock cavity, or as a stalactite in a cave, however, were never a sediment. Neither was limestone precipitated directly from seawater as a solid rock by corals.

Organic sedimentary rocks are rocks that accumulate from the remains of organisms. *Coal* is an organic rock that forms from the compression of plant remains, such as moss, leaves, twigs, roots, and tree trunks. A limestone formed from the accumulation of clam shells on the sea floor might also be called an organic rock.

Appendix B describes and helps you identify the common sedimentary rocks. The geologic symbols for these rocks (such as dots for sandstone, and a "brick-wall" symbol for limestone) are shown in Appendix F and will be used in the remainder of the book.

Clastic Rocks

Breccia and Conglomerate

Sedimentary breccia is a coarse-grained sedimentary rock formed by the cementation of coarse, angular fragments of rubble (figure 6.8). Because grains are rounded so rapidly during transport, it is unlikely that the angular fragments within breccia have moved very far from their source. Sedimentary breccia might form, for example, from fragments that have accumulated at the base of a steep slope of rock that is being mechanically weathered. Landslide deposits also might lithify into sedimentary breccia. This type of rock is not particularly common.

Conglomerate is a coarse-grained sedimentary rock formed by the cementation of rounded gravel. It can be distinguished from breccia by the definite roundness of its particles (figure 6.9). Because conglomerates are coarse-grained, the particles may not have traveled far; some transport, however, was necessary to round the particles. Angular fragments that fall from a cliff and then are carried a few miles by a river or by waves are quickly rounded. Gravel that is transported down steep submarine canyons, or carried by glacial ice as *till,* however, can be transported tens or even hundreds of miles before deposition.

Figure 6.10

Types of sandstone. (*A*) Quartz sandstone; more than 90% of the grains are quartz. (*B*) Arkose; the grains are mostly feldspar and quartz. (*C*) Graywacke; the grains are surrounded by dark, fine-grained matrix. (Small scale divisions are 1 millimeter; most of the sand grains are about 1 mm in diameter.)

Sandstone

Sandstone is a medium-grained sedimentary rock formed by the cementation of sand grains (figure 6.10). Any deposit of sand can lithify to sandstone. Rivers deposit sand in their channels, and wind piles up sand into dunes. Waves deposit sand on beaches and in shallow water. Deep-sea currents spread sand over the sea floor. As you might imagine, sandstones show a great deal of variation in mineral composition, degree of sorting, and degree of rounding.

Quartz sandstone is a sandstone in which more than 90% of the grains are quartz (figure 6.10*A*). Because quartz is not subject

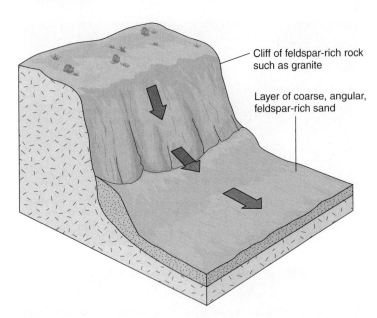

Figure 6.11

Feldspar-rich sand (arkose) may accumulate from the rapid erosion of feldspar-containing rock such as granite. Steep terrain accelerates erosion rates so that feldspar may be eroded before it is completely chemically weathered into clay minerals.

to chemical weathering, it tends to concentrate residually in sand as less resistant minerals such as feldspar are weathered away, probably in a low-lying humid region that allows chemical weathering to continue for a long time. The quartz grains in a quartz sandstone are usually well-sorted and well-rounded because they have been transported for great distances. Most quartz sandstone was deposited as beach sand or dune sand.

A sandstone with more than 25% of the grains consisting of feldspar is called *arkose* (figure 6.10*B*). Because feldspar grains are preserved in the rock, the original sediment obviously did not undergo severe chemical weathering, or the feldspar would have been destroyed. Cliffs of granite in a desert could be a source for such a sediment, for the rapid erosion associated with rugged terrain would allow feldspar to be eroded before it is chemically weathered (a dry climate slows chemical weathering). Most arkoses contain coarse, angular grains, so transportation distances were probably short. An arkose may have been deposited at or near the base of a cliff, as shown in figure 6.11.

Sandstones may contain a substantial amount of **matrix,** fine-grained silt and clay found in the space between larger sand grains (figure 6.12). Matrix usually consists of fine-grained quartz and clay minerals. A matrix-rich sandstone is poorly sorted and often dark in color. It is sometimes called a "dirty sandstone."

Graywacke (pronounced "gray-whacky") is a type of sandstone in which more than 15% of the rock's volume consists of fine-grained matrix (figure 6.10*C*). Graywackes are often tough and dense and are generally dark gray or green. The sand grains may be so coated with matrix that they are hard to see, but they typically consist of quartz, feldspar, and sand-sized fragments of fine-grained sedimentary, volcanic, and metamorphic rocks.

Most graywackes probably formed from sediments transported by **turbidity currents,** dense masses of sediment-laden

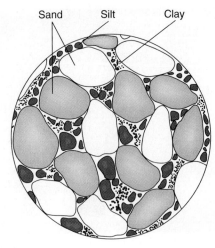

Figure 6.12

A poorly sorted sediment of sand grains surrounded by a matrix of silt and clay grains. Lithification of such a sediment would produce a "dirty sandstone."

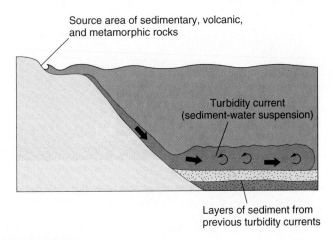

Figure 6.13

A turbidity current flow downslope along the sea floor. Dense sediment-laden water is heavier than the clear water that it flows beneath.

water that flow downslope along the sea floor. The sediment-water mixture is heavier than clear water, so it is pulled downslope by gravity until it comes to rest on the flat sea floor at the base of the slope (figure 6.13). Turbidity currents may be generated by underwater landslides (perhaps triggered by earthquakes) or by violent surface storms such as hurricanes, which stir up bottom sediment. Sediment-laden rivers discharging directly into the sea may also cause turbidity currents.

The Fine-Grained Rocks

Rocks consisting of fine-grained silt and clay are called *shale, siltstone, claystone,* and *mudstone.*

Shale is a fine-grained sedimentary rock notable for its splitting capability (called *fissility*). Splitting takes place along the surfaces of very thin layers (called *laminations*) within the shale (figure 6.14). Most shales contain both silt

B

A

Figure 6.14

(*A*) An outcrop of shale in the Grand Canyon. Note how this fine-grained rock tends to break into small, flat pieces. (*B*) Shale pieces; note the very fine grain (scale in centimeters), very thin layers (laminations) on the edge of the large piece, and tendency to break into small, flat pieces (fissility).

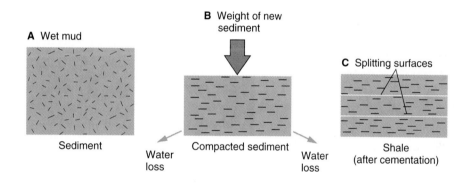

A Wet mud

Sediment

B Weight of new sediment

Water loss   Compacted sediment   Water loss

C Splitting surfaces

Shale (after cementation)

Figure 6.15

Lithification of shale from the compaction and cementation of wet mud. (*A*) Randomly oriented silt and clay particles in wet mud. (*B*) Particles reorient, water is lost, and pore space decreases during compaction caused by the weight of new sediment deposited on top of the wet mud. (*C*) Splitting surfaces in cemented shale form parallel to the oriented mineral grains.

and clay (averaging $^2/_3$ clay-sized clay minerals, $^1/_3$ silt-sized quartz) and are so fine-grained that the surface of the rock feels very smooth. The silt and clay deposits that lithify as shale accumulate on lake bottoms, at the ends of rivers in deltas, beside rivers in flood, and on quiet parts of the deep ocean floor.

Fine-grained rocks such as shale typically undergo pronounced compaction as they lithify. Figure 6.15 shows the role of compaction in the lithification of shale from wet mud. Before compaction, as much as 80% of the volume of the wet mud may have been pore space, the pores being filled with water. The flakelike clay minerals were randomly arranged within the mud. Pressure from overlying material packs the sediment grains together and reduces the overall volume by squeezing water out of the pores. The clay minerals are reoriented perpendicular to the pressure, becoming parallel to one

another like a deck of cards. The fissility of shale is due to splitting between these parallel clay flakes.

Compaction by itself does not generally lithify sediment into sedimentary rock. It does help consolidate clayey sediments by pressing the microscopic clay minerals so closely together that attractional forces at the atomic level tend to bind them together. Even in shale, however, the primary method of lithification is cementation.

A rock consisting mostly of silt grains is called *siltstone*. Somewhat coarser-grained than most shales, siltstones lack the fissility and laminations of shale. *Claystone* is a rock composed predominantly of clay-sized particles, but lacking the fissility of shale. *Mudstone* contains both silt and clay, having the same grain size and smooth feel of shale but lacking shale's laminations and fissility. Mudstone is massive and blocky, while shale is visibly layered and fissile.

Sediments and Sedimentary Rocks   **129**

Figure 6.16

Corals precipitate calcium carbonate to form limestone in a reef. Water depth about 25 feet (8 meters), San Salvador Island, Bahamas.

Chemical Sedimentary Rocks

Chemical sedimentary rocks have been precipitated from an aqueous environment. Chemical sedimentary rocks are either precipitated directly by inorganic processes or by the actions of organisms. Chemical rocks include *carbonates, chert,* and *evaporates.*

Carbonate Rocks

Carbonate rocks contain CO_3 as part of their chemical composition. The two main types of carbonates are limestone and dolomite.

Limestone

Limestone is a sedimentary rock composed mostly of calcite ($CaCO_3$). Limestones are either precipitated by the actions of organisms or are precipitated directly as the result of inorganic processes. The two major types of limestone can be classified as either *biochemical* or *inorganic limestone.*

Biochemical limestones are precipitated through the actions of organisms. Most biochemical limestones are formed on continental shelves in warm, shallow water. Biochemical limestone may be precipitated directly as a solid rock in the core of a reef by corals and encrusting algae (figure 6.16). Such a rock would have a crystalline texture and would contain the fossil remains of organisms still in growth position.

The great majority of limestones are biochemical limestones formed of wave-broken fragments of algae, corals, and shells. The fragments may be of any size (gravel, sand, silt, and clay) and are often sorted and rounded as they are transported by waves and currents across the sea floor (figure 6.17). The action of these waves and currents and subsequent cementation of these fragments into rock give these limestones a clastic texture. These *bioclastic* (or *skeletal*) *limestones* take a great variety of appearances. They may be relatively coarse-grained with recognizable fossils (figure 6.18) or uniformly fine-grained and dense from the accumulation of microscopic fragments of coralline algae (figures 6.18 and 6.19). A variety of limestone called *coquina* forms from the cementation of shells that accumulated on the shallow sea floor near shore (figure 6.20). It has a clastic texture and is usually coarse-grained, with easily recognizable shells and shell frag-

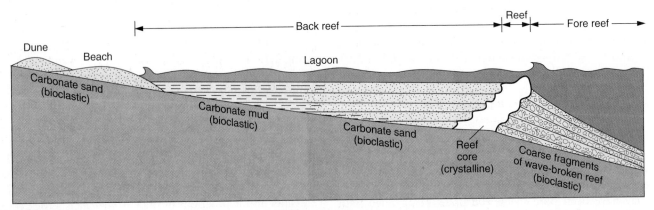

Figure 6.17

A living coral-algal reef sheds bioclastic sediment into the fore-reef and back-reef environments. The fore reef consists of coarse, angular fragments of reef. Coralline algae are the major contributors of carbonate sand and mud in the back reef. Beaches and dunes are often bioclastic sand. The sediments in each environment can lithify to form highly varied limestones.

Figure 6.18

Bioclastic limestones. The two on the left are coarse-grained and contain visible fossils of corals and shells. The limestone on the right consists of fine-grained carbonate mud formed by coralline algae.

Figure 6.19

Coralline algae on the sea floor in 10 feet (3 meters) of water on the Bahama Banks. The "shaving brush" alga is Penicillus, which produces great quantities of fine-grained carbonate mud.

Figure 6.20

Coquina, a variety of bioclastic limestone, is formed by the cementation of coarse shells.

Figure 6.21

Chalk is a fine-grained variety of bioclastic limestone formed of the remains of microscopic marine organisms that live near the sea surface.

A

B

Figure 6.22

(*A*) Aerial photo of underwater dunes of oolites (ooids) chemically precipitated from seawater on the shallow Bahama Banks, south of Bimini. Tidal currents move the dunes. (*B*) An oolitic limestone formed by the cementation of oolites (small spheres). Small divisions on scale are one millimeter wide.

ments in it. *Chalk* is a light-colored, porous, very fine-grained variety of bioclastic limestone that forms from the sea-floor accumulation of microscopic marine organisms that drift near the sea surface (figure 6.21).

Inorganic limestones are precipitated directly as the result of inorganic processes. *Oolitic limestone* is a distinctive variety of inorganic limestone formed by the cementation of sand-sized *oolites* (or *ooids*), small spheres of calcite inorganically precipitated in warm, shallow seawater (figure 6.22). Strong tidal currents roll the oolites back and forth daily, allowing them to maintain a spherical shape as they grow. Wave action may also contribute to their growth.

Oolitic limestone has a clastic texture. *Tufa* and *travertine* are inorganic limestones that form from fresh water. Tufa is precipitated from solution in the water of a continental spring, lake, or from percolating ground water. Travertine forms in caves when droplets of carbonate-rich water lose CO_2 to the cave atmosphere. Tufa and travertine have a crystalline texture.

Limestones are particularly susceptible to **recrystallization,** the process by which new crystals, often of the same composition as the original grains, develop in a rock. Calcite grains recrystallize easily, particularly in the presence of water and under the weight of overlying sediment. The new crystals that form are often large and can be easily seen in a rock as light reflects off their broad, flat faces. Because recrystallization often destroys the original clastic texture and fossils of a rock, replacing them with a new crystalline texture, the geologic history of such a rock can be very difficult to determine.

Table 6.2 — Chemical Sedimentary Rocks

Inorganic Sedimentary Rocks

Rock	Composition	Texture	Origin
Limestone	$CaCO_3$	Crystalline	Cementation of oolites (ooids) precipitated chemically from warm shallow seawater (*oolitic limestone*). May be precipitated directly from seawater. Also forms in caves as *travertine* and in springs, lakes, or percolating ground water as *tufa*.
Dolomite	$CaMg(CO_3)_2$	Crystalline	Alteration of limestone by Mg-rich solutions (usually)
Evaporites			Evaporation of seawater or a saline lake.
Rock salt	$NaCl$	Crystalline	
Rock gypsum	$CaSO_4 \cdot 2H_2O$	Crystalline	

Biochemical Sedimentary Rocks

Rock	Composition	Texture	Origin
Limestone	$CaCO_3$ (calcite)	Clastic or crystalline	Cementation of fragments of shells, corals, and coralline algae (*bioclastic limestone* such as coquina and chalk). Also precipitated directly by organisms in reefs.
Chert	SiO_2 (silica)	Crystalline (usually)	Cementation of microscopic marine organisms; rock usually recrystallized.

Dolomite

The term **dolomite** (table 6.2) is used to refer to both a sedimentary rock and the mineral that composes it, $CaMg(CO_3)_2$. (Some geologists use *dolostone* for the rock.) Dolomite often forms from limestone as the calcium in calcite is partially replaced by magnesium, usually as water solutions move through the limestone.

$$Mg^{++} \quad + \quad 2\,CaCO_3 \quad \rightarrow \quad CaMg(CO_3)_2 \quad + \quad Ca^{++}$$

magnesium   calcite   dolomite   calcium
in solution   in solution

Regionally extensive layers of dolomite are thought to form in one of two ways:

1. As magnesium-rich brines created by solar evaporation of seawater trickle through existing layers of limestone.
2. As chemical reactions take place at the boundary between fresh underground water and seawater; this boundary could migrate through layers of limestone as sea level rises or falls. This replacement process tends to cause recrystallization of the preexisting limestone, so evidence of the rock's origin is often obscured.

Chert

A hard, compact, fine-grained sedimentary rock formed almost entirely of silica, **chert** occurs in two principal forms—as irregular, lumpy nodules within other rocks and as layered deposits like other sedimentary rocks (figure 6.23). The nodules, often found in limestone, probably formed from inorganic precipitation as underground water replaced part of the original rock with silica. The layered deposits may also have formed from inorganic precipitation or from the accumulation of hard, shell-like parts of microscopic marine organisms on the sea floor, or from a combination of both.

Microscopic fossils composed of silica are abundant in some cherts. But because chert is susceptible to recrystallization, the original fossils are easily destroyed, and the origin of many cherts remains doubtful.

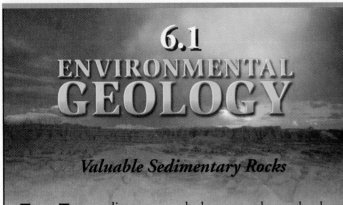

6.1 ENVIRONMENTAL GEOLOGY

Valuable Sedimentary Rocks

Many sedimentary rocks have uses that make them valuable. *Limestone* is widely used as building stone and is also the main rock type quarried for crushed rock for road construction. Pulverized limestone is the main ingredient of cement and is also used as a soil conditioner in the humid regions of the United States. *Coal* is a major fuel, used widely for generating electrical power and for heating. Plaster and plasterboard for home construction are manufactured from *gypsum,* which is also used to condition soils in some areas. Huge quantities of *rock salt* are consumed by industry, primarily for the manufacture of hydrochloric acid. More familiar uses of rock salt are for table salt and for melting ice on roads.

Some *chalk* is used in the manufacture of blackboard chalk, although most classroom chalk is now made from pulverized limestone. The filtering agent for beer brewing and for swimming pools is likely to be made of *diatomite,* an accumulation of the siliceous remains of microscope diatoms.

Clay from *shale* and other deposits supplies the basic material for ceramics of all sorts, from hand-thrown pottery and fine porcelain to sewer pipe. *Sulfur* is used for matches, fungicides, and sulfuric acid; and *phosphates* and *nitrates* for fertilizers are extracted from natural occurrences of special sedimentary rocks (although other sources also are used). Potassium for soap manufacture comes largely from *evaporites,* as does boron for heat-resistant cookware and fiberglass, and sodium for baking soda, washing soda, and soap. *Quartz sandstone* is used in glass manufacturing.

Many *metallic ores,* such as the most common iron ores, have a sedimentary origin. The pore space of sedimentary rocks acts as a reservoir for ground water (chapter 11), crude oil, and natural gas. In chapter 21 we take a closer look at these resources and other useful earth materials.

A

B

Figure 6.23

(*A*) Chert nodules in Redwall Limestone, Grand Canyon. (*B*) Bedded chert from the Coast Ranges, California. Camera lens cap (5.5 cm) for scale.

Evaporites

Rocks formed from crystals that precipitate during evaporation of water are called **evaporites.** They form from the evaporation of seawater or a saline lake (figure 6.24), such as Great Salt Lake in Utah. *Rock gypsum,* formed from the mineral gypsum ($CaSO_4 \cdot 2H_2O$), is a common evaporite. *Rock salt,* composed of the mineral halite (NaCl), may also form if evaporation continues. Other less common evaporites include the borates, potassium salts, and magnesium salts. All evaporites have a crystalline texture.

Figure 6.24

Salt (and mud) deposited on the floor of a dried-up desert lake, Bonneville salt flats, Utah.

Figure 6.25

A bed of coal near Trinidad, Colorado. Hammer at bottom for scale.

Organic Sedimentary Rocks

Coal

Coal is a sedimentary rock that forms from the compaction of plant material that has not completely decayed (figure 6.25). Rapid plant growth and deposition in water with a low oxygen content are needed, so shallow swamps or bogs in a temperate or tropical climate are likely environments of deposition. The plant fossils in coal beds include leaves, stems, tree trunks, and stumps with roots often extending into the underlying shales, so apparently most coal formed right at the place where the plants grew. Coal usually develops from *peat,* a brown, light-weight, unconsolidated or semiconsolidated deposit of moss and other plant remains that accumulates in wet bogs. Peat is transformed into coal largely by compaction after it has been buried by sediments.

Partial decay of the abundant plant material uses up any oxygen in the swamp water, so the decay stops and the remaining organic matter is preserved. Burial by sediment compresses the plant material, gradually driving out any water or other volatile compounds. The coal changes from brown to black as the amount of carbon in it increases. Several varieties of coal are recognized on the basis of the type of original plant material and the degree of compaction (chapter 21).

The Origin of Oil and Gas

Oil and natural gas seem to originate from organic matter in marine sediment. Microscopic organisms, such as diatoms and other single-celled algae, settle to the sea floor and accumulate in marine mud. The most likely environments for this are restricted basins with poor water circulation, particularly on continental shelves. The organic matter may partially decompose, using up the dissolved oxygen in the sediment. As soon as the oxygen is gone, decay stops and the remaining organic matter is preserved.

Continued sedimentation buries the organic matter and subjects it to higher temperatures and pressures, which convert the organic matter to oil and gas. As muddy sediments compact, the gas and small droplets of oil may be squeezed out of the mud and may move into more porous and permeable sandy layers nearby. Over long periods of time large accumulations of gas and oil can collect in the sandy layers. Both oil and gas are less dense than water, so they generally tend to rise upward through water-saturated rock and sediment.

Sedimentary Structures

Sedimentary structures are features found within sedimentary rock. They usually form during or shortly after deposition of the sediment, but before lithification. Structures found in sedimentary rocks are important because they provide clues that help geologists determine the means by which sediment was transported and also its eventual resting place, or environment of deposition. Sedimentary structures may also reveal the original stratigraphic top, or upward direction of deposition, which helps geologists unravel the geometry of rocks that have been folded and faulted in tectonically active regions.

One of the most prominent structures, seen in most large bodies of sedimentary rock, is **bedding,** a series of visible layers within rock (figure 6.26). Most bedding is horizontal because the sediments from which the sedimentary rocks formed were originally deposited as horizontal layers. The principle of **original horizontality** states that most water-laid sediment is deposited in horizontal or near-horizontal layers that are essentially parallel to the earth's surface. In many cases this is also true for sediments deposited by ice or wind. If each new layer of sediment buries previous layers, a stack of horizontal layers

Figure 6.26

Bedding in sandstone and shale, Utah. The horizontal layers formed as one type of sediment buried another in the geologic past. The layers get younger upwards.

will develop with the oldest layer on the bottom and the layers becoming younger upward. Sedimentary rocks formed from such sediments preserve the horizontal layering in the form of beds (figure 6.26). A **bedding plane** is a nearly flat surface of deposition separating two layers of rock. A change in the grain size or composition of the particles being deposited, or a pause during deposition, can create bedding planes.

A specialized type of bedding that is not horizontal is **cross-bedding,** a series of thin, inclined layers within a larger bed of rock. The cross-beds form a distinct angle to the horizontal (or near-horizontal) bedding planes of the larger rock unit (figure 6.27). Cross-bedding is found most often in sandstone. It develops as sand is deposited on steep, local slopes (figure 6.28). The distinctive patterns of cross-bedding can form in many places—in sand dunes deposited by wind; in sand ridges deposited by ocean currents on the sea floor; in sediment bars and dunes deposited by rivers in their channels; and in deltas that form at the mouths of rivers.

A **graded bed** is a layer with a vertical change in particle size, usually from coarse grains at the bottom of the bed to progressively finer grains toward the top (figure 6.29). A single

bed may have gravel at its base and grade upward through sand and silt to fine clay at the top. A graded bed may build up as sediment is deposited by a gradually slowing current. This seems particularly likely to happen during deposition from a *turbidity current* on the deep sea floor (figure 6.13). Figure 6.30 shows the development of a graded bed by turbidity-current deposition.

Mud cracks are a polygonal pattern of cracks formed in very fine-grained sediment as it dries (figure 6.31). Because drying requires air, mud cracks form only in sediment exposed above water. Mud cracks may form in lake-bottom sediment as the lake dries up; in flood-deposited sediment as a river level drops; or in marine sediment exposed to the air, perhaps temporarily by a falling tide. Cracked mud can lithify to form shale, preserving the cracks. The filling of mud cracks by sand can form casts of the cracks in an overlying sandstone.

Ripple marks are small ridges formed on the surface of a sediment layer by moving wind or water. The ridges form perpendicular to the motion. Ripple marks, found in gravel, sand, or silt, can be caused by either waves or currents of water or wind (figure 6.32*A,B*). Wave-caused ripple marks are symmetric

Figure 6.27

Cross-bedded sandstone in Zion National Park, Utah. Note how the thin layers have formed at an angle to the more extensive bedding planes (also tilted) in the rock. This cross-bedding was formed in sand dunes deposited by the wind.

Figure 6.28

The development of cross-bedding in wind-blown sand (*A* and *B*) and current-deposited sand (*C* and *D*). (*A*) Sand is deposited in inclined layers on the downwind side of a dune. (*B*) Second dune covers first. Cross-bedding may change orientation if wind direction shifts. (*C*) A water current fills in a depression on river bottom or sea floor with sediment. (*D*) Continued sedimentation may cover first set of cross-beds with another set.

Figure 6.29

A graded bed has coarse grains at the bottom of the bed and progressively finer grains toward the top. Coin for scale.

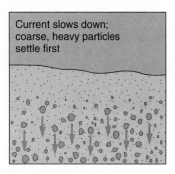

Sediment-laden turbidity current flows beneath clear water

Current slows down; coarse, heavy particles settle first

Main body of current comes to rest

Fine-grained "tail" of turbidity current continues to flow, adding fine-grained sediment to top of deposit

Progressively finer sediments settle on top of coarse particles

A graded bed

Figure 6.30

Development of a graded bed of sediment deposited by a turbidity current.

A

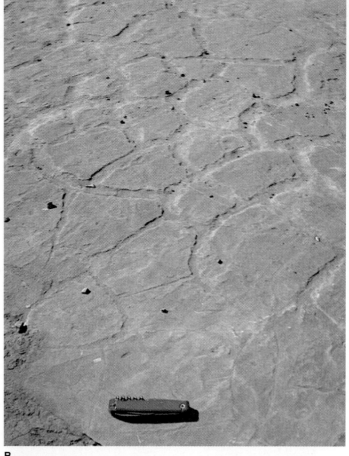

B

Figure 6.31

(*A*) Mud cracks in recently dried mud. (*B*) Mud cracks preserved in shale; they have been partially filled with sediment.

Figure 6.32

Development of ripple marks in loose sediment. (*A*) Symmetric ripple marks form beneath waves. (*B*) Asymmetric ripple marks, forming beneath a current, are steeper on their down-current sides. (*C*) Ripple marks on a bedding plane in sandstone, Capitol Reef National Park, Utah. Scale in centimeters. (*D*) Current ripples in wet sediment of a tidal flat, Baja California. Current flowed to right.

Photo © by Frank M. Hanna

C

A

B

D

ridges; current-caused ripple marks are asymmetric, with steeper sides in the down-current direction. Either type can be preserved in rock (figure 6.32*C,D*).

Fossils, traces of plants or animals preserved in rock, are relatively common sedimentary structures. The hard parts of organisms are most likely to be fossilized. If a bone or shell is covered by sediment before it decays, it may be preserved in the rock that forms from the sediment (figure 6.33). The original bone or shell material is seldom preserved unaltered; calcite or silica may fill the pore spaces of the fossil or may completely replace the original material. Sometimes fossils dissolve entirely, leaving an open mold in a rock. A mold may be filled in later by silica or calcite from underground water, forming a cast of the original fossil. Sometimes flat fossils of fish or leaves are formed as the weight of overburden presses most of the original material out of the fossil, leaving a thin film of black carbon behind (figure 6.34). Preserved footprints, trails, and burrows are also fossils.

Because many limestones are composed of shell and coral debris, fossils are particularly common in limestones, but they also occur in shales and sandstones. Because coal, crude oil, and natural gas form from ancient organic matter, they are sometimes called *fossil fuels.*

Formations

A **formation** is a body of rock of considerable thickness with characteristics that distinguish it from adjacent rock units. Although a formation is usually composed of one or more beds of sedimentary rock, units of metamorphic and igneous rock are also called formations. It is a convenient unit for mapping, describing, or interpreting the geology of a region.

Figure 6.33

Fossil shells of clams in sandstone, California. Some of the fossils retain the original white shells; in others the shells have dissolved to form open molds.

A

B

Figure 6.34

(*A*) Fossil fish in a rock from western Wyoming. (*B*) Dinosaur footprint in shale, Tuba City, Arizona. Scale in centimeters.

Photo © by U.S. Geological Survey

Formations are often based on rock type. A formation may be a single thick bed of sandstone. A sequence of several thin sandstone beds could also be called a formation, as could a sequence of alternating limestone and shale beds.

The main criterion for distinguishing and naming a formation is some visible characteristic that makes it a recognizable unit. This characteristic may be rock type or sedimentary structures or both. For example, a thick sequence of shale may be overlain by basalt flows and underlain by sandstone. The shale, the basalt, and the sandstone are each a different formation. Or a sequence of thin limestone beds, with a total thickness of hundreds of feet, may have recognizable fossils in the lower half and distinctly different fossils in the upper half. The limestone sequence is divided into two formations on the basis of its fossil content.

Formations are given proper names: the first name is often a geographic location where the rock is well exposed, and the second the name of a rock type, such as Navajo Sandstone, Austin Chalk, Baltimore Gneiss, Onondaga Limestone, or Chattanooga Shale. If the formation has a mixture of rock types, so that one rock name does not accurately describe it, it is called simply "formation," as in the Morrison Formation or the Martinsburg Formation.

A **contact** is the boundary surface between two different rock types or ages of rocks. In sedimentary rock formations, the contacts are usually bedding planes.

Figure 6.35 shows the three formations that make up the upper part of the canyon walls in Grand Canyon National Park in Arizona. The contacts between formations are also shown.

Interpretation of Sedimentary Rocks

Sedimentary rocks are important in interpreting the geologic history of an area. Geologists examine sedimentary formations to look for clues such as fossils; sedimentary structures; grain shape, size, and composition; and the overall shape and extent of the formation. These clues are useful in determining the source area of the sediment, environment of deposition, and the possible plate tectonic setting at the time of deposition.

Source Area

The **source area** of a sediment is the locality that eroded and provided the sediment. The most important things to determine about a source area are the type of rocks that were exposed in it and its location and distance from the site of eventual deposition.

The *rock type* exposed in the source area determines the character of the resulting sediment. The composition of a sediment can indicate the source area rock type, even if the source area has been completely eroded away. A conglomerate may contain cobbles of basalt, granite, and chert; these rock types were obviously in its source area. An arkose containing coarse feldspar, quartz, and biotite may have come from a granitic source area. Furthermore, the presence of feldspar indicates the

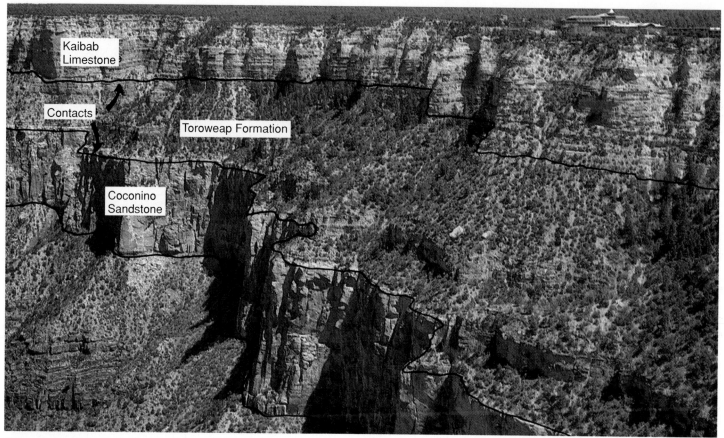

Figure 6.35

The upper three formations in the cliffs of the Grand Canyon, Arizona. The Kaibab Limestone and the Coconino Sandstone are resistant in the dry climate and form cliffs. The Toroweap Formation contains some shale and is less resistant, forming slopes.

source area was not subjected to extensive chemical weathering and that erosion probably took place in an arid environment with high relief. A quartz sandstone containing well-rounded quartz grains, on the other hand, probably represents the reerosion and redeposition of quartz grains from preexisting sandstone. Quartz is a hard, tough mineral very resistant to rounding by abrasion, so if quartz grains are well-rounded they have undergone many cycles of erosion, transportation, and deposition, probably over tens of millions of years.

Sedimentary rocks are also studied to determine the *direction* and *distance* to the source area. Figure 6.36 shows how several characteristics of sediment may vary with distance from a source area. Many sediment deposits get thinner away from the source, and the sediment grains themselves usually become finer and (for coarse grains) more rounded.

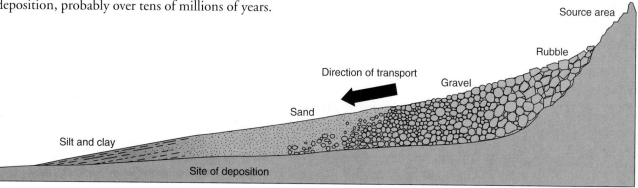

Figure 6.36

Sediment deposits often become thinner away from the source area, and sediment grains usually become finer and more rounded. The rocks that form from these sediments would change with distance from the source area from breccia to conglomerate to sandstone to shale. See Appendix F for rock symbols.

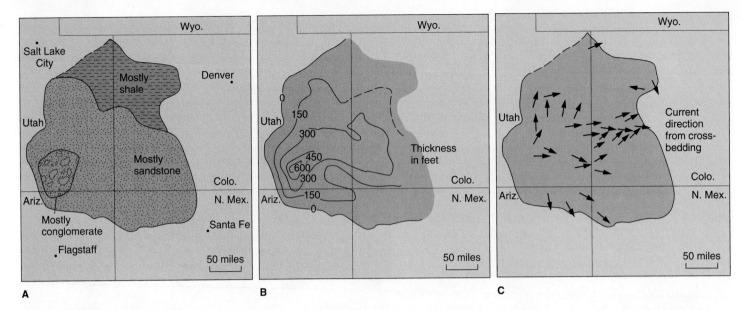

Figure 6.37

Characteristics of the Salt Wash Member of the Morrison Formation that help locate its source area. (*A*) The sediment grains become coarser to the southwest. (*B*) The deposit becomes thicker to the southwest. The contour lines show the thickness of the Salt Wash Member in feet. (*C*) Cross-bedding shows that the depositing currents came mostly from the southwest (arrows point down current).

Redrawn and simplified from Craig and others, 1955, U.S. Geological Survey Bulletin

Sedimentary structures often give clues about the directions of old currents (*paleocurrents*) that deposited sediments. Refer back to figure 6.27 and notice how cross-beds slope downward in the direction of current flow. Old current direction can also be determined from asymmetric ripple marks (figure 6.32).

Figure 6.37 shows how three of these characteristics were used to determine the location of the source area for a particular rock unit in the southwestern United States. The unit is the Salt Wash Member of the Morrison Formation (a *member* is a subdivision of a formation). It is an important rock unit, for it contains a great deal of uranium, deposited within the rock by ground water long after the rock formed. The unit thickens and coarsens to the southwest, and cross-beds show that the old currents that deposited the sediment came from the southwest. These three facts strongly suggest that the source area was to the southwest. This information helps exploration geologists search for uranium within the Salt Wash.

Environment of Deposition

Figure 6.38 shows the common environments in which sedimentary rocks are deposited. Geologists study modern environments in great detail so that they can interpret ancient rocks. Clues to the ancient environment of deposition come from a rock's composition, the size and sorting and rounding of the grains, the sedimentary structures present (including fossils), and the shape and vertical sequence of the sedimentary layers.

Continental environments include alluvial fans, river channels, flood plains, lakes, and dunes. Sediments deposited on land are subject to reerosion, so they often are destroyed. The great bulk of sedimentary rocks comes from the more easily preserved shallow marine environments, such as deltas, beaches, lagoons, shelves, and reefs. The characteristics of major environments are covered in detail in later chapters (10, 12–14, 18). In this section we describe the main sediment types and sedimentary structures found in each environment.

Glacial Environments

Glacial ice deposits narrow ridges and layers of sediment in valleys and widespread sheets of sediment on plains. Glacial sediment (*till*) is an unsorted mix of unweathered boulders, cobbles, pebbles, sand, silt, and clay. The boulders and cobbles may be scratched from grinding over one another under the great weight of the ice.

Alluvial Fan

As rivers emerge from mountains onto flatter plains, they deposit broad, fan-shaped piles of sediment. The sediment often consists of coarse, arkosic sandstones and conglomerates, marked by coarse cross-bedding and lens-like channel deposits (figure 6.39).

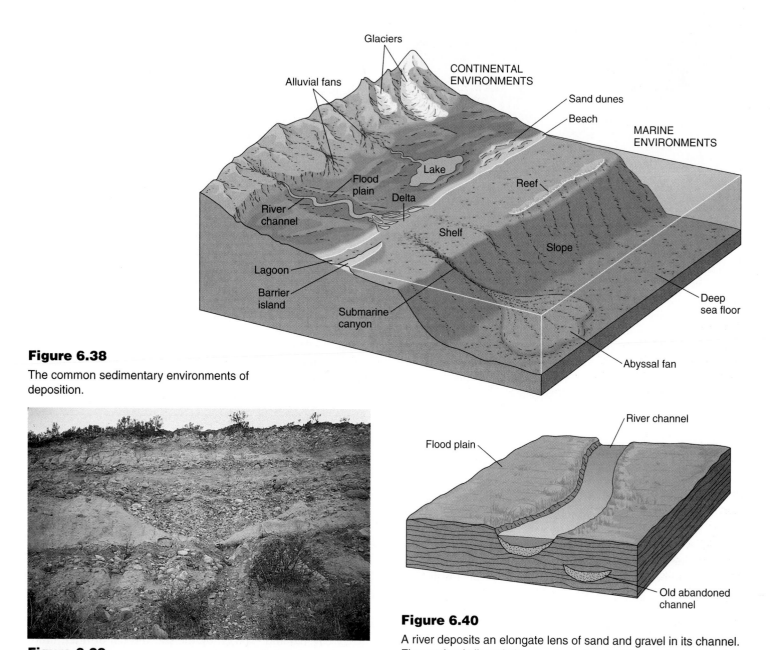

Figure 6.38

The common sedimentary environments of deposition.

Figure 6.39

Alluvial fan deposits, Baja California. A channel deposit of conglomerate occurs within the coarse-grained sequence.

Figure 6.40

A river deposits an elongate lens of sand and gravel in its channel. Fine-grained silt and clay are deposited beside the channel on the river's flood plain.

River Channel and Flood Plain

Rivers deposit elongate lenses of conglomerate or sandstone in their channels (figure 6.40). The sandstones may be arkoses or may consist of sand-sized fragments of fine-grained rocks. River channel deposits typically contain cross-beds and current ripple marks. Broad, flat flood plains beside channels are covered by periodic floodwaters, which deposit thin-bedded shales characterized by mud cracks and fossil footprints of animals. Hematite staining may color flood plain deposits red.

Lake

Thin-bedded shale, perhaps containing fish fossils, is deposited on lake bottoms. If the lake periodically dries up, the shales will be mud-cracked and perhaps interbedded with evaporites such as gypsum or rock salt.

6.2

IN GREATER DEPTH

Transgressions and Regressions

Sea level is not stable. Sea level has risen and fallen many times in the geologic past, flooding and exposing much of the land of the continents as it did so.

On a very broad, shallow, marine shelf several types of sediments may be deposited. On the beach and near shore, waves will deposit sand, which is usually derived from land. Farther from shore, in deeper, quieter water, land-derived silt and clay will be deposited. If the shelf is broad enough and covered with warm water, corals and algae may form carbonate sediments still farther seaward, beyond the reach of land-derived sediment. These sediments can lithify to form a seaward sequence of sandstone, shale, and limestone (box figure 1*A*).

If sea level rises or the land sinks (subsides), large areas of land will be flooded and these three environments of rock deposition will migrate across the land (box figure 1*B*). This is a *transgression* of the sea as it moves across the land, and it can result in a bed of sandstone overlain by shale, which in turn is overlain by limestone. Note that different parts of a single rock bed are deposited at different times—the seaward edge of the sandstone bed, for example, is older than the landward edge.

In a *regression* the sea moves off the land and the three rock types are arranged in a new vertical sequence—limestone is overlain by shale, and shale by sandstone (box figure 1*C*). A drop in sea level alone will not preserve this regression sequence. The land must usually subside rapidly to preserve these rocks, so that they are not destroyed by continental erosion.

The angles shown in the figure are exaggerated—the rocks often appear perfectly horizontal. Geologists use these two contrasting vertical sequences of rock to identify ancient transgressions and regressions.

Delta

A delta is a body of sediment deposited when a river flows into standing water, such as the sea or a lake. Deltas contain a great variety of subenvironments but are generally made up of thick sequences of siltstone and shale, marked by low-angle cross-bedding and cut by coarser channel deposits. Delta sequences may contain beds of peat or coal, as well as marine fossils such as clam shells.

Beach, Barrier Island, Dune

A barrier island is an elongate bar of sand built by wave action. Well-sorted quartz sandstone with well-rounded grains is deposited on beaches, barrier islands, and dunes. Beaches and barrier islands are characterized by cross-bedding (often low-angle) and marine fossils. Dunes have both high-angle and low-angle cross-bedding and occasionally contain fossil footprints of land animals such as lizards. All three environments can also contain carbonate sand in tropical regions, thus yielding cross-bedded clastic limestones.

Lagoon

A semi-enclosed, quiet body of water between a barrier island and the mainland is a lagoon. Fine-grained dark shale, cut by tidal channels of coarse sand and containing fossil oysters and other marine organisms, is formed in lagoons. Limestones may also form in lagoons adjacent to reefs (figure 6.17).

Shallow Marine Shelves

On the broad, shallow shelves adjacent to most shorelines, sediment grain size decreases offshore. Widespread deposits of sandstone, siltstone, and shale can be deposited on such shelves. The sandstone and siltstone contain symmetrical ripple marks, low-angle cross-beds, and marine fossils such as clams and snails. If fine-grained *tidal flats* near shore are alternately covered and exposed by the rise and fall of tides, mud-cracked marine shale will result.

Reefs

Massive limestone is deposited in reef cores, with steep beds of limestone breccia forming seaward of the reef, and horizontal beds of sand-sized and finer-grained limestones forming landward (figure 6.17). All these limestones are full of fossil fragments of corals, coralline algae, and numerous other marine organisms.

Deep Marine Environments

On the deep sea floor are deposited shale and graywacke sandstones. The graywackes are deposited by turbidity currents

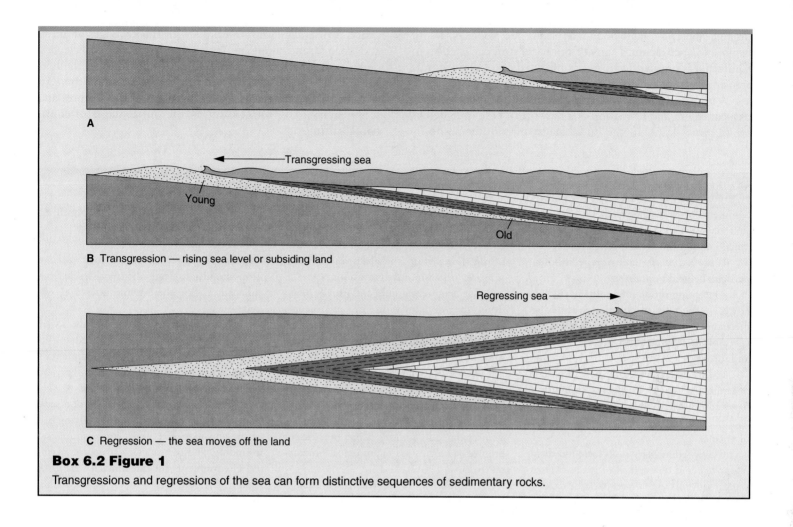

A

Transgressing sea ←

Young

Old

B Transgression — rising sea level or subsiding land

Regressing sea →

C Regression — the sea moves off the land

Box 6.2 Figure 1

Transgressions and regressions of the sea can form distinctive sequences of sedimentary rocks.

(figure 6.13) and typically contain graded bedding and current ripple marks.

Plate Tectonics and Sedimentary Rocks

The dynamic forces that move plates on Earth are also responsible for the distribution of many sedimentary rocks. As such, the distribution of sedimentary rocks often provides information that helps geologists reconstruct past plate tectonic settings.

In tectonically active areas, particularly along *convergent plate boundaries,* erosion is typically quite rapid resulting in thick accumulations of clastic sediments that record the uplift and erosion of mountain ranges. Coarse-grained clastic sediments are transported mainly by streams and turbidity currents and are usually deposited in low-lying basins adjacent to mountains. For example, uplift of the ancestral Sierra Nevada and Klamath mountain ranges in California is recorded by the thick accumulation of turbidite deposits preserved in basins to the west of the mountains. There, graywacke sandstone deposited by turbidity currents contains mainly volcanic clasts

in the lower part of the sedimentary sequence and abundant feldspar clasts in the upper part of the sequence. This indicates that a cover of volcanic rocks was first eroded from the ancestral mountains and then, as uplift and erosion continued, the underlying plutonic rocks were unroofed and eroded. Other eroded mountains, such as the Appalachians, have left similar records of uplift and erosion in the sedimentary record.

It is not uncommon for rugged mountain ranges such as the Canadian Rockies, European Alps, and Himalayas, that stand several thousand meters above sea level, to contain sedimentary rocks of marine origin that were originally deposited below sea level. The presence of marine sedimentary rocks such as limestone, chert, and shale containing marine fossils at high elevations, attests to the tremendous uplift associated with mountain building at convergent plate boundaries (see chapter 20).

Transform plate boundaries are also characterized by rapid rates of erosion and deposition of sediments as fault-bounded basins open and subside rapidly with continued plate motion. Because of the rapid rate of deposition and the rapid burial of organic material, fault-bounded basins are good places to explore for petroleum. Many of the petroleum occurrences in California are related to basins that formed as the San Andreas transform fault developed.

A *divergent plate boundary* may result in the splitting apart of a continent and formation of a new ocean basin (see figures 19.22 and 19.26). In the initial stages of continental divergence, a rift valley forms and fills with thick wedges of gravel and coarse sand along its fault-bounded margins; lake bed deposits and associated evaporite rocks may form in the bottom of the rift valley. The Red Sea is located along the East African Rift Zone and is a good example of the features and sedimentary rocks formed during the initial stages of continental rifting.

Summary

Sediment forms by the weathering and erosion of preexisting rocks and by chemical precipitation, sometimes by organisms.

Gravel, sand, silt, and *clay* are sediment particles defined by grain size.

The composition of sediment is governed by the rates of chemical weathering, mechanical weathering, and erosion. During transportation, grains can become rounded and sorted.

Sedimentary rocks form by *lithification* of sediment, by *crystallization* from solution, or by consolidation of remains of organisms. Sedimentary rocks may be *clastic, chemical,* or *organic.*

Clastic sedimentary rocks form mostly by *compaction* and *cementation* of grains. *Matrix* can partially fill the *pore space* of clastic rocks.

Conglomerate forms from coarse, rounded sediment grains that often have been transported only a short distance by a river or waves. *Sandstone* forms from sand deposited by rivers, wind, waves, or turbidity currents. *Shale* forms from river, lake, or ocean mud.

Limestone consists of calcite, formed either as a chemical precipitate in a reef or, more commonly, by the cementation of shell and coral fragments or of oolites. *Dolomite* usually forms from the alteration of limestone by magnesium-rich solutions.

Chert consists of silica and usually forms from the accumulation of microscopic marine organisms. *Recrystallization* often destroys the original texture of chert (and some limestones).

Evaporites, such as rock salt and gypsum, form as water evaporates. *Coal,* a major fuel, is consolidated plant material.

Sedimentary rocks are usually found in *beds* separated by *bedding planes* because the original sediments are deposited in horizontal layers.

Cross-bedding forms where sediment is deposited on a sloping surface in a sand dune, delta, or river bar.

A *graded bed* forms as coarse particles fall from suspension before fine particles, perhaps in a turbidity current.

Mud cracks form in drying mud. *Ripple marks* form beneath waves or currents.

Fossils are the traces of an organism's hard parts or tracks preserved in rock.

A *formation* is a convenient rock unit for mapping and describing rock. Formations are lithologically distinguishable from adjacent rocks; their boundaries are *contacts.*

Geologists try to determine the *source area* of a sedimentary rock by studying its grain size, composition, and sedimentary structures. The source area's rock type and location are important to determine.

The *environment of deposition* of a sedimentary rock is determined by studying bed shape and sequence, grain composition and rounding, and sedimentary structures. Typical environments include alluvial fans, river channels, flood plains, lakes, dunes, deltas, beaches, shallow marine shelves, reefs, and the deep sea floor.

Plate tectonics plays an important role in the distribution of sedimentary rocks; the occurrence of certain types of sedimentary rocks is used by geologists to construct past plate-tectonic settings.

Terms to Remember

bedding 135
bedding plane 136
cement 125
cementation 125
chemical sedimentary rock 126
chert 133
clastic sedimentary rock 126
clastic texture 125
clay 122
coal 135
compaction 125
conglomerate 126

contact 140
cross-bedding 136
crystalline texture 125
crystallization 125
deposition 123
dolomite 133
environment of deposition 124
evaporite 134
formation 139
fossil 139
graded bed 136
gravel 122

limestone 130
lithification 125
matrix 128
mud crack 136
organic sedimentary rock 126
original horizontality 135
pore space 125
recrystallization 132
ripple mark 136
rounding 122
sand 122
sandstone 127

sediment 122
sedimentary breccia 126
sedimentary rock 126

sedimentary structure 135
shale 128
silt 122

sorting 123
source area 140
turbidity current 128

Testing Your Knowledge

Use the questions below to prepare for exams based on this chapter.

1. Quartz is a common mineral in sandstone. Under certain circumstances, feldspar is common in sandstone, even though it normally weathers rapidly to clay. What conditions of climate, weathering rate, and erosion rate could lead to a feldspar-rich sandstone? Explain your answer.

2. Describe with sketches how wet mud compacts before it becomes shale.

3. What do mud cracks tell about the environment of deposition of a sedimentary rock?

4. How does a graded bed form?

5. List the clastic sediment particles in order of decreasing grain size.

6. How does a sedimentary breccia differ in appearance and origin from a conglomerate?

7. Describe three different origins for limestone.

8. How does dolomite usually form?

9. What is the origin of coal?

10. Sketch the cementation of sand to form sandstone.

11. How do evaporites form? Name two evaporites.

12. Name the three most common sedimentary rocks.

13. What is a formation?

14. Sketch two ways that cross-bedding can form.

15. Particles of sediment from 1/16 mm to 2 mm in diameter are of what size? (a) gravel (b) sand (c) silt (d) clay

16. Rounding is (a) the rounding of a grain to a spherical shape (b) the grinding away of sharp edges and corners of rock fragments during transportation (c) a type of mineral (d) none of the above

17. Compaction and cementation are two common processes of (a) erosion (b) transportation (c) deposition (d) lithification

18. Which is not a chemical or organic sedimentary rock? (a) rock salt (b) shale (c) limestone (d) gypsum

19. The major difference between breccia and conglomerate is (a) size of grains (b) rounding of the grains (c) composition of grains (d) all of the above

20. Which is not a type of sandstone? (a) quartz sandstone (b) arkose (c) graywacke (d) coal

21. Shale differs from mudstone in that (a) shale has larger grains (b) shale is visibly layered and fissile; mudstone is massive and blocky (c) shale has smaller grains (d) there is no difference between shale and mudstone

22. The chemical element found in dolomite not found in limestone is (a) Ca (b) Mg (c) C (d) O (e) Al

23. In a graded bed the particle size decreases (a) upward (b) downward (c) in the direction of the current (d) particle size stays the same

24. A body or rock of considerable thickness with characteristics that distinguish it from adjacent rock units is called a/an (a) formation (b) contact (c) bedding plane (d) outcrop

25. If sea level drops or the land rises, what is likely to occur? (a) a flood (b) a regression (c) a transgression (d) no geologic change will take place

26. Thick accumulations of graywacke and volcanic sediments can indicate an ancient (a) divergent plate boundary (b) convergent boundary (c) transform boundary

27. A sedimentary rock made of fragments of preexisting rocks is (a) organic (b) chemical (c) clastic

28. Clues to the nature of the source area of sediment can be found in (a) the composition of the sediment (b) sedimentary structures (c) rounding of sediment (d) all of the above

Expanding Your Knowledge

1. How might graded bedding be used to determine the tops and bottoms of sedimentary rock layers in an area where sedimentary rock is no longer horizontal? What other sedimentary structures can be used to determine the tops and bottoms of tilted beds?

2. Which would weather faster in a humid climate, a quartz sandstone or an arkose? Explain your answer.

3. A cross-bedded quartz sandstone may have been deposited as a beach sand or as a dune sand. What features could you look for within the rock to tell if it had been deposited on a beach? On a dune?

4. Why is burial usually necessary to turn a sediment into a sedimentary rock?

5. Why are most beds of sedimentary rock formed horizontally?

6. Discuss the role of sedimentary rocks in the rock cycle, diagramming the rock cycle as part of your answer. What do sedimentary rocks form from? What can they turn into?

Exploring Resources

 Blatt, H. 1992. *Sedimentary petrology.* 2d ed. New York: W. H. Freeman.

Boggs, S. 1987. *Principles of sedimentology and stratigraphy.* Columbus, Ohio: Charles E. Merrill.

Boggs, S. 1992. *Petrology of sedimentary rocks.* New York: Macmillan.

Busby, C. J., and R. V. Ingersoll. 1995. *Tectonics of sedimentary basins.* Palo Alto, Calif.: Blackwell.

Davis, R. A. 1992. *Depositional systems.* 2d ed. Englewood Cliffs, N.J.: Prentice-Hall.

Pettijohn, F. J. 1975. *Sedimentary rocks.* 3d ed. New York: Harper & Row.

Pettijohn, F. J., and P. E. Potter. 1964. *Atlas and glossary of primary sedimentary structures.* New York: Springer-Verlag.

Prothero, D. R., and F. Schwab. 1996. *Sedimentary geology: An introduction to sedimentary rocks and stratigraphy.* New York: W. H. Freeman.

Reading, H. G., ed. 1986. *Sedimentary environments and facies.* 2d ed. Oxford: Blackwell.

Scholle, P. A., D. G. Debout, and C. H. Moore, eds. 1983. *Carbonate depositional environments.* Tulsa: Amer. Assoc. of Petroleum Geologists Memoir 33.

Scholle, P. A., and D. Spearing, eds. 1982. *Sandstone depositional environments.* Tulsa: Amer. Assoc. of Petroleum Geologists Memoir 31.

Selley, R. C. 1982. *An introduction to sedimentology.* 2d ed. New York: Academic Press.

Tucker, M. E. 1991. *Sedimentary petrology: An introduction.* 2d ed. Oxford: Blackwell.

http://www-marine.wr.usgs.gov/docs/seds/sedimentology.html

Visit the U.S. Geological Survey Bedform and Sedimentology site for computer and photographic images and movies of sedimentary structures.

http://zircon.geology.union.edu/Gildner/stack.html

For a virtual field trip of the Ordovician-age rocks in the Mohawk Valley of New York state, visit Union College's geology website.

http://www.unistuttgart.de/geologie/Jurassic_Reef_Park/

For a virtual field trip of the Jurassic Park Reef in Germany.

Interacting with Journey Through Geology CD-ROM

Expand your knowledge of the concepts presented in this chapter by using the CD-ROM to answer the following question.

1. Stories in Rocks, Sandstone: How can geologists use the size of the sediment in a sedimentary rock as a clue to the environment in which the sediment was deposited?

Photo taken through a polarizing microscope of a metamorphic rock that was once shale. Micas (colored minerals) grew while the rock was being folded during regional metamorphism.

Photo by C. C. Plummer

CHAPTER
7

Metamorphism, Metamorphic Rocks, and Hydrothermal Rocks

This chapter on metamorphic rocks, the third major category of rocks in the rock cycle, completes our description of earth materials (rocks and minerals). The information on igneous and sedimentary processes in previous chapters should help you understand metamorphic rocks, which form from *preexisting* rocks.

After reading the chapter on weathering, you know how rocks are altered when exposed at the earth's surface. *Metamorphism* (a word from Latin and Greek that means literally "changing of form") also involves alterations, but the changes are due to deep burial, tectonic forces, and/or high temperature rather than surface conditions.

As you study this chapter, try to keep clearly in mind how the chemical composition of a rock and the temperature, pressure, and water present each contribute to the metamorphic process and the resultant metamorphic rock.

We also discuss hydrothermally deposited rocks and minerals, which are found in association with both igneous and metamorphic rocks. Hydrothermal ore deposits, while not volumetrically significant, are of great importance to the world's supply of metals.

Because nearly all metamorphic rocks form deep within the earth's crust, they provide geologists with many clues about conditions at depth. Therefore, understanding metamorphism will help you when we consider geologic processes involving the earth's internal forces. Metamorphic rocks are a feature of the oldest exposed rocks of the continents and of major mountain belts. They are especially important in providing evidence of what happens during subduction and plate convergence.

From your study so far of earth materials and the rock cycle, you know that rocks change, given enough time, when their physical environment changes radically. In chapter 3, you saw how deeply buried rocks melt (or partially melt) to form magma when temperatures are high enough. What happens to rocks that are deeply buried but are not hot enough to melt? They become metamorphosed. **Metamorphism** refers to changes to rocks that take place in the earth's interior. The changes may be new textures, new mineral assemblages, or both. Transformations occur in the solid state (meaning the rock does not melt).

The new rock, the **metamorphic rock,** in nearly all cases has a texture clearly different from that of the *original* rock, or **parent rock.** When limestone is metamorphosed to marble, for example, the fine grains of calcite coalesce and recrystallize into larger calcite crystals. The calcite crystals are interlocked in a mosaic pattern that gives marble a texture distinctly different from that of the parent limestone. If the limestone is composed entirely of calcite, then metamorphism into marble involves no new minerals, only a change in texture.

More commonly, the various elements of a parent rock react chemically and crystallize into new minerals, thus making the metamorphic rock distinct both mineralogically and texturally from the parent rock. This is because the parent rock is unstable in its new environment. The old minerals recrystallize into new ones that are at *equilibrium* in the new environment. For example, clay minerals form at the earth's surface (see chapter 5). Therefore, they are stable at the low temperature and pressure conditions both at and just below the earth's surface. When subjected to the temperatures and pressures deep within the earth's crust, the clay minerals of a shale can recrystallize into coarse-grained mica. Another example is that under appropriate temperature and pressure conditions, a quartz sandstone with a calcite cement metamorphoses as follows:

$$CaCO_3 + SiO_2 \rightarrow CaSiO_3 + CO_2$$

| calcite | quartz | wollastonite (a mineral) | carbon dioxide gas |

No one has observed metamorphism taking place, just as no one has ever seen a granite pluton form. What, then, leads us to believe that metamorphic rocks form in a solid state (i.e., without melting) at high pressure and temperature? Many metamorphic rocks found on the earth's surface exhibit contorted banding (figure 7.1). Commonly, banding in metamorphic rocks can be demonstrated to have originally been flat-lying sedimentary layering (even though the rock has since recrystallized). These rocks, now hard and brittle, would shatter if smashed with a hammer. But they must have been **plastic** (capable of being bent and molded under stress) to have been folded into such contorted patterns. Because high temperature and pressure are necessary to make rocks plastic, a reasonable conclusion is that these rocks formed at considerable depth, where such conditions exist. Moreover, crystallization of a magma would not produce contorted layering.

Figure 7.1

Metamorphic rock from Greenland. Metamorphism took place 3,700 million years ago—it is one of the oldest rocks on earth.

Photo by C. C. Plummer

Factors Controlling the Characteristics of Metamorphic Rocks

A metamorphic rock owes its characteristic texture and particular mineral content to several factors, the most important being (1) the composition of the parent rock before metamorphism, (2) temperature and pressure during metamorphism, (3) the effects of tectonic forces, and (4) the effects of fluids, such as water.

Composition of the Parent Rock

Usually no new elements or chemical compounds are added to the rock during metamorphism, except perhaps water. (Metasomatism, discussed later in this chapter, does involve the addition of other elements.) Therefore, the mineral content of the metamorphic rock is controlled by the chemical composition

7.1
ENVIRONMENTAL
GEOLOGY

Metamorphic Exhalations May Have Changed the Climate

The world's highest mountain chain, the Himalaya, may have been responsible for a period of global warming some 60 million years ago. As explained in chapter 20, collision between India and Asia created the Himalaya. Before the collision, a great thickness of sedimentary rocks built up on the ocean floor that separated the two landmasses. Limestones were especially abundant. Upon collision, the layers of sedimentary rocks were crumpled, as if caught in a giant vise. Rocks that were deeply buried by the process were metamorphosed.

Derrill Kerrick and his colleagues at Pennsylvania State University have proposed a hypothesis that links the global warming of the time to the release of CO_2 during metamorphism. If limestone contains significant amounts of quartz or clay in addition to the prevailing calcite, chemical reactions during metamorphism produce new minerals as well as carbon dioxide gas. The scientists estimated that several hundred million tons of carbon dioxide per year may have been released into the atmosphere over a period of around ten million years. This amount of CO_2 added to the atmosphere and contributing to the greenhouse effect would account for the warmer climate inferred for that part of the earth's history.

If global warming is taking place at present (as is believed by many) due to increasing CO_2 levels, then our cars, factories and cutting down of forests can in a few decades do what metamorphism took millions of years to accomplish.

---

of the parent rock. For example, a basalt always metamorphoses into a rock in which the new minerals can collectively accommodate the approximately 50% silica and relatively high amounts of the oxides of iron, magnesium, calcium, and aluminum in the original rock. On the other hand, a limestone, composed essentially of calcite ($CaCO_3$), cannot metamorphose into a silica-rich rock.

Temperature

Heat, necessary for metamorphic reactions, comes primarily from the outward flow of geothermal energy from the earth's deep interior. The deeper a rock is beneath the surface, the hotter it will be. The particular temperature for rock at a given depth depends on the local *geothermal gradient* (described in chapter 3). Additional heat could be derived from magma, if magma bodies are locally present, and by friction. The frictional heat is from the grinding of rock against rock in response to tectonic forces.

A mineral is said to be *stable* if, given enough time, it does not react with another substance or convert to a new mineral or substance. Any mineral is stable only within a given temperature range. The stability temperature range of a mineral varies with factors such as pressure and the presence or absence of other substances. Some minerals are stable over a wide temperature range. Quartz, if not mixed with other minerals, is stable at atmospheric pressure (i.e., at the earth's surface) up to about 800°C. At higher pressures, quartz remains stable to even higher temperatures. Other minerals are stable over a temperature range of only 100° or 200°C.

By knowing (from results of laboratory experiments) the particular temperature range in which a mineral is stable, a geologist may be able to deduce the temperature of metamorphism for a rock that includes that mineral.

Minerals stable at higher temperatures tend to be less dense (or have a lower specific gravity) than chemically identical minerals stable at lower temperatures. As temperature increases, the ions vibrate more within their sites in the crystal structure. A more open (less tightly packed) crystal structure, such as high-temperature minerals tend to have, allows greater vibration of ions. (If the heat and resulting vibrations become too great, the bonds between atoms in the crystal break and the substance becomes liquid.)

The upper limit on temperature in metamorphism overlaps the temperature of partial melting of a rock. If partial melting takes place, the component that melts becomes a magma; the solid residue remains a metamorphic rock. Temperatures at which the igneous and metamorphic realms can coexist vary considerably. For an ultramafic rock (containing only ferromagnesian silicate minerals) the temperature will be over 1200°C; for a granite or rhyolite under high water pressure, it could be as low as 650°C.

Pressure

Usually, when we talk about pressure, we mean **confining pressure**; that is, pressure applied equally on all surfaces of a substance as a result of burial or submergence. A diver senses confining pressure proportional to the weight of the overlying water (figure 7.2). The pressure uniformly squeezes the diver's entire body surface. Likewise, an object buried deeply within

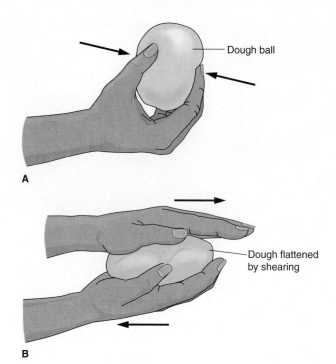

Figure 7.3

(*A*) Compressive stress. More force is exerted in the direction of the arrows than elsewhere on the ball. (*B*) Shearing. Parallel arrows indicate the direction of forces.

Figure 7.2

Confining pressure. The diver's suit is pressurized to counteract hydrostatic pressure. Object (cube) has a greater volume at low pressure than at high pressure.

the earth's crust is compressed by strong confining pressure, called *lithostatic pressure,* which forces grains closer together and eliminates pore space.

Any new mineral that has crystallized under high-pressure conditions tends to occupy less space than did the mineral or minerals from which it formed. The new mineral is denser than its low-pressure counterparts because the pressure forces atoms closer together into a more closely packed crystal structure.

Differential Stress

Most metamorphic rocks show the effects of tectonic forces. When forces are applied to an object, the object is under **stress.**

If the forces on a body are stronger or weaker in different directions, a body is subjected to **differential stress.** Differential stress tends to deform objects into oblong or flattened forms. If you squeeze a rubber ball between your thumb and forefinger, the ball is under differential stress (figure 7.3*A*). If you squeeze a ball of dough, it will remain flattened after you stop squeezing, because dough is plastic. To illustrate the difference between confining pressure and differential stress, visualize a drum filled with water. If you place a ball of putty underwater in the bottom of the drum, the ball will not change its shape (its volume will decrease slightly due to the weight of the overlying water). Now take the putty ball out of the water and place it under the drum. The putty will be flattened into the shape of a pancake due to the differential stress. In this case, the putty is subjected to *compressive differential stress* or, more simply, **compressive stress** (as is the dough ball shown in figure 7.3*A*).

Differential stress is also caused by **shearing,** which causes parts of a body to move or slide relative to one another across a plane. An example of shearing is when you spread out a deck of cards on a table with your hand moving parallel to the table. Figure 7.3*B* shows that initially dough is flattened at a low angle to the shear force (the moving hands). As shearing continues, the flattened dough is rotated toward *parallelism* to the shear force. In contrast, compressive stress flattens objects *perpendicular* to the applied force.

Foliation

Differential stress has a very important influence on the texture of a metamorphic rock because it forces the constituents of the

Figure 7.4

Metamorphosed conglomerate in which the pebbles have been flattened (sometimes called a stretched pebble conglomerate). Compare to figure 6.9.

Photo by C. C. Plummer

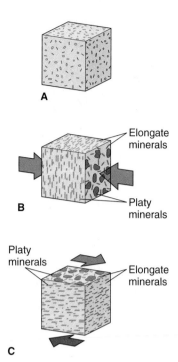

Figure 7.5

Orientation of platy and elongate minerals in metamorphic rock. (*A*) Platy minerals randomly oriented (e.g., clay minerals before metamorphism). No differential stress involved. (*B*) Platy minerals (e.g., mica) and elongate minerals (e.g., amphibole) have crystallized under the influence of compressive stress. (*C*) Platy and elongate minerals developed under the influence of shearing.

Figure 7.6

Schistose texture.

rock to become parallel to one another. For instance, the pebbles in the metamorphosed conglomerate shown in figure 7.4 were originally more spherical but have been flattened by differential stress. When a rock has a planar texture, it is said to be **foliated.** Foliation is manifested in various ways. If a platy mineral (such as mica) is crystallizing within a rock that is undergoing differential stress, the mineral grows in such a way that it remains parallel to the direction of shearing or perpendicular to the direction of compressive stress (figure 7.5). Any platy mineral attempting to grow against shearing is either ground up or forced into alignment. Minerals that crystallize in needlelike shapes (for example, hornblende) behave similarly, growing with their long axes parallel to the plane of shearing or perpendicular to compressive stress. The three very different textures described below (from lowest to highest degree of metamorphism) are all variations of foliation and are important in classifying metamorphic rocks:

1. If the rock splits easily along nearly flat and parallel planes, indicating that preexisting, microscopic, platy minerals were pushed into alignment during metamorphism, we say the rock is **slaty,** or that it possesses **slaty cleavage.**
2. If visible platy or needle-shaped minerals have grown essentially parallel to a plane due to differential stress, the rock is **schistose** (figure 7.6).
3. If the rock became very plastic and the new minerals separated into distinct (light and dark) layers or lenses, the rock has a layered or **gneissic** texture, such as in figure 7.14.

Fluids

Hot water (as vapor) is the most important fluid involved in metamorphic processes, although other gases, such as carbon dioxide, sometimes play a role. The water may have been trapped in a parent sedimentary rock or given off by a cooling pluton.

Water is thought to help trigger metamorphic chemical reactions. Water is a sort of intra-rock rapid transit for ions. Under high pressure, it moves between grains, dissolves ions from one mineral, and then carries these ions elsewhere in the rock where

A

B

Figure 7.7

Photomicrographs of metamorphic rocks in thin sections of (*A*) nonfoliated rock and (*B*) foliated rock.

Photo by C. C. Plummer

they can react with the ions of a second mineral. The new mineral that forms is stable under the existing conditions.

Time

The effect of time on metamorphism is hard to comprehend. Most metamorphic rocks are composed predominantly of silicate minerals, and silicate compounds are notorious for their sluggish chemical reaction rates. Recently, garnet crystals taken from a metamorphic rock collected in Vermont were analyzed and scientists calculated a growth rate of 1.4 millimeters per million years. The garnets' growth was sustained over a 10.5 million year period. Many laboratory attempts to duplicate metamorphic reactions believed to occur in nature have been frustrated by the time element. The several million years during which a particular combination of temperature and pressure may have prevailed in nature are impossible to duplicate.

Classification of Metamorphic Rocks

As we noted before, the kind of metamorphic rock that forms is determined by the metamorphic environment (primarily the particular combination of pressure, stress, and temperature) and by the chemical constituents of the parent rock. Many kinds of metamorphic rocks exist because of the many possible combinations of these factors. These rocks are classified based on broad similarities. (Appendix B contains a systematic procedure for identifying common metamorphic rocks.)

First consider the texture of a metamorphic rock. Is it *foliated* or *nonfoliated* (figure 7.7)? If the rock is nonfoliated, it is named on the basis of its composition. For example, a nonfoliated quartz-rich metamorphic rock is a *quartzite;* one composed almost entirely of calcite is a *marble*.

If the rock is foliated, you must determine the type of foliation. For example, a schistose rock is called a *schist*. But this name tells us nothing about what minerals are in this rock; so we add adjectives to describe the composition. Thus, a *garnet-mica schist* (whose parent was shale) is easily distinguishable from an *amphibole schist* (a metamorphic product of basalt). The relationship of texture to rock name is summarized in table 7.1.

Types of Metamorphism

The two most common types of metamorphism are contact metamorphism and regional metamorphism. These are discussed next. Hydrothermal metamorphism, in which hot water plays a major role, is discussed later in this chapter.

Contact Metamorphism

Contact metamorphism (also known as *thermal* metamorphism) is metamorphism in which high temperature is the dominant factor. Confining pressure may influence which new minerals crystallize. However, the confining pressure is usually relatively low. This is because contact metamorphism mostly takes place not too far beneath the earth's surface (less than 10 kilometers). Contact metamorphism occurs adjacent to a plu-

Table 7.1 — Classification and Naming of Metamorphic Rocks (Based Primarily on Texture)

Nonfoliated
Name Based on Mineral Content of Rock

Usual Parent Rock	Rock Name	Predominant Minerals	Identifying Characteristics
Limestone Dolomite	Marble Dolomite marble	Calcite Dolomite	Coarse interlocking grains of calcite (or, less commonly, dolomite). Calcite (or dolomite) has rhombohedral cleavage; hardness intermediate between glass and fingernail. Calcite effervesces in weak acid.
Quartz sandstone	Quartzite	Quartz	Rock composed of interlocking small granules of quartz. Has a sugary appearance and vitreous luster; scratches glass.
Shale Basalt	Hornfels Hornfels	Fine-grained micas Fine-grained ferro-magnesian minerals, plagioclase	A fine-grained, dark rock that generally will scratch glass. May have a few coarser minerals present.

Foliated
Name Based Principally on Kind of Foliation Regardless of Parent Rock. Adjectives Describe the Composition (e.g., biotite-garnet schist)

Texture	Rock Name	Identifying Characteristics
Slaty	Slate	A very fine-grained rock with an earthy luster. Splits easily into thin, flat sheets.
Intermediate between slaty and schistose	Phyllite	Fine-grained rock with a silky luster. Generally splits along wavy surfaces.
Schistose	Schist	Composed of visible platy or elongated minerals that show planar alignment. A wide variety of minerals can be found in various types of schist (e.g., garnet mica schist, hornblende schist, etc.).
Gneissic	Gneiss	Light and dark minerals are found in separate, parallel layers or lenses. Commonly, the dark layers include biotite and hornblende; the light-colored layers are composed of feldspars and quartz. The layers may be folded or appear contorted.

ton when a body of magma intrudes relatively cool country rock. The process can be thought of as the "baking" of country rock adjacent to an intrusive contact; hence the term *contact metamorphism.* The zone of contact metamorphism (also called an *aureole*) is usually quite narrow—generally from 1 to 100 meters wide. Differential stress is rarely significant; therefore, these rocks typically are nonfoliated.

During contact metamorphism, shale is changed into the very fine-grained metamorphic rock **hornfels.** Characteristi-cally, only microscopically visible micas develop. Sometimes a few minerals grow large enough to be seen with the naked eye; these are minerals that are especially capable of crystallizing under the particular temperature attained during metamor-phism. Hornfels can also form from basalt, in which case amphibole, rather than mica, is the predominant fine-grained mineral produced.

Limestone recrystallizes during metamorphism into **mar-ble,** a coarse-grained rock composed of interlocking calcite

7.1 Astrogeology

Impact Craters and Shock Metamorphism

Astrogeology Box 7.1 Figure 1

Meteor crater in Arizona.

Photo by Frank M. Hanna

The spectacular collision of the comet Shoemaker-Levy with Jupiter in 1994 served to remind us that asteroids and comets occasionally collide with a planet. Earth is not exempt from collisions. Large meteorites have produced impact craters when they have collided with Earth's surface. One well-known meteorite crater is Meteor Crater in Arizona, which is a little more than a kilometer in diameter (box figure 1). Many much larger craters are known in Canada, Germany, Australia, and other places.

Impact craters display an unusual type of metamorphism called *shock metamorphism*. The sudden impact of a large extraterrestrial body results in brief, but extremely high pressures. Quartz may recrystallize into the rare SiO_2 mineral coesite. Quartz that is not as intensely impacted suffers damage (detectable under a microscope) to its crystal lattice.

The impact of a meteorite also may generate enough heat to locally melt rock. Molten blobs of rock are thrown into the air and become streamlined in the Earth's atmosphere before solidifying into what are called *tektites*. Tektites may be found hundreds of kilometers from the point of meteorite impact.

A large meteorite would also throw large quantities of dust high into the atmosphere. According to one theory, the change in global climate due to a meteorite impact around 65 million years ago caused extinctions of many varieties of creatures (see Box 8.2 on the extinction of dinosaurs). Evidence for this impact includes finding tiny fragments of shock metamorphosed quartz and tektites in sedimentary rock that is 65 million years old.

Shock metamorphosed rock fragments are much more common on the Moon than on Earth. There may be as many as 400,000 craters larger than a kilometer in diameter on the Moon. Mercury's surface is remarkably similar to that of the Moon. Our two neighboring planets, Venus and Mars, are not as extensively cratered as is the Moon.

---

crystals (figure 7.8). Dolomite, less commonly found, recrystallizes into a *dolomitic marble*. Marble has long been valued as a building material and as a material for sculpture, partly because it is easily cut and polished and partly because it reflects light in a shimmering pattern, a result of the excellent cleavage of the individual calcite crystals. Marble is, however, highly susceptible to chemical weathering (see chapter 5).

Quartzite is produced when grains of quartz in sandstone are welded together while the rock is subjected to high temperature. This makes it as difficult to break along grain boundaries as through the grains. Therefore quartzite, being as hard as a single quartz crystal, is difficult to crush or break. It is the most durable of common rocks used for construction, both because of its hardness and because quartz is not susceptible to chemical weathering.

Marble and quartzite also form under conditions of regional metamorphism. When grains of calcite or quartz recrystallize, they tend to be equidimensional, rather than elongate or platy. For this reason, marble and quartzite do not usually exhibit foliation, even though subjected to differential stress during metamorphism.

Regional Metamorphism

The great majority of the metamorphic rocks found on the earth's surface are products of **regional metamorphism** (also known as *dynamothermal* metamorphism) which is metamorphism that takes place at considerable depth underground (generally greater than 5 kilometers). Regional metamorphic rocks are almost always foliated, indicating differential stress during recrystallization. Metamorphic rocks are prevalent in the most intensely deformed portions of mountain ranges. They are visible where once deeply buried cores of mountain ranges are exposed by erosion. Furthermore, large regions of the continents are underlain by metamorphic rocks, thought to be the roots of ancient mountains long since eroded down to plains or rolling hills.

Temperatures during regional metamorphism vary widely. Usually, the temperatures are in the range of 300 to 800°C. Temperature at a particular place depends to a large extent on depth of burial and the geothermal gradient of the region (see Box 7.3). Locally, temperature may also increase because of heat from friction (due to shearing) or from heat

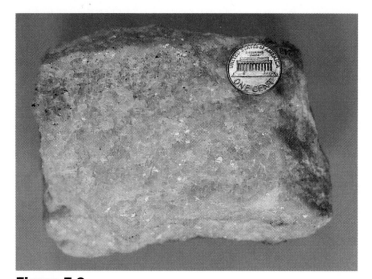

Figure 7.8

Marble.

Photo by C. C. Plummer

Table 7.2	Regional Metamorphic Rocks that Form under Approximately Similar Pressure and Temperature Conditions	
Parent Rock	**Rock Name**	**Predominant Minerals**
Basalt	Amphibole schist (amphibolite)	Hornblende, plagioclase, garnet
Shale	Mica schist	Biotite, muscovite, quartz, garnet
Quartz sandstone	Quartzite	Quartz
Limestone or dolomite	Marble	Calcite or dolomite

given off by nearby magma bodies. The high confining pressure is due to burial under 5 or more kilometers of rock. The differential stress is due to tectonism; that is, the constant movement and squeezing of the crust during mountain-building episodes.

Depending on the pressure and temperature conditions during metamorphism, a particular parent rock may recrystallize into one of several metamorphic rocks. For example, if basalt is metamorphosed at relatively low temperatures and pressures, it will recrystallize into a *greenschist,* a schistose rock containing chlorite (a green sheet-silicate), actinolite (a green amphibole), and sodium-rich plagioclase. At higher temperatures and pressures, the same basalt would recrystallize into an *amphibole schist* (also called *amphibolite*), a rock composed of hornblende, plagioclase feldspar, and, perhaps, garnet. Metamorphism of other parent rocks under conditions similar to those that produce amphibole schist from basalt should produce the metamorphic rocks shown in table 7.2.

Progressive Metamorphism

To show how rocks are changed by regional metamorphism, we look at what happens to shale during *progressive metamorphism*—that is, as progressively greater pressure and temperature act on a rock type with increasing depth in the earth's crust (figure 7.9).

Shale, the parent rock, is formed largely of submicroscopic, platy, clay minerals. The metamorphic rock associated with the lowest pressure and temperature conditions of regional metamorphism is **slate,** a very fine-grained rock that splits easily along flat, parallel planes (figures 7.10 and 7.11). Slate develops under temperatures and pressures only slightly greater than those found in the sedimentary realm. The temperatures are not high enough for the rock to thoroughly recrystallize. The important controlling factor is differential stress. The original clay minerals partially recrys-

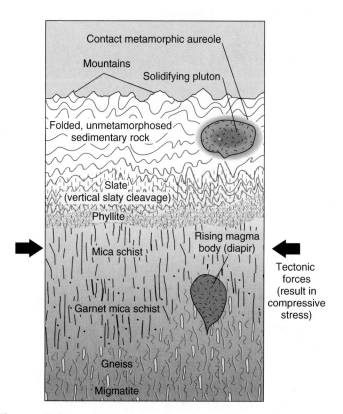

Figure 7.9

Schematic cross section representing an approximately 30-kilometer portion of the earth's crust during metamorphism. Rock names given are those produced from shale.

tallize into equally fine-grained micas. Under differential stress, the old and new platy minerals are aligned, creating slaty cleavage in the rock. A slate indicates that a relatively cool and brittle rock has been subjected to intense tectonic activity.

Metamorphism, Metamorphic Rocks, and Hydrothermal Rocks   **159**

Figure 7.10

Slate outcrop in Antarctica.

Photo by P. D. Rowley, U.S. Geological Survey

Figure 7.11

Slate.

Photo by C. C. Plummer

Because of the ease with which it can be split into thin, flat sheets, slate is used for making chalkboards, pool tables, and roofs.

Phyllite is a rock in which the newly formed micas are larger than in slate, but still cannot be seen with the naked eye.

Figure 7.12

Phyllite, exhibiting a crinkled, silky-looking surface.

Photo by C. C. Plummer

This requires a further increase in temperature over that needed for slate to form. The very fine-grained mica imparts a silky sheen to the rock, which may otherwise closely resemble slate (figure 7.12). However, the slaty cleavage may be crinkled in the process of conversion of slate to phyllite.

A **schist** is characterized by megascopically visible, approximately parallel-oriented minerals. Platy or elongate minerals that crystallize from the parent rock are clearly visible to the naked eye. Shale may recrystallize into several mineralogically distinct varieties of schist. Which minerals form depends on the particular combination of temperature and pressure prevailing during recrystallization. For instance, if the rock is a *mica schist,* metamorphism probably took place at only slightly higher temperatures and pressures than those at which a phyllite forms. A *garnet-mica schist* (figure 7.13) indicates that the temperature and pressure were somewhat greater than necessary for a mica schist to form (see Box 7.2).

In a **gneiss,** a rock consisting of light and dark mineral layers or lenses, the highest temperatures and pressures have changed the rock so that minerals have separated into layers. Platy or elongate minerals (such as mica or amphibole) in dark layers alternate with layers of light-colored minerals of no particular shape. Within the light-colored layers coarse feldspars have crystallized. In composition, a gneiss may resemble granite or diorite, but it is distinguishable from those plutonic rocks by its foliation (figure 7.14).

Temperature conditions under which a gneiss develops approach those at which granite solidifies. It is not surprising, then, that the same minerals are found in gneiss and in granite. In fact, a previously solidified granite can be converted to a gneiss under appropriate pressure and temperature conditions and if the rock is under differential stress.

If the temperature is high enough, partial melting of rock may take place, and a magma is sweated out into layers within the foliation planes of the solid rock. After the magma solidifies, the rock becomes a **migmatite,** a mixed igneous and

7.2
IN GREATER DEPTH

Index Minerals

Box 7.2 Figure 1

Phase diagram showing the stability relationships for the Al_2SiO_5 minerals.

M. J. Holdaway, 1971, *American Journal of Science*, v. 271. Reprinted by permission of *American Journal of Science*.

Certain minerals can only form under a restricted range of pressure and temperature. Stability ranges of these minerals have been determined in laboratories. When found in metamorphic rocks, these minerals can help us infer, within limits, what the pressure and temperature conditions were during metamorphism. For this reason, they are known as *index minerals*. Among the best known are *andalusite, kyanite,* and *sillimanite*. All three have an identical chemical composition (Al_2SiO_5), but different crystal structures. They are found in metamorphosed shales that have an abundance of aluminum. Box figure 1 is a phase diagram showing the pressure/temperature fields in which each is stable.

If andalusite is found in a rock, this indicates that pressures and temperatures were relatively low. Andalusite is often found in contact metamorphosed shales (hornfels). Kyanite, when found in schists, is regarded as an indicator of high pressure; but note that the higher the temperature of

the rock, the greater the pressure needed for kyanite to form. Sillimanite is an indicator of high temperature and can be found in some contact metamorphic rocks adjacent to very hot intrusions as well as in regionally metamorphosed schists and gneisses that formed at considerable depths.

Note that if you find all three minerals in the same rock, and could determine that they were mutually stable, you could infer that the temperature was close to 500°C and the confining pressure was almost 4 kilobars during metamorphism.

Figure 7.13

Garnet-mica schist. Small, subparallel flakes of mica reflect light. Garnet crystals give the rock a "raisin bread" appearance.

Photo by C. C. Plummer

Figure 7.14

Gneiss.

Photo by C. C. Plummer

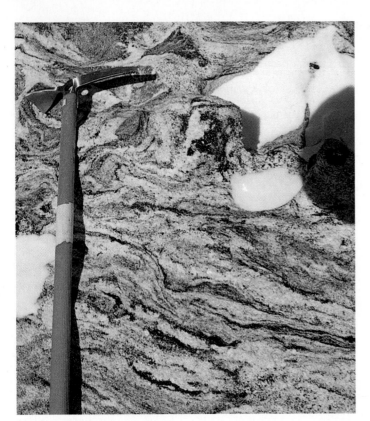

Figure 7.15

Migmatite in the Daniels Range, Antarctica.

Photo by C. C. Plummer

metamorphic rock (figure 7.15). A migmatite can be thought of as a "twilight zone" rock that is neither fully igneous nor entirely metamorphic.

Plate Tectonics and Metamorphism

The model of the earth's crust and upper mantle that we get from plate tectonic theory allows us to explain many of the observed characteristics of metamorphic rocks. To demonstrate the relationship between regional metamorphism and plate tectonics, we will look at what is believed to take place at a converging boundary in which oceanic lithosphere is subducted beneath continental lithosphere, as shown in figure 7.16.

Differential stress, which is responsible for foliation, occurs wherever rocks are being squeezed between the two plates or wherever rocks are sliding past one another. Shearing is expected in the subduction zone where the oceanic crust slides beneath the continental lithosphere (figure 7.16). In the chapter on mountains (chapter 20), we describe the new concept of *gravitational collapse and spreading,* in which the central part of a mountain belt becomes too high after plate convergence and is gravitationally unstable. This forces rock downward and outward as shown by the arrows in figure 7.16 (see also figure 20.20). At deeper levels the plastic rock flows during metamorphism and folia-

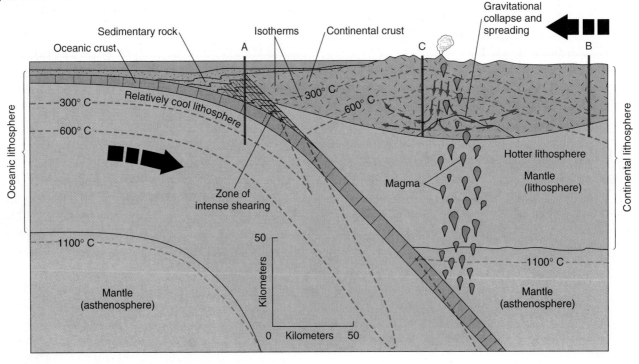

Figure 7.16

Metamorphism across a converging plate boundary.

From W. G. Ernst. *Metamorphism and Plate Tectonic Regimes.* Stroudsberg, Pa.: Dowden, Hutchinson, & Ross, 1975; p. 425. Reprinted by permission of the publisher.

tion should develop parallel to the direction of flowage. Also, vertical foliation may develop throughout much of the region due to the compressive stress in which the crust is caught, as if in a vise (see figure 7.9).

Confining pressure is directly related to depth. For this reason, we would expect the same pressure at a depth of 20 kilometers beneath a volcanic area as beneath the relatively cool rocks of a plate's interior.

Metamorphosed rocks indicate wide ranges of temperatures for a given pressure. This is understandable if we realize that the geothermal gradient is not the same everywhere in the world. Each of the three places (A, B, and C) in figure 7.16 would have a different geothermal gradient. If you were somehow able to push a thermometer through the lithosphere, you would find the rock is hotter at shallower depths in areas with higher geothermal gradients than at places where the geothermal gradient is low. As indicated in figure 7.16, the geothermal gradient is higher progressing downward through an active volcanic-plutonic complex (for instance, the Cascade Mountains of Washington and Oregon) than it is in the interior of a plate (beneath the Great Plains of North America, for example).

Heat, then, is the most variable of the controlling factors of metamorphism. Figure 7.16 shows the temperatures calculated for a converging boundary. A line connecting points that have the same temperature is called an **isotherm.** Note how the 300°C, 600°C, and 1100°C isotherms change depth radically across the subduction zone. This is because rocks that were cool because they were relatively near the earth's surface have been transported rapidly to depth. ("Rapidly" in the geologic sense—movement being around 1 cm/year.) The oceanic lithosphere along a subduction zone has not had time to heat up and come into thermal equilibrium with the relatively hot rocks elsewhere at these depths. On the top of the subducted oceanic slab, the extra heat provided by friction along with the heat that is normal for the overlying continental lithosphere and asthenosphere cause the isotherms to rise sharply toward the surface. The isotherms are bowed upward in the region of the volcanic-plutonic complex because magma created along the lower levels of the subduction zone works its way upward and brings heat from the asthenosphere into the mantle and crust of the continental lithosphere.

To understand why metamorphic rocks can form under such a wide variety of temperatures and pressures, study figure 7.16. You may observe that the bottom of line A is at a depth of about 50 kilometers, and if a hypothetical thermometer were here, it would read just over 300° because it would be just below the 300° isotherm. Compare this to vertical line C in the volcanic-plutonic complex. The confining pressure at the base of this line would be the same as at the base of line A, yet the temperature at the base of line C would be well over 600°. The minerals that could form at the base of line A would not be the same as those that could form at line C. Therefore, we would expect quite different metamor-

phic rocks in the two places, even if the parent rock had been the same (Box 7.3).

Hydrothermal Processes

Rocks that have precipitated from hot water or have been altered by hot water passing through are hard to classify. As described earlier, hot water is involved to some extent in most metamorphic processes. Beyond metamorphism, hot water also plays an important role creating new rocks and minerals. These form entirely by precipitation of ions derived from hydrothermal solutions. *Hydrothermal minerals* can form in void spaces or between the grains of a host rock. An aggregate of hydrothermal minerals, a **hydrothermal rock,** may crystallize within a preexisting fracture in a rock to form a hydrothermal **vein.**

Hydrothermal processes are summarized in table 7.3. The "dry" metamorphism referred to is relatively rare. Most of the examples given so far in this chapter probably took place with water present and so would be classed as "wet" metamorphism.

Hydrothermal Activity at Diverging Plate Boundaries

Hydrothermal processes are particularly important at mid-oceanic ridges (which are also diverging plate boundaries). As shown in figure 7.17, cold seawater moves downward through cracks in the basaltic crust and is cycled upward by heat from magma beneath the ridge crest. Very hot water returns to the ocean at submarine hot springs. Hot water traveling through the basalt, gabbro, and ultramafic rocks of the oceanic lithosphere helps metamorphose these rocks while they are close to the diverging boundary.

Table 7.3	Hydrothermal Processes
Role of Water	**Name of Process or Product**
No water	"Dry" metamorphism
Water merely transporting ions between grains in a rock	"Wet" metamorphism
Water brings ions from outside the rock, and they are added to the rock during metamorphism. Other ions may be dissolved and removed.	Metasomatism
Water passes through cracks or pore spaces in rock and precipitates minerals on the walls of cracks and within pore spaces.	Hydrothermal rocks

Metamorphic Facies and the Relationship to Plate Tectonics

During the early part of the twentieth century, geologists in Scandinavia introduced the concept of metamorphic facies. They noted that metamorphic rocks that were once basalts had one set of minerals in some parts of Europe, but in other areas, metabasalts had quite different mineral assemblages. As these rocks were *chemically* similar, the different *mineral assemblages* were regarded as indicating significantly different pressure and temperature conditions during metamorphism. Rocks having the same mineral assemblage are regarded as belonging to the same **metamorphic facies,** implying that they formed under broadly similar pressure and temperature conditions. The name for each facies is based on the assemblage of minerals or the name of a rock common to that facies. For instance, a schistose metabasalt composed mostly of the minerals chlorite, actinolite and epidote (all of which are green minerals) belongs to the *greenschist facies.* On the other hand, rocks of the same chemical composition (metabasalts) belonging to the *amphibolite facies* are largely made up of hornblende and garnet. (Do not try to remember the names of the facies or their compositions, your aim should be to understand the concept.)

Based on the geologic setting, early workers inferred that the temperature conditions during metamorphism were

Box 7.3 Figure 1

The metamorphic facies. Facies are named after minerals (prehnite, zeolite, pumpellyite) or rock types (e.g., blueschist, granulite). Boundaries between facies are approximate. The arrows represent increases in temperature with depth for the three lines labeled *A, B, C* in figure 2 and in figure 7.16.

From W. G. Ernst, *Metamorphism and Plate Tectonic Regimes.* Stroudsberg, Pa.: Dowden, Hutchinson, & Ross, 1975; p. 425. Reprinted by permission of the publisher.

lower for the greenschist facies than for the amphibolite facies. Laboratory work has since confirmed this as well as determined the pressure and temperature *stability fields* for each of the facies (box figure 1).

Ore Deposits at Divergent Plate Boundaries

As the sea water moves through the crust, it dissolves metals and sulfur from the crustal rocks and magma. When the hot, metal-rich solutions contact cold sea water, metal sulfides are precipitated in a mound around the hot spring. This process has been filmed in the Pacific, where some springs spew clouds of fine-grained ore minerals that look like black smoke (figure 7.18).

The metals in rift-valley hot springs are predominantly iron, copper, and zinc, with smaller amounts of manganese, gold, and silver. Although the mounds are nearly solid metal

Figure 7.17

Cross section of a mid-oceanic ridge (diverging plate boundary). Water descends through fractures in the oceanic lithosphere, is heated by magma and hot igneous rocks, and rises.

Box 7.3 Figure 2

Schematic representation of the distribution of facies across a convergent plate boundary.

From W. G. Ernst, *Metamorphism and Plate Tectonic Regimes.* Stroudsberg, Pa.: Dowden, Hutchinson, & Ross, 1975; p. 426. Reprinted by permission of the publisher.

The concept of metamorphic facies is analogous to defining climatic zones by the combinations of plants found in each zone. A place where ferns, palm trees, and vines flourish corresponds to a climate with warm temperatures and abundant rainfall. On the other hand, a combination of palm trees, cactus, and sagebrush implies a hot, dry climate.

By identifying the metamorphic facies of rocks presently cropping out on the surface, geologists can infer, within broad limits, the depth at which metamorphism took place. They may also (again, within broad limits) be able to determine the corresponding temperature.

The concept of metamorphic facies preceded plate tectonic theory by several decades. Although earlier geologists were able to relate the individual facies to pressure and temperature combinations, they had no satisfactory explanation for the environments that produced the various combinations. Figure 7.16, which relates the temperatures of regional metamorphism to plate tectonics, may be used to infer the environment for each of the metamorphic facies shown in box figure 1. Box figure 2 shows the likely distribution of metamorphic facies across the same converging boundary as in figure 7.16. To understand the relationship, study box figures 1 and 2 as well as figure 7.16.

If one were to determine the geothermal gradient represented by the three vertical lines marked *A, B,* and *C* on figure 7.16 and box figure 2, the temperatures for particular depths should plot on the corresponding arrows shown in box figure 1. Follow arrow *A* in box figure 1. This means that if you were able to drill vertically downward along line *A,* you would find, beneath unmetamorphosed rocks, rocks of the zeolite facies. At a greater depth would be the boundary between the zeolite and prehnite-pumpellyite facies. You would reach this boundary at a depth where the pressure is about 4 kilobars and the temperature is approximately 150°C. Your drill would penetrate rocks of the prehnite-pumpellyite facies until you reached the blueschist facies. In hole *B,* in the interior of the plate, the progression would be from zeolite facies to prehnite-pumpellyite facies to greenschist facies to amphibolite facies to granulite facies. Hole *C,* in the volcanic-plutonic complex, would not pass through the prehnite-pumpellyite facies but would go from zeolite facies to greenschist facies to amphibolite facies to granulite facies.

sulfide, they are small and widely scattered on the sea floor, so commercial mining of them may not be practical.

Metasomatism

Metasomatism is metamorphism coupled with the introduction of *ions* from an external source. The ions are brought in by water from outside the immediate environment and are incorporated into the newly crystallizing minerals. At the same time, the hot water may dissolve minerals that were part of the rock and carry them away.

When metasomatism takes place during regional metamorphism very hot water travels through a rock while gneiss or schist is crystallizing. Ions (typically K^+, Na^+, Si^{+4}, and O^{-2}) are carried by the water and participate in metamorphic reactions. Large feldspar crystals may grow in schist due to the addition of potassium or sodium ions.

If metasomatism is associated with contact metamorphism, the ions are introduced from a cooling magma. Some important commercially mined deposits of metals such as iron, tungsten, copper, lead, zinc, and silver are attributed to metasomatism. Figure 7.19 shows how magnetite (iron oxide) ore bodies have formed through metasomatism. Ions of the metal are transported by water and react with minerals in the host rock. Elements within the host rock are simultaneously dissolved out of the host rock and replaced by the metal ions brought in by the fluid. Because of the solubility of calcite, marble commonly serves as a host for metasomatic ore deposits.

Figure 7.18

"Black smoker" or submarine hot spring on the crest of the mid-oceanic ridge in the Pacific Ocean near 21° North latitude. The "smoke" is a hot plume of metallic sulfide minerals being discharged into cold sea water from a chimney 0.5 meters high. The large mounds around the chimney are metal deposits. The instruments in the foreground are attached to the small submarine from which the picture was taken.

Photo by U.S. Geological Survey

Hydrothermal Rocks and Minerals

Quartz veins (figure 7.20) are especially common where igneous activity has occurred. These can form from hot water given off by a cooling magma; but probably they are mostly produced by ground water heated by a pluton and circulated by convection, as shown in figure 7.21. Where the water is hottest, the most material (notably silica) is dissolved. As water vapor continues upward through increasingly cooler rocks during its journey toward the earth's surface, pressure decreases and heat is lost. Fewer ions can be carried in solution, and so

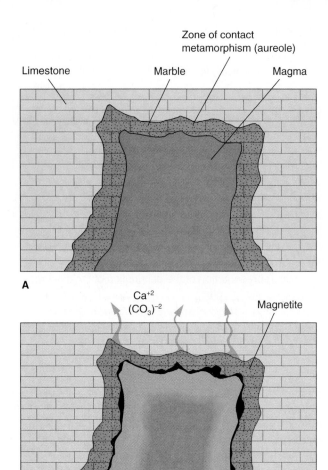

A

B

Figure 7.19

Development of a contact metasomatic deposit of iron (magnetite). (*A*) Magma intrudes country rock (limestone), and marble forms along contact. (*B*) As magma solidifies, gases bearing ions of iron leave the magma, dissolve some of the marble, and deposit iron as magnetite.

Figure 7.20

Ore-bearing veins in a mine.

Photo by C. C. Plummer

Figure 7.21

How veins form. Cold water descends, is heated, dissolves material, ascends, and deposits material as water cools and pressure drops upon ascending.

Figure 7.22

Water at a converging boundary. Seawater trapped in the oceanic crust is carried downward and released upon heating at various depths within the subduction zone.

the silicon and oxygen leave the water and cake onto the walls of the crack as silica (SiO_2), forming a quartz vein.

Veins consisting only of quartz are the most widespread, although some quartz veins contain other minerals. Veins with no quartz are not as common and are composed of calcite or a wide range of other minerals.

Hydrothermal veins are very important economically. They form most of the world's great deposits of zinc, lead, silver, gold, tungsten, tin, mercury, and, to some extent, copper. Ore minerals containing these metals are usually found in quartz veins. Veins containing commercially extractable amounts of metals are by no means common (and might even be regarded as freaks of nature).

Some ore-bearing solutions percolate upward between the grains of the rock and deposit very fine grains of ore mineral throughout. These are called *disseminated ore deposits* and account for the largest copper mines in the world (see Box 7.4) as well as some very large gold mines.

Sources of Water

Where does the water come from? The following is a logical explanation. *Ground water* seeps downward from the earth's surface through pores and fractures in rocks; however, the depth to which surface-derived water can penetrate is quite limited.

Plate tectonics can account for water at deeper levels in the lithosphere as seawater trapped in the oceanic crust can be carried to considerable depths through subduction (figure 7.22). Water trapped in sediment and in sedimentary rocks lying on

7.4
ENVIRONMENTAL GEOLOGY

The World's Largest Human-made Hole—The Bingham Canyon Copper Mine

The biggest man-made hole in the world is located at Bingham Canyon near Salt Lake City, Utah. The 800 meter ($^1/_2$ mile) deep open pit mine is 4 km ($2^1/_2$ miles) wide at the top and continues to be enlarged. The reason for this hole is copper. About half a million tons of rock are blasted apart and moved each day.

About 40,000 kilograms of explosives are used per day to blast apart over 60,000 tons of ore (copper-bearing rock) and an equal amount of waste rock. A conveyor belt system moves up to 10,000 tons of crushed rock per hour through a tunnel out of the pit for processing.

Mining began here as a typical underground operation in the last century. The shafts and tunnels of the mine followed a series of veins. Originally ores of silver and lead were mined. Later it was discovered that fine-grained, copper-bearing minerals (chalcopyrite and other copper sulfide minerals) were disseminated in tiny pore spaces throughout a granite stock rather than restricted to the veins. Although the percentage of copper in the rock was small, the total volume of copper was recognized as huge. With efficient earth-moving techniques, large volumes of ore-bearing rock can be moved and processed. Today mining is still going on, and the company is able to make a profit even though less than 0.4% of the rock being mined is copper. Since 1904, 12 million tons of copper have been mined, processed, and sold. The mine has also produced impressive amounts of gold, silver, and other metals.

Obviously, such an operation is not without environmental problems. Some people regard the huge hole in the mountains as an eyesore (but it is a popular tourist attraction). Disposing of the waste—over 99% of the rock material mined—creates problems. Wind stirs up dust storms from the piles of finely crushed waste rock unless it is kept wet. The smelters that extract the pure copper from the sulfide minerals emit a toxic smoke containing sulfuric acid fumes. The prevailing winds blow the smoke away from Salt Lake City for all but a few days each year. A 360-meter-high smokestack was built as part of an emission control program that cost the company $280 million to meet federal health-related air standards.

Box 7.4 Figure 1

Bingham Canyon copper mine in Utah.
Courtesy Kennecott Copper Company.

basalt may be carried down with the descending crust. However, recent studies indicate that most of the water is carried by hydrous minerals (amphibole, for example) in the basaltic crust. When the rocks get hot enough the hydrous minerals recrystallize, releasing water. The water vapor works its way upward through the overlying continental lithosphere through fissures. In the process of ascending, water assists in the metamorphism of rocks, dissolves minerals, and carries the ions to interact during metasomatism, or it deposits quartz and other minerals in fissures as veins. The water can also lower the melting points of rocks at depth, allowing magma to form (as described in the chapter on igneous rocks).

Summary

Metamorphic rocks form from other rocks that are subjected to high temperature generally accompanied by high confining pressure. Recrystallization takes place in the solid state although water, usually present, aids metamorphic reactions. Foliation in metamorphic rocks is due to *differential stress* (either *compressive stress* or *shearing*). Slate, phyllite, schist, and gneiss are foliated and distinguished from one another by the type of foliation.

Contact metamorphic rocks are produced during metamorphism usually without significant differential stress but with high temperature. Contact metamorphism occurs in rocks immediately adjacent to intruded magmas.

Regional metamorphism, which involves heat, confining pressure, and differential stress, has created most of the metamorphic rock of the earth's crust. Different parent rocks as well as widely varying combinations of pressure and temperature result in a large variety of metamorphic rocks. Combinations of minerals in a rock can indicate what the pressure and temperature conditions were during metamorphism. Extreme metamorphism, where the rock partially melts, can result in *migmatites.*

Hydrothermal veins form when hot water precipitates material that crystallizes into minerals. During *metasomatism,* hot water introduces ions into a rock being metamorphosed, changing the chemical composition of the metasomatized rock from that of the parent rock.

Plate tectonic theory accounts for the features observed in metamorphic rocks and relates their development to other activities in the earth. In particular, plate tectonics explains (1) the deep burial of rocks originally formed at or near the earth's surface; (2) the intense squeezing necessary for the differential stress, implied by foliated rocks; (3) the presence of water deep within the lithosphere; and (4) the wide variety of pressures and temperatures believed to be present during metamorphism.

Terms to Remember

compressive stress 154

confining pressure 153

contact metamorphism 156

differential stress 154

foliation 155

gneiss 160

gneissic 155

hornfels 157

hydrothermal rock 163

isotherm 163

marble 157

metamorphic facies 164

metamorphic rock 152

metamorphism 152

metasomatism 165

migmatite 160

parent rock 152

phyllite 160

plastic 152

quartzite 158

regional metamorphism 158

schist 160

schistose 155

shearing 154

slate 159

slaty 155

slaty cleavage 155

stress 154

vein 163

Testing Your Knowledge

Use the questions below to prepare for exams based on this chapter.

1. What are the effects on metamorphic minerals and textures of temperature, confining pressure, and differential stress?

2. What are the various sources of heat for metamorphism?

3. How do regional metamorphic rocks differ in texture from contact metamorphic rocks?

4. Why is such a variety of combinations of pressure and temperature environments possible during metamorphism?

5. How would you distinguish

    a. schist and gneiss?   b. slate and phyllite?

    c. quartzite and marble?   d. granite and gneiss?

6. Why is an edifice built with blocks of quartzite more durable than one built of marble blocks?

7. Metamorphism of limestone may contribute to global warming by the release of (a) oxygen (b) sulfuric acid (c) nitrogen (d) CO_2

8. Shearing is a type of (a) compressive stress (b) confining pressure (c) lithostatic pressure (d) differential stress

9. Metamorphic rocks with a planar texture (the constituents of the rock are parallel to one another) are said to be (a) concordant (b) foliated (c) discordant (d) nonfoliated

10. Metamorphic rocks are classified primarily on (a) texture—the presence or absence of foliation (b) mineralogy—the presence or absence of quartz (c) environment of deposition (d) chemical composition

11. Which is not a foliated metamorphic rock? (a) gneiss (b) schist (c) quartzite (d) slate

12. Limestone recrystallizes during metamorphism into (a) hornfels (b) marble (c) quartzite (d) schist

13. Quartz sandstone is changed during metamorphism into (a) hornfels (b) marble (c) quartzite (d) schist

14. The correct sequence of rocks that are formed when shale undergoes progressive metamorphism is (a) slate, gneiss, schist, phyllite (b) phyllite, slate, schist, gneiss (c) slate, phyllite, schist, gneiss (d) schist, phyllite, slate, gneiss

15. The major difference between metamorphism and metasomatism is (a) temperature at which each takes place (b) the minerals involved (c) the area or region involved (d) metasomatism is metamorphism coupled with the introduction of ions from an external source

16. Ore bodies at divergent plate boundaries can be created through (a) contact metamorphism (b) regional metamorphism (c) hydrothermal processes

17. Metamorphic rocks with the same mineral assemblage belong to the same (a) metamorphic facies (b) progressive metamorphism (c) schistosity

18. A metamorphic rock which has undergone partial melting to produce a mixed igneous-metamorphic rock is a (a) gneiss (b) hornfels (c) schist (d) migmatite

Expanding Your Knowledge

1. Should ultramafic rocks in the upper mantle be regarded as metamorphic rocks rather than igneous rocks?

2. Where were the metals before they were concentrated in hydrothermal vein ore deposits?

3. What happens to originally horizontal layers of sedimentary rock when they are subjected to the deformation associated with regional metamorphism?

4. Where in the earth's crust would you expect most migmatites to form?

Exploring Resources

Blatt, H., and R. C. Tracy. 1996. *Petrology: Igneous, Sedimentary, and Metamorphic.* 2d Ed. New York: W. H. Freeman.

Hibbard, M. J. 1995. *Petrography to petrogenesis.* Englewood Cliffs, New Jersey: Prentice-Hall.

Kerrick, D. M., ed. 1991. *Contact metamorphism.* Reviews in Mineralogy,

vol. 26. Washington, D.C.: Mineralogical Society of America.

Mason, R. 1989. *Petrology of the metamorphic rocks.* 2d ed. London: Unwin Hyman.

Yardley, B. W. D. 1989. *An introduction to metamorphic petrology.* Essex, England: Longman.

 http://www.science.ubc.ca/~geol202/meta/metamorphic.html

University of British Columbia's Metamorphic Rocks Home Page.

Expand your knowledge of the concepts presented in this chapter by using the CD-ROM to answer the following questions.

1. Spreading Center module, Action at Spreading Ridges, Visit a Black Smoker: Where does the heat fueling the black smoker come from? What kind of metamorphism would you expect to find in the rock of the seafloor underlying the black smoker?

2. Changing Earth: Watch as Pangaea breaks up and the continents reassemble. Identify areas of the world where you would expect to find rock metamorphosed either by continental collision or by the collision (accretion) of volcanic island chains (island arcs) with the continents.

Grand Canyon, Arizona. Horizontal Paleozoic beds (top of photo) overlie tilted
Precambrian beds (bottom of photo) and older, Precambrian metamorphic
rock (at far left of photo).

Photo © John Shelton

CHAPTER
8

Time and Geology

*T*he immensity of geologic time is hard for humans to perceive. It is unusual for someone to live a hundred years, but a person would have to live 10,000 times that long to observe a geologic process that takes a million years. In this chapter we try to help you develop a sense of the vast amounts of time over which geologic processes have been at work.

To determine how many years ago a rock formed requires the special techniques of radioactive isotope dating. To most geologists, however, being able to assign years (absolute age) to a process, significant as it is, is less important than establishing relative time and the sequence in which events have occurred. This chapter explains how to apply several basic principles to decipher a sequence of events responsible for geologic features. These principles can be applied to many aspects of geology—as, for example, in understanding geologic structures (chapter 15). Understanding the complex history of mountain belts (chapter 20) also requires knowing the techniques for determining relative ages of rocks.

Determining age relationships between geographically widely separated rock units is necessary for understanding the geologic history of a region, a continent, or the whole earth. Substantiation of the plate tectonics theory depends on intercontinental correlation of rock units and geologic events, piecing together evidence that the continents were once one great body.

Widespread use of fossils led to the development of the standard geologic time scale. Originally based on relative age relationships, the subdivisions of the standard geologic time scale have now been assigned absolute ages through isotopic dating. Think of the geologic time scale as a sort of calendar to which events and rock units can be referred. Its major subdivisions are referred to elsewhere in this book.

The Key to the Past

We who are part of present-day Western civilization are not so resistant to the concept of immense geologic time as were those who lived two or three centuries ago. Then it was commonly believed in the Western world that the earth was only a few thousand years old (although Chinese and Hindu cultures had a better concept of the vast age of the universe). In early Christendom, geologic events were placed within a biblical chronology, and catastrophic happenings were blamed for features of the landscape. When rocks several thousand meters above sea level seemed to have been formed of sediment deposited in water (even containing fossils somewhat like modern marine organisms), the explanation was that a worldwide inundation had simply drowned all the earth's mountains in a matter of days. Because no known physical laws could account for such events, they were attributed to divine intervention. In the eighteenth century, however, James Hutton, a Scotsman who is regarded as the father of modern geology, realized that geologic features could be explained through present-day processes. He recognized that our mountains are not permanent, but have been carved into their present shapes and will be worn down by the slow agents of erosion now working on them. He realized that the great thicknesses of sedimentary rock we find on the continents are products of sediment removed from land and deposited as mud and sand in seas. The time required for these processes to take place had to be incredibly long. Hutton upset conventional thinking (the world was believed to be less than 6,000 years old) by writing in 1788, "We find no sign of a beginning—no prospect for an end." Hutton's writings, which were somewhat murky, received scant notice until his ideas were given widespread attention by Charles Lyell in a landmark book, *Principles of Geology.* Lyell referred to Hutton's concept that geologic processes operating at present are the same processes that operated in the past as the principle of **uniformitarianism.** The principle is stated more succinctly as "The present is the key to the past."

The term uniformitarianism is a bit unfortunate, because it suggests that changes take place at a uniform *rate.* Hutton recognized that sudden, violent events, such as a major, short-lived volcanic eruption, also influence the earth's history. In some countries actualism is used in place of uniformitarianism. The term **actualism** comes closer to conveying Hutton's principle that the same processes and natural laws that operated in the past are those we can actually observe or infer from observation as operating at present. Under present usage, uniformitarianism has the same meaning as actualism for most geologists.

We now realize that geology involves time periods much greater than a few thousand years. But how long? For instance, were rocks near the bottom of Grand Canyon (chapter opening photo) formed closer to 10,000 or 100,000 or 1,000,000 or 1,000,000,000 years ago? What geologists needed was some "clock" that began working when rocks formed. Such a "clock" was found when radioactivity was discovered. Dating based on radioactivity (discussed later in this chapter) allows us to deter-

mine a rock's **absolute age** (age given in years or some other unit of time). Geologists, however, usually are more concerned with **relative time,** the *sequence* in which events took place, than with the number of years involved.

These statements show the difference between absolute age and relative time:

"Uncle George planted the plum tree sometime between the time little Oscar was born and the time when the house across the street burned down." This statement gives the time of an event (planting a tree) relative to other events.

But speaking in terms of absolute age: "Let's see, Uncle George planted that plum tree about four and a half, maybe five, years ago." Note that an absolute age does not have to be an *exact* age, merely age given in units of time. Because most geologic problems are concerned with the sequence of events, we discuss relative time first.

Relative Time

The geology of an area may seem, at first glance, to be hopelessly complex. A nongeologist might think it impossible to decipher the sequence of events that created such a geologic pattern. However, a geologist has learned to approach seemingly formidable problems by breaking them down to a number of simple problems. As an example, the geology of Grand Canyon, shown diagrammatically in figure 8.15 and in the chapter opening photo, can be analyzed in four parts: (1) horizontal layers of rock; (2) inclined layers; (3) rock underlying the inclined layers (plutonic and metamorphic rock); and (4) the canyon itself, carved into these rocks.

After you have studied the following section, return to the photo of the Grand Canyon and see if you can determine the sequence of geologic events that took place.

Principles Used to Determine Relative Age

Most of the individual parts of the larger problem are solved by applying several simple principles (or laws, as some prefer to call them) while studying the exposed rock. In this way the sequence of events or the relative time involved can be determined. Contacts are particularly useful for deciphering the geologic history of an area. (**Contacts,** as described in previous chapters, are the surfaces separating two different rock types or rocks of different ages.) To explain various principles, we will use a fictitious place that bears some resemblance to Grand Canyon. We will call this place, represented by the block diagram of figure 8.1, Minor Canyon. The formation names are also fictitious. (**Formations,** as described in chapter 6, are bodies of rock of considerable thickness with recognizable characteristics that make each distinguishable from adjacent rock units. They are named after local geographic features, such as towns.) Note the contacts between the tilted formations, the horizontal formations, the granite, and the dike. What sequence of events might be responsible for the geology of

Figure 8.1

Block diagram representing the Minor Canyon area.

Minor Canyon? (You might briefly study the block diagram and see how much of the geologic history of the area you can decipher before reading further.)

To determine the relationship of geologic events to one another, four basic principles are applied. These are the principles of (1) original horizontality, (2) superposition, (3) lateral continuity, and (4) cross-cutting relationships. These principles will be used in interpreting figure 8.1.

Original Horizontality

The principle of **original horizontality** states that beds of sediment deposited in water formed as horizontal or nearly horizontal layers (as described in chapter 6). (All the layers in figure 8.1 represent sedimentary rock originally deposited in a marine environment.)

Note in figure 8.1 that the Larsonton Formation and overlying rock units (Foster City Formation, Hamlinville Formation, and Skinner Gulch Limestone) are horizontal. Evidently their original horizontal attitude has not changed since they were deposited. However, the Lutgrad, Birkland, Tarburg, and Leet Junction Formations, must have been tilted after they were deposited as horizontal layers. By applying the principle of original horizontality, we have determined that a geologic event—tilting of bedrock—occurred after the Leet Junction, Tarburg, Birkland, and Lutgrad Formations were deposited on a sea floor. We can also see that the tilting event did not affect the Larsonton and overlying formations. (A reasonable conclusion is that tilting was accompanied by uplift and erosion, all before renewed deposition of younger sediment.)

Superposition

The principle of **superposition** states that within a sequence of undisturbed sedimentary or volcanic rocks, the layers get younger going from bottom to top.

Obviously, if sedimentary rock is formed by sediment settling onto the sea floor, then the first (or bottom) layer must be there before the next layer can be deposited on top of it. The principle of superposition also applies to layers formed by multiple lava flows, where one lava flow is superposed on a previously solidified flow.

Applying the principle of superposition, we can determine that the Skinner Gulch Limestone is the youngest layer of sedimentary rock in the Minor Canyon area. The Hamlinville Formation is the next oldest formation, and the Larsonton Formation is the oldest of the still horizontal sedimentary rock units. Similarly, we assume that the inclined layers were originally horizontal (by the first principle). By mentally restoring them to their horizontal position (or "untilting" them), we can see that the youngest formation of the sequence is the Leet Junction Formation and that the Tarburg, Birkland, and Lutgrad Formations are progressively older.

Principle of Lateral Continuity

The principle of **lateral continuity** states that an original sedimentary layer extends laterally until it tapers or thins at its edges. This is what we expect at the edges of a depositional environment, or where one type of sediment interfingers laterally with another type of sediment as environments change. In

Table 8.1	Relative Ages of Features in Figure 8.1 Determinable by Cross-cutting Relationships	
Feature	**Is Younger Than**	**But Older Than**
Valley (canyon)	Skinner Gulch Ls.	
Foster City Fm.	Dike	Hamlinville Fm.
Dike	Larsonton Fm.	Foster City Fm.
Larsonton Fm.	Leet Junction Fm. and granite	Dike
Granite	Tarburg Fm.	Larsonton Fm.

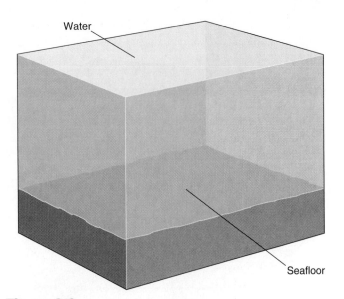

Figure 8.2

The area during deposition of the initial sedimentary layer of the Lutgrad Formation.

figure 8.1 the bottom bed of the Hamlinville Formation tapers as we would expect from this principle. We are not seeing any other layers taper, either because we are not seeing their full extent within the diagram or because they have been truncated (cut off abruptly) due to later events.

Cross-cutting Relationships

The fourth principle can be applied to determine the remaining age relationships at Minor Canyon. The principle of **cross-cutting relationships** states that a disrupted pattern is older than the cause of disruption. A layer cake (the pattern) has to be baked (established) before it can be sliced (the disruption).

To apply this principle, look for disruptions in patterns of rock. Note that the valley in figure 8.1 is carved into the horizontal rocks as well as into the underlying tilted rocks. The sedimentary beds on either side of the valley appear to have been sliced off, or *truncated,* by the valley. (The principle of lateral continuity tells us that sedimentary beds normally become thinner toward the edges rather than stop abruptly. So the event that caused the valley must have come after the sedimentation responsible for deposition of the Skinner Gulch Limestone and underlying formations. That is, the valley is younger than these layers. We can apply the principle of cross-cutting relationships to contacts elsewhere in figure 8.1 with the results shown in table 8.1.

We can now describe the geological history of the Minor Canyon area represented in figure 8.1 on the basis of what we have learned through applying the principles. Figures 8.2 through 8.11 show how the area changed over time, progressing from oldest to youngest events.

By *superposition,* we know that the Lutgrad Formation, the lowermost rock unit in the tilted sequence, must be the oldest of the sedimentary rocks as well as the oldest rock unit in the diagram. From the principle of *original horizontality,* we infer that these layers must have been tilted after they formed. Figure 8.2 shows initial sedimentation of the Lutgrad Formation taking place. If the entire depositional basin were shown, the

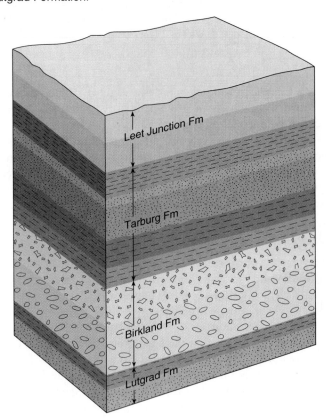

Figure 8.3

The area before intrusion of the granite.

layer would be tapered at its edges, according to the principle of *lateral continuity.*

Superposition indicates that the Birkland Formation was deposited on top of the Lutgrad Formation. Deposition of the Tarburg and Leet Junction Formations followed in turn (figure 8.3).

The truncation of bedding in the Lutgrad, Birkland, and Tarburg Formations by the granite tells us that the granite intruded sometime after the Tarburg Formation was formed (this is an *intrusive contact*). Although figure 8.4 shows that the granite was emplaced before tilting of the layered rock, we cannot determine from looking at figure 8.1 whether the granite intruded the sedimentary rocks before or after tilting. We can, however, determine through *cross-cutting relationships* that tilting and intrusion of the granite occurred before deposition of the Larsonton Formation. Figure 8.5 shows the rocks in the area have been tilted and erosion has taken place. Sometime later, sedimentation was renewed and the lowermost layer of the Larsonton Formation was deposited on the erosion surface, as shown in figure 8.6. Contacts representing buried erosion surfaces such as these are called *unconformities* and are discussed in more detail later in this chapter.

After the Larsonton Formation was deposited, an unknown additional thickness of sedimentary layers was deposited, as shown in figure 8.7. This can be determined through application of cross-cutting relationships. The dike is truncated by the Foster City Formation; therefore, it must have extended into some rocks that are no longer present, such as shown in figure 8.8. Figure 8.9 shows the area after the erosion that truncated the dike took place.

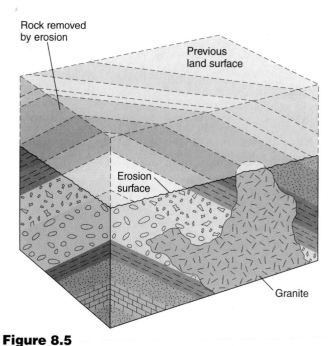

Figure 8.5

The area before deposition of the Larsonton Formation. Dashed lines show rock probably lost through erosion.

Figure 8.4

The area before layers were tilted.

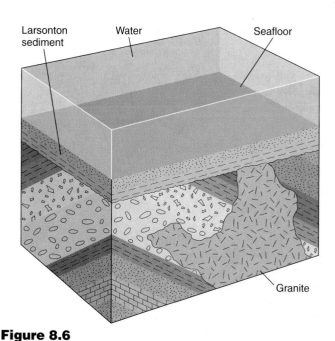

Figure 8.6

The area at the time the Larsonton Formation was being deposited.

Figure 8.7

Area before intrusion of dike. Thickness of layers above the Larsonton Formation is indeterminable.

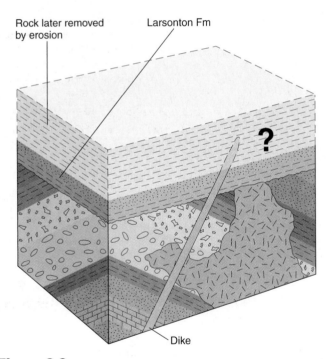

Figure 8.8

Dike intruded into the Larsonton Formation and preexisting overlying layers of indeterminable thickness.

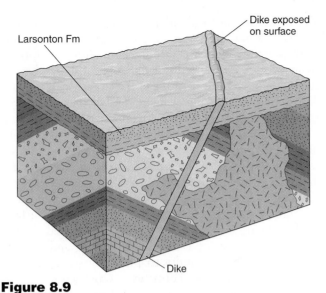

Figure 8.9

The area after rock overlying the Larsonton Formation, along with part of the dike, was removed by erosion.

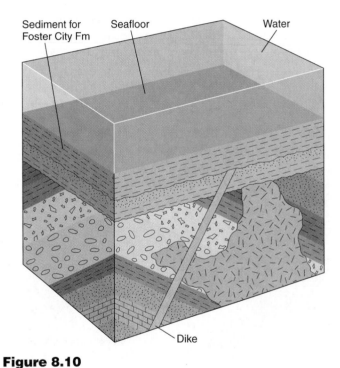

Figure 8.10

Sediment being deposited that will become part of the Foster City Formation.

Once again, sedimentation took place as the lowermost layer of the Foster City Formation blanketed the erosion surface (figure 8.10). Sedimentation continued until the uppermost layer (top of the Skinner Gulch Limestone) was deposited. At some later time, the area was raised above sea level and the stream began to carve the canyon (figure 8.11). Because the valley sides truncated the youngest layers of rock, we can determine from figure 8.1 that the last event was the carving of the valley.

Note that there are limits on how precisely we can determine the relative age of the granite body. It definitely intruded *before* the Larsonton Formation was deposited and *after* the Tarburg Formation was deposited. As no contacts can be observed between the Leet Junction Formation and

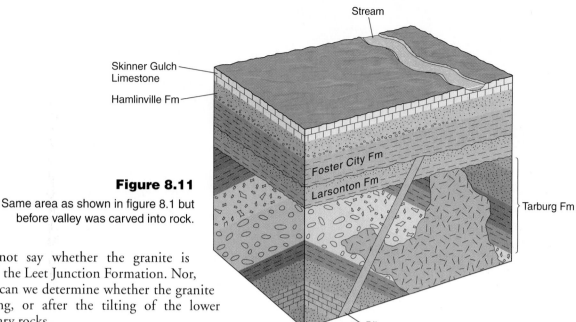

Figure 8.11

Same area as shown in figure 8.1 but before valley was carved into rock.

the granite, we cannot say whether the granite is younger or older than the Leet Junction Formation. Nor, as mentioned earlier, can we determine whether the granite formed before, during, or after the tilting of the lower sequence of sedimentary rocks.

We must also point out some special circumstances under which these principles do not apply. For example, intense deformation by tectonic forces can overturn or disrupt beds so much that the principle of superposition cannot be used (see chapter 15). A geologist must avoid being dogmatic in applying principles.

Now, if you take another look at the chapter opening photo of Grand Canyon (and figure 8.15), you should be able to determine the sequence of events. The sequence (going from older to younger) is as follows. Regional metamorphism took place resulting in the schists of the lower part of Grand Canyon (you cannot tell these are schists from the photograph). Erosion followed and leveled the land surface. Sedimentation followed. These sedimentary layers were subsequently tilted (they were also faulted). Once again erosion took place. The lowermost of the presently horizontal layers of sedimentary rock was deposited. Subsequently, each of the layers progressively higher up the sequence formed. Finally, the stream (the Colorado River) eroded its way through the rock, carving the Grand Canyon.

Other Time Relationships

Other characteristics of geology can be applied to help determine relative ages. For instance, we would expect the portion of the Tarburg Formation (the tilted layers) immediately adjacent to the granite body to have been *contact metamorphosed* (or "baked"). This would indicate that the Tarburg Formation had to be there before intrusion of the hot, granite magma. The base of the Larsonton Formation in contact with the granite would not be contact metamorphosed because it was deposited after the granite had cooled (and exposed by erosion). (The relationships are shown in figure 8.12, but, to keep them simple, not in the previous diagrams.)

The principle of **inclusions** states that fragments included in a host rock are older than the host rock. In figure 8.12, the granite contains inclusions of the tilted sedimentary rock. Therefore, the granite is younger than the tilted rock. The rock

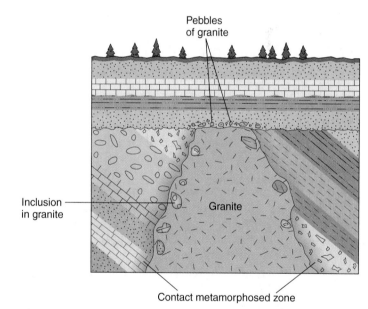

Figure 8.12

Age relationships indicated by contact metamorphism, inclusions (zenoliths) in granite, and pebbles of granite.

overlying the granite has granite pebbles in it. Therefore, the granite is older than the horizontal sedimentary rock.

Correlation

In geology **correlation** usually means determining time equivalency of rock units. Rock units may be correlated within a region, a continent, and even between continents. Various methods of correlation are described below along with examples of how the principles we described earlier in this chapter are used to determine whether rocks in one area are older or younger than rocks in another area.

Time and Geology   **179**

Figure 8.13

The upper layers of sedimentary rock in Grand Canyon. The Coconino Sandstone forms the white cliff above the red rocks.

Photo by C. C. Plummer

Figure 8.14

Sandstone in Zion National Park, Utah.

Photo by C. C. Plummer

Physical Continuity

Finding **physical continuity**—that is, being able to trace physically the course of a rock unit—is one way to correlate rocks between two different places. In figure 8.13, the prominent white layer of cliff-forming rock, the Coconino Sandstone, exposed along the upper part of Grand Canyon can be seen all the way across the photograph. You can physically follow this unit for several tens of kilometers, thus verifying that, wherever it is exposed in Grand Canyon, it is the same rock unit. (Incidentally, the Coconino Sandstone is also shown on figure 6.36.) Grand Canyon is an ideal location for correlating rock units by physical continuity. However, it is not possible to follow this rock unit from the Grand Canyon into another region because it is not continuously exposed. We usually must use other methods to correlate rock units between regions.

Similarity of Rock Types

Under some circumstances, correlation between two regions can be made by assuming that similar rock types in two regions formed at the same time. This method must be used with extreme caution, especially if the rocks being correlated are common ones.

To show why correlation by similarity of rock type does not always work, we can try to correlate the white, cliff-forming Coconino Sandstone in Grand Canyon (shown in figure 8.13) with a rock unit of similar appearance in Zion National Park about 100 kilometers away (figure 8.14). Both units are white sandstone. Cross-bedding indicates that both were once a series of sand dunes. It is tempting to correlate them and conclude that both formed at the same time. But if you were to drive or walk from the rim of Grand Canyon (where the Coconino Sandstone is *below* you), you would get to Zion by ascending a series of layers of sedimentary rock stacked on one another. In other words, you would be getting into progressively younger rock as shown diagrammatically in figure 8.15. In short, you have shown through *superposition* that the sandstone in Zion (called the Navajo Sandstone) is younger than the Coconino Sandstone.

Correlation by similarity of rock types is more reliable if a very unusual sequence of rocks is involved. If you find in one area a layer of green shale on top of a red sandstone that, in turn, overlies basalt of a former lava flow, and then find the same sequence in another area, you probably would be correct in concluding that the two sequences formed at essentially the same time.

When the hypothesis of continental drift was first proposed (see chapter 1) important evidence was provided by correlating a sequence of rocks (figure 8.16) consisting of glacially deposited sedimentary rock (tillites, described in the chapter on glaciation), overlain by continental sandstones, shales, and coal beds. These strata are in turn overlain by basalt flows. The sequence is found in parts of South America, Australia, Africa, Antarctic and India. It is very unlikely that an identical sequence of rocks could have formed on each of the continents if they were widely separated, as they are at present. Therefore, the continents on which the sequence is found is likely to have been part of a single, super-continent on which the rocks were deposited. Fossils found in these rocks further strengthened the correlation.

Correlation by Fossils

Fossils are common in sedimentary rock, and their presence is important for correlation. Plants and animals that lived at the time the rock formed were buried by sediment, and their fossil remains are preserved in sedimentary rock. Most of the fossil

Figure 8.15

Schematic cross section through part of the Colorado Plateau.

species found in rock layers are now extinct—99.9% of all species that ever lived are extinct. (The concept of *species* in geology is similar to that in biology.)

In a thick sequence of sedimentary rock, the fossils nearer the bottom (that is, in the older rock) are more unlike today's plants and animals than are those near the top. As early as the beginning of the nineteenth century, biologists began to notice that fossils in one layer of sedimentary rock often differ markedly from fossils in the layer above (or below). It became clear that many species of plants or animals existed only during the time that certain sedimentary layers were being deposited. Some species survived through the time interval represented by rock layers thousands of meters thick. Other species, confined to thin sequences of rock, apparently became extinct after a relatively short existence on earth. These observations, verified around the world, became formalized into the principle of **faunal succession:** fossil species succeed one another in a definite and recognizable order. In general, fossils in progressively older rock show increasingly greater differences from species now living. Plants and animals have tended to evolve from simpler to more complex species through geologic time. The regularity of changes in species of fossils through successive layers is one of the very strong lines of evidence supporting the theory of evolution.

Historical geology is concerned with the study of life and changes in life-forms throughout geologic time as well as with the physical evolution of the earth. *Paleontologists,* specialists in the study of fossils, have patiently and meticulously over the years identified thousands of species of fossils and determined which plant and animal species lived during which subdivision of geologic time. They have also been able to describe the physical environment in which the fossil species lived.

No matter where on earth they are found, individual fossil species always occur in the same sequence relative to one another. By comparing fossils found in a layer of rock in one area with similar fossils in another area, we can correlate the two rock units. More accurately, we can say that both rock units formed during the span of time that the species existed on the earth.

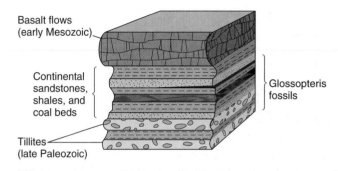

Figure 8.16

Rock sequences similar to this are found in India, Africa, South America, Australia, and Antarctica. The rocks in each of these localities contain the fossil plant *Glossopteris.*

Ideally, a geologist hopes to find an **index fossil,** a fossil from a very short-lived, geographically widespread species known to exist during a specific period of geologic time. A single index fossil allows the geologist to correlate the rock in which it is found with all other rock layers in the world containing that fossil.

Many fossils are of little use in time determination because the species thrived during too large a portion of geologic time. Sharks, for instance, have been in the oceans for a long time, so discovering a shark's tooth in a rock is not very helpful in determining the rock's relative age.

A geologist is likely to find a **fossil assemblage,** several different fossil species in a rock layer. A fossil assemblage is generally more useful for dating rocks than a single fossil is, because the sediment must have been deposited at a time when all the species represented existed (figure 8.17).

Some fossils are restricted in geographic occurrence, representing organisms adapted to special environments. However, many former organisms apparently lived over most of the earth, and fossil assemblages from these may be used for worldwide correlation. Fossils in the lowermost horizontal layers of Grand Canyon are comparable to ones collected in Wales,

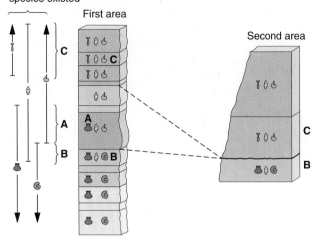

Time intervals
over which
species existed

First area

Second area

Figure 8.17

The use of fossil assemblages for determining relative ages. Rock A contains 🐚◊🐌. Therefore, it must have formed during time interval A. Rock B contains 🐚◊🐚. Therefore, it must have formed during time interval B. Rock C contains 🐚◊🐌. Therefore, it must have formed sometime during time interval C. In the second area, sequence A is missing. Therefore, there is an unconformity between sequence B and sequence C.

Era	Period	Epoch
Cenozoic	*Quaternary*	Recent (Holocene)
		Pleistocene
	Tertiary	Pliocene
		Miocene
		Oligocene
		Eocene
		Paleocene
Mesozoic	*Cretaceous*	
	Jurassic	
	Triassic	
Paleozoic	*Permian*	
	Pennsylvanian	
	Mississippian	
	Devonian	
	Silurian	
	Ordovician	
	Cambrian	
Precambrian Time		

Table 8.2   Geologic Time Scale

Figure 8.18

Trilobite fossils from the Cambrian Period.

Courtesy Ward's Natural Science Est. Inc., Rochester, N.Y.

Great Britain, and many other places in the world (the trilobites in figure 8.18 are an example). We can therefore correlate these rock units and say they formed during the same general span of geologic time.

The Standard Geologic Time Scale

Geologists can use fossils in rock to refer the age of the rock to the **standard geologic time scale,** a worldwide relative time scale. Based on fossil assemblages, the geologic time scale sub-divides geologic time. On the basis of fossils found, a geologist can say, for instance, that the rocks of the lower portion of horizontal layers in Grand Canyon formed during the *Cambrian Period.* This implicitly correlates these rocks with certain rocks in Wales (in fact, the period takes its name from Cambria, the Latin name for Wales) and elsewhere in the world where similar fossils occur.

The geologic time scale, shown in a somewhat abbreviated form in table 8.2, has had tremendous significance as a unifying concept in the physical and biological sciences. The working out of the evolutionary chronology by successive generations of geologists and other scientists has been a remarkable human achievement. The geologic time scale, representing an extensive fossil record, consists of three **eras,** which are subdivided into **periods,** which are, in turn, divided into **epochs.** (Remember that this is a relative time scale.)

Precambrian denotes the vast amount of time that preceded the Paleozoic Era (which begins with the Cambrian Period). The **Paleozoic Era** (meaning "old life") began with the appearance of abundant and complex life, as indicated by fossils. Rocks older than Paleozoic contain few fossils. This is because creatures with shells or other hard parts, which are easily preserved in fossils, did not evolve until the beginning of the Paleozoic.

The **Mesozoic Era** (meaning "middle life") followed the Paleozoic. On land, dinosaurs became the dominant animals of the Mesozoic. We live in the **Recent** (or **Holocene**) **Epoch** of the **Quaternary Period** of the **Cenozoic Era** (meaning "new life"). The Quaternary also includes the most recent ice ages, which were part of the **Pleistocene Epoch.**

The oldest fossils found are prokaryotes—microscopic, single-celled organisms that lack a nucleus. These date back to around 3.5 billion years (b.y.) ago, so life on earth is at least that old. It is likely that even more primitive organisms, similar to viruses, date back further in time but are not preserved in the fossil record. Fossils of much more complex, single-celled organisms that contained a nucleus (eukaryotes) are found in rocks as old as 1.4 b.y. These are the earliest living creatures to have reproduced sexually. Colonies of unicellular organisms likely evolved into multicellular organisms. Multicellular algae fossils date back at least a billion years.

Imprints of larger multicellular creatures appear in rocks of late Precambrian age, about 700 to 650 million years ago (m.y.). These resemble jellyfish and worms.

Sedimentary rocks from the Paleozoic, Mesozoic, and Cenozoic eras have abundant fossils. Large numbers of fossils mark the beginning (545 m.y. ago) of the Cambrian period. The reason for the sudden abundance of fossils is that organisms with shells and other hard parts had evolved and flourished. Animals with hard parts are much more likely to be preserved as fossils than soft-bodied organisms. Trilobites (figure 8.18) evolved into many species and were particularly abundant during the Cambrian. They became less significant later in the Paleozoic and finally all trilobites became extinct by the end of the Paleozoic.

The most primitive fish, the first vertebrates, date back to late in the Cambrian. Fish similar to presently living species of fish (including sharks) flourished during the Devonian (named after Devonshire, England). The Devonian is often called the "age of fish." Amphibians evolved from fish that had developed lungs late in the Devonian. These were the first land animals. Land plants, however, date back to the Ordovician. Reptiles evolved from amphibians in Pennsylvanian time or perhaps earlier.

The Paleozoic ended with the greatest mass extinction ever to occur on earth. Over 95% of species that existed died out.

During the Mesozoic (which began around 245 m.y. ago) new creatures evolved to occupy ecological domains left vacant by extinct creatures. Dinosaurs and mammals evolved from reptiles that survived the great extinction. Dinosaurs became the dominant group of land animals. Birds likely evolved from dinosaurs in the Mesozoic. Large, now extinct, marine reptiles, such as ichthysaurs, lived in the Mesozoic seas while flying reptiles, pterosaurs, soared through the air.

The Cretaceous period (and Mesozoic era) ended with the second largest mass extinction (around 75% of species were wiped out).

The Cenozoic, which began around 66 m.y. ago, is often called the "age of mammals." Mammals, which were small, insignificant creatures during the Mesozoic, evolved into the many groups of mammals (whales, bats, canines, cats, elephants, primates, and so forth) that occupy the earth at present. Many species of mammals evolved and became extinct throughout the Cenozoic. Hominids (modern humans and our extinct ancestors) have a fossil record dating back 4 m.y. and likely evolved from a now extinct ancestor common to hominids, chimpanzees, and other apes.

Fossils have been used to determine ages of the horizontal rocks in Grand Canyon. All are Paleozoic. The lowermost horizontal formations (figure 8.15) are Cambrian, above which are Mississippian, Pennsylvanian, and Permian rock units. By referring to the geologic time scale (table 8.2), we can see that Ordovician, Silurian, and Devonian rocks are not represented. Thus an unconformity (buried erosion surface) is present within the horizontally layered rocks of Grand Canyon.

Unconformities

In this and earlier chapters, we noted the importance of *contacts* for deciphering the geologic history of an area. In chapters 3 and 6 we described intrusive contacts and sedimentary contacts. Faults (described in chapter 15) are a third type of contact. The final important type of contact is an *unconformity.*

Each type of contact has very different implication about what took place in the geologic past.

An **unconformity** is a surface (or contact) that represents a *gap in the geologic record,* with the rock unit immediately above the contact being considerably younger than the rock beneath. Most unconformities are buried erosion surfaces. Unconformities are classified into three types—disconformities, angular unconformities, and nonconformities—with each type having important implications for the geologic history of the area in which it occurs.

Disconformities

In a **disconformity,** the contact representing missing rock strata separates beds that are parallel to one another. Probably what has happened is that older rocks were eroded away paral-

8.2

IN GREATER DEPTH

Demise of the Dinosaurs—Was It Extraterrestrial?

Dinosaurs dominated the continents during the Mesozoic Era. Now they prey on the imaginations of children of all ages. It's hard to accept that beings as powerful and varied as dinosaurs existed and were wiped out. But the fossil record is clear—when the Mesozoic came to a close, dinosaurs became extinct. Not a single of the numerous dinosaur species survived into the Cenozoic Era. Not only did the dinosaurs go, but about 75% of all plant and animal species, marine as well as terrestrial, were extinguished. This was one of the Earth's "great dyings"—an even "greater dying" was when the Paleozoic Era ended with the extinction of over 95% of all species. Most major extinctions have been gradual and scientists usually have attributed them to climate changes.

A decade ago, geologist Walter Alvarez, his father, physicist Luis Alvarez, and two other scientists proposed a hypothesis that the dinosaur extinction was caused by the impact of an asteroid. This was based on the chemical analysis of a thin layer of clay marking the boundary between the Mesozoic and Cenozoic eras (usually referred to as the K-T boundary—it separates the Cretaceous [K] and Tertiary [T] periods). The K-T boundary clay was found to have about 30 times the amount of the rare element iridium as is normal for crustal rocks. Iridium is relatively abundant in meteorites and other extraterrestrial objects such as comets and the scientists suggested that the iridium was brought in by an extraterrestrial body.

The scientists proposed a scenario in which an asteroid 10 kilometers in diameter struck the earth. This generated a gigantic dust cloud, which darkened the earth long enough to disrupt plant life growth and reproduction as well as drop the temperature of the earth worldwide. Creatures perished because of the disruption in their food supply or because of their inability to withstand the sudden climate change.

Other scientists proposed an alternative hypothesis blaming exceptional volcanic activity for the extinction as well as for the high iridium in the K-T boundary layer. (Some Hawaiian eruptions have a high iridium content.)

Debate between proponents of the two alternate hypotheses became heated and further evidence to support each was sought. K-T layers throughout the world were found to have grains of quartz that have been subjected to shock metamorphism (see Astrogeology Box 7.1), supporting the asteroid hypothesis. Concentrations of some elements common in volcanic rocks but not in meteorites were found in the K-T boundary layers, supporting the volcanic hypothesis. Critics of the asteroid hypothesis also pointed out that the dinosaurs had been on the decline for millions of years before the end of the Cretaceous. On the other hand, organisms in the oceans apparently died out very rapidly.

The asteroid hypothesis advocates predicted that a large meteorite crater should be found someplace on earth that could be dated as having formed around 66 million years ago when the Mesozoic ended.

In 1990, evidence was found suggesting a large (120 km in diameter) crater centered off the coast of Mexico's Yucatan peninsula at a place called Chicxulub. The alleged crater at Chicxulub, now buried beneath sediment, would be the right size to have been formed by a 10-kilometer asteroid. Geologists were quick to search for and present evidence in support of the Chicxulub crater. Among evidence cited was sediment that appeared to have been deposited in various locations by giant sea waves. Also cited were widespread microscopic spheres of glass (now weathered into clays) that formed when rock melted from the impact and droplets were thrown high into the air.

Geologists checking Mexican records compiled during drilling offshore for oil find there are breccias in the Chicxulub area. Breccias, due to meteorite impact, are common at known meteorite craters. Also, the age of the impact seems to be correct for the K-T boundary. The evidence for an asteroid impact is overwhelming. However, the impact may not be entirely to blame for dinosaur extinction.

We do know that climates were changing toward the end of the Cretaceous and dinosaur species were decreasing. Perhaps exceptional volcanic activity led to climate changes that were not favorable to dinosaur survival and the asteroid dealt the final unfortunate blow to dinosaurs.

"Unfortunate" is from the perspective of dinosaurs, not humans. The only mammals in the Cretaceous were inconsequential, mouse-sized creatures. They survived the K-T extinction and, with dinosaurs no longer dominating the land, evolved into the many mammal species that populate the earth today, including humans.

Related Web Resources

http://ericir.syr.edu/projects/Newton/12/Lessons/dinoextn.html

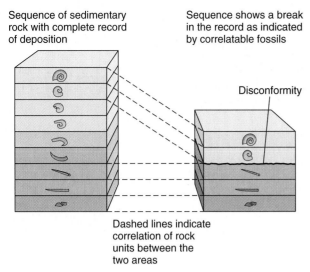

Figure 8.19
Schematic representation of a disconformity. The disconformity is in the block on the right.

Sequence of sedimentary rock with complete record of deposition

Sequence shows a break in the record as indicated by correlatable fossils

Disconformity

Dashed lines indicate correlation of rock units between the two areas

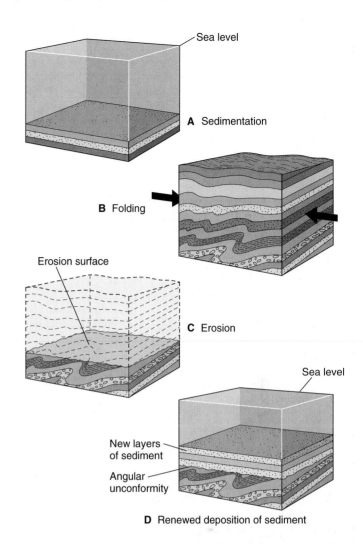

A Sedimentation

B Folding

Erosion surface

C Erosion

Sea level

New layers of sediment

Angular unconformity

D Renewed deposition of sediment

lel to the bedding plane; renewed deposition later buried the erosion surface (figure 8.19).

Because it often appears to be just another sedimentary contact (or bedding plane) in a sequence of sedimentary rock, a disconformity is the hardest type of unconformity to detect in the field. Rarely, a telltale weathered zone is preserved immediately below a disconformity. Usually the disconformity can be detected only by studying fossils from the beds in a sequence of sedimentary rocks. If certain fossil beds are absent, indicating that a portion of geologic time is missing from the sedimentary record, it can be inferred that a disconformity is present in the sequence. Although it is most likely that some rock layers are missing because erosion followed deposition, in some instances possibly neither erosion nor deposition took place for a significant amount of geologic time.

Angular Unconformities

An **angular unconformity** is a contact in which younger strata overlie an erosion surface on tilted or folded layered rock. It implies the following sequence of events, from oldest to youngest: (1) deposition and lithification of sedimentary rock (or solidification of successive lava flows if the rock is volcanic); (2) uplift accompanied by folding or tilting of the layers; (3) erosion; (4) renewed deposition (usually preceded by subsidence) on top of the erosion surface (figure 8.20).

If fossils are abundant in the rocks above and below an angular unconformity, the laws of *faunal succession* and *cross-cutting relationships* can be used to determine the relative time of folding and tilting. For example, if the youngest fossils identified in the folded sequence are late Paleozoic and the fossils in the horizontal layer just over the unconformity contain middle Mesozoic fossils, the folding must have occurred between late Paleozoic and middle Mesozoic times.

E

Figure 8.20
Development of an angular unconformity. (*A*) Deposition of sediment. (*B*) Folding. (*C*) Erosion. (*D*) Renewed deposition of sediment. (*E*) An angular unconformity at Cody, Wyoming.
Photo by C. C. Plummer

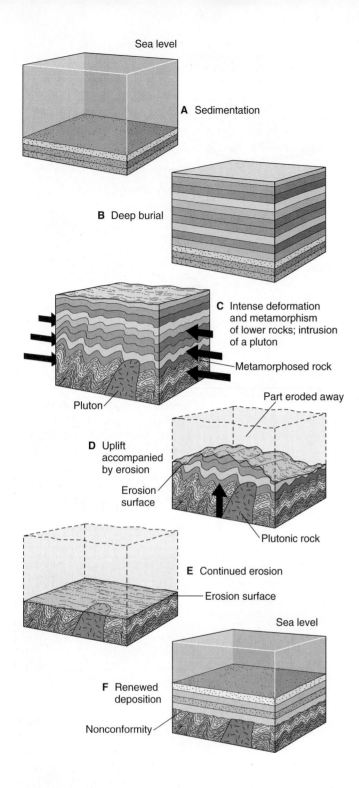

A Sedimentation

B Deep burial

C Intense deformation and metamorphism of lower rocks; intrusion of a pluton

Metamorphosed rock

Pluton

Part eroded away

D Uplift accompanied by erosion

Erosion surface

Plutonic rock

E Continued erosion

Erosion surface

Sea level

F Renewed deposition

Nonconformity

G

Figure 8.21

(*A–F*) Sequence of events implied by a nonconformity underlain by metamorphic and plutonic rock. (*G*) A nonconformity in Grand Canyon, Arizona. Horizontal Paleozoic sedimentary rocks overlie vertically foliated Precambrian metamorphic rocks.

Photo by Frank M. Hanna

Nonconformities

A **nonconformity** is a contact in which an erosion surface on plutonic or metamorphic rock has been covered by younger sedimentary or volcanic rock (figure 8.21). A nonconformity generally indicates deep or long-continued erosion before subsequent burial, because metamorphic or plutonic rocks form at considerable depths in the earth's crust.

The geologic history implied by a nonconformity, shown in figure 8.21, is (1) crystallization of igneous or metamorphic rock at depth; (2) erosion of at least several kilometers of overlying rock (the great amount of erosion further implies considerable uplift of this portion of the earth's crust); (3) deposition of new sediment, which eventually becomes sedimentary rock, on the ancient erosion surface.

Absolute Age

Counting annual growth rings in a tree trunk will tell you how old a tree is. Similarly, layers of sediment deposited annually in glacial lakes can be counted to determine how long those lakes existed (*varves,* as these deposits are called, are explained in chapter 12). But only within the few decades since the discovery of radioactivity have scientists been able to determine absolute ages of rock units. We have subsequently been able to assign absolute values to the geologic time scale and determine how many years ago the various eras, periods, and epochs began and ended. We can now state that the Cenozoic Era began some 66 million years ago, the Mesozoic Era started about 245 million years ago, and the Precambrian ended (or the Paleozoic began) about 545 million years ago (until recently, the beginning of the Paleozoic was regarded as 570 million years old). The Precambrian includes most of geologic time, because the age of the earth is commonly regarded as about 4.5 to 4.6 billion years. The oldest rocks found on earth are from northwestern Canada and have been dated at 4.03 billion years old.

Isotopic Dating

Radioactivity provides a "clock" that begins working when radioactive elements are sealed into newly crystallized miner-

ENVIRONMENTAL GEOLOGY

Radon, a Radioactive Health Hazard

Radon is an odorless, colorless gas. Every time you breathe outdoors you inhale a harmless, minute amount of radon. If the concentration of radon that you breathe in a building is too high, however, you could, over time, develop lung cancer. It is one of the intermediate daughter products in the radioactive disintegration of U-238 to Pb-206. It has a half-life of only 3.8 days.

Concentrations of radon are highest in areas where the bedrock is granite, gneiss, limestone, black shale, or phosphate-rich rock—rocks in which uranium is relatively abundant. Concentrations are also high where glacial deposits are made of fragments of these rocks. Even in these areas, radon levels are harmless in open, freely circulating air. Radon may dissolve in ground water or build up to high concentrations in confined air spaces.

The United States Environmental Protection Agency (EPA) regards 5 million American homes to have unacceptable radon levels in the air. Scientists outside of EPA have concluded that the standards the EPA is using are too stringent. They think that a more reasonably defined danger level means that only 50,000 homes have radon concentrations that pose a danger to their occupants.

Radon was first recognized in the 1950s as a health hazard in uranium mines, where the gas would collect in poorly ventilated air spaces. Radon lodges in the respiratory system of an individual and, as it deteriorates into daughter products, the subatomic particles given off cause damage to lung tissue. Three-quarters of the uranium miners studied were smokers. Thus, it is difficult to determine the extent to which smoking or radon induced lung cancer. (All studies show, however, that smoking and exposure to high radon levels are more likely to cause lung cancer than either alone.)

Interpolating the high rates of cancer incidence from the uranium miners to the population exposed to the very much lower radium levels in homes, as the EPA has done, is scientifically questionable.

What should you do if you are living in a high radon area? First have your house checked to see what the radon level is. Then read up on what acceptable standards should be. In most buildings with a high radon level, the gas seeps in from the underlying soil through the building's foundation. If a building's windows are kept open and fresh air circulates freely, radon concentrations cannot build up. However, houses are often kept sealed for air conditioning during the summer and heating during the winter. Air circulation patterns are such that a slight vacuum sucks the gases from the underlying soil into the house. Thus, radon concentrations might build up to dangerous levels.

The problem may be solved in several ways (aside from leaving windows open winter and summer). Basements can be made air tight so that gases cannot be sucked into the house from the soil. Air circulation patterns can be altered so that gases are not sucked in from underlying soil, or they are mixed with sufficient fresh, outside air.

If you are purchasing a new house, it would be a good idea to have it tested for radon before buying, particularly if the house is in an area of high-uranium bedrock or soil.

Related Web Resources
Radon in earth, air, and water.
http://sedwww.cr.usgs.gov:8080/radon/radonhome.html

---

als. The rates at which radioactive elements decay can be measured and duplicated in many different laboratories. Therefore, if we can determine the ratio of a particular radioactive element and its decay products in a mineral, we can calculate how long ago that mineral crystallized.

Determining the age of a rock through its radioactive elements is known as **isotopic dating** (previously, and somewhat inaccurately, called *radiometric dating*).

Isotopes and Radioactive Decay

As discussed in chapter 2, every atom of a given element possesses the same number of protons in its nucleus. The number of neutrons, however, need not be the same in all atoms of the same element. The **isotopes** of a given element have different numbers of neutrons, but the same number of protons.

Uranium, for example, commonly occurs as two isotopes, U-238 and U-235. The former has 238 protons and neutrons in its nucleus, whereas the latter has 235 (figure 8.22). Of these particles, 92 (the atomic number of uranium) must be protons and the rest neutrons. U-238 weighs slightly more than U-235. Because of the difference in mass, the two isotopes may be separated with an apparatus called a mass spectrometer.

Radioactive decay is the spontaneous nuclear disintegration of isotopes with unstable nuclei. As protons and neutrons leave these atoms, energy is produced. The particles departing from radioactive elements can be detected by a Geiger counter or similar device, and, in high concentrations, can damage or kill humans (see Box 8.3).

Figure 8.22

Nuclei of isotopes of U-238 and U-235.

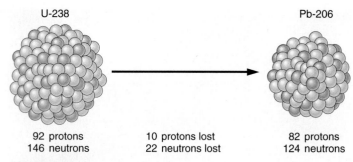

Figure 8.23

U-238 disintegrates radioactively to Pb-206 (intermediate daughter products are not shown).

When protons are lost during radioactive decay, the atom becomes a different element. When a uranium-238 atom disintegrates radioactively, it loses a total of 32 protons and neutrons, in a complex series of steps (beyond the scope of this book; however it should be described in this book's website.), before becoming a nonradioactive, stable atom. Having lost 32 subatomic particles (out of 238), the now-stable atom has 206 neutrons and protons left. Ten of the lost particles are protons, so the atomic number has changed from 92 to 82, and the atom is now the element lead (abbreviated Pb) (figure 8.23). Specifically, it is the stable Pb-206 isotope. Lead-206 is the **daughter product,** the isotope produced by radioactive decay, of U-238.

An isotope of potassium, K-40, converts to argon, Ar-40, through a different process. A K-40 nucleus captures one of its orbiting electrons. The negative charge of the electron neutralizes a single positively charged proton. The proton becomes a neutron. Because there is one less proton in the nucleus, that atom no longer is a potassium atom (atomic number 19) but an argon atom (atomic number 18).

Radioactive elements are useful for determining age because in a large number of atoms of a radioactive element, the *proportion* of atoms that will radioactively decay is constant. Even a rock possessing a small fraction of a percent of uranium contains millions of uranium atoms. The proportional amount of atoms which periodically decay is apparently unaffected by chemical reactions or by the high pressures and high temperatures of the earth's interior.

The rate of proportional decay rate for isotopes is expressed as **half-life,** the time it takes for a given amount of a radioactive isotope to be reduced by one-half. (The other half disintegrates into daughter products and energy.) The half-lives of some isotopes created in nuclear reactors are in fractions of a second. Naturally occurring isotopes that we find in rocks have very long half-lives (table 8.3). K-40 has a half-life of 1.3 billion years. If you began with one milligram of K-40, 1.3 billion years later one-half milligram of K-40 would remain. After another 1.3 billion years, there would be one-fourth of a milligram, and after another half-life only one-eighth of a milligram. (Note that two half-lives do not equal a whole life.)

To determine the age of a rock by using K-40, the amount of K-40 in that rock must first be determined by chemical analysis. The amount of Ar-40 (the daughter product) must also be determined and then used to calculate how much K-40 was present when the rock formed. By knowing how much K-40 was originally present in the rock and how much is still there, we can calculate the age of the rock on the basis of its half-life mathematically (see Box 8.4). The graph in figure 8.24 demonstrates the mathematical relationship between a radioactively decaying isotope and time.

Table 8.3	**Radioactive Isotopes Commonly Used for Determining Ages of Rocks**		
Isotope	**Half-Life**	**Product Daughter**	**Effective Dating Range (years)**
K-40	1.25 billion years	Ar-40	100,000–4.6 billion
U-238	4.5 billion years	Pb-206	10 million–4.6 billion
U-235	713 million years	Pb-207	10 million–4.6 billion
Th-232	14.1 billion years	Pb-208	10 million–4.6 billion
Rb-87	49 billion years	Sr-87	10 million–4.6 billion
C-14	5,730 years	N-14	100–40,000

8.4 IN GREATER DEPTH

Calculating the Age of a Rock

The relationship between time and radioactive decay of an isotope is expressed by the following equation (which is used to plot curves such as shown in figure 8.24).

$$N = N_0 e^{-\lambda t}$$

N is the number of atoms of the isotope at time t, the time elapsed. N_0 is the number of atoms of that isotope present when the "clock" was set. The mathematical constant e has a value of 2.718. λ is a decay constant—that portion of the isotope that decays per unit time.

The relationship between λ and the half-life (t_{hl}) is

$$\lambda = \frac{\ln 2}{t_{hl}} = \frac{0.693}{t_{hl}}$$

Replacing λ in the first equation and converting that equation to natural logarithmic (to the base e) form we get

$$t = \frac{t_{hl}}{.693} \ln \frac{N}{N_0}$$

N/N_0 is the ratio of parent atoms at present to the original number of parent atoms.

As an example, we will calculate the age of a mineral using U-235 decaying to Pb-207. Table 8.3 indicates that the half-life is 713 million years. A laboratory determines that, at present, there are 440,000 atoms of U-235 and that the amount of Pb-207 indicates that when the mineral crystallized there were 1,200,000 atoms of U-235. (We assume that there was no Pb-207 in the mineral at the time the mineral crystallized.) Plugging these values into the formula we get

$$t = \frac{713,000,000}{.693} \ln \frac{440,000}{1,200,000}$$

Solving this gives us 1,032,038,250 years. The technique is not that precise, but rounded off, we can say the mineral formed 1.032 billion years ago.

A

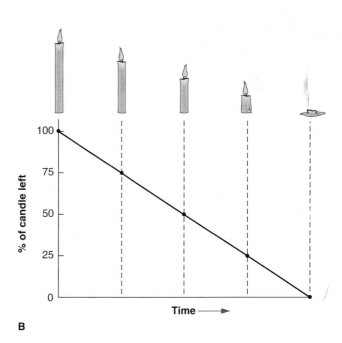

B

Figure 8.24

(*A*) The curve used to determine the age of a rock by comparing the percentage of radioactive isotope remaining in time to the original amount. Darker bars show the amount left after each half-life. Dashed bars show the amount disintegrated into daughter product and lost nuclear particles. The numbers of dots in the squares above the graph are proportional to the numbers of atoms. (*B*) For comparison, a candle burns at a linear rate.

Radiocarbon Dating

Because of its short half-life of 5,730 years, radiocarbon dating is useful only in dating things and events accurately back to about 40,000 years—about seven half-lives. (However, some people think it can be used for dating up to 70,000 years ago.) The technique is most useful in archaeological dating and for very young geologic events (Recent, or Holocene, volcanic and glacial features for instance). It has even been used to date historical artifacts, such as the Shroud of Turin. The shroud was shown not to be old enough to have been the cloth that covered Christ's face, as claimed by believers in Italy.

Carbon (atomic number 6) is in the air as part of CO_2. It is mostly the stable isotope C-12. However, C-14 is created in the atmosphere when cosmic radiation bombards nitrogen (N), atomic number 7. A neutron strikes and is captured by an N-14 atom. A proton is expelled from the nucleus and the atom becomes C-14. The nucleus of the newly created carbon atom is unstable and will, sooner or later, revert to N-14 when a neutron splits into an electron plus a proton. The electron is emitted from the atom as radiation. The rate of production of C-14 approximately balances the rate at which C-14 reverts to N-14 so that the level of C-14 remains approximately constant in the atmosphere.

Living matter incorporates C-12 and C-14 into its tissues; the ratios of C-12 and C-14 in the new tissues are usually the same as in the atmosphere. On dying, the plant or animal ceases to build new tissue. The C-14 disintegrates radioactively at the fixed rate of its half-life (5,730 years). The radioactive emissions per gram of carbon from the plant or animal remains are determined. The greater the radioactivity, the younger the sample. By comparing the radioactivity per gram to previously determined standards, we can indirectly determine the ratio of C-12 to radioactive C-14 in organic remains, and we can determine the time elapsed since the death of the organism.

Uses of Isotopic Dating

Some techniques determine isotopic ratios for a whole rock, others use single minerals within a rock. Exactly what is being dated depends on the isotope analyzed. Usually, an isotopic date determines how long ago the rock or mineral became a closed system. That is how long ago it was sealed off so that neither parent nor daughter isotopes could enter or leave the mineral or rock. Different isotopic pairs have different closure temperatures; when a rock cools below that temperature the system is closed and the "clock" starts. For instance, the K-40–Ar-40 isotopic pair has closure temperatures ranging from 150°C to 550°C, depending on the mineral. (Ar is a gas and gets trapped in different crystal structures at different temperatures.)

The best dates are obtained from igneous rocks. For a lava flow, which cools and solidifies rapidly, the age determined gives us the time of eruption and solidification. On the other hand, plutonic rocks, which may take over a million years to solidify, will not necessarily yield the time of intrusion, but the time at which a mineral cooled below the closure temperature.

Dating metamorphic rocks usually means determining the time of cooling following the end of a metamorphic episode. Sedimentary rocks are difficult to date reliably.

In order for an isotopic age determination to be accurate, several conditions must be met. To ensure that the isotopic system has remained closed, the rock collected must show no signs of weathering or hydrothermal alteration. Second, one should be able to infer there were no daughter isotopes in the system at the time of closure or make corrections for probable amounts of daughter isotopes present before the "clock" was set. Third, there must be sufficient parent and daughter atoms to be measurable by the mass spectrometer being used. And, of course, technicians and geochronologists (geologists specializing in radiometric dating), must be highly skilled at working sophisticated equipment and collecting and processing rock specimens.

Whenever possible, geochronologists will use more than one isotope pair for a rock. The two U-Pb systems (table 8.3) can usually be used together and provide an internal cross-check on the age determinate.

How Reliable is Isotopic Dating?

Half-lives of radioactive isotopes, whether short-lived, such as used in medicine, or long-lived such as used in isotopic dating, have been found not to vary beyond statistical expectations. The half-life of each of the isotopes we use for dating rocks has not changed with physical conditions or chemical activity nor could the rates have been different in the distant past. It would violate laws of physics for decay rates (half-lives) to have been different in the past. Moreover, when several isotopic dating systems are painstakingly done on a single ancient lava flow, the same age is obtained, confirming that the decay constants for each system are indeed constant.

Comparing isotopic ages with relative age relationships confirms the reliability of isotopic dating. For instance, a dike that crosscuts rocks containing Cenozoic fossils gives us a relatively young isotopic age (less than 66 million years old) whereas a pluton truncated by overlying sedimentary rocks with Paleozoic fossils yields a relatively old age (greater than 245 million years). Many thousands of similar determinations have confirmed the reliability of radiometric dating.

Combining Relative and Absolute Ages

Radiometric dating can provide absolute time brackets for events whose relative ages are known. Figure 8.25 adds radiometric dates for each of the two igneous bodies in the fictitious Minor Canyon area of figure 8.1. The date obtained for the granite is 540 million years B.P. (before present), while the dike formed 78 million years ago. We can now state that the Tarburg Formation and older tilted layers formed before 540 million years ago (though we cannot say how much older they are). We still do not know whether the

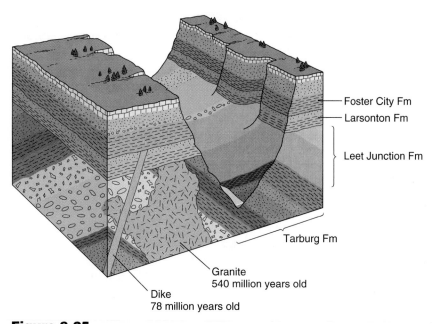

Foster City Fm

Larsonton Fm

Leet Junction Fm

Tarburg Fm

Granite
540 million years old

Dike
78 million years old

Figure 8.25

The Minor Canyon area as shown in figure 8.1 but with isotopic dates for igneous rocks indicated.

Leet Junction Formation is older or younger than the granite because of the lack of cross-cutting relationships. The Larsonton Formation's age is bracketed by the age of the granite and the age of the dike. That is, it is between 540 and 78 million years old. The Foster City and overlying formations are younger than 78 million years old; how much younger we cannot say.

Isotopic dates from volcanic ash layers or lava flows interlayered between fossiliferous sedimentary rocks have been used to assign absolute ages to the geologic time scale (figure 8.26). Isotopic dating has also allowed us to extend the time scale back into the Precambrian. There is, of course, a margin of uncertainty in each of the given dates. The beginning of the Paleozoic, for instance, was regarded until recently to be 570 million years ago, but with an uncertainty of ±30 million years. Recent work has fixed the age as closer to 545 million years. There are inherent limitations on the dating techniques as well as problems in finding the ideal rock for dating. For instance, if you wanted to obtain the date for the end of the Paleozoic Era and the beginning of the Mesozoic Era, the ideal rock would be found where there is no break in deposition of sediments between the two eras, as indicated by fossils in the rocks. But the difficulties in dating sedimentary rock mean you would be unlikely to date such rocks. Therefore, you would need to date volcanic rocks interlayered with sedimentary rocks found as close as possible to the transitional sedimentary strata. Alternatively, isotopically dated intrusions, such as dikes, whose cross-cutting relationships indicate that the age of intrusion is close to that of the transitional sedimentary layers, could be used to approximate the absolute age of the transition.

Isotopic dating has shown that the Precambrian took up most of geologic time (87%). Obviously, the Precambrian needed to be subdivided. The three major subdivisions of the Precambrian are the **Prearchean** (or **Hadean**), the **Archean** and the **Proterozoic** (Greek for beginning life). Each is regarded as an **Eon,** the largest unit of geological time. A fourth, and youngest, eon is the **Phanerozoic** (Greek for visible life). The Phanerozoic eon is all of geologic time with an abundant fossil record; in other words, it is made up of the three eras that followed the Precambrian.

Age of the Earth

In 1625 Archbishop James Ussher determined that the earth was created in the year 4004 B.C. His age determination was made by counting back generations in the Bible. This would make the earth 6,000 years old at present. That very young age of the earth was largely taken for granted by Western countries. By contrast, Hindus at the time regarded the earth as very old. According to an ancient Hindu calendar, our year 2000 A.D. would be year 1,972,949,101.

With the popularization of uniformitarianism in the early 1800s, earth scientists began to realize that the earth must be very old—at least in the hundreds of millions of years. They were dealt a setback by the famous English physicist, Lord Kelvin. Kelvin, in 1866, calculated from the rate at which the earth loses heat that the earth must have been entirely molten between 20 and 100 million years ago. He later refined his estimate to between 20 and 40 million years. He was rather arrogant in scoffing at earth scientists that believed that uniformitarianism indicated a much older age for the earth. The discovery of radioactivity in 1896 invalidated Kelvin's claim because it provided a heat source that he had not known about. When radioactive elements decay, heat is given off and that heat is added to the heat already in the earth. The amount of radioactive heat given off at present approximates the heat the earth is losing. So, for all practical purposes, the earth is not getting cooler.

The discovery of radioactivity also provided the means to determine how old the earth is. In 1905, the first crude isotopic dates were done and indicated an age of 2 billion years. Isotopic dating of meteorites has shown that they are between 4.5 and 4.6 billion years old. Some of the meteorites dated are planetary fragments. It is probable that all of the planets formed at the same time.

Comprehending Geologic Time

The vastness of geologic time (sometimes called deep time) is difficult for us to comprehend. One way of visualizing deep time is to visualize driving from Los Angeles to New York, a

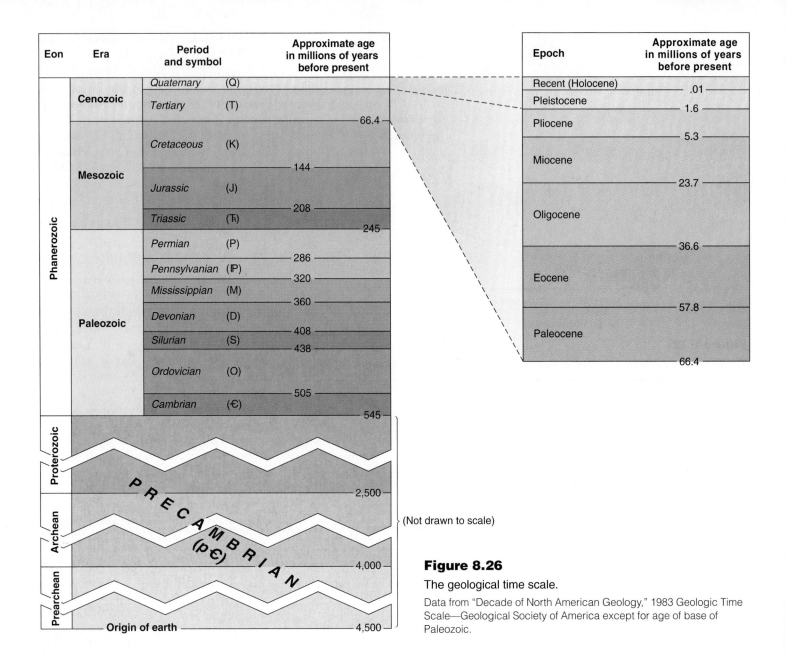

Eon	Era	Period and symbol		Approximate age in millions of years before present
Phanerozoic	Cenozoic	Quaternary	(Q)	
		Tertiary	(T)	66.4
	Mesozoic	Cretaceous	(K)	
		Jurassic	(J)	144
		Triassic	(Ŧ)	208 / 245
	Paleozoic	Permian	(P)	286
		Pennsylvanian	(IP)	320
		Mississippian	(M)	360
		Devonian	(D)	408
		Silurian	(S)	438
		Ordovician	(O)	505
		Cambrian	(€)	545
Proterozoic		PRECAMBRIAN (pЄ)		2,500
Archean				4,000
Prearchean		Origin of earth		4,500

(Not drawn to scale)

Epoch	Approximate age in millions of years before present
Recent (Holocene)	.01
Pleistocene	1.6
Pliocene	5.3
Miocene	23.7
Oligocene	36.6
Eocene	57.8
Paleocene	66.4

Figure 8.26

The geological time scale.

Data from "Decade of North American Geology," 1983 Geologic Time Scale—Geological Society of America except for age of base of Paleozoic.

distance of approximately 4,500 kilometers. Each kilometer represents 1 million years—this is a very, very slow trip. The highlights of the trip corresponding to the earth's history are shown in figure 8.27. Note that if you live to be a hundred, your life is represented by less than the width of a curb at the edge of a sidewalk.

Another way to get a sense of geologic time is to compare it to a motion picture. A movie is projected at a rate of 32 frames per second; that is, each image is flashed on the screen for only 1/32 of a second, giving the illusion of continuous motion. But suppose that each frame represented 100 years. If you lived 100 years, one frame would represent your whole lifetime.

If we were able to show the movie on a standard projector, each 100 years would flash by in 1/32 of a second. It would take only one-sixteenth of a second to go back to the signing of the Declaration of Independence. The 2,000-year-old Christ-

ian era would be on screen for 3/4 of a second. A section showing all time back to the last major ice age would only be less than seven seconds long. However, you would have to sit through almost six hours of film to view a scene at the close of the Mesozoic Era (perhaps you would see the last dinosaur die). And to give a complete record from the beginning of the Paleozoic Era, this epic film would have to run continuously for two days. You would have to spend over two weeks (sixteen days) in the theater, without even a popcorn break between reels, to see a movie entitled "The Complete Story of Earth, from Its Birth to Modern Civilization."

Thinking of our lives as taking less than a frame of such a movie can be very humbling. From the perspective of being stuck in that one last frame, geologists would like to know what the whole movie is like or, at least, get a synopsis of the most dramatic parts of the film.

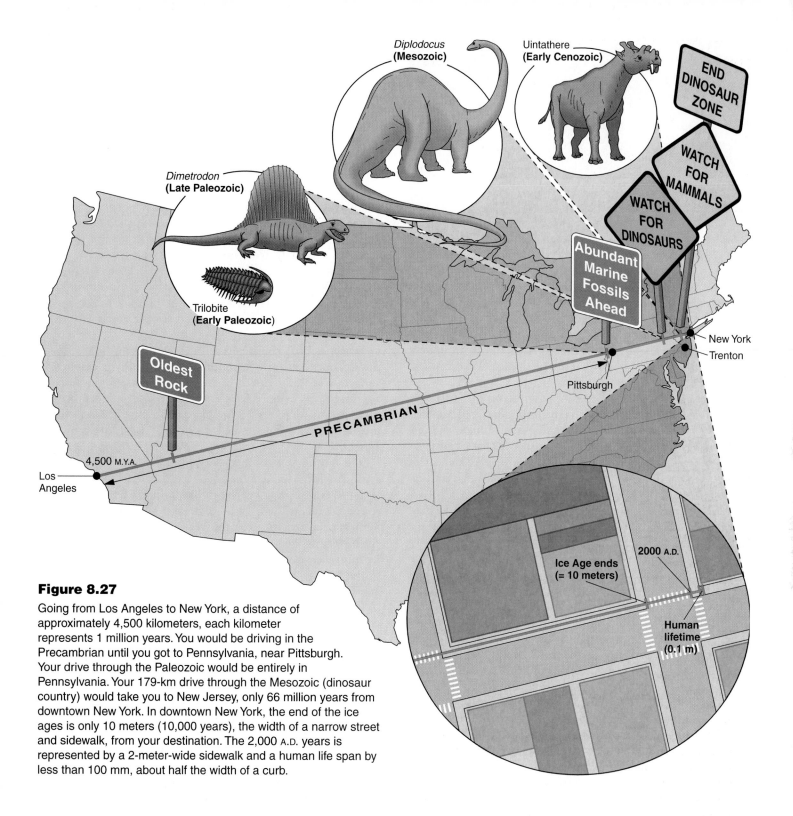

Figure 8.27

Going from Los Angeles to New York, a distance of approximately 4,500 kilometers, each kilometer represents 1 million years. You would be driving in the Precambrian until you got to Pennsylvania, near Pittsburgh. Your drive through the Paleozoic would be entirely in Pennsylvania. Your 179-km drive through the Mesozoic (dinosaur country) would take you to New Jersey, only 66 million years from downtown New York. In downtown New York, the end of the ice ages is only 10 meters (10,000 years), the width of a narrow street and sidewalk, from your destination. The 2,000 A.D. years is represented by a 2-meter-wide sidewalk and a human life span by less than 100 mm, about half the width of a curb.

8.1
Astrogeology

The Origin and History of the Solar System

The solar system began with the condensation of a large volume of interstellar gas and dust called a *nebula.* Much of the material in the nebula collected in the center to form the Sun, but some of it remained farther out. Small groupings of particles formed randomly, and because of the combined gravitational pull of the particles in the group, they attracted other particles. The groupings continued to increase in size until, on the average, they were probably several kilometers in diameter.

These larger bodies sometimes joined together by collision and sometimes became fragmented. The largest bodies tended to increase in size by capturing smaller ones, and the smaller bodies tended to be fragmented or were captured. The bodies that grew larger became the planets and satellites; those that were fragmented became interstellar dust or asteroids.

Variations in composition among the planets can be partially explained by differences in their sizes. Larger planets were able to attract and hold larger amounts of lighter gases, such as hydrogen and helium. Another important factor in determining planetary composition was the location within the nebula where the planet formed. Those bodies nearest the center (Mercury, Venus, Earth, the Moon, and Mars) would have experienced higher temperatures and so retained smaller percentages of materials that are easily vaporized. The outer planets and their satellites were formed at a great enough distance from the Sun for large quantities of gases and ices to collect around their rocky cores.

Heat from building of planets along with heat derived from the decay of radioactive elements caused partial melting of the planets' interiors. Heavier materials such as iron-nickel alloys would sink to the planets' centers, while lighter materials such as silicate magmas would rise to form crusts. On Earth, the original crust was mostly basalt, with some granitic rocks. The areas of granitic rocks would have formed the first continents. Convection in the mantle probably would have begun at the time of crustal differentiation and would have set in motion plate tectonic processes.

On Venus, which is about the same size as the Earth, the presence of possible folded mountains (the parallel ridges that have been observed) and chains of volcanoes may indicate that differentiation and some kind of lithospheric motions have occurred. However, no surface features have been found yet that could be interpreted as diverging plate boundaries similar to the Earth's oceanic ridges. The observation that volcanic activity appears to be concentrated in one area at the equator has led to the suggestion that Venus' tectonic activity is controlled by at least one giant mantle plume area. The absence of folded mountain chains and the presence of large numbers of undeformed craters on Mars indicate that plate tectonic activity has not occurred on Mars. It has been noted that the very large size of volcanoes such as Olympus Mons could be explained by assuming a stationary lithosphere over a mantle plume. Mercury and the Moon show no evidence of plate tectonic processes having occurred.

Space "debris" continued to be swept up by the planets throughout their early histories, producing heavily cratered surfaces and multi-ringed basins. The Moon and Mercury are still heavily cratered. The Earth, too, was subjected to an early period of intensive bombardment. But erosion and tectonic activity have erased most of those early-formed craters.

The heating up of planetary interiors would produce great quantities of magma that would rise to the surface along fractures produced by meteorite impact or by some tectonic process. Vast areas of the Moon, Mercury, and Mars are covered with volcanic rocks, most formed early in the history of the solar system. Similar early extrusive activity must have occurred on Venus and on the Earth. Volcanic activity on a much smaller scale still occurs on the Earth and Venus and may still be occurring on Mars.

Gases would escape from planetary interiors as a result of volcanic activity, later condensing to form atmospheres, oceans, ice shells, and polar caps. Lighter elements would have escaped quickly from objects as small as the Moon and Mercury but could have been partially retained by the Earth, Venus, and Mars. The distances of these planets from the Sun might determine whether these light elements exist mostly as atmospheres (as on Venus); as atmospheres, oceans, underground water, and ice caps (as on Earth); or mostly as ice caps and underground ice (as on Mars). On planets with appropriate conditions and temperature ranges, life may have formed, as it did on the Earth. The presence of life on the Earth has certainly extensively modified the Earth's atmosphere (by producing oxygen, for example).

The presence of atmospheres, oceans, underground ice, and so on would have determined the extent to which weathering and erosion modified a planet's surface.

Summary

The principle of *uniformitarianism,* a fundamental concept of geology, states that the present is the key to the past.

Relative time, or the sequence in which geologic events occur in an area, can be determined by applying the principles of *original horizontality, superposition, lateral continuity,* and *cross-cutting relationships.*

Rocks can be correlated by determining the physical continuity of rocks between the two areas (generally this works only for a short distance). A less useful means of correlation is similarity of rock types (which must be used cautiously).

Unconformities are buried erosion surfaces that help geologists determine the relative sequence of events in the geologic past. Beds above and below a *disconformity* are parallel, generally indicating less intense activity in the earth's crust. An *angular unconformity* implies that folding or tilting of rocks took place before or around the time of erosion. A *nonconformity* implies deep erosion because metamorphic or plutonic rocks have been exposed and subsequently buried by younger rock.

Fossils are used for worldwide correlation of rocks. Sedimentary rocks are assigned to the various subdivisions of the *geologic time scale* on the basis of fossils they contain, which are arranged according to the principle of *faunal succession.*

Absolute age—how many years ago a geologic event took place—is generally obtained by using *isotopic dating* techniques. Isotopic dating is accomplished by determining the ratio of the amount of a radioactive isotope presently in a rock or mineral being dated to the amount originally present. The time it takes for a given amount of an isotope to decay to half that amount is the *half-life* for that isotope. Absolute ages have been determined for the subdivisions of the geologic time scale.

Terms to Remember

absolute age 174
actualism 174
angular unconformity 185
Archean Eon 191
Cenozoic Era 182
contacts 174
correlation 179
cross-cutting relationship 176
daughter product 188
disconformity 183
eon 191
epoch 182
era 182
faunal succession 181

formation 174
fossil assemblage 181
half-life 188
inclusion 179
index fossil 181
isotope 187
isotopic dating 187
lateral continuity 175
Mesozoic Era 182
nonconformity 186
original horizontality 175
Paleozoic Era 182
period 182
Phanerozoic Eon 191

physical continuity 180
Pleistocene Epoch 182
Prearchean (Hadean) Eon 191
Precambrian 182
Proterozoic Eon 191
Quaternary Period 182
radioactive decay 187
Recent (Holocene) Epoch 182
relative time 174
standard geologic time scale 182
superposition 175
unconformity 183
uniformitarianism 174

Testing Your Knowledge

Use the questions below to prepare for exams based on this chapter.

1. Why is it desirable to find an index fossil in a rock layer? In the absence of index fossils, why is it desirable to find several fossils in a rock unit to determine relative age?

2. Suppose you had a radioactive isotope X whose half-life in disintegrating to daughter product Y is 120,000 years. By calculating how much it took to make the present amount of $Y,$ you determine that, originally, the rock contained 8 grams of isotope $X.$ At present only 1/4 gram of X is in the rock. How many half-lives have gone by? How old is the rock?

3. By applying the various principles, draw a cross section of an area in which the following sequence of events occurred:

a. Several layers of sedimentary rocks were deposited in the Cambrian through Devonian periods on a much older Precambrian basement of metamorphic rock.

b. No record exists of Mississippian through Triassic rocks.

c. A stock intruded in the Jurassic Period.

d. Tilting and erosion to a flat plain preceded deposition of a sedimentary layer in late Cretaceous.

e. In the Tertiary Period, more and steeper tilting affected the entire area.

f. Erosion during the Quaternary Period created slightly hilly terrain.

g. Following erosion, a volcano with a feeder dike erupted in recent time.

4. Name as many types of contacts (e.g., intrusive contact) as you can.

5. Using information from Box 8.4, calculate the age of a feldspar. At present, there are 1.2 million atoms of K-40. The amount of Ar-40 in the mineral indicates that originally there were 1.9 million K-40 atoms in the rock. Use a half-life of 1.3 billion years. (Hint: The answer is 862 million years.)

6. "Geological processes operating at present are the same processes that have operated in the past" is the principle of (a) correlation (b) catastrophism (c) uniformitarianism (d) none of the above

7. "Within a sequence of undisturbed sedimentary rocks, the layers get younger going from bottom to top" is the principle of (a) original horizontality (b) superposition (c) cross-cutting (d) none of the above

8. If rock A cuts across rock B, then rock A is _____ rock B. (a) younger than (b) the same age as (c) older than

9. Which is a method of correlation? (a) physical continuity (b) similarity of rock types (c) fossils (d) all of the above

10. Eras are subdivided into (a) periods (b) eons (c) ages

11. Periods are subdivided into (a) eras (b) epochs (c) ages

12. Which division of geologic time was the longest? (a) Precambrian (b) Paleozoic (c) Mesozoic (d) Cenozoic

13. Which is a useful radioactive decay scheme? (a) U-238/Pb-206 (b) U-235/Pb-207 (c) K-40/Ar-40 (d) Rb-87/Sr-87 (e) all of the above

14. C-14 dating can be used on all of the following except (a) wood (b) shell (c) the Shroud of Turin (d) granite (e) bone

15. Concentrations of radon are highest in areas where the bedrock is (a) granite (b) gneiss (c) limestone (d) black shale (e) phosphate-rich rock (f) all of the above

16. Which is not a type of unconformity? (a) disconformity (b) angular unconformity (c) nonconformity (d) triconformity

17. A geologist could use the principle of inclusions to determine the relative age of (a) fossils (b) metamorphism (c) shale layers (d) xenoliths

18. The oldest abundant fossils of complex multicellular life date from the (a) Precambrian (b) Paleozoic (c) Mesozoic (d) Cenozoic

19. A contact between parallel sedimentary rock which records missing geologic time is (a) a disconformity (b) an angular unconformity (c) a nonconformity

Expanding Your Knowledge

1. How much of the uranium-238 originally in the earth is still present?

2. As indicated by fossil records, why have some ancient organisms survived through very long periods of time whereas others have been very short-lived?

3. To what extent would a composite volcano (chapter 4) be subject to the three principles described in this chapter?

4. Suppose a sequence of sedimentary rock layers was tilted into a vertical position by tectonic forces. How might you determine (a) which end was originally up and (b) the relative ages of the layers?

5. Note that in table 8.3 the epochs are given only for the Cenozoic Era (as is commonly done in geology textbooks). Why are the epochs for the Mesozoic and Paleozoic considered less important and not given?

6. Why would you not be able to use the principle of superposition to determine the age of a sill (defined in chapter 3)?

7. Moon rocks have been dated that are considerably older than any dated on earth. Give a hypothesis to explain this.

Exploring Resources

Cowen, R. 1994. *History of life.* 2d ed. Cambridge, Massachusetts: Blackwell Scientific Publications.

Dalrymple, G. B. 1991. *The age of the Earth.* Palo Alto, California: Stanford University Press.

Faure, G. 1986. *Principles of isotope geology.* 2d ed. New York: John Wiley & Sons.

Geyh, M. A., and H. Schleicher. 1990. *Absolute age determination.* New York: Springer-Verlag.

Gould, S. J. 1988. *Time's arrow/time's cycle.* Boston: Harvard University Press.

Raup, D. M. 1991. *Extinction: Bad genes or bad luck?* New York: W. W. Norton & Company.

The earth has a history. Geological Society of America, Boulder, Colorado.

Iso-topics: Using isotopes to study our planet. Berkeley Geochronology Center, 2455 Ridge Road, Berkeley, CA 94709

http://www.ucmp.berkeley.edu/exhibit/exhibits.html
Paleontology without Walls. University of California Museum of Paleontology virtual exhibit. Click on Geologic Time.

http://www.talkorigins.org/origins/faqs-youngearth.html

Isotopic dating and the age of the universe are covered in detail in The Talk. Origins archive, Exploring the Creation/Evolution Controvers

Expand your knowledge of the concepts presented in this chapter by using the CD-ROM to answer the following question.

1. Changing Earth: Using the geological time scale in fig 8.26 and the various buttons on this screen, determine the geologic period in which each of these events occurred:

- North America separated from the rest of Pangaea
- the great North American inland sea formed
- North and South America become connected
- India and Australia separated from Antarctica
- India collided with Asia

Earthflow at Stone Canyon, Los Angeles, California, 1980.
©Tom McHugh/Photo Researchers, Inc.

CHAPTER

9

Mass Wasting

When material on a hillside has weathered (the process described in chapter 5), it is likely to move downslope because of the pull of gravity. Soil or rock moving in bulk at the earth's surface is called mass wasting. Mass wasting is one of several surficial processes. Other processes of erosion, transportation, and deposition—involving streams, glaciers, wind, and ocean waves—are discussed in following chapters.

Landsliding is the best known type of mass wasting. Landslides destroy towns and kill people. While these disasters involve relatively rapid movement of debris and rock, mass wasting can also be very slow. Creep is a type of mass wasting too slow to be called a landslide.

In this chapter we describe how different types of mass wasting shape the land and alter the environment and what factors control the rapidity or slowness of the process. Understanding mass wasting and its possible hazards is particularly important in hilly or mountainous regions.

Type of Movement	Slowest ⟶	Increasing Velocities ⟶		Fastest
	Less than 1 cm/year	1 mm/day to 1 km/hr	1 to 5 km/hr	Velocities generally greater than 4 km/hour
Flow	⟵——————— Debris Flow ———————⟶			
	Creep (Debris)	Earthflow	Mudflow (Water-saturated debris)	Debris avalanche (Debris) Rock avalanche (Bedrock)
Slide		⟵————— Debris Slide —————⟶		
		⟵———— Rockslide (Bedrock) ————⟶		
Fall				Rockfall (Bedrock) Debris fall (Debris)
	⟵————— "Landslides" —————⟶			

1. The type of material at the start of movement is shown in parentheses. Rates given are typical velocities for each type of movement.

You may recall from previous chapters that mountains are products of tectonic forces. Most mountains are associated with present or past converging plate boundaries. If tectonism were not at work, the surfaces of the continents would long ago have been reduced to featureless plains due to weathering and erosion. We consider the material on mountain slopes or hillsides to be out of equilibrium with respect to gravity. Because of the force of gravity, the various agents of erosion (moving water, ice, and wind) work to make slopes gentler and therefore increasingly more stable. The process of erosion discussed in this chapter is mass wasting.

Mass wasting (also called mass movement) is movement in which bedrock, rock debris, or soil moves downslope in bulk, or as a mass, because of the pull of gravity. Mass wasting includes movement so slow that it is almost imperceptible (called *creep*) as well as **landslides,** a general term for the slow to very rapid descent of rock or soil.

Mass wasting affects humans in many ways. Its effects range from the devastation of a killer landslide (see Box 9.1) to the nuisance of having a fence slowly pulled apart by soil creep. The cost in lives and property from landslides is surprisingly high. According to the U.S. Geological Survey, more people in the United States died from landslides during the last three months of 1985 than were killed during the last twenty years by all other geologic hazards, such as earthquakes and volcanic eruptions. Over time, landslides have cost Americans triple the combined costs of earthquakes, hurricanes, floods, and tornadoes. On average, the annual cost of landslides in the United States has been 1.5 billion dollars and 25 lost lives. In many cases of mass wasting, a lit-tle knowledge of geology, along with appropriate preventive action, could have averted destruction.

Classification of Mass Wasting

A number of systems are used by geologists, engineers, and others for classifying mass wasting, but none has been universally accepted. Some are very complex and useful only to the specialist.

The classification system used here and summarized in table 9.1 is based on (1) rate of movement, (2) type of material, and (3) nature of the movement.

Rate of Movement

A landslide like the one in Peru (Box 9.1) clearly involves rapid movement. Just as clearly, movement of soil at a rate of less than a centimeter a year is slow movement. Between these extremes is a wide range of velocities.

Type of Material

Mass wasting processes are usually distinguished on the basis of whether the descending mass started as bedrock (as in a rockslide) or as debris. The term **debris,** as applied to mass wasting processes, means any unconsolidated material at the earth's surface, such as soil and rock fragments (weathered or unweathered) of any size.

The amount of water (or ice and snow) in a descending mass strongly influences the rate and type of movement.

9.1
ENVIRONMENTAL GEOLOGY

Disaster in the Andes

As a result of a tragic combination of geological conditions, one of the most devastating landslides in history destroyed the town of Yungay in Peru in 1970. Yungay was one of the most picturesque towns in the Santa River Valley, which runs along the base of the highest peaks of the Peruvian Andes. Heavily glaciated Nevado Huascarán, 6,663 meters (21,860 feet) above sea level, rises steeply above the populated narrow plains along the Santa River.

In May 1970 a sharp earthquake occurred. The earthquake was centered offshore from Peru about 100 kilometers from Yungay. Although the tremors in this part of the Andes were no stronger than those that have done only light damage to cities in the United States, many poorly constructed homes collapsed. Because of the steepness of the slopes, thousands of small rockfalls and rockslides were triggered.

The greatest tragedy began when a slab of glacier ice about 800 meters wide, perched near the top of Huascarán, was dislodged by the shaking. (A few years earlier American climbers returning from the peak had warned that the ice looked highly unstable. The Peruvian press briefly noted the danger to the towns below, but the warning was soon forgotten.)

The mass of ice rapidly avalanched down the extremely steep slopes, breaking off large masses of rock debris, scooping out small lakes and loose rock that lay in its path. Eyewitnesses described the mass as a rapidly moving wall the size of a ten-story building. The sound was deafening. More than 50 million cubic meters of muddy debris traveled 3.7 kilometers (12,000 feet) vertically and 14.5 kilometers (9 miles) horizontally in less than four minutes, attaining speeds between 200 and 435 kilometers per hour (125 to 270 miles per hour). The main mass of material traveled down a steep valley until it came to rest blocking the Santa River and burying about 1,800 people in the small village of Ranrahirca (box figure 1). A relatively small part of the mass of mud and debris that was moving especially rapidly shot up the valley sidewall at a curve and overtopped a ridge. The mass was momentarily airborne before it fell on the town of Yungay, completely burying it under several meters of mud and loose rock. Only the top of the church and tops of palm trees were visible, marking where the town center was

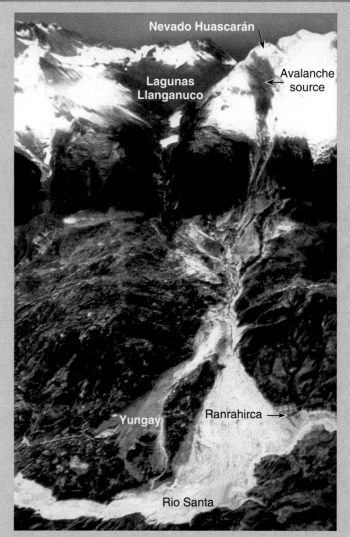

Box 9.1 Figure 1

Air photo showing the 1970 debris avalanche in Peru, which buried Yungay. The main mass of debris destroyed the small village of Ranrahirca.

Photo by Servicio Aerofotografico de Peru, courtesy of U.S. Geological Survey

buried (box figure 2). Ironically, the cemetery was not buried because it occupied the high ground. The few survivors were people who managed to run to the cemetery.

The estimated death toll at Yungay was 17,000. This was considerably more than the town's normal population, because it was Sunday, a market day, and many families had come in from the country.

For several days after the slide the debris was too muddy for people to walk on, but within three years grass had grown over the site. Except for the church steeple and the tops of palm trees that still protrude above the ground, and the crosses

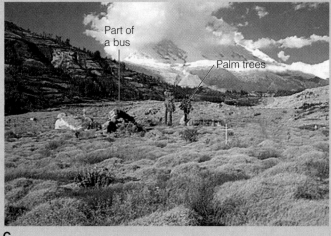

Box 9.1 Figure 2

(*A*) Yungay is completely buried, except for the cemetery and a few houses on the small hill in the lower right of the photograph. (*B*) Behind the palm trees is the top of a church buried under five meters of debris at Yungay's central plaza. (*C*) Three years later.

Photos *A* and *B* by George Plafker, U.S. Geological Survey; photo *C* by C. C. Plummer

erected by families of those buried in the landslide, the former site of Yungay appears to be a scenic meadow overlooking the Santa River. The U.S. Geological Survey and Peruvian geologists found evidence that Yungay itself had been built on top of debris left by an even bigger slide in the recent geologic past. More slides will almost surely occur here in the future. Forewarned, the Peruvian government will not allow the building of a new town in the danger area.

Further Reading

Ericksen, G. E., G. Plafker, and J. Fernandez Concha. 1970. *Preliminary report on the geologic events associated with the May 31, 1970, Peru earthquake.* U.S. Geological Survey Circular 639.

Type of Movement

In general, the type of movement in mass wasting can be classified as mainly flow, slide, or fall (figure 9.1). A **flow** implies that the descending mass is moving downslope as a viscous fluid. **Slide** means the descending mass remains relatively coherent, moving along one or more well-defined surfaces. A **fall** occurs when material free-falls or bounces down a cliff.

Two kinds of slip are shown in figure 9.1. In a **translational slide**, the descending mass moves along a plane approx-

Figure 9.1

Flow, slide, and fall.

imately parallel to the slope of the surface. A **rotational slide** (also called a *slump*) involves movement along a curved surface, the upper part moving downward while the lower part moves outward.

Controlling Factors in Mass Wasting

Table 9.2 summarizes the factors that influence the likelihood and the rate of movement of mass wasting. The table makes apparent some of the reasons why the landslide (a debris avalanche) in Peru (Box 9.1) occurred and why it moved so rapidly. (1) The slopes were exceptionally steep, and (2) the **relief** (the vertical distance between valley floor and mountain summit) was great, allowing the mass to pick up speed and momentum. (3) Water and ice not only added weight to the mass of debris but also acted as lubricants. (4) Abundant loose rock and debris were available in the course of the slide. (5) Where the slide began, there were no plants with roots to anchor loose material on the slope. Finally, (6) the area is earthquake prone. Although the slide would have occurred eventually even without an earthquake, it was triggered by an earthquake.

Other factors influence susceptibility to mass wasting as well as its rate of movement. The orientation of planes of weakness in bedrock (bedding planes, foliation planes, etc.) is important if the movement involves bedrock rather than debris. Fractures or bedding planes oriented so that slabs of rock can slide easily along these surfaces greatly increase the likelihood of mass wasting.

| Table 9.2 | Summary of Controls of Mass Wasting |

Driving Force: Gravity

Contributing Factors	Most Stable Situation	Most Unstable Situation
Slope angle	Gentle slopes or horizontal surface	Steep or vertical
Local relief	Low	High
Thickness of debris over bedrock	Slight thickness (usually)	Great thickness
Orientation of planes of weakness in bedrock	Planes at right angles to hillside slopes	Planes parallel to hillside slopes
Climatic factors:		
Ice	Temperature stays above freezing	Freezing and thawing for much of the year
Water in soil or debris	Film of water around fine particles	Saturation of debris with water
Precipitation	Frequent but light rainfall or snow	Long periods of drought with rare episodes of heavy precipitation
Vegetation	Heavily vegetated	Sparsely vegetated

Triggering Mechanisms: (1) earthquakes; (2) weight added to upper part of a slope; (3) undercutting of bottom of slope.

Climatic controls inhibit some types of mass wasting and aid others (Table 9.2). Climate influences how much and what kinds of vegetation grow in an area and what type of weathering occurs. Infrequent but heavy rainfall aids mass wasting because it quickly saturates debris that lacks the protective vegetation found in wetter climates. By contrast, rain that drizzles intermittently much of the year results in vegetation that tends to inhibit mass wasting. In cold climates, freezing and thawing contribute to downslope movement.

Gravity

Gravity is the driving force for mass wasting. Figure 9.2*A* and *B* show gravity acting on a block on a slope. The length of the vertical arrow is proportional to the force—the heavier the material, the longer the arrow. The effect of gravity is resolvable into two component forces, indicated by the black arrows. One, the *normal force,* is perpendicular to the slope and its value indicates the block's ability to stay in place because of frictional considerations. The other, called the **shear force,** is parallel to the slope, and indicates the block's ability to move. The length of the arrows is proportional to the strength of each force. The steeper the slope (and the heavier the block), the greater the shear force and the greater the tendency of the block to slide. Friction counteracts the shear force. If friction is greater than the shear force, the block will not move. If the force of friction is reduced (for instance, with water) so that it is less than the shear force, the block will slide. Similar forces act on debris on a hillside (figure 9.2*C*). The resistance to movement or deformation of that debris is its **shear strength.** Shear strength is controlled by factors such as the cohesiveness of the material, friction between particles and the anchoring effect of plant roots. Shear strength is also related to the normal force. The larger the normal force, the greater the shear strength. If the shear strength is greater than the shear force, the debris will not move or be deformed. On the other hand, if shear strength is less than shear force, the debris will flow or slide.

Building a heavy structure high on a slope demands special precautions. To prevent movement of both the slope and the building, pilings may have to be sunk through the debris, perhaps even into bedrock. Developers may have to settle for fewer buildings than planned if the weight of too many structures will make the slope unsafe.

Water

Water is a critical factor in mass wasting. When debris is saturated with water (as from heavy rain or melting snow), it becomes heavier and is more likely to flow downslope. The added gravitational shear force from the increased weight, however, is probably less important than the reduction in shear strength. This is due to increased *pore pressure* in which water forces grains of debris apart.

Paradoxically, a small amount of water in soil can actually prevent downslope movement. When water does not completely fill the pore spaces between the grains of soil, it forms a thin film around each grain (as shown in figure 9.3). Loose grains adhere to one another because of the *surface tension* created by the film of water and shear strength increases. Surface tension of water between sand grains is what allows you to build a sand castle. The sides of the castle can be steep or even vertical because surface tension holds the moist sand grains in place. Dry sand cannot be shaped into a sand castle because the sand grains slide back into a pile that generally slopes at an angle of about 30° to 35° from the horizontal. On the other hand, an experienced sand castle builder also knows that it is impossible to build anything with sand that is too wet. In this case the water completely occupies the pore space between sand grains, forcing them apart and allowing them to slide easily past one another. When the tide comes in, or someone pours a pail of water on your sand castle, all you have is a puddle of wet sand.

Similarly, as the amount of water in debris increases, rate of movement tends to increase. Damp debris may not

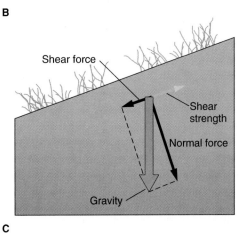

Figure 9.2

Relationship of shear force and normal force to gravity. (*A*) For a block on a gently inclined surface. (*B*) For a block on a steep surface. (*C*) Forces acting at a point in debris. Shear strength is represented by a yellow arrow. If that arrow is longer than the one represented by shear force, debris at that point will not slide or be deformed.

move at all, whereas moderately wet debris moves slowly downslope. Slow types of mass wasting, such as creep, are generally characterized by a relatively low ratio of water to debris. Mudflows always have high ratios of water to debris. A mudflow that continues to gain water eventually becomes a muddy stream.

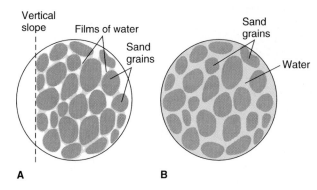

Figure 9.3

The effect of water in sand. (*A*) Unsaturated sand held together by surface tension of water. (*B*) Saturated sand grains forced apart by water; mixture flows easily.

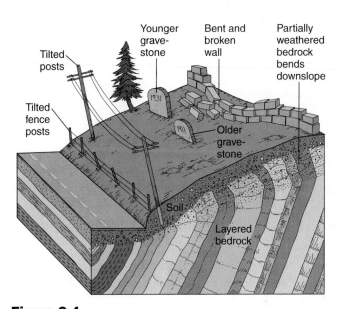

Figure 9.4

Indicators of creep.

After C. F. S. Sharpe.

Common Types of Mass Wasting

The common types of mass wasting are shown in table 9.1. Here we will describe each type in detail.

Creep

Creep is very slow, downslope movement of soil or unconsolidated debris. Shear forces, over time, are only slightly greater than shear strengths. The rate of movement is usually less than a centimeter per year and can be detected only by observations taken over months or years. When conditions are right, creep can take place along nearly horizontal slopes. Some indicators of creep are illustrated in figures 9.4 and 9.5.

A

B

Figure 9.5

(*A*) Tilted gravestones in a churchyard at Lyme Regis, England (someone probably straightened the one upright gravestone). Grassy slope is inclined gently to the left. (*B*) Soil and partially weathered, nearly vertical sedimentary strata have crept downslope.

Photo *A* by C. C. Plummer; photo *B* by Frank M. Hanna

A

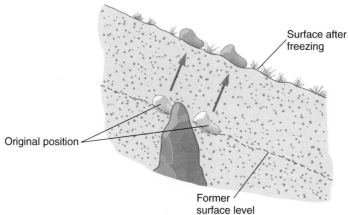

B

C

Figure 9.6

Downslope movement of soil, illustrated by following two sand grains (each less than a millimeter in size) during a freeze-thaw cycle.

Two factors that contribute significantly to creep are water in the soil and daily cycles of freezing and thawing. As we have said, water-saturated ground facilitates movement of soil downhill. What keeps downslope movement from becoming more rapid in most areas is the presence of abundant grass or other plants that anchor the soil. (Understandably, overgrazing can severely damage sloping pastures.)

Although creep does take place in year-round warm climates, the process is more active where the soil freezes and thaws during part of the year. During the winter in regions such as the northeastern United States, the temperature may rise above and fall below freezing once a day. When there is moisture in the soil, each freeze-thaw cycle moves soil particles a minute amount downhill, as shown in figure 9.6.

Debris Flow

The general term **debris flow** is used for mass wasting in which motion is taking place throughout the moving mass (flow). The common varieties earthflow, mudflow, and debris avalanche are described in this section.

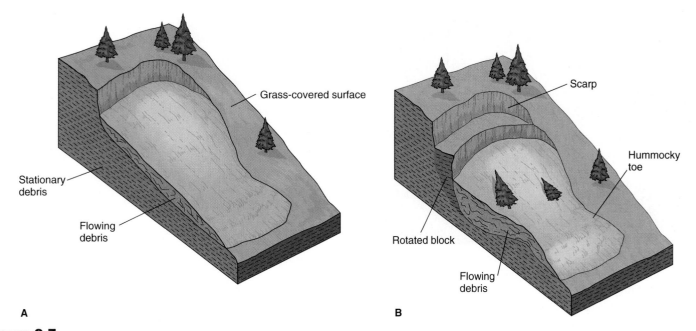

Figure 9.7

Earthflow. (*A*) Earthflow involving only flow. (*B*) Earthflow and rotational sliding.

Earthflow

In an **earthflow,** debris moves downslope as a viscous fluid; the process can be slow or rapid. Earthflows usually occur on hillsides that have a thick cover of debris, often after heavy rains have saturated the soil. Typically, the flowing mass remains covered by a blanket of vegetation, with a *scarp* (steep cut) developing where the moving debris has pulled away from the stationary upper slope.

A landslide may be entirely an earthflow, as in figure 9.7*A,* with debris particles moving past one another roughly parallel to the slope. Commonly, however, rotational sliding takes place above the earthflow as in figures 9.7*B* and 9.8. This example is a *debris slide* (upper part) and an *earthflow* (lower part). In such cases, debris remains in a relatively coherent block or blocks that rotate downward and outward, forcing the debris below to flow.

A hummocky lobe usually forms at the toe or front of the earthflow where debris has accumulated. An earthflow can be active over a period of hours, days, or months; in some earthflows intermittent slow movement continues for years.

Humans can trigger earthflows by adding too much water to soil from septic tank systems or by overwatering lawns. In one case, in Los Angeles, a man departing on a long trip forgot to turn off the sprinkler system for his hillside lawn. The soil became saturated, and both house and lawn were carried downward on an earthflow whose lobe spread out over the highway below.

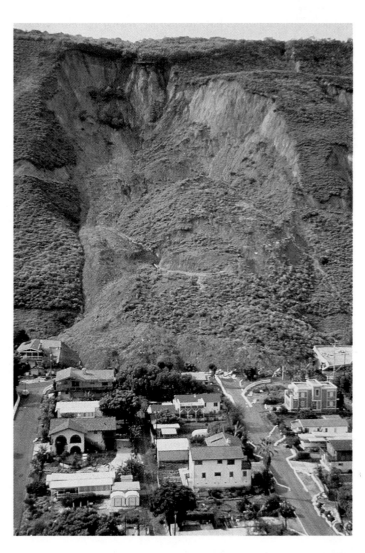

Figure 9.8

Earthflow that destroyed several houses in March, 1995 at La Conchita, California.

Photo by Robert L. Schuster, U.S. Geological Survey

Figure 9.9

Landslide with houses and roads moved on slide blocks. Point Fermin, Los Angeles, California.

Photo by Frank M. Hanna

Ice-saturated debris

A Winter

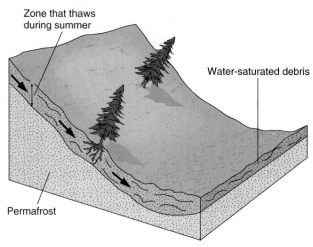

Zone that thaws during summer

Water-saturated debris

Permafrost

B Summer

Figure 9.10

Solifluction due to thawing of ice-saturated debris.

Earthflows, like other kinds of landslides, can be triggered by undercutting at the base of a slope. The undercutting can be caused by waves breaking along shorelines or streams eroding and steepening the base of a slope. Along coastlines, mass wasting commonly destroys buildings (figure 9.9). Entire housing developments and expensive homes built for a view of the ocean are lost. A home buyer who knows nothing of geology may not realize that the sea cliff is there because of the relentless erosion of waves along the shoreline. Nor is the person likely to be aware that a steepened slope creates the potential for landslides.

Bulldozers can undercut the base of a slope more rapidly than wave erosion, and such oversteepening of slopes by human activity has caused many landslides. Unless careful engineering measures are taken at the time a cut is made, roadcuts or platforms carved into hillsides for houses may bring about disaster (figure 9.18).

Solifluction and Permafrost

One variety of earthflow is usually associated with colder climates. **Solifluction** is the flow of water-saturated debris over impermeable material. Because the impermeable material beneath the debris prevents water from draining freely, the debris between the vegetation cover and the impermeable material becomes saturated (figure 9.10). Even a gentle slope is susceptible to movement under these conditions (figure 9.11).

The impermeable material beneath the saturated soil can be either impenetrable bedrock or, as is more common, **permafrost,** ground that remains frozen for many years. Most

Figure 9.11

A railroad built on permafrost terrain in Alaska.

Photo by Lynn A. Yehle, U.S. Geological Survey

9.2
ENVIRONMENTAL GEOLOGY

Los Angeles, A Mobile Society

The following satirical newspaper column was written by humorist Art Buchwald in 1978, a year, like the "El Niño year" of 1998, in which southern California had many landslides because of unusually wet weather.

Los Angeles—I came to Los Angeles last week for rest and recreation, only to discover that it had become a rain forest.

I didn't realize how bad it was until I went to dinner at a friend's house. I had the right address, but when I arrived there was nothing there. I went to a neighboring house where I found a man bailing out his swimming pool.

I beg your pardon, I said. Could you tell me where the Cables live?

"They used to live above us on the hill. Then, about two years ago, their house slid down in the mud, and they lived next door to us. I think it was last Monday, during the storm, that their house slid again, and now they live two streets below us, down there. We were sorry to see them go—they were really nice neighbors."

I thanked him and slid straight down the hill to the new location of the Cables' house. Cable was clearing out the mud from his car. He apologized for not giving me the new address and explained, "Frankly, I didn't know until this morning whether the house would stay here or continue sliding down a few more blocks."

Cable, I said, you and your wife are intelligent people, why do you build your house on the top of a canyon, when you know that during a rainstorm it has a good chance of sliding away?

"We did it for the view. It really was fantastic on a clear night up there. We could sit in our Jacuzzi and see all of Los Angeles, except of course when there were brush fires.

"Even when our house slid down two years ago, we still had a great sight of the airport. Now I'm not too sure what

kind of view we'll have because of the house in front of us, which slid down with ours at the same time."

But why don't you move to safe ground so that you don't have to worry about rainstorms?

"We've thought about it. But once you live high in a canyon, it's hard to move to the plains. Besides, this house is built solid and has about three more good mudslides in it."

Still, it must be kind of hairy to sit in your home during a deluge and wonder where you'll wind up next. Don't you ever have the desire to just settle down in one place?

"It's hard for people who don't live in California to understand how we people out here think. Sure we have floods, and fire and drought, but that's the price you have to pay for living the good life. When Esther and I saw this house, we knew it was a dream come true. It was located right on the tippy top of the hill, way up there. We would wake up in the morning and listen to the birds, and eat breakfast out on the patio and look down on all the smog.

"Then, after the first mudslide, we found ourselves living next to people. It was an entirely different experience. But by that time we were ready for a change. Now we've slid again and we're in a whole new neighborhood. You can't do that if you live on solid ground. Once you move into a house below Sunset Boulevard, you're stuck there for the rest of your life.

"When you live on the side of a hill in Los Angeles, you at least know it's not going to last forever."

Then, in spite of what's happened, you don't plan to move out?

"Are you crazy? You couldn't replace a house like this in L.A. for $500,000."

What happens if it keeps raining and you slide down the hill again?

"It's no problem. Esther and I figure if we slide down too far, we'll just pick up and go back to the top of the hill, and start all over again; that is, if the hill is still there after the earthquake."

Reprinted by permission of the author.

Further Reading
John McPhee's *The control of nature* (see Exploring Resources) contains a factual, and highly readable, account of 1978 landslides in southern California.

---

solifluction takes place in areas of permanently frozen ground, such as in Alaska and northern Canada. Permafrost occurs at depths ranging from a few centimeters to a few meters beneath the surface. The ice in permafrost is a cementing agent for the debris. Permafrost is as solid as concrete.

Above the permafrost is a zone that, if the debris is saturated, is frozen during the winter and indistinguishable from

the underlying permafrost. When this zone thaws during the summer, the water, along with water from rain and runoff, cannot percolate downward through the permafrost, and so the slopes become susceptible to solifluction.

As solifluction movement is not rapid enough to break up the overlying blanket of vegetation into blocks, the water-saturated debris flows downslope, pulling vegetation along

9.1 Astrogeology

Permafrost Ice on Mars

On Earth most underground water is in the liquid form, with the only abundant underground ice occurring as permafrost in the polar regions. In these areas underground ice is responsible for producing several distinctive landforms; most notable are surfaces with polygonal cracks called patterned ground (box figure 1*A*).

On Mars, because of its cooler temperatures, ice is the normal state of underground water and is apparently quite abundant. Many features of the Martian landscape have been attributed to this underground ice, including polygonally patterned ground (box figure 1*B*), rampart craters, and chaotic terrain.

Rampart craters are Martian meteorite craters that are surrounded by material that appears to have flowed from the point of impact. It has been suggested that the underground ice in the surface layers of Mars melted as a result of the impact and, along with rock fragments, sloshed outward from the impact area.

A

Astrogeology Box 9.1 Figure 1

(*A*) Patterned ground in Alaska. Coarse rock forms the edges of polygons; fine-grained material in center supports vegetation. (*B*) Patterned ground on Mars. The polygons in photo *B* are over 100 times larger than those in photo *A*. Note the impact craters.

Photo *A* by C. C. Plummer; Photo *B* NASA

Patches of jumbled and broken angular slabs and blocks called *chaotic terrain* occur in some places on Mars (box figure 2). Some channels originate in these areas, and it is believed that this terrain may be caused by melting of

with it and forming a wrinkled surface. Gradually the debris collects at the base of the slope, where the vegetated surface bulges into a hummocky lobe.

Solifluction is not the only hazard associated with permafrost. Great expanses of flat terrain in arctic and subarctic climates become swampy during the summer because of permafrost, making overland travel very difficult. Building and maintaining roads is an engineering headache. In the preliminary stages of planning the Alaska pipeline, a road was bulldozed across permafrost terrain during the winter, removing the vegetation from the rock-hard ground. It was an excellent truck route during the winter, but when summer came, the road became a quagmire several hundred kilometers long (see Box 1.2, figure 1). The strip can never be used by vehicles as planned, nor will the vegetation return for many decades. Building structures on permafrost terrain presents serious problems. For instance, heat from a build-

ing can melt underlying permafrost; the building then sinks into the mud.

Mudflow

A **mudflow** is a flowing mixture of debris and water, usually moving down a channel (figure 9.12). It can be visualized as a stream with the consistency of a thick milkshake. Most of the solid particles in the slurry are clay and silt (hence the muddy appearance), but coarser sediment commonly is part of the mixture. Usually after a heavy rainfall a slurry of debris and water forms and begins moving down a slope. Most mudflows quickly become channeled into valleys. They then move downvalley like a stream except that, because of the heavy load of debris, they are more viscous. Mud moves more slowly than a stream but, because of its high viscosity, can transport boulders, automobiles, and even locomotives.

B

Astrogeology Box 9.1 Figure 2

Chaotic terrain with associated channels on Mars. Photo is a composite of several images. White rectangles are where images were not taken.

Photo by NASA

permafrost and consequent collapse of the ground. Subsidence due to withdrawal of magma has also been suggested as the cause of chaotic terrain.

Some astrogeologists think that there is a good chance that Mars is the only other planet besides Earth that has life. According to their hypothesis, primitive organisms evolved during an earlier time when Mars had liquid water. The microorganisms might be dormant in the permafrost and perhaps could be revived by melting the ice. This hypothesis received a boost in 1996 when scientists studying a meteorite from Mars reported that it revealed possible signs of former life (see box 12.4). Hardly the alien creatures pictured in science fiction movies or books (e.g., figure 2.16), but organic life nevertheless.

Houses in the path of a mudflow will be filled with mud, if not broken apart and carried away.

Mudflows are most likely to occur in places where debris is not protected by a vegetative cover. For this reason, mudflows are more likely to occur in arid regions than in wet climates. A hillside in a desert environment, where it may not have rained for many years, may be covered with a blanket of loose material. With sparse desert vegetation offering little protection, a sudden thunderstorm with drenching rain can rapidly saturate the loose debris and create a mudflow in minutes.

Mudflows frequently occur on young volcanoes that are littered with ash. Water from heavy rains mixes with pyroclastic debris as at Mount Pinatubo in 1991. Or the water can come from glaciers that are melted by lava or hot pyroclastic

Figure 9.12

A dried mudflow in the Peruvian Andes.

Photo by C. C. Plummer

Figure 9.13

Man examining a 75-meter-long bridge on Washington state highway 504, across the North Fork of the Toutle River. The bridge was washed out by mudflow during the May 18, 1980 eruption of Mount St. Helens. The steel structure was carried about 0.5 kilometers downstream and partially buried by the mudflow.

Photo by Robert L. Schuster, U.S. Geological Survey

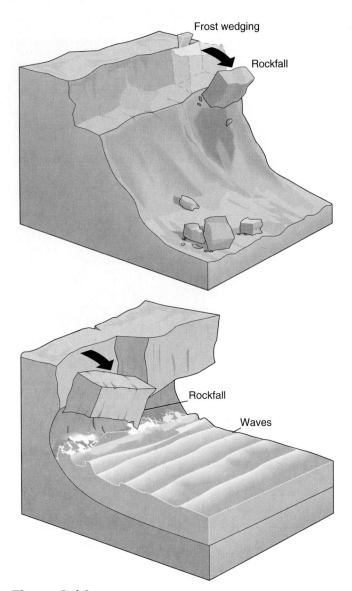

Figure 9.14

Two examples of rockfall.

debris, as occurred at Mount St. Helens in 1980 (figure 9.13) and at Colombia's Nevado del Ruiz in 1985, which cost 23,000 lives. Mudflows also occur after forest fires destroy slope vegetation that normally anchors soil in place. Burned-over slopes are extremely vulnerable to mudflow if heavy rains fall before the vegetation is restored.

The year 1978 was particularly bad for debris flows in southern California. One mudflow roared through a Los Angeles suburb carrying almost as many cars as large boulders. A sturdily built house withstood the onslaught but began filling with muddy debris. Two of its occupants were pinned to the wall of a bedroom and could do nothing as the room filled slowly with mud. The mud stopped rising just as it was reaching their heads. Hours later they were rescued. (John McPhee's *The Control of Nature,* listed at the end of this chapter, is a highly readable account of this and other debris flows in southern California.)

Debris Avalanche

The fastest variety of debris flow is a **debris avalanche,** a very rapidly moving, turbulent mass of debris, air, and water. The best modern example is the one that buried Yungay (see Box 9.1, pgs. 201–202).

Some geologists have suggested that in very rapidly moving rock avalanches, air trapped under the rock mass creates an air cushion that reduces friction. This could explain why some landslides reach speeds of several hundred kilometers per hour. But other geologists have contended that the rock mass is too turbulent to permit such an air cushion to form.

Rockfalls and Rockslides

Rockfall

When a block of bedrock breaks off and falls freely or bounces down a cliff, it is a **rockfall** (figure 9.14). Cliffs may form naturally by the undercutting action of a river, wave action, or glacial erosion. Highway or other construction projects may also oversteepen slopes. Bedrock commonly has cracks (joints) or other planes of weakness such as foliation (in metamorphic rocks) or sedimentary bedding planes. Blocks of rock will break off along these planes. In colder climates rock is effectively broken apart by frost wedging (as explained in chapter 5).

Commonly, an apron of fallen rock fragments, called **talus,** accumulates at the base of a cliff (figure 9.15).

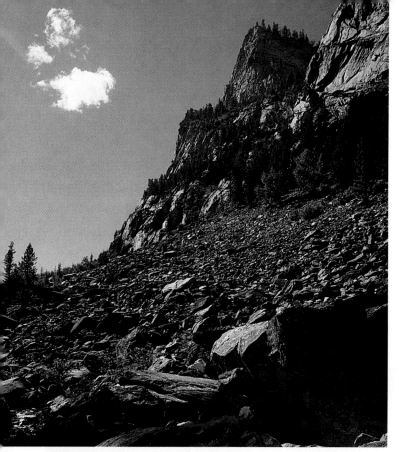

Figure 9.15
Talus.
Photo by C. C. Plummer

Figure 9.16
Small dust clouds linger high above Yosemite Valley where rock slabs broke loose and fell to the valley floor which, upon impact, created the debris-laden blast of air climbing up the other side of the valley. The photo was taken by a rock climber on a nearby cliff.
Photo by Ed Youmans

A spectacular rockfall took place in Yosemite National Park in the summer of 1996, killing one man and injuring several other people. The rockfall originated from near Glacier Point (the place where the photo for figure 12.1 was taken). Two huge slabs (weighing approximately 80,000 tons) of an overhanging arch broke loose just seconds apart. (The arch was a product of exfoliation, and broke loose along a sheet joint—see chapter 5.) The slabs slid a short distance over steep rock from which they were launched outward, as if from a ski jump, away from the vertical cliffs. The slabs fell free for around 500 meters (1700 feet) and hit the valley floor 30 meters out from the base of the cliff (you would not have been hit if you were standing at the base of the cliff). They shattered upon impact and created a dust cloud (figure 9.16) that obscured visibility for hours. A powerful air blast was created as air between the rapidly falling rock and the ground was compressed. The debris-laden wind felled a swath of trees between the newly deposited talus and a nature center building.

Rockslide

A **rockslide** is, as the term suggests, the rapid sliding of a mass of bedrock along an inclined surface of weakness, such as a bedding plane (figure 9.17), a major fracture in the rock, or a foliation plane (box 9.3). Once sliding begins, a rock slab usually breaks up into rubble. Like rockfalls, rockslides can be caused by undercutting at the base of the slope from erosion or construction.

Some rockslides travel only a few meters before halting at the base of a slope. In country with high relief, however, a rockslide may travel hundreds or thousands of meters before reaching a valley floor. If movement becomes very rapid, the rockslide may break up and become a rock avalanche. A **rock avalanche** is a very rapidly moving, turbulent mass of broken-up bedrock. Movement in a rock avalanche is flowage on a grand scale. The only difference between a rock avalanche and a debris avalanche is that a rock avalanche begins its journey as bedrock.

Ultimately, a rockslide or rock avalanche comes to rest as the terrain becomes less steep. Sometimes the mass of rock fills the bottom of a valley and creates a natural dam. If the rock mass suddenly enters a lake or bay, it can create a huge wave that destroys lives and property far beyond the area of the original landslide.

An example is a disastrous landslide that took place in northern Italy in 1963. A huge layer of limestone broke loose parallel to its bedding planes. The translational slide involved around 250 million cubic meters that slid into the Vaiont Reservoir creating a giant wave. The 175-meter (almost two football fields) high wave overtopped the Vaiont Dam (it was the world's highest dam, rising 265 meters above the valley floor). Three thousand people were killed in the villages that it flooded in the valleys below. The dam was not destroyed, a tribute to excellent engineering, but the men in charge of the building project were convicted of criminal negligence for ignoring the landslide hazards.

A

B

Figure 9.17

(*A*) and (*B*) Diagram of the Gros Ventre, Wyoming, slide. (*C*) Photo of Gros Ventre slide.

A and *B* after W. C. Alden. U.S. Geological Survey. Photo C by D. A. Rahm, courtesy of Rahm Memorial Collection, Western Washington University

As in slower mass movements, water can play an important role in causing a rockslide. In 1925, exceptionally heavy rains in the Gros Ventre Mountains of Wyoming caused water to seep into a layer of sandstone, wetting the underlying layer of shale, greatly reducing its shear strength (figure 9.17). The layers of sedimentary rock were inclined roughly parallel to the hillside. With the wet shale acting as a lubricant, the overlying sedimentary rock and its soil cover slid into the valley, blocking the river. The slide itself merely created a lake, but the natural dam broke two years later and the resulting flood destroyed the small town of Kelly several kilometers downstream. Several res-

idents who were standing on a bridge watching the floodwaters come down the valley were killed.

Debris Slides and Debris Falls

As the names suggest, debris slides and debris falls behave similarly to rockslides and rockfalls, except that they involve debris that moves as a coherent mass (at least initially). A **debris fall** is a free-falling mass of debris.

A **debris slide** is a coherent mass of debris moving along a well-defined surface (or surfaces). If the movement is along a curved surface the landslide is a *rotational debris slide.* Debris slides were mentioned earlier with earthflows, with which they are commonly associated (figure 9.7). Debris may slide, however, without an earthflow taking place.

Preventing Landslides

Preventing Mass Wasting of Debris

Usually mass movements of debris can be prevented. Proper engineering is essential when the natural environment of a hillside is altered by construction. As shown in figure 9.18, construction generally makes a slope more susceptible to mass wasting of debris in several ways: (1) the base of the slope is undercut, removing the natural support for the upper part of the slope; (2) vegetation is removed during construction; (3) buildings constructed on the upper part of a slope add weight to the potential slide; and (4) extra water may be allowed to seep into the debris.

Some preventive measures can be taken during construction. A retaining wall is usually built where a cut has been made in the slope, but this alone is seldom as effective a

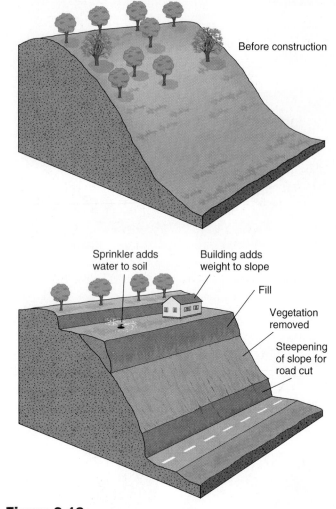

Figure 9.18

A hillside becomes vulnerable to mass wasting due to construction activities.

Before construction

Sprinkler adds water to soil

Building adds weight to slope

Fill

Vegetation removed

Steepening of slope for road cut

Water trapped in soil causes movement, pushing down retaining wall.

A

Water drains through pipe, allowing wall to keep slope from moving.

B

Figure 9.19

Use of drains to help prevent mass wasting.

deterrent to down-slope movement as people hope. If, in addition, drain pipes are put through the retaining wall and into the hillside, water can percolate through and drain away rather than collecting in the debris behind the wall (figure 9.19). Without drains, excess water results in decreased shear strength and the whole soggy mass can easily burst through the wall.

Another practical preventive measure is to avoid over-steepening the slope. The hillside can be cut back in a series of terraces rather than in a single steep cut. This not only reduces the slope angle but also reduces the shear force by removing much of the overlying material. It also prevents loose material (such as boulders dislodged from the top of the cut) from rolling to the base. Road cuts constructed in this way are usually reseeded with rapidly growing grass or plants whose roots help anchor the slope. A vegetation cover also minimizes erosion from running water.

Preventing Rockfalls and Rockslides on Highways

Rockslides and rockfalls are a major problem on highways built through mountainous country. Steep slopes and cliffs are created when road cuts are blasted and bulldozed into mountain sides. If the bedrock has planes of weakness (such as joints, bedding planes, or foliation planes), the orientation of these planes relative to the road cut determines whether there is a rockslide hazard (as in figure 9.20). If the planes of weakness are inclined into the hill, there is no chance of a rockslide. On the other hand, where the planes of weakness are approximately parallel to the slope of the hillside, a rock-slide may occur.

Various techniques are used to prevent rockslides. By doing a detailed geologic study of an area before a road is built,

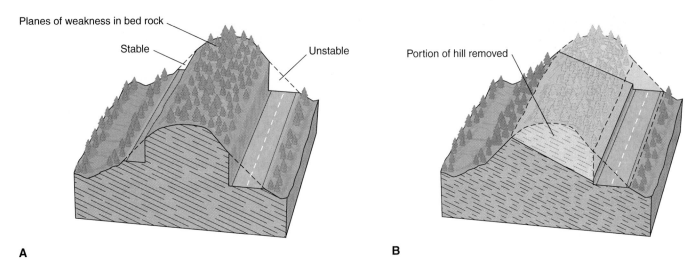

A **B**

Figure 9.20

(*A*) Cross section of a hill showing a relatively safe road cut on the left and a hazardous road cut on the right. (*B*) The same hazardous road cut after removal of rock that might slide.

builders might avoid a hazard by choosing the least dangerous route for the road. If a road cut must be made through bedrock that appears prone to sliding, all of the rock that might slide could be removed (sometimes at great expense), as shown in figure 9.20*B*.

In some instances, slopes prone to rock sliding have been "stitched" in place by the technique shown in figure 9.21.

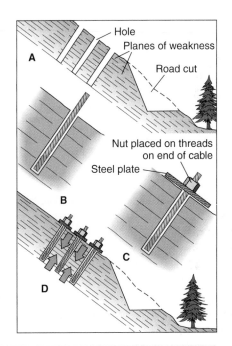

Figure 9.21

"Stitching" a slope to keep bedrock from sliding along planes of weakness. (*A*) Holes are drilled through unstable layers into stable rock. (*B*) Expanded view of one hole. A cable is fed into the hole and cement is pumped into the bottom of the hole and allowed to harden. (*C*) A steel plate is placed over the cable and a nut tightened. (*D*) Tightening all the nuts pulls unstable layers together and anchors them in stable bedrock. (*E*) Stabilized road cut along Richardson Highway near Valdez, Alaska.

Photo by Paul G. Bauer

E

Summary

Mass wasting is the movement of a mass of debris (soil and loose rock fragments) or bedrock toward the base of a slope. Movement can take place as a flow, slide, or fall. Gravity is the driving force. The component of gravitational force that propels mass wasting is the *shear force,* which occurs parallel to the slopes. The resistance to that force is the *shear strength* of rock or debris. If shear force exceeds shear strength, mass wasting takes place. Water is an important factor in mass wasting.

A number of other factors determine whether movement will occur and, if it does, the rate of movement.

The slowest type of movement, *creep,* occurs mostly on relatively gentle slopes, usually aided by water in the soil. In colder climates, repeated freezing and thawing of water within the soil contributes to creep. *Landsliding* is a general term for more rapid mass wasting of rock, debris, or both. *Debris flows* include earthflows, mudflows, and debris avalanches. *Earthflows* vary greatly in velocity although they are not as rapid as *debris avalanches,* which are turbulent masses of debris, water, and air. *Solifluction,* a special variety of earthflow, usually takes place in arctic or subarctic climates where ground is permanently frozen (*permafrost*). A *mudflow* is a slurry of debris, and water. Most mudflows flow in channels much as streams do.

Rockfall is the fall of broken rock down a vertical or near-vertical slope. A *rockslide* is a slab of rock sliding down a less-than-vertical surface. *Debris falls* and *debris slides* involve unconsolidated material rather than bedrock.

Terms to Remember

creep 205
debris 200
debris avalanche 212
debris fall 214
debris flow 206
debris slide 214
earthflow 207
fall 202

flow 202
landslide 200
mass wasting 200
mudflow 210
permafrost 208
relief 203
rock avalanche 213
rockfall 212

rockslide 213
rotational slide (slump) 203
shear force 204
shear strength 204
slide 202
solifluction 208
talus 212
translational slide 202

Testing Your Knowledge

Use the questions below to prepare for exams based on this chapter.

1. Describe the effect on shear strength of the following: (a) thickness of debris; (b) orientation of planes of weakness; (c) water in debris; and (d) vegetation.

2. Compare the shear force to the force of gravity (drawing diagrams similar to figure 9.2) for the following situations: (a) a vertical cliff; (b) a flat horizontal plane; and (c) a 45° slope.

3. How does a rotational slide differ from a translational slide?

4. What role does water play in each of the types of mass wasting?

5. Why is solifluction more common in colder climates than in temperate climates?

6. List and explain the key factors that control mass wasting.

7. What is the slowest type of mass wasting process? (a) debris flow (b) rockslide (c) creep (d) rockfall (e) avalanche

8. Any unconsolidated material at the earth's surface of any size is called (a) debris (b) sediment (c) soil (d) talus

9. A descending mass moving downslope as a viscous fluid is referred to as a (a) fall (b) landslide (c) flow (d) slide

10. The driving force behind all mass wasting processes is (a) gravity (b) slope angle (c) type of bedrock material (d) presence of water (e) vegetation

11. The resistance to movement or deformation of debris is its (a) mass (b) shear strength (c) shear force (d) density

12. Flow of water-saturated debris over impermeable material is called (a) solifluction (b) flow (c) slide (d) fall

13. A flowing mixture of debris and water, usually moving down a channel is called a (a) mudflow (b) slide (c) fall (d) debris flow

14. An apron of fallen rock fragments that accumulates at the base of a cliff is called (a) debris (b) sediment (c) soil (d) talus

15. How does construction destabilize a slope? (a) adds weight to the top of the slope (b) decreases water content of the slope (c) adds weight to the bottom of the slope (d) increases the shear strength of the slope

16. How can landslides be prevented during construction? (choose all that apply) (a) retaining walls (b) cut steeper slopes (c) install water drainage systems (d) add vegetation

Expanding Your Knowledge

1. Why do people fear earthquakes, hurricanes, and tornadoes more than they fear landslides?
2. If you were building a house on a cliff, what would you look for to ensure that your house would not be destroyed through mass wasting?
3. Why isn't the land surface of the earth flat after millions of years of erosion by mass wasting as well as by other erosional agents?
4. Can any of the indicators of creep be explained by processes other than mass wasting?

Exploring Resources

Brabb, E. E., and B. L. Harrod. 1989. *Landslides: Extent and economic significance.* Brookfield, Vermont: A. A. Balkema.

Dennen, W. H., and B. R. Moore. 1986. *Geology and engineering.* Dubuque, Iowa: Wm. C. Brown Publishers.

McPhee, J. 1989. *The control of nature.* New York: Noonday Press.

Nuhfer, E. B., R. J. Proctor, and P. H. Moser. 1993. *The citizens guide to geologic hazards.* Arvada, Colorado: American Institute of Professional Geologists.

Ritter, D. F., R. C. Kochel, and J. R. Miller. 1995. *Process geomorphology.* 3d ed. Dubuque, Iowa: Wm. C. Brown Publishers.

Turner, A. K., and R. L. Schuster, eds. 1996. *Landslides: investigation and mitigation.* Transport Research Board Special Report 247. Washington, D.C.: National Academy of Sciences.

In the path of a killer volcano. Films for the Humanities & Sciences, Princeton, N.J. Contains footage of volcanic mudflow associated with the eruption of Mount Pinatubo, Philippines.

http://geohazards.cr.usgs.gov/html_files/nlicsun.html
USGS, National Landslide Information Center. Includes images of landslides and how to get information on landslides for a particular state.

http://www.NRCan.gc.ca/gsc/cpdnew/lndsld_e.html
Landslide hazards in Canada.

Interacting with *Journey Through Geology* CD-ROM

Expand your knowledge of the concepts presented in this chapter by using the CD-ROM to answer the following question.

1. Convergent Margins, Alaska, Earthquake Footage: Earthquakes often trigger landslides. Look for evidence of mass wastage in this footage of the 1964 Alaskan earthquake. What type of mass wastage happened here?

Grand Forks, North Dakota during the 1997 flooding of the Red River.
Photo by Brooks Kraft/Sygma

CHAPTER
10

Streams and Floods

*R*unning water, aided by mass wasting, is the most important geologic agent in eroding, transporting, and depositing sediment. Almost every landscape on earth shows the results of stream erosion or deposition. Although other agents—ground water, glaciers, wind, and waves—can be locally important in sculpturing the land, stream action and mass wasting are the dominant processes of landscape development.

The first part of this chapter deals with the various ways that streams erode, transport, and deposit sediment. The second part describes landforms produced by stream action, such as valleys, flood plains, deltas, and alluvial fans, and shows how each of these is related to changes in stream characteristics. The chapter also includes a discussion of the causes and effects of flooding, and various measures used to control flooding.

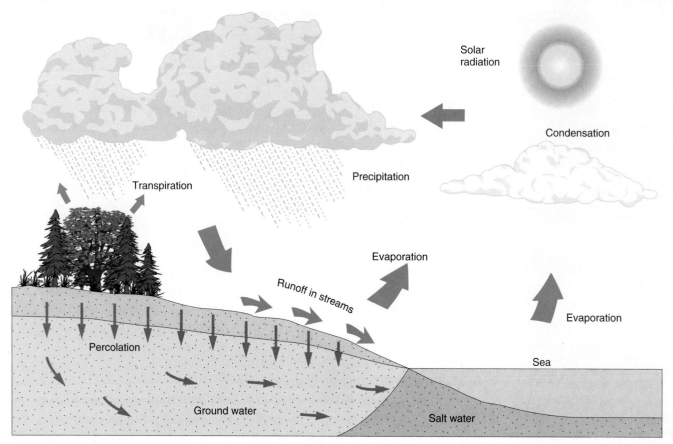

Figure 10.1

The hydrologic cycle. Water vapor evaporates from the land and sea, condenses to form clouds, and falls as precipitation (rain and snow). Water falling on land runs off over the surface as streams, or percolates into the ground to become ground water. It returns to the atmosphere again by evaporation and transpiration (the loss of water to the air by plants).

The Hydrologic Cycle

The movement of water and water vapor from the sea to the atmosphere, to the land, and back to the sea and atmosphere again is called the **hydrologic cycle** (figure 10.1). When rain (or snow) falls on the land surface, more than half the water returns rather rapidly to the atmosphere by evaporation or by transpiration from plants. The remainder either flows over the land surface as runoff in streams or percolates down into the ground to become ground water.

Only about 15% to 20% of rainfall normally ends up as surface runoff in rivers, although the amount of runoff can range from 2% to more than 25% with variations in climate, steepness of slope, soil and rock type, and vegetation. Steady, continuous rains can saturate the ground and the atmosphere, however, and lead to floods as runoff approaches 100% of rainfall.

Channel Flow and Sheet Flow

A **stream** is a body of running water that is confined in a channel and moves downhill under the influence of gravity. In some parts of the country, *stream* implies size: rivers are large, streams somewhat smaller, and brooks or creeks even smaller.

Geologists, however, use *stream* for any body of running water, from a small trickle to a huge river.

Figure 10.2*A* shows a *longitudinal profile* of a typical stream viewed from the side. The stream begins in steep mountains and flows out across a gentle plain into the sea. The *headwaters* of a stream are the upper part of the stream near its source in the mountains. The *mouth* is the place where a stream enters the sea, a lake, or a larger stream. A *cross section* of a stream in steep mountains is usually a V-shaped valley cut into solid rock, with the stream channel occupying the narrow bottom of the valley; there is little or no flat land next to the stream on the valley bottom (figure 10.2*B*). Near its mouth a stream usually flows within a broad, flat-floored valley. The stream channel is surrounded by a flat *flood plain* of sediment deposited by the stream (figure 10.2*C*).

A stream normally stays in its **stream channel,** a long narrow depression eroded by the stream into rock or sediment. The stream *banks* are the sides of the channel; the stream *bed* is the bottom of the channel. During a flood the waters of a stream may rise and spill over the banks onto the flat flood plain of the valley floor (figure 10.3).

Not all water that moves over the land surface is confined to channels. Sometimes, particularly during heavy rains, water

A Longitudinal profile (dark blue line) of a stream beginning in mountains and flowing across a plain into the sea.

B Cross section of the stream at point B. The channel is at the bottom of a V-shaped valley cut into rock.

C Cross section at point C. The channel is surrounded by a broad flood plain of sediment.

Figure 10.2

Longitudinal and cross sections of a typical stream.

A B

Figure 10.3

A stream normally stays in its channel, but during a flood it can spill over its banks onto the adjacent flatland. Todd Valley in New Zealand during normal discharge (*A*) and also in flood stage when 15 centimeters (6 inches) of rain fell in 3 hours (*B*).

Photos © G.R. 'Dick' Roberts

runs off as **sheetwash,** a thin layer of unchanneled water flowing downhill. Sheetwash is particularly common in deserts, where the lack of vegetation allows rainwater to spread quickly over the land surface. It also occurs in humid regions during heavy thunderstorms when water falls faster than it can soak into the ground. A series of closely spaced storms can also promote sheetwash; as the ground becomes saturated, more water runs over the surface.

Sheetwash, along with the violent impact of raindrops on the land surface, can produce considerable *sheet erosion,* in which a thin layer of surface material, usually topsoil, is removed by the flowing sheet of water. This gravity-driven

Figure 10.4

The drainage basin of the Mississippi River is the land area drained by the river and all its tributaries, including the Ohio and Missouri Rivers; it covers more than 1 million square miles. Heavy rain in any part of the basin can cause flooding on the lower Mississippi River in the states of Mississippi and Louisiana. The Continental Divide separates rivers that flow into the Pacific from rivers that flow into the Atlantic and the Gulf of Mexico.

movement of sediment is a process intermediate between mass wasting and stream erosion.

Overland sheetwash becomes concentrated in small channels, forming tiny streams called *rills*. Rills merge to form small streams, and small streams join to form larger streams. Most regions are drained by networks of coalescing streams.

Drainage Basins

Each stream, small or large, has a **drainage basin,** the total area drained by a stream and its tributaries (a **tributary** is a small stream flowing into a larger one). A drainage basin can be outlined on a map by drawing a line around the region drained by all the tributaries to a river (figure 10.4). The Mississippi River's drainage basin, for example, includes all the land area drained by the Mississippi River itself and by all its tributaries, including the Ohio and Missouri Rivers. This great drainage system includes more than one-third the land area of the contiguous 48 states.

A ridge or strip of high ground dividing one drainage basin from another is termed a **divide** (figure 10.4). The best known in the United States is the Continental Divide, a line separating streams that flow to the Pacific Ocean from those that flow to the Atlantic and the Gulf of Mexico. The Continental Divide, which extends from the Yukon Territory down into Mexico, crosses Montana, Idaho, Wyoming, Colorado, and New Mexico in the United States. Signs indicating the crossing of the Continental Divide have been placed at numerous points where major highways intersect the divide.

Drainage Patterns

The arrangement, in map view, of a river and its tributaries is a **drainage pattern.** A drainage pattern can in many cases reveal the nature and structure of the rocks underneath it.

Most tributaries join the main stream at an acute angle, forming a V (or Y) pointing downstream. If the pattern resembles branches of a tree or veins in a leaf, it is called **dendritic** (fig-

A Dendritic

B Radial

C Rectangular

Ridge — Valley

Fractures

D Trellis

Figure 10.5

Drainage patterns can reveal something about the rocks underneath. (*A*) Dendritic pattern develops on uniformly erodible rock. (*B*) A radial pattern develops on a conical mountain or dome. (*C*) A rectangular pattern develops on regularly fractured rock. (*D*) A trellis pattern develops on alternating ridges and valleys caused by the erosion of resistant and nonresistant tilted rock layers.

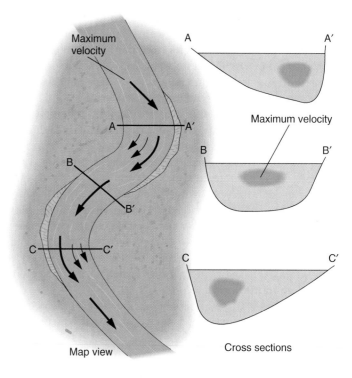

Maximum velocity

Maximum velocity

Map view

Cross sections

Figure 10.6

Regions of maximum velocity in a stream. Arrows on the map show how the maximum velocity shifts to the outside of curves. Sections show maximum velocity on outside of curves and in the center of the channel on a straight stretch of stream.

where tiled layers of resistant rock such as sandstone alternate with nonresistant rock such as shale. Erosion of such a region results in a surface topography of parallel ridges and valleys.

Factors Affecting Stream Erosion and Deposition

Stream erosion and deposition are controlled primarily by a river's *velocity* and, to a lesser extent, by its *discharge*. Velocity is largely controlled by the stream *gradient,* channel shape, and channel roughness.

Velocity

The distance water travels in a stream per unit time is called the **stream velocity.** A moderately fast river flows at about 5 kilometers per hour (3 miles per hour). Rivers flow much faster during flood, sometimes exceeding 25 kilometers per hour (15 miles per hour).

The cross-sectional views of a stream in figure 10.6 show that a stream reaches its maximum velocity near the middle of the channel. Near the stream's banks and bed, friction between the water and the stream channel slows the water. When a stream goes around a curve, the region of maximum velocity is

ures 10.4 and 10.5*A*). Dendritic drainage patterns develop on uniformly erodible rock or regolith, and are the most common type of pattern. A **radial pattern,** in which streams diverge outward like spokes of a wheel, forms on high conical mountains, such as composite volcanoes and domes (figure 10.5*B*). A **rectangular pattern,** in which tributaries have frequent 90° bends and tend to join other streams at right angles, develops on regularly fractured rock (figure 10.5*C*). A network of fractures meeting at right angles forms pathways for streams because fractures are eroded more easily than unbroken rock. A **trellis pattern** consists of parallel main streams with short tributaries meeting them at right angles (figure 10.5*D*). A trellis pattern forms in a region displaced by centrifugal force toward the outside of the curve. Velocity is the key factor in a stream's ability to erode, transport, and deposit. High velocity (meaning greater energy) generally results in erosion and transportation; low velocity causes sediment deposition. Slight changes in velocity can cause great changes in the sediment load carried by the river.

Figure 10.7 shows the stream velocities at which sediments are eroded, transported, and deposited. For each grain size, these velocities are different. The upper curve represents the minimum velocity needed to erode sediment grains. This curve shows the velocity at which previously stationary grains are first picked up by moving water. The lower curve represents the velocity below which deposition occurs, when moving grains come to rest. Between the two curves the water is moving fast enough to transport grains that have already been eroded. Note that it takes a higher stream velocity to erode grains (set them in motion) than to transport grains (keep them in motion).

Point A on figure 10.7 represents fine sand on the bed of a stream that is barely moving. The vertical red arrows represent a flood with gradually increasing stream velocity. No sediment moves until the velocity is high enough to intersect the *upper* curve and move into the area marked "erosion." As the flood recedes, the velocity drops below the upper curve and into the transportation area. Under these conditions the sand that was already eroded continues to be transported, but no new sand is eroded. As the velocity falls below the lower curve, all the sand is deposited again, coming to rest on the stream bed.

The right half of the diagram shows that coarser particles require progressively higher velocities for erosion and transportation, as you might expect (boulders are harder to move than sand grains). The erosion curve also rises toward the left of the diagram, however. This shows that fine-grained silt and clay are actually harder to erode than sand. The reason is that molecular forces tend to bind silt and clay into a smooth, cohesive mass that resists erosion. Once silt and clay are eroded, however, they are easily transported. As you can see from the lower curve, the silt and clay in a river's suspended load are not deposited until the river virtually stops flowing.

Gradient

One factor that controls a stream's velocity is the **stream gradient,** the downhill slope of the bed (or of the water surface, if the stream is very large). A stream gradient is usually measured in feet per mile in the United States, because these units are used on U.S. maps (elsewhere, gradients are expressed in meters per kilometer). A gradient of 5 feet per mile means that the river drops 5 feet vertically for every mile that it travels horizontally. Mountain streams may have gradients as steep as 50 to 200 feet per mile (10 to 40 m/km). The lower Mississippi River has a very gentle gradient, 0.5 foot/mile (0.1 m/km) or less.

A stream's gradient usually decreases downstream. Typically, the gradient is greatest in the headwater region and decreases toward the mouth of the stream (see figure 10.2). Local increases in the gradient of a stream are usually marked by rapids.

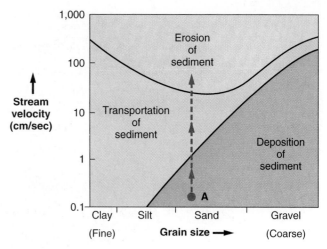

Figure 10.7

Logarithmic graph showing the stream velocities at which erosion and deposition of sediment occur. These velocities vary with the grain size of the sediment. See text for a discussion of point *A* and the dashed red line above it.

Channel Shape and Roughness

The *shape of the channel* also controls stream velocity. Flowing water drags against the stream banks and bed, and the resulting friction slows the water down. In figure 10.8, the streams in *A* and *B* have the same cross-sectional area, but stream *B* flows slower than *A* because the wide, shallow channel in *B* has more surface for the moving water to drag against.

A stream may change its channel width as it flows across different rock types. Hard, resistant rock is difficult to erode, so a stream may have a relatively narrow channel in such rock. As a result it flows rapidly (figure 10.9*A*). If the stream flows onto a softer rock that is easier to erode, the channel may widen, and the river will slow down because of the increased surface area dragging on the flowing water. Sediment may be deposited as the velocity decreases.

The width of a stream may be controlled by factors external to the stream. A landslide may carry debris onto a valley floor, partially blocking a stream's channel (figure 10.9*B*). The constriction causes the stream to speed up as it flows past the slide, and the increased velocity may quickly erode the landslide debris, carrying it away downstream. Human interference with a river can promote erosion and deposition. Construction of a culvert or bridge can partially block a channel, increasing the stream's velocity (figure 10.9*C*). If the bridge was poorly designed, it may increase velocity to the point where erosion may cause the bridge to collapse.

The *roughness of the channel* also controls velocity. A stream can flow rapidly over a smooth channel, but a rough, boulder-strewn channel floor creates more friction and slows the flow (see figure 10.8*C*). Coarse particles increase the roughness more than fine particles, and a rippled or wavy sand bottom is rougher than a smooth sand bottom.

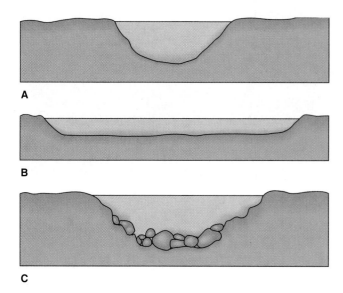

Figure 10.8

Channel shape and roughness influence stream velocity.
(A) Semi-circular channel allows stream to flow rapidly. (B) Wide, shallow channel increases friction, slowing river down. (C) Rough, boulder-strewn channel slows river.

Discharge

The **discharge** of a stream is the volume of water that flows past a given point in a unit of time. It is found by multiplying the cross-sectional area of a stream by its velocity (or width × depth × velocity). Discharge can be reported in cubic feet per second (cfs), which is standard in the United States, or in cubic meters per second (m³/sec).

$$\text{Discharge (cfs)} = \text{average stream width (ft)}$$
$$\times \text{ average depth (ft)}$$
$$\times \text{ average velocity (ft/sec)}$$

A stream 100 feet wide and 15 feet deep flowing at 4 miles per hour (6 ft/sec) has a discharge of 9,000 cubic feet per second (cfs). In streams in humid climates, discharge increases downstream for two reasons: (1) water flows out of the ground into the river through the stream bed; and (2) small tributary streams flow into a larger stream along its length, adding water to the stream as it travels.

To handle the increased discharge, these streams increase in width and depth downstream. Some streams surprisingly increase slightly in velocity downstream, as a result of the increased discharge (the increase in discharge and channel size, and the typical downstream smoothness of the channel, override the effect of a lessening gradient).

During floods a stream's discharge and velocity increase, usually as a result of heavy rains over the stream's drainage basin. Flood discharge may be 50 to 100 times normal flow. Stream erosion and transportation generally increase enormously as a result of a flood's velocity and discharge. Swift

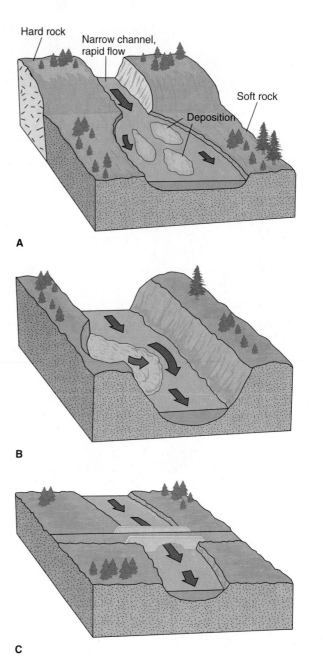

Figure 10.9

Channel width variations caused by rock type and obstructions. Length of arrow indicates velocity. (A) A channel may widen in soft rock. Deposition may result as stream velocity drops.
(B) Landslide may narrow a channel, increasing stream velocity. Resulting erosion usually removes landslide debris. (C) Bridge piers (or other obstructions) will increase velocity and sometimes erosion next to the piers.

mountain streams in flood can sometimes move boulders the size of automobiles (figure 10.10). Flooded areas may be intensely scoured, with river banks and adjacent lawns and fields washed away. As floodwaters recede, both velocity and discharge decrease, leading to the deposition of a blanket of sediment, usually mud, over the flooded area.

Figure 10.11

Hydraulic action can loosen, roll and lift grains from the stream bed.

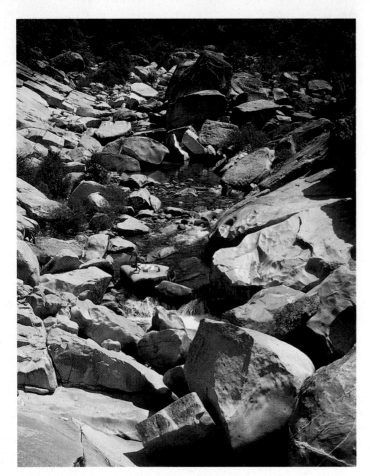

Figure 10.10

These large boulders of granite in a mountain stream are moved only during floods. Note the rounding of the boulders and the scoured high-water mark of floods on the valley walls. Note people for scale.

In a dry climate, a river's discharge can decrease in a downstream direction as river water evaporates into the air and soaks into the dry ground (or is used for irrigation). As the discharge decreases, the load of sediment is gradually deposited.

Stream Erosion

A stream usually erodes the rock and sediment over which it flows. In fact, streams are one of the most effective sculptors of the land. Streams cut their own valleys, deepening and widening them over long periods of time and carrying away the sediment that mass wasting delivers to valley floors. The particles of rock and sediment that a stream picks up are carried along to be deposited farther downstream. Streams erode rock and sediment in three ways—*hydraulic action, solution,* and *abrasion.*

Hydraulic action refers to the ability of flowing water to pick up and move rock and sediment (figure 10.11). The force of running water swirling into a crevice in a rock can crack the rock and break loose a fragment to be carried away by the

stream. Hydraulic force can also erode loose material from a stream bank on the outside of a curve. The pressure of flowing water can roll or slide grains over a stream bed, and a swirling eddy of water may exert enough force to lift a rock fragment above a stream bed. The great force of falling water makes hydraulic action particularly effective at the base of a waterfall (figure 10.12), where it may erode a deep plunge pool. You may be able to hear the results of hydraulic action by standing beside a swift mountain stream and listening to boulders and cobbles hitting one another as they tumble along downstream.

From what you have learned about weathering, you know that some rocks can be dissolved by water. **Solution,** although ordinarily slow, can be an effective process of weathering and erosion (weathering because it is a response to surface chemical conditions; erosion because it removes material). A stream flowing over limestone, for example, gradually dissolves the rock, deepening the stream channel. A stream flowing over other sedimentary rocks, such as sandstone, can dissolve calcite cement, loosening grains that can then be picked up by hydraulic action.

The erosive process that is usually most effective on a rocky stream bed is **abrasion,** the grinding away of the stream channel by the friction and impact of the sediment load. Sand and gravel tumbling along near the bottom of a stream wear away the stream bed much as moving sandpaper wears away wood. The abrasion of sediment on the stream bed is generally much more effective in wearing away the rock than hydraulic action alone. The more sediment a stream carries, the faster it is likely to wear away its bed.

The coarsest sediment is the most effective in stream erosion. Sand and gravel strike the stream bed frequently and with great force, while the finer-grained silt and clay particles weigh so little that they are easily suspended throughout the stream and have little impact when they hit the channel.

Potholes are depressions that are eroded into the hard rock of a stream bed by the abrasive action of the sediment load (figure 10.13). During high water when a stream is full, the swirling water can cause sand and pebbles to scour out smooth, bowl-shaped depressions in hard rock. Potholes tend

Figure 10.12

Hydraulic action can erode a deep plunge pool at the base of a waterfall. Niagara Falls.

Natural Science Est., Inc., Rochester, N.Y.

to form in spots where the rock is a little weaker than the surrounding rock. Although potholes are fairly uncommon, you can see them on the beds of some streams at low water level. Potholes may contain sand or an assortment of beautifully rounded pebbles.

Stream Transportation of Sediment

The sediment load transported by a stream can be subdivided into *bed load, suspended load,* and *dissolved load.* Most of a stream's load is carried in suspension and in solution.

The **bed load** is the large or heavy sediment particles that travel on the stream bed (figure 10.14). Sand and gravel, which form the usual bed load of streams, move by either *traction* or *saltation.*

Large, heavy particles of sediment, such as cobbles and boulders, may never lose contact with the stream bed as they move along in the flowing water. They roll or slide along the stream bottom, eroding the stream bed and each other by abrasion. Movement by rolling, sliding, or dragging is called **traction.**

Sand grains move by traction, but they also move downstream by **saltation,** a series of short leaps or bounces off the bottom (see figure 10.14). Saltation begins when sand grains are momentarily lifted off the bottom by *turbulent* water (eddying, swirling flow). The force of the eddying water temporarily counteracts the downward force of gravity, suspending the grains in water above the stream bed. The water soon slows down because the velocity of water in an eddy is not constant; then gravity overcomes the lift of the water, and the sand grain

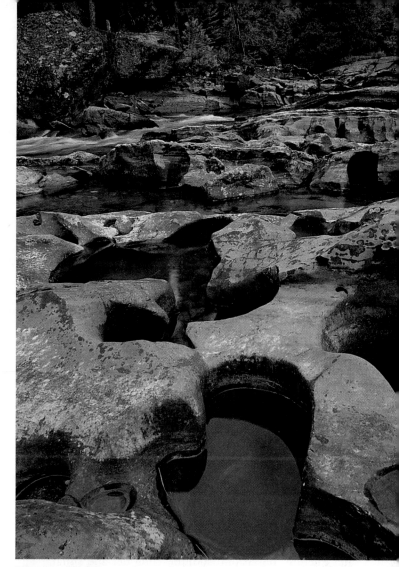

Figure 10.13

Potholes scoured along bed of McDonald River in Glacier National Park, Montana.

Photo © Joe McDonald/Visuals Unlimited

once again falls to the bed of the stream. While it is suspended, the grain moves downstream with the flowing water. After it lands on the bottom, it may be picked up again if turbulence increases, or it may be thrown up into the water by the impact of another falling sand grain. In this way sand grains saltate downstream in leaps and jumps, partly in contact with the bottom and partly suspended in the water.

The **suspended load** is sediment that is light enough to remain lifted indefinitely above the bottom by water turbulence (see figure 10.14). The muddy appearance of a stream during a flood or after a heavy rain is due to a large suspended load. Silt and clay usually are suspended throughout the water, while the coarser bed load moves on the stream bottom. Suspended load has less effect on erosion than the

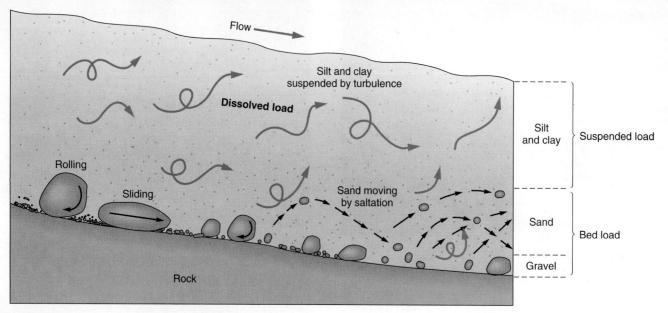

Figure 10.14

A stream's bed load consists of sand and gravel moving on or near the stream bed by traction and saltation. Finer silt and clay form the suspended load of the stream. The dissolved load of soluble ions is invisible.

less visible bed load, which causes most of the abrasion of the stream bed. Vast quantities of sediment, however, are transported in suspension.

Soluble products of chemical weathering processes can make up a substantial **dissolved load** in a stream. Most streams contain numerous ions in solution, such as bicarbonate, calcium, potassium, sodium, chloride, and sulfate. The ions may precipitate out of water as evaporite minerals if the stream dries up, or they may eventually reach the ocean. Very clear water may in fact be carrying a large load of material in solution, for the dissolved load is invisible. Only if the water evaporates does the material become visible as crystals begin to form.

One estimate is that rivers in the United States carry about 250 million tons of solid load and 300 million tons of dissolved load each year. (It would take a freight train eight times as long as the distance from Boston to Los Angeles to carry 250 million tons.)

Stream Deposition

The sediments transported by a stream are often deposited temporarily along the stream's course (particularly the bed load sediments). Such sediments move sporadically downstream in repeated cycles of erosion and deposition, forming *bars* and *flood-plain deposits*. At or near the end of a stream, sediments may be deposited more permanently in a *delta* or an *alluvial fan.*

Bars

Stream deposits may take the form of a **bar,** a ridge of sediment, usually sand or gravel, deposited in the middle or along the banks of a stream (figure 10.15). Bars are formed by deposition when a stream's discharge or velocity decreases. During a flood, a river can move all sizes of sediment, from silt and clay up to huge boulders, because the greatly increased volume of water is moving very rapidly. As the flood begins to recede, the water level in the stream falls and the velocity drops. With the stream no longer able to carry all its sediment load, the larger boulders drop down on the stream bed, slowing the water locally even more. Finer gravel and sand are deposited between the boulders and downstream from them. In this way, deposition builds up a sand or gravel bar that may become exposed as the water level falls.

The next flood on the river may erode all the sediment in this bar and move it farther downstream. But as the flood slows, it may deposit new gravel in approximately the same place, forming a new bar (figure 10.16). After each flood, river fishermen and boat operators must relearn the size and posi-

Figure 10.15

Gravel bars along the banks and in the middle of a stream. Note the sand deposited on the boulders at the right.

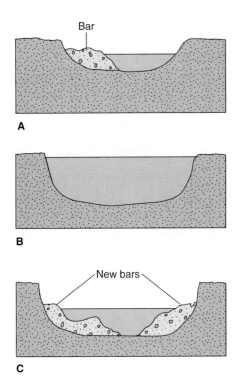

Figure 10.16

A flood can wash away bars in a stream, depositing new bars as the water recedes. (*A*) Normal water flow with sand and gravel bar. (*B*) Increased discharge and velocity during flood moves all sediment downstream. Channel deepens and widens if banks erode easily. (*C*) New bars are deposited as water level drops and stream slows down.

tion of the bars. Sometimes gold panners discover fresh gold in a mined-out river bar after a flood has shifted sediment downstream. A dramatic example of the shifting of sand bars occurred during the planned flood on the Colorado River downstream from the Glen Canyon Dam (Box 10.1).

Placer Deposits

Placer deposits are found in streams where the running water has mechanically concentrated heavy sediment. The heavy sediment is concentrated in the stream where the velocity of the water is high enough to carry away lighter material but not the heavy sediment. Such places include river bars on the inside of meanders, plunge pools below waterfalls and depressions on a stream bed (figure 10.17). Grains concentrated in this manner include gold nuggets and dust, native platinum, diamonds and other gemstones, and worn pebble or sand grains composed of the heavy oxides of titanium and tin.

Braided Streams

Deposition of a bar in the center of a stream (a *midchannel bar*) diverts the water toward the sides where it washes against the stream banks with greater force, eroding the banks and widening the stream (figure 10.18). A stream heavily loaded with sediment may deposit many bars in its channel, causing the stream to widen continually as more bars are deposited. Such a stream typically goes through many stages of deposition, erosion, redeposition, and reerosion, especially if its

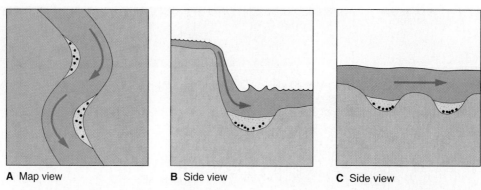

Figure 10.17

Types of placer deposits. (*A*) Stream bar. (*B*) Below waterfall. (*C*) Depressions on stream bed. Valuable grains shown in black.

Figure 10.18

(*A*) A midchannel bar can divert a stream around it, widening the stream. (*B*) Braided stream. Bars split main channel into many smaller channels, greatly widening the stream.

Figure 10.19

A braided river with a heavy suspended load of silt and clay, Alaska.

Photo © Steve McCutcheon/Visuals Unlimited

discharge fluctuates. The stream may lose its main channel and become a **braided stream,** flowing in a network of interconnected rivulets around numerous bars (figures 10.18 and 10.19). A braided stream characteristically has a wide, shallow channel.

A stream tends to become braided when it is heavily loaded with sediment (particularly bed load) and has banks that are easily eroded. The braided pattern develops in deserts as a sediment-laden stream loses water through evaporation and through percolation into the ground. In meltwater streams flowing off glaciers, braided patterns tend to develop when the discharge from the melting glaciers is low relative to the great amount and ranges of size of sediment the stream has to carry.

10.1
ENVIRONMENTAL GEOLOGY

A Controlled Flood in the Grand Canyon: A Bold Experiment to Restore Sediment Movement in the Colorado River

On March 26, 1996, one of the largest experiments ever conducted on a river took place along the Colorado River below the Glen Canyon Dam (box figure 1). For six days the discharge from the Glen Canyon Dam was increased from 8,000 cfs to 45,000 cfs (a spike flow) to emulate the effects of a flood on the Colorado River (box figure 2). One of the main goals of this controlled flooding experiment was to determine whether the higher flows would result in bed scour and redeposition of sand bars and beaches along the sides of the channel. Another goal was to measure and observe how rocks move along the bed of the river bed with increasing discharge and velocity of floodwaters.

The Colorado River had not experienced its usual summertime floods since the Glen Canyon Dam was completed in 1963. The construction of the dam controlled peak discharges or flows on the Colorado River, which resulted in sand being deposited mainly along the bed or bottom of the river and erosion of beaches along the banks of the river. The Glen Canyon Dam cuts off a significant percentage of the sand supply to the lower Colorado River such that most of the downstream sand is supplied by two tributary streams, the Paria and Little Colorado Rivers. In August 1992 the Paria River flooded and deposited 330,000 tons of sand into the Colorado River, and in January 1993 a flood on the Little Colorado River deposited 10 million tons of sediment below its confluence with the Colorado River. The influx of sediment, coupled with the relatively low discharges from the dam (8–20,000 cfs), resulted in sand being concentrated along the bed of the Colorado River.

One of the main predictions of the experiment proved true. That is, sand caught in deep pools in the bottom of the main channel was scoured and carried in suspension downstream where it was redeposited along the river banks as beaches (box figure 3). Deposition of sand along the banks occurs due to back eddies that upwell and move upstream along the river banks and decrease the velocity of the downstream flow so that deposition can occur. Most of the scouring and deposition occurred in the first three days of the experiment; however, when flows were reduced

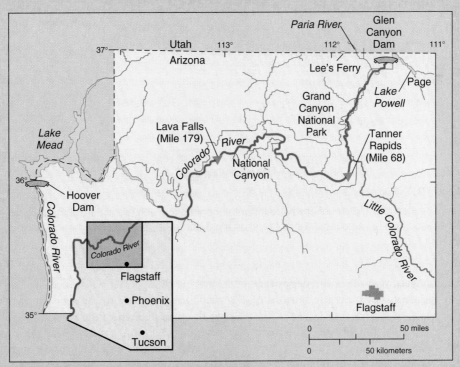

Box 10.1 Figure 1

Location map of the Grand Canyon controlled-flood experiment.

U.S. Geological Survey

Box 10.1 Figure 2

Graph of annual peak discharges before (black) and after Glen Canyon Dam (purple) and the 1996 controlled high-flow release.

U.S. Geological Survey

Box 10.1 Figure 3

Cross section of the channel downstream of the confluence with the Little Colorado River. Increased flows have scoured the bottom sediment and redeposited it as a beach along the river bank.

U.S. Geological Survey

back to 8,000 cfs beaches began to erode and redeposition occurred once again in the deep pools on the bottom of the river.

Downstream at Lava Falls, an experiment was set up to determine how and if large boulders deposited in the main channel from a debris flow would move with the increased discharge and velocity of the floodwater. Holes were drilled into 150 basalt boulders and radio tags were inserted (box figure 4) so their movement could be monitored and correlated with the increase in discharge and velocity of the river. Surface velocity measurements were taken by kayaking the river and charting the speed at which floating balls moved. The surface velocities were used to calculate the velocity of the water close to the river bed where the boulders were positioned. Dye was also injected into the river at peak flows to determine the average velocity of the water. The dye indicated that the velocity of the water increased downstream, particularly at the Lava Falls debris flow. This is because the floodwater accelerated as it flowed downstream, pushing the river water in front of it, which increased the downstream velocity. The first crest of water actually arrived behind Hoover Dam at Lake Mead a day ahead of the floodwater marked with a red dye.

A **B**

Box 10.1 Figure 4

(A) Hole being drilled into a basalt boulder and (B) radio tag installed to track the movement of boulders as the discharge and velocity of the Colorado River increases at the Lava Falls debris flow locality.

Photos courtesy of KUAT-TV, University of Arizona, photo by Dan Duncan

curring flood (box figure 2), showed that beaches could be restored below a dam and that boulders could be moved out of rapids much like that which occurs on an undammed river during a seasonal flood. It is proposed that other dammed rivers would benefit from periodic floods to help restore their natural conditions and thus minimize the adverse effects of damming a river.

Additional Information:
Flooding in Grand Canyon. *Scientific American*, January 1997, pp. 82–89.
Grand Canyon Flood! *NOVA Video*, 1997.

Related Web Resources:
For an overview and details of the specific experiments conducted during the planned flood, visit the following websites:
http://www.daztcn.wr.usgs.gov/flood.html
http://www-marine.wr.usgs.gov/docs/projects/
    grandcan/flood.html
http://www.pbs.org/kuat/grandcanyonflood

The experiment was deemed a success and for the first time a flood was studied as it happened. The experimental flood, even though smaller than the size of a naturally oc-

Meandering Streams and Point Bars

Rivers that carry fine-grained silt and clay in suspension tend to be narrow and deep and to develop pronounced, sinuous curves called **meanders** (figure 10.20). In a long river, sediment tends to become finer downstream, so meandering is common in the lower reaches of a river.

You have seen in figure 10.16 that a river's velocity is higher on the outside of a curve than on the inside. This high velocity can erode the river bank on the outside of a curve, often quite rapidly (figure 10.21).

The low velocity on the inside of a curve promotes sediment deposition. The sand bars in figure 10.22 have been deposited on the inside of curves because of the lower velocity there. Such a bar is called a **point bar** and usually consists of a series of arcuate ridges of sand or gravel.

The simultaneous erosion on the outside of a curve and deposition on the inside can deepen a gentle curve into a hairpin-like meander (see figure 10.22). Meanders are rarely fixed in position. Continued erosion and deposition cause them to migrate back and forth across a flat valley floor, as well as downstream, leaving scars and arcuate point bars to mark their former positions (figure 10.23).

At times, particularly during floods, a river may form a **meander cutoff,** a new, shorter channel across the narrow neck of a meander (figure 10.24). The old meander may be abandoned as sediment separates it from the new, shorter channel. The cutoff meander becomes a crescent-shaped **oxbow lake** (figure 10.25). With time, an oxbow lake may fill with sediment and vegetation (see figure 10.23).

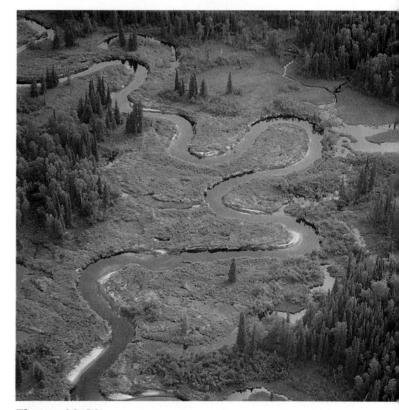

Figure 10.20

Meanders in a stream. These sinuous curves develop because a stream's velocity is highest on the outside of curves, promoting erosion there.

Photo © Glenn M. Oliver/Visuals Unlimited

A

B

Figure 10.21

River erosion on the outside of a curve, Newaukum River, Washington. Pictures were taken in (*A*) January and (*B*) March 1965.

Photos by P. A. Glancy, U.S. Geological Survey

B

Erosion   Deposition

Cross section

A

Corkscrew water motion on a curve helps cause erosion and deposition.

Figure 10.22

(*A*) Development of river meanders and point bars by erosion and deposition on curves. (*B*) A point bar of gravel has been deposited on the inside of this curve.

Photo *B* by Frank Hanna

Figure 10.23

The migration of river meanders across this valley has left arcuate point bar deposits and visible scars marking old meander positions. Sacramento River, California.

Photo © Martin G. Miller/Visuals Unlimited

Figure 10.24

Creation of an oxbow lake by a meander neck cutoff. Old channel is separated from river by sediment deposition.

Figure 10.25

An oxbow lake marks the old position of a river meander, Sweetwater River, Wyoming.

Photo by W. R. Hansen, U.S. Geological Survey

Flood Plains

A **flood plain** is a broad strip of land built up by sedimentation on either side of a stream channel. During floods, flood plains may be covered with water carrying suspended silt and clay (figure 10.26A). When the floodwaters recede, these fine-grained sediments are left behind as a horizontal deposit on the flood plain (figure 10.26B).

Some flood plains are constructed almost entirely of horizontal layers of fine-grained sediment, interrupted here and there by coarse-grained channel deposits (figure 10.27A). Other flood plains are dominated by meanders shifting back and forth over the valley floor and leaving sandy point bar deposits on the inside of curves. Such a river will deposit a characteristic fining-upward sequence of sediments: coarse channel deposits are gradually covered by medium-grained point bar deposits, which in turn are overlain by fine-grained flood deposits (figure 10.27B).

As a flooding river spreads over a flood plain, it slows down. The velocity of the water is abruptly decreased by the increase in friction as the water leaves the deep channel and moves in a thin sheet over the flat valley floor. The sudden decrease in velocity of the water causes the river to deposit most of its sediment near the main channel, with progressively less sediment deposited away from the channel (figure 10.28). A series of floods may build up **natural levees**—low ridges of flood-deposited sediment that form on either side of a stream channel and thin away from the channel. The sediment near the river is coarsest, often sand and silt, while the finer clay is carried farther from the river into the flat, lowland area (the backswamp).

A

B

Figure 10.26

River flood plains. (A) Flooded flood plain of the Animas River, Colorado. (B) Sediment deposited on a flood plain by the Trinity River in California during a 1964 flood.

Photo A by D. A. Rahm, courtesy Rahm Memorial Collection, Western Washington University; photo B by A. O. Waananen, U.S. Geological Survey

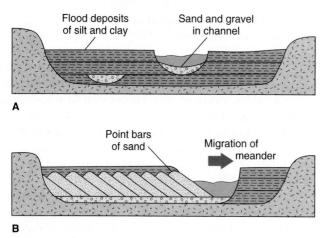

Figure 10.27

Flood plains. (A) Horizontal layers of fine-grained flood deposits with lenses of coarse-grained channel deposits. (B) A fining-upward sequence deposited by a migrating meander. Channel gravel is overlain by sandy point bars, which are overlain by fine-grained flood deposits.

A

B

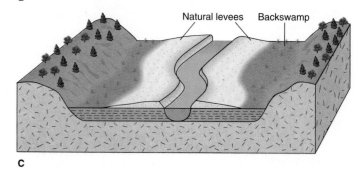

C

Figure 10.28▸

Natural levee deposition during a flood. Levees are thickest and coarsest next to the river channel and build up from many floods, not just one. (Relief of levees is exaggerated.) (A) Normal flow. (B) Flood. (C) After flood.

Deltas

Most streams ultimately flow into the sea or into large lakes. A stream flowing into standing water usually builds a **delta,** a body of sediment deposited at the mouth of a river when the river's velocity decreases (figure 10.29).

The surface of most deltas is marked by **distributaries**—small, shifting channels that carry water away from the main river channel and distribute it over the surface of the delta (figure 10.30). Sediment deposited at the end of a distributary tends to block the water flow, causing distributaries and their sites of sediment deposition to shift periodically.

The shape of a marine delta in map view depends on the balance between sediment supply from the stream and the erosive power of waves and tides (figure 10.31). Some deltas, like that of the Nile river, are broadly triangular; this delta's resemblance to the Greek letter *delta* (Δ) is the origin of the name.

The Nile delta is a *wave-dominated delta* that contains barrier islands along its oceanward side (figure 10.31*A*); the barrier islands form by waves actively reworking the deltaic sediments. Some deltas form along a coast that is dominated by strong tides and the sediment is reshaped into tidal bars that are aligned parallel to a tidal current (figure 10.31*B*). The Ganges-Brahmaputra delta in Bangladesh is a good example of a *tide-dominated* delta. Other deltas, including that of the Mississippi River, are created when very large amounts of sediment are carried into relatively quiet water. Partly because dredging has kept the major distributary channels (locally called "passes") fixed in position for many decades, the Mississippi's distributaries have built long fingers of sediment out into the sea. The resulting shape has been termed a *birdfoot delta.* Because of the dominance of stream sedimentation that forms the finger-like distributaries, birdfoot deltas like the Mississippi are also referred to as *stream-dominated* deltas.

Figure 10.29

Delta of Tongariro River, New Zealand. Note the sediment-laden water, and how the land is being built outward by river sedimentation. A river typically divides into several channels (distributaries) on a delta.

Photo © G. R. 'Dick' Roberts

Many deltas, particularly small ones in freshwater lakes, are built up from three types of deposits, shown in the diagram in figure 10.30. *Foreset beds* form the main body of the delta. They are deposited at an angle to the horizontal. This angle can be as great as 20° to 25° in a small delta where the

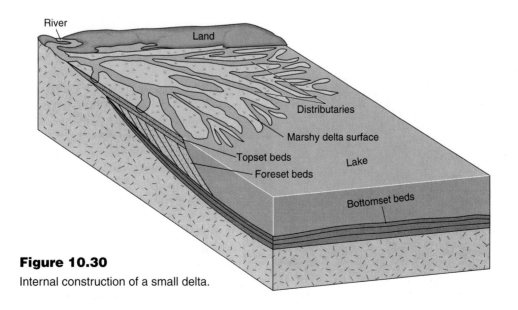

Figure 10.30

Internal construction of a small delta.

A Wave-dominated delta

B Tidal-dominated delta

Figure 10.31

The shape of a delta depends on the amount of sediment being carried by the river and on the vigor of waves and tides in the sea. (*A*) The Nile delta is a wave-dominated delta with prominent barrier islands. (*B*) The Ganges-Brahmaputra delta in Bangladesh contains tidal sand bars formed by strong tidal currents. (*C*) is a Landsat image of the Mississippi River delta. Note how sediment (Light yellow) is carried by both river and ocean currents.

Source: After R. G. Walker, "Facies Models," *Geoscience Canada,* fig. 7, p. 109, 1979; Photo by NASA

C Stream-dominated delta

foreset beds are sandy, or less than 5° in large deltas with fine-grained sediment. On top of the foreset beds are the *topset beds,* nearly horizontal beds of varying grain size formed by distributaries shifting across the delta surface. Out in front of the foreset beds are the *bottomset beds,* deposits of the finest silt and clay that are carried out into the lake by the river water flow or by sediments sliding downhill on the lake floor. Many of the world's great deltas in the ocean are far more complex than the simplified diagram shown in the figure. Shifting river mouths, wave energy, currents, and other factors produce many different internal structures.

The persistence of large deltas as relatively "dry" land depends upon a balance between the rate of sedimentation and the rates of tectonic subsidence and of compaction of water-saturated sediment. Many deltas are sinking, with seawater encroaching on once-dry land. The Mississippi River delta in Louisiana is sinking, as upstream dams catch sediment, reducing the delta's supply, and as extraction of oil and gas from beneath the delta accelerates subsidence. The flat surface of a delta is a risky place to live or farm, particularly in regions threatened by the high waves and raised sea level of hurricanes, such as the U.S. Gulf Cost and the country of Bangladesh (the Ganges-Brahmaputra delta) on the Indian Ocean.

Alluvial Fans

Some streams, particularly in dry climates, do not reach the sea or any other body of water. They build alluvial fans instead of deltas. An **alluvial fan** is a large, fan- or cone-shaped pile of sediment that usually forms where a stream's velocity decreases as it emerges from a narrow mountain canyon onto a flat plain (figures 10.32 and 10.33). Alluvial fans are particularly well developed and exposed in the southwestern desert of the United States and in other desert regions; but they are by no means limited to arid regions.

An alluvial fan builds up its characteristic fan shape gradually as streams shift back and forth across the fan surface and deposit sediment, usually in a braided pattern. Deposition on an alluvial fan in the desert is discontinuous because streams typically flow for only a short time after the infrequent rainstorms. When rain does come, the amount of sediment to be moved is often greater than the available water. When the rain stops, the stream may end up as a debris flow and deposit its load.

The sudden loss of velocity when a stream flows from steep mountains onto a broad plain causes the sediment to deposit on an alluvial fan. The loss of velocity is due to the

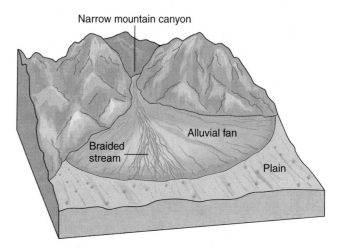

Figure 10.32

An alluvial fan at the mouth of a desert canyon.

A

B

Figure 10.33

(*A*) This alluvial fan formed on the salt-encrusted floor of Death Valley, California, as sediment was washed out of the canyon by thunderstorms. (*B*) Giant alluvial fans in the distance joined together as they grew at the base of the mountains. The surface of another fan forms the foreground. Death Valley.

Photo by Frank M. Hanna

widening or branching of the channel as it leaves the narrow mountain canyon. The gradual loss of water as it infiltrates into the fan also promotes sediment deposition. On large fans, deposits are graded in size within the fan, with the coarsest sediment dropped nearest the mountains and the finer material deposited progressively father away. Small fans do not usually show such grading.

Flooding

Many of the world's cities, such as Pittsburgh, St. Louis, and Portland, Oregon, are built beside rivers and therefore can be threatened by floods. Rivers are important transportation routes for ships and barges, and flat flood plains have excellent agricultural soil and offer attractive building sites for houses and industry.

Flooding does not occur every year on every river, but flooding is a natural process on all rivers and must be prepared for by river cities and towns. Heavy rains and the rapid melting of snow in the springtime are the usual causes of floods. The rate and volume of rainfall, and the geographic path of rainstorms, often determine whether flooding will occur.

Floods are described by *recurrence interval*, the average time between floods of a given size. A "100-year flood" is one that can occur, on the *average*, every 100 years (box 10.2). A 100-year flood has a 1-in-100, or one percent, chance of occurring in any given year. It is perfectly possible to have two 100-year floods in successive years—or even in the same year. If a 100-year flood occurs this year on the river you live beside, you should not assume that there will be a 99-year period of safety before the next one.

Flood erosion is caused by the high velocity and large volume of water in a flood. Although relatively harmless on an uninhabited flood plain, flood erosion can be devastating to a city. As a river undercuts its banks, particularly on the outside of curves where water velocity is high, buildings and piers may fall into the river. As sections of flood plain are washed away, highways and railroads are cut.

High water covers streets and agricultural fields and invades buildings, shorting out electrical lines and backing up sewers. Water-supply systems may fail or be contaminated. A foot of water in your living room will be drawn upward in your walls by capillary action in wall plasterboard and insulation, creating a soggy mess that has to be torn out and replaced. High water on flat flood plains often drains away very slowly; street travel may be by boat for weeks. If floodwaters are deep enough, houses may float away (figure 10.34).

Flood deposits are usually silt and clay. A new layer of wet mud on a flood plain in an agricultural region can be beneficial in that it renews the fields with topsoil from upstream. The same mud in a city will destroy lawns, furniture, and machinery. Cleanup is slow; imagine shoveling four inches of worm-filled mud that smells like sewage out of your house.

Figure 10.34

January 1997 flood waters were so deep in the Arboga, California area that a house floated off its foundation.

Source: Andrea Booher, FEMA

Figure 10.35

The presence of a city can increase the chance of floods. The blue curve shows the normal increase in a river's discharge following a rainstorm (black bar). The red curve shows the great increase in runoff rate and amount caused by pavement and storm sewers in a city.

Urban Flooding

Urbanization enhances flooding. Paved areas and storm sewers increase the amount and rate of surface runoff of water, making river levels higher during storms (figure 10.35). Such rapid increases in runoff or discharge to a river is called a "flashy" discharge. Storm sewers are usually designed for a 100-year storm; however, large storms that drop a lot of rain in a short period of time (cloudburst) may overwhelm sewer systems and cause localized flooding. Rising river levels may block storm sewer outlets and also add to localized flooding problems.

Bridges, docks and buildings built on flood plains can also constrict the flow of flood waters, increasing the water height and velocity and promoting erosion.

10.2 IN GREATER DEPTH

Estimating the Size and Frequency of Floods

Because people have encroached on the flood plains of many rivers, flooding is one of the most universally experienced geologic hazards. To minimize flood damage and loss of life, it is useful to know the potential size of large floods and how often they might occur. This is often a difficult task because of the lack of long-term records for most rivers. The U.S. Geological Survey monitors the stage (water elevation) and discharge of rivers and streams throughout the U.S. in order to collect data that can be used to attempt to predict the size and frequency of flooding and to make estimates of water supply.

Hydrologists designate floods based on their *recurrence interval,* or *return period.* For example, a 100-year flood is the largest flood expected to occur within a period of 100 years. This does not mean that a 100-year flood occurs once every century, but that there is 1-in-100 chance, or a one percent probability, each year that a flood of this size will occur. Usually flood control systems are built to accommodate a 100-year flood because that is the minimum margin of safety required by the federal government if an individual wants to obtain flood insurance subsidized by the Federal Emergency Management Agency (FEMA).

To calculate the recurrence interval of flooding for a river, the annual peak discharges (largest discharge of the year) are collected and ranked according to size (box figure 1 and table 1). The largest annual peak discharge is assigned a rank (m) of one, the second a two, and so on until all of the discharges are assigned a rank number. The *recurrence interval* (**R**) of each annual peak discharge is then calculated by adding one to the *number of years of record* (**n**) and dividing by its *rank* (**m**).

$$R = \frac{n + 1}{m}$$

For example, the Cosumnes River in California has 90 years of record (**n = 90**), and in 1907 the second largest peak discharge (**m = 2**) of 71,000 cfs occurred. The recurrence interval (**R**), or expected frequency of occurrence, for a discharge this large is 45.5 years:

$$R = \frac{90 + 1}{2} = 45.5 \text{ years}$$

That is, there is a 1-in-45.5, or 2 percent, chance each year of a peak discharge of 71,000 cfs or greater occurring on the Cosumnes River.

The flood of record (largest recorded discharge) occurred on January 2, 1997 when heavy, unseasonably warm rains rapidly melted snow in the Sierra Nevada and caused flooding in much of northern California. A peak discharge of 93,000 cfs in the Cosumnes River resulted in levee breaks and widespread flooding of homes and agricultural areas (box figure 2). The recurrence interval for the 1997 flood (93,000 cfs) is 91 years:

$$R = \frac{90 + 1}{1} = 91 \text{ years}$$

Box 10.2 Figure 1

Annual peak discharge for the Cosumnes River.

After U.S. Geological Survey Water-Data Report, CA-97-3.

Box 10.2 Figure 2

Levee break along the Cosumnes River.

Box 10.2 Figure 3

Flood-frequency curves for the Cosumnes River.

Preliminary data from Richard Hunrichs, hydrologist, U.S. Geological Survey and
U.S. Geological Survey Water-Data Report, CA-97-3.

A *flood-frequency curve* can be useful in providing an estimate of the discharge and the frequency of floods. The flood-frequency curve is generated by plotting the annual peak discharges against the calculated recurrence intervals (box figure 3). Because most of the data points defining the curve plot in the lower range of discharge and recurrence interval, there is some uncertainty in projecting larger flood events. Two flood-frequency curves are drawn in box figure 3; the red line represents the best-fit curve for all of the data, whereas the dashed blue line excludes the 1997 flood of record. Notice that the curve has a steeper slope when the 1997 data is included, and that the size of the 100-year flood has increased from 73,000 cfs to 82,000 cfs based on the one additional year of record. Because large floods do not occur as often as small floods, the rare large flood can have a dramatic effect on the shape of the flood-frequency curve and the estimate of a 100-year event. This is particularly true for a river like the Cosumnes that has had only

Box 10.2 Table 1	Annual Peak Discharges and Recurrence Intervals in Rank Order for the Cosumnes River at Michigan Bar, California.		
Year	Peak Discharge (cfs)	Magnitude Rank (m)	Recurrence Interval
1997	93,000	1	91.0
1907	71,000	2	45.5
1986	45,100	3	30.33
1956	42,000	4	22.75
1963	39,400	5	18.42
1909	28,400	10	9.20
1943	22,900	20	4.60
1970	16,800	30	3.07
1960	11,200	40	2.30
1971	8,590	50	1.84
1991	6,670	60	1.53

Source: Preliminary data from Richard Hunrichs, hydrologist, U.S. Geological Survey and U.S. Geological Survey Water-Data Report, CA-97-3.

two large events, one in 1907 and the other ninety years later in 1997.

The 100-year flood plain is based on the estimate of the discharge of the 100-year flood and on careful mapping of the flood plain. Changes in the estimated size of the 100-year flood could result in property that no longer has 100-year flood protection. In this case, property owners may be prevented from getting flood insurance or money to rebuild from FEMA unless new flood control structures are built to provide additional protection or houses are raised or even relocated out of the flood plain.

Related Web Resources
To find data sets to calculate the recurrence interval for rivers throughout the United States, access the U.S. Geological Survey Water Data Retrieval Website: http://h2o.usgs.gov/swr/

Further Readings
Water Resources Data for California, Water Year 1997. U.S. Geological Survey Water-Data Report CA-97-3.
Riggs, H.C. 1968. *Frequency Curves.* Techniques of Water-Resources Investigations of the U.S. Geological Survey. Book 4, Hydrologic Analysis and Interpretation.

Flash Floods

Some floods occur rapidly and die out just as quickly. *Flash floods* are local, sudden floods of large volume and short duration, often triggered by heavy thunderstorms. A startling example occurred in 1976 in north-central Colorado along the Big Thompson River (figure 10.36). Strong winds from the east pushed moist air up the front of the Colorado Rockies, causing thunderstorms in the steep mountains. The storms were unusually stationary, allowing as much as 30 cm of rain to fall in two days. Some areas received 19 cm in just over an hour. Little of this torrential rainfall could soak into the ground. The volume of water in the Big Thompson River swelled to four times the previously recorded maximum, and the river's velocity rose to an impressive 25 kilometers per hour (15 miles per hour) for a few hours on the night of July 31. By the next morning the flood was over, and the appalling toll became apparent—139 people dead, 5 missing, and more than $35 million in damages (figure 10.37).

On July 29, 1997, just two days before the 21st anniversary of the Big Thompson River flood, Fort Collins, Colorado was struck by a flash flood when 20 cm (8 inches) of rain fell in only 5 hours. A 4-meter (12 foot) wall of water rushed down Spring Creek, a tributary to the Cache la Poudre River, nearly devastating two trailer parks. Five people lost their lives and 40 were injured when a 5-meter (15 foot) high railroad embankment that had temporarily dammed Spring Creek, broke and sent the wall of

water into the trailer park (figure 10.37*B*). Unlike the Big Thompson Canyon flood, the heavy rains fell over the city of Fort Collins rather than upstream in the steep mountain canyons.

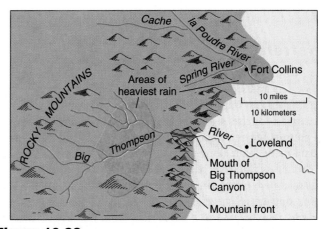

Figure 10.36
Location map of the 1976 flash flood on the Big Thompson River in Colorado and the 1997 flash flood in Fort Collins.

A

B

Figure 10.37

(*A*) A cabin sits crushed against a bridge following the Big Thompson Canyon flash flood of 1976. (*B*) Devastation along Spring Creek after Fort Collins flash flood.

Photos by W.R. Hansen, U.S. Geological Survey

Controlling Floods

Flood-control structures can partially reduce the dangers of floodwaters and sedimentation to river cities (figure 10.38). Upstream dams can trap water and release it slowly after the storm. (A dam also catches sediment, which eventually fills its reservoir and ends its life as a flood-control structure.) Artificial levees are embankments built along the banks of a river channel to contain floodwaters within the channel. Protective walls of stone (riprap) or concrete are often constructed along river banks, particularly on the outside of curves, to slow erosion. Floodwalls, walls of concrete, may be used to protect cities from flooding. However, these flood-control structures may constrict the channel and cause water to flow faster with more erosive power downstream. Bypasses are also used along the Mississippi and other rivers to re-

duce the discharge in the main channel by diverting water into designated basins in the flood plain. The bypasses serve to give part of the natural flood plain back to the river.

Dams and levees are designed to control certain specified floods. If the flood-control structures on your river were designed for 75-year floods, then a much larger 100-year flood will likely overtop these structures, and may destroy your home. The disastrous floods along the Missouri and Mississippi Rivers and their tributaries north of Cairo, Illinois in 1993 resulted from many such failures in flood control.

Wise land-use planning and zoning for flood plains should go hand in hand with flood control. Wherever possible, buildings should be kept out of areas that might someday be flooded by 100-year floods.

Figure 10.38

Examples of flood control structures.

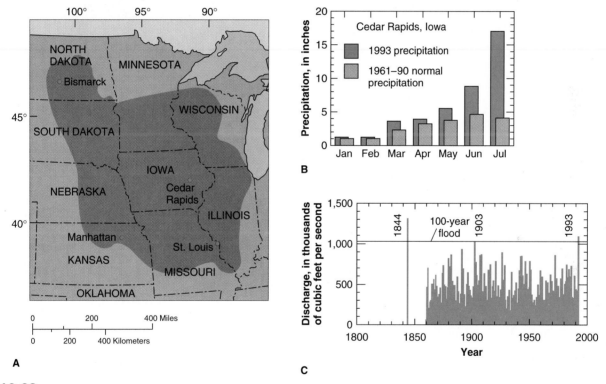

A

B

C

Figure 10.39

The Great Flood of 1993. (*A*) Area of flood. (*B*) 1993 rainfall at Cedar Rapids, Iowa, compared to normal rainfall. (*C*) Discharge of the Mississippi River at St. Louis compared to 100-year flood.

Photo by U.S. Geological Survey

The Great Flood of 1993

In the Spring of 1993, heavy rains soaked the ground in the upper Midwestern United States. In June and July a stationary weather pattern created a shifting band of thunderstorms that dumped as much as four inches of rain in a single day on localities such as Bismarck, North Dakota; Cedar Rapids, Iowa; and Manhattan, Kansas (figure 10.39). These torrential rains falling on already-saturated ground created the worst flood disaster in U.S. history, as swollen rivers flooded 6.6 million acres in 9 states, killing 38 and causing $12 billion in damage to houses and crops.

River discharges exceeded 100-year discharges on many rivers including the Mississippi, the Missouri, the Iowa, the Platte, and the Raccoon. At St. Louis, the Mississippi River discharge was greater than 1,000,000 cubic feet per second on

Figure 10.40

Mississippi River water pours through a broken levee onto a farm near Columbia, Illinois, 1993.

Photo © St. Louis Post-Dispatch

August 1 (six times normal flow), and the river crested 20 feet above flood stage (and 23 feet above flood stage farther downstream at Chester, Illinois). At Hannibal, Missouri, where the 500-year flood height had been determined to be 30 feet, the new flood-control levee 31 feet high was completed in April. The river crested at 32 feet. Some rivers remained above flood stage for months.

The high water in major rivers such as the Mississippi and the Missouri caused many smaller tributary streams to back up, flooding numerous small towns. Many levees broke as water overtopped them or just physically pushed the saturated levee sediment aside (figure 10.40). After the rains stopped, in some places the floodwaters took weeks or even months to recede. Some entire towns have been relocated to drier ground.

Stream Valley Development

Valleys, the most common landforms on the earth's surface, are usually cut by streams. By removing rock and sediment from the stream channel, a stream deepens, widens, and lengthens its own valley.

Downcutting and Base Level

The process of deepening a valley by erosion of the stream bed is called **downcutting.** If a stream removes rock from its bed, it can cut a narrow *slot canyon* down through rock (figures 10.41

Figure 10.41

Downcutting, mass wasting, and sheet erosion shape canyons and valleys. (*A*) Downcutting can create slot canyons in resistant rock, particularly where downcutting is rapid and fractures in the rock are favorably oriented. (*B*) Downslope movement of rock and soil on valley walls widens most canyons into V-shaped valleys.

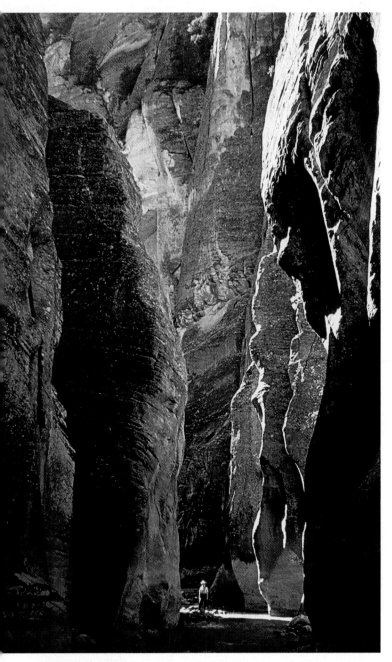

Figure 10.42

Stream erosion has cut this unusual slot canyon through porous sandstone, Zion National Park, Utah.

Photo by Allen Hagood, Zion Natural History Association

Figure 10.43

The waterfall and rapids on the Yellowstone River in Wyoming indicate that the river is ungraded and actively downcutting. Note the V-shaped cross-profile and lack of flood plain.

and 10.42). Such narrow canyons do not commonly form because mass wasting and sheet erosion usually remove rock from the valley walls. These processes widen the valley from a narrow, vertical-walled canyon to a broader, open, V-shaped canyon (figures 10.41*B* and 10.43). Slot canyons persist, however, in very resistant rock with favorably oriented fractures or in regions where downcutting is rapid.

Downcutting cannot continue indefinitely because the headwaters of a stream cannot cut below the level of the stream bed at the mouth. If a river flows into the ocean, sea level becomes the lower limit of downcutting. The river cannot cut below sea level, or it would have to flow uphill to get the sea. For most streams, sea level controls the level to which the land can be eroded.

The limit of downcutting is known as **base level;** it is a theoretical limit for erosion of the earth's surface (figure 10.44). Downcutting will proceed until the stream bed reaches base level. If the stream is well above base level, downcutting can be quite rapid; but as the stream approaches base level, the rate of downcutting slows down. For streams that reach the ocean, base level is close to sea level, but since streams need at least a gentle gradient in order to flow, base level slopes gently upward in an inland direction.

During the glacial fluctuations of the Pleistocene Epoch (see chapter 12), sea level rose and fell as water was removed from the sea to form the glaciers on the continents and returned to the sea when the glaciers melted. This means that

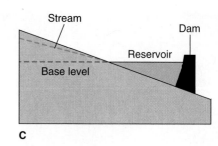

Figure 10.44

Base level is the lowest level of downcutting.

base level rose and fell for streams flowing into the sea. As a result, the lower reaches of such rivers alternated between erosion (caused by low sea level) and deposition (caused by high sea level). Since the glaciers advanced and retreated several times, the cycle of erosion and deposition was repeated many times, resulting in a complex history of cutting and filling near the mouths of most old rivers.

Base levels for streams that do not flow into the ocean are not related to sea level. In Death Valley in California (figure 10.44B), base level for in-flowing streams corresponds to the lowest point in the valley, 86 meters (282 feet) *below* sea level (the valley has been dropped below sea level by tectonic movement along faults). On the other hand, base level for a stream above a high reservoir or a mountain lake can be hundreds of feet above sea level. The surface of the lake or reservoir serves as base level for all the water upstream (figure 10.44C). The base level of a tributary stream is governed by the level of its junction with the main stream. A ledge of resistant rock may act as a temporary base level if a stream has difficulty eroding through it.

The Concept of a Graded Stream

As a stream begins downcutting into the land, its longitudinal profile is usually irregular, with rapids and waterfalls along its course (see figure 10.43). Such a stream, termed *ungraded*, is using most of its erosional energy in downcutting to smooth out these irregularities in gradient.

As the stream smoothes out its longitudinal profile to a characteristic concave-upward shape (figure 10.45), it becomes graded. A **graded stream** is one that exhibits a delicate balance between its transporting capacity and the sediment load available to it. This balance is maintained by cutting and filling any irregularities in the smooth longitudinal profile of the stream.

Earlier in this chapter you learned how changes in a stream's gradient can cause changes in its sediment load. An increase in gradient causes an increase in a stream's velocity, allowing the stream to erode and carry more sediment. A balance is maintained—the greater load is a result of the greater transporting capacity caused by the steeper gradient.

Figure 10.45

An ungraded stream has an irregular longitudinal profile with many waterfalls and rapids. A graded stream has smoothed out its longitudinal profile to a smooth, concave-upward curve.

The relationship also works in reverse—a change in sediment load can cause a change in gradient. For example, a decrease in sediment load may bring about erosion of the stream's channel, thus lowering the gradient. Because dams trap sediment in the calm reservoirs behind them, most streams are sediment-free just downstream from dams. In some streams this loss of sediment has caused severe channel erosion below a dam, as the stream adjusts to its new, reduced load.

A river's energy is used for two things—transporting sediment and overcoming resistance to flow. If the sediment load decreases, the river has more energy for other things. It may use this energy to erode more sediment, deepening its valley. Or it may change its channel shape or length, increasing resistance to flow, so that the excess energy is used to overcome friction. Or the river may increase the roughness of its channel, also increasing friction. The response of a river is not always predictable, and construction of a dam can sometimes have unexpected and perhaps harmful results.

Figure 10.46

Lateral erosion can widen a valley by undercutting and eroding valley walls.

Lateral Erosion

A graded stream can be deepening its channel by downcutting while part of its energy is also widening the valley by **lateral erosion,** the erosion and undercutting of a stream's banks and valley walls as the stream swings from side to side across its valley floor. The stream channel remains the same width as it moves across the flood plain, but the valley widens by erosion, particularly on the outside of curves and meanders where the stream impinges against the valley walls (figure 10.46). The valley widens as its walls are eroded by the

Figure 10.47

Headward erosion is lengthening this gully and destroying the field. The head of the gully is extending uphill, largely by mass wasting.

Courtesy Ward's Natural Science Est., Inc., Rochester, N.Y.

stream and as its walls retreat by mass wasting triggered by stream undercutting. As a valley widens, the stream's flood plain increases in width also.

Headward Erosion and Stream Piracy

Building a delta or alluvial fan at its mouth is one way a river can extend its length. A stream can also lengthen its valley by **headward erosion,** the slow uphill growth of a valley above its original source through gullying, mass wasting, and sheet erosion (figure 10.47). This type of erosion is particularly difficult to stop. When farmland is being lost to gullies that are eroding headward into fields and pastures, farmers must divert sheet flow and fill the gully heads with brush and other debris to stop, or at least retard, the loss of topsoil.

Headward erosion sometimes results in a stream capturing the drainage of a nearby stream, a process known as *stream piracy* (or stream capture). If the two streams flow downhill in different directions, one stream may erode faster than the other, perhaps because it is on the rainy side of the mountain and has more water in it or because it is flowing over softer rock or a steeper slope than the other stream. The rapidly eroding stream will try to become graded by extending its length by headward erosion and may eventually capture the drainage of the slower stream (figure 10.48). The faster stream grows at the expense of the slower one, and the water that used to flow in the slow stream has now been captured in the new, larger stream.

A

B

Figure 10.48

The stages of stream piracy. (*A*) A divide separates two streams. (*B*) Headward erosion by stream flowing on steeper slope captures drainage of stream with the lower gradient.

Figure 10.49

Stream terraces near Jackson Hole, Wyoming. The stream has cut downward into its old flood plain.

Stream Terraces

Stream terraces are steplike landforms found above a stream and its flood plain (figure 10.49). Terraces may be benches cut in rock (sometimes sediment-covered), or they may be steps formed in sediment by deposition and subsequent erosion.

Figure 10.50 shows how one type of terrace forms as a river cuts downward into a thick sequence of its own flood-plain deposits. Originally the river deposited a thick section of flood-plain sediments. Then the river changed from deposition to erosion, and cut into its old flood plain, parts of which remain as terraces above the river.

A

B

C

Figure 10.50

Terraces formed by a stream cutting downward into its own flood-plain deposits. (*A*) Stream deposits thick, coarse, flood-plain deposits. (*B*) Stream erodes its flood plain by downcutting. Old flood-plain surface forms terraces. (*C*) Lateral erosion forms new flood plain below terraces.

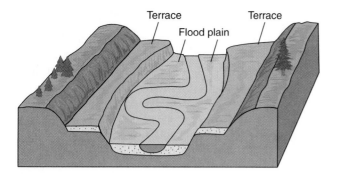

Figure 10.51

Terraces may be erosional benches cut in rock.

Figure 10.52

Incised meanders of the Colorado River ("The Loop"), Canyonlands National Park, southwestern Utah.

Photo by Frank M. Hanna

Why might a river change from deposition to erosion? One reason might be regional uplift, raising a river that was once meandering near base level to an elevation well above base level. Uplift would steepen a river's gradient, causing the river to speed up and begin erosion. But there are several other reasons why a river might change from deposition to erosion. A change from a dry to a wet climate may increase discharge and cause a river to begin eroding. A drop in base level (such as lowering of sea level) can have the same effect. A situation like that shown in figure 10.50 can develop in a recently glaciated region. Thick valley fill such as glacial outwash (see chapter 12) may be deposited in a stream valley and later, after the glacier stops producing large amounts of sediment, be dissected into terraces by the river.

Terraces can also develop from erosion of a bedrock valley floor (figure 10.51). Bedrock benches are usually capped by a veneer of flood-plain deposits.

Incised Meanders

Incised meanders are meanders that retain their sinuous curves as they cut vertically downward below the level at which they originally formed. The result is a meandering *valley* with essentially no flood plain, cut into the land as a steep-sided canyon (figure 10.52).

Some incised meanders may be due to the profound effects of a change in base level. They may originally have been formed as meanders in a laterally eroding river flowing over a flat flood plain, perhaps near base level. If regional uplift elevated the land high above base level, the river would begin downcutting and might be able to maintain its characteristic meander pattern while deepening its valley (figure 10.53). A drop in base level without land uplift (possibly because of a lowering of sea level) could bring about the same result.

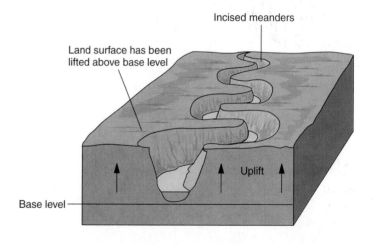

Figure 10.53

Incised meanders can form by uplift.

10.1 Astrogeology

Stream Features on the Planet Mars

There is no liquid water on the surface of Mars today. With the present surface temperatures, atmospheric pressures, and water content in the Martian atmosphere, any liquid water would immediately evaporate. There are some indications, however, that conditions may have been different in the past and that liquid water existed on Mars, at least temporarily. Certain features on Mars, called *channels,* closely resemble certain types of stream channels on Earth. They have tributary systems and meanders and are sometimes braided. The channels trend downslope and tend to get wider toward their mouths. These Martian channels are restricted to certain areas and appear to have been formed by intermittent episodes of erosion.

One type of stream channel on Mars appears to have formed by large flooding events and is similar in appearance to those observed in the Channeled Scablands of Washington state. The Channeled Scablands were formed by extensive flooding during the Pleistocene glacial ages when a naturally formed ice dam broke and released water from a large lake. The mouth of Ares Vallis, an ancient Martian flood channel similar to those observed in the Channeled Scablands, was selected for the July 4, 1997 landing of the *Pathfinder* spacecraft and Sojourner Rover. It was postulated that a variety of rock types should be present in the mouth of what appeared to be an ancient flood channel. The first photos from the Mars Pathfinder Lander and Sojourner Rover, a "robotic field geologist", revealed a variety of rock types (box figure 1) and also an imbricate or overlapping arrangement of the rocks similar to that seen in stream beds here on Earth.

A second kind of Martian channel (box figure 2) occurs mostly on the sides of some volcanoes and some meteorite impact craters, and is more typical of the gullies and dendritic drainage patterns seen on Earth. Channels of this type occur on the older surfaces of Mars (more than 3.5 billion years old) and may indicate that early in the history of Mars temperature and atmospheric conditions were such that rainfall could have occurred and long-lived river systems could have existed. Possibly as the climate of Mars cooled, the water froze and formed the extensive deposit of underground ice known to exist in the surface layers of Mars today. (See Astrogeology Box 9.1 for a look at underground ice.)

Astrogeology Box 10.1 Figure 2

Viking Orbiter view of a portion of channel Nirgal Vallis and its tributary system.

Photo by Dr. Michael H. Carr/National Space Science Data Center

Astrogeology Box 10.1 Figure 1

View from the Mars Pathfinder Lander showing the Sojourner Rover and a variety of rocks from Ares Vallis.

Photo Courtesy of Jet Propulsion Laboratory

Although uplift may be a key factor in the formation of many incised meanders, it may not be *required* to produce them. Lateral erosion certainly seems to become more prominent as a river approaches base level, but some meandering can occur as soon as a river develops a graded profile. A river flowing on a flat surface high above base level may develop meanders early in its erosional history, and these meanders may become incised by subsequent downcutting. In such a case uplift is not necessary.

Superposed Streams

Some narrow mountain ranges have steep-sided river valleys slicing directly across them. The origin of these valleys is often puzzling. Why did the river erode *through* the mountain range rather than flow around it?

The usual reason is that the river has been **superposed** on the range from a gently sloping plain that used to exist above it (figure 10.54). If sediment once buried the range, a river could develop on the smooth surface of the sediment layer. As the sediment gradually was eroded away by the river, the river itself would be let down onto the once-buried range and could cut a canyon through the more resistant rocks in the underlying range.

The folded rocks of the Appalachians were worn away by erosion and then partially covered by flat sediments during the Mesozoic. Gradual uplift in the Cenozoic has caused rivers such as the Potomac and Susquehanna to cut down into the once-buried range.

Figure 10.54

Development of a superposed stream. (*A*) Stream initially cuts through the horizontal sediment. (*B*) Continued erosion removes horizontal strata and stream cuts through underlying rock forming narrow valleys in resistant rock.

Summary

Normally stream *channels* are eroded and shaped by the streams that flow in them. Unconfined sheet flow can cause significant erosion.

Drainage basins are separated by *drainage divides.*

A river and its tributaries form a *drainage pattern*. A *dendritic* drainage pattern develops on uniform rock, a *rectangular* pattern on regularly jointed rock. A *radial* pattern forms on conical mountains, while a *trellis* pattern usually indicates erosion of folded sedimentary rock.

Stream *velocity* is the key factor controlling sediment erosion, transportation, and deposition. Velocity is in turn controlled by several factors.

An increase in a stream's *gradient* increases the stream's velocity. *Channel shape* and *roughness* affect velocity by increasing or lessening friction. As tributaries join a stream, the stream's *discharge* increases downstream. Floods increase stream discharge and velocity.

Streams erode by *hydraulic action, abrasion,* and *solution.* They carry coarse sediment by *traction* and *saltation* as *bed load.* Finer-grained sediment is carried in *suspension.* A stream can also have a substantial *dissolved load.*

Streams create features by erosion and deposition. *Potholes* form by abrasion of hard rock on a stream bed. *Bars* form in the middle of streams or on stream banks, particularly on the inside of curves where velocity is low (*point bars*). A *braided pattern* can develop in streams with a large amount of bed load.

Meanders are created when a laterally eroding stream shifts across the flood plain, sometimes creating cutoffs and oxbow lakes.

A *flood plain* develops by both lateral and vertical deposition. *Natural levees* are built up beside streams by flood deposition.

A *delta* forms when a stream flows into standing water. The shape and internal structure of deltas are governed by river deposition and wave and current erosion. *Alluvial fans* form, particularly in dry climates, at the base of mountains as a stream's channel widens and its velocity decreases.

Rivers deepen their valleys by *downcutting* until they reach *base level,* which is either sea level or a local base level.

A *graded stream* is one with a delicate balance between its transporting capacity and its available load.

Lateral erosion widens a valley after the stream has become graded.

A valley is lengthened by both *headward erosion* and sediment deposition at the mouth.

Stream terraces can form by erosion of rock benches or by dissection of thick valley deposits during downcutting.

Incised meanders form as (1) river meanders are cut vertically downward following uplift or (2) lateral erosion and downcutting proceed simultaneously.

Superposed streams form as a stream cuts down through a sediment layer into the underlying, more resistant rock of a once-buried mountain range.

Terms to Remember

abrasion 228
alluvial fan 241
bar 230
base level 249
bed load 229
braided stream 232
delta 239
dendritic pattern 224
discharge 227
dissolved load 230
distributary 239
divide 224
downcutting 248
drainage basin 224
drainage pattern 224

flood plain 237
graded stream 250
headward erosion 251
hydraulic action 228
hydrologic cycle 222
incised meander 253
lateral erosion 251
meander 235
meander cutoff 235
natural levee 237
oxbow lake 235
point bar 235
pothole 228
radial pattern 225
rectangular pattern 225

saltation 229
sheetwash 223
solution 228
stream 222
stream channel 222
stream gradient 226
stream terrace 252
stream velocity 225
superposed stream 255
suspended load 229
traction 229
trellis pattern 225
tributary 224

Testing Your Knowledge

Use the questions below to prepare for exams based on this chapter.

1. What factors control a stream's velocity?
2. Describe how bar deposition creates a braided stream.
3. In what part of a large alluvial fan is the sediment the coarsest? Why?
4. What does a trellis drainage pattern tell about the rocks underneath it?
5. Describe one way that incised meanders form.
6. How does a meander neck cutoff form an oxbow lake?
7. How does a natural levee form?
8. Describe how stream terraces form.
9. Describe three ways in which river erodes its channel.
10. Name and describe the three main ways in which a stream transports sediment.
11. How does a stream widen its valley?
12. What is base level?
13. The total area drained by a stream and its tributaries is called the (a) hydrologic cycle (b) tributary area (c) drainage divide (d) drainage basin
14. Stream erosion and deposition are controlled primarily by a river's (a) velocity (b) discharge (c) gradient (d) channel shape (e) channel roughness
15. What is the gradient of a stream that drops 10 vertical feet over

a 2-mile horizontal distance? (a) 20 feet per mile (b)10 feet per mile (c) 5 feet per mile (d) 2 feet per mile
16. What are typical units of discharge? (a) miles per hour (b) cubic meters (c) cubic feet per second (d) meters per second
17. Hydraulic action, solution, and abrasion are all examples of stream (a) erosion (b) transportation (c) deposition
18. Cobbles are more likely to be transported in a stream's (a) bed load (b) suspended load (c) dissolved load (d) all of the above
19. A river's velocity is _____ on the outside of a meander curve compared to the inside. (a) higher (b) equal (c) lower
20. Sand bars deposited on the inside of meander curves are called (a) dunes (b) point bars (c) cutbanks (d) none of the above
21. Which is not a drainage pattern? (a) dendritic (b) radial (c) rectangular (d) trellis (e) none of the above
22. The broad strip of land built up by sedimentation on either side of a stream channel is (a) a floodplain (b) a delta (c) an alluvial fan (d) a meander
23. The average time between floods of a given size is (a) the discharge (b) the gradient (c) the recurrence interval (d) the magnitude
24. A platform of sediment formed where a stream flows into standing water is (a) an alluvial fan (b) a delta (c) a meander (d) a flood plain

Expanding Your Knowledge

1. Several rivers have been set aside as "wild rivers" on which dams cannot be built. Give at least four arguments against building dams on rivers. Give at least four arguments in favor of building dams.

2. Discuss the similarities between deltas and alluvial fans. Describe the differences between them.

3. What conditions are necessary to erode a region down to base level?

Exploring Resources

 Beven, K., and P. Carling, eds. 1989. *Floods.* New York: John Wiley and Sons.

Bloom, A. L. 1978. *Geomorphology: A systematic analysis of late Cenozoic landforms.* Englewood Cliffs, N.J.:Prentice-Hall.

Clark, C. 1982. *Planet earth: Flood.* Alexandria, Va.: Time-Life Books.

Czaya, E. 1981. *Rivers of the World.* New York: Van Nostrand Reinhold.

Easterbrook, D. J. 1993. *Surface processes and landforms.* New York: Macmillan.

Knighton, D. 1984. *Fluvial forms and processes.* London: Edward Arnold.

Leopold, L. B. 1974. *Water, a primer.* New York: W. H. Freeman.

———. 1994. *A view of the river.* Cambridge, Massachusetts: Harvard University Press.

Leopold, L. B., and W. B. Langbein. 1966. *River meanders. Scientific American* 214(6), 60–70.

McPhee, J. 1989. *The control of nature.* New York: Farrar, Straus, & Giroux.

Morisawa, M. 1968. *Streams: Their dynamics and morphology.* New York: McGraw-Hill.

Nuhfer, E. B., R. J. Proctor, and P. H. Moser. 1993. *The citizens' guide to geologic hazards.* Arvada, Colorado: American Institute of Professional Geologists.

Ritter, D. F., R. C. Kochel, and J. R. Miller. 1995. *Process geomorphology.* 3d ed. Dubuque, Iowa: Wm. C. Brown Publishers.

St. Louis Post Dispatch. 1993. *High and mighty: the flood of 1993.* Kansas City, Missouri: Andrews and McMeel.

Schumm, S. A. 1977. *The fluvial system.* New York: John Wiley & Sons.

Tuttle, S. D. 1980. *Landforms and landscapes.* 3d ed. Dubuque, Iowa: Wm. C. Brown Publishers.

 http://h2o.usgs.gov/ Contains information on surface water issues and specific data for stream flow throughout the U.S.

Interacting with Journey Through Geology CD-ROM

Expand your knowledge of the concepts presented this chapter by using the CD-ROM to answer the following question.

1. Stories in Rocks, Sandstone: How does the velocity of a stream affect patterns of sediment transport and deposition in streams?

Groundwater gushes from a large spring in the Redwall Limestone, Vasey's Paradise, Grand Canyon National Park, Arizona.

Photo © Kerrick James

CHAPTER
11

Ground Water

Surprisingly, water underground is about 60 times as plentiful as fresh water in lakes and rivers on the land surface (not including water stored as ice in glaciers). Ground water is a tremendously important resource. How it gets underground, where it is stored, how it moves while underground, how we look for it, and, perhaps most important of all, why we need to protect it are the main topics of this chapter.

Also important is how ground water is related to surface rivers and springs. Ground water can form distinctive geologic features, such as caves, sinkholes, and petrified wood. It also can appear as hot springs and geysers. Hot ground water can be used to generate power.

World Distribution of Water (%)	
Ocean	97.2
Glaciers and other ice	2.15
Ground water	.61
Lakes—Fresh	.009
—saline	.008
Soil moisture	.005
Atmosphere	.001
Rivers	.0001

Table 11.1   Porosity and Permeability of Sediments and Rocks

Sediment	Porosity (%)	Permeability
Gravel	25 to 40	excellent
Sand (clean)	30 to 50	good to excellent
Silt	35 to 50	moderate
Clay	35 to 80	poor
Glacial till	10 to 20	poor to moderate
Rock		
Conglomerate	10 to 30	moderate to excellent
Sandstone		
Well-sorted, little cement	20 to 30	good to very good
Average	10 to 20	moderate to good
Poorly sorted, well-cemented	0 to 10	poor to moderate
Shale	0 to 30	very poor to poor
Limestone, dolomite	0 to 20	poor to good
Cavernous limestone	up to 50	excellent
Crystalline rock		
Unfractured	0 to 5	very poor
Fractured	5 to 10	poor
Volcanic rocks	0 to 50	poor to excellent

Many communities obtain the water they need from rivers, lakes, or reservoirs, sometimes using aqueducts or canals to bring water from distant surface sources. Another source of water lies directly beneath most towns. This resource is **ground water,** the water that lies beneath the ground surface, filling the pore space between grains in bodies of sediment and clastic sedimentary rock, and filling cracks and crevices in all types of rock.

Ground water is a major economic resource, particularly in the dry western areas of the United States and Canada where surface water is scarce. Many towns and farms pump great quantities of ground water from drilled wells. Even cities next to large rivers may pump their water from the ground because ground water is commonly less polluted and more economical to use than surface water.

The source of ground water is rain and snow that falls to the ground. A portion of this precipitation percolates down into the ground to become ground water (see figure 10.1 and the Hydrologic Cycle discussion in chapter 10). How much precipitation soaks into the ground is influenced by climate, land slope, soil and rock type, and vegetation. In general, approximately 15% of the total precipitation ends up as ground water, but that varies locally and regionally from 1% to 20%.

Porosity and Permeability

Porosity, the percentage of rock or sediment that consists of voids or openings, is a measurement of a rock's ability to hold water. Most rocks can hold some water. Some sedimentary rocks, such as sandstone, conglomerate, and many limestones, tend to have a high porosity and therefore can hold a considerable amount of water. A deposit of loose sand may have a porosity of 30% to 50%, but this may be reduced to 10% to 20% by compaction and cementation as the sand lithifies (table 11.1). A sandstone in which pores are nearly filled with cement and fine-grained matrix material may have a porosity of 5% or less. Crystalline rocks, such as granite, schist, and some limestones, do not have pores but may hold some water in joints and other openings.

Although most rocks can hold some water, they vary a great deal in their ability to allow water to pass through them. **Permeability** refers to the capacity of a rock to transmit a fluid such as water or petroleum through pores and fractures. In other words, permeability measures the relative ease of water flow and indicates the degree to which openings in a rock interconnect. The distinction between porosity and permeability is important. A rock that holds much water is called *porous;* a rock that allows water to flow easily through it is described as *permeable.* Most sandstones and conglomerates are both porous and permeable. An *impermeable* rock is one that does not allow water to flow through it easily. Unjointed granite and schist are impermeable. Shale can have substantial porosity, but it has low permeability because its pores are too small to permit easy passage of water.

The Water Table

Responding to the pull of gravity, water percolates down into the ground through the soil and through cracks and pores in the rock. Several kilometers down in the crust percolation stops. With increasing depth, sedimentary rock pores tend to be closed by increasing amounts of cement and by the weight of the overlying rock. Moreover, sedimentary rock overlies igneous and metamorphic crystalline basement rock, which

B

Figure 11.1

(*A*) The water table marks the top of the saturated zone, in which water completely fills the rock pore space (inset figure). Above the water table is the vadose zone in which rock openings typically contain both air and water. (*B*) Ground water fills this abandoned limestone quarry in Indiana that is cut below the water table.

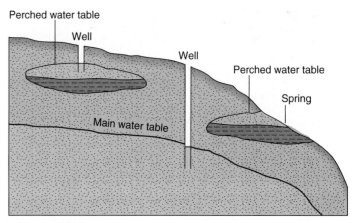

Figure 11.2

Perched water tables above lenses of less permeable shale within a large body of sandstone. Downward percolation of water is impeded by the less permeable shale.

The subsurface zone in which all rock openings are filled with water is called the **saturated zone** (figure 11.1*A*). If a well were drilled downward into this zone, ground water would fill the lower part of the well. The water level inside the well marks the upper surface of the saturated zone; this surface is the **water table.**

Most rivers and lakes intersect the saturated zone. Rivers and lakes occupy low places on the land surface, and ground water flows out of the saturated zone into these surface depressions. The water level at the surface of most lakes and rivers coincides with the water table. Ground water also flows into mines and quarries cut below the water table (figure 11.1*B*).

Above the water table there is a zone that is generally unsaturated and is referred to as the **vadose zone** (figure 11.1*A*). Within the vadose zone, capillary action causes water to be held above the water table. The *capillary fringe* is a transition zone with higher moisture content at the base of the vadose zone just above the water table. Some of the water in the capillary fringe has been drawn or wicked upward from the water table (much like water rising up a paper towel if the corner is dipped in water), whereas most of the capillary fringe water is due to fluctuations in the level of the water table. The capillary fringe is generally less than a meter thick, but may be much thicker in fine-grained sediments and thinner in coarse-grained sediments such as sand and gravel.

Plant roots generally obtain their water from the belt of soil moisture near the top of the vadose zone, where fine-grained clay minerals hold water and make it available for plant growth. Most plants "drown" if their roots are covered by water in the saturated zone; plants need both water and air in soil pores to survive. (The water-loving plants of swamps and marshes are an exception.)

A **perched water table** is the top of a body of ground water separated from the main water table beneath it by a zone that is not saturated (figure 11.2). It may form as ground water collects above a lens of less permeable shale within a more permeable rock, such as sandstone. If the perched water table intersects the land

usually has very low porosity. The depth where downward percolation stops is generally about 5 kilometers below the surface, although in thick wedges of sedimentary rock, such as those found along the Gulf Coast of the United States, percolation may reach a depth of 10 kilometers or more.

11.1
IN GREATER DEPTH

Darcy's Law and Fluid Potential

I n 1856 Henry Darcy, a French engineer, found that the velocity at which ground water moves depends upon the *hydraulic head* of the water and upon the permeability of the material that the water is moving through.

The hydraulic head of a drop of water is equal to the elevation of the drop plus the water pressure on the drop:

Hydraulic head = elevation + pressure.

In box figure 1*A* the points A and B are both on the water table, so the pressure at both points is zero (there is no water above points A and B to create pressure). Point A is at a higher elevation than B, so A has a higher hydraulic head than B. The difference in elevation is equal to the difference in head, which is labeled **h**. Water will move from point A to point B (as shown by the dark blue arrow), because water moves from a region of high hydraulic head to a region of low head. The distance the water moves from A to B is labeled **L**. The *hydraulic gradient* is the difference in head between two points divided by the distance between the two points:

$$\text{Hydraulic gradient} = \frac{\text{difference in head}}{\text{distance}} = \frac{h}{L}$$

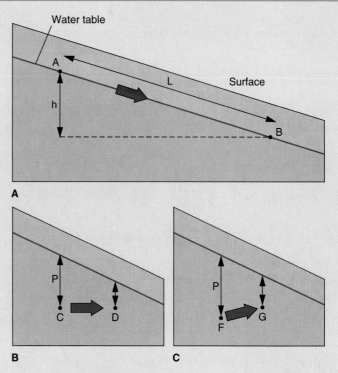

Box 11.1 Figure 1

Ground water moves in response to hydraulic head (elevation plus pressure). Water movement shown by dark blue arrows. (*A*) Points A and B have the same pressure, but A has a higher elevation, therefore water moves from A to B. (*B*) Point C has a higher pressure (arrow marked P) than D, therefore water moves from C to D at the same elevation. (*C*) Pressure also moves water upward from F to G.

In figure 1*B* the two points have equal elevation, but the pressure on point C is higher than on point D (there is more water to create pressure above point C than point D). The head is higher at point C than at point D, so the water

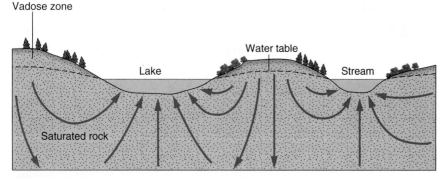

Figure 11.3

Movement of ground water beneath a sloping water table in uniformly permeable rock. Near the surface the ground water tends to flow parallel to the sloping water table.

surface, a line of springs can form along the upper contact of the shale lens. The water perched above a shale lens can provide a limited water supply to a well; it is an unreliable long-term supply.

The Movement of Ground Water

Compared to the rapid flow of water in surface streams, most ground water moves relatively slowly through rock underground. Because it moves in response to differences in water pressure and elevation, water within the upper part of the saturated zone tends to move downward following the slope of the water table (figure 11.3). See box 11.1 for Darcy's Law.

The circulation of ground water in the saturated zone is not confined to a shallow layer beneath the water table. Ground water may move hundreds of feet vertically downward before rising again to discharge as a spring or to seep into the beds of rivers and lakes at the surface (figure 11.3) due to the combined effects of gravity and the slope of the water table.

Box 11.1 Figure 2

Dark blue arrows are flow lines, which show direction of ground-water flow. Flow is perpendicular to equipotential lines (black lines with numbers), which show regions of equal fluid potential. Ground water generally flows from hilltops toward valleys, emerging from the ground through stream beds and banks (and as springs, lakes, and swamps).

moves from C to D. In figure 1*C* point F has a lower elevation than point G, but F also has a higher pressure than G. The difference in pressure is greater than the difference in elevation, so F has a higher head than G, and water moves from F to G. Note that underground water may move downward, horizontally, or upward in response to differences in head, but that it always moves in the direction of the downward slope of the water table above it. One of the first goals of ground-water geologists, particularly in pollution cases, is to find the slope of the local water table in order to determine the direction (and velocity) of ground-water movement.

Darcy found that the velocity of ground-water flow is controlled by both the permeability of the sediment or rock and the hydraulic gradient of the water. *Darcy's Law* states that ground-water

$$\text{velocity} = \text{permeability} \times \text{hydraulic gradient}$$

$$V = K\frac{h}{L}$$

(Darcy called **K** the hydraulic conductivity; it is a measure of permeability).

Most ground-water literature today, including most of the chapter-end supplementary readings, relate ground-water movement to *fluid potential*. Potential is found by multiplying the acceleration of gravity (**g**) times the hydraulic head (**h**):

$$\textbf{potential = acceleration of gravity} \times \textbf{hydraulic head}$$
$$\textbf{= gh}$$

For ground water the acceleration of gravity is constant, so potential is essentially identical to head. Modern figures show ground-water movement in relation to *equipotential lines* (lines of equal potential). Ground water moves from regions of high potential to regions of low potential (from high to low head). Figure 2 shows how *flow lines,* which show ground-water movement, cross equipotential lines at right angles as water moves from high to low potential.

The *slope of the water table* strongly influences ground-water velocity. The steeper the slope of the water table, the faster ground water moves. Water-table slope is controlled largely by topography—the water table roughly parallels the land surface (particularly in humid regions), as you can see in figure 11.3. Even in highly permeable rock, ground water will not move if the water table is flat.

How fast ground water flows also depends on the *permeability* of the rock or other materials through which it passes. If rock pores are small and poorly connected, water moves slowly. When openings are large and well connected, the flow of water is more rapid. One way of measuring ground-water velocity is to introduce a tracer, such as a dye, into the water and then watch for the color to appear in a well or spring some distance away. Such experiments have shown that the velocity of ground water varies widely, averaging a few inches to many feet a day. Nearly impermeable rocks may allow water to move only a few inches a *year,* but highly permeable materials, such as

unconsolidated gravel or cavernous limestone, may permit flow rates of hundreds or even thousands of feet per day.

Aquifers

An **aquifer** is a body of saturated rock or sediment through which water can move easily. Aquifers are both highly permeable and saturated with water. A well must be drilled into an aquifer to reach an adequate supply of water (figure 11.4). Good aquifers include sandstone, conglomerate, well-jointed limestone, bodies of sand and gravel, and some fragmental or fractured volcanic rocks such as columnar basalt (table 11.1). These favorable geologic materials are sought in "prospecting" for ground water or looking for good sites to drill water wells.

Wells drilled in shale beds are not usually very successful because shale, although sometimes quite porous, is relatively impermeable (figure 11.4). Wet mud may have a porosity of 80% to 90% and even when compacted to form shale may still

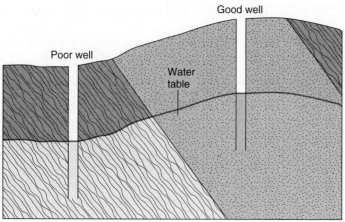

Figure 11.4

A well must hit an aquifer to obtain water. The saturated part of the highly permeable sandstone is an aquifer, but the less permeable shale is not. Although the shale is saturated, it will not readily transmit water.

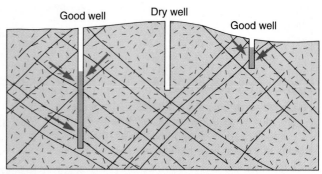

Figure 11.5

Wells can obtain some water from fractures in crystalline rock. Wells must intersect fractures to obtain water.

have a high porosity of 30%. Yet the extremely small size of the pores, together with the electrostatic attraction that clay minerals have for water molecules (see chapter 5), prevents water from moving readily through the shale into a well.

Because they are not very porous, crystalline rocks such as granite, gabbro, gneiss, schist, and some types of limestone are not good aquifers. The porosity of such rocks may be 1% or less. (Shale and crystalline rocks are sometimes called *aquitards* because they retard the flow of ground water.) Crystalline rocks that are highly fractured, however, may be porous and permeable enough to provide a fairly dependable water supply to wells (figure 11.5).

Figure 11.6 shows the difference between an **unconfined aquifer,** which has a water table because it is only partly filled

with water, and a **confined aquifer,** which is completely filled with water under pressure, and which is usually separated from the surface by a relatively impermeable confining bed, or aquitard, such as shale. An unconfined aquifer is recharged rapidly by precipitation, has a rising and falling water table during wet and dry seasons, and has relatively rapid movement of ground water through it (figure 11.7). A confined aquifer is recharged slowly through confining shale beds. With very slow movement of ground water, a confined aquifer may have no response at all to wet and dry seasons.

Wells

A **well** is a deep hole, generally cylindrical, that is dug or drilled into the ground to penetrate an aquifer within the saturated zone (figure 11.4). Usually water that flows into the well

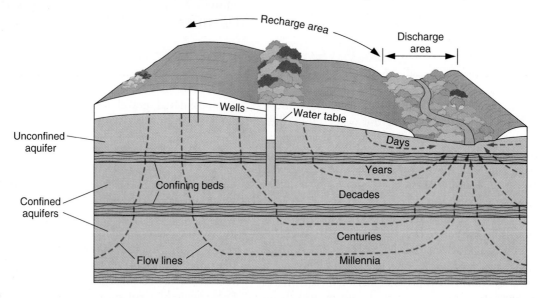

Figure 11.6

An unconfined aquifer is exposed to the surface and is only partly filled with water; water in a shallow well will rise to the level of the water table. A confined aquifer is separated from the surface by a confining bed, and is completely filled with water under pressure; water in wells rises above the aquifer. Flow lines show direction of ground-water flow. Days, years, decades, centuries, and millennia refer to the time required for ground water to flow from the recharge area to the discharge area. Water enters aquifers in recharge areas, and flows out of aquifers in discharge areas.

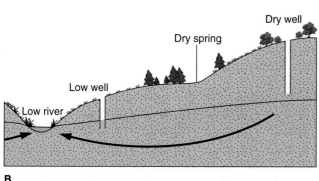

Figure 11.7

The water table in an unconfined aquifer rises in wet seasons and falls in dry seasons as water drains out of the saturated zone into rivers. (*A*) Wet season: water table and rivers are high; springs and wells flow readily. (*B*) Dry season: water table and rivers are low; some springs and wells dry up.

from the saturated rock must be lifted or pumped to the surface. As figure 11.7 shows, a well dug in a valley usually has to go down a shorter distance to hit water than a well dug on a hilltop. During dry seasons the water table falls as water flows out of the saturated zone into springs and rivers. Wells not deep enough to intersect the lowered water table go dry, but the rise of the water table during the next rainy season normally returns water to the dry wells. The addition of new water to the saturated zone is called **recharge.**

When water is pumped from a well, the water table is typically drawn down around the well into a depression shaped like an inverted cone known as a **cone of depression** (figure 11.8). This local lowering of the water table, called **drawdown,** tends to change the direction of ground-water flow by changing the slope of the water table. In lightly used wells that are not pumped, drawdown does not occur and a cone of depression does not form. In a simple, rural well with a bucket lowered on the end of a rope, water cannot be extracted rapidly enough to significantly lower the water table. A well of this type is shown in figure 11.1*A*.

In unconfined aquifers, water rises in shallow wells to the level of the water table. In confined aquifers, the water is under pressure and rises in wells to a level above the top of the aquifer (figure 11.6). Such a well is

called an **artesian well** (and confined aquifers are also called *artesian aquifers*).

In some artesian wells the water rises above the land surface, producing a flowing well that spouts continuously into the air unless it is capped (figure 11.9). Flowing wells used to occur in South Dakota, when the extensive Dakota Sandstone aquifer was first tapped (figure 11.10), but continued use has lowered the water pressure surface below the ground surface in most parts of the state. Water still rises above the aquifer, but does not reach the land surface.

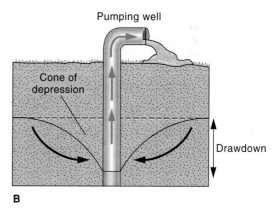

Figure 11.8

Pumping well lowers the water table into a cone of depression.

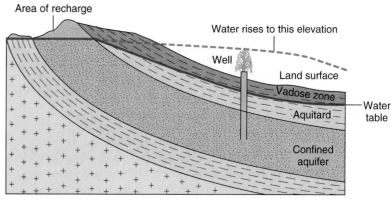

Figure 11.9

The Dakota Sandstone in South Dakota is a relatively unusual type of confined aquifer because it is tilted and exposed to the surface by erosion. Water in most wells rose above the land surface when the aquifer was first tapped in the 1800s.

Figure 11.10

Artesian well spouts water above land surface in South Dakota, early 1900s. Heavy use of this aquifer has reduced water pressure so much that spouts do not occur today.

Photo by N. H. Darton, U.S. Geological Survey

Springs and Streams

A **spring** is a place where water flows naturally from rock onto the land surface (figure 11.11). Some springs discharge where the water table intersects the land surface, but they also occur where water flows out from caverns or along fractures, faults, or rock contacts that come to the surface (figure 11.12).

Climate determines the relationship between stream flow and the water table. In rainy regions most streams are **gaining streams;** that is, they receive water from the saturated zone (figure 11.13). The surface of these streams coincides with the water table. Water from the saturated zone flows into the stream through the stream bed and banks that lie below the water table. Because of the added ground water, the discharge of these streams increases downstream. Where the water table intersects the land surface over a broad area, ponds, lakes, and swamps are found.

In drier climates rivers tend to be **losing streams;** that is, they are losing water to the saturated zone (figure 11.13). The channels of losing streams lie above the water table. The water percolating into the ground beneath a losing stream may cause the water table below the stream to rise. This ground-water mound remains beneath the stream even when the stream bed is dry, and in a desert the nearest source of water may lie a short distance under a dry stream bed.

A

B

Figure 11.11

(*A*) A large spring issuing from a cavern in limestone, Jasper National Park, Alberta, Canada. (*B*) A line of springs seeps from the ground at the contact between less permeable shale and the overlying permeable sandstone. Southern Utah.

Pollution of Ground Water

Ground water in its natural state tends to be relatively free of contaminants in most areas. Because it is a widely used source of drinking water, pollution of ground water can be a very serious problem.

Pesticides and *herbicides* (such as DDT and 2,4-D) applied to agricultural crops (figure 11.14*A*) can find their way into ground water when rain or irrigation water leaches the poisons downward into the soil. *Fertilizers* are also a concern. Nitrate, one of the most widely used fertilizers, is harmful in even small quantities in drinking water.

Figure 11.12

Springs can form in many ways. (*A*) Water moves along fractures in crystalline rock and forms springs where the fractures intersect the land surface. (*B*) Water enters caves along joints in limestone and exits as springs at the mouths of caves. (*C*) Springs form at the contact between a permeable rock such as sandstone and an underlying less permeable rock such as shale. (*D*) Springs can form along faults when permeable rock has been moved against less permeable rock. Arrows show relative motion along fault.

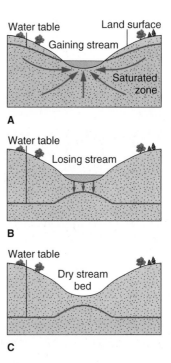

Figure 11.13

Gaining and losing streams. (*A*) Stream gaining water from saturated zone. (*B*) Stream losing water through stream bed to saturated zone. (*C*) Water table can be close to the land surface beneath a dry stream bed.

Rain can also leach pollutants from city dumps into ground-water supplies (figure 11.14*B*). Consider for a moment some of the things you threw away last year. A partially empty aerosol can of ant poison? The can will rust through in the dump, releasing the poison into the ground and into the saturated zone below. A broken thermometer? The toxic mercury may eventually find its way to the ground-water supply. A half-used can of oven cleaner? The dried-out remains of a can of lead-base paint? *Heavy metals* such as mercury, lead, chromium, copper, and cadmium, together with household chemicals and poisons, can all be concentrated in ground-water supplies beneath dumps (figure 11.15).

Liquid and solid wastes from septic tanks, sewage plants, and animal feedlots and slaughterhouses may contain *bacteria, viruses,* and *parasites* that can contaminate ground water (figure 11.14*C*). Liquid wastes from industries (figure 11.14*D*) and military bases can be highly toxic, containing high concentrations of heavy metals and compounds such as cyanide and *PCBs* (polychlorinated biphenyls), which are widely used in industry. A degreaser called *TCE* (trichloroethylene) has been increasingly found to pollute both surface and underground water in numerous regions. Toxic liquid wastes are often held in surface ponds or pumped down deep disposal wells. If the

Figure 11.14

Some sources of ground-water pollution. (*A*) Pesticides. (*B*) Household garbage. (*C*) Animal waste. (*D*) Industrial toxic waste.

Photo *A* by Michael Stimmann; photo *B* by Frank M. Hanna; photos *C* and *D* from USDA-Soil Conservation Service

ponds leak, ground water can become polluted. Deep wells may be safe for liquid waste disposal if they are deep enough, but contamination of drinking water supplies and even surface water has resulted in some localities from improper design of the disposal wells.

Acid mine drainage from coal and metal mines can contaminate both surface and ground water. It is usually caused by sulfuric acid formed by the oxidation of sulfur in pyrite and other sulfide minerals when they are exposed to air by mining activity. Fish and plants are often killed by the acid waters draining from long-abandoned mines.

Radioactive waste is both an existing and a very serious potential source of ground-water pollution. The shallow burial of *low-level* solid and liquid radioactive wastes from the nuclear power industry has caused contamination of ground water, particularly as liquid waste containers leak into the saturated zone and as the seasonal rise and fall of the water table at some sites periodically covers the waste with ground water. The search for a permanent disposal site for solid, *high-level* radioactive waste (now stored temporarily on the surface) is a major national concern for the United States. The permanent site will be deep underground and must be isolated from ground-water circulation for thousands of years. Salt beds, shale, glassy tuffs, and crystalline rock deep beneath the surface have all been studied, particularly in arid regions where the water table is hundreds of feet below the land surface. The likely site for disposal of high-level waste, primarily spent fuel from nuclear reactors, is Yucca Mountain, Nevada, 180 km (110 miles) northwest of Las Vegas. The site would be deep underground in volcanic tuff well above the current (or predicted future) water table, and in a region of very low rainfall. The U.S. Congress,

A Cross section

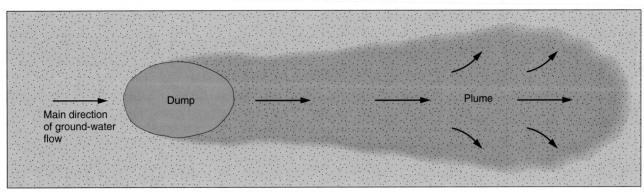

B Map view of contaminant plume. Note how it grows in size with distance from the pollution source.

Figure 11.15

Dump waste piled on the land surface creates a ground-water mound beneath it because the dump forms a hill, and because the waste material is more porous and permeable than the surrounding soil and rock. Rain leaches pollutants into the saturated zone. A plume of polluted water will spread out in the direction of ground-water flow.

under intense political pressure from other candidate states who did not want the site, essentially chose Nevada in late 1988 by eliminating the funding for the study of all alternative sites, but the final decision regarding the safety of Yucca Mountain will not be made until after much additional study. Even if the site is deemed safe, it could not open before the year 2010. It could be delayed much later than this: in 1992, a 5.6-magnitude aftershock of the Landers, California, earthquake occurred 19 kilometers (12 miles) from the proposed disposal site. The quake caused $1 million damage to a U.S. Energy Department office building near the site, and may indicate that the region is too seismically active for the site to be built here at all.

Not all ground-water pollutants form plumes within the saturated zone as shown in figure 11.15. *Gasoline,* which leaks from gas station storage tanks at tens of thousands of U.S. locations, is less dense than water, and floats upon the water table (figure 11.16). Some liquids are heavier than water such as TCE and sink to the bottom of the saturated zone, perhaps traveling in unpredicted directions upon the surface of an impermeable layer (figure 11.16). Determining the extent and flow direction of ground-water pollution is a lengthy process requiring the drilling of tens, or even hundreds, of costly wells for each pollution site.

Not all sources of ground-water pollution are manmade. Naturally occurring *minerals within rock and soil* may

11.2 ENVIRONMENTAL GEOLOGY

Prospecting for Ground Water

Many wells are drilled or dug without any effort to locate a good aquifer. Many of these wells are successful (especially if only small amounts of water are needed) because most rocks hold some water, which flows into wells that intersect the water table. However, if a large and dependable supply of water is needed—as for a city water system—specialists in ground-water geology (hydrogeologists) may be called in to locate a promising well site. Hydrogeologists use many methods to locate aquifers. A detailed knowledge of the local rocks is necessary. Therefore, a geologist may map the surface rocks and use electrical, magnetic, and seismic surveys to study subsurface rocks to determine the presence of possible aquifers. Sometimes a small-diameter test well is drilled before the larger, more expensive supply well is sunk. Hydrogeologists look for potentially high-producing aquifers rather than searching for water directly, which would be much more difficult. In some regions, however, the presence of certain plants may be a useful guide to locating water, particularly the depth of the water table.

Some people search for water by water witching, or *dowsing*, with a divining rod (also sometimes used to search for metals or lost objects). Usually the dowser holds a forked stick horizontally in the hands while walking over an area. The stick is supposed to deflect or twist downward of its own accord when the dowser passes over water. This method has been tried for centuries, the only modification being that a twisted metal rod, often made of a coat hanger, now may be substituted for the stick. Carefully controlled tests conducted by workers in psychic research have shown that water witchers' "success" is equal to or less than pure chance, while geologists' results are superior both to witching and to chance. Records kept on thousands of wells in Australia in the early 1900s show that of wells that were not divined, more than 83% produced flows of 100 gallons per hour and 7.4% were failures, finding no water at all. Of wells that were divined by water witchers, only about 70% produced more than 100 gallons per hour, and 14.7% were dry. In the early part of this century the U.S. Geological Survey concluded that any future testing of the results of water witching would be a misuse of public funds.

Despite such findings, many people believe strongly in dowsing. Water witchers themselves devoutly believe they can find water, and in some regions of the United States almost no wells are drilled without a witcher's advice. Dowsers are helped in locating water by the fact that most rocks hold some water, and dowsers often have a long-standing knowledge of a particular region and its potential water resources. This is not to say that dowsers are deliberate frauds. Many are convinced that they perform a valuable public service, and some do not charge for their services. Scientists see no reason for dowsing to work and are skeptical about dowsers' "success." Geologists would almost unanimously urge you not to pay for a dowser's service if a fee were charged.

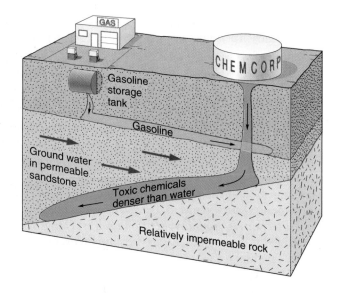

contain elements such as arsenic, selenium, mercury, and other toxic metals. Circulating ground water can leach these elements out of the minerals and raise their concentrations to harmful levels within the water. Not all spring water is safe to drink. Like a "bad waterhole" depicted in a Western movie, some springs contain such high levels of toxic elements that the water can sicken or kill humans and animals

Figure 11.16

Not all pollutants move within the saturated zone as shown in figure 11.15. Gasoline floats on water; many dense chemicals move along impermeable rock surfaces below the saturated zone.

11.3
ENVIRONMENTAL GEOLOGY

Hard Water and Soapsuds

"Hard water" is water that contains relatively large amounts of dissolved calcium (often from the chemical weathering of calcite or dolomite) and magnesium (from the ferromagnesian minerals or dolomite). Water taken from the ground-water supply or from a stream for home use may contain enough of these ions to prevent soap from lathering. Calcium and magnesium ions in hard water form gray curds with soap. The curd continues to form until all the calcium and magnesium ions are removed from the water and bound up in the curd. Only then will soap lather and clean laundry. Cleaning laundry in hard water therefore takes an excessively large amount of soap.

Hard water may also precipitate a scaly deposit inside teakettles and hot-water tanks and pipes (box figure 1). The entire hot-water piping system of a home in a hard-water area eventually can become so clogged that the pipes must be replaced.

"Soft water" may carry a substantial amount of ions in solution but not the ions that prevent soap from lathering. Water softeners in homes replace calcium and magnesium ions with sodium ions, which do not affect

Box 11.3 Figure 1

Scale caused by hard water coats the inside of this hot-water pipe.

Photo by Hauser Water Conditioning, Inc.

lathering or cause scale. However, water containing a large amount of sodium ions, whether from a softener or from natural sources, may be harmful if used as drinking water by persons on a "salt-free" (low sodium) diet for health reasons.

that drink it. Many desert springs contain such high concentrations of sodium chloride or other salts that their water is undrinkable.

Soil and rock filter some contaminants out of ground water. This filtering ability depends on the permeability and mineral composition of rock and soil. Under ideal conditions, human sewage can be purified by only 30 to 45 meters (100 to 150 feet) of travel through a sandy loam soil (a mixture of clay minerals, sand, and organic humus). The sewage is purified by filtration, ion absorption by clay minerals and humus, and decomposition by soil organisms (figure 11.17). On the other

hand, extremely permeable rock, such as highly fractured granite or cavernous limestone, has little purifying effect on sewage. Ground water flows so rapidly through such rocks that it is not purified even after hundreds of feet of travel. Some pesticides and toxic chemicals are not purified by passage through rock and soil at all, even soil rich in humus and clay minerals.

Polluted ground water is extremely difficult to clean up. Networks of expensive wells may be needed to pump contaminated water out of the ground and replace it with clean water. Because of the slow movement and large volume of ground

Figure 11.17

Rock type and distance control possible sewage contamination of neighboring wells. (*A*) As little as 30 meters (100 feet) of movement can effectively filter human sewage in sandstone and some other rocks and sediments. (*B*) If the rock has large open fractures, contamination can occur many hundreds of meters away.

water, the clean-up process for a large region can take decades and tens of millions of dollars to complete.

Ground-water pollution can be largely prevented with careful thought and considerable expense. A city dump can be sited high above the water table and possible flood levels, or located in a region of ground-water discharge rather than recharge. A site can be sealed below by impermeable (and expensive) clay barriers and plastic liners, and sealed off from rainfall by an impermeable cover. Dikes can prevent surface runoff through or from the site. Although nonpolluting waste disposal is now expensive, it is much cheaper than ground-water cleanup.

Pumping wells can cause or aggravate ground-water pollution (figure 11.18). Well drawdown can increase the slope of the water table locally, thus increasing the rate of ground-water flow and giving the water less time to be purified underground before it is used (figure 11.18*A*). Drawdown can even reverse the original slope of the water table, perhaps contaminating wells that were pure before pumping began (figure 11.18*B*). Heavily pumped wells near a coast can be contaminated by *saltwater intrusion* (figure 11.18*C* and *D*).

Balancing Withdrawal and Recharge

A local supply of ground water will last indefinitely if it is withdrawn for use at a rate equal to or less than the rate of recharge to the aquifer. If ground water is withdrawn faster than it is being recharged, however, the supply is being reduced and will one day be gone.

Heavy use of ground water causes a regional water table to drop. In parts of western Texas and eastern New Mexico the pumping of ground water has caused the water table to drop 30 meters (100 feet) over the past few decades. The lowering of the water table means that wells must be deepened and more electricity must be used to pump the water to the surface. Moreover, as water is withdrawn, the ground surface may settle because the water no longer supports the rock and sediment. Mexico City has subsided more than 7 meters (23 feet) and portions of California's Central Valley 9 meters (30 feet) because of extraction of ground water (figure 11.19). Such *subsidence* can crack building foundations, roads, and

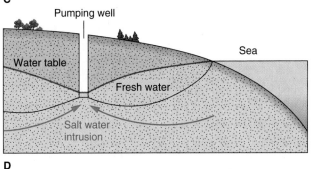

Figure 11.18

Ground-water pollution problems caused or aggravated by pumping wells. (*A*) Water table steepens near a dump, increasing the velocity of ground-water flow and drawing pollutants into a well. (*B*) Water-table slope is reversed by pumping, changing direction of the ground-water flow, and polluting the well. (*C*) Well near a coast (before pumping). Fresh water floats on salt water. (*D*) Well in *C* begins pumping, thinning the freshwater lens and drawing salt water into the well.

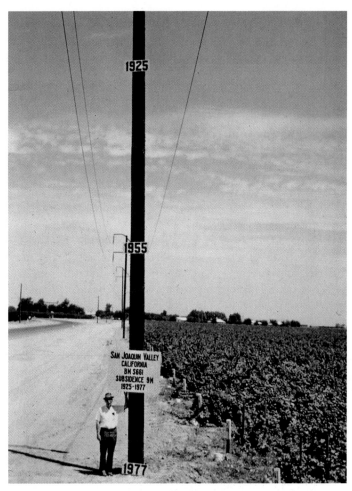

Figure 11.19

Subsidence of the land surface caused by the extraction of ground water, near Mendota, San Joaquin Valley, California. Signs on the pole indicate the positions of the land surface in 1925, 1955, and 1977. The land sank 9 meters (30 feet) in 52 years.

Photo by Richard O. Ireland, U.S. Geological Survey

pipelines. Overpumping of ground water also causes compaction and porosity loss in rock and soil, and can permanently ruin good aquifers.

To avoid the problems of falling water tables, subsidence, and compaction, many towns use *artificial recharge* to increase the natural rate of recharge. Natural floodwaters or treated industrial or domestic wastewaters are stored in infiltration ponds on the surface to increase the rate of water percolation into the ground. Reclaimed, clean water from sewage treatment plants is commonly used for this purpose. In some cases, especially in areas where ground water is under confined conditions, water is actively pumped down into the ground to replenish the ground-water supply. This is more expensive than filling surface ponds, but it reduces the amount of water lost through evaporation.

Effects of Ground-Water Action

Caves, Sinkholes, and Karst Topography

Caves (or **caverns**) are naturally formed underground chambers. Most caves develop when slightly acidic ground water dissolves limestone along joints and bedding planes, opening up cavern systems as calcite is carried away in solution (figure 11.20). Natural ground water is commonly slightly acidic because of dissolved carbon dioxide (CO_2) from the atmosphere or from soil gases (see chapter 5).

Geologists disagree whether limestone caves form above, below, or at the water table. Most caves probably are formed by ground water circulating below the water table, as shown in figure 11.20. If the water table drops or the land is elevated above the water table, the cave may begin to fill in again by calcite precipitation. Read the equation below from left to right for calcite solution, and from right to left for the calcite precipitation reaction (see also table 5.1).

Ground water with a high concentration of calcium (Ca^{++}) and bicarbonate (HCO_3^-) ions may drip slowly from the ceiling of an air-filled cave. As a water drop hangs on the ceiling of the cave, some of the dissolved carbon dioxide (CO_2) may be lost into the cave's atmosphere. The CO_2 loss causes a small amount of calcite to precipitate out of the water onto the cave ceiling. When the water drop falls to the cave floor, the impact may cause more CO_2 loss, and another small amount of calcite may precipitate on the cave floor. A falling water drop, therefore, can precipitate small amounts of calcite on both the cave ceiling and the cave floor and each subsequent drop adds more calcite to the first deposits.

Deposits of calcite (and, rarely, other minerals) built up in caves by dripping water are called *dripstone.* **Stalactites** are icicle-like pendants of dripstone hanging from cave ceilings (figure 11.20*B*). They are generally slender and are commonly aligned along cracks in the ceiling, which act as conduits for ground water. **Stalagmites** are cone-shaped masses of dripstone formed on cave floors, generally directly below stalactites. Splashing water precipitates calcite over a large area on the cave floor, so stalagmites are usually thicker than the stalactites above them. As a stalactite grows downward and a stalagmite grows upward, they may eventually join to form a *column* (figure 11.20*B*). Figure 11.21 shows some of the intriguing features formed in caves.

$$H_2O \quad + \quad CO_2 \quad + \quad CaCO_3 \quad \rightleftarrows \quad Ca^{++} \quad + \quad 2HCO_3^-$$

water · carbon dioxide · calcite in limestone · calcium ion · bicarbonate ion

development of caves (solution) →

← development of flowstone and dripstone (precipitation)

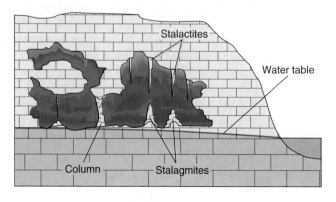

In parts of some caves, water flows in a thin film over the cave surfaces rather than dripping from the ceiling. Sheetlike or ribbonlike *flowstone* deposits develop from calcite that is precipitated by flowing water on cave walls and floors.

The floors of most caves are covered with sediment, some of which is *residual clay,* the fine-grained particles left behind as insoluble residue when a limestone containing clay dissolves. (Some limestone contains only about 50% calcite.) Other sediment, including most of the coarse-grained material found on cave floors, may be carried into the cave by streams, particularly when surface water drains into a cave system from openings on the land surface.

Solution of limestone underground may produce features that are visible on the surface. Extensive cavern systems can undermine a region so that roofs collapse and form depressions in the land surface above. **Sinkholes** are closed

Figure 11.20

Solution of limestone to form caves. (*A*) Water moves along fractures and bedding planes in limestone, dissolving the limestone to form caves below the water table. (*B*) Falling water table allows cave system, now greatly enlarged, to fill with air. Calcite precipitation forms stalactites, stalagmites, and columns above the water table.

depressions found on land surfaces underlain by limestone (figure 11.22). They form either by the *collapse* of a cave roof or by *solution* as descending water enlarges a crack in limestone. Limestone regions in Florida, Missouri, Indiana, and Kentucky are heavily dotted with sinkholes. Sinkholes can also form in regions underlain by gypsum or rock salt, which are also soluble in water.

An area with many sinkholes and with cave systems beneath the land surface is said to have **karst topography** (figure 11.23). Karst areas are characterized by a lack of surface streams, although one major river may flow at a level lower than the karst area.

Streams sometimes disappear down sinkholes to flow through caves beneath the surface. In this specialized instance, a true *underground stream* exists. Such streams are quite rare, however, as most ground water flows very slowly through pores and cracks in sediment or rock. You may hear people with wells describe the "underground stream" that their well penetrates, but this is almost never the case. Wells tap ground water in the rock pores and crevices, not underground streams. If a well did tap a true underground river in a karst region, the water would probably be too polluted to drink, especially if it had washed down from the surface into a cavern without being filtered through soil and rock.

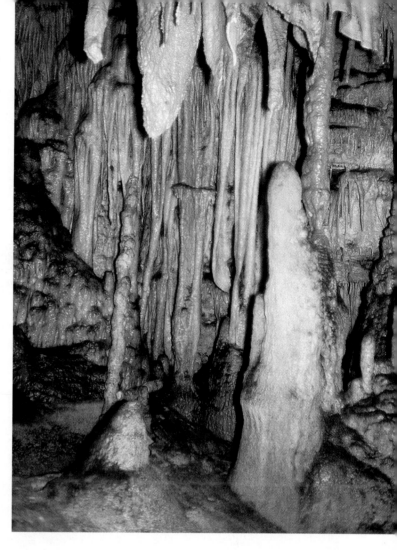

Figure 11.21

Stalactites, stalagmites, and flowstone in Great Onyx Cave, Kentucky.

Photo courtesy Stanley Fagerlin

A

B

Figure 11.22

(*A*) Trees grow in a sinkhole formed in limestone near Mammoth Cave, Kentucky. (*B*) A collapse sinkhole that formed suddenly in Winter Park, Florida, in 1981.

Photo *A* courtesy Ward's Natural Science. Photo *B* by Richard A. Davis, Jr.

Ground Water   275

Figure 11.23

Karst topography is marked by underground caves and numerous surface sinkholes. A major river may cross the region, but small surface streams generally disappear down sinkholes.

Figure 11.24

Petrified log in the Painted Desert, Arizona. Small amounts of iron and other elements color the silica in the log.

Photo © Eric & David Hosking/Corbis Media

Figure 11.25

Concretions that have weathered out of shale. Concretions contain more cement than the surrounding rock and therefore are very resistant to weathering.

Other Effects

Ground water is important in the preservation of *fossils* such as **petrified wood,** that develops when porous buried wood is either filled in or replaced by inorganic silica carried in by ground water (figure 11.24). The result is a hard, permanent rock, commonly preserving the growth rings and other details of the wood. Calcite or silica carried by ground water can also replace the original material in marine shells and animal bones.

Sedimentary rock *cement,* usually silica or calcite, is carried into place by ground water. When a considerable amount of cementing material precipitates locally in a rock, a hard rounded mass called a **concretion** develops, typically around an organic nucleus such as a leaf, tooth, or other fossil (figure 11.25).

Geodes are partly hollow, globe-shaped bodies found in some limestones and locally in other rocks. The outer shell is amorphous silica, and well-formed crystals of quartz, calcite, or other minerals project inward toward a central cavity (figure 11.26). The origin of geodes is complex but clearly related to ground water. Crystals in geodes may have filled original cavities or have replaced fossils or other crystals.

In arid and semiarid climates, *alkali soil* may develop because of the precipitation of great quantities of sodium salts by evaporating ground water. Such soil is generally unfit for plant growth. Alkali soil generally forms at the ground surface in low-lying areas. (See chapter 5.)

Hot Water Underground

Hot springs are springs in which the water is warmer than human body temperature. Water can gain heat in two ways while it is underground. First, and more commonly, ground water may circulate near a magma chamber or a body of cooling igneous rock. In the United States most hot springs are found in the western states where they are associated with relatively recent volcanism. The hot springs and pools of Yellowstone National Park in Wyoming are of this type.

Ground water can also gain heat if it circulates unusually deeply in the earth, perhaps along joints or faults. As discussed in chapter 3, the normal geothermal gradient (the increase in temperature with depth) is 25°C/kilometer (about 75°F/mile). Water circulating to a depth of 2 or 3 kilometers is warmed substantially above normal surface water temperature. The famous springs at Warm Springs, Georgia, have been warmed by deep circulation. Warm water, regardless of its origin, is lighter than cold water and readily rises to the surface.

A **geyser** is a type of hot spring that periodically erupts hot water and steam. The water is generally near boiling (100°C). Eruptions may be caused by a constriction in the underground "plumbing" of a geyser, which prevents the water from rising and cooling. The events thought to lead to a geyser eruption are illustrated in figure 11.27. Water gradually

Figure 11.26

Geodes. Concentric layers of amorphous silica are lined with well-formed quartz crystals growing inward toward a central cavity. (Scale is in centimeters.)

Figure 11.27

Eruptive history of a typical geyser in A through D. Photo shows the eruption of Old Faithful geyser in Yellowstone National Park, Wyoming. See text for explanation.

Photo ©Hal Beral/Visuals Unlimited

Figure 11.28

Precipitation of calcite in the form of travertine terraces around a hot spring (Mammoth Hot Springs, Yellowstone National Park). Algae living in the hot water provide the color.

Photo by Diane Carlson

seeps into a partially emptied geyser chamber and heat supplied from below slowly warms the water. Bubbles of water vapor and other gases then begin to form as the temperature of the water rises. The bubbles may clog the constricted part of the chamber until the upward pressure of the bubbles pushes out some of the water above in a gentle surge, thus lowering the pressure on the water in the lower part of the chamber. This drop in pressure causes the chamber water, now very hot, to flash into vapor. The expanding vapor blasts upward out of the chamber, driving hot water with it and condensing into visible steam. The chamber, now nearly empty, begins to fill again and the cycle is repeated. The entire cycle may be quite regular, as it is in Yellowstone's Old Faithful Geyser, which averages about 65 minutes between eruptions (though it varies from about 30 to 95 minutes). Many geysers, however, erupt irregularly, some with weeks or months between eruptions.

As hot ground water comes to the surface and cools, it may precipitate some of its dissolved ions as minerals. *Travertine* is a deposit of *calcite* that often forms around hot springs (figure 11.28), while dissolved *silica* precipitates as *sinter* (called *geyserite* when deposited by a geyser, as shown in figure 11.29). The composition of the subsurface rocks generally determines which type of deposit forms, although sinter can indicate higher subsurface temperatures than travertine because silica is harder to dissolve than calcite. Both deposits can be stained by the pigments of algae living in the hot water. The algae can be used to estimate water temperature because their color changes from green to brown to orange to yellow as the temperature rises.

A *mudpot* is a special type of hot spring that contains thick, boiling mud. Mudpots are usually marked by a small amount of water and strongly sulfurous gases, which combine

Figure 11.29

Geyserite deposits around the vent of Castle Geyser, Yellowstone National Park.

to form strongly acidic solutions. The mud probably results from intense chemical weathering of the surrounding rocks by these strong acids (see figure 5.16).

Geothermal Energy

Electricity can be generated by harnessing naturally occurring steam and hot water in areas that are exceptionally hot underground. In such a *geothermal area,* wells can tap steam (or superheated water that can be turned into steam) that is then piped to a powerhouse where it turns a turbine that spins a generator, creating electricity.

Geothermal energy production requires no burning of fuel, so the carbon dioxide emissions of power plants that burn coal, oil, or natural gas are eliminated (as are the nuclear waste and dangers of nuclear power plants). Although geothermal energy is relatively clean, it has some

Figure 11.30

Geothermal power plant at The Geysers, California. Underground steam, piped from wells to the power plant, is being discharged from the cooling towers in the background.

Photo by M. Smith, U.S. Geological Survey

environmental problems. Workers need protection from toxic hydrogen sulfide gas in the steam, and the hot water commonly contains dissolved ions and metals, such as lead and mercury, that can kill fish and plants if discharged on the surface. Geothermal fluids are often highly corrosive to equipment, and their extraction can cause land subsidence. Pumping wastewater underground can help reduce subsidence problems.

Geothermal fields can be depleted. The largest field in the world is at The Geysers in California (figure 11.30), 120 kilometers (80 miles) north of San Francisco. The Geysers field increased its capacity in recent years to 2,000 megawatts of electricity (enough for 2 million people), but production has declined, and the field may soon run out of steam.

Nonelectric uses of geothermal energy include space heating (in Boise, Idaho; Klamath Falls, Oregon; and Reykjavik, the capital of Iceland), as well as paper manufacturing, ore processing, and food preparation.

Summary

About 15 percent of the water that falls on land percolates underground to become ground water. Ground water fills pores and joints in rock, creating a large reservoir of usable water in most regions.

Porous rocks can hold water. *Permeable* rocks permit water to move through them.

The *water table* is the top surface of the *saturated zone* and is overlain by the *vadose zone*.

Local variations in rock permeability may develop a *perched water table* above the main water table.

Ground-water velocity depends on rock permeability and the slope of the water table.

An *aquifer* is porous and permeable and can supply water to wells. A *confined aquifer* holds water under pressure, which can create *artesian wells*.

Gaining streams, springs, and lakes form where the water table intersects the land surface. *Losing streams* contribute to the ground water in dry regions.

Ground water can be polluted by city dumps, agriculture, industry, or sewage disposal. Some pollutants can be filtered out by passage of the water through moderately permeable geologic materials.

A pumped well causes a *cone of depression* that in turn can cause or aggravate ground-water pollution. Near a coast, it can cause *saltwater intrusion.*

Artificial *recharge* can help create a balance between withdrawal and recharge of ground-water supplies, and help prevent subsidence.

Solution of limestone by ground water forms *caves, sinkholes,* and *karst topography.* Calcite precipitating out of ground water forms *stalactites* and *stalagmites* in caves.

Precipitation of material out of solution by ground water helps form petrified wood, other fossils, sedimentary rock cement, concretions, geodes, and alkali soils.

Geysers and *hot springs* occur in regions of hot ground water. *Geothermal energy* can be tapped to generate electricity.

Terms to Remember

aquifer 263
artesian well 265
cave (cavern) 274
concretion 276
cone of depression 265
confined (artesian) aquifer 264
drawdown 265
gaining stream 266
geode 276
geyser 277

ground water 260
hot spring 277
karst topography 275
losing stream 266
perched water table 261
permeability 260
petrified wood 276
porosity 260
recharge 265
saturated zone 261

sinkhole 274
spring 266
stalactite 274
stalagmite 274
unconfined aquifer 264
vadose zone 261
water table 261
well 264

Testing Your Knowledge

Use the questions below to prepare for exams based on this chapter.

1. What conditions are necessary for an artesian well?

2. What distinguishes a geyser from a hot spring? Why does a geyser erupt?

3. What is karst topography? How does it form?

4. What chemical conditions are necessary for caves to develop in limestone? For stalactites to develop in a cave?

5. What causes a perched water table?

6. Describe several ways in which ground water can become polluted.

7. Discuss the difference between porosity and permeability.

8. What is the water table? Is it fixed in position?

9. Sketch four different origins for springs.

10. What controls the velocity of ground-water flow?

11. Name several geologic materials that make good aquifers. Define *aquifer*.

12. How does petrified wood form?

13. What happens to the water table near a pumped well?

14. How does a confined aquifer differ from an unconfined aquifer?

15. Porosity is (a) the percentage of a rock's volume that is openings (b) the capacity of a rock to transmit a fluid (c) the ability of a sediment to retard water (d) none of the above

16. Permeability is (a) the percentage of a rock's volume that is openings (b) the capacity of a rock to transmit a fluid (c) the ability of a sediment to retard water (d) none of the above

17. The subsurface zone in which all rock openings are filled with water is called the (a) saturated zone (b) water table (c) vadose zone

18. An aquifer is (a) a body of saturated rock or sediment through which water can move easily (b) a body of rock that retards the flow of ground water (c) a body of rock that is impermeable

19. Which rock type would make the best aquifer? (a) shale (b) mudstone (c) sandstone (d) all of the above

20. Which of the following determines how quickly ground water flows? (a) elevation (b) water pressure (c) permeability (d) all of the above

21. Ground water flows (a) always downhill (b) from areas of high hydraulic head to low hydraulic head (c) from high elevation to low elevation (d) from high permeability to low permeability

22. The drop in the water table around a pumped well is the (a) drawdown (b) hydraulic head (c) porosity (d) fluid potential

Expanding Your Knowledge

1. Describe any difference between the amounts of water that would percolate downward to the saturated zone beneath a flat meadow in northern New York and beneath a rocky hillside in southern Nevada. Discuss the factors that control the amount of percolation in each case.

2. Where should high-level nuclear waste from power plants be stored? If your state or community uses nuclear power, where is your local waste stored?

3. Should all polluted ground water be cleaned up? How much money has been set aside by the federal government for cleaning polluted ground water? Who should pay for ground-water cleanup if the company that polluted the water no longer exists? Should some aquifers be deliberately left contaminated if there is no current use of the water, or if future use could be banned?

4. Why are most of North America's hot springs and geysers in the western states and provinces?

Exploring Resources

Baldwin, H. L., and C. I. McGuinness. 1963. *A primer on ground water.* Washington, D.C.: U.S. Geological Survey.

Bouwer, H. 1978. *Groundwater hydrology.* New York: McGraw-Hill.

Davis, S. N., and R. J. M. De Wiest. 1966. *Hydrogeology.* New York: John Wiley & Sons.

Driscoll, F. G. 1986. *Groundwater and wells.* 2d ed. St. Paul, Minnesota: Johnson Division.

Fetter, C. W. 1993. *Applied hydrogeology.* 3d ed. New York: Macmillan Publishing Company, Inc.

———. 1993. *Contaminant hydrogeology.* New York: Macmillan Publishing Company, Inc.

Freeze, R. A., and J. A. Cherry. 1979. *Groundwater.* Englewood Cliffs, N.J.: Prentice-Hall.

Heath, R. C. 1983. *Basic ground-water hydrology.* Washington, D.C.: U.S. Geological Survey Water-Supply Paper 2220.

———1984. *Ground-water regions of the United States.* Washington, D.C.: U.S. Geological Survey Water-Supply Paper 2242.

Jennings, J. N. 1983. Karst landforms. *American Scientist* 71(6): 578–86.

Keefer, W. R. 1971. *The geologic story of Yellowstone National Park.* U.S. Geological Survey Bulletin 1347.

Leopold, L. B. 1974. *Water: A primer.* San Francisco: W. H. Freeman & Co.

Leopold, L. B., and W. B. Langbein. 1960. *A primer on water.* Washington, D.C.: U.S. Geological Survey.

Moore, G. W., and G. Nicholas. 1964. *Speleology: The study of caves.* Boston: D. C. Heath.

Palmer, A. N. 1991. *Origin and morphology of limestone caves.* Geological Society of America Bulletin, v. 103, pp. 1–21.

Palmer, C. M. 1992. *Principles of contaminant hydrogeology.* Chelsea, Michigan: Lewis Publishers, Inc.

Pye, V. I., R. Patrick, and J. Quarles. 1983. *Groundwater contamination in the United States.* Philadelphia: Univ. of Pennsylvania.

Ritter, D. F., R. C. Kochel, and J. R. Miller. 1995. *Process geomorphology.* 3d ed. Dubuque, Iowa: Wm. C. Brown Publishers.

Swenson, H. A., and H. L. Baldwin. 1965. *A primer on water quality.* Washington, D.C.: U.S. Geological Survey.

Todd, D. K. 1980. *Ground water hydrology.* 2d ed. New York: John Wiley & Sons.

Waller, R. M. 1988. *Ground water and the rural homeowner.* Washington, D.C.: U.S. Geological Survey General Interest Publication.

Waltham, T. 1975. *Caves.* New York: Crown Publishers.

 http://toxics.usgs.gov/toxics/
Various sites and information about clean up of toxics in surface and ground water.

http://water.usgs.gov/public/wid/html/bioremed.html
Information about using bioremediation to clean up toxics in the soil, surface and ground water.

http://water.wr.usgs.gov/gwatlas/index.html
Ground Water Atlas for the United States. Good general information about aquifers.

http://h2o.usgs.gov/
Good general web site that has a lot of links to water topics in the United States from the USGS.

http://www.goodearth.com/virtcave.html
Contains a virtual field trip of caves and many pictures of cave formations.

http://www.marcusj.com/cave.htm and http://www.goodearth.com/under_earth.html
Virtual tour and pictures from Jewell Cave in South Dakota and Lechuigilla Cave in New Mexico.

Interacting with Journey Through Geology CD-ROM

Expand your knowledge of the concepts presented in this chapter by using the CD-ROM to answer the following questions.

Hot Spots module, Yellowstone, Geyser, and Hot Springs Animation:

1. What is the difference between the conduits of geysers and hot springs?

2. What creates the heat that powers the geysers and hot springs of Yellowstone?

Icebergs and glacier in Antarctica.
Photo © Gregory Dimijian/Photo Researchers, Inc.

CHAPTER
12

Glaciers and Glaciation

*I*n the preceding three chapters you have seen how the surface of the land is shaped by mass wasting, running water, and, to some extent, ground water. Running water is regarded as the erosional agent most responsible for shaping the earth's land surface. Where glaciers exist, however, they are far more effective agents of erosion, transportation, and deposition. Geologic features characteristic of glaciation are distinctly different from the features formed by running water. Once recognized, they lead one to appreciate the great extent of glaciation during the recent geologic past (that age popularly known as the Ice Age).

Immense and extensive glaciers, covering as much as a third of the earth's land surface, had a profound effect on the landscape and on our present civilization. Moreover, worldwide climatic changes during the glacial ages distinctively altered landscapes in areas far from the glacial boundaries. For instance, water stored as ice in glaciers lowered the levels of the world's oceans, exposing more land than is presently above sea level.

These episodes of glaciation took place within only the last few million years, ending about 10,000 years ago. Preserved in the rock record, however, is evidence of extensive older glaciations. The chapter on plate tectonics shows how the record of these ancient glaciations supports the theory of plate tectonics.

To understand how glacial erosion and deposition could have created the features regarded as evidence for past glaciation, you must first appreciate how present-day glaciers erode, transport, and deposit material. In other words, you must apply the principle of uniformitarianism to your study of glaciation.

Figure 12.1

Yosemite Valley, as seen from Glacier Point, Yosemite National Park, California. Its U-shaped cross profile is typical of glacially carved valleys.

Photo by C. C. Plummer

A **glacier** is a large, long-lasting mass of ice, formed on land that moves under the influence of gravity and its own weight. It develops as snow is compacted and recrystallized. A glacier can develop anyplace where, over a period of years, more snow accumulates than melts away or is otherwise lost.

There are two types of *glaciated* terrain on the earth's surface. **Alpine glaciation** is found in mountainous regions, while **continental glaciation** exists where a large part of a continent (thousands of square kilometers) is covered by glacial ice. In both cases the moving masses of ice profoundly and distinctively change the landscape.

The Theory of Glacial Ages

In the early 1800s the hypothesis of past extensive continental glaciation of Europe was proposed. Among the many people who regarded the hypothesis as outrageous was the Swiss naturalist Louis Agassiz. But, after studying the evidence in Switzerland, he changed his mind. In 1837, he published a discourse that eventually led to wide acceptance of the theory. Later, Agassiz traveled widely throughout Europe and North America promoting and extending the theory. Agassiz and his colleagues had observed the characteristic erosional and depositional features of present glaciers in the Alps and had compared these with similar features found in northern Europe and the British Isles, well beyond the farthest extent of the Alpine glaciers. Based on these observations, they proposed that very large glaciers had covered most of Europe while a colder climate prevailed during the past. Agassiz later came to North America and worked with American geologists who had found similar indications of large-scale past glaciation on this continent.

As more evidence accumulated, the hypothesis became accepted as a theory that today is seldom questioned. The **theory of glacial ages** states that at times in the past, colder climates prevailed during which much more of the land surface of the earth was glaciated than at present.

Because the last episode of glaciation was at its peak only about 18,000 years ago, its record has remained largely undestroyed by subsequent erosion and so provides abundant evidence to support the theory. This most recent glacial episode was the last of several glacial ages that alternated with periods of warmer climate (only slightly warmer than today's climate) around the world.

The glacial ages are not just a scientific curiosity. Our lives and environment today have been profoundly influenced by their effects. For example, much of the fertile soil of the northern Great Plains of the United States developed on the loose debris transported and deposited by glaciers that moved southward from Canada. The spectacularly scenic areas in many North American national parks owe much of their beauty to glacial action. Yosemite Valley in California might have been another nondescript valley if glaciers had not carved it into its present shape (figure 12.1).

Before we can understand how a continental glacier was responsible for much of the soil in the Midwest, or how a glacier confined to a valley could carve a Yosemite, we must learn something about present-day glaciers.

Glaciers—Where They Are, How They Form and Move

Distribution of Glaciers

Glaciers occur both in polar regions, where there is little melting during the summer, and in temperate climates that have heavy snowfall during the winter months. Certain climatic conditions are necessary for glaciation.

The coastal mountains of Washington have more glaciers than any state other than Alaska, even though the climate of Washington is considerably warmer than that of the Rocky Mountain states. Glaciers can flourish in these coastal mountains because Washington has very high precipitation during the winter months, and more snow accumulates in higher elevations than melts away during the summer. Glaciers are common even near the equator in the very high mountains of South America because of the low temperatures at high altitudes.

Figure 12.2

Valley glacier on the flanks of Mount Logan, Canada's highest mountain.

Photo by C. C. Plummer

Glaciation is most extensive in polar regions, where little melting takes place at any time of year. At present about one-tenth of the land surface of the earth is covered by glaciers (compared with about one-third during the peak of the glacial ages). Approximately 85% of the present-day glacier ice is on the Antarctic continent, covering an area larger than the combined areas of western Europe and the United States; 10% is in Greenland. All the remaining glaciers of the world amount to only about 5% of the earth's freshwater ice. This means that Antarctica is in fact storing most of the earth's fresh water in the form of ice. Some have suggested that ice from the Antarctic could be brought to areas of dry climate to alleviate water shortages. While no one has suggested large-scale melting of the Antarctic glaciers, it is worth noting that if all that continent's ice were to melt, sea level around the world would rise

over 60 meters (200 feet). This would flood the world's coastal cities and significantly decrease the land surface available for humans to live on.

Types of Glaciers

A simple criterion—whether or not a glacier is restricted to a valley—is the basis for classifying glaciers by form. A **valley glacier** is a glacier that is confined to a valley and flows from a higher to a lower elevation. Like streams, small valley glaciers may be tributaries to a larger trunk system. Valley glaciers are prevalent in areas of alpine glaciation. As might be expected, most glaciers in the United States and Canada, being in mountains, are of the valley type (figure 12.2).

In contrast, an **ice sheet** is a mass of ice that is not restricted to a valley but covers a large area of land (over 50,000 square kilometers). Ice sheets are associated with continental glaciation. Only two places on earth now have ice sheets, Greenland and Antarctica. A similar but smaller body is called an **ice cap.** Ice caps (and valley glaciers as well) are found in a few mountain highlands and on islands in the Arctic Ocean, off Canada, Russia, and Siberia. An ice cap or ice sheet flows downward and outward from a central high point, as figure 12.3 shows.

Formation and Growth of Glaciers

Snow converts to glacier ice in somewhat the same way that sediment turns into a sedimentary rock and then into metamorphic rock; figure 12.4 shows the process. A snowfall can be compared to sediment settling out of water. A new snowfall may be in the form of light "powder snow," which consists mostly of air trapped between many six-pointed snowflakes. In a short time the snowflakes settle by compaction under their own weight and much of the air between them is driven out. Meanwhile, the sharp points of the snowflakes are destroyed as flakes reconsolidate into granules. In warmer climates, partial thawing and refreezing results in coarse granules—the "corn

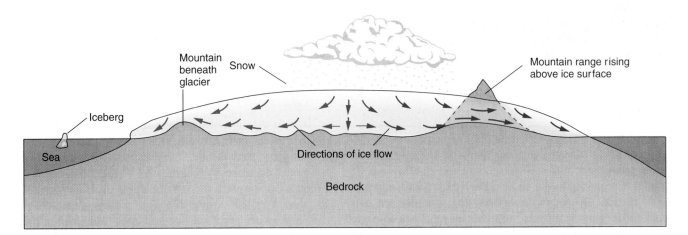

Figure 12.3

Diagrammatic cross section of an ice sheet. Vertical scale is highly exaggerated.

Figure 12.4

(*A*) Conversion of snow to glacier ice. (*B*) Thin slice of glacier ice in polarized light. In polarized light, the colors of individual ice grains vary depending on their crystallographic orientation.

Photo by C. C. Plummer

Figure 12.5

An iceberg off the coast of Antarctica. Its height is that of a 5-story building.

Photo by C. C. Plummer

snow" of spring skiing. In colder climates where little or no melting takes place, the snowflakes will recrystallize into fine granules. After the granular snow is buried by a new snowpack, usually during the following winter, the granules are compacted and weakly "cemented" together by ice. The compacted mass of granular snow, transitional between snow and glacier ice, is called *firn*. Firn is analogous to a sedimentary rock such as sandstone.

Through the years, the firn becomes more deeply buried as more snow accumulates. More air is expelled, the remaining pore space is greatly reduced and granules forced together recrystallize into the tight, interlocking mosaic of *glacier ice* (figure 12.4*B*). The recrystallization process involves little or no melting and is comparable to metamorphism. Glacier ice is texturally similar to the metamorphic rock, quartzite.

Under the influence of gravity, glacier ice moves downward and is eventually **wasted,** or lost. (An alternate term for wastage is **ablation.**) For glaciers in all but the coldest parts of the world, wastage is due mostly to melting, although some ice evaporates directly into the atmosphere. If a moving glacier reaches a body of water, blocks of ice break off (or *calve*) and float free as **icebergs** (figure 12.5). In most of the Antarctic, wastage takes place largely through calving of icebergs and direct evaporation. Only along the coast does melting take place, and there for only a few weeks of the year.

Glacial Budgets

If, over a period of time, the amount of snow a glacier gains is greater than the amount of ice and water it loses, the glacier's budget is *positive* and it expands. If the opposite occurs, the glacier decreases in volume and is said to have a *negative budget*. Glaciers with positive budgets push outward and downward at their edges; they are called **advancing glaciers.** Those with negative budgets grow smaller and their edges melt back; they are **receding glaciers.** Bear in mind that the glacial ice moves downvalley, as shown in figure 12.6, whether the glacier is advancing or receding. In a receding

12.1
ENVIRONMENTAL GEOLOGY

Glaciers as a Water Resource

Few people think of glaciers as frozen reservoirs supplying water for irrigation, hydroelectric power, recreation, and industrial and domestic use. In the state of Washington, however, streamflow from the approximately 800 glaciers there amounts to about 470 billion gallons of water during a summer, according to the U.S. Geological Survey. More water is stored in glacier ice in Washington than in all of the state's lakes, reservoirs, and rivers.

One important aspect of glacier-derived water is that it is available when needed most. Snow accumulates on glaciers during the wet winter months. During the winter, streams at lower elevations, where rain falls rather than snow, are full and provide plenty of water. During the summer, however, the climate in the Pacific Northwest is hotter and drier. Because of this, demand for water increases, especially for irrigation of crops. Streams that were fed by rainwater may have dried up. Yet, in the heat of summer, the period of peak demand, snow and ice on glaciers are melting, and streams draining glaciers are at their highest level.

Paradoxically, the greater the snowfall on a glacier during a winter, the smaller the amount of meltwater during the summer. A larger blanket of white snow reflects the sun's radiation more effectively than the darker, bare glacier ice, which absorbs more of the heat of the sun. Experiments have shown that melting can be greatly increased by darkening the snow surface, for instance, by sprinkling coal dust on it. Similarly, the melting of a glacier can be slowed artificially by covering it with highly reflective material. Such means of controlling glacial meltwater have been proposed to benefit power generating stations or to provide additional irrigation.

These ideas are appealing from a shortsighted point of view. However, the long-term effect of tampering with a glacier's natural regime can adversely affect the overall environment. It is conceivable, for example, that we could melt a glacier out of existence. Because most of the United States' glaciers are in national parks or wilderness areas, glaciers have not been tampered with to control the output of meltwater.

Further Reading
U.S. Geological Survey. 1973. *Glaciers, a water resource.* U.S. Geological Survey Information Pamphlet.

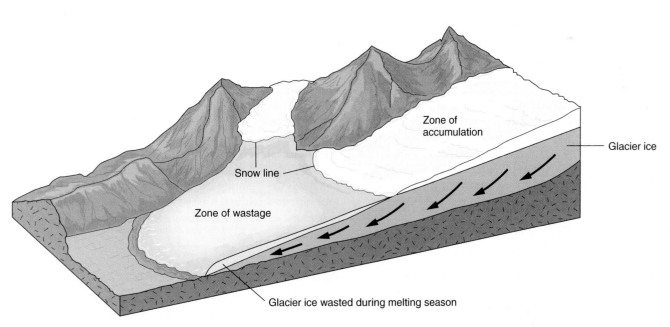

Figure 12.6
A valley glacier as it would appear at the end of a melt season. Below the snow line, glacier ice and snow have been lost during the melting season. In the zone of accumulation above that line, firn is added to the glacier from the previous winter snowfall.

A

B

Figure 12.7

South Cascade Glacier, Washington. If the photos were taken at the end of the melt season, the snow line would be the boundary between white snow and darker glacier ice. These two photos were taken 23 years apart. Photo (*A*) was taken in 1957; note that the glacier extended into the lake and that small icebergs calved from it. Photo (*B*) was taken in 1980; notice that the glacier has shrunk as well as retreated. During the interval between photos, the glacier lost approximately 7.5 meters of water averaged over its surface or the equivalent of 18.7 million cubic meters of water for the entire glacier.

Photos by U.S. Geological Survey

glacier, however, the rate of flow of ice is insufficient to replace all of the ice lost in the lower part of the glacier. If the amount of snow retained by the glacier equals the amount of ice and water lost, the glacier has a *balanced budget* and is neither advancing nor receding.

The upper part of a glacier, called the **zone of accumulation,** is the part of the glacier with a perennial snow cover (figure 12.6). The lower part is the **zone of wastage,** for there ice is lost, or wasted, by melting, evaporation, and calving.

The boundary between these two altitudinal zones of a glacier is an irregular line called the **snow line,** which marks the highest point at which the glacier's winter snow cover is lost during a melt season (figure 12.7).

The snow line may shift up or down from year to year, depending on whether there has been more accumulation or more wastage. Its location therefore indicates whether a glacier has a positive or negative budget. A snow line migrating upglacier over a period of years is a sign of a negative budget, whereas a snow line migrating downglacier indicates that the glacier has a positive budget. If a snow line remains essentially in the same place year after year, the glacier has a balanced budget.

The **terminus,** the lower edge of a glacier, moves farther downvalley when a valley glacier has a positive budget. In a receding glacier the terminus melts back upvalley. Because most glaciers move slowly, migration of the terminus tends to lag several years behind a change in the budget.

An ice sheet with a positive budget increases in volume, advancing its outer margins. If the expanding ice sheet extends into the ocean, an increasing number of icebergs break off and float away in the open sea.

Advancing or receding glaciers are significant and sensitive indicators of climatic change. However, an advancing glacier does not necessarily indicate that the climate is getting colder. It may mean that the climate is getting wetter, or that more precipitation is falling during the winter months, or that the summers are cloudier. It is estimated that a worldwide decrease in the mean annual temperature of about 5°C could bring about a new ice age.

Movement of Valley Glaciers

Valley glaciers move downslope under the influence of gravity and their own weight, the rate being variable, ranging from less than a few millimeters a day to more than 15 meters a day. The upper part of a glacier—where the volume of ice is greater and slopes tend to be steeper—generally moves faster than ice farther down or on gentler slopes. In this way ice from the higher altitudes keeps replenishing ice lost in the zone of wastage.

12.1 Astrogeology

Ice Elsewhere in the Solar System

The polar regions of Mars are covered with ice caps, which are only a few meters thick and composed mostly of frozen carbon dioxide (dry ice). During the summer on each hemisphere, the ice caps shrink markedly as the carbon dioxide vaporizes; however, a small cap remains. This small residual cap (400 kilometers in diameter) is probably composed of water ice.

Two distinctive types of terrain can be observed in the Martian polar regions. *Laminated terrain* is the name given to areas where series of alternating light and dark layers can be seen. The layers are essentially horizontal, and each is about 15 to 35 meters thick. As many as fifty layers have been counted in one location. The layers are thought to represent alternating beds of high dust content and high ice content, and their alternation may be due to some kind of climatic change. That the layers are stratified outwash from glaciers has also been suggested. Near the margins of the polar caps are large troughs and ridges that could be glacial valleys and moraines.

Underlying the laminated terrain is another terrain, which is characterized by small pits. The pits of this *etch-pitted terrain* may be due to wind erosion, and, if so, they would be deflation basins (blowouts). They may also be glacial kettles or collapse pits caused by the loss of underground ice.

Comets (such as Hale-Bopp, which graced nighttime skies in 1997) are essentially chunks of water ice, frozen carbon dioxide (dry ice), methane, ammonia, and rock dust that travel through space. Although they are usually only a kilometer or so in diameter, their tails of evaporating gases and dust extend over 100 million kilometers. A small comet probably landed in Siberia in 1908, felling trees up to 30 kilometers away and rattling windows 450 kilometers away.

Some of the satellites for the large outer planets have abundant ice and some may have liquid water (raising the possibility that life could exist on these satellites). Europa, one of Jupiter's moons (box figure 1) is covered

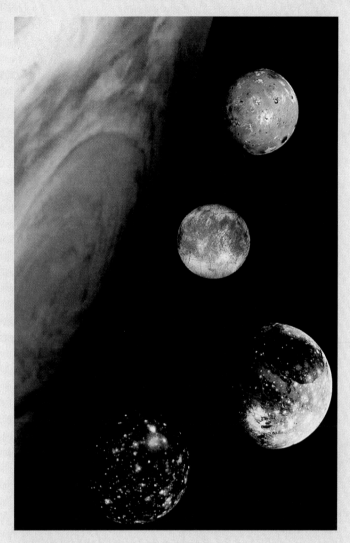

Astrogeology Box 12.1 Figure 1

Composite image of Jupiter and its moons. From top to bottom: Io, Europa, Ganymede, and Callisto.

JPL/NASA

with ice. Images (received in 1997) from the spacecraft Galileo show that the ice has been cracked, bulged, and disrupted, indicating movement from below. Some scientists think that a deep ocean underlies the ice and that motion in the ocean causes disruption of the ice. An alternate hypothesis is that the ice layer is at least 100 kilometers thick and warm enough to move and churn the surface ice.

Figure 12.8

Movement of a glacier. Markers on the glacier indicate the center of the glacier moves faster than its side. Cross-sectional view shows movement within the glacier.

Glaciers in temperate climates—where the temperature of the glacier is at or near the melting point for ice—tend to move faster than those in colder regions—where the ice temperature stays well below freezing.

Velocity also varies within the glacier itself (figure 12.8). The central portion of a valley glacier moves faster than the sides (as water does in a stream), and the surface moves faster than the base. How ice moves within a valley glacier has been demonstrated by studies in which holes are drilled through the glacier ice and flexible pipes inserted. Changes in the shape and position of the pipes are measured periodically. The results of these studies are shown diagrammatically in figure 12.8.

Note in the diagram that the base of the pipe has moved downglacier. This indicates **basal sliding,** which is the sliding of the glacier as a single body over the underlying rock. A thin film of meltwater that develops along the base from the pressure of the overlying glacier facilitates basal sliding. Think of a large bar of wet soap sliding down an inclined board.

Note that the lower portion of the pipe is bent in a downglacier direction. This indicates **plastic flow** of ice, movement that occurs within the glacier due to the plastic or "deformable" nature of the ice itself. Visualize two neighboring grains of ice within the glacier, one over the other. Both are moving—carried along by the ice below them, however, the higher of our two ice grains slides over its underlying neighbor a bit further. The reason the pipe is bent more sharply near the base of the glacier is that pressure from overlying ice results in greater flowage with increasing depth. Deep in the glacier, ice grains are sliding past their underlying neighbors further than similar ice grains higher up where the pipe is less bent. (We

should point out that a glacier flows not only because ice grains slide past one another but also because ice grains deform and recrystallize.)

In the **rigid zone,** or upper part of the glacier, the pipe has been moved downglacier; however, it has remained unbent. The ice nearer the top apparently rides along passively on the plastically moving ice closer to the base. In the rigid zone grains of ice do not move relative to their neighbors.

Crevasses

Along its length, a valley glacier moves at different rates in response to changes in the steepness of the underlying rock. Typically a valley glacier rides over a series of rock steps. Where the glacier passes over a steep part of the valley floor, it moves faster. The upper rigid zone of ice, however, cannot stretch to move as rapidly as the underlying plastic-flowing ice. Being brittle, the ice of the rigid zone is broken by the tensional forces. Open fissures, or **crevasses,** develop (figures 12.9 and 12.10). Crevasses also form along the margins of glaciers in places where the path is curved, as shown in part of figure 12.9. This is because ice (like water) flows faster toward the outside of the curve. For glaciers in temperate climates, a crevasse should be no deeper than about 40 meters, the usual thickness of the rigid zone. If you are falling down a crevasse, it is some compensation to realize on the way down that you will not fall more than 40 meters (130 feet).

After the ice has passed over a steep portion of its course, it slows down, and compressive forces close the crevasses. If a glacier descends a long, very steep slope, the ice breaks into a chaotic jumble of blocks and crevasses called an *icefall* (figure 12.11).

Figure 12.9

Crevasses on a glacier, looking down from Mount Logan, Yukon Territory, Canada.

Photo by C. C. Plummer

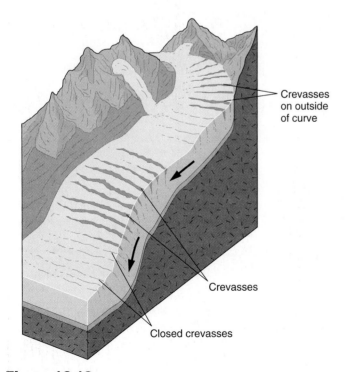

Figure 12.10

Crevasses along the course of a glacier.

Figure 12.11

A glacier in the Peruvian Andes. Crevasses in the icefall (right) close at its base where the glacier's gradient flattens out. The small crevasse in the foreground is at the top of the icefall.

Photo by C. C. Plummer

12.2 ENVIRONMENTAL GEOLOGY

Water Beneath Glaciers: Floods, Giant Lakes, and Galloping Glaciers

A Galloping Glacier

Glacial motion is often used as metaphor for slowness ("The trial proceeded at a glacial pace"). However, some glaciers will *surge,* that is, move very rapidly for short periods following years of barely moving at all. The most extensively documented surge (or "galloping glacier") was that of Alaska's Bering Glacier in 1993–94. The Bering Glacier is the largest glacier in continental North America and it surges on a 20–30 year cycle. After its previous surge in 1967, its terminus had retreated 10 kilometers. In August 1993, the latest surge began. Ice traveled at velocities up to 100 meters per day for short periods of time and sustained velocities of 35 meters per day over a period of several months. The terminus advanced nine kilometers by the time the surge ended in November 1994. When glaciers surge, the previously slow moving, lower part of a glacier breaks into a chaotic mass of blocks (box figure 1). Surges are usually attributed to a buildup of water beneath part of a glacier floating it above its bed. In July 1994, a large flood of water burst from Bering Glacier's terminus, carrying with it blocks of ice up to 25 meters across.

A Flood

Glacial outburst floods are not always associated with surges. In October 1996, a volcano erupted beneath a glacier in Iceland. The glacier, which is up to 500 meters thick, covers one-tenth of Iceland. Emergency teams prepared for the flood that geologists predicted would follow the eruption. The expected flood took place early in November with a peak flow of 45,000 cubic meters per second (over 1.5 million cubic feet per second)! The flood lasted only a few hours; however, it caused between 10 and 15 million dollars worth of damage. Three major bridges were destroyed or damaged and 10 kilometers of roads were washed away. Because people had been kept away from the expected flood path there were no casualties.

A Giant Lake

One of the world's largest lakes was only recently discovered. But don't expect to take a dip or go windsurfing on it. It lies below the thickest part of the East Antarctic Ice Sheet and is named after the Russian research station, Vostok, which is 4,000 meters above the lake at the coldest and most remote part of Antarctica. It was discovered in the 1970s through ice-penetrating radar. However, its extent was unknown until 1996 when satellite-borne radar revealed how large a lake it is. Studies indicate that the lake is 200 km long and 50 km wide—about the size of Lake Ontario. At its deepest, it is 510 meters, placing it among the ten deepest lakes in the world.

The lake has been sealed off from the rest of the world for around a million years and it is likely that it contains organisms, such as microbes, dating back to that time. These organisms (and their genes) would not have been affected by modern pollution or nuclear bomb fallout. By coincidence, the world's deepest ice hole (over 3 kilometers) was being drilled from Vostok Station above the lake when the size of Lake Vostok was being determined. The ice core from this hole should add to the findings from the Greenland drilling projects (box 12.3) and provide an even greater picture of earth's climate during the ice ages. When the hole was completed in 1997, drilling was halted short of reaching the lake due to fear of contaminating it and compromising whatever living organisms might be in the very old water. Study of the lake and its organisms is curtailed until a future generation of scientists can devise means of sampling the waters without altering its ecosystem.

Box 12.2 Figure 1

Part of a glacier after a surge (lower part of photo). The debris-covered ice has been broken up into a chaotic mass of blocks. In the background is a small glacier that has retreated up its valley. Photo taken near the Canada-Alaska border.

Photo by C. C. Plummer

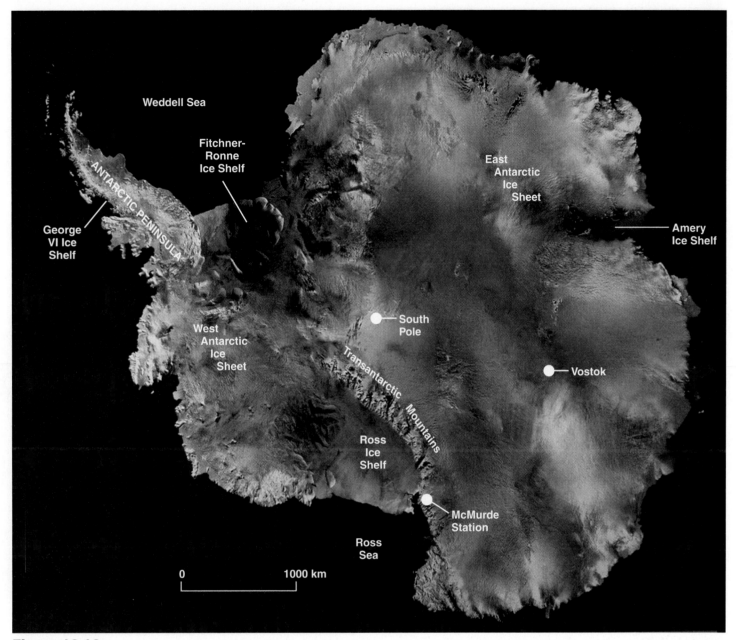

Figure 12.12

The Antarctic continent and its ice sheets. Vostok is at the highest part of the East Atlantic Ice Sheet. The distance between the South Pole and Vostok is approximately 500 kilometers.

U.S. Geological Survey/NASA

Movement of Ice Sheets

An ice sheet or ice cap moves like a valley glacier except that it moves downward and outward from a central high area toward the edges of the glacier (as shown in figure 12.3).

Glaciological research in Antarctica has determined how ice sheets grow and move. Antarctica has two ice sheets, the West Antarctic Ice Sheet is separated by the Transantarctic Mountains from the much larger East Antarctic Ice Sheet (fig-ure 12.12). The two ice sheets join in the low areas between mountain ranges. Both are nearly completely within a zone of accumulation because so little melting takes place (wastage is largely by calving of icebergs) and because occasional snowfalls nourish their high central parts. The ice sheets mostly overlie interior lowlands, but also completely bury some mountain ranges. Much of the base of the West Antarctic Ice Sheet is on

Figure 12.13

The South Pole. Actually, the true South Pole is several kilometers from here. The moving ice sheet has carried the striped pole away from the site of the true South Pole, where the pole was erected in 1956.

Photo by C. C. Plummer

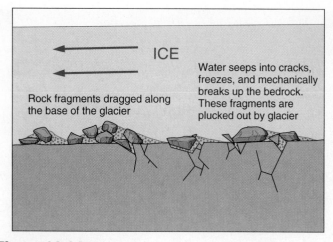

Figure 12.14

Plucking and abrasion beneath a glacier.

bedrock that is below sea level. At least one active volcano underlies the West Antarctic Ice Sheet (resulting in a depression in the ice sheet). Where mountain ranges are higher than the ice sheet, the ice flows through as valley glaciers.

At the South Pole (figures 12.12 and 12.13)—neither the thickest part nor the center of the East Antarctic Ice Sheet—the ice is 2,700 meters thick. The thickest part of the East Antarctic Ice Sheet is 4,776 meters (15,670 feet).

Most of the movement of the East Antarctic Ice Sheet is by means of plastic flow. It has been thought that most of the ice sheet is frozen to the underlying rocks and basal sliding takes place only locally. However, the recent discovery of a giant lake beneath the thickest part of the Antarctic Ice Sheet (box 12.2) may indicate that liquid water at its base is more widespread and basal sliding might be more important than previously regarded.

Glacial Erosion

Wherever basal sliding takes place, the rock beneath the glacier is abraded and modified. As meltwater works into cracks in bedrock and refreezes, pieces of the rock are broken loose and frozen into the base of the moving glacier, a process known as *plucking*. While being dragged along by the moving ice, the rock within the glacier grinds away at the underlying rock (figure 12.14). The thicker the glacier, the more pressure on the rocks and the more effective the grinding and crushing.

Pebbles and boulders that are dragged along are *faceted,* that is, given a flat surface by abrasion. The bedrock, as well as the ice-carried rocks, is *polished* by the grinding. Sharp corners of rock fragments dragged along make grooves and **striations,** or scratches, in the rock, in the direction of ice movement (figure 12.15).

The grinding of rock across rock produces a powder of fine fragments called **rock flour.** Rock flour is composed

Figure 12.15

Striated and polished bed rock surface in South Australia. Unlike glacial striations commonly found in North America, these were caused by late Paleozoic glaciation.

Photo by C. C. Plummer

largely of very fine (silt- and clay-sized) particles of unaltered minerals (pulverized from chemically unweathered bedrock). When *meltwater* washes rock flour from a glacier, the streams draining the glacier appear milky.

Not all glacier-associated erosion is caused directly by glaciers. Mass wasting takes place on steep slopes created by downcutting glaciers. Frost wedging breaks up bedrock ridges and cliffs above a glacier, causing frequent rockfalls. Snow avalanches bring down loose rocks onto the glacier surface, where they ride on top of the ice. Debris may also fall into crevasses to be transported within or at the base of a glacier, as shown in figure 12.21.

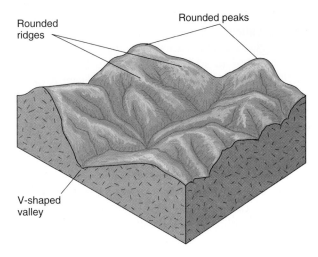

Rounded
ridges

Rounded peaks

V-shaped
valley

A

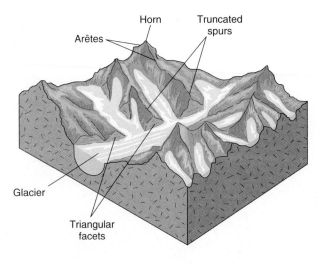

Horn

Truncated
spurs

Arêtes

Glacier

Triangular
facets

B

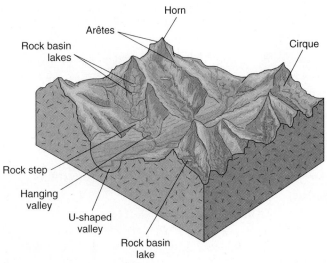

Horn

Arêtes

Rock basin
lakes

Cirque

Rock step

Hanging
valley

U-shaped
valley

Rock basin
lake

C

Figure 12.16

(*A*) A stream-carved mountain landscape before glaciation.
(*B*) The same area during glaciation. Ridges and peaks become
sharper due to frost wedging. (*C*) The same area after glaciation.

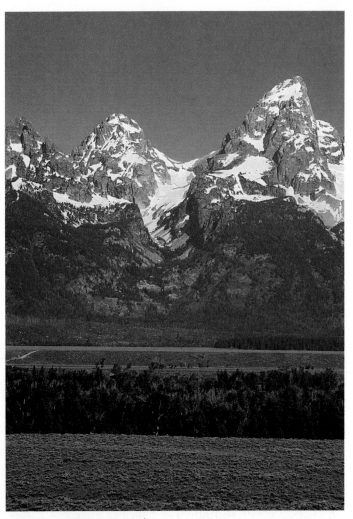

Figure 12.17

Glacially carved valley in Grand Teton National Park, Wyoming.
Rounded knobs produced from glacial erosion are visible in the
lower slopes, in front of the U-shaped valley.

Photo by C. C. Plummer

Erosional Landscapes Associated with Alpine Glaciation

We are in debt to glaciers for the rugged and spectacular
scenery of high mountain ranges. Figure 12.16 shows how
glaciation has radically changed a previously unglaciated
mountainous region. The striking and unique features associated with mountain glaciation are due to the erosional effects
of glaciers as well as frost wedging on exposed rock.

Glacial Valleys

Glacially carved valleys are easy to recognize. A **U-shaped valley**
(in cross profile) is characteristic of glacial erosion (figure 12.17),
just as a **V**-shaped valley is characteristic of stream erosion.

The thicker a glacier is, the more erosive force it exerts on the
underlying valley floor, and the more bedrock is ground away.
For this reason, a large trunk glacier erodes downward more
rapidly and carves a deeper valley than do the smaller tributaries

Figure 12.18

A hanging valley in Yosemite National Park, California.

Photo by C. C. Plummer

that join it. After the glaciers disappear, these tributaries remain as **hanging valleys** high above the main valley (figure 12.18).

Valley glaciers, which usually occupy valleys formerly carved by streams, tend to straighten the curves formed by running water. This is because the mass of ice of a glacier is too sluggish and inflexible to move easily around the curves. In the process of carving the sides of its valley, a glacier erodes or "truncates" the lower ends of ridges that extended to the valley. **Truncated spurs** are ridges that have *triangular facets* produced by glacial erosion at their lower ends (figure 12.16*B*).

Although a glacier tends to straighten and smooth the side walls of its valley, ice action often leaves the surface of the underlying bedrock carved into a series of steps. This is due to the variable resistance of bedrock to glacial erosion. Figure 12.19 shows what happens when a glacier abrades a relatively weak rock with closely spaced fractures. Water seeps into cracks in the bedrock, freezes there, and enlarges fractures or makes new ones. Rock frozen into the base of the glacier grinds and loosens more pieces. After the ice has melted back, a chain of **rock-basin lakes** (also known as **tarns**) may occupy the depressions carved out of the

Figure 12.19

Development of rock steps. (*A*) Valley floor before glaciation.
(*B*) During glaciation. (*C*) Rock steps and rock-basin lakes, Sierra Nevada, California.

A and *B* after F. E. Matthes, 1930, U.S. Geological Survey; Photo by C. C. Plummer

Figure 12.20

A cirque occupied by a small glacier in the Canadian Rocky Mountains. The glacier was much larger during the ice ages.

Photo by C. C. Plummer

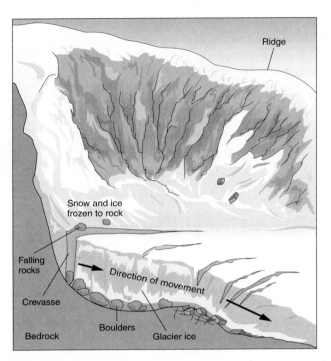

Figure 12.21

Cutaway view of a cirque.

weaker rock. A series of such lakes, reminiscent of a string of prayer beads, is sometimes called *paternoster lakes.*

Areas where the bedrock is more resistant to erosion stand out after glaciation as *rounded knobs* (see figure 12.17), usually elongated parallel to the direction of glacier flow. These are also known as *rouche moutonnées* (French for "sheeplike rocks"). On stronger, less fractured rock, glacial erosion works in a similar way but is less effective because of the bedrock's resistance to grinding and crushing.

Cirques, Horns, and Arêtes

A glacial **cirque** is a steep-sided, half-bowl shaped recess carved into a mountain at the head of a glacial valley (figure 12.20). In this unique, often spectacular, topographic feature, a large percentage of the snow accumulates that eventually converts to glacier ice and spills over the threshold as the valley glacier starts its downward course.

A cirque is not entirely carved by the glacier itself but is also shaped by the weathering and erosion of the rock walls above the surface of the ice. Frost wedging and avalanches break up the rock and steepen the slopes above the glacier. Broken rock tumbles onto the valley glacier and becomes part of its load, and some rock may fall into a crevasse that develops where the glacier is pulling away from the cirque wall (figure 12.21).

The headward erosional processes that enlarge a cirque also help create the sharp peaks and ridges characteristic of glaciated mountain ranges. A **horn** is the sharp peak that remains after cirques have cut back into a mountain on several sides (figure 12.22).

Frost wedging works on the rock exposed above the glacier, steepening and cutting back the side walls of the valley.

Figure 12.22

The Matterhorn in Switzerland. Engraving by Edward Whymper, who, in 1865, became the first person to climb the Matterhorn.

On July 1, 1993, a drill penetrating the thickest part of Greenland's ice sheet struck rock. This ended drilling for an American project that retrieved just over 3 kilometers of ice core—a complete cross-sectional sampling of the ice sheet. Thirty-two kilometers away, European scientists had finished drilling through the ice sheet in a similar project a year earlier. Field work began for both projects in 1989. The two projects were closely coordinated. It was decided that data from the two cores would complement one another and determine possible influences from local flow patterns of the glacier.

The holes were drilled using a hollow bit from which a cylindrical core of ice 10 cm in diameter was retrieved in segments around 3 m long. As the cores were retrieved, a portion of each was analyzed by scientists at each site. The remaining ice was shipped to laboratories for more detailed studies.

The reason for drilling at the highest part of the ice sheet is that here the snow converts to glacier ice and moves straight downward as it becomes buried (as shown on figure 12.3). Thus, drilling here assures a complete record of past snowfall and eliminates complications from lateral flow. The ice record here extends back 250,000 years.

Among other investigations, chemical analyses were made to determine precisely the composition of air in bubbles within the ice. The bubbles in each layer contain air trapped at the time of snowfall. Other analyses determined the composition of dust and minute particles that settled out of the atmosphere with the snow.

Results from these cores added to information obtained from previous drilling near the edges of the Greenland ice sheet and from cores obtained by Russians at Vostok, the central and thickest part of the East Antarctic Ice Sheet (see figure 12.12).

Layers representing annual snowfall are well-defined for the upper part of the glacier. A high quality, continuous record of annual snowfall and clues to the climate for each year can be determined for at least as far back as 13,000 years ago. This means we have a precise record of the transition from the last ice age to the present warmer climate. Climate changes are inferred in part by determining variations in annual snowfall as represented by the thicknesses of ice layers. Other techniques for determining climatic conditions involve correlating particular oxygen and hydrogen isotopic ratios to variations in global climate.

One surprising result is how quickly the climate changed at the end of the last ice age. Research on Greenland cores indicates that about 11,700 years ago the temperature increased by 6°C within just a few years' time. Scientists had previously thought the warming that ended an ice age was gradual. This and other discoveries should give us new insight on what factors are most significant for climate changes.

Moreover, researchers have also been able to correlate major volcanic eruptions with specific past years based on the concentration of sulfates in layers of ice. The sulfates are fallout from SO_2 blasted into the high atmosphere by volcanoes (e.g., as at Mount Pinatubo, described in chapters 1 and 4). For cores dating from the present to 2,000 years ago, most of the ice layers having relatively high concentrations of sulfates can be correlated to known historical eruptions. Going further back in time, the ice cores give us a detailed record of when eruptions occurred before significant records were kept by humans. The ice cores provide evidence that increased volcanic activity 8,000 to 9,000 years ago correlates with cooler temperatures for that period, indicating that heavy volcanic activity is a significant factor in changing the earth's climate.

Sharp ridges called **arêtes** separate adjacent glacially carved valleys (figure 12.23).

Erosional Landscapes Associated with Continental Glaciation

In contrast to the rugged and angular nature of glaciated mountains, an ice sheet tends to produce rounded topography. The rock underneath an ice sheet is eroded in much the same way as the rock beneath a valley glacier. However, the weight and thickness of the ice sheet may produce more pronounced effects. Grooved and striated bedrock is common. Some grooves are actually channels several meters deep and many kilometers long. The orientation of grooves and striations indicates the direction of movement of a former ice sheet.

An ice sheet may be thick enough to bury mountain ranges, rounding off the ridges and summits and perhaps streamlining them in the direction of ice movement. Much of northeastern Canada, with its rounded mountains and grooved and striated bedrock surface, shows the erosional effects of ice sheets that formerly covered that part of North America (figure 12.24).

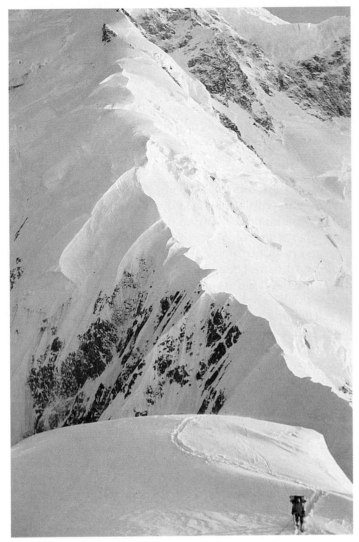

Figure 12.23

An arête on Mount Logan, Yukon Territory, Canada.

Photo by C. C. Plummer

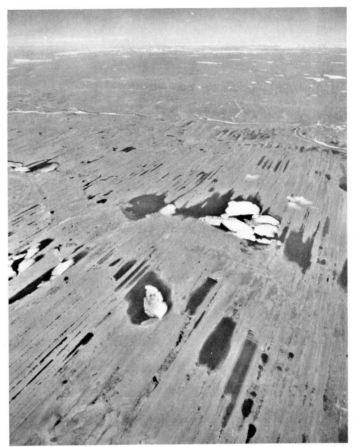

Figure 12.24

Air view of glacially scoured terrain in Canada. Ice moved from upper right to lower left.

National Air Photo Library of Canada

Glacial Deposition

The rock fragments scraped and plucked from the underlying bedrock and carried along at the base of the ice make up most of the load carried by an ice sheet, but only part of a valley glacier's load. Much of a valley glacier's load comes from rocks broken from the valley walls.

Most of the rock fragments carried by glaciers are angular, as the pieces have not been tumbled around enough for the edges and corners to be rounded. The debris is unsorted, and clay-sized to boulder-sized particles are mixed together (figure 12.25). The unsorted and unlayered rock debris carried or deposited by a glacier is called **till.**

Glaciers are capable of carrying virtually any size of rock fragment, even boulders as large as a house. An **erratic** is an ice-transported boulder that has not been derived from the nearby

Figure 12.25

Till transported on top of and alongside a glacier in Peru. View is downglacier. The lake is dammed up by an end moraine at its far end.

Photo by C. C. Plummer

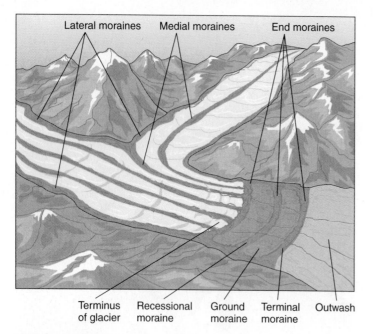

Figure 12.26
Moraines associated with valley glaciers.

bedrock. If its bedrock source can be found, the erratic indicates the direction of movement of the glacier that carried it.

Moraines

When till occurs as a body of unsorted debris either on a glacier or left behind by a glacier, the body is regarded as one of several types of **moraines. Lateral moraines** are elongate mounds of till which form along the sides of a valley glacier (figures 12.25, 12.26, 12.27, and 12.28). Rockfall debris from the steep cliffs that border valley glaciers accumulates along the edges of the ice to form lateral moraines.

Where tributary glaciers come together, the adjacent lateral moraines join and are carried downglacier as a single long mound of till known as a **medial moraine.** In a large trunk glacier that has formed from many tributaries, the numerous medial moraines give the glacier the appearance from the air of a multilane highway (figures 12.26 and 12.27).

An actively flowing glacier brings debris to its terminus. If the terminus remains stationary for a few years or advances, a distinct **end moraine,** a curved ridge of till, piles up along the front edge of the ice. Valley glaciers build end moraines that are crescent-shaped or sometimes horseshoe-shaped (figures 12.26 and 12.28). The end moraine of an ice sheet takes a similar lobate form, but is much longer and more irregular than that of a valley glacier (figure 12.29).

Geologists distinguish two special kinds of end moraines. A *terminal moraine* is the end moraine marking the farthest advance of a glacier. A *recessional moraine* is an end moraine built while the terminus of a receding glacier remains temporarily stationary. A single receding glacier can build several recessional moraines (as in figures 12.26, 12.28, and 12.29).

As ice melts, rock debris that has been carried by a glacier is deposited to form a **ground moraine,** a fairly thin, extensive layer or blanket of till (figures 12.26 and 12.29). Very large areas that were once covered by an ice sheet now have the gently rolling surface characteristic of ground moraine deposits.

Figure 12.27
Medial and lateral moraines on valley glaciers, Yukon Territory, Canada. Ice is flowing toward viewer and to lower right.
Photo by C. C. Plummer

Figure 12.28

In the foreground, end moraines (recessional moraines) curve into two long lateral moraines. The two lateral moraines extend back to a glacially-carved valley in the Sierra Nevada, California.

Photo by C. C. Plummer

Figure 12.29

Depositional features in front of a receding ice sheet.

Figure 12.30

Drumlin in New York state.

Photo by Ward's Natural Science Est., Inc., Rochester, N.Y.

Figure 12.31

An esker in northeastern Washington.

Photo by D. A. Rahm, courtesy Rahm Memorial Collection, Western Washington University

In some areas of past continental glaciation there are bodies of till shaped into streamlined hills called **drumlins** (figures 12.29 and 12.30). A drumlin is shaped like an inverted spoon aligned parallel to the direction of ice movement of the former glacier. Its gentler end points in the downglacier direction. Because we cannot observe drumlins forming beneath present ice sheets there is uncertainty how till becomes shaped into these streamlined hills.

Outwash

In the zone of wastage, large quantities of meltwater usually run over, beneath, and away from the ice. The material deposited by the debris-laden meltwater is called **outwash.** Because it has the characteristic layering and sorting of stream-deposited sediment, outwash can be distinguished easily from the unlayered and unsorted deposits of till. Because outwash is fairly well sorted and the particles generally are not chemically weathered, it is an excellent source of aggregate for building roadways and for mixing with cement to make concrete.

An outwash feature of unusual shape associated with former ice sheets and some very large valley glaciers is an **esker,** a long, sinuous ridge of water-deposited sediment (figures 12.29 and 12.31). Eskers can be up to 10 meters high and are formed of cross-bedded and well-sorted sediment. Evidently eskers are deposited in tunnels within or under glaciers, where meltwater loaded with sediment flows under and out of the ice.

As meltwater builds thick deposits of outwash alongside and in front of a retreating glacier, blocks of stagnant ice may be surrounded and buried by sediment. When the ice block

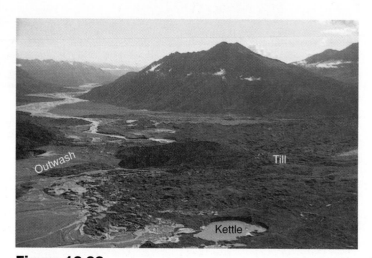

Figure 12.32

A kettle (*foreground*) and outwash (*background* and *left*) from a glacier. Stagnant ice underlies much of the till. Yukon Territory, Canada.

Photo by C. C. Plummer

finally melts (sometimes years later), a depression called a **kettle** forms (figures 12.29 and 12.32). Many of the small scenic lakes in the upper Middle West of the United States are kettle lakes.

The streams that drain glaciers tend to be very heavily loaded with sediment, particularly during the melt season. As they come off the glacial ice and spread out over the outwash deposits, the streams form a braided pattern (see chapter 10 on streams).

Figure 12.33

Varves from a former glacial lake. Each pair of light and dark layers represents a year's deposition. Gradations on ruler are centimeters.

Photo by Brian Atwater, U.S. Geological Survey

The large amount of rock flour that these streams carry in suspension settles out in quieter waters. In dry seasons or drought, the water may dry up and the rock flour deposits may be picked up by the wind and carried long distances. Some of the best agricultural soil in the United States has been formed by rock flour that has been redeposited by wind. Such fine-grained, wind-blown deposits of dust are called *loess* (see chapter 13).

Glacial Lakes and Varves

Lakes often occupy depressions carved by glacial erosion but can also form behind dams built by glacial deposition. Commonly a lake forms between a retreating glacier and an earlier end moraine (see figure 12.25).

In the still water of the lake, clay and silt settle on the bottom in two thin layers—one light-colored, one dark—that are characteristic of glacial lakes. Two layers of sediment representing one year's deposition in a lake are called a **varve** (figure 12.33). The light-colored layer consists of slightly coarser sediment (silt) deposited during the warmer part of the year when the nearby glacier is melting and sediment is transported to the lake. The silt settles within a few weeks or so after reaching the lake. The dark layer is finer sediment (clay)—material that sinks down more slowly during the winter after the lake surface freezes and the supply of fresh, coarser sediment stops due to

lack of meltwater. The dark color is attributed to fine organic matter mixed with the clay.

Because each varve represents a year's deposit, varves are like tree rings and indicate how long a glacial lake existed.

Effects of Past Glaciation

As the glacial theory gained general acceptance during the latter part of the nineteenth century, it became clear that much of northern Europe and the northern United States as well as most of Canada had been covered by great ice sheets during the so-called Ice Age. It also became evident that even areas not covered by ice had been affected because of the changes in climate and the redistribution of large amounts of water.

We now know that the last of the great North American ice sheets melted away from Canada less than 10,000 years ago. In many places, however, till from that ice sheet overlies older tills, deposited by earlier glaciations. The older till is distinguishable from the newer till because the older till was deeply weathered during times of warmer climate between glacial episodes.

The Glacial Ages

Geologists can reconstruct with considerable accuracy the last episode of extensive glaciation, which covered large parts of North America and Europe and was at its peak about 18,000 years ago. There has not been enough time for weathering and erosion to alter significantly the effects of glaciation. Less evidence is preserved for each successively older glacial episode, because (1) weathering and erosion occurred during warm interglacial periods and (2) later ice sheets and valley glaciers overrode and obliterated many of the features of earlier glaciation. However, from piecing together the evidence, geologists can see that earlier glaciers covered approximately the same region as the more recent ones.

Until a few years ago geologists thought the Pleistocene Epoch (see chapter 8) included all the glacial ages, but recent work indicates that worldwide climate changes necessary for continental glaciation probably began late in the Cenozoic Era, at least a million years before the Pleistocene. Antarctica has been glaciated for at least 20 million years, and continental glaciers elsewhere probably existed at least 3 million years ago.

The significance of this is that the earth has undergone episodic changes in climate during the last 2 to 3 million years. Actually, the climate changes necessary for a glacial age to occur are not so great as one might imagine. During the height of a glacial age, the worldwide average of annual temperatures was probably only about 5°C cooler than at present. The intervening interglacial periods were probably a bit warmer worldwide than present-day average annual temperatures.

12.4
IN GREATER DEPTH

Mars on a Glacier

Meteorites are extraterrestrial rocks—fragments of material from space that have managed to penetrate the earth's atmosphere and land on the earth's surface. They are of interest not only to astronomers but to geologists, for they help us date the earth (chapter 8) and give us clues to what the earth's interior is like (see chapter 17) because many of the meteorites are thought to represent fragments of destroyed minor planets. Meteorites are rarely found; for one reason, they usually do not look very different from the earth's rocks with which they are mixed.

In recent years, a bonanza of meteorites has been found on the Antarctic ice sheets. Over a thousand meteorites have been collected from one small area where the ice sheet abuts against the Transantarctic Mountains. The reason for this heavy concentration is that meteorites landing on the surface of the ice over a vast area have been incorporated into the glacier and transported to where wastage takes place. The process is illustrated in box figure 1.

A few of the meteorites are especially intriguing. Some almost certainly are rocks from the moon, while several others apparently came from Mars. Their chemistry and physical properties match what we would expect of a Martian rock. But how could a rock escape Mars and travel to earth? Scientists suggest that a meteorite hit Mars with such force that fragments of that planet were launched into space. Eventually, some of the fragments reached Earth.

In 1997, researchers announced that they found what could be signs of former life on Mars in one of the meteorites collected 13 years earlier in Antarctica. The evidence included carbon-containing molecules that might have been produced by living organisms as well as microscopic blobs that could be fossil alien bacteria. But there are alternate explanations for each line of evidence and a hot debate has ensued between scientists with opposing viewpoints. The ongoing space missions to Mars will be looking closely for evidence of primitive life on Mars, past as well as present.

Box 12.4 Figure 1

Diagram showing the way in which meteorites are concentrated in a narrow zone of wastage along the Transantarctic Mountains. Two meteorites are shown as well as the paths they would have taken from the time they hit the ice sheet until they reached the zone of wastage. The vertical scale is greatly exaggerated.

Source: Antarctic Journal of the United States.

Figure 12.34

Extent of maximum glaciation in North America during the Pleistocene Epoch. Arrows show direction of ice movement.

After C. S. Denny, U.S. Geological Survey National Atlas of the United States

Direct Effects of Past Glaciation in North America

Moving ice abraded vast areas of northern and eastern Canada during the growth of the North American ice sheets (figure 12.34). Most of the soil was scraped off and bedrock was scoured. Many thousands of future lake basins were gouged out of the bedrock.

The directions of ice flow can be determined from the orientation of striations and grooves in the bedrock. The largest ice sheet (the Laurentide Ice Sheet) moved outward from the general area now occupied by Hudson Bay, which is where the ice sheet was thickest. The present generally barren surface of the Hudson Bay area contrasts markedly with the Great Plains surface of southern Canada and northern United States, where vast amounts of till were deposited.

12.5 IN GREATER DEPTH

Causes of Glacial Ages

What caused the glacial ages? This question has been asked by scientists since the theory of glacial ages was accepted over a century ago. Only in the last couple of decades have climatologists thought they were beginning to provide acceptable answers.

The question is of more than academic importance. Understanding climate changes could provide the key to accurate, long-range weather predictions. Even minor climate changes affect crops and the problem of feeding humanity. If we knew the causes of fluctuations between glacial and interglacial episodes, we would probably know whether another glacial age is imminent or whether we are in a cooling or warming stage in the earth's history.

A number of theories and hypotheses have been proposed to account for glacial ages. Each explains some of our observations, but none alone satisfactorily accounts for all of the data. We shall review some of the more plausible ideas.

Variations in the Earth's Orbit and Inclination to the Sun

The amount of heat from solar radiation received by any particular portion of the earth is related to the angle of the incoming sun's rays and, to a lesser degree, the distance to the sun. The dimensions of the earth's orbit around the sun change slightly over a period of thousands of years. The angle of the earth's poles relative to the plane of the earth's orbit about the sun also changes periodically. Recent analysis of data has added strong support to the theory (now known as the *Milankovitch theory*) that variations in orbital relationships and "wobble" of the earth's axis are largely responsible for the glacial and interglacial episodes. In the 1920s, Milutin Milankovitch, a Serbian mathematician, calculated that incoming solar radiation

varies in cycles of 21,000, 41,000 and 100,000 years. Proof for the Milankovitch cycles came from oceanographic data. Cores of deep sea sediment have provided a fairly precise record of climatic variations over the past few hundred thousand years. The cycles of cooling and warming determined from the marine sediments are 23,000, 42,000 and 100,000 years, thus closely matching the times predicted by Milankovitch.

The orbital variation theory fails to explain the absence of glaciation over most of geologic time. Thus, one or more of the other mechanisms described below may have contributed to glacial ages.

Changes in the Atmosphere

A set of hypotheses seeks to explain glacial ages by changes in the ability of the atmosphere to filter solar radiation. Much of the solar energy reaching our planet is either reflected back out to space or absorbed by the atmosphere. If the composition of the atmosphere changes, the amount of the sun's heat reaching and retained by the earth's surface also changes.

One hypothesis regards carbon dioxide as responsible for major climate changes. According to this proposal, if there is an increase in the carbon dioxide content in the atmosphere, the earth warms up. This is because of a "greenhouse effect" in which solar energy penetrates the atmosphere and heat is retained or trapped at the earth's surface. Carbon dioxide in the atmosphere reduces the amount of heat that can radiate from earth back out into space. This results in a warming trend until, for some reason, the carbon dioxide in the atmosphere decreases. High carbon dioxide content would coincide with warm episodes between glacial ages, low carbon dioxide content with periods of glaciation.

A major problem in the carbon dioxide hypothesis is explaining a cyclic change in the concentration of carbon dioxide in the atmosphere. One suggestion has been that when vegetation is abundant, there is less carbon dioxide in the atmosphere. The cooling this causes results in an ice age, and much of the earth's vegetation dies, releasing carbon dioxide to the atmosphere to begin the warming trend. However, there is no evidence of profound changes of vegetation. An additional difficulty comes in explaining why the process is effective for such a small percentage of geologic time.

Another hypothesis involves volcanoes. Worldwide temperature dropped about 2°C following the major vol-

Why is this so? The ice sheet's erosive ability may have been greater in northern Canada, where it was thickest. On the other hand, the erodibility of the bedrock may have been a more significant factor. The scoured bedrock of northern Canada is igneous and metamorphic rock, resistant to erosion. The bedrock beneath most of the till in midwestern Canada and northern United States is easily eroded sedimentary rock.

Most of the till was deposited as ground moraine, which, along with outwash deposits, has partially weathered canic eruption of Krakatoa in Indonesia in 1883. The SO_2 and dust carried by high atmospheric winds reduced solar energy penetrating the atmosphere for a few years (the opposite of the "greenhouse effect"). Presumably a closely spaced series of large eruptions could cause enough of a temperature drop to trigger glaciation and begin a glacial age.

Changing of the Positions of Continents

Another hypothesis is that glacial ages occur when land masses move closer to polar regions. Plate tectonics provides the mechanism for motion of landmasses. But the fluctuations in late Cenozoic climates are not explained by plate tectonics. All evidence indicates that the positions of the continents have not changed significantly with respect to the poles during the late Cenozoic Era. Certainly the continents have not shuffled back and forth during the recent glacial and interglacial ages.

However, the earlier movement of continents from positions closer to the equator into more northerly latitudes might have placed the present northern continents in a position favorable to glaciation.

Changes in Circulation of Sea Water

Our present climates are very much affected by circulation patterns of sea water. Landmasses block the worldwide free circulation of ocean water, thus some oceans are warmer than others.

According to one hypothesis, a glacial episode begins when Atlantic water circulates freely with Arctic Ocean water. At present the warmer waters of the Atlantic Ocean cannot mix freely with water in the Arctic Ocean. For this reason the surface of the Arctic Ocean is frozen for much of the year. Continental glaciation begins, according to this hypothesis, when warmer Atlantic water flows through a shallow channel between Greenland and Canada. This would keep the Arctic Ocean from completely freezing over during the winter. The moisture picked up by winds blowing over the Arctic Ocean would precipitate heavy amounts of snow on the northern continental landmasses. Ice sheets would grow so long as ice-free seas supplied the moisture. But as glaciation continued, sea level would drop due to the loss of water to the ice sheets. Sea level would eventually drop below the floor of the shallow channel between Greenland and Canada, shutting off the supply of warm Atlantic

waters to the Arctic Ocean. Because of this, the surface of the Arctic Ocean would freeze, reducing the moist air supplied to the ice sheets. The ice sheets would recede and disappear.

Some scientists question the adequacy of this mechanism to explain large-scale glaciation. Others point out that this hypothesis cannot explain the very warm interglacial climates. Also, snow and ice reflect much solar radiation. Therefore, a frozen Arctic Ocean coupled with glacial ice covering up to one-third of the world's landmass should tend to perpetuate a glacial age by reflecting more solar radiation into space.

Sliding of the West Antarctic Ice Sheet over the Ocean

One intriguing although highly speculative hypothesis regards changes in the West Antarctic Ice Sheet as responsible for glacial ages. A large segment of the ice sheet, lubricated by water along its base, slides as a mass onto the ocean surface. Before the large floating slab of ice can be broken up by wave action, the ice reflects a significant amount of solar heat back into space. This results in a worldwide cooling sufficient to trigger a glacial age.

There is some evidence that parts of the Antarctic Ice Sheet have slid rapidly and suddenly in the past. Whether the mechanism is capable of cooling the world's climate so much has yet to be demonstrated.

Summary

Scientists do not fully understand what causes glacial ages and intervening warm episodes. Only recently has the Milankovitch theory become widely accepted. This, at least, seems to explain what controls the cycles of climatic variation. But one or more of the other postulated mechanisms must also contribute significantly.

Further Reading

Imbrie, J. I., and K. P. Imbrie. 1986. *Ice Ages: Solving the mystery.* Boston: Harvard University Press.

Related Web Resources
Goddard Space Center Global Change
http://daac.gsfc.nasa.gov/

to give excellent soil for agriculture. In many areas along the southern boundaries of land covered by ground moraines, broad and complex end moraines extend for many kilometers, indicating that the ice margin must have been virtually stationary for a long period of time. Numerous drumlins are preserved in areas such as New England and upstate New York. Kettle lakes dot the landscape in large sections of Wisconsin and Minnesota. New York's Long Island was built up as morainal debris, most of it probably scoured out of New England.

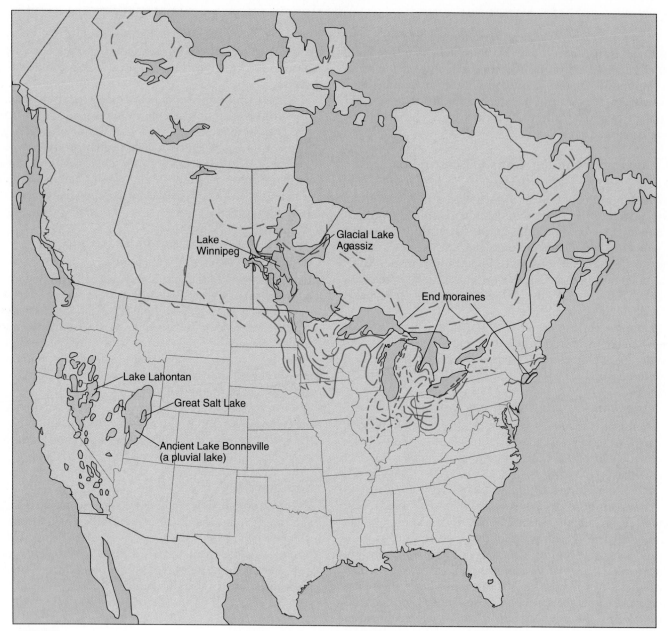

Figure 12.35

End moraines in the contiguous United States and Canada (shown by red lines), Glacial Lake Agassiz and pluvial lakes in the western United States (purple).

After C. S. Denny, U.S. Geological Survey, and the Geological Map of North America, Geological Society of America, and The Geological Survey of Canada.

The Great Lakes are, at least in part, a legacy of continental glaciation. Former stream valleys were widened by the ice sheet eroding weak layers of sedimentary rock into the present lake basins. End moraines border the Great Lakes, as shown in figure 12.35. Large regions of Manitoba, Saskatchewan, North Dakota, and Minnesota were covered by ice-dammed lakes (Lake Agassiz is the name for the largest of these). The former lake beds are now rich farmland.

Alpine glaciation was much more extensive throughout the world during the glacial ages than it is now. For example, small glaciers in the Rocky Mountains that now barely extend beyond their cirques were then valley glaciers 10, 50, or 100 kilometers in length. Yosemite Valley, which is no longer glaciated, was filled by a glacier about a kilometer thick. Its terminus was at an elevation of about 1,300 meters above sea level. Furthermore, cirques and other features

12.6
IN GREATER DEPTH

The Channeled Scablands

I n chapter 4 we described how the Columbia plateau in the Pacific Northwest (see figures 1.5 and 4.26) was built by a series of successive lava floods. The northeastern part of the plateau features a unique landscape, known as the channeled scablands, where the basalt bedrock has been carved into a series of large, interweaving valleys. From the air, the pattern looks like that of a giant braided stream. The channels, however, which range up to 30 kilometers wide and are usually 15 to 30 meters deep, are mostly dry.

The scablands are believed to have been carved by gigantic floods of water. Huge ripples in gravel bars (box figure 1) support this idea. To create these ripples, a flood would have to be about 100 times that of the Mississippi River in flood stage. This is much larger than any flood in recorded history.

What seems to have occurred is that, during the ice ages, a lobe of the ice sheet extended southward into northern Washington, Montana, and Idaho, blocking the head of the valley occupied by the Clark Fork River. The ice provided a natural dam for what is now known as Glacial Lake Missoula. Lake Missoula drowned a system of valleys in western Montana that extended hundreds of kilometers into the Rocky Mountains.

Ice is not ideal for building dams. Upon failure of the glacial dam, the contents of Lake Missoula became the torrential flood that scoured the Columbia plateau. There were at least four giant floods. Advancing ice from Canada would reestablish the dam, only to be destroyed after the reservoir refilled.

Box 12.6 Figure 1

Giant ripples of gravel from the draining of Lake Missoula, Montana. Note farm buildings for scale.

Photo by P. Weiss, U.S. Geological Survey

typical of valley glaciers can be found in regions that at present have no glaciers, such as the northern Appalachians—notably in the White Mountains of New Hampshire. Apparently, alpine glaciers were present in parts of New England after the main ice sheet retreated.

Indirect Effects of Past Glaciation

As the last continental ice sheet wasted away, what effects did the tremendous volume of meltwater have on American rivers? Rivers that now contain only a trickle of water were huge in the glacial ages. Other river courses were blocked by the ice sheet or clogged with morainal debris. Large dry stream channels have been found that were preglacial tributaries to the Mississippi and other river systems.

Pluvial Lakes

During the glacial ages the climate in North America, even beyond the glaciated parts, was more humid than it is now. Most of the presently arid regions of the western United States had moderate rainfall, as traces or remnants of numerous lakes indicate. These **pluvial lakes** (formed in a period of abundant rainfall) once existed in Nevada, Utah, and eastern California (figure 12.35). Some may have been fed by meltwater from mountain glaciers, but most were simply the result of a wetter climate.

Great Salt Lake in Utah is but a small remnant of a much larger body of fresh water called Lake Bonneville, which, at its maximum size, was nearly as large as Lake Michigan is today. Ancient beaches and wave-cut terraces on hillsides indicate the depth and extent of ancient Lake Bonneville. As the climate became more arid, lake levels lowered, outlets were cut off, and the water became salty, eventually leaving behind the Bonneville salt flats and the present very saline Great Salt Lake (see figure 6.24).

Even Death Valley in California—now the driest and hottest place in the United States—was occupied by a deep lake during the Pleistocene. The salt flats that were left when this lake dried include rare boron salts that were mined during the pioneer days of the American West.

Lowering of Sea Level

All of the water for the great glaciers had to come from somewhere. It is reasonable to assume that the water was "borrowed" from the oceans and that sea level worldwide was lower than it is today—at least 130 meters lower, according to scientific estimates.

What is the evidence for this? Stream channels have been charted in the present continental shelves, the gently inclined, now submerged edges of the continents (described in the chapter about the sea floor). These submerged channels are continuations of today's major rivers and had to have been above sea

Figure 12.36

Fiord in Alaska.

Photo by D. A. Rahm, courtesy Rahm Memorial Collection, Western Washington University

level for stream erosion to take place. Bones and teeth from now-extinct mammoths and mastodons have been dredged up from the Atlantic continental shelf, indicating that these relatives of elephants roamed over what must have been dry land at the time.

A **fiord** (also spelled fjord) is a coastal inlet that is a drowned glacially carved valley (figure 12.36). Fiords are common along the mountainous coastlines of Alaska, British Columbia, Chile, New Zealand, and Norway. They are evidence that valleys eroded by past glaciers were later partly submerged by rising seas.

Crustal Rebound

The weight of an ice sheet several thousand meters thick depresses the crust of the earth much as the weight of a person depresses a mattress. A land surface bearing the weight of a continental ice sheet may be depressed several hundred meters.

Once the glacier is gone, the land begins to rebound slowly to its previous height (see chapter 17 and figures 17.13 and 17.14). Uplifted and tilted shorelines along lakes are an indication of this process. The Great Lakes region is still rebounding as the crust slowly adjusts to the removal of the last ice sheet.

Evidence for Older Glaciation

Throughout most of geologic time, the climate has been warmer and more uniform than it is today. We think that the late Cenozoic Era is unusual because of the periodic fluctuations of climate and the widespread glaciations. However, glacial ages are not restricted to the late Cenozoic.

The evidence of older glaciation comes from rocks called tillites. A **tillite** is lithified till. Unsorted rock particles, including angular, striated, and faceted boulders, have been consolidated into a sedimentary rock. In some places, tillite layers overlie surfaces of older rock that have been polished and striated. Tillites of the late Paleozoic and tillites representing a minor part of the late Precambrian crop out in parts of the southern continents. (The striated surface in Australia, shown in figure 12.15, is overlain in places by late Paleozoic tillite.)

The oldest glaciation, for which we have evidence, appears to have taken place in what is now Ontario around 2.3 billion years ago.

The late Paleozoic tillites in the southern continents (South Africa, Australia, Antarctica, South America) have been used as evidence that these landmasses were once joined (see chapter 19 on plate tectonics). Directions of striations indicate that an ice sheet flowed onto South America from what is now the South Atlantic Ocean. Because an ice sheet can build up only on land, it is reasonable to conclude that the former ice sheet was centered on the ancient supercontinent before it broke up into the present continents.

Summary

A *glacier* is a large, long-lasting mass of ice that forms on land and moves under the influence of gravity and its own weight. A glacier can form wherever more snow accumulates than is lost. *Ice sheets* and *valley glaciers* are the two most important types of glaciers. Glaciers move downward from where the most snow accumulates toward where the most ice is wasted.

A glacier moves both by basal sliding and by internal flow. The upper portion of a glacier tends to remain rigid and is carried along by the ice moving beneath it.

Glaciers advance and recede in response to changes in climate. A receding glacier has a *negative budget* and an advancing one has a *positive budget*. A glacier's budget for the year can be determined by noting the relative position of the *snow line*.

Snow recrystallizes into firn, which eventually becomes converted to glacier ice. Glacier ice is lost (or wasted) by melting, by breaking off as icebergs, and by direct evaporation of the ice into the air.

A glacier erodes by plucking and the grinding action of the rock it carries. The grinding produces rock flour and faceted and polished rock fragments. Bedrock over which a glacier moves is generally polished and grooved.

A mountain area showing the erosional effects of alpine glaciation possesses relatively straight valleys with U-shaped cross profiles. The floor of a glacial valley usually has a *cirque* at its head and descends as a series of rock steps. Small *rock-basin lakes* are commonly found along the steps and in cirques. A *hanging valley* indicates that a smaller tributary joined the main glacier. A *horn* is a peak between several cirques. *Arêtes* usually separate adjacent glacial valleys.

A glacier deposits unsorted rock debris or *till*, which contrasts sharply with the sorted and layered deposits of glacial *outwash*. Till forms various types of *moraines*.

Fine silt and clay may settle as *varves* in a lake in front of a glacier, each pair of layers representing a year's accumulation.

Multiple till deposits and other glacial features indicate several major episodes of glaciation during the late Cenozoic Era. During each of these episodes, large ice sheets covered most of northern Europe and northern North America, and glaciation in mountain areas of the world was much more extensive than at present. At the peak of glaciation about a third of the earth's land surface was glaciated (in contrast to the 10% of the land surface presently under glaciers). Warmer climates prevailed during interglacial episodes.

The glacial ages also affected regions never covered by ice. Because of wetter climate in the past, large lakes formed in now-arid regions of the United States. Sea level was considerably lower.

Glacial ages also occurred in the more distant geologic past, as indicated by late Paleozoic and Precambrian tillites.

Terms to Remember

advancing glacier 286
alpine glaciation 284
arête 298
basal sliding 290
cirque 297
continental glaciation 284
crevasse 290
drumlin 302

end moraine 300
erratic 299
esker 302
fiord 310
glacier 284
ground moraine 300
hanging valley 296
horn 297

iceberg 286
ice cap 285
ice sheet 285
kettle 302
lateral moraine 300
medial moraine 300
moraine 300
outwash 302

plastic flow 290
pluvial lake 310
receding glacier 286
rigid zone 290
rock-basin lake (tarn) 296
rock flour 294
snow line 288

striations 294
terminus 288
theory of glacial ages 284
till 299
tillite 311
truncated spur 296
U-shaped valley 295

valley glacier 285
varve 303
wastage (ablation) 286
zone of accumulation 288
zone of wastage 288

Testing Your Knowledge

Use the questions below to prepare for exams based on this chapter.

1. How do erosional landscapes formed beneath glaciers differ from those that developed in rock exposed above glaciers?

2. How do features caused by stream erosion differ from features caused by glacial erosion?

3. How does material deposited by glaciers differ from material deposited by streams?

4. Why is the North Pole not glaciated?

5. How do arêtes, cirques, and horns form?

6. How does the glacial budget control the migration of the snow line?

7. How do recessional moraines differ from terminal moraines?

8. Alpine glaciation (a) is found in mountainous regions (b) exists where a large part of a continent is covered by glacial ice (c) is a type of glacier (d) none of the above

9. Continental glaciation (a) is found in mountainous regions (b) exists where a large part of a continent is covered by glacial ice (c) is a glacier found in the subtropics of continents (d) none of the above

10. At present about _____% of the land surface of the earth is covered by glaciers. (a) 1/2 (b) 1 (c) 2 (d) 10 (e) 33 (f) 50

11. Which is not a type of glacier? (a) valley glacier (b) ice sheet (c) ice cap (d) sea ice

12. The boundary between the zone of accumulation and the zone of wastage of a glacier is called the (a) firn (b) snow line (c) ablation zone (d) moraine

13. The ice caps on Mars are composed mostly of frozen (a) water (b) carbon dioxide (c) nitrogen (d) helium

14. Recently geologists have been drilling through ice sheets for clues about (a) ancient mammals (b) astronomical events (c) extinctions (d) past climates

15. Glacially carved valleys are usually _____ shaped. (a) V (b) U (c) Y (d) all of the above

16. Which is not a type of moraine? (a) medial (b) end (c) terminal (d) recessional (e) ground (f) esker

17. The last episode of extensive glaciation in North America was at its peak about _____ years ago. (a) 2,000 (b) 5,000 (c) 10,000 (d) 18,000

18. How fast does the central part of a valley glacier move compared to the sides of the glacier? (a) faster (b) slower (c) at the same rate

19. During the Ice Ages, much of Nevada, Utah, and Eastern California were covered by (a) ice (b) huge lakes (c) deserts (d) the sea

Expanding Your Knowledge

1. How might a warming trend cause increased glaciation?

2. How do, or do not, the Pleistocene glacial ages fit in with the principle of uniformitarianism?

3. Is ice within a glacier a mineral? Is a glacier a rock?

4. Could a rock that looks like a tillite have been formed by any agent other than glaciation?

5. What is the likelihood of a future glacial age? What effect might human activity have on causing or preventing a glacial age?

Exploring Resources

Bennett, M. R., and N. F. Glasser. 1996. *Glacial geology: Ice sheets and Landforms.* New York: J. Wiley.

Easterbrook, D. J. 1993. *Surface processes and landforms.* New York: McMillan Publishing Company.

Ehlers, J. 1996. *Quaternary and glacial geology.* New York: J. Wiley.

Hambrey, M. J. 1995. *Glacial environments.* Vancouver, B.C.: UBC Press.

Post, A., and E. R. LaChappelle. 1971. *Glacier ice.* Seattle: University of Washington Press. (This is a book of photographs.)

Ritter, D. F., R. C. Kochel, and J. R. Miller. 1995. *Process geomorphology.* 3d ed. Dubuque, Iowa: Wm. C. Brown Publishers.

Sharp, R. P. 1989. *Living ice: Understanding glaciers and glaciation.* New York: Cambridge University Press.

 http://Yahoo.com/Science/Earth_Sciences/Geology_and_Geophysics/Glaciology

Glaciers and climate change. Numerous websites may be accessed on this excellent site from the University of Illinois.

http://geochange.er.usgs.gov/pub/info/facts/atlas/index.html

U.S.G.S. Satellite Image Atlas of Glaciers of the World

Interacting with *Journey Through Geology* CD-ROM

Expand your knowledge of the concepts presented in this chapter by using the CD-ROM to answer the following questions.

1. Changing Earth, Atlantic Ocean: As this animation approaches the modern day, watch carefully the continental shelves surrounding North America. Why do they suddenly increase in size just before the animation stops?

Mitten Buttes in Monument Valley Tribal Park, Arizona.

Photo by © Doug Sherman, photographer

CHAPTER 13

Deserts and Wind Action

*D*eserts have a distinctive appearance because a dry climate controls erosional and depositional processes and the rates at which they operate. Although it seldom rains in the desert, running water is the dominant agent of land sculpture. Flash floods cause most desert erosion and deposition, even though they are rare events.

In chapters 9 through 12 you have seen how the land is sculptured by mass wasting, streams, ground water, and glaciers. Here we discuss the fifth agent of erosion and deposition: wind. Deserts and wind action are discussed together because of the wind's particular intensity in dry regions. But wind erosion and deposition can be very significant in other climates as well.

The word *desert* may suggest shifting sand dunes. Although moviemakers usually film sand dunes to represent deserts, only small portions of most deserts are covered with dunes. Actually a **desert** is any region with low rainfall. A region is usually classified as a desert if it has a dry or *arid climate* with less than 25 centimeters (10 inches) of rain per year. Few plants can tolerate low rainfall, so most deserts have a barren look.

Some specialized types of plants, however, grow well in desert climates despite the dryness. These plants are generally salt-tolerant; and they have extensive root systems to conserve water, so they often are widely spaced (figure 13.1). The leaves are usually very small, minimizing water loss by transpiration; they may even drop off the plants between rainstorms. During much of the year many desert plants look like dead, dry sticks. When rain does fall on the desert, the plants become green, and many will bloom.

Distribution of Deserts

The location of most deserts is related to descending air. The global pattern of air circulation is shown in simplified form in figure 13.2. The equator receives the sun's heat more directly than the rest of the earth. Air warms and rises at the equator, then moves both northward and southward to sink near 30° North latitude and 30° South latitude. The world's best-known deserts lie beneath these two belts of sinking air (figure 13.3).

Air sinking down through the atmosphere is compressed by the weight of the air above it. As air compresses, it warms up; and as it warms, it is able to hold more water vapor. Evaporation of water from the land surface into the warm, dry air is so great under belts of sinking air that moisture seldom falls back to earth in the form of rain. The two belts at 30° North and South latitude characteristically have clear skies, much sunshine, little rain, and high evaporation.

In contrast to the belts at 30°, the equator is marked by rising air masses which expand and cool as they rise. In cooling, the air loses its moisture, causing cloudy skies and heavy precipitation. Thus a belt of high rainfall at the equator separates the two major belts of deserts.

Not all deserts lie on the 30° latitude belts. Some of the world's deserts are the result of the **rain shadow** effect of mountain ranges (figure 13.4). As moist air is forced up to pass over a mountain range, it expands and cools, losing moisture as it rises. The dry air coming down the other side of the mountain compresses and warms, bringing high evaporation with little or no rainfall to the downwind side of the range. This dry region downwind of mountains is the *rain shadow zone*. Parts of the southwestern United States desert in Nevada and northern Arizona are largely the result of the rain shadow effect of the Sierra Nevada range in eastern California.

Great distance from the ocean is another factor that can create deserts, since most rainfall comes from water evaporated from the sea. The dry climate of the large arid regions in China, well north of 30° North latitude, is due to their location in a continental interior and to the rain shadow effect of mountains such as the Himalayas.

Deserts also tend to develop on tropical coasts next to *cold ocean currents*. Cold currents run along the western edges of con-

Figure 13.1

Typical appearance of the southwestern United States desert. Widely spaced plants have adapted to less than 25 centimeters (10 inches) of rain per year.

Figure 13.2

Global air circulation. Red arrows show surface winds. Blue arrows show vertical circulation of air. Air sinks at 30°N and 30°S latitude (and at the poles).

tinents, cooling the air above them. The cold marine air warms up as it moves over land, causing high evaporation and little rain on the coasts. This effect is particularly pronounced on the Pacific coast of South America and the Atlantic coast of Africa.

Not all deserts are hot. The cold, descending air at the North and South Poles (figure 13.2) creates *polar deserts* that have an arid climate along with a snow or ice cover. The entire continent of Antarctica is a desert, as are most of Greenland and the northernmost parts of Alaska, Canada, and Siberia.

Some Characteristics of Deserts

Because of their low rainfall, deserts have characteristic drainage and topography that differ from those of humid regions. Desert

Figure 13.3

World distribution of nonpolar deserts. Most deserts lie in two bands near 30°N and 30°S.

From map by U.S. Department of Agriculture.

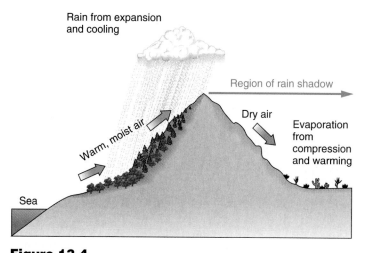

Figure 13.4

Rain shadow causes deserts on the downwind side of mountain ranges. Prevailing winds are from left to right.

streams usually flow intermittently. Water runs over the surface after storms, but during most of the year stream beds are dry. As a result, most deserts *lack through-flowing streams.* The Colorado River in the southwestern United States and the Nile River in Egypt are notable exceptions. Both are fed by heavy rainfall in distant mountains. The runoff is great enough to sustain stream flow across dry regions with high evaporation.

Many desert regions have *internal drainage;* the streams drain toward landlocked basins instead of toward the sea. The surface of an enclosed basin acts as a local base level. Because each basin is generally filled to a different level than the neighboring basins, desert erosion may be controlled by many different *local base levels.* As a basin fills with sediment, its surface rises, leading to a *rising base level,* which is a rare situation in humid (wet) regions.

The limited rainfall that does occur in deserts often comes from violent thunderstorms, with a high volume of rain falling in a very short time. Desert thunderstorms may dump more than 13 centimeters (5 inches) of rain in one hour. Such a large amount of rain cannot soak readily into the sun-baked caliche soil, so the water runs rapidly over the land surface, particularly where vegetation is sparse. This high runoff can create sudden local floods of high discharge and short duration called **flash floods.** Flash floods are more common in arid regions than in humid regions. They can turn normally dry stream beds into raging torrents for a short time after a thunderstorm. Because soil particles are not held in place by plant roots, these occasional floods can effectively erode the land surface in a desert region. As a result, desert streams normally are very heavily laden with sediment. Flash floods can easily erode enough sediment to become *mudflows* (see chapter 9).

Desert stream channels are distinctive in appearance because of the great erosive power of flash floods and the intermittent nature of streamflow. Most stream channels are normally dry and covered with sand or gravel that is moved only during occasional flash floods. Rapid downcutting by sediment-laden floodwaters tends to produce narrow canyons with vertical walls and flat, gravel-strewn floors (figure 13.5A). Such channels are often called *arroyos* or *dry washes.*

Deserts and Wind Action   **317**

Expanding Deserts

Many geologists and geographers use a two-part definition of *desert*. A desert must not only have less than 25 centimeters (10 inches) of rain per year, but it also must be so devoid of vegetation that few people can live there. Many dry regions have supported marginally successful agriculture and moderate human populations in the past but are being degraded into barren deserts today by overgrazing and overpopulation. The expansion of barren deserts into once-populated regions is called *desertification*.

Limited numbers of people can exist in dry regions through careful agricultural practices that protect water sources and limit grazing of sparse vegetation. Overuse of the land by livestock and humans, however, can strip it bare and make it uninhabitable. The large desert in northern Africa, shown in figure 13.3, is the Sahara Desert, and the semiarid region (10 to 20 inches of rain per year) to the south of it is the Sahel. In the early 1960s a series of abnormally wet years encouraged farmers in the Sahel to expand their herds and grazing lands.

A severe drought throughout the 1970s and 1980s caused devastation of the plant life of the region as starving livestock searched desperately for food, and humans gathered the last remaining sticks for firewood (box figure 1). Vast areas that were once covered with trees and sparse grass became totally barren, and an acute famine began, killing more than 100,000 people. The desert expanded southward, advancing in some places as much as 50 kilometers (30 miles) per year.

Box 13.1 Figure 1

Desertification in Africa after a period of drought has left inhabitants desperately searching for water.

Photo © Walt Anderson/Visuals Unlimited

The denuded soil in many regions became susceptible to wind erosion, leading to choking dust storms and new, advancing dune fields (some even migrating into cities).

Some of the same problems afflicted the midwestern United States in the 1930s, as intense land cultivation coupled with a prolonged drought produced the barren Dust Bowl during the time of the Great Depression. Renewed rains and improved soil-conservation practices have reversed the trend in the United States, but the area is still vulnerable to a future drought, and the possibility of future dust storms in prairie states is very real.

Drought accelerates desertification but is not necessary for it to occur. Overloading the land with livestock and humans can strip marginal regions of vegetation even in wet years.

Related Web Resources
http://pubs.usgs.gov/gip/deserts/desertification.html
Good overview of desertification around the world. Includes photographs.

Newcomers to deserts sometimes get into serious trouble in desert canyons in rainy weather. Imagine for a moment that you have camped on the canyon floor in figure 13.5*A* to get out of strong desert winds. Later that night a towering thunderhead cloud forms, and heavy rain falls on the mountains several miles away from you. Although no rain has fallen on you, you are awakened several minutes later by a distant roar. The roar grows louder until a 3-meter (10-foot) "wall" of water rounds a bend in the canyon, heading straight for you at the speed of a galloping horse. Boulders, brush, and tree trunks are being swept along in this raging flash flood. The walls of the canyon are too steep to climb. Several hikers died during the summer of 1997 when such a wall of water roared down a side canyon in Grand Canyon National Park in Arizona. Stay out of desert canyons if there is any sign of rain; sleeping in such canyons is particularly dangerous.

The resistance of some rocks to weathering and erosion is partly controlled by climate. In a humid (wet) climate limestone dissolves easily, forming low places on the earth's surface. In a desert climate the lack of water makes limestone resistant, so it stands up as ridges and cliffs in the desert just as sandstone and conglomerate do. Lava flows and most igneous and metamorphic rock are also resistant. Shale is the least resistant rock in a desert so it usually erodes more deeply than other rock types and forms gentler slopes (figure 10.52) or badland topography (figure 13.5*B*).

Although intersecting joints form angular blocks of rock in all climates, desert topography characteristically looks more angular than the gently rounded hills and valleys of a humid region. This may be due indirectly to the low rainfall in deserts. Shortage of water slows chemical weathering processes to the point where few minerals break down to form fine-

A

B

Figure 13.5

Desert features. (*A*) Desert stream channel showing dry, gravel-covered floor and steep, vertical sides (cut in sandstone and conglomerate) in Death Valley, California. (*B*) Badland topography (sharp ridges and V-shaped channels) eroded on shale in a dry region where plants are scarce. Badlands National Park, South Dakota.

Photo *B* © Dick Scott, Visuals Unlimited

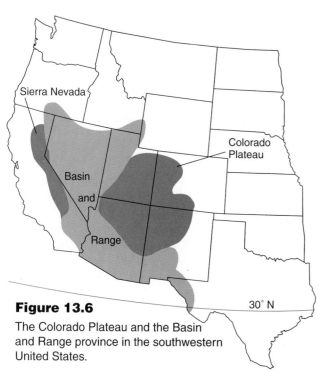

Figure 13.6

The Colorado Plateau and the Basin and Range province in the southwestern United States.

grained clay minerals. Soils are coarse and rocky with few chemically weathered products. Plants, which help bind soil into a cohesive layer in humid climates, are rare in deserts, and so desert soils are easily eroded by wind and rainstorms. Downhill creep of thick, fine-grained soil is partly responsible for softening the appearance of jointed topography in humid climates. With thin, rocky soil and slow rates of creep, desert topography remains steep and angular.

As you learned in chapter 10, climate is only one of many things that determine the shape and appearance of the land. Rock structure is another. As an example, in the next section we will look closely at two different structural regions within the desert of the southwestern United States.

Desert Features in the Southwestern United States

Much of the southwestern United States has an arid (or semiarid) climate, partly because it is close to 30° North latitude and partly because of the rain shadow effect of the Sierra Nevada and other mountain ranges. Within this region of low rainfall are two areas of markedly different geologic structure. One area is the Colorado Plateau and the other is the Basin and Range province, a mountainous region centered on the state of Nevada. The boundaries of these two areas are shown in figure 13.6 (see also figure 1.5).

Deserts and Wind Action   **319**

Figure 13.7

Characteristic landforms of the Colorado Plateau. (*A*) Erosional retreat of a cliff at the edge of a plateau can leave behind mesas and buttes as erosional remnants of the plateau. This is an example of parallel retreat of slopes, which was discussed in chapter 10. (*B*) Mesas and buttes in Monument Valley, Arizona, an area of eroded, horizontal, sedimentary rocks.

The *Colorado Plateau* centers roughly on the spot known as the Four Corners, where the states of Utah, Colorado, Arizona, and New Mexico meet at a common point. The rocks near the surface of the Colorado Plateau are mostly flat-lying beds of sedimentary rock over 1500 meters above sea level. These rocks are well exposed at the Grand Canyon in Arizona.

Because the rock layers are well above sea level, they are vulnerable to erosion by the little rain that does fall in the region. Flat-lying layers of resistant rock, such as sandstone, limestone, and lava flows, form **plateaus**—broad, flat-topped areas elevated above the surrounding land and bounded, at least in part, by cliffs. As erosion removes the rock at its base,

the cliff is gradually eroded back into the plateau (figure 13.7*A*). Remnants of the resistant rock layer may be left behind, forming flat-topped mesas or narrow buttes (figure 13.7*B*). A **mesa** is a broad, flat-topped hill bounded by cliffs and capped with a resistant rock layer. A **butte** is a narrow hill of resistant rock with a flat top and very steep sides. Most buttes form by continued erosion of mesas. (The term *butte* is also used in other parts of the country for any isolated hill.)

The Colorado Plateau is also marked by peculiar steplike folds (bends in rock layers) called *monoclines.* Erosion of monoclines (and other folds) leaves resistant rock layers protruding above the surface as ridges (figure 13.8). A steeply tilted resis-

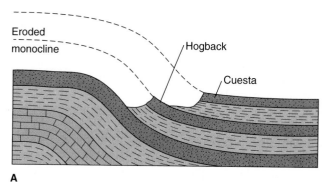

Eroded
monocline
Hogback
Cuesta

A

B

Figure 13.8

(A) Steplike monocline folds often erode so that resistant rock layers form hogbacks and cuestas (these features are not unique to deserts). (B) Monocline near the Big Horn Mountains, Wyoming.

Figure 13.9

Basin and Range topography in Death Valley, California. In the distance the fault-bounded Panamint Mountains rise more than 3 kilometers (11,000 feet) above Death Valley. Giant alluvial fans at the base of the mountains show a braided stream pattern. Fine-grained sediments and salt deposits underlie the playa in the foreground.

tant layer erodes to form a *hogback,* a sharp ridge that has steep slopes. A gently tilted resistant layer forms a *cuesta,* with one steep side and one gently sloping side.

Note that plateaus, monoclines, hogbacks, and cuestas are not unique to deserts. They are surface features found in all cli-

mates, but are particularly well exposed in deserts because of thin soil and sparse vegetation.

The *Basin and Range province* is characterized by rugged mountain ranges separated by flat valley floors (figure 13.9). The blocks of rock that form the mountain ranges and the

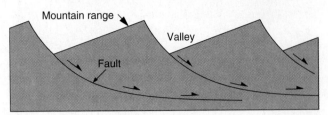

Figure 13.10

Two possible origins of the mountains and valleys of the Basin and Range province. Movement along the faults is shown by arrows.

valley floors are bounded by **faults,** cracks in the earth along which some rock movement has taken place. (Chapter 15 discusses faults in more detail.) In the Basin and Range province, movement on the faults has dropped the valleys down relative to the adjacent mountain ranges to accommodate crustal thinning and extension (figure 13.10). Fault-controlled topography is found throughout the Basin and Range province, which covers almost all of Nevada and portions of bordering states as well as southern Arizona and New Mexico. The numerous ranges in this province create multiple rain shadow zones and therefore a very dry climate (see figure 1.5).

Heavy rainfall from occasional thunderstorms in the mountain ranges causes rapid erosion of the steep mountain fronts and resulting deposition on the valley floors (figure 13.11). Rock debris from the mountains, picked up by flash floods and mudflows, is deposited at the base of the mountain ranges in the form of alluvial fans. Alluvial fans (described in chapter 10) build up where stream channels widen as they flow out of narrow canyons onto the open valley floors causing a decrease in velocity and rapid deposition of sediment.

Although most of the sediment carried by runoff is deposited in alluvial fans, some fine clay may be carried in suspension onto the flat valley floor. If no outlet drains the valley, runoff water may collect and form a **playa lake** on the valley floor. Playa lakes are usually very shallow and temporary, lasting for only a few days after a rainstorm. After the lake evaporates, a thin layer of fine mud may be left on the valley floor. The mud dries in the sun, forming a **playa,** a very flat surface underlain by hard, mud-cracked clay (figures 13.12 and 13.13). If the runoff contained a large amount of dissolved salt or if seeping ground water brings salt to the surface, the flat playa surface may be underlain by a bright white layer of dried salt instead of mud, as on the Bonneville Salt Flats in Utah (see figure 6.24).

Continued deposition near the base of the mountains may form a **bajada,** a broad, gently-sloping depositional surface formed by the coalescing of individual alluvial fans (figure 13.11). A bajada is much more extensive than a single alluvial fan and may have a gently rolling surface resulting from the merging of the cone-shaped fans.

Erosion of the mountain can eventually form a **pediment,** which is a gently sloping surface, commonly covered with a veneer of gravel, cut into the solid rock of the mountain (figure 13.11). A pediment develops uphill from a bajada as the mountain front retreats. It can be difficult to distinguish a pediment from the surface of the bajada downhill, because both have the same slope and gravel cover. The pediment, however, is an erosional surface, usually underlain by solid rock, while the bajada surface is depositional and may be underlain by hundreds of meters of sediment.

An abrupt change in slope marks the upper limit of the pediment, where it meets the steep mountain front. Many geologists who have studied desert erosion believe that as this steep mountain front erodes, it retreats uphill, maintaining a relatively constant angle of slope. (This is *parallel retreat* of a slope—see figure 10.50.)

Notice that rock structure, not climate, largely controls the fact that plateaus and cliffs are found in the Colorado Plateau, while mountain ranges, broad valleys, alluvial fans, and pediments are found in the Basin and Range province. Features such as plateaus, mesas, and alluvial fans can also be found in humid climates wherever the rock structure is favorable to their development; they are not controlled by climate. Features such as steep canyons, playa lakes, thin soil, and sparse vegetation, however, *are* typically controlled by climate.

Wind Action

Wind can be an important agent of erosion and deposition in any climate, as long as sediment particles are loose and dry. Wind differs from running water in two important ways. Because air is less dense than water, wind can erode only fine sediment—sand, silt, and clay. But wind is not confined to channels as running water is, so wind can have a widespread effect over vast areas.

In general, the faster the wind blows, the more sediment it can move. Wind velocity is affected by air temperature. As air warms and cools, it changes density, and these density changes create air pressure differences that cause wind. Wet climates and cloud cover help buffer changes in air temperature, but in dry climates daily temperature changes can be extreme. In a desert, the temperature may range from 10°C (50°F) at night to more than 40°C (100°F) in the daytime. Because of these temperature fluctuations, wind is generally stronger in deserts than in humid regions, typically exceeding 100 kilometers per hour (60 miles per hour). The scarcity of vegetation in deserts to slow wind velocity by friction increases the effectiveness of desert winds.

Although strong winds are also associated with rainstorms and hurricanes, these winds seldom erode sediment because

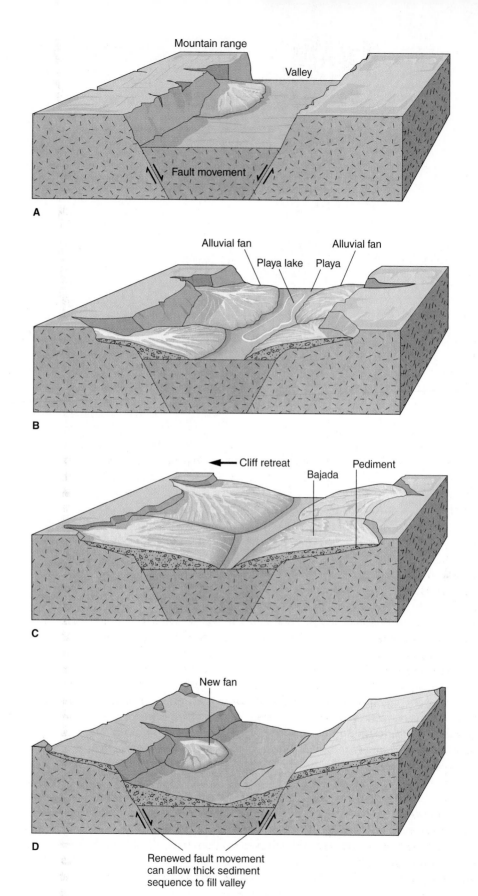

Figure 13.11

Development of landforms associated with Basin and Range topography.

Figure 13.12

Alluvial fans at the base of mountain canyons, Death Valley, California. The white salt flat in the foreground is part of a playa.

Photo by Frank M. Hanna

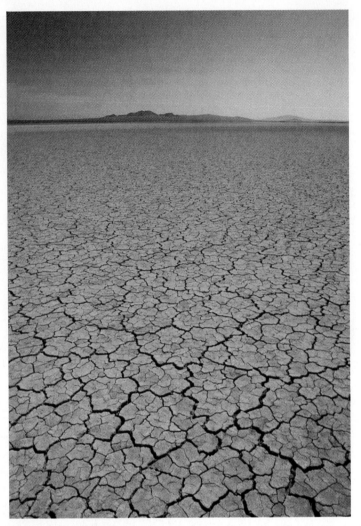

Figure 13.13

Mud-cracked playa surface.

Photo © Bill Ross/Westlight

rain wets the surface sediment. Wet sediment is heavy and cohesive and will not be blown away. Strong winds in the desert, however, blow over loose, dry sediment, so wind is an effective erosional agent in dry climates. (As we said earlier in the chapter, running water in the form of flash floods is a far more important erosive agent than wind, even though the wind can be very strong.)

Wind Erosion and Transportation

Thick, choking *dust storms* are one example of wind action (figure 13.14). "Dust Bowl" conditions in the 1930s in the agricultural prairie states lasted for several years due to droughts and poor soil-conservation practices. Loose silt and clay are easily picked up from barren dry soil, such as in a cultivated field. Wind erosion is even greater if the soil is disturbed by animals or vehicles. Silt and clay can remain suspended in turbulent air for a long time, so a strong wind may carry a dust cloud hundreds of meters upward and hundreds of kilometers horizontally. Dust storms of the 1930s frequently blacked out the midday sun, fertile soil was lost over vast regions ruining many farms, and streets and rivers downwind were filled with thick dust deposits.

Wind-blown sediment is sometimes picked up on land and carried out to sea. Particles from the Sahara Desert in Africa have been collected from the air over the islands of the West Indies after having been carried across the entire Atlantic Ocean. A substantial amount of the fine-grained sediment that settles to the sea floor is land-derived sediment that the wind has deposited on the sea surface. Ships 800 kilometers (500 miles) offshore have reported dustfalls a few millimeters thick covering their decks.

Volcanic ash can be carried by wind for very great distances. An explosive volcanic eruption can blast ash more than 15 kilometers (10 miles) upward into the air. Such ash may be caught in the high-altitude *jet streams,* narrow belts of strong winds with velocities sometimes greater than 300 kilometers (200 miles) per hour. Following the 1980 eruption of Mount St. Helens in western Washington, a visible ash layer blanketed parts of Washington, Idaho, and Montana to the east. At high altitudes, St. Helens ash could be detected blowing over New York and out over the Atlantic Ocean, 5,000 kilometers (3,000 miles) from the volcano. But the St. Helens eruption was a relatively small one. Ash from the 1883 Krakatoa eruption in Indonesia circled the globe for

A

B

Figure 13.14

(*A*) Wind erosion of soil from a cultivated field. (*B*) Approaching dust storm, Prowers County, Colorado, late afternoon (1930s). The storm lasted almost three hours, with wind speeds of approximately 50 kilometers (30 miles) per hour.

Photos *A* and *B* by U.S. Department of Agriculture, Soil Conservation Service

A

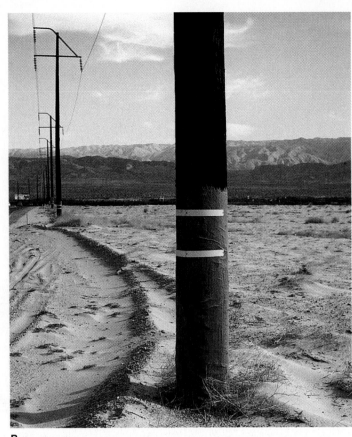

B

Figure 13.15

(*A*) Wind erosion near the ground has sandblasted the lower 1 meter (3 feet) of this basalt outcrop, Death Valley, California. Hammer for scale. (*B*) Power pole with its base wrapped in an abrasion-resistant material to minimize wind erosion.

Photo *B* courtesy Paul Bauer

two years, causing spectacular sunsets and a slight, but measurable, drop in temperature as the ash reflected sunlight back into space. The 1815 eruption of Tambora, also in Indonesia, put so much ash into the air that there were summer blizzards, crop failures, and famine in New England and northern Europe in 1816, "the year without a summer." Lower temperatures (1°F) and brilliant sunsets also marked the 1991 eruption of Pinatubo in the Philippines.

Because sand grains are heavier than silt and clay, sand moves close to the ground in the leaping pattern called *saltation* (as does some sediment in streams). High-speed winds can cause *sandstorms,* clouds of sand moving rapidly near the land surface. The high-speed sand in such a storm can sandblast smooth surfaces on hard rock and scour the windshields and paint of automobiles. Because of the weight of the sand grains, however, sand rarely rises more than 1 meter (3 feet) above a flat land surface, even under extremely strong winds. Therefore, most of the sandblasting action of wind occurs close to the ground (figure 13.15). Telephone poles in regions of wind-driven sand often are severely abraded near the ground. To prevent such abrasion, desert residents pile small stones or wrap sheet metal around the base of the poles.

Wind seldom moves particles larger than sand grains, but wind-blown sand may sculpture isolated pebbles, cobbles, or

Figure 13.16

Ventifacts eroded by blowing sand, Death Valley, California. Most show two flat sides joined by a sharp ridge. Attached sand shows original depth of burial.

boulders into **ventifacts**—rocks with flat, wind-abraded surfaces (figure 13.16). If the wind direction shifts, or if the stone is turned, more than one flat face may develop on the ventifact.

Figure 13.17

Deflation by wind erosion can form a blowout in loose, dry sediment. Deflation stops at the water table. A pillar, or erosional remnant, may be found in the center of a blowout.

Deflation

The removal of clay, silt, and sand particles from the land surface by wind is called **deflation.** If the sediment at the land surface is made up only of fine particles, the erosion of these particles by the wind can lower the land surface substantially. A **blowout** is a depression on the land surface caused by wind erosion (figure 13.17). A *pillar,* or erosional remnant of the former land, may be left at the center of a blowout (figures 13.17 and 13.18).

Blowouts are common in the Great Plains states (figure 13.18). One in Wyoming measures 15 by 5 kilometers (9 by 3 miles) and is 45 meters (150 feet) deep. The enormous Qattara Depression in western Egypt, more than 250 kilometers (150 miles) long and more than 100 meters (300 feet) *below* sea level, has been attributed to wind deflation. Deflation can continue to deepen a blowout in fine-grained sediment until it reaches wet, cohesive sediment at the water table.

Wind Deposition

Loess

Loess is a deposit of wind-blown silt and clay composed of unweathered, angular grains of quartz, feldspar, and other minerals weakly cemented by calcite. Loess has a high porosity, typically near 60%. Deposits of loess may blanket hills and valleys downwind of a source of fine sediment, such as a desert or a region of glacial outwash.

China has extensive loess deposits (figure 13.19), more than 100 meters (300 feet) thick in places. Wind from the Gobi Desert carried the silt and clay that formed these deposits. Loess is easy to dig into and has the peculiar ability to stand as a vertical cliff without slumping (figure 13.20), perhaps because of its cement or perhaps because the fine, angular, sediment grains interlock with one another. For centuries the Chinese have dug cavelike homes in loess cliffs. When a large earthquake shook China in 1920, however, many of these cliffs collapsed, burying alive about 100,000 people.

During the glacial ages of the Pleistocene Epoch, the rivers that drained what is now the midwestern United States transported and deposited vast amounts of glacial outwash. Later,

Figure 13.18

Large blowout near Harrison, Nebraska. Pillar top is the original level of land, before wind ersosion lowered the land surface by more than 3 meters (10 feet). The pillar is the erosional remnant at the center of the blowout.

Photo by N. H. Darton, U.S. Geological Survey

winds eroded silt and clay (originally glacial rock flour) from the flood plains of these rivers and blanketed large areas of the Midwest with a cover of loess (figure 13.20). Soils that have developed from the loess are usually fertile and productive. The grain fields of much of the Midwest (and in the Palouse area in eastern Washington) are planted on these rich soils. Wind erosion of cultivated, loess-covered hills in the Palouse region is a serious problem that has locally removed fertile soil from the hilltops.

Sand Dunes

Sand dunes are mounds of loose sand grains heaped up by the wind. Dunes are most likely to develop in areas with strong winds that generally blow in the same direction. Patches of dunes are scattered throughout the southwestern United States desert. More extensive dune fields occur on some of the other

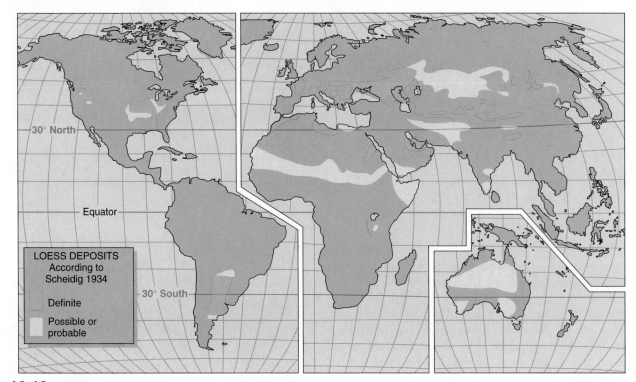

Figure 13.19

Major loess-covered areas in the world.

Source: Data from map from U.S. Department of Agriculture; Smalley, I. J., Loess Lithology and Genesis, fig. 1, p. 768, Halsted Press, 1975.

Figure 13.20

Vertical road cuts in loess, Vicksburg, Mississippi.

Photo by E. W. Shaw, U.S. Geological Survey

Figure 13.21

Coastal dunes formed from beach sand blowing inland, Pismo Beach, California.

Photo by Frank M. Hanna

deserts of the world, such as the Sahara Desert of Africa, which contains vast *sand seas.* Dunes are also commonly found just landward of beaches (figure 13.21), where sand is blown inland. Beach dunes are common along the shores of the Great Lakes and along both coasts of the United States. Braided rivers (see chapter 10) can also be sources of sand for dune fields.

The mineral composition of the sand grains in sand dunes depends on both the character of the original sand source and the intensity of chemical weathering in the region. Many dunes, particularly those near beaches in humid regions, are composed largely of quartz grains because quartz is so resistant to chemical weathering. Inland dunes, such as the Great Sand Dunes National Monument in Colorado, often contain unstable feldspar and rock fragments in addition to quartz. Some dunes are formed mostly of carbonate grains, particularly those near tropical beaches.

Desert Pavement and Desert Varnish

Two intriguing features can be seen in many deserts, particularly on the surface of old alluvial fans no longer receiving new sediment.

Desert pavement is a thin, surface layer of closely packed pebbles (box figure 1). The pebbles were once thought to be residual, left behind as strong winds blew away all the fine grains of a rocky soil. The pebbles are now thought to be brought to the surface by cycles of wetting and drying, which cause the soil to swell and shrink as water is absorbed and lost by soil particles. Swelling soil lifts pebbles slightly; drying soil cracks, and fine grains fall into the cracks. In this way pebbles move up, while fine grains move down. The surface layer of

pebbles protects the land from wind erosion and deflation. When the pavement is disturbed (as in the 1991 Gulf War), dust storms and new sand dunes may result.

Many rocks on the surface of deserts are darkened by a chemical coating known as *desert varnish.* Although the interior of the rocks may be light colored, a hard, often shiny, coating of dark iron and manganese oxides and clay minerals can build up on the rock surface over long periods of time (box figure 2). Although no one is quite certain how this coating develops, it seems to be added to the rocks from the outside, for even white quartzite pebbles with no internal source of iron, manganese or clay minerals can develop desert varnish. One current hypothesis is that the clay is windblown, perhaps sticking to rocks dampened by dew. A film of clay on a rock may draw iron and manganese-containing solutions upward from the soil by capillary action, and the presence of the clay minerals may help deposit the dark manganese oxide that cements the clay to the rock. Another hypothesis is that the oxide is deposited biologically by microbes. Regardless of how the varnish forms, the longer a rock is exposed on a desert land surface, the darker it becomes.

Box 13.2 Figure 1

Desert pavement on an old alluvial fan surface. The surface pebbles are closely packed; fine sand underlies the pebbles.

Photo by Frank M. Hanna

Box 13.2 Figure 2

Petroglyphs carved on this rock cut through the dark desert varnish to show the lighter color of the interior of the rock, Valley of Fire, Nevada.

Photo by J. Freeberg, U.S. Geological Survey

At White Sands, New Mexico, dunes are made of gypsum grains, eroded by wind from playa lake beds.

Sand grains found in dunes are commonly well-sorted and well-rounded because wind is very selective as it moves sediment. Fine-grained silt and clay are carried much farther than sand, and grains coarser than sand are left behind when sand moves. The result is a dune made solely of sand grains, commonly all very nearly the same size. The prevalence of well-rounded grains in many dunes also may be due to selective sorting by the wind. Rounded grains roll more easily than

angular grains, and so the wind may remove only the rounded grains from a source to form dunes. Wind will often selectively roll oolitic grains from a carbonate beach of mixed oolitic and skeletal grains.

Most sand dunes are asymmetric in cross section, with a gentle slope facing the wind and a steeper slope on the downwind side. The steep downwind slope of a dune is called the **slip face** (figure 13.22). It forms from loose, cascading sand that generally keeps the slope at the *angle of repose,* which is about 34° for loose, dry sand. Sand is blown

Figure 13.22

(*A*) A sand dune forms with a gentle upwind slope and a steeper slip face on the downwind side. Sand eroded from the upwind side of the dune is deposited on the slip face, forming cross-beds. Movement of sand causes the dune to move slowly downwind. (*B*) Strong desert winds (60 miles per hour) blowing to the right remove sand from the gently sloping upwind side of this dune. The sand settles onto the steep slip face on the right.

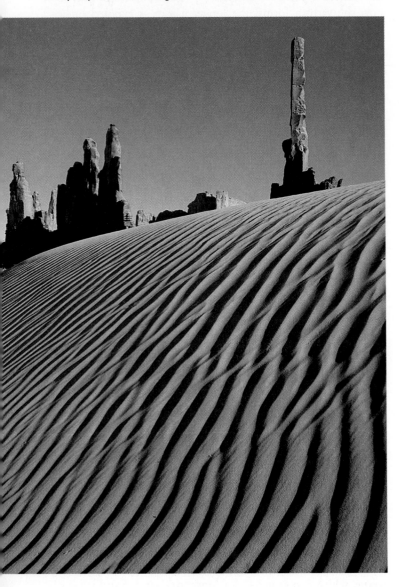

up the gentle slope and over the top of the dune. Sand grains fall like snow onto the slip face when they encounter the calm air on the downwind side of the dune. Loose sand settling on the top of the slip face may become oversteepened and slide as a small avalanche down the slip face. These processes form high-angle cross-bedding within the dune. When found in sandstone, such cross-bedding strongly suggests deposition as a dune (see figure 6.27).

In passing over a dune, the wind erodes sand from the gentle upwind slope and deposits it downwind on the slip face. As a result, the entire dune moves slowly in a downwind direction. The rate of dune motion is much slower than the speed of the wind, of course, because only a thin layer of sand on the surface of the dune moves at any one time. The dune may move only 10 to 15 meters (30 to 50 feet) per year. Over many years, however, the movement of dunes can be significant, a fact not always appreciated by people who build homes close to moving sand dunes.

If a dune becomes overgrown with grass or other vegetation, movement stops. The Sand Hills of north-central Nebraska are large dunes, formed during the Pleistocene or Holocene Epochs, that have become stabilized by vegetation. The migration of many beach dunes toward beach homes and roads has been stopped by planting a cover of beach grass over the dunes. Dune-buggy tires can trample and kill the grass, however, and start the dunes moving again.

Sand moving over a dune surface typically forms *wind ripples*—small, low ridges of sand produced by saltation of the grains (figure 13.23). The ripples are similar to those formed in sediment by a water current (see chapter 6). Because sand moves perpendicularly to the long dimension

Figure 13.23

Wind ripples on sand surface, Monument Valley, Utah.

Photo by © Doug Sherman

A Barchans

B Transverse dunes

C Parabolic dunes

D Longitudinal dunes (seifs)

Figure 13.24

Types of sand dunes.

of the ripples, a rippled sand surface indicates the direction of sand movement.

Types of Dunes

As figure 13.24 shows, dunes tend to develop certain characteristic shapes, depending on (1) the wind's velocity and direction (that is, whether constant or shifting); (2) the sand

Figure 13.25

These barchan dunes are advancing as much as 15 meters (50 feet) a year over this barren valley floor in southern California.

Photo by John S. Shelton

supply available; and (3) how the vegetation cover, if any, is distributed.

Where the sand supply is limited, a type of dune called a **barchan** generally develops. The barchan is crescent-shaped with a steep slip face on the inward or concave side. The horns on a barchan dune point in the downwind direction (figure 13.24*A*). Barchan dunes are usually separated from one another and move across a barren surface (figure 13.25). If more sand is available, the wind may develop a **transverse dune,** a relatively straight, elongate dune oriented perpendicular to the wind direction (figures 13.24*B* and 13.26).

Figure 13.26

Transverse dunes, Oregon.

Photo by D. A. Rahm, courtesy Rahm Memorial Collection, Western Washington University

Figure 13.27

Parabolic dunes near Pismo Beach, central California. Wind blows from left to right. The ocean and a sand beach are just to the left of the photo.

Photo by Frank M. Hanna

Figure 13.28

Longitudinal dunes in the Sahara Desert, Algeria. Photo from Gemini spacecraft at an altitude of about 100 kilometers (65 miles).

Photo by NASA

A **parabolic dune** is somewhat similar in shape to a barchan dune, except that it is deeply curved and is convex in the downwind direction. The horns point upwind and are commonly anchored by vegetation (figure 13.24C). The parabolic dune requires abundant sand and commonly forms around a blowout. Because they require abundant sand and strong winds, parabolic dunes are typically found inland from an ocean beach (figure 13.27). All three of these dune shapes develop in areas having steady wind direction, and all three have steep slip faces on the downwind side.

One of the largest types of dunes is the **longitudinal dune** or *seif* (figure 13.27D), which is a symmetrical ridge of sand that forms parallel to the prevailing wind direction. Longitudinal dunes occur in long, parallel ridges that are exceptionally straight and regularly spaced. They are typically separated by barren ground or desert pavement. Longitudinal dunes in the Sahara Desert (figure 13.28) are as high as 200 meters (650 feet) and more than 120 kilometers (75 miles) in length. Numerous hypotheses have been proposed to explain the development of longitudinal dunes, but none can ade-

quately explain their spectacular size and regular spacing. It appears that crosswinds are important in piling up sand, which adds to the height of longitudinal dunes, whereas the more constant prevailing wind direction redistributes the sand down the length of the dunes. Smoke bomb experiments to analyze airflow have shown that the wind spirals down the intervening troughs between longitudinal dunes and may control the regularity of their spacing.

Not all dunes can be classified by an easily recognizable shape. Many of them are quite irregular.

Summary

Deserts are located in regions where less than 25 cm (10 inches) of rain falls in a year. Such regions are found primarily in belts of descending air at 30° North and South latitude. Arid regions also may be due to the *rain shadow* of a mountain range, great distance from the sea, and proximity to a cold ocean current. Descending air forms cold deserts at the poles.

Desert landscapes differ from those of humid regions of lacking through-flowing streams and in having internal drainage and many local, rising base levels. *Flash floods* caused by desert thunderstorms are effective agents of erosion despite the low rainfall. Limestone is resistant in deserts. Thin soil and slow rates of creep may give desert topography an angular look.

Parts of the southwestern United States are desert, the topography determined primarily by rock structure. Flat-lying sedimentary rocks of the Colorado Plateau are sculptured into cliffs, *plateaus, mesas,* and *buttes.* The fault-controlled topography of the Basin and Range province is marked by *alluvial fans, bajadas, playas,* and *pediments.*

Although wind erosion can be intense in regions of low moisture, streams are usually more effective than wind in sculpturing landscapes, regardless of climate.

Fine-grained sediment can be carried long distances by wind, even across entire continents and oceans.

Sand moves by *saltation* close to the ground, occasionally carving *ventifacts.*

Wind can *deflate* a region, creating a *blowout* in fine sediment.

Sand dunes move slowly downwind as sand is removed from the gentle upwind slope and deposited on the steeper *slip face* downwind.

Dunes are classified as *barchans, transverse dunes, parabolic dunes,* and *longitudinal dunes,* but many dunes do not resemble these types. Dune type depends on wind strength and direction, sand supply, and vegetation.

Terms to Remember

bajada 322
barchan 331
blowout 327
butte 320
deflation 327
desert 316
fault 322

flash flood 317
loess 327
longitudinal dune (seif) 332
mesa 320
parabolic dune 332
pediment 322
plateau 320

playa 322
playa lake 322
rain shadow 316
sand dune 327
slip face 329
transverse dune 331
ventifact 326

Testing Your Knowledge

Use the questions below to prepare for exams based on this chapter.

1. What are two reasons why parts of the southwestern United States have an arid climate?

2. Sketch a cross section of an idealized dune, labeling the slip face and indicating the wind direction. Why does the dune move?

3. Describe the geologic structure and sketch the major landforms of

   a.   the Colorado Plateau;

   b.   the Basin and Range province.

4. How does a flash flood in a dry region differ from most floods in a humid region?

5. Give two reasons why wind is a more effective agent of erosion in a desert than in a humid region.

6. Name four types of sand dunes and describe the conditions under which each forms.

7. The defining characteristic of a desert is (a) shifting sand dunes (b) high temperatures (c) low rainfall (d) all of the above (e) none of the above

8. Which is characteristic of deserts? (a) internal drainage (b) limited rainfall (c) flash floods (d) slow chemical weathering (e) all of the above

9. The major difference between a mesa and a butte is one of (a) shape (b) elevation (c) rock type (d) size

10. The Basin and Range province covers almost all of (a) Utah (b) Nevada (c) Texas (d) Colorado

11. A very flat surface underlain by a dry lake bed of hard, mud-cracked clay is called a (a) ventifact (b) plateau (c) playa (d) none of the above

12. Rocks with flat, wind-abraded surfaces are called (a) ventifacts (b) pediments (c) bajadas (d) none of the above

13. The removal of clay, silt, and sand particles from the land surface by wind is called (a) deflation (b) depletion (c) deposition (d) abrasion

14. Which is not a type of dune? (a) barchan (b) transverse (c) parabolic (d) longitudinal (e) all of the above are dunes

15. Much of the Southwestern U.S. is desert because (choose as many as apply) (a) it is near 30 degrees North (b) the western mountains create a rain shadow (c) cold ocean currents in the Pacific cause high evaporation rates in the land (d) they are a great distance from the ocean

16. A broad ramp of sediment formed at the base of mountains when alluvial fans merge is (a) a playa (b) a bajada (c) a pediment (d) an arroyo

17. A surface layer of closely packed pebbles is called (a) desert varnish (b) deflation (c) a blowout

Expanding Your Knowledge

1. Study the photos of sand dunes in this chapter. Which way does the prevailing wind blow in each case?

2. Can deserts be converted into productive agricultural regions? How? Are there any environmental effects from such conversion?

3. At what relative depth is ground water likely to be found in a desert? Why? Is the water likely to be drinkable? Why?

Exploring Resources

 Bagnold, R. A. 1941. *The physics of blown sand and desert dunes.* New York: William Morrow, 1954 (reprinted).

Blackwelder, E. 1954. Geomorphic processes in the desert. *Bulletin of California Division of Mines* 170:11–20. Sacramento: Department of Natural Resources.

Brookfield, M. E., and T. S. Ahlbrandt. 1983. *Eolian sediments and processes.* New York: Elsevier.

Cooke, R. U., and A. Warren. 1973. *Geomorphology in deserts.* Berkeley: University of California Press.

Easterbrook, D. J. 1993. *Surface processes and landforms.* New York: Macmillan.

Glennie, K. W. 1970. *Desert sedimentary environments.* New York: Elsevier.

Greeley, R., and J. Iversen. 1985. *Wind as a geological process.* Cambridge: Cambridge University Press.

Mabbutt, J. A. 1977. *Desert landforms.* Cambridge, Mass.: MIT Press.

McGinnies, W. G., B. J. Goldman, and P. Paylore, eds. 1968. *Deserts of the world.* Tucson, Arizona: University of Arizona Press.

McKee, E. D., ed. 1979. *A study of global sand seas.* U.S. Geological Survey Professional Paper 1052.

Pewe, T. L., ed. 1981. *Desert dust: Origin, characteristics, and effect on man.* Boulder, Colorado: Geological Society of America Special Paper 186.

Pye, K., and H. Tsoar. 1990. *Aeolian sand and sand dunes.* New York: Harper Collins Academic.

Ritter, D. F., R. C. Kochel, and J. R. Miller.

1995. *Process geomorphology.* 3d ed. Dubuque, Iowa: Wm. C. Brown Publishers.

Sheridan, D. 1981. *Desertification of the United States.* Washington, D.C.: Council on Environmental Quality.

Thornbury, W. D. 1969. *Principles of geomorphology.* 2d ed. New York: John Wiley & Sons.

Walker, A. S. 1982. Deserts of China. *American Scientist* 70(4): 366–76.

Wells, S. G., and D. R. Haragan, eds. 1983. *Origin and evolution of deserts.* Albuquerque: Univ. of New Mexico Press.

 http://pubs.usgs.gov/gip/deserts/contents/
Good overview of deserts, processes, and mineral resources.

Expand your knowledge of the concepts presented in this chapter by using the CD-ROM to answer the following question.

1. Changing Earth: Watch as Pangaea breaks up. How would the movement of the continents affect the distribution of the deserts? Predict where major deserts might have formed at different times in Earth's history.

A wave-eroded coast at Rialto Beach, Olympic National Park, Washington

Photo © Doug Sherman

CHAPTER 14

Waves, Beaches, and Coasts

*T*he previous five chapters have dealt with the sculpturing of the land by mass wasting, streams, ground water, glaciers, and wind. Water waves are another agent of erosion, transportation, and deposition of sediment. Along the shores of oceans and lakes, waves break against the land, building it up in some places and tearing it down in others.

The energy of the waves comes from the wind. This energy is used to a large extent in eroding and transporting sediment along the shoreline. Understanding how waves travel and move sediment can help you see how easily the balance of supply, transportation, and deposition of beach sediment can be disturbed. Such disturbances can be natural or human-made, and the changes that result often destroy beachfront homes and block harbors with sand.

Beaches have been called "rivers of sand" because breaking waves, as they sort and transport sediment, tend to move sand parallel to the shoreline. In this chapter we look at how beaches are formed and also examine the influence of wave action on such coastal features as sea cliffs, barrier islands, and terraces.

Figure 14.1

These beach homes in North Carolina were built too close to the ocean, and are being destroyed as sand is removed from the beach by storm waves.

Photo © Bruce Roberts/Photo Researchers, Inc.

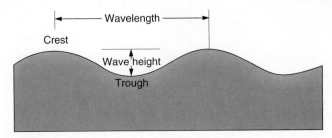

Figure 14.2

Wave height is the vertical distance between the wave crest and the wave trough. Wavelength is the horizontal distance between two crests.

If you spend a week at the shore during the summer, you may not notice any great change in the appearance of the beach while you are there. Even if you spend the whole summer at the seaside, nothing much seems to happen to the beach during those months. Tides rise and fall every day and waves strike the shore, but the sand that you walk on one day looks very much like the sand that you walk on the next day. The shape of the beach does not appear to change, nor does the sand seem to move very much.

On most beaches, however, the sand is moving, in some cases quite rapidly. The beach looks the same from day to day only because new sand is being supplied at about the same rate that old sand is being removed.

Where is the sand going? Some sand is carried out to deep water. Some is piled up and stored high on the beach. But on most shores much more of the sand moves along parallel to the beach in relatively shallow water. In this way, loose sand grains travel hundreds of feet per day along some coasts, especially those subject to strong waves.

On some beaches, sand is being removed faster than it is being replenished. When this happens, beaches become narrower and less attractive for swimming. Where erosion is severe, buildings close to the beach can be undermined and destroyed by waves as the beach disappears (figure 14.1). The sand moved from the beach may be redeposited in inconvenient places, such as across the mouth of a harbor, where it must be dredged out periodically. Because moving sand can create many problems for people in coastal towns and cities, it is important to understand something of how and why the sand moves.

Water Waves

The energy that moves sand along a beach comes from the wind-driven water waves that break upon the shore. As wind blows over the surface of an ocean or a lake, some of the wind's energy is transferred to the water surface, forming the waves that move through the water. The height of waves (and their length and speed) are controlled by the wind speed, the length of time that the wind blows, and the distance that the wind blows over the water.

Wave shapes can vary. Short, choppy *seas* in and near a storm create a confused sea surface, often with considerable white foam as strong winds blow the tops off of waves. Long, rolling *swells* form a regular series of similar-sized waves on shores that may be thousands of miles from the storms that generated the waves. (Summer surfing waves in southern California are usually generated by large storms south of Australia in the southern hemisphere winter.) When waves break against the shore as *surf*, a large portion of their energy is spent moving sand along the beach.

The height of waves is the key factor in determining wave energy. **Wave height** is the vertical distance between the **crest**, which is the high point of a wave, and the **trough**, which is the low point (figure 14.2). In the open ocean, normal waves have heights of about 0.3 to 5 meters (1 to 15 feet), although during violent storms, including hurricanes, waves can be more than 15 meters (50 feet) high. The highest wind wave ever measured was 34 meters (112 feet) by the anxious crew of a ship in the north Pacific in 1933. (The highest tsunami ever measured, caused by a submarine earthquake rather than wind, was 85 meters (278 feet), in the Ryukyu island chain south of Japan in 1971. See chapter 16.)

Wavelength is the horizontal distance between two wave crests (or two troughs). Most ocean wind waves are between 40 to 400 meters (130 and 1,300 feet) in length, and move at speeds of 25 to 90 kilometers per hour (15 to 55 miles per hour) in deep water.

The movement of water in a wave is like the movement of wheat in a field when wind blows across it. You can see the ripple caused by wind blowing across a wheat field, but the wheat does not pile up at the end of the field. Each stalk of wheat bends over when the wind strikes it and then returns to its

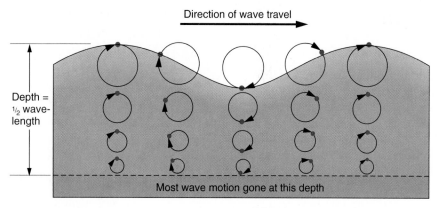

Figure 14.3

Orbital motion of water in waves dies out with depth. At the surface the diameter of the orbits is equal to the wave height.

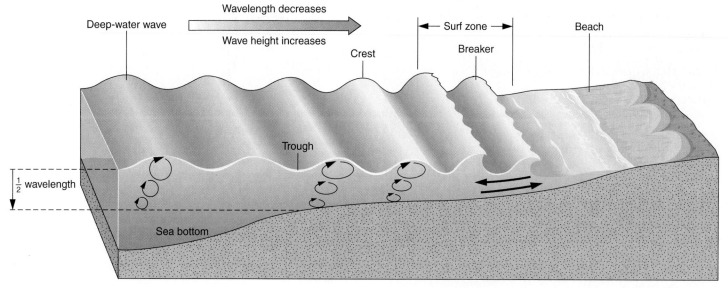

Figure 14.4

As a deep-water wave approaches shore, it begins to "feel" the sea bottom and slow down. Circular water orbits flatten and the wave peaks and breaks. In the foamy surf zone, water moves back and forth rather than in orbits.

original position. A particle of water moves in an *orbit,* a nearly circular path, as the wave passes (figure 14.3); the particle returns to its original position after the wave has passed. In deep water, when a wave moves across the water surface, energy moves with the wave; but the water, like the wheat, does not move with the wave.

At the surface, the diameter of the orbital path of a water particle is equal to the height of the wave (figure 14.3). Below the surface, the orbits decrease in size until the motion is essentially gone at a depth equal to half the wavelength. This is why a submarine can cruise in deep, calm water beneath surface ships that are being tossed by the orbital motion of large waves.

Surf

As waves move from deep water to shallow water near shore, they begin to be affected by the ocean bottom. A wave first

begins to "feel bottom" at the level of lowest orbital motion—that is, when the depth to the bottom equals half the wavelength. For example, a wave 150 meters (500 feet) long will begin to be influenced by the bottom at a water depth of 75 meters (250 feet).

In shallow water the presence of the bottom interferes with the circular orbits, which flatten into ovals (figure 14.4). The waves slow down and their length decreases. Meanwhile, the sloping bottom wedges the moving water upward, increasing the wave height. Because the height is increasing while the length is decreasing, the waves become steeper and steeper until they break. A **breaker** is a wave that has become so steep that the crest of the wave topples forward, moving faster than the main body of the wave. The breaker then advances as a turbulent, often foamy, mass. Breakers collectively are called **surf.** Water in the surf zone has lost its orbital motion and moves back and forth, alternating between onshore and offshore flow.

Nearshore Circulation

Wave Refraction

Most waves do not come straight into shore. A wave crest usually arrives at an angle to the shoreline (figure 14.5A). One end of the wave breaks first, and then the rest breaks progressively along the shore.

This angled approach of a wave toward shore can change the direction of wave travel. One end of the wave reaches shallow water first. This end of the wave "feels bottom" and slows

A

B

Figure 14.5

(A) These waves are arriving at an angle to the shoreline. They break progressively along the shore, from the upper right of the photo to the lower left. (B) Wave refraction changes the wave direction, bending the wave so it becomes more parallel to shore. The angled approach of waves to shore sets up a longshore current parallel to the shoreline.

A

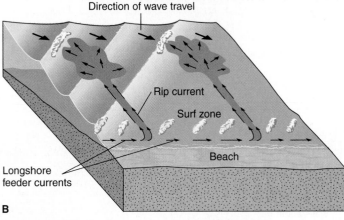

B

Figure 14.6

Rip currents and their feeder currents can develop regardless of the angle of approach of waves. (A) Waves approach parallel to shore; feeder currents on both sides of rip currents. (B) Waves approach at an angle to the shore; feeder current on only one side of rip current.

down while the rest of the wave continues at its deep-water speed (figure 14.5B). As more and more of the wave comes into contact with the bottom, more of the wave slows down. As the wave slows progressively along its length, the wave crest changes direction and becomes more nearly parallel to the shoreline. This bending of waves is called **wave refraction** (figure 14.5B).

Longshore Currents

Although most wave crests become nearly parallel to shore as they are refracted, waves do not generally strike *exactly* parallel to shore. Even after refraction, a small angle remains between the wave crest and the shoreline. As a result, the water in the wave is pushed both *up* the beach toward land and *along* the beach parallel to shore.

Each wave that arrives at an angle to the shore pushes more water parallel to the shoreline. Eventually a moving mass of water called a **longshore current** develops parallel to the shoreline (figure 14.6). The width of the longshore current is

c

Figure 14.6 Continued

(*C*) Rip currents carry dirty water and foam seaward; they can cause incoming waves to break early.

Photo © Sanford Berry/Visuals Unlimited

equal to the width of the surf zone. The seaward edge of the current is the outer edge of the surf zone, where waves are just beginning to break; the landward edge is the shoreline. A longshore current can be very strong, particularly when the waves are large. Such a current can carry swimmers hundreds of yards parallel to shore before they are aware that they are being swept along. It is these longshore currents that transport most of the beach sand parallel to shore.

Rip Currents

Rip currents are narrow currents that flow straight out to sea in the surf zone, returning water seaward that breaking waves have pushed ashore (figure 14.6). Rip currents travel at the water surface and die out with depth. They pulsate in strength, flowing most rapidly just after a set of large waves has carried a large amount of water onto shore. Rip currents can be important transporters of sediment, as they carry fine-grained sediment out of the surf zone into deep water.

As a single wave comes toward shore, its height varies from place to place. Rip currents tend to develop locally where wave height is low. Rip currents that are fixed in position are apt to be found over channels on the sea floor, because depressions on the bottom reduce wave height. Complex wave interactions can also lower wave height, and rip currents that form because of wave interactions tend to shift position along the shore. Such shifting rip currents are usually spaced at regular intervals along the beach.

Rip currents are fed by water within the surf zone. They flow rapidly out through the surf zone and then die out quickly. Where waves are nearly parallel to a shoreline, longshore feeder currents of equal strength develop in the surf zone on either side of a rip current (figure 14.6*A*). Where waves strike the shore at an angle and set up a strong, unidirectional, longshore current, a rip current is fed from one side by the longshore current, which increases in strength as it nears the rip current (figure 14.6*B*). Rip currents are also found alongside points of land and engineered structures such as jetties and piers, which can deflect longshore currents seaward.

You can easily learn to spot rip currents at a beach. Look for discoloration in the water where sediment is being picked up in the surf zone and moved seaward (figure 14.6*C*). Another sign is incoming waves breaking early within a rip current as they meet the opposing flow. The diffuse heads of rip currents outside the surf zone may be marked at the edge with foam lines. Even on very calm days rips can often be identified by subtle changes in the water surface, such as a different pattern of water ripples or light reflection off the water.

Getting caught in a rip current and being carried out to sea can panic an inexperienced swimmer—even though the trip will stop some distance beyond the surf zone as the rip dies out. A swimmer frightened by being carried away from land and into breaking waves can grow exhausted fighting the current to get back to shore. The thing to remember is that rip currents are narrow. Therefore, you can get out of a rip easily by swimming *parallel* to the beach instead of struggling against the current.

Surfers, on the other hand, often look for rip currents and paddle intentionally into them to get a quick ride out into the high breakers.

Beaches

A **beach** is a strip of sediment (usually sand or gravel) that extends from the low-water line inland to a cliff or a zone of permanent vegetation. Waves break on beaches, and rising and falling tides may regularly change the amount of beach sediment that is exposed above water (figure 14.7*A*).

The steepest part of a beach is the **beach face,** which is the section exposed to wave action, particularly at high tide. Offshore from the beach face there is usually a **marine terrace,** a broad, gently sloping platform that may be exposed at low tide if the shore has significant tidal action. Marine terraces may be

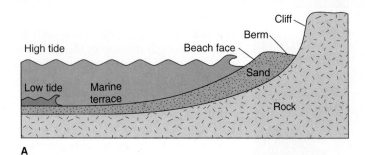

Figure 14.7

(*A*) Parts of a beach. (*B*) The beach face (on the left) and berm (on the right) on a northern California beach.

Photo B by Diane Carlson

A Winter beach

B Summer beach

Figure 14.8

Seasonal cycle of a beach caused by differing wave types.
(*A*) Narrow winter beach. Waves may break once on the winter sandbar, then re-form and break again on the beach face.
(*B*) Wide summer beach.

wave-built terraces constructed of sediment carried away from the shore by waves, or they may be *wave-cut* rock benches or platforms, perhaps thinly covered with a layer of sediment.

The upper part of the beach, landward of the usual high-water line is the **berm,** a wave-deposited sediment platform that is flat or slopes slightly landward (figure 14.7*B*). It is usually dry, being covered by waves only during severe storms.

Beach sediment is usually sand, typically quartz-rich because of quartz's resistance to chemical weathering. Heavy metallic minerals ("black sands") can also be concentrated on some beaches as less dense minerals such as quartz and feldspar are carried away by waves or wind (titanium-bearing sands are mined on some beaches in Florida and Australia). Tropical beaches may be made of bioclastic carbonate grains from offshore corals, algae, and shells. Some Hawaiian beaches are made of sand-sized fragments of basalt. Gravel beaches are found on coasts attacked by the high energy of

B

large waves (*shingle* is a regional name for disk-shaped gravel). Gravel beaches have a steeper face slope than sand beaches.

In seasonal climates, beaches often go through a summer-winter cycle (figure 14.8). This is due to the greater frequency of storms with strong winds during the winter months, which tend to produce high waves with short wavelengths. These high energy waves tend to crash onshore and erode sand from the beach face and narrow the berm. Offshore, in less turbulent water, the sand settles to the bottom and builds an underwater sandbar (parallel to the beach) that serves as a "storage facility" for the next summer's sand supply. The following summer, or during calmer weather, lower energy waves with long wavelengths break over the sandbar and gradually push the sand back onto the beach face to widen the berm. Each season the beach changes in shape until it comes into equilibrium with the prevailing wave type.

Many winter beaches can be dangerous because of high waves and narrowed beaches. Several beaches along the Pacific coast of the United States are nearly free of accidents in the summer, when they are heavily used, but are regularly marked by drownings in the winter as beachwalkers are swept off narrower beaches out to sea by large storm waves.

Longshore Drift of Sediment

Longshore drift is the movement of sediment parallel to shore when waves strike the shoreline at an angle. Figure 14.9 shows the two ways in which this movement of sediment (usually sand) occurs. Some longshore drift takes place directly on the beach face when waves wash up on land. A wave washing up on the beach at an angle tends to wash sand along at the same angle. After the wave has washed up as far as it can go, the water returns to the sea by running down the beach face by the shortest possible route, that is, straight downhill to the shoreline, not back along the oblique route it came up. (Wave run-up is known as *swash,* the return as *backwash.*) The net effect of this motion is to move the sand in a series of arcs along the beach face.

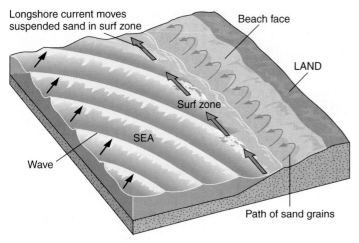

Figure 14.9

Longshore drift of sand on the beach face and by a longshore current within the surf zone.

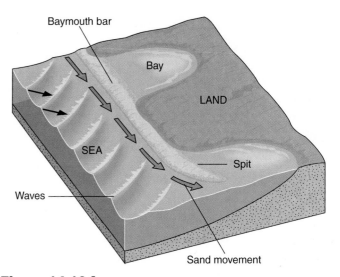

Figure 14.10A

Longshore drift of sand can form spits and baymouth bars.

Much more sand is moved by longshore transport in the surf zone, where waves are breaking into foam. The turbulence of the breakers erodes sand from the sea bottom and keeps it suspended. Even a weak longshore current can move the suspended sand parallel to the shoreline. The sand in the longshore current moves in the same direction as the sand drift on the beach face (figure 14.9).

Vast amounts of sand can be moved by longshore transport. The U.S. Army Corps of Engineers estimates that 436,000 cubic yards of sand per year are moved northward by waves at Sandy Hook, New Jersey, and 1,000,000 cubic yards of sand per year are moved southward at Santa Monica, California.

Eventually the sand that has moved along the shore by these processes is deposited. Sediment may build up off a point of land to form a **spit,** a fingerlike ridge of sediment that extends out into open water (figures 14.10*A,B*). A **baymouth bar,** a ridge of sediment that cuts a bay off from the ocean, is formed by sediment migrating across what was earlier an open bay (figures 14.10*A,C*). Off the western coast of the United States, a considerable amount of drifting sand is carried into the heads of underwater canyons, where the sediments slide down into deep, quiet water.

A striking, but rare, feature formed by longshore drift is a **tombolo,** a bar of sediment connecting a former island to the mainland. As shown in figure 14.11, waves are refracted around an island in such a way that they tend to converge behind the island. The waves sweep sand along the mainland (and from the island) and deposit it at this zone of convergence, forming a bar that grows outward from the mainland and eventually connects to the island.

Figure 14.10B

Curved spit near Victoria, British Columbia.

Photo by D. A. Rahm, courtesy Rahm Memorial Collection, Western Washington University

Figure 14.10C

A baymouth bar has sealed off this bay from the ocean as sand migrated across the mouth of the Russian River in northern California.

Photo by Diane Carlson

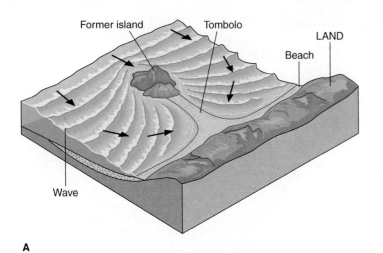

A

Figure 14.11

(*A*) Formation of a tombolo. Wave refraction around an island interrupts the longshore current and creates a sand bar that connects the island with the mainland. (*B*) A tombolo has connected this rock, once an island, to the shore. Note the waves bending around the two sides of the rock. Near Santa Cruz, California.

B

Human Interference with Sand Drift

Several engineered features can interrupt the flow of sand along a beach (figure 14.12). *Jetties,* for example, are rock walls designed to protect the entrance of a harbor from sediment deposition and storm waves. Usually built in pairs, they protrude above the surface of the water. Figure 14.12*A* shows how sand piles up against one jetty while the beach next to the other, deprived of a sand supply, erodes back into the shore.

Groins are sometimes built in an attempt to protect beaches that are losing sand from longshore drifting. These short walls are built perpendicular to shore to trap moving sand and widen a beach (figure 14.12*B*).

Sand deposition also occurs when a stretch of shore is protected from wave action by a *breakwater,* an offshore structure built to absorb the force of large breaking waves and provide quiet water near shore. When the city of Santa Monica in California built a rock breakwater parallel to the shore to create a protected small-boat anchorage, the lessening of wave action on the shore behind the breakwater allowed sand to build up there (figure 14.12*C*), threatening eventually to fill in the anchorage. The city had to buy a dredge to remove the sand from the protected area and redeposit it farther along the shore where the waves could resume moving sediment.

A beach attempts to come into equilibrium with the waves that strike it. The type and amount of sediment, the position of the sediment, and especially the movement of the sediment, adjust to the incoming wave energy. Whenever human activity interferes with sand drift or wave action, the beach responds by changing its configuration, usually through erosion or deposition in a nearby part of the beach.

Sources of Sand on Beaches

Some beach sand comes from the erosion of local rock, such as points of land or cliffs nearby. On a few beaches replenishment comes from sand stored outside the surf zone in the deeper water offshore. Bioclastic carbonate beaches are formed from the remains of marine organisms offshore. But the greater part of the sand on most beaches comes from river sediment brought down to the ocean. Waves pick up this sediment and move it along the beach by longshore drift.

What happens to a beach if all the rivers contributing sand to it are dammed? Although damming a river may be desirable for many reasons (flood control, power generation, water supply, recreation), when a river is dammed, its sediment load no longer reaches the sea. The sand that supplied the beach in the past now comes to rest in the quiet waters of the reservoir behind the dam. Longshore drift, however, continues to remove sand from beaches even though little new sand is being supplied, and the result is a net loss of sand from beaches. Beaches without a sand supply eventually disappear. To prevent this, some coastal communities have set up expensive programs of draining reservoirs and trucking the trapped sand down to the beaches.

Coasts and Coastal Features

A beach is just a small part of the **coast,** which is all the land near the sea, including the beach and a strip of land inland from it. Coasts can be rocky, mountainous, and cliffed, as in northern New England and on the Pacific shore of North America; or they can be broad, gently sloping plains, as along much of the southeastern United States. Wave erosion and deposition can greatly modify coasts from their original shapes. Many coasts have been drowned during the past 15,000 years by the rise in sea level caused by the melting of the Pleistocene glaciers (see chapter 12). Other coasts have been lifted up by tectonic forces at a rate greater than the rise in sea level so that sea-floor features are now exposed on dry land.

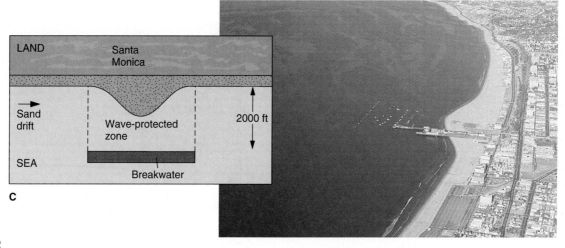

Figure 14.12

Sand piles up against obstructions and in areas deprived of wave energy. (*A*) Jetties at Manasquan Inlet, New Jersey. Sand drift to the right has piled sand against the left jetty, and removed sand near the right jetty. (*B*) Groins at Ocean City, New Jersey. Sand drift is to the right. (*C*) Breakwater at Santa Monica, California has caused deposition of sand in the wave-protected zone.

Photos *A* and *B* by S. Jeffress Williams; photo *C* © John Shelton

Erosional Coasts

A great many steep, rocky coasts have been visibly changed by wave erosion. Soluble rocks such as limestone dissolve as waves wash against them, and more durable rocks such as granite are fractured by the enormous pressures caused by waves slamming into rock (wave impact pressures have been measured as high as 6 atmospheres—575 kPa or 6 tons per square foot).

Figure 14.13

(*A*) Wave refraction on an irregular coast. Black arrows show transport direction of wave energy, concentrated on headlands, spread out in bays. (*B*) Wave refraction around a point of land, Rincon Point, California. Note waves at right center have been bent 90 degrees from their original direction (parallel to the bottom of the photo).

Photo by Frank M. Hanna

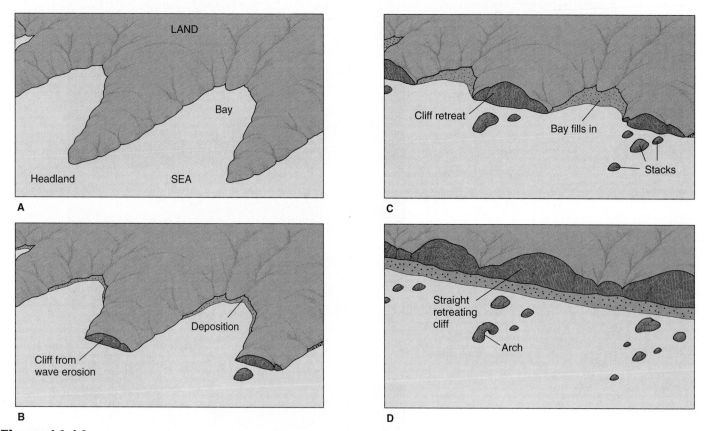

Figure 14.14

Coastal straightening of an irregular coastline by wave erosion of headlands and wave deposition of sediment in bays. Continued erosion produces a straight, retreating cliff.

An irregular coast with bays separated by rocky **headlands** (points of land) can be gradually straightened by wave action. Because wave refraction bends waves approaching such a coast until they are nearly parallel to shore, most of the waves' energy is concentrated on the headlands, while the bays receive smaller, diverging waves (figure 14.13). Rocky cliffs form from wave erosion on the headlands. The eroded material is deposited in the quieter water of nearby bays, forming broad beaches. **Coastal straightening** of an irregular shore gradually takes place through wave erosion of headlands and wave deposition in bays (figure 14.14).

Figure 14.15

Retreating wave-cut cliff, north of Bodega Bay, Sonoma County, California. A concrete seawall has been built at the base of the cliff to slow wave erosion and help protect the cliff-edge homes. Note fragments of wave-destroyed structures near the seawall. Seawalls usually increase the erosion of sand beaches.

Wave erosion of headlands produces **sea cliffs,** steep slopes that retreat inland by mass wasting as wave erosion undercuts them (figure 14.15). At the base of sea cliffs are sometimes found *sea caves,* cavities eroded by wave action along zones of weakness in the cliff rock. As headlands on irregular coasts are eroded landward, sea cliffs enlarge until the entire coast is marked by a retreating cliff (figure 14.14). On some exposed coasts the rate of cliff retreat can be quite rapid, particularly if the rock is weakly consolidated. Some sea cliffs north of San Diego, California, and at Cape Cod

National Seashore in Massachusetts, are retreating at an average rate of 1 meter (3 feet) per year. Because sea-cliff erosion in weak rock is often in the form of large, infrequent slumps (see chapter 9), some portions of these coasts may retreat 10 to 30 meters (30 to 100 feet) in a single storm. Some of these cliffs have "ocean-view" homes and hotels at their very edges. Sea cliffs in hard, durable rock such as granite and schist retreat much more slowly.

Seawalls may be constructed along the base of retreating cliffs to prevent wave erosion (figure 14.15). Seawalls of giant pieces of broken stone (riprap) or concrete tetrahedrons are designed to absorb wave energy rather than allow it to erode cliff rock. Vertical or concave seawalls of concrete are designed to reflect wave energy seaward rather than allow it to impact the shore. Reflection of waves from a seawall will increase the amount of wave energy just offshore, often increasing the amount of sand erosion offshore. Thus a seawall designed to protect a sea cliff (and the buildings at its edge) may in some cases destroy a sand beach at the base of the cliff. Seawalls are difficult and expensive to build and maintain, and they may destroy beaches, but political pressure to build more of them will increase as the sea level rises in the future.

Wave erosion produces other distinctive features in association with sea cliffs. A **wave-cut platform** (or *terrace*) is a horizontal bench of rock formed beneath the surf zone as a coast retreats by wave erosion (figure 14.16). The platform widens as the sea cliffs retreat. The depth of water above a wave-cut platform is generally 6 meters (20 feet) or less, coinciding with the depth at which turbulent breakers actively erode the sea bottom. **Stacks** are erosional remnants of headlands left behind as the coast retreats inland (figure 14.17). They form small, rocky

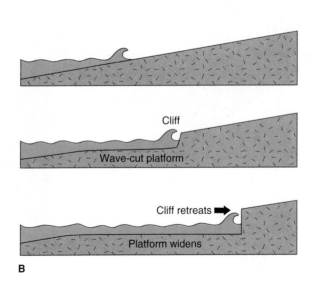

Figure 14.16

(*A*) A wave-cut platform (the wide, horizontal bench of dark rock at the base of the cliffs) is exposed at low tide. La Jolla, California.
(*B*) A wave-cut platform widens as a sea cliff retreats.

Figure 14.17

Stacks and an arch mark old headland positions on this retreating, wave-eroded coast. Northern California.

islands off retreating coasts, often directly off headlands (figure 14.14). **Arches** (or *sea arches*) are bridges of rock left above openings eroded in headlands or stacks by waves. The openings are eroded in spots where the rock is weaker than normal, perhaps because of closely spaced fractures.

Depositional Coasts

Many coasts are gently sloping plains and show few effects of wave erosion. Such coasts are found along most of the Atlantic Ocean and Gulf of Mexico shores of the United States. These coasts are primarily shaped by sediment deposition, particularly by longshore drift of sand.

Coasts such as these are often marked by **barrier islands**—ridges of sand that parallel the shoreline and extend above sea level (figure 14.18). These barrier islands may have formed from sand eroded by waves from deeper water offshore, or they may be greatly elongated sand spits formed by longshore drift. The slowly rising sea level associated with the melting of the Pleistocene glaciers may have been a factor in their development. A protected lagoon separates barrier islands from the mainland. Because the lagoon is protected from waves, it provides a quiet waterway for boats. A series of such lagoons stretches almost continuously from New York to Florida, and many also exist along the Gulf Coast, forming an important route for barge traffic. As tides rise and fall, strong tidal currents may wash in and out of gaps between barrier islands, distributing sand in submerged *tidal deltas* both landward and seaward of the gaps.

Some barrier islands along the Atlantic and Gulf coasts are densely populated. Atlantic City (New Jersey), Ocean City (Maryland), Miami Beach (Florida), and Galveston (Texas) are examples of cities built largely on barrier islands. In some of

Figure 14.18

(*A*) A barrier island on a gently sloping coast. A lagoon separates the barrier island from the mainland, and tidal currents flowing in and out of gaps in the barrier island deposit sediment as submerged tidal deltas. (*B*) A barrier island near Pensacola, Florida. Open ocean to left, lagoon to right, mainland Florida on far right. Light-colored lobes of sand within lagoon were eroded from the barrier island by hurricane waves.

Photo by Frank M. Hanna

Figure 14.19

Hotels built upon the loose sand of a barrier island, Miami Beach, Florida. The lagoon and mainland Florida are visible at the upper left.

Photo by Florida Division of Tourism

these cities, houses, luxury hotels, and condominiums are clustered near the edge of the sea; many are built upon the loose sand of the island (figure 14.19). These developed areas are vulnerable to late-summer hurricanes that sooner or later bring huge storm waves onto these coasts, eroding the sand and undermining the building foundations at the water's edge.

Nonmarine deposition may also shape a coast. Rapid sedimentation in *deltas* by rivers can build a coast seaward

(see figure 10.31). *Glacial deposition* can form shoreline features. Several islands off the New England coast were glacially deposited; Long Island, New York, formed from two end moraines.

Drowned Coasts

Drowned (or *submergent*) coasts are common because sea level has been rising worldwide for the past 15,000 years. During the glacial ages of the Pleistocene, sea level was 130 meters (400 feet) below its present level. The shallow sea floor near the continents was then dry land, and rivers flowed across it, cutting valleys. As the great ice sheets melted, sea level began to rise, drowning the river valleys. These drowned river mouths, called **estuaries,** mark many coasts today (figure 14.20). They extend inland as long arms of the sea. Fresh water from rivers mixes with the sea water to make most estuaries brackish. The quiet, protected environment of estuaries makes them very rich in marine life, particularly the larval forms of numerous species. Unfortunately, cities and factories built on many estuaries to take advantage of quiet harbors are severely polluting the water and the sediment of the estuaries. The poor circulation that characterizes most estuaries hinders the flushing away of this pollution, and estuary shellfish are sometimes not safe to eat as a result.

Figure 14.20

(*A*) An estuary formed as rising sea level drowned a river valley, Malibu, California. (*B*) Landsat satellite photo of estuaries, Albemarle and Pamlico Sounds, North Carolina. Barrier islands are visible in upper right. Infrared image shows vegetation as red.

Photo *A* by Frank M. Hanna; photo *B* by NASA

B

Waves, Beaches, and Coasts   **349**

14.1
ENVIRONMENTAL
GEOLOGY

The Effects of Rising Sea Level

Long Term

Sea level has risen about 130 meters (400 feet) in the past 15,000 years as the Pleistocene glaciers melted, adding water to the oceans. During this period, sea level initially rose at the rapid rate of 1.3 meters (4 feet) per century, but the rate of rise gradually slowed down, so that for the past 3,000 years the rise has only been about 4 cm (1.5 inches) per century. Since about 1930, however, the rate of sea-level rise has increased six-fold, to 23 cm (9 inches) per century along the Atlantic and Gulf coasts (the reasons for this increase are not clear). When sea level rose this rapidly a few thousand years ago, the coasts were rapidly eroded.

On low-lying coasts a rise in sea level causes ocean waves (particularly storm waves) to extend much farther inland than before, flooding the land (box figure 1*A*). Barrier islands migrate landward. In the United States from New Jersey south to Florida, and along the Gulf of Mexico to southern Texas, the coastal land is very flat. A small rise in sea level can send seawater many miles inland.

On steep coasts a rise in sea level can accelerate the erosion of coastal cliffs, destroying oceanfront property (box figure 1*B*).

Some scientists predict that the rise in sea level may accelerate in the next century as a result of global warming by the "greenhouse effect." Burning of coal and oil has increased the amount of carbon dioxide in the atmosphere by more than 10 percent in this century. The carbon dioxide and other gases may trap more of the sun's energy on earth, warming the air and the sea. Warm air would accelerate the melting of glaciers on Greenland and Antarctica. Warm seawater expands. Together these two effects could raise sea level.

Predictions on the amount of sea level rise by the year 2100 vary widely. Early predictions of a rise of 1 to 2 meters (4 to 7 feet) have recently been revised downward to .3 to .6 meter (1 to 2 feet) as computer models were refined (by considering cloud cover, for example). A rise of even 1

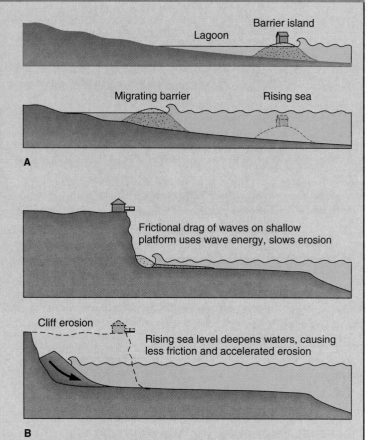

A

B

Box 14.1 Figure 1

Rising sea level can cause erosion of both gentle and steep coasts, leading to destruction of buildings.

meter on a low-lying coast could destroy thousands of beachfront houses and hotels.

Global warming is an emotional and controversial subject. The recent rise in atmospheric carbon dioxide is indisputable, but temperature data, while convincing some scientists that the earth is warming, indicate to other scientists that the earth is not changing in temperature, or is even cooling. The possibility of future warming is widely accepted, but evidence of warming to date is debatable. The debate over global warming does not change the fact that glaciers are now melting (as they have for 15,000 years) and sea level is now rising. The scientific argument is about whether the present rise of 23 centimeters per century will continue or will accelerate.

Short Term

Hurricanes, which are common along the Atlantic and Gulf coasts, tend to cause short-term rises in sea level called *storm*

Box 14.1 Figure 2

Strong onshore winds in a hurricane pile water against the shore, forming a storm surge (high sea level) that may cause severe flooding on a low-lying coast.

surges. Hurricanes consist of strong winds rotating counterclockwise (in the northern hemisphere) around a region of very low air pressure. They may be 500 kilometers (300 miles) in diameter and contain winds up to 300 kilometers (180 miles) per hour. The low air pressure in the center of a hurricane literally sucks up the sea surface into a broad dome, and this rise in sea level is accentuated by strong onshore winds, which pile water against shore (box figure 2). Storm surges can easily raise sea level 5 meters (15 feet). A storm surge 8 meters (26 feet) high struck Galveston, Texas, in 1900, completely covering the barrier island that Galveston is built upon. High storm waves on top of the high sea level destroyed countless buildings, and 6,000 people died, many by drowning. Hurricane-tracking programs cut the death tolls of storm surges today by providing advance warning to coastal communities. In September, 1989, hurricane Hugo hit South Carolina with 135 mile-per-hour winds and a 5-meter storm surge north of Charleston, causing $10 billion in damage. Because more than 500,000 people were evacuated along the low-lying coast, however, the death toll was only 29.

Careful planning for the use of coastal land in your lifetime will be necessary to lessen destruction caused by storm

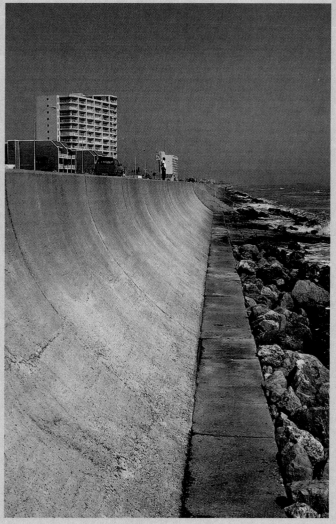

Box 14.1 Figure 3

Seventeen-foot seawall of concrete and granite blocks at Galveston, Texas, built to protect the barrier island from storm waves and storm surges. Note the lack of a sand beach seaward of the seawall.

surges and long-term sea-level rise. Developed land at the ocean's edge may face a grim choice in the future—abandonment of existing buildings or expensive "armoring" of the coast with structures such as seawalls in hopes of protecting the buildings from wave damage (box figure 3).

Further Reading

Williams, S. J., K. Dodd, and K. K. Gohn. 1990. *Coasts in crisis.* U.S. Geological Survey Circular 1075.

Figure 14.21

Uplifted marine terrace, northern California. The flat land surface at the top of the sea cliff was eroded by wave action, then raised above sea level by tectonic uplift. The rock knob on the terrace was once a stack.

Drowned coasts may be marked by **fiords,** glacially cut valleys flooded by rising sea level (see figure 12.35). They form in the same way as estuaries, except they were cut by glacial ice rather than rivers during low sea-level stands.

Uplifted Coasts

Uplifted (or *emergent*) coasts have been elevated by deep-seated tectonic forces. The land has risen faster than sea level, so parts of the old sea floor are now dry land.

Marine terraces form just offshore from the beach face, as described earlier in this chapter. These terraces can be wave-cut platforms caused by erosion of rock associated with cliff retreat, or they can be wave-built terraces caused by deposition of sediment. If the shore is elevated by tectonic uplift, these flat surfaces will become visible as *uplifted marine terraces* (figure 14.21). They formed below the ocean surface but are visible now because of uplift. The tectonically unstable Pacific coast of the United States and Canada has many areas marked by uplifted terraces, along with the erosional coast features described earlier.

Coasts Shaped by Organisms

The growth of coral and algal *reefs* offshore can shape the character of a coast. The reefs act as a barrier to strong waves, protecting the shoreline from most wave erosion (see figure 18.22).

Carbonate sediments blanket the sea floor on both sides of a reef and usually form a carbonate sand beach on land (see figure 6.18). Southernmost Florida has a coast of this type.

Branching *mangrove roots* dominate many parts of the southeastern United States coast. The roots dam wave and current action, creating a quiet environment that provides a haven for the larval forms of many marine organisms, and may trap fine-grained sediment. Mangroves also deposit layers of organic peat on low-lying coasts.

Summary

Wind blowing over the sea surface forms waves, which transfer some of the wind's energy to shorelines. Orbital water motion extends to a depth equal to half the wavelength.

As a wave moves into shallow water, the ocean bottom flattens the orbital motion and causes the wave to slow and peak up, eventually forming a *breaker* whose crest topples forward. The turbulence of *surf* is an important agent of sediment erosion and transportation.

Wave refraction bends wave crests and makes them more parallel to shore. Few waves actually become parallel to the shore, and so *longshore currents* develop in the surf zone. *Rip currents* carry water seaward from the surf zone.

A beach consists of a *berm, beach face,* and *marine terrace.* Summer beaches have a wide berm and a smooth offshore profile. Winter beaches are narrow, with offshore bars.

Longshore drift of sand is caused by the waves hitting the beach face at an angle and also by longshore currents.

Deposition of sand that is drifting along the shore can form *spits* and *baymouth bars.* Drifting sand may also be deposited against jetties or groins or inside breakwaters.

Rivers supply most sand to beaches, although local erosion may also contribute sediment. If the river supply of sand is cut off by dams, the beaches gradually disappear.

Coasts may be erosional or depositional, drowned or uplifted, or shaped by organisms

such as corals and mangroves.

Coastal straightening by waves is caused by headland erosion and by deposition within bays.

A coast retreating under wave erosion can be marked by *sea cliffs,* a *wave-cut platform, stacks,* and *arches.*

Waves can form *barrier islands* off gently sloping coasts. River and glacial deposition can also shape coasts.

Drowned coasts are marked by *estuaries* and *fiords. Uplifted marine terraces* characterize coasts that have risen faster than the recent rise in sea level.

Terms to Remember

arch (sea arch) 348
barrier island 348
baymouth bar 343
beach 341
beach face 341
berm 342
breaker 339
coast 344
coastal straightening 346

crest (of wave) 338
estuary 349
fiord 352
headland 346
longshore current 340
longshore drift 342
marine terrace 341
rip current 341
sea cliff 347

spit 343
stack 347
surf 339
tombolo 343
trough (of wave) 338
wave-cut platform 347
wave height 338
wavelength 338
wave refraction 340

Testing Your Knowledge

Use the questions below to prepare for exams based on this chapter.

1. Show in a sketch how longshore drift of sand can form a baymouth bar.

2. In a sketch, show how and why sand moves along a beach face when waves approach a beach at an angle.

3. How are summer beaches different from winter beaches? Discuss the reasons for these differences.

4. What would happen to the beaches of most coasts if all the rivers flowing to the sea were dammed? Why?

5. What does the presence of an estuary imply about the recent geologic history of a region?

6. Describe how waves can straighten an irregular coastline.

7. Describe the transition of deep-water waves into surf.

8. Show in a sketch the refraction of waves approaching a straight coast at an angle. Explain why refraction occurs.

9. What is a longshore current? Why does it occur?

10. What is a rip current? Why does it occur? How do you get out of a rip current?

11. The path a water particle makes as a wave passes in deep water is best described as (a) elliptical (b) orbital (c) spherical (d) linear

12. The easiest method of escaping a rip current is to (a) swim toward shore (b) swim parallel to the shore (c) swim away from the shore

13. Why is beach sediment typically quartz-rich sand? (a) other minerals are not deposited on beaches (b) quartz is the only mineral that can be sand-sized (c) quartz is resistant to chemical weathering (d) none of the above

14. Longshore drift (a) is the movement of sediment parallel to shore when waves strike the shoreline at an angle (b) a type of rip current (c) a type of tide (d) the movement of waves

15. Which structure would interfere with longshore drift? (a) jetties (b) groins (c) breakwaters (d) all of the above

16. What is the most common source of sand on beaches? (a) sand from river sediment brought down to the ocean (b) land next to the beach (c) offshore sediments

17. Which would characterize an erosional coast? (a) headlands (b) sea cliffs (c) stacks (d) arches (e) all of the above

18. Which would characterize a depositional coast? (a) headlands (b) sea cliffs (c) stacks (d) arches (e) barrier islands

19. A glacial valley drowned by rising sea level is (a) a fiord (b) an estuary (c) a tombolo (d) a headland

20. The surf zone is (a) the region in which waves break (b) water less than one half wavelength in depth (c) where the longshore current flows (d) all of the above

21. The storm surge of a hurricane is (a) the highest winds (b) the tallest waves (c) the dome of high water in the center of the hurricane (d) the area of high pressure within the storm

Expanding Your Knowledge

1. Sea level would rise by about 60 meters (200 feet) if *all* the glacial ice on the earth's surface melted. How many U.S. cities would this affect?

2. What happens to a coast if its offshore reef dies?

3. Is a beach a good place to mine sand for construction? Explain your answer carefully.

4. The seaward tip of a headland may be the most rapidly eroding locality on a coast, yet may also be the most expensive building site on a coast. Why is this so?

Bascom, W. 1980. *Waves and beaches.* Rev. ed. New York: Doubleday Anchor Books.

Bird, E. C. F. 1985. *Coastline changes: A global review.* New York: John Wiley & Sons.

Bird, E. C. F., and M. L. Schwartz. 1985. *The world's coastlines.* New York: Van Nostrand Reinhold.

Davis, R. A., Jr., and R. L. Ethington. 1976. *Beach and nearshore sedimentation.* Tulsa: Society of Economic Paleontologists and Mineralogists, Special Publication 24.

Dolan, R., B. Hayden, and H. Lins. 1980. Barrier islands. *American Scientist* 68: 16–25.

Dolan, R., and H. Lins. 1987. Beaches and barrier islands. *Scientific American* 257(7): 68–77.

Easterbrook, D. J. 1993. *Surface processes and landforms.* New York: Macmillan.

Hardisty, J. 1990. *Beaches, form and process.* New York: Harper Collins Academic.

Inman, D. L., and B. M. Brush. 1973. The coastal challenge. *Science* 180 (4094): 20–32.

Kaufman, W., and O. H. Pilkey, Jr. 1983. *The beaches are moving: The drowning of America's shoreline.* Durham, N.C.: Duke University Press.

Komar, P. D. 1976. *Beach processes and sedimentation.* Englewood Cliffs, N.J.: Prentice-Hall.

Mack, W. N., and E. A. Leitikow. 1996. Sands of the world. *Scientific American.* August: 62–67.

Nuhfer, E. B., R. J. Proctor, and P. H. Moser. 1993. *The citizens' guide to geologic hazards.* Arvada, Colorado: American Institute of Professional Geologists.

Shepard, F. P. 1973. *Submarine geology.* 3d ed. New York: Harper & Row.

Shepard, F. P., and H. R. Wanless. 1971. *Our changing coastlines.* New York: McGraw-Hill.

Snead, R. E. 1982. *Coastal landforms and surface features: A photographic atlas and glossary.* Stroudsburg, Pennsylvania: Hutchinson Ross.

Williams, S. J., K. Dodd, and K. K. Gohn. 1990. *Coasts in crisis.* U.S. Geological Survey Circular 1075.

Rocks of Paleozoic age that were once horizontal have been contorted into folds during mountain building. Peter Lougheed Provincial Park, Alberta, Canada.

Photo © Doug Sherman

CHAPTER
15

Geologic Structures

*I*n previous chapters we have discussed how rock at the earth's surface is affected by erosional agents such as wind and water. We now shift our focus to changes in bedrock caused by powerful forces originating deep within the earth. In this chapter we explain how rocks respond to these tectonic forces.

The main purpose of this chapter is to help you recognize certain geologic structures, understand the forces that caused them, and thus determine the geologic history of an area.

Some principles discussed in chapter 8 should help you interpret the way structures develop in an area and the sequence. Recognition of unconformities as well as the principles of original horizontality, superposition, and cross-cutting relationships are as important to structural geology as they are to determining relative time.

Subsequent chapters will require an understanding and knowledge of structural geology as presented in this chapter. To understand earthquakes, for instance, one must know about faults. Appreciating how major mountain belts and the continents have evolved (chapter 20) calls for a comprehension of faulting and folding. Understanding plate-tectonic theory as a whole (chapter 19) also requires a knowledge of structural geology. (Plate-tectonic theory developed primarily to explain certain structural features.) In areas of active tectonics, the location of geologic structures is important in the selection of suitable sites for schools, hospitals, dams, bridges, and nuclear power facilities.

Also, understanding structural geology can help us more fully appreciate the problem of finding more of the earth's dwindling natural resources. Chapter 21 discusses the association of certain geologic structures with petroleum deposits and other valuable resources.

Figure 15.1

Folded and faulted sedimentary beds exposed in a road cut near Palmdale, California.

Photo by C. C. Plummer

In a broad sense structural geology can be thought of as the study of the architecture of the earth's crust, its deformational features, and their mutual relations and origins. For our purposes, **structural geology** can be defined as the branch of geology concerned with the shapes, arrangement, and interrelationships of bedrock units and the forces that cause them.

Tectonic Forces at Work

Stress and Strain in the Earth's Crust

Tectonic forces move and deform parts of the earth's crust, particularly along plate margins. When studying deformed rocks, structural geologists typically refer to **stress,** a force per unit area. Where stress can be measured, it is expressed as the force per unit area at a particular point. However, it is difficult to measure stress in rocks that are currently buried. We can observe the effects of past stress (caused by tectonic forces and confining pressure from burial) when bedrock is exposed after uplift and erosion. From our observations we may be able to infer the principal directions of stress that prevailed. We also can observe in exposed bedrock the effect of forces on a rock that was stressed. **Strain** is the change in size (volume) or shape, or both, while an object is undergoing stress. In figure 15.1, originally horizontal rock layers have changed in shape

and are strained into wave-like folds that are broken by faults. The layers have been strained, probably by a horizontal stress that pushed or compressed the layers together until they were shortened by buckling.

The relationship between stress and strain can be illustrated by deforming a piece of Silly Putty (figure 15.2). If the Silly Putty is pushed together or squeezed from opposite directions, we say the stress is **compressive.** Compressive stress is common along convergent plate boundaries and typically results in rocks being deformed by a *shortening strain.* In figure 15.2*A*, an elongate piece of Silly Putty may shorten by bending, or folding, whereas a ball of Silly Putty will shorten in the direction parallel to the compressive stress and elongate or stretch in the direction perpendicular to it. Rocks subjected to compressive stresses, particularly along convergent plate boundaries, behave in the same way and are typically shortened in the horizontal direction and elongated in the vertical direction.

A **tensional stress** is caused by forces pulling away from one another in opposite directions (figure 15.2*B*) and results in a *stretching* or *extensional strain.* If we apply a tensional stress on a ball of Silly Putty, it will elongate or stretch parallel to the applied stress. If the tensional stress is applied rapidly, the Silly Putty will first stretch and then break apart (figure 15.2*B*). Tensional stresses are quite rare in the crust; in fact, most stresses directly measured in the Earth are compressive.

A

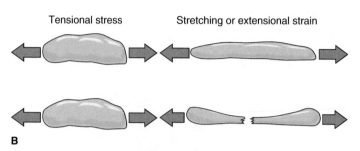

B

Figure 15.2

The effects of compressional and tensional stresses on Silly Putty. (A) Compressing Silly Putty results in shortening either by folding or flattening. (B) Pulling (tensional stress) Silly Putty causes stretching or extension; if pulled (strained) too fast, or chilled, the Silly Putty will break after first stretching.

Figure 15.3

Shear stress can be modeled by shearing a deck of cards.

The third type of stress is a **shear stress** that is due to movement parallel to but in opposite directions along a fault or other boundary. It is much like putting a deck of cards in your hands and shearing the deck by moving your hands in opposite directions (figure 15.3). A shear stress results in a *shear strain* parallel to the direction of the stresses. Shear stresses are notable along transform plate boundaries and along other actively moving faults.

Behavior of Rocks to Stress and Strain

Rocks behave as plastic, elastic, or brittle materials depending on the amount and rate of stress applied, the type of rock, and the temperature and pressure under which the rock is strained.

Figure 15.4

Graph shows the behavior of rocks with increasing stress and strain. Elastic behavior occurs along the straight line portions (shown in blue) of the graph. At stresses greater than the elastic limit (red points), the rock will either deform as a plastic material or break, as shown in the deformed rock cylinders.

If a deformed body recovers its original shape after the stress is reduced or removed, the behavior is **elastic.** For example, if a tensional stress is applied to a rubber band it will stretch as long as the stress is applied; but, once the stress is removed the rubber band returns or recovers to its original shape and its behavior is elastic. Silly Putty will also behave elastically if molded into a ball and bounced. Most rocks can behave in an elastic way at very low stresses (a few kilobars), however, once the stress applied exceeds the **elastic limit** (figure 15.4) the rock will deform in a permanent way.

A rock that behaves in a ductile or **plastic** manner will bend while under stress and does not return to its original shape after relaxation of the stress. Silly Putty behaves as a plastic material unless the rate of strain is rapid. Rocks exposed to elevated pressure and temperature during regional metamorphism also behave in a plastic manner. As shown in figure 15.4, material behaving in a plastic manner does not require much of an increase in stress to continue to strain (relatively flat curve). A plastic response results in rocks that are permanently deformed mainly by folding or bending of rock layers (figure 15.1). Silly Putty is a good example of a plastic solid, like glacial ice or the asthenosphere.

A rock exhibiting **brittle** behavior will break or fracture at stresses higher than its elastic limit, much like a rubber band will break if stretched too far. Rocks typically exhibit brittle behavior at or near the earth's surface where pressure and temperatures are low. Under these conditions, rocks favor breaking rather than bending. Faults and joints are examples of structures that form by brittle behavior of the crust.

A sedimentary rock exposed at the earth's surface is brittle; it will fracture if you hit it with a hammer. How then do sedimentary rocks, such as those shown in figure 15.1, become bent (or deformed plastically)? The answer is that either stress increased very slowly or that the rock was deformed under considerable confining pressure (buried under more rock).

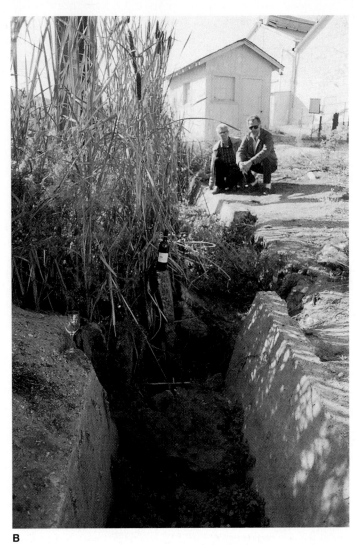

A

B

Figure 15.5

Ditch at a winery near Hollister, California, displaced by the San Andreas fault. Photo *A* was taken in 1975, *B* in 1992.

Photos by C. C. Plummer

Note, however, that there are some fractures (faults) disrupting the bent layers in figure 15.1. This tells us that although the rock was plastic initially, the stress became too intense or the rate of strain increased and the rock fractured.

Present Deformation of the Crust

Geologists often say the crust of the earth is "mobile" or "restless" because bedrock is moving and being deformed in many parts of the world. Bedrock may be displaced suddenly during earthquakes. By contrast, much of the movement of rock within the earth's crust is continuous and very slow, at rates of less than a millimeter per year.

Compared with most geologic processes, the present-day movement of the crust in much of California is very rapid. A large slice of the coastal portion of the state is moving relentlessly northward relative to the rest of North America and has

apparently been doing so for millions of years. Some of the movement is jerky and associated with earthquakes (discussed in the following chapter), but elsewhere it is essentially continuous and smooth.

Around Hollister, California, ditches, curbs, homes, and buildings are being slowly torn apart because they straddle an active fault (figure 15.5). A **fault** is a fracture in bedrock along which movement has taken place. The fault that goes through Hollister (part of the San Andreas fault system) is one of several major cracks extending deep into the earth's lithosphere, separating the northward-moving coastal portion of California from the rest of the state. Hollister residents whose homes "ride" the fault report hearing almost constant creaking, evidently due to motion along the fault. Yet the movement—about one centimeter per year (roughly the same rate at which a fingernail grows)—is slow enough that walls can be patched with plaster as they crack.

Geologically rapid movement of the crust can be observed in young, developing mountain regions such as the California Coast Ranges, which have been forming throughout the Cenozoic Era (the last 65 million years). In other parts of the world, however, the continents and sea floors are shifting very slowly up and down as well as moving laterally. Some of these motions can be detected only by precise, repeated surveying.

Structures as a Record of the Geologic Past

Some geologic structures that give us clues to the past have been described in earlier chapters. Batholiths, stocks, dikes, and sills, for example, are keys to past igneous activity (see chapter on intrusive activity). In this chapter we are mainly concerned with types of structures that can provide a record of crustal deformation that is no longer active. Very old structures that are now visible at the earth's surface were once buried and are exposed through erosion.

The study of geologic structures is of more than academic interest. The petroleum and mining industries, for example, employ geologists to look for and map geologic structures associated with oil and metallic ore deposits. Understanding and mapping geologic structures is also important for evaluating problems related to engineering decisions and environmental planning, such as the siting of dams, large bridges, or nuclear reactors, and even the building of houses, schools, and hospitals.

Geologic Maps and Field Methods

In an ideal situation, a geologist studying structures would be able to fly over an area and see the local and regional patterns of bedrock from above. Sometimes this is possible, but more often soil and vegetation conceal the bedrock. Therefore, geologists ordinarily use observations from a number of individual *outcrops* (exposures of bedrock at the surface) in determining the patterns of geologic structures. The characteristics of rock at each outcrop in an area are plotted on a map by means of appropriate symbols. With the data that can be collected, a geologist can make inferences about those parts of the area he or she cannot observe. A **geologic map,** which uses standardized symbols and patterns to represent rock types and geologic structures, is typically produced from the field map for a given area (for example see the geologic map of North America inside the front cover). On such a map are plotted the distribution of rock units, the occurrence of structural features (folds, faults, joints, etc.), ore deposits, and so forth. Sometimes surficial features, such as deposits by former glaciers, are included, but these may be shown separately on a different type of geologic map.

Anyone trained in the use of geologic maps can find on them considerable information about local geologic structures because standard symbols and terms are used on the maps and the accompanying reports. For example, the symbol ⊕ on a geologic map denotes horizontal bedding in an outcrop. Different colors or patterns on a geologic map represent distinct rock units.

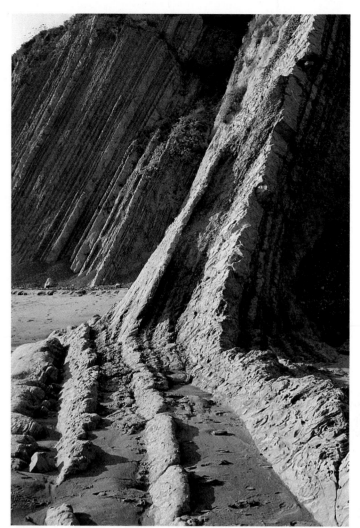

Figure 15.6

Tilted sedimentary beds along the coast of northern California near Pt. Arena. Here, the strike is the line formed by the intersection of the tilted sedimentary beds and the sand in the foreground. The direction of dip is toward the left.

Photo by Diane Carlson

Strike and Dip

According to the principle of *original horizontality,* sedimentary rocks and some lava flows and ash falls are deposited as horizontal beds or strata. Where these originally horizontal rocks are found tilted, it indicates that tilting must have occurred after deposition and lithification (figure 15.6). Someone studying a geologic map of the area would want to know the extent and direction of tilting. By convention, this is determined by plotting the relationship between a surface of an inclined bed and an imaginary horizontal plane. You can understand the relationship by looking carefully at figure 15.7, which represents sedimentary beds cropping out alongside a lake (the lake surface provides a convenient horizontal plane for this discussion).

Strike is the compass direction of a line formed by the intersection of an inclined plane with a horizontal plane. In

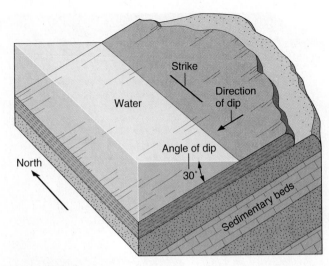

Figure 15.7

Strike, angle of dip, and direction of dip.

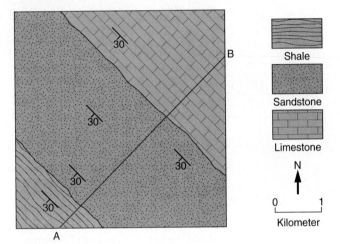

Figure 15.8

A geologic map of an area with three sedimentary formations. (Each formation may contain many individual sedimentary layers, as explained in chapter 6.) Beds strike northwest and dip 30° to the southwest.

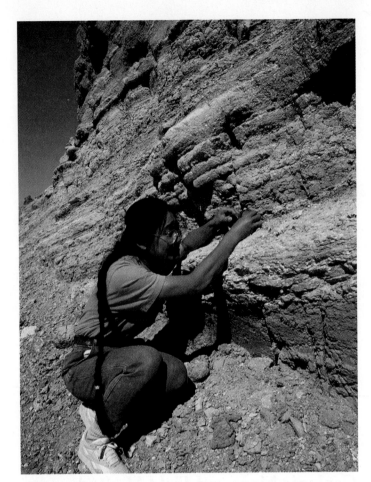

Figure 15.9

Geologist determining the strike and dip of inclined beds in Nevada.

Photo by C. C. Plummer

this example, the inclined plane is a bedding plane. You can see from figure 15.7 that the beds are striking from north to south. Customarily only the northerly direction (of the strike line) is given, so we simply say that these beds strike north.

Observe that the **angle of dip** is measured downward from the horizontal plane to the bedding plane (an inclined plane). Note that the angle of dip (30° in the figure) is measured within a vertical plane that is perpendicular to both the bedding and the horizontal planes.

The **direction of dip** is the compass direction in which the angle of dip is measured. If you could roll a ball down a bedding surface, the compass direction in which the ball rolled would be the direction of dip.

The dip angle is always measured at a right angle to the strike, that is, perpendicular to the strike line as shown in fig-

ure 15.7. Because the beds could dip away from the strike line in either of two possible directions, the general direction of dip is also specified—in this example, west.

Besides recording strike and dip measurements in a field notebook, a geologist who is mapping an area draws strike and dip symbols on the field map, such as ⌐ or ⌐ for each outcrop with dipping or tilted beds. The long line of the symbol is aligned with the compass direction of the strike. The small tick, which is always drawn perpendicular to the strike line, is put on one side or the other, depending on which of the two directions the beds actually dip toward. The angle of dip is given as a number next to the appropriate symbol on the map. Thus, ^{25}⌐ indicates that the bed is dipping 25° from the horizontal (and is striking northeast, assuming that the top of the page is north). Figure 15.8 is a geologic map that shows all the sedimentary layers striking northwest and dipping 30° to the southwest.

On a map the intersection of the two lines at the center of each strike and dip symbol represents the location of the outcrop where the strike and dip of the bedrock were measured. A specially designed instrument called a Brunton pocket transit (after the inventor) is used by geologists for this purpose (figure 15.9). The Brunton pocket transit contains a compass, a level, and a device for measuring angles of inclination.

Figure 15.10

Geologic cross section for line *A–B* on the geologic map in figure 15.8.

Beds with vertical dip require a unique symbol because they dip neither to the left nor the right of the direction of strike. The symbol used is ⤬, which indicates that the beds are striking northeast and that they are vertical.

Geologic Cross Sections

A **geologic cross section** represents a vertical slice through a portion of the earth. It is much like a roadcut (see figure 15.1) or the wall of a quarry in that it shows the orientation of rock units and structures in the vertical dimension. Geologic cross sections are constructed from geologic maps by projecting the dip of rock units into the subsurface, and are quite useful in helping visualize geology in three dimensions. They are used extensively throughout this book as well as in professional publications. Figure 15.10 shows a geologic cross section constructed between points *A* and *B* on the geologic map shown in figure 15.8.

Folds

Folds are bends or wave-like features in layered rock. Folded rock can be compared to several layers of rugs or blankets that have been pushed into a series of arches and troughs. Oftentimes folds in rock can be seen in roadcuts or other exposures (figure 15.11). When the arches and troughs of folds are concealed (or when they exist on a grand scale), geologists can still determine the presence of folds by noticing repeated reversals in the direction of dip taken on outcrops in the field or shown on a geologic map.

The fact that the rock is folded shows that it was strained plastically rather than by elastic or brittle strain. Yet the rock exposed in outcrops is generally brittle and shatters when struck with a hammer. The rock is not metamorphosed (most metamorphic rock is intensely folded because it is plastic under the high pressure and temperature environment of deep burial and tectonic stresses). Perhaps folding took place when the rock was buried at a moderate depth where high confining pressure favors plastic behavior. Alternatively, folding could have taken place close to the surface under a very low rate of strain. (When we strike a rock with a hammer, the strain rate is very high at the point of impact.)

Geometry of Folds

Determining the geometry or shape of folds may have important economic implications because many oil and gas deposits

Figure 15.11

Folded rock, Calico Hills, California.

Photo by C. C. Plummer

and also some metallic mineral deposits are localized in folded rocks (see chapter 21). The geometry of folds is also important in unraveling how a rock was strained and how it might be related to the movement of tectonic plates. Folds are usually associated with compressive stresses along convergent plate boundaries, but are also commonly formed where rock has been sheared along a fault.

Because folds are wave-like forms that usually form by the shortening of rock layers, two basic fold geometries are common—anticlines and synclines (figure 15.12).

An **anticline** is an upward arching fold. Usually the rock layers dip away from the **hinge line** (or *axis*) of the fold. The downward-arching counterpart of an anticline is a **syncline,** a troughlike fold. The layered rock usually dips toward the syncline's hinge line. In the series of folds shown in figure 15.12, two anticlines are separated by a syncline. Each anticline and adjacent syncline share a **limb.** Note the hinge lines on the crests of the two anticlines and bottom of the syncline. Similar hinge lines could be located in the hinge areas at the contacts between any two adjacent folded layers. For each anticline and the syncline, the hinge lines are contained within the shaded vertical planes. Each of these planes is an **axial plane,** a plane containing all of the hinge lines of a fold.

It is important to remember that anticlines are not necessarily related to ridges nor synclines to valleys, because valleys and ridges are nearly always erosional features. In an area that has been eroded to a plain, the presence of underlying anticlines and synclines is determined by the direction of dipping

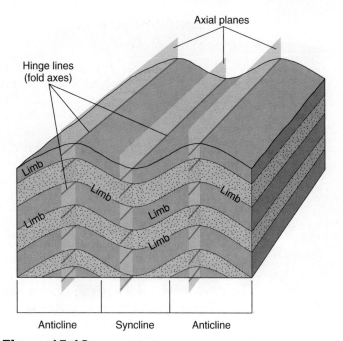

Figure 15.12

Two anticlines and a syncline.

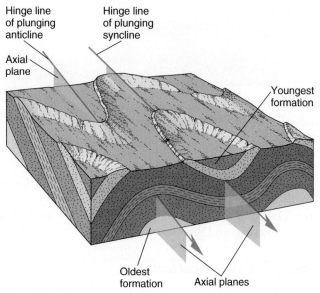

Figure 15.14

Plunging folds: anticline on left and right, syncline in center. The hinge lines are at an angle to the block diagram, penetrating the surface and emerging from the front cross section.

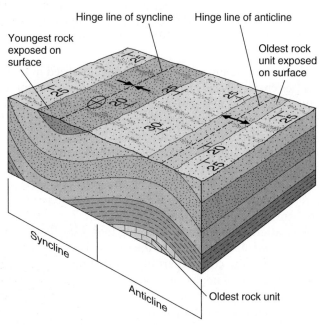

Figure 15.13

Folded rock. This view is a *block diagram.* Its top represents the land surface and its two visible sides are vertical cross sections. The surface has been eroded to a nearly horizontal plain. Side views are interpretations based on what the geologist notices on the surface.

beds in exposed bedrock, as shown in figure 15.13. (In the field, of course, the cross sections are not exposed to view as they are in the diagram.)

Figure 15.13 also illustrates how determining the relative ages of the rock layers, or beds, can tell us whether a structure is an anticline or a syncline. Observe that the oldest exposed rocks are along the hinge line of the anticline. This is because lower layers in the originally flat-lying sedimentary or volcanic rock were moved upward and are now in the core of the anticline. The youngest rocks, on the other hand, which were originally in the upper layers, were folded downward and are now exposed along the synclinal hinge line.

Plunging Fold

The examples shown so far have been of folds with horizontal hinge lines. These are the easiest to visualize. In nature, however, anticlines and synclines are apt to be **plunging folds**—that is, folds in which the hinge lines are not horizontal. On a surface leveled by erosion, the patterns of exposed strata (beds) resemble V's or horseshoes (figures 15.14 and 15.15) rather than the striped patterns of nonplunging folds. However, plunging anticlines and synclines are distinguished from one another in the same way as are nonplunging folds—by directions of dip or by relative ages of beds.

A plunging syncline contains the youngest rocks in its center or core, and the V or horseshoe points in the direction opposite of the plunge. Conversely, a plunging anticline contains the oldest rocks in its core and the V points in the same direction as the plunge of the fold.

Figure 15.15

Plunging folds. Anticline in Utah plunging in the direction of the upper part of the photo.

Photo by Frank M. Hanna

Structural Domes and Structural Basins

A **structural dome** is a structure in which the beds dip away from a central point. In cross section, a dome resembles an anticline. In a **structural basin,** the beds dip toward a central point; in cross section, it is comparable to a syncline (figure 15.16). A structural basin is like a set of nested bowls. If the set of bowls is turned upside down, it is analogous to a structural dome.

Domes and basins tend to be features on a grand scale (some are more than a hundred kilometers across), formed by uplift somewhat greater (for domes) or less (for basins) than that of the rest of a region. Most of Michigan and parts of adjoining states and Ontario are on a large structural basin (see map on the inside front cover). Domes of similar size are found in other parts of the Middle West. Smaller domes are found in the Rocky Mountains (figure 15.17).

Domes and anticlines (as well as some other structures) are important to the world's petroleum resources, as described in chapter 21 (see figure 21.3).

Oldest formation exposed on the surface

Youngest formation

A

Youngest formation

Oldest formation exposed on the surface

B

Figure 15.16

(*A*) Structural dome. (*B*) Structural basin.

Figure 15.17

Dome near Casper, Wyoming. The ridges are sedimentary layers that are resistant to erosion. Beds dip away from the center of the dome.

Photo by D. A. Rahm, courtesy of Rahm Memorial Collection, Western Washington University

A Strata before folding

Buckling   Bending

B Open folds–the two diagrams show alternate ways that stresses may have been distributed to have caused the folding.

C Isoclinal ("hairpin") folds

Figure 15.18

Various types of folds. The length of the arrows in A through E is proportional to the amount and direction of stresses that caused folding. (A) Strata before folding. (B) Open folds in Spain (they are plunging away from the people). (C) Isoclinal folds from northeastern Nevada.

Photo *B* by C. C. Plummer, photo *C* by Martin G. Miller/Visuals Unlimited

Interpreting Folds

Folds occur in many varieties and sizes. Some are studied under the microscope, while others can have adjacent hinge lines tens of kilometers apart. Some folds are a kilometer or more in height. Figure 15.18 shows several of the more common types of folds. **Open folds** (figure 15.18*B*) have limbs that dip gently. All other factors being equal, the more open the fold, the less intense the stress involved. By contrast, an **isoclinal fold,** one in which limbs are parallel to one another, implies intense compressive or shear stress (figure 15.18*C*).

Figure 15.18 (Continued)

(*D*) Overturned anticline from northern California. (*E*) Recumbent folds in the Alps.

Photo *D* by Diane Carlson; photo *E* courtesy of Professor John Ramsay, from *"The Techniques of Modern Structural Geology, Vol. 2,"* J. G. Ramsay & M. J. Huber, © 1987 Academic Press.

Folds described thus far have vertical axial planes and are referred to as upright folds. However, where the axial plane of a fold is not vertical but is inclined, the fold is classified as *asymmetric*. If the axial plane is inclined to such a degree that the fold limbs dip in the same direction, the fold is classified as an **overturned fold.** Overturned folds imply that unequal compressive stresses or even a shearing stress caused the upper limb of the fold to override the lower limb (figure 15.18*D*). Looking at an outcrop where only the overturned limb of a fold is exposed, you would probably conclude that the youngest bed is at the top. However, the principles of *superposition* (see chapter 8) cannot be applied to determine top and bottom for overturned beds. You must either see the rest of the fold or find primary sedimentary structures within the beds such as mudcracks that indicate the original top or upward direction.

Recumbent folds (figure 15.18*E*) are overturned to such an extent that the limbs are essentially horizontal. Recumbent folds are found in the cores of mountain ranges such as the Canadian Rockies, Alps, and Himalayas and indicate compressive and/or shear stresses were more intense in one direction and probably record shortening of the crust associated with plate convergence.

Fractures in Rock

If a rock is brittle, or if the strain rate is too great for deformation to be accommodated by plastic behavior, the rock fractures. Commonly there is some movement or displacement. If essentially no displacement occurs, a fracture or crack in bedrock is called a **joint.** If the rock on either side of a fracture

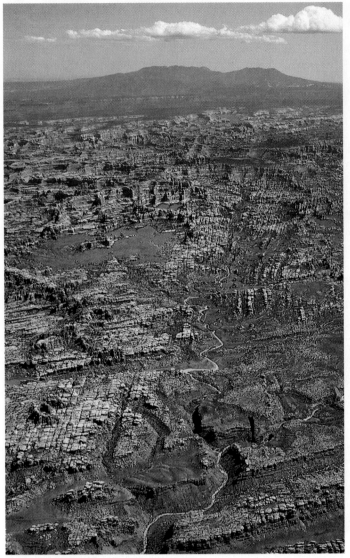

Figure 15.19

Vertical joints in sedimentary rock of the Colorado Plateau formed in response to tectonic uplift of the region.

Photo by Frank M. Hanna

moves, the fracture is a *fault* (as defined earlier). Most rock at or near the surface is brittle, so nearly all exposed bedrock is jointed to some extent.

Joints

In discussing volcanoes, we described *columnar jointing,* in which hexagonal columns form as the result of tension and contraction of a cooling, solidified lava flow. *Sheet jointing,* a type of jointing due to expansion (discussed along with weathering in chapter 5) is caused by tension. The pressure release due to removal of overlying rock has the effect of creating tensional stress perpendicular to the land surface.

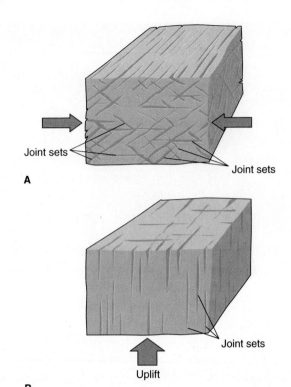

Figure 15.20

Joint sets. Arrows indicate directions of principal stress. (*A*) Three joint sets associated with a horizontal compressional stress. (*B*) Two joint sets associated with uplift and a vertical compressional stress.

Columnar and sheet joints are examples of fractures that form from nontectonic stresses and are therefore referred to as primary joints. In this chapter we are concerned with joints that form not from cooling or unloading but from tectonic stresses.

Joints are one of the most commonly observed structures in rocks (figure 15.19). Where joints are oriented approximately parallel to one another, a **joint set** can be defined. Joints are important because they often indicate the direction of compressive stress operative during their formation. For example, in figure 15.20*A* three sets of joints have formed in response to a horizontal compressive stress. Vertical joint sets (figure 15.20*B*) are often associated with tectonic uplift of a region.

Geologists sometimes find valuable ore deposits by studying a joint system. For example, gold-bearing hydrothermal solutions may migrate upward through a set of joints and deposit quartz and gold in the cracks. Accurate information about joints also is important in the planning and construction of large engineering projects, particularly dams and reservoirs. If the bedrock at a proposed location is intensely jointed, the possibility of dam failure or reservoir leakage may make that site too hazardous.

15.1
IN GREATER DEPTH

Structures Associated with Salt Domes

Most folding is the product of horizontal compression; what we would expect if rock layers were placed in a giant vise. However, we get vertical compressive stresses when bodies of salt rise through the crust. Domes and normal faults are the characteristic structures that form above the rising bodies of salt.

Blobs (more correctly *diapirs,* similar to magma diapirs described in chapter 3) of rock salt rise through layers of sedimentary rock. Sometimes these bodies, which are a kilometer (or more) wide, travel upward 10 kilometers. They originate from buried layers of evaporites. Because rock salt is easily deformed plastically, it squeezes out of the original beds. It works its way upward because salt is less dense than the overlying rocks (box figure 1). Where it breaks through rock layers, those layers are upturned adjacent to the rock salt (box figure 2). The rising salt pushes the overlying rock into a dome or, less commonly, an anticline. The compressive stresses caused by the rising salt diapirs may also create normal faults that trap oil (box figure 2).

Many important oil fields along the Gulf coast of Texas and Louisiana are located at salt domes. A few salt-created anticlines are also found in the Colorado plateau.

If the rock salt breaks through the earth's surface in a dry climate, it will ooze very slowly over the land surface and continue to move, like a glacier, but much more slowly. Layering within the mass is highly contorted from the squeezing that takes place (box figure 3).

| 1 Kilometer |

Box 15.1 Figure 2

A salt dome. Oil and gas are trapped in folds and along faults above the dome and within upturned sandstones along the flanks of the dome. Insoluble cap rock may contain recoverable sulfur.

Box 15.1 Figure 1

Salt domes may form as a bed of rock salt is loaded unevenly by a thick wedge of sediment. The salt flows toward the thin part of the sediment wedge and also flows upward.

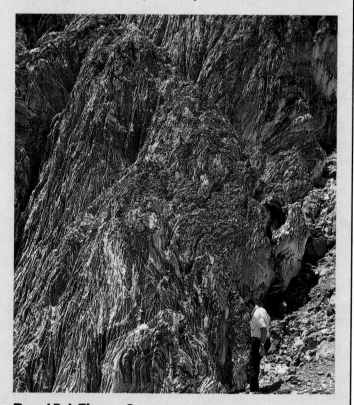

Box 15.1 Figure 3

Rock salt body in Spain.

Photo by C. C. Plummer

Related Web Resources

Visit the Applied Geodynamics Laboratory web site at the University of Texas–Austin for interesting three dimensional images and animation of salt domes: http://www.utexas.edu/research/beg/giovanni/

Figure 15.21

Fault in Big Horn Mountains, Wyoming is marked by broken, red-stained rocks, and displaced rock layers.

Photo by Diane Carlson

Reverse fault

Normal fault

A Dip-slip faults

B Strike-slip faults

C Oblique-slip fault

Figure 15.22

Three types of faults illustrated by displaced blocks. Heavier arrows show direction in which block to the left moved. (*A*) Dip-slip movement. (*B*) Strike-slip movement. (*C*) Oblique-slip movement. Black arrows show dip-slip and strike-slip components of movement.

Faults

Faults were defined earlier as fractures in bedrock along which movement has taken place. The displacement may be only several centimeters or may involve hundreds of kilometers. For many geologists, an active fault is regarded as one along which movement has taken place during the last 11,000 years. Most faults, however, are no longer active.

The nature of past movement ordinarily can be discerned where a fault is exposed in an outcrop (figure 15.21). The geologist looks for dislocated beds or other features of the rock that might show how much displacement has occurred and the relative direction of movement. In some faults the contact between the two displaced sides is a crack. In others the rock has been broken or ground to a fractured or pulverized mass sandwiched between the displaced sides.

Geologists describe fault movement in terms of direction of slippage: dip-slip, strike-slip, or oblique-slip (figure 15.22). In a **dip-slip fault,** movement is parallel to the dip of the fault surface. A **strike-slip fault** indicates *horizontal* motion parallel to the strike of the fault surface. An **oblique-slip fault** has both strike-slip and dip-slip components.

Dip-slip Faults

Normal and reverse faults, the most common types of dip-slip faults, are distinguished from each other on the basis of the relative movement of the *footwall block* and the *hanging-wall block.* The **footwall** is the underlying surface of an inclined fault plane, whereas the overlying surface is the **hanging wall.** These old mining terms brought into geology are illustrated in figure 15.23. If a miner is tunneling along the strike of a fault, his feet are on the footwall and his lantern is hung on the hanging wall.

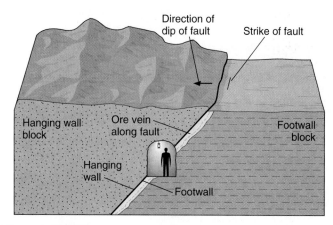

Figure 15.23

Relationship between the hanging wall block and footwall block of a fault.

In a **normal fault** (figures 15.24 and 15.25), the hanging-wall block has moved downward relative to the footwall block. The relative movement is represented on a geological cross section by a pair of arrows, because we cannot generally tell which block actually moved. As shown in figure 15.24, a normal fault is caused by vertical compressive stresses and results in extension or lengthening of the crust. When there is extension of the crust, the hanging-wall block moves downward along the fault to compensate for the pulling apart of the rocks. Sometimes a block bounded by normal faults will drop down, creating a *graben,* as shown in figure 15.24C. (*Graben* is the German word for "ditch.") *Rifts* are grabens associated with diverging plate boundaries, either along mid-oceanic ridges or on continents (see chapters 18 and 19).

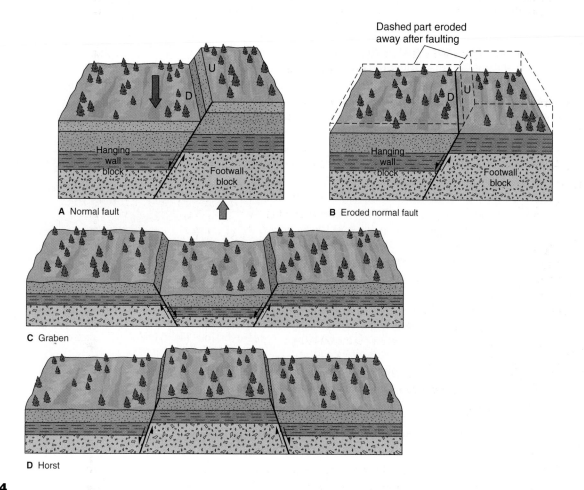

Figure 15.24

Normal faults. (*A*) Diagram shows the fault before erosion and the geometric relationships of the fault and how it might have been caused by vertical compressive stresses. (*B*) The same area after erosion. (*C*) A graben. (*D*) A horst.

A

B

Box 15.2 Figure 1

(*A*) Oil and gas are concentrated or trapped in hinge of anticline. Gas and oil float on water in porous and permeable reservoir rock (sandstone). (*B*) Eroded anticline forms trap in Lander oil field, Wyoming.

Photo B by Diane Carlson

An "oil pool" can exist only under certain conditions. Crude oil does not fill caves underground as the term *pool* may suggest; rather, it simply occupies the pore spaces of certain sedimentary rocks, such as poorly cemented sandstone, in which void space exists between grains. Natural gas (being lighter) often occupies the pore spaces above the crude oil, while water (being heavier) is generally found saturating the rock below the oil pool (box figure 1).

A **source rock,** which is always a sedimentary rock, must be present for oil to form. The sediment of the source rock has to include remains of organisms buried during sedimentation. This organic matter partially decomposes into petroleum and natural gas. Once formed, the droplets of petroleum tend to migrate, following fractures and interconnecting pore spaces. Being lighter than the rock, the petroleum usually migrates upward, although horizontal migration does occur.

If it is not blocked by impermeable rock, the oil may migrate all the way to the surface, where it is dissipated and permanently lost for human use. Natural oil seeps, where leakage of petroleum is taking place, exist both on land and offshore. Where impermeable rock blocks the oil droplets' path of migration, an oil pool may accumulate below the rock, much like helium-filled balloons might collect under a domed ceiling. For any significant amount of oil to collect, the rock below the impermeable rock must be porous as well as permeable. Such a rock, when it contains oil, is called a **reservoir rock.**

Another necessary condition is that the geologic structure must be one that favors the accumulation and retention of petroleum. An "anticlinal trap" is one of the best structures for holding oil. As oil became a major energy source and the demand for it increased, most of the newly discovered wells penetrated anticlinal traps. Geologists discovered

A Fault

B Uncomformity

C Sedimentary facies

these by looking for indication of anticlines exposed at the surface. As time went on, other types of structures were also found to be oil traps. Many of these were difficult to find because of the lack of telltale surface patterns indicating favorable underground structures. Box figure 2 illustrates some structures other than anticlinal traps that might have a potential for oil production.

At present, oil companies rely on detailed and sophisticated geologic studies of an area they hope may have the potential for an "oil strike." The petroleum industry also depends heavily on geophysical techniques (see chapter 17) for determining, by indirect means, the subsurface structural geology.

Even when everything indicates that conditions are excellent for oil to be present underground, there is no guarantee that oil will be found. Eventually an oil company must commit a million dollars or more to drill a deep test well, or "wildcat" well. Statistics indicate that the chance of a test well yielding commercial quantities of oil is much less than 1 in 10. As more and more of the world's supply of petroleum is used up, what is left becomes increasingly harder—and costlier—to find.

Box 15.2 Figure 2

Structures other than anticlines that trap oil.

Figure 15.25

Normal faults with prominent horst block offsets volcanic ash layers in southern Oregon.

Photo by Diane Carlson

If a block bounded by normal faults is uplifted sufficiently, it becomes a fault-block mountain range. (This is also called a *horst,* the opposite of a graben.) The Basin and Range province (described in other chapters) of Nevada and portions of adjoining states is characterized by numerous mountain ranges separated from adjoining valleys by normal faults.

In a **reverse fault,** the hanging-wall block has moved upward relative to the footwall block (figures 15.26 and 15.27). As shown in figure 15.26, horizontal compressive stresses cause reverse faults. (The faults in figure 15.1 are reverse faults.) Reverse faults tend to shorten the crust.

A **thrust fault** is a reverse fault in which the dip of the fault plane is at a low angle to horizontal (figures 15.26*C* and 15.28). In some mountain regions it is not uncommon for the upper plate (or hanging-wall block) of a thrust fault to have overridden the lower plate (footwall block) for several tens of kilometers. Thrust faults typically move or thrust older rocks on top of younger rocks (figure 15.28), and result in an extreme shortening of the crust. Thrust faults commonly form at convergent plate boundaries to accommodate shortening during collision.

Figure 15.26

(*A*) A reverse fault. The fault is unaffected by erosion. Arrows indicate compressive stress. (*B*) Diagram shows area after erosion; dashed lines indicate portion eroded away. (*C*) Thrust fault due to horizontal compression.

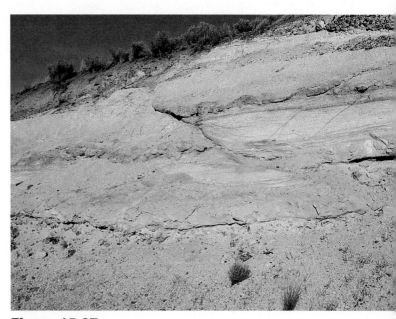

Figure 15.27

Reverse fault in volcanic ash beds, southern Oregon.

Photo by Diane Carlson

A

Chief
Mountain

Precambrian

Cretaceous shale

B

Figure 15.28

(*A*) Chief Mountain in Glacier National Park, Montana, is an erosional remnant of a major thrust fault. (*B*) Cross section of the area. Older (Precambrian) rocks have been thrust over younger (Cretaceous) rocks. Dashed lines show where the fault has been eroded away.

Photo by Frank M. Hanna

Strike-slip Faults

A fault where the movement (or *slip*) is predominantly horizontal and parallel to the strike of the fault, is called a **strike-slip fault.** The displacement along a strike-slip fault is either left-lateral or right-lateral and can be determined by looking across the fault. For instance, if a recent fault displaced a stream (figure 15.29), a person walking along the stream would stop where it is truncated by the fault. If the person looks across the fault and sees the displaced stream to the right, it is a **right-lateral fault.** A **left-lateral fault** is one in which displacement is to the right as seen from across the fault. Again, we cannot tell which side actually moved, so pairs of arrows are used to indicate relative movement.

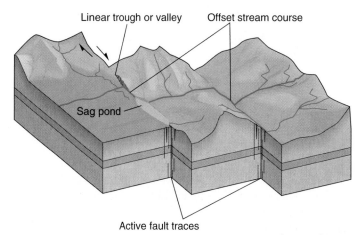

Linear trough or valley   Offset stream course

Sag pond

Active fault traces

Figure 15.29

Right-lateral strike-slip fault showing offset stream channels, sag ponds, and linear valleys and ridges.

15.3
IN GREATER DEPTH

California's Greatest Fault—
The San Andreas

The San Andreas fault in California is the best known geologic structure in the United States, but the geologists and seismologists who study it admit that our knowledge of its activity and its history is far from complete. Actually, the San Andreas is the longest of several, subparallel faults that transect western California (box figure 1). Collectively, these right-lateral faults are known as the San Andreas fault system. The system is in a belt approximately 100 kilometers wide and 1,300 kilometers long that extends into Mexico, ending at the Gulf of California.

Los Angeles is slowly moving toward San Francisco because of San Andreas fault motion. At an average rate of movement of about 2 centimeters per year, Los Angeles could be a western suburb of San Francisco (or San Francisco an eastern suburb of Los Angeles) in some 25 million years. Earthquakes are produced by sudden movement within the fault system as explained in chapter 16. Bedrock along the San Andreas fault shifted as much as 4.5 meters in association with the 1906 quake that destroyed much of San Francisco.

The San Andreas fault is not a simple crack, but a belt of broken and ground-up rock, usually a hundred meters or more wide. Its presence is easy to determine throughout most of its length. Along the fault trace are long, straight valleys (formed by erosion and subsidence) that show quite different terrain on either side. Stream channels follow much of the fault zone because the weak, ground-up material along the fault is easily eroded. Locally, elongate lakes (called sag ponds) are found where the ground-up material has settled more than the surface of adjacent parts of the fault zone. The fault was named after one of these ponds, San Andreas Lake, just south of San Francisco (box figure 2).

One can visually follow the fault northward from San Andreas Lake into the southwestern suburbs of San Francisco. There the fault zone is hidden by recently built housing tracts. Apparently the builders and residents have chosen to ignore the hazards of living on the nation's most famous fault.

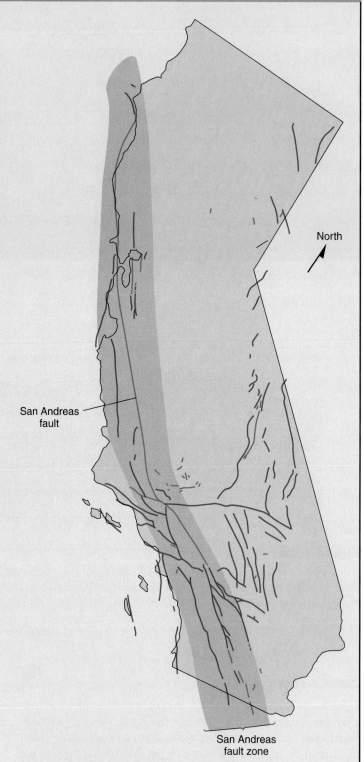

North

San Andreas fault

San Andreas fault zone

Box 15.3 Figure 1

California has its faults. Red lines indicate fault movement within the last 200 years. Blue lines indicate movement within the last 2 million years.

From California Division of Mines and Geology.

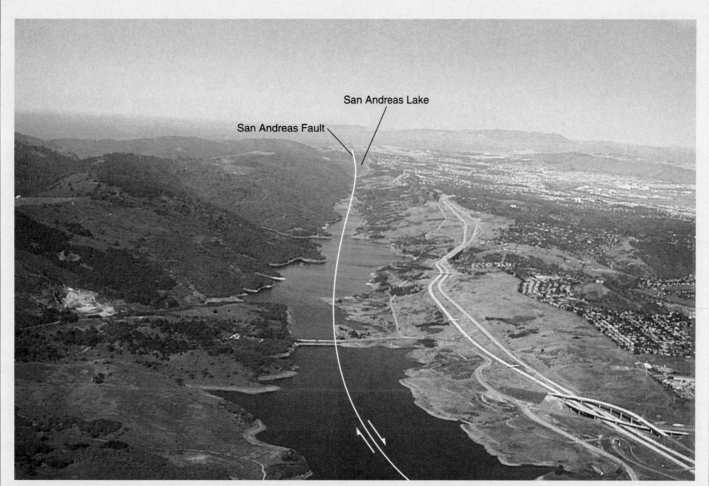

Box 15.3 Figure 2

Part of the San Andreas fault. View northward toward San Francisco. Lakes occupy the fault zone. Hills to the left of the fault are moving northward.

Photo by B. Amundson

Geologists have been unable to agree on the total displacement of the fault or on how long it has been active. Some believe movement began in the Mesozoic Era (over 65 million years ago); most think that it began later. The difficulty in establishing an age for the inception of the faulting lies in finding clear evidence of displaced bedrock. What geologists would like to find, if it exists, is a rock unit that can be isotopically dated and that was formed across the fault zone just about the time faulting was beginning. Now displaced rock on both sides of the fault zone would have to be clearly identifiable as having been the same unit.

Geologically young features that cross the fault, such as displaced stream channels (box figure 3), are common. Similarly, ancient rocks that undoubtedly were there before faulting began are recognized as having been displaced. Many California geologists believe that the belt of granitic rock just west of the fault was once the southern continuation of the granitic batholiths of the Sierra Nevada (box figure 4A), which are more than 80 million years old. But these extremes tell us only that the age of the San Andreas is somewhere between approximately 80 million years and a few thousand years, when the stream channel in box figure 3 carved its course across the fault.

The strongest evidence for long-term faulting comes from almost identical volcanic sequences now 315 kilometers apart. The volcanic activity took place 23.5 million years ago. Using these figures, we can calculate the average rate of motion as 1.3 centimeters per year for the San Andreas fault. (However, movement along other faults means the rate of motion is higher for the fault system.) Older rocks that appear to have been offset 560 kilometers have been correlated with less certainty, suggesting that the total offset for the San Andreas fault is at least 560 kilometers.

Box 15.3 Figure 3

Stream channel displaced by the San Andreas fault. The arrows on either side of the fault trace indicate relative motion.

Photo by C. C. Plummer

How long ago faulting began remains controversial. According to plate-tectonic theory, the San Andreas fault is a transform boundary that separates the North American plate from the Pacific plate. One hypothesis places the beginning of strike-slip movement at about 30 million years ago. According to this hypothesis, the Baja California peninsula split from mainland Mexico as seafloor spreading began (box figure 4*B*). As the Gulf of California widens, the block of crust west of the San Andreas is pushed northward.

Additional Resources

Wallace, R. E., ed. 1990. The San Andreas fault system, California: U.S. Geological Survey Professional Paper 1515.

Related Web Resources

Internet version of the San Andreas fault system: http://pubs.usgs.gov.gip/earthq3/

For additional photographs of the San Andreas Fault: du/cgi-bin/illus_query?table-char_table=photocd&like-char-subject=SAN+ANDREAS

Box 15.3 Figure 4

(*A*) Reconstruction of California and Mexico as they may have been before faulting. (*B*) Continuous opening of the Gulf of California creates motion along the San Andreas fault.

After Tanya Atwater, 1970, *Geological Society of America Bulletin.*

Large strike-slip faults, like the San Andreas fault in California, typically define a zone of faulting that may be several kilometers wide and hundreds of kilometers long (see box 15.3). The surface trace of an active strike-slip fault is usually defined by a prominent linear valley that has been more easily eroded where the rock has been ground up along the fault during movement. The linear valley may contain lakes or sag ponds (figure 15.29) where the highly permeable fault rock allows ground water to freely flow to the surface. The trace of the fault may also be marked by offset surface features such as streams, fences, and roads or by distinctive rock units.

Strike-slip faults accommodate shearing stress along transform plate boundaries where plates slide past one another. One of the most famous examples of a transform fault is the San Andreas Fault. The San Andreas fault is a right-lateral strike-slip fault that forms the boundary between the North American and Pacific plates (see box 15.3 figure 4).

Summary

Tectonic forces result in deformation of the earth's crust. *Stress* (force per unit area) is a measure of the tectonic force and confining pressure acting on bedrock. Stress can be *compressive, tensional,* or *shearing. Strained* (changed in size or shape) rock records past stresses, usually as joints, faults, or folds.

A geologic map shows the structural characteristics of a region. *Strike* and *dip* symbols on geologic maps indicate the attitudes of inclined surfaces such as bedding planes. The strike and dip of a bedding surface indicate the relationship between the inclined plane and a horizontal plane.

If rock layers bend (plastic behavior) rather than break, they become folded.

Rock layers are folded into *anticlines* and *synclines* and recumbent folds. If the hinge line of a fold is not horizontal, the fold is *plunging.* Older beds exposed in the core of a fold indicate an anticline, whereas younger beds in the center of the structure indicate a syncline. In places where folded rock has been eroded to a plain, an anticline can usually be distinguished from a syncline by whether the beds dip toward the center (syncline) or away from the center (anticline).

Fractures in rock are either *joints* or *faults.* A joint indicates that movement has not occurred on either side of the fracture; displaced rock along a fracture indicates a fault. *Dip-slip* faults are either *normal* or *reverse,* depending on the motion of the hanging wall block relative to the footwall block. A reverse fault with a low angle of dip for the fault plane is a *thrust fault.* Reverse faults accommodate horizontal shortening of the crust whereas normal faults accommodate horizontal stretching or extension.

In a *strike-slip* fault, which can be either left-lateral or right-lateral, horizontal movement has occurred.

Terms to Remember

angle of dip 362
anticline 363
axial plane 363
brittle 359
compressive stress 358
dip-slip fault 370
direction of dip 362
elastic 359
elastic limit 359
fault 360
fold 363
footwall 370
geologic cross section 363
geologic map 361
hanging wall 370

hinge line 363
isoclinal fold 366
joint 367
joint set 368
left-lateral fault 375
limb 363
normal fault 371
oblique-slip fault 370
open fold 366
overturned fold 367
plastic 359
plunging fold 364
recumbent fold 367
reservoir rock 372
reverse fault 374

right-lateral fault 375
shear stress 359
source rock 372
strain 358
stress 358
strike 361
strike-slip fault 370, 375
structural basin 365
structural dome 365
structural geology 358
syncline 363
tensional stress 358
thrust fault 374

Testing Your Knowledge

Use the questions below to prepare for exams based on this chapter.

1. Most anticlines have both limbs dipping away from their hinge lines. For which kind of fold is this not the case?

2. What are the four main types of contacts and how would you distinguish between them if you were a geologist doing field work?

3. On a geologic map, if no cross sections were available, how could you distinguish an anticline from a syncline?

4. If you locate a dip-slip fault while doing field work, what kind of evidence would you look for to determine whether the fault is normal or reverse?

5. Name several geologic structures described in earlier chapters.

6. What is the difference between strike, direction of dip, and angle of dip?

7. Draw a simple geologic map, using strike and dip symbols for a syncline plunging to the west.

8. How does a structural dome differ from a plunging anticline?

9. Which of the statements is true? (a) when forces are applied to an object, the object is under stress (b) strain is the change in size (volume) or shape, or both, while an object is undergoing stress (c) stresses can be compressive, tensional or shear (d) all of the above

10. The compass direction of a line formed by the intersection of an inclined plane with a horizontal plane is called (a) strike (b) direction of dip (c) angle of dip

11. Folds in a rock show that the rock behaved in a _____ way. (a) plastic (b) elastic (c) brittle (d) all of the above

12. An anticline is (a) any fold (b) overturned fold (c) an upward-arched fold (d) a downward-arched fold

13. A syncline is (a) an upward-arched fold (b) overturned fold (c) a downward-arched fold (d) horizontal beds

14. A structure in which the beds dip away from a central point is called a (a) basin (b) anticline (c) structural dome (d) syncline

15. Which is not a type of fold? (a) open (b) isoclinal (c) overturned (d) recumbent (e) thrust

16. Fractures in bedrock along which movement has taken place are called (a) joints (b) faults (c) cracks (d) crevasses

17. In a normal fault, the hanging-wall block has moved _____ relative to the footwall block. (a) upward (b) downward (c) sideways

18. Normal faults accommodate what kind of strain? (a) shortening (b) extensional (c) plastic

19. Faults that typically move older rock on top of younger rock are (a) normal faults (b) thrust faults (c) strike-slip faults

Expanding Your Knowledge

1. Can a fault always be distinguished from an unconformity?

2. In what parts of North America would you expect to find the most intensely folded rock?

3. A subduction zone can be regarded as a very large example of what type of fault?

4. Why do some horizontal compressive stresses cause thrust faults while others cause strike-slip faults?

5. What features in sedimentary or volcanic rock layers would you look for to tell you that the rock was part of the overturned limb of a fold?

6. Can you identify and name the various geologic structures shown in the figures in chapter 8?

Exploring Resources

Davis, G. H., and S. J. Reynolds 1996. *Structural geology of rocks and regions.* 2d ed. New York: John Wiley & Sons.

Hatcher, R. D. 1995. *Structural geology—Principles, concepts, and problems.* 2d ed. Englewood Cliffs, N.J.: Prentice-Hall.

Keller, E. A., and N. Pinter. 1996. *Active tectonics: Earthquakes, uplift, and landscapes.* Upper Saddle River, N.J.: Prentice-Hall.

Lisle, R. J. 1988. *Geological structures and maps: A practical guide.* New York: Pergamon.

Marshak, S., and G. Mitra. 1988. *Basic methods of structural geology.* Englewood Cliffs, N.J.: Prentice-Hall.

McClay, K. 1987. *The mapping of geologic structures.* Geological Society of London Handbook. Open University Press.

Roberts, J. L. 1982. *Introduction to geologic maps and structures.* Oxford: Pergamon Press.

Rowland, S. M., and E. M. Duebendorfer. 1994. *Structural analysis and synthesis.* 2d ed. Boston: Blackwell Scientific Publications.

Suppe, J. 1985. *Principles of structural geology.* Englewood Cliffs, N.J.: Prentice-Hall.

Twiss, R. J., and E. M. Moores. 1992. *Structural geology.* New York: W. H. Freeman.

van der Pluijm, B. A., and S. Marshak. 1997. *Earth structure.* Dubuque, Iowa: WCB/McGraw-Hill.

 http://home.earthlink.net/~ schimmrich/structure/structure.html
Structural geology site maintained by S. H. Schimmrich at Kuztown University contains many links to on-line courses, computer software, bibliographies, and research projects dealing with structural geology.

http://www.geo.cornell.edu/geology/classes/ geol326/326.html
Web site for structural geology course taught by the Department of Geological Sciences at Cornell University contains images showing structural features and models of thrust-fault movement.

http://craton.geol.brocku.ca/ctg.html
Canadian Tectonics Group web site contains structural geology images, computer software, and a newsletter outlining research projects.

Interacting with Journey Through Geology CD-ROM

Expand your knowledge of the concepts presented in this chapter by using the CD-ROM to answer the following questions.

1. Changing Earth: Watch as the continents move and collide. Identify when and where you see compression or tension at work. From this, predict where in the world you would expect to find compressional features such as folds and thrust faults, and where you would expect to find tensional features such as normal faults.

2. Rocks Deform: Cite examples of ductile and brittle behavior in the everyday objects shown in the CD video sequence. Do any of these objects behave elastically?

3. Rocks Deform: What evidence of ductile behavior in rocks can be seen on the microscopic scale? On the scale of mountain ranges?

4. Rocks Deform: What variables determine whether a rock will deform in a ductile or plastic manner?

These train tracks have been severely bent by earthquake waves that rippled across the earth's surface during the January, 1995 earthquake in Kobe, Japan. The earthquake killed more than 5,000 people and injured 27,000. Damage exceeded $400 billion; the earthquake destroyed or severely damaged more than 88,000 buildings.

Photo © Patrick Robert/SYGMA

CHAPTER
16

Earthquakes

*T*his chapter will help you understand the nature and origin of earthquakes. We discuss the seismic waves created by earthquakes and how the quakes are measured and located by studying these waves. We also describe some effects of earthquakes, such as ground motion and displacement, damage to buildings, and quake-caused fires, landslides, and tsunamis ("tidal waves").

Earthquakes are largely confined to a few narrow belts on earth. This distribution was once puzzling to geologists, but here we show how the concept of plate tectonics neatly explains it.

As geologists learn more about earthquake behavior, there is the intriguing possibility that we will be able to predict—and even control—earthquakes. We conclude the chapter with a look at this developing branch of earth study.

On April 18, 1906, at 5:12 in the morning, part of California slid abruptly past the rest of the state during a great earthquake. A visible scar 450 kilometers (280 miles) long was left where the earth was torn along coastal northern California. Rock was displaced horizontally as much as 4.5 meters (15 feet); soil above the rock was displaced up to 6.5 meters (21 feet). The quake, located on a segment of the San Andreas fault near San Francisco, shook the ground for one full minute. Buildings toppled in San Francisco, and broken gas mains fed fires that raged for three days. Broken water mains hampered fire fighting. The fires were finally extinguished when buildings were dynamited to create a firebreak. Terrified and homeless people moved to refugee camps set up in city parks. Looters were shot on sight. As the city gradually recovered from the shock of the devastation, it was found that at least 3,000 people had died and $400 million (in 1906 dollars) of damage had been done. Perhaps 90% of the destruction was caused by the fires.

At 5:04 P.M. on October 17, 1989, San Francisco was again severely shaken for 15 seconds by the Loma Prieta earthquake located to the south on the San Andreas fault near Santa Cruz. Although the quake did not tear the ground surface, it collapsed some buildings and freeway overpasses built upon the soft "bay fill" sediment in San Francisco and Oakland. A section of the Bay Bridge collapsed. Just as in 1906, raging fires were fed by broken gas mains in the Marina district of San Francisco, and were hard to fight because of broken water mains; fireboats helped extinguish them. Very severe damage occurred in small towns near the center of the quake. The death toll was 63, and damage was $6 billion.

At 5:30 P.M. on March 27, 1964, southern Alaska was rocked by an earthquake that lasted for three minutes. Although the force of this earthquake was twice as strong as the 1906 San Francisco earthquake, loss of life and property was relatively low because of Alaska's small population—15 people died as a direct result of the shaking, and damage amounted to slightly over $300 million (in 1964 dollars). The tremor was felt over an area of more than 1 million square kilometers (350,000 square miles). A section of the earth's surface 50 by 200 kilometers (30 by 125 miles) was raised as much as 13 meters (40 feet), and a similar block of land sank 1 to 2 meters (3 to 6 feet). Horizontal movement was slight. In Anchorage, 150 kilometers (90 miles) from the center of the earthquake, landslides wrecked parts of the city. The greatest loss of life was caused by large sea waves (or tsunamis) generated by land movement associated with the earthquake—almost 100 people drowned in Alaska, and a few people drowned as far away as Oregon and northern California, as the waves spread over the Pacific Ocean.

On January 17, 1994, at 4:31 A.M., the Northridge earthquake rocked the San Fernando Valley just north of Los Angeles, California, for 40 seconds. The quake, about 2 miles from California State University, Northridge, damaged or destroyed all 53 CSUN buildings, and seriously damaged 300 other schools. Numerous freeway overpasses collapsed (including some that had previously collapsed in a nearby 1971 quake), closing four Interstates and seven other highways for months. The two upper stories of the Northridge Meadows Apartments

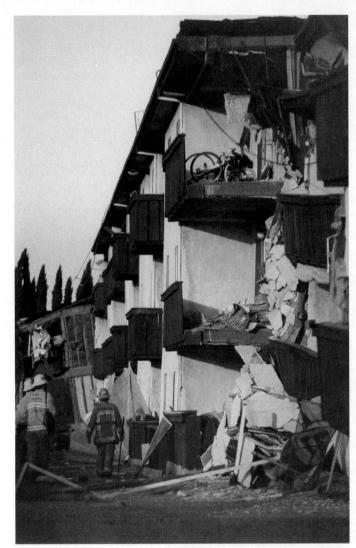

Figure 16.1

The collapse of the lower story of the Northridge Meadows Apartments killed 16 people in the 1994 earthquake. San Fernando Valley, southern California.

Photo © Joel Lugavere/Los Angeles Times

collapsed onto the lower story, killing 16 people (figure 16.1). A 3-story concrete parking garage at a shopping center pancaked down, trapping a worker for hours under 3 meters of concrete. Gas and water mains were broken, triggering several hard-to-fight fires; about 100 homes burned at a Sylmar mobile-home park. Fifteen thousand newly homeless people had to live in tents, and tens of thousands of people had no water, electricity, or gas for several days. The death toll of 61 was very low because the quake occurred early in the morning on the Martin Luther King holiday, so very few commuters were on the collapsed freeways. Damage exceeded $10 billion.

Causes of Earthquakes

What causes earthquakes? An **earthquake** is a trembling or shaking of the ground caused by the sudden release of energy stored in the rocks beneath the earth's surface. As described in rock breaks suddenly, causing an earthquake. Two masses of rock move past one another along a fault. The movement may be vertical, horizontal, or both (figure 16.3). The strain on the rock is released; the energy is expended by moving the rock into new positions and by creating seismic waves.

Recently some modifications have been suggested for the sequence of events shown in figure 16.2. The classic model implies that existing faults are strong; a very large stress must act to break rocks along a fault. The new idea is that faults are weak, and only need a small stress to cause rupture and an earthquake. The evidence for the new idea is suggestive, but not yet conclusive, so we currently have two models for fault behavior. The weak-fault model poses serious problems for earthquake prediction, as you will see later in the chapter.

The brittle behavior of breaking rock is characteristic only of rocks near the earth's surface. Rocks at depth are subject to increased temperature and pressure, which tend to reduce brittleness. Deep rocks deform plastically (*ductile* behavior) instead of breaking (*brittle* behavior); hence, there is a limit to the depth where faults can occur.

Most earthquakes are associated with movement on faults, but in some quakes the connection with faulting may be difficult to establish. Four recent California quakes, including the 1994 Northridge quake, occurred on buried thrust faults, some of which were unknown and none of which involved

Figure 16.2

The elastic rebound theory of the cause of earthquakes. (*A*) Rock with stress acting on it. (*B*) Stress has caused strain in the rock. Strain builds up over a long period of time. (*C*) Rock breaks suddenly, releasing energy, with rock movement along a fault. Horizontal motion is shown; rocks can also move vertically or diagonally. (*D*) Fence offset nearly 3 meters after 1906 San Francisco earhquake.

Photo by G. K. Gilbert, U.S. Geological Survey

chapter 15, great forces acting deep in the earth may put a *stress* on the rock, which may bend or change in shape (*strain*). If you bend a stick of wood, your hands put a stress (the force per unit area) on the stick; its bending (a change in shape) is the strain.

Like a bending stick, rock can deform only so far before it breaks. When a rock breaks, waves of energy are released and sent out through the earth. These are **seismic waves,** the waves of energy produced by an earthquake. It is the seismic waves that cause the ground to tremble and shake during an earthquake.

The sudden release of energy when rock breaks may cause one huge mass of rock to slide past another mass of rock into a different relative position. As you know from chapter 15 the break between the two rock masses is a *fault*. The classic explanation of why earthquakes take place is called the **elastic rebound theory** (figure 16.2). It involves the sudden release of progressively stored strain in rocks, causing movement along a fault. Deep-seated internal forces (*tectonic forces*) act on a mass of rock over many decades. Initially the rock bends, but does not break. More and more energy is stored in the rock as the bending becomes more severe. Eventually the energy stored in the rock exceeds the breaking strength of the rock, and the

Figure 16.3

Horizontal offset of rows in a lettuce field, 1979, El Centro, California.

Photo by Univ. of Colorado; courtesy National Geophysical Data Center, Boulder, Colorado

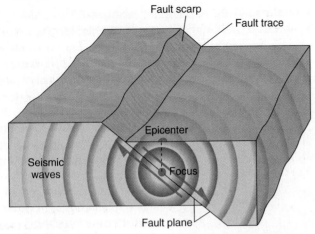

Figure 16.4

The focus of an earthquake is the point where rocks first break along a fault; seismic waves radiate from the focus. The epicenter is the point on the earth's surface directly above the focus.

surface displacement. Most earthquakes in the eastern United States are also not associated with surface displacement. Earthquakes also occur during explosive volcanic eruptions and as magma forcibly fills underground magma chambers prior to many eruptions; these quakes may not be associated with fault movement at all.

Another cause has been recently postulated for deep earthquakes (100 to 670 kilometers below the surface), essentially all of which are found on cold, subducting plates sliding down into the mantle. Although the downgoing plates are colder than the surrounding rock, the high temperature and pressure at depth suggests to some geologists that the rock in the plates should behave plastically rather than breaking in the brittle manner of near-surface rocks. The suggested cause of deep quakes is mineral transformations within the downgoing rock, as pressure collapses one mineral into a denser form. Lab experiments have shown these transformations occur suddenly, forming lens-shaped bodies of the new, denser minerals along fractures. Whether the process occurs on a large scale to produce large quakes is unknown. Similar suggestions for the cause of deep quakes include the dehydration of water-containing serpentine and the conversion of serpentine into glass. Both these processes occur suddenly on small fractures in lab experiments.

Seismic Waves

The point within the earth where seismic waves originate is called the **focus** (or *hypocenter*) of the earthquakes (figure 16.4). This is the center of the earthquake, the point of initial breakage and movement on a fault. Rupture begins at the focus and then spreads rapidly along the fault plane. The point on the earth's surface directly above the focus is the **epicenter.**

Two types of seismic waves are generated during earthquakes. **Body waves** are seismic waves that travel through the earth's interior, spreading outward from the focus in all direc-

tions, like sound waves moving through air. **Surface waves** are seismic waves that travel on the earth's surface away from the epicenter, like water waves spreading out from a pebble thrown into a pond. Rock movement associated with seismic surface waves dies out with depth into the earth, just as water movement in ocean waves dies out with depth.

Body Waves

There are two kinds of body waves, both shown in figure 16.5. A **P wave** is a compressional (or longitudinal) wave in which rock vibrates back and forth *parallel* to the direction of wave propagation. Because it is a very fast wave, traveling through near surface rocks at speeds of 4 to 7 kilometers per second (9,000 to more than 15,000 miles per hour), a P wave is the first (or *primary*) wave to arrive at a recording station following an earthquake.

The second kind of body wave is called an **S wave** (*secondary*) and is a slower, transverse wave that travels at 2 to 5 kilometers per second. An S wave is propagated by a shearing motion much like that in a stretched, shaken rope. The rock vibrates *perpendicular* to the direction of wave propagation, that is, crosswise to the direction the waves are moving.

Both P waves and S waves pass easily through solid rock. A P wave can also pass through a fluid (gas or liquid), but an S wave cannot. We discuss the importance of this fact in chapter 17.

Surface Waves

Surface waves are the slowest waves set off by earthquakes. In general, surface waves cause more property damage than body waves because surface waves produce more ground movement and travel more slowly, so they take longer to pass. The two most important kinds of surface waves are Love waves and Rayleigh waves, named after the geophysicists who discovered them.

Love waves are most like S waves that have no vertical displacement. The ground moves side to side in a horizontal plane that is perpendicular to the direction the wave is traveling or propagating (figure 16.5*C*). Like S waves, Love waves do not travel through liquids and would not be felt on a body of water. Because of the horizontal movement, Love waves tend to knock buildings off their foundations.

Rayleigh waves behave like rolling ocean waves. Rayleigh waves cause the ground to move in an elliptical path as the wave passes (figure 16.5*D*). Rayleigh waves tend to be incredibly destructive to buildings because they produce more ground movement and take longer to pass.

Locating and Measuring Earthquakes

The invention of instruments that could accurately record seismic waves was an important scientific advance. They measure the amount of ground motion and can be used to find the location, depth, and size of an earthquake.

A Primary wave

C Love wave

B Secondary wave

D Rayleigh wave

Figure 16.5

Particle motion in seismic waves. (*A*) A P wave is illustrated by a sudden push on the end of a stretched spring or Slinky. The particles vibrate *parallel* to the direction of wave propagation. (*B*) An S wave is illustrated by shaking a loop along a stretched rope. The particles vibrate *perpendicular* to the direction of wave propagation. (*C*) Love waves behave like S waves in that the particle motion is perpendicular to the direction of wave travel along the earth's surface. (*D*) Rayleigh waves are like ocean waves and cause a rolling motion on the earth's surface. The particle motion is elliptical in the direction of wave propagation.

The instrument used to detect seismic waves is a *seismometer*. The principle of the seismometer is to keep a heavy suspended mass as motionless as possible—suspending it by springs or hanging it as a pendulum from the frame of the instrument (figure 16.6). When the ground moves, the frame of the instrument moves with it. However, the inertia of the heavy mass suspended inside keeps the mass motionless to act as a point of reference in determining the amount of ground motion.

A seismometer by itself cannot record the motion that it measures. A **seismograph** is a recording device that produces a permanent record of earth motion detected by a seismometer, usually in the form of a wiggly line drawn on a moving strip of paper (figure 16.7). The paper record of earth vibration is called a **seismogram.**

A network of seismograph stations is maintained all over the world to record and study earthquakes (and nuclear bomb explosions). Within minutes after an earthquake

Figure 16.6

A simple seismograph for detecting vertical rock motion. The pen records the ground motion on the seismogram as the spring stretches and compresses with up and down movement of the spring. Frame and recording drum move with the ground. Inertia of the weight keeps it and the needle relatively motionless.

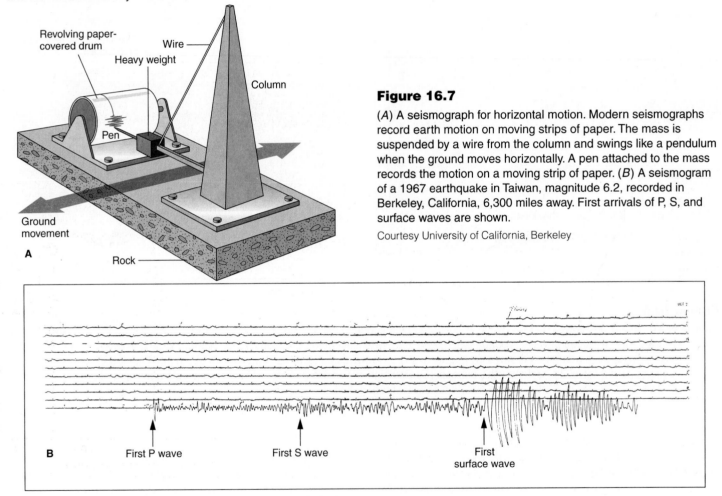

Figure 16.7

(*A*) A seismograph for horizontal motion. Modern seismographs record earth motion on moving strips of paper. The mass is suspended by a wire from the column and swings like a pendulum when the ground moves horizontally. A pen attached to the mass records the motion on a moving strip of paper. (*B*) A seismogram of a 1967 earthquake in Taiwan, magnitude 6.2, recorded in Berkeley, California, 6,300 miles away. First arrivals of P, S, and surface waves are shown.

Courtesy University of California, Berkeley

occurs, distant seismographs begin to pick up seismic waves. A large earthquake can be detected by seismographs all over the world.

Because the different types of seismic waves travel at different speeds, they arrive at seismograph stations in a definite order, first the P waves, then the S waves, and finally the surface waves. These three different waves can be distinguished on the seismograms. By analyzing these seismograms, geologists can learn a great deal about an earthquake, including its location and size.

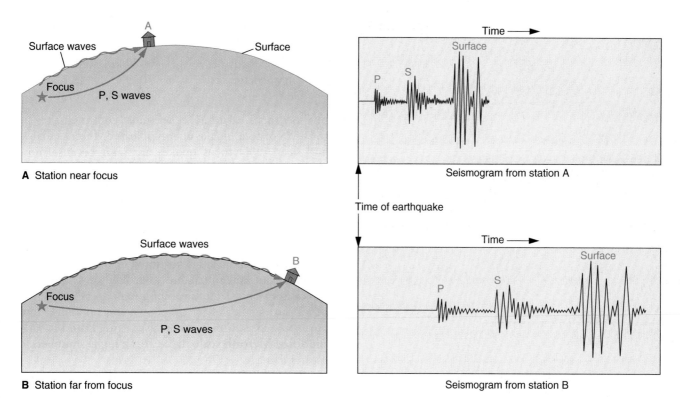

Figure 16.8

Intervals between P waves, S waves, and surface waves increase with distance from the focus.

Determining the Location of an Earthquake

P and S waves start out from the focus of an earthquake at essentially the same time. As they travel away from the quake, the two kinds of body waves gradually separate because they are traveling at different speeds. On a seismogram from a station close to the earthquake, the first arrival of the P wave is separated from the first arrival of the S wave by a short distance on the paper record (figure 16.8). At a recording station far from the earthquake, however, the first arrivals of these waves will be recorded much farther apart on the seismogram. The farther the seismic waves travel, the longer the time intervals between the arrivals of P and S waves and the more they are separated on the seismograms.

Because the time interval between the first arrivals of P and S waves increases with distance from the focus of an earthquake, this interval can be used to determine the distance from the seismograph station to a quake. The increase in the P-S interval is regular with increasing distance for several thousand kilometers and so can be graphed in a **travel-time curve,** which plots seismic-wave arrival time against distance (figure 16.9).

In practice, a station records the P and S waves from a quake, then a seismologist matches the interval between the waves to a standard travel-time curve. By reading directly from the graph, one can determine, for example, that an earthquake has occurred 5,500 kilometers (3,400 miles) away.

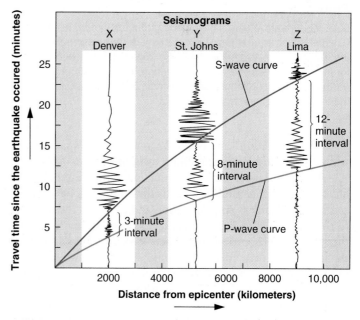

Figure 16.9

A travel-time curve is used to determine the distance to an earthquake. Note that the time interval between the first arrival of P and S waves increases with distance from the epicenter. Seismogram X has a 3-minute interval between P and S waves corresponding to a distance of 2,000 km from the epicenter, Y has an interval of 8 minutes, so the earthquake occurred 5,200 kilometers away, and Z an interval of 12 minutes, and is a distance of 9,000 km from the epicenter.

Figure 16.10

Locating an earthquake. The distance from each of three stations (Denver, Quebec, and Lima) is determined from seismograms and the travel-time curves shown in figure 16.9. Each distance is used for the radius of a circle about the station. The location of the earthquake is just offshore of Vancouver, British Columbia, where the three circles intersect.

This determination can often be made very rapidly, even while the ground is still trembling from the quake.

A single station can determine only the distance to a quake, not the direction. A circle is drawn on a globe with the center of the circle being the station and its radius the distance to the quake (figure 16.10). The scientists at the station know that the quake occurred somewhere on that circle, but from the information recorded they are not able to tell where. With information from other stations, however, they can pinpoint the location of the quake. If three or more stations have determined the distance to a single quake, a circle is drawn for each station. If this is done on a map, the intersection of the circles locates the quake.

Analyses of seismograms can also indicate at what depth beneath the surface the quake occurred. Most earthquakes occur relatively close to the earth's surface, although a few occur much deeper. The maximum **depth of focus**—the distance between focus and epicenter—for earthquakes is about

670 kilometers. Quakes are classified into three groups according to their depth of focus:

Shallow focus	0–70 km deep
Intermediate focus	70–350 km deep
Deep focus	350–670 km deep

Shallow focus earthquakes are most common; they account for 85% of total quake energy released. Intermediate (12%) and deep (3%) focus quakes are rarer because most deep rocks flow plastically when stressed or deformed; they are unable to store and suddenly release energy as brittle surface rocks do.

Measuring the Size of an Earthquake

The size of earthquakes is measured in two ways. One method is to find out how much and what kind of damage the quake has caused. This determines the **intensity,** which is a measure of an earthquake's effect on people and buildings. Intensities are expressed as Roman numerals ranging from I to XII on the **modified Mercalli scale** (table 16.1); higher numbers indicate greater damage.

Although intensities are widely reported at earthquake locations throughout the world, using intensity as a measure of earthquake strength has a number of drawbacks. Because damage generally lessens with distance from a quake's epicenter, different locations report different intensities for the same earthquake (figure 16.11). Moreover, damage to buildings and other structures depends greatly on the type of geologic material on which a structure was built as well as the type of construction. Houses built on solid rock normally are damaged far less than houses built upon loose sediment, such as delta mud or bay fill. Brick and stone houses usually suffer much greater damage than wooden houses, which are somewhat flexible. Damage estimates are also subjective: people may exaggerate damage reports consciously or unconsciously. Intensity maps can be drawn for a single earthquake to show the approximate damage over a wide region (figure 16.11). But such maps cannot be drawn for uninhabited areas (the open ocean, for instance), so not all quakes can be assigned intensities. The one big advantage of intensity ratings is that no instruments are required, which allows seismologists to estimate the size of earthquakes that occurred before seismographs were available.

The second method of measuring the size of a quake is to calculate the amount of energy released by the quake. This method is usually done by measuring the height (amplitude) of one of the wiggles on a seismogram. The larger the quake, the more the ground vibrates and the larger the wiggle. After measuring a specific wave on a seismogram, and correcting for the type of seismograph and for the distance from the quake, scientists can assign a number called the **magnitude.** It is a measure of the energy released during the earthquake.

For the past several decades magnitude has been reported on the **Richter scale,** a numerical scale of magnitudes. The Richter scale is open ended, meaning there are no earthquakes

Table 16.1	Modified Mercalli Intensity Scale of 1931 (Abridged)

I. Not felt except by a very few under especially favorable circumstances.

II. Felt only by a few persons at rest, especially on upper floors of buildings. Delicately suspended objects may swing.

III. Felt quite noticeably indoors, especially on upper floors of buildings, but many people do not recognize it as an earthquake. Standing motor cars may rock slightly. Vibration like passing of truck. Duration estimated.

IV. During the day felt indoors by many, outdoors by few. At night some awakened. Dishes, windows, doors disturbed; walls made cracking sound. Sensation like heavy truck striking building. Standing motor cars rocked noticeably.

V. Felt by nearly everyone; many awakened. Some dishes, windows, etc., broken; a few instances of cracked plaster; unstable objects overturned. Disturbance of trees, poles, and other tall objects sometimes noticed. Pendulum clocks may stop.

VI. Felt by all; many frightened and run outdoors. Some heavy furniture moved; a few instances of fallen plaster or damaged chimneys. Damage slight.

VII. Everybody runs outdoors. Damage *negligible* in buildings of good design and construction; *slight* to moderate in well-built ordinary structures; *considerable* in poorly built or badly designed structures; some chimneys broken. Noticed by persons driving motor cars.

VIII. Damage *slight* in specially designed structures; *considerable* in ordinary substantial buildings with partial collapse; *great* in poorly built structures. Panel walls thrown out of frame structures. Fall of chimneys, factory stacks, columns, monuments, walls. Heavy furniture overturned. Sand and mud ejected in small amounts. Changes in well water. Persons driving motor cars disturbed.

IX. Damage *considerable* in specially designed structures; well-designed frame structures thrown out of plumb; *great* in substantial buildings, with partial collapse. Buildings shifted off foundations. Ground cracked conspicuously. Underground pipes broken.

X. Some well-built wooden structures destroyed; most masonry and frame structures destroyed with foundations; ground badly cracked. Rails bent. Considerable landslides from river banks and steep slopes. Shifted sand and mud. Water splashed (slopped) over banks.

XI. Few, if any, (masonry) structures remain standing. Bridges destroyed. Broad fissures in ground. Underground pipe lines completely out of service. Earth slumps and land slips in soft ground. Rails bent greatly.

XII. Damage total. Waves seen on ground surface. Lines of sight and level distorted. Objects thrown upward into the air.

From Wood and Neumann, 1931, Bulletin Seismological Society of America.

Figure 16.11

Zones of different intensity from the 1886 Charleston, South Carolina, earthquake. The map illustrates the general decrease in intensity with increasing distance from the epicenter, and also the effect of different types of earth materials.

U.S. Geological Survey

too large or too small to fit on the scale. The higher numbers indicate larger earthquakes. Very small earthquakes can have negative magnitudes, but these are seldom reported. The largest Richter magnitude measured so far is 8.6. Smaller earthquakes are much more common than large ones:

Richter Magnitude		Number Per Year (World)
2	Just felt	More than 100,000
4.5	Damage usually newsworthy	A few thousand
7	"Major"	16 to 18
8	"Great"	1 or 2
8.6	Maximum recorded	

There are several methods of measuring magnitude, however. The original Richter scale applied only to shallow earthquakes in southern California. Different seismic waves (body or surface) can be measured to make the scale more useful over larger areas, so several different magnitudes are sometimes reported for a single quake. A further complication is that magnitudes calculated from seismograms tend to be inaccurate (usually too low) above magnitude 7.

A new method of calculating magnitude involves the use of the *seismic moment* of a quake, which is determined from the strength of the rock, surface area of the rupture, and the amount of rock displacement along the fault. A **moment magnitude** can be higher than 9. The 1964 Alaska quake is estimated to have a moment magnitude of 9.2, and the 1960 Chile quake 9.5. Unfortunately the media rarely indicate which type of magnitude they are reporting, and scientists typically revise magnitudes for several weeks after a quake as they receive more information, so trying to find out the "real" magnitude of a recent quake can be confusing. Table 16.2 lists Richter magnitudes and moment magnitudes for many earthquakes of interest. Richter magnitudes are the most popularly quoted.

Because the Richter scale is logarithmic, the difference between two consecutive whole numbers on the scale means an increase of 10 times in the amplitude of the earth's vibrations, particularly below magnitude 5. This means that if the measured amplitude of vibration for certain rocks is 1 inch during a magnitude-4 quake, these rocks will move 10 inches during a magnitude-5 quake occurring at the same location.

It has been estimated that a tenfold increase in the size of the earth vibrations is caused by an increase of roughly 32 times in terms of energy released. A quake of magnitude 5, for example, releases approximately 32 times more energy than one of magnitude 4. A magnitude-6 quake is about 1,000 times (32×32) more powerful in terms of energy released than a magnitude-4 quake.

Table 16.2	Earthquake Magnitudes	Richter Magnitude	Moment Magnitude
1811–12	New Madrid, Missouri area	7.5, 7.3, 7.8	7.7, 7.6, 7.9
1857	Fort Tejon, S. Calif.	7.6	7.9
1872	Lone Pine, Calif.	7.3	7.8
1886	Charleston, South Carolina	6.7	7.0
1906	San Francisco, N. Calif.	8.25	7.7
1915	Pleasant Valley, Nevada	7.7	7.1
1933	Long Beach, S. Calif.	6.3	6.2
1952	Kern County, S. Calif.	7.2	7.5
1954	Dixie Valley, Nevada	7.2, 7.1	7.3, 6.9
1957	Aleutian Islands, Alaska	8.1	8.8
1958	Southeastern Alaska	7.9	8.3
1959	Hebgen Lake, Montana	7.7	7.3
1960	Chile	8.5	9.5
1964	near Anchorage, Alaska	8.6	9.2
1965	Aleutian Islands, Alaska	8.2	8.7
1970	Peru	7.75	7.9
1971	San Fernando Valley, S. Calif.	6.4	6.6
1975	Hawaii	7.2	7.5
1976	China	7.6	7.5
1980	Humboldt County, N. Calif.	6.9	7.2
1983	Coalinga, Calif	6.7	6.2
1983	Challis, Idaho	7.2	7.0
1983	Adirondack Mountains, New York	5.1	4.9
1983	Hawaii	6.6	
1985	Ixtapa, Mexico		8.1, 7.5
1987	Lawrenceville, Illinois	5.0	5.0
1987	Whittier, S. Calif.	5.9	5.9
1988	Quebec	6.0	
1989	Loma Prieta, N. Calif.	7.0	7.2
1989	Hawaii	6.1	6.4
1992	Humboldt County, N. Calif.	7.1, 6.6, 6.7	
1992	Landers, S. Calif.	7.6, 6.7	7.25, 6.4
1994	Northridge, S. Calif.	6.4	6.7
1995	Kobe, Japan (7.2 on Japanese scale)	6.8	6.9

U.S. Geological Survey and other sources.

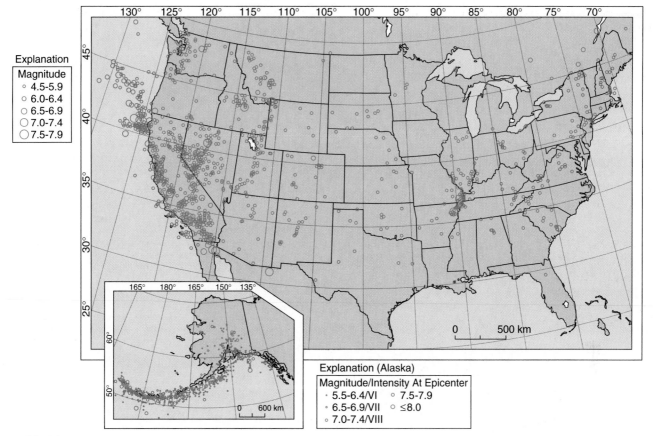

Figure 16.12

Locations of damaging earthquakes (magnitude 4.5 and greater) in the United States and southern Canada, 1568–1989 in the east, 1769–1989 in the west.

U.S. Geological Survey.

Although a seismograph is usually required to measure magnitude, this measure has many advantages over intensity as an indicator of earthquake strength. A worldwide network of standard seismograph stations now makes determining magnitude a routine matter; and the press reports magnitudes for all earthquakes of interest to the United States. Eventually, a single magnitude number can be assigned to a single earthquake, whereas intensity varies for a single earthquake, depending on the amount and kind of local damage. Magnitudes can be reported for all quakes, even those in distant uninhabited areas where there is no property to affect.

Location and Size of Earthquakes in the United States

Figure 16.12 shows the locations of all damaging earthquakes that have occurred in the United States through 1989 (small quakes, which are very common, have been omitted). Note that only a few localities are relatively free of earthquakes.

Most of the large earthquakes occur in the western states. Quakes in California, Nevada, Utah, Idaho, Montana, Washington, and other western states are related to known faults

and usually (but not always) involve surface rupture of the ground. Earthquakes in Alaska occur mainly below the Aleutian Islands where the Pacific plate is converging with and being subducted under the North American plate.

Earthquakes east of the Rocky Mountains are more rare, and are generally smaller and deeper than earthquakes in the western United States. They are not associated with surface rupture. The quakes may be occurring on the deeply buried, relatively inactive faults of old *divergent plate boundaries* and *failed rifts* (*aulacogens*), both of which are described in chapter 19.

Although large quakes are extremely rare in the central and eastern United States, when they do occur they can be very destructive and widely felt, because the earth's crust is older, cooler, and more brittle in the east than in the west and seismic waves travel more efficiently. The Saint Lawrence River Valley along the Canadian border has had several intensity IX and X earthquakes, most recently in 1944. Plymouth, Massachusetts, had an intensity IX quake in 1638, and a quake of intensity VIII occurred in 1775 near Cambridge, Massachusetts. In 1929 in Attica, New York, an earthquake of intensity IX knocked over 250 chimneys. A series of quakes (intensity XI)

Peak Horizontal Ground Acceleration

- ☐ 4 % of gravity
- ☐ 10% of gravity
- ▨ greater than 20% of gravity
- ▨ greater than 30% of gravity

Figure 16.13

Map of seismic risk for Canada and the United States. The risk is shown as the expected maximum horizontal acceleration of the ground during an earthquake, shown as a percent of gravity. Damage usually begins at 10% of gravity; accelerations of 250% of gravity have been measured in some locations. Acceleration shown has only a 1-in-10 chance of being exceeded in a 50-year period. Maximum acceleration (in California) is 80% of gravity.

U.S. Geological Survey and Canadian Earthquake Hazards Program.

that occurred near New Madrid, Missouri, in the winter of 1811–1812 were the most widely felt earthquakes to occur in North America in recorded history. The quakes knocked over chimneys as far away as Richmond, Virginia, and rang church bells in Boston, 1100 miles away.

The 1886 quake in Charleston, South Carolina, (intensity X) was felt throughout almost half the United States (figure 16.11) and killed sixty people; it was sharply felt in New York City. Moderate quakes hit Arkansas and New Hampshire in 1982, and in 1983 a quake of 5.1 magnitude rocked New York's Adirondack Mountains. A 5.0-magnitude quake near Lawrenceville, Illinois, was felt from Kansas to South Carolina to Ontario in 1987. In 1988 a 6.0-magnitude quake north of Quebec City was felt as far away as Indiana and Washington, D.C.

Geologists have mapped regions of seismic risk in the United States (figure 16.13) primarily on the assumption that large earthquakes will occur in the future in places where they have occurred in the past.

A

B

C

Figure 16.14

Earthquake damage to structures. (*A*) Fallen statue at Stanford University, San Francisco earthquake, 1906. (*B*) Cars crushed by a collapsed apartment building, Northridge earthquake, southern California, 1994. (*C*) Damage to overpasses of Interstate 5 near Castaic Junction, north of Northridge, 1994. Note stranded and fallen vehicles. (*D*) Collapse of the upper portion of Anaheim Stadium, Northridge quake, 1994.

Photo *A* by W. C. Mendenhall, U.S. Geological Survey; photo *B* © Rolando Otero/Los Angeles Times; photo *C* © Michael Edwards/Los Angeles Times; photo *D* © Rod Boren/Los Angeles Times

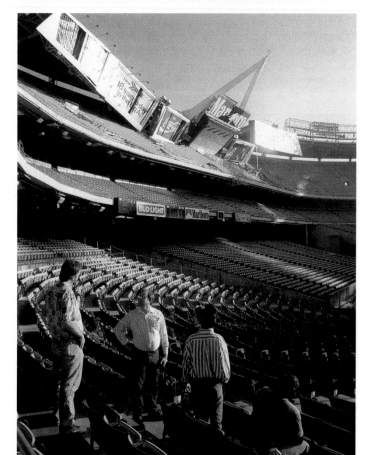

D

Effects of Earthquakes

Ground motion is the trembling and shaking of the land that can cause buildings to vibrate. During small quakes, windows and walls may crack from such vibration. In a very large quake the ground motion may be visible. It can be strong enough to topple large structures such as bridges and office buildings (figure 16.14). Most people injured or killed in an earthquake are hit by falling debris from buildings. Because proper building construction can greatly reduce the dangers, building codes need to be both strict and strictly enforced in earthquake-prone areas. As we have seen, the location of buildings also needs to be controlled; buildings built on soft sediment are damaged more than buildings on hard rock.

Figure 16.15

Almost 100 homes burned at a Sylmar mobile-home park following the Northridge earthquake, southern California, 1994.

Photo © Ken Lubas/Los Angeles Times

Fire is a particularly serious problem just after an earthquake because of broken gas and water mains and fallen electrical wires (figure 16.15). Although fire was the cause of most of the damage to San Francisco in 1906, changes in building construction and improved fire-fighting methods have reduced (but not eliminated) the fire danger to modern cities. The stubborn Marina district fires in San Francisco in 1989 attest to modern dangers of broken gas and water mains.

Landslides can be triggered by the shaking of the ground (figure 16.16). The 1959 Madison Canyon landslide in Montana was triggered by a nearby quake of magnitude 7.7. Landslides and subsidence caused extensive damage in downtown and suburban Anchorage during the 1964 Alaskan quake (magnitude 8.6). The 1970 Peruvian earthquake (magnitude 7.75) set off thousands of landslides in the steep Andes Mountains, burying more than 17,000 people (see Box 9.1). In 1920 in China over 100,000 people living in hollowed-out caves in cliffs of loess (described in chapter 13) were killed when a quake collapsed the cliffs.

A special type of ground failure caused by earthquakes is *liquefaction.* This occurs when a water-saturated soil or sediment turns from a solid to a liquid as a result of earthquake shaking. Liquefaction may occur several minutes after an earthquake, causing buildings to sink and underground tanks to float as once-solid sediment flows like water (figure 16.16*B*). Liquefaction was responsible for much of the damage in the 1989 Loma Prieta quake, and contributed to the damage in the 1906 San Francisco, the 1964 Alaska, and the 1995 Kobe, Japan, quakes.

Figure 16.16

(*A*) Landslide in Pacific Palisades triggered by the Northridge earthquake, 1994. (*B*) Liquefaction of soil by a 1964 quake in Niigata, Japan, caused earthquake-resistant apartment buildings to topple over intact.

Photo *A* © Al Seib/Los Angeles Times; photo *B* by National Geophysical Data Center

A

B

A

B

C

D

Figure 16.17

Varieties of ground displacement caused by earthquakes.
(*A*) Sixteen-foot scarp (cliff) formed by vertical ground motion,
Alaska, 1964. (*B*) Tearing of the ground near Olema, California,
1906. (*C*) Fence compressed by ground movement, Gallatin
County, Montana, 1959. (*D*) Compression of concrete freeway,
San Fernando Valley, California, 1971.

Photo *A* by U.S. Geological Survey; photo *B* by G. K. Gilbert, U.S.
Geological Survey; photo *C* by I. J. Witkind, U.S. Geological Survey

Permanent displacement of the land surface may be the result of movement along a fault. Rocks can move vertically, those on one side of a fault rising while those on the other side drop. Rocks can also move horizontally, those on one side of a fault sliding past those on the other side. Diagonal movement with both vertical and horizontal components can also occur during a single quake. Such movement can affect huge areas, although the displacement in a single earthquake seldom exceeds 8 meters (25 feet). The trace of a fault on the earth's surface may appear as a low cliff, called a *scarp,* or as a closed tear in the ground (figure 16.17). In rare instances small cracks open during a quake (but not to the extent that Hollywood

16.1
IN GREATER DEPTH

The Mexico City Earthquake of 1985

A

B

Box 16.1 Figure 1

Mexico City earthquake, 1985. (*A*) This 15-story building collapsed completely, crushing all its occupants as its reinforced-concrete floors "pancaked" together. (*B*) The upper floors of the left portion of this hotel collapsed. The lower floors and the entire right portion of the hotel remained intact. Note similar destruction in the left rear, and undamaged buildings nearby.

Photos by M. Celebi, U.S. Geological Survey

On September 19, 1985, central Mexico was rocked by an earthquake measuring 8.1 in magnitude. Thirty-six hours later, a magnitude-7.5 aftershock occurred nearby. The first quake was larger than the 1906 San Francisco quake. The second was fifteen times bigger than the 1994 Northridge earthquake north of Los Angeles. This one-two punch in Mexico caused some severe damage in Mexico City, 350 kilometers (200 miles) from the epicenter. At least 5,000 people died and 30,000 were injured in a population of 18 million. Damage totaled $5 billion.

The ground shaking was relatively mild in most parts of Mexico, but it was amplified in some buildings and by the lake-bed sediments beneath Mexico City. Most buildings are not rigidly stiff but slightly flexible; they sway gently like large pendulums when struck by wind or seismic waves. The time necessary for a single back-and-forth oscillation is called a period, and it varies with a building's height and mass. Natural bodies of rock and sediment vibrate the same way, with periods that vary with the body's size and density.

When earthquake waves struck Mexico City, many were vibrating with a two-second period. The body of lake sediment beneath the city has a natural period of two seconds also, so the wave motion was amplified by the sediments. This type of amplification, or resonance, occurs when you push a child on a swing—if you push gently in time with the swing's natural period, the swing goes higher and higher. The water-saturated sediment began to move like a sloshing waterbed. The ground moved back and forth by 40 centimeters (16 inches) every two seconds, and it did this fifteen to twenty times.

This shaking was devastating to buildings with a natural two-second period (generally those five to twenty stories high), and several hundred buildings collapsed as they further resonated with the shaking (box figure 1). Many structures weakened by the main shock collapsed during the large aftershock. Most shorter and taller buildings had other periods of vibration and rode out the shaking with little damage. Although 800 buildings collapsed or were seriously damaged, most of the city's 600,000 structures survived with little or no damage. The Mexico City quake vividly illustrates the fact that proper building design can greatly lessen earthquake damage.

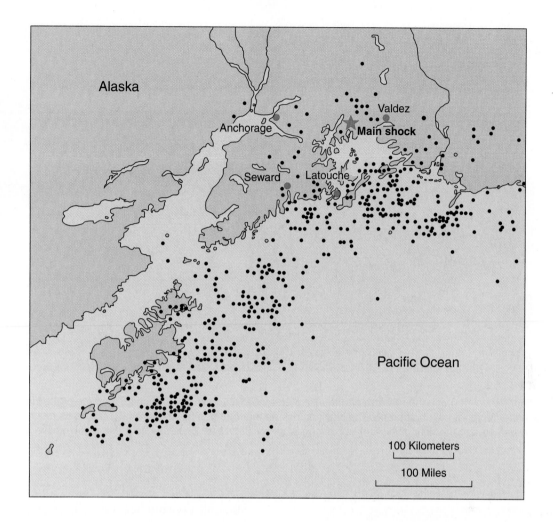

Figure 16.18

Aftershocks that followed the 8.6 (Richter magnitude) Alaska earthquake, March 27 to December 31, 1964. The aftershocks ranged from 4.4 to 6.7 magnitude (numerous smaller aftershocks are not mapped).

From U.S. Geological Survey.

films often portray). Ground displacement during quakes can tear apart buildings, roads, and pipelines that cross faults. Sudden subsidence of land near the sea can cause flooding and drownings.

Aftershocks are small earthquakes that follow the main shock (figure 16.18). In the months after the Alaskan earthquake of 1964 there were several thousand aftershocks, tapering off in frequency. Although aftershocks are smaller than the main quake, they can cause considerable damage, particularly to structures previously weakened by the powerful main shock. A long period of aftershocks can be extremely unsettling to people who have lived through the main shock. *Foreshocks* are

small quakes that precede a main shock. They are usually less common and less damaging than aftershocks, but can sometimes be used to help predict large quakes (although not all large quakes have foreshocks).

Tsunamis

The sudden movement of the sea floor upward or downward during a submarine earthquake can generate very large sea waves, popularly called "tidal waves." Because the ocean tides have nothing to do with generating these huge waves, the Japanese term **tsunami** is preferred by geologists. Tsunamis are

Figure 16.19

Uplift of the sea floor associated with the Alaskan earthquake, 1964, near the town of Latouche. The trace of the Hanning Bay fault runs diagonally through the center of the photograph. The block of rock on the right has been displaced upward as much as 5 meters (16 feet). The large, light-colored area to the right of the fault is bleached reef rock that was once sea floor but is now above sea level.

Photo by George Plafker, U.S. Geological Survey

A Before earthquake

B Sudden displacement of sea floor causes sea level to drop momentarily

C Water rushes into depression and overcorrects, raising sea level slightly

D Sea level oscillates before coming to rest; long, low waves (tsunamis) are sent out over sea surface

also called **seismic sea waves.** They usually are caused by great earthquakes (magnitude 8⁺) that disturb the sea floor, but they also result from submarine landslides or volcanic explosions. When a large section of sea floor suddenly rises or falls during a quake (figure 16.19), all the water over the moving area is lifted or dropped for an instant. As the water returns to sea level, it sets up long, low waves that spread very rapidly over the ocean (figure 16.20).

Tsunamis are unlike ordinary water waves on the sea surface. A large wind-generated wave may have a wavelength of 400 meters (1,300 feet) and be moving in deep water at a speed of 90 kilometers per hour (55 miles per hour). The wave height when it breaks on shore may be only 0.6 to 3 meters (2 to 10 feet), although in the middle of hurricanes the waves can be more than 15 meters (50 feet) high. A tsunami, however, may have a wavelength of 160 kilometers (100 *miles*), and may

Figure 16.20

Generation of tsunamis by a submarine earthquake (rock and water displacements are not drawn to scale).

A　　　　　　　　　　　　　　　　　　　　**B**

Figure 16.21

Tsunami damage in Alaskan earthquake, 1964. (*A*) Fishing boat carried inland in Resurrection Bay at Seward. (*B*) Spruce trees 0.7 m (2 feet) thick broken by a wave at an elevation of 30 m (100 feet) above sea level near Valdez.

Photo *A* by National Geophysical Data Center; photo *B* by U.S. Geological Survey

be moving at 725 kilometers per hour (450 miles per hour). In deep water the wave height may be only 0.6 to 2 meters (2 to 6 feet), but near shore the tsunami may peak up to heights of 15 to 30 meters (50 to 100 feet). This great increase in wave height near shore is caused by bottom topography; only a few localities have the combination of gently sloping offshore shelf and funnel-shaped bay that force tsunamis to awesome heights (the record height was 85 meters, or 278 feet in 1971 in the Ryukyu islands south of Japan). Along most coastlines tsunami height is very small.

Although the speed of the wave slows drastically as it moves through shallow water, a tsunami can still hit some shores as a very large, very fast wave. Because of its extremely long wavelength, a tsunami does not withdraw quickly as normal waves do. The water keeps on rising for five to ten minutes, causing great flooding before the wave withdraws. A tsunami's long duration and great height can bring widespread destruction to the entire shore zone (figure 16.21).

A tsunami formed by the 1960 Chilean quake crossed the Pacific Ocean and did extensive damage in Japan. One of the most destructive tsunamis of modern times was generated April 1, 1946, by a 7.3-magnitude earthquake offshore from Alaska. It devastated the city of Hilo, Hawaii, causing 159 deaths. The tsunami following the 1964 Alaskan quake drowned twelve people in Crescent City, California, a small coastal town near the Oregon border. Wave damage near an epicenter can be awesome. The 1946 Alaska tsunami destroyed the Scotch Cap lighthouse on nearby Unimak Island, sweeping

it off its concrete base, which was 10 meters (35 feet) above sea level, and killing its five occupants. The wave also swept away a radio tower, whose base was 31 meters (103 feet) above sea level.

World Distribution of Earthquakes

Most earthquakes are concentrated in narrow geographic belts (figure 16.22*A*), although *some* earthquakes have occurred in most regions on earth. The boundaries of plates in the plate-tectonic theory are defined by these earthquake belts (figure 16.22*B*). The most important concentration of earthquakes by far is in the **circum-Pacific belt,** which encircles the rim of the Pacific Ocean. Within this belt occur approximately 80% of the world's shallow-focus quakes, 90% of the intermediate-focus quakes, and nearly 100% of the deep-focus quakes (figure 16.23).

Another major concentration of earthquakes is in the **Mediterranean-Himalayan belt,** which runs through the Mediterranean Sea, crosses the Mideast and the Himalayas, and passes through the East Indies to meet the circum-Pacific belt north of Australia.

A number of shallow-focus earthquakes occur in two other significant locations on earth. One is along the summit or crest of the *mid-oceanic ridge,* a huge underwater mountain range that runs through all the world's oceans (see figure 1.11 and chapter 18). A few earthquakes have also

Figure 16.22

(*A*) World distribution of earthquakes recorded over a six-year period with focal depths between 0 and 700 kilometers. (*B*) The major plates of the world in the theory of plate tectonics. Compare the locations of plate boundaries with earthquake locations shown in figure 16.22*A*. Double lines show diverging plate boundaries, single lines show transform boundaries. Heavy lines with triangles show converging boundaries; triangles point down subduction zone.

A modified from Barazangi and Dorman, *Bulletin of Seismological Society of America,* 1969. *B* modified from W. Hamilton, U.S. Geological Survey.

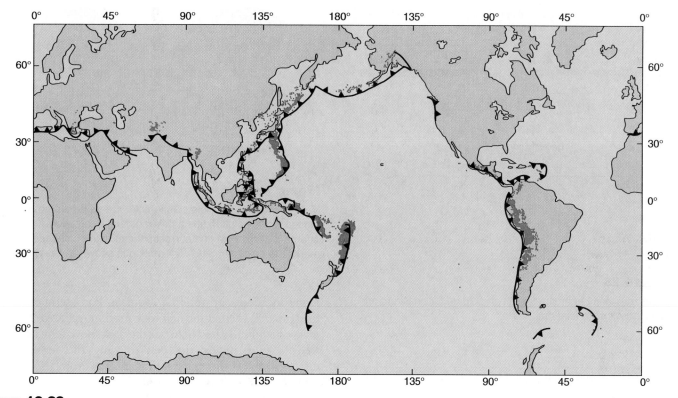

Figure 16.23

Selected earthquakes with depth of focus between 100 and 670 kilometers. These quakes define Benioff zones.

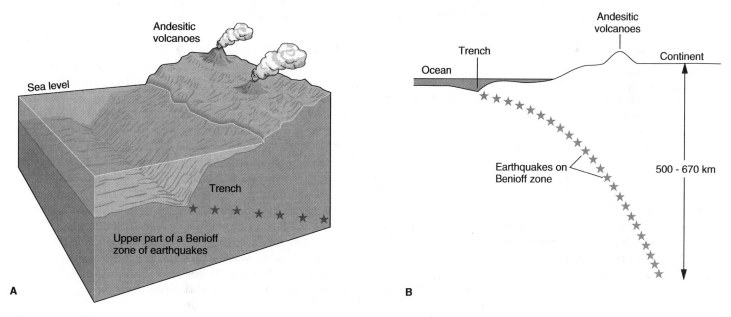

A

B

Figure 16.24

A Benioff zone of earthquakes begins at an oceanic trench and dips under a continent (such as South America). Andesitic volcanoes form near the edge of the continent. (*A*) Upper part of Benioff zone. (*B*) A Benioff zone may extend to a depth of 670 kilometers.

been recorded in isolated spots usually associated with basaltic volcanoes, such as those of Hawaii.

In most parts of the circum-Pacific belt, earthquakes, andesitic volcanoes, and *oceanic trenches* (see chapter 18) appear to be closely associated. Careful determination of the locations and depths of focus of earthquakes has revealed the existence of distinct *earthquake zones* that begin at oceanic trenches and slope landward and downward into the earth at an angle of about 30° to 60° (figures 16.23 and 16.24). Such zones are called **Benioff zones** after the man who first recognized them.

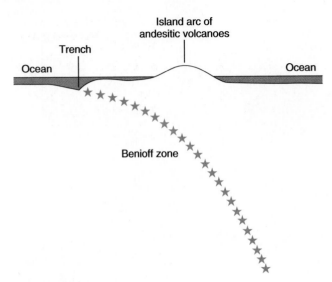

Figure 16.25

A Benioff zone can begin at a trench and dip under an island arc (such as the Aleutian Islands).

Figure 16.26

Seismograms showing how first horizontal motions of rocks along a fault are determined. (*A*) If the first motion is a push (from the epicenter to the seismograph station), the seismogram trace is deflected upward. (*B*) If it is a pull (away from the station), the deflection is downward.

Benioff zones slope under a continent or a curved line of islands called an **island arc** (figure 16.25). Andesitic volcanoes may form the islands of the island arc, or they may be found near the edge of a continent that overlies a Benioff zone.

Most of the circum-Pacific belt is made up of Benioff zones associated in this manner with oceanic trenches and andesitic volcanoes. Parts of the Mediterranean-Himalayan belt represent Benioff zones, too, notably in the eastern Mediterranean Sea and in the East Indies. Essentially all the world's intermediate- and deep-focus earthquakes occur in Benioff zones.

First-Motion Studies of Earthquakes

By studying seismograms of an earthquake on a distant fault, geologists can tell which way rocks moved along that fault. Rock motion is determined by examining seismograms from many locations surrounding a quake. Each seismograph station can tell whether the first rock motion recorded there was a push or a pull (figure 16.26). If the rock moved toward the station (a push), then the pen drawing the seismogram is deflected up. If the first motion is away from the station (a pull), then the pen is deflected down.

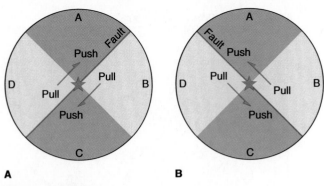

Figure 16.27

Map view of two possible solutions for the same pattern of first motion. Each solution has a different fault orientation. If the fault orientation is known, the correct solution can be chosen. The star marks the epicenter, and rock motion is shown by the arrows.

If an earthquake occurs on a fault as shown in figure 16.27*A*, large areas around the fault will receive a push as first motion, while different areas will receive a pull. Any station within the green area marked *A* will receive a push, for the rock is moving from the epicenter toward those stations, as shown by the arrows in the figure. All stations in area *C* will also receive a push, but areas *B* and *D* will record a pull as the first motion.

In figure 16.27*B* you can see the same pattern of pushes and pulls, but in this case the pattern is caused by a fault with a different orientation. Either fault can cause the same pattern, if the rock moves in the direction shown by the arrows. In other words, there are two possible solutions to any pattern of first motions.

If the orientation of the fault trace on the earth's surface is known, as it is for most faults on land (and for a few on the ocean floor), the correct choice of the two solutions can be made. But if the orientation of the fault is not known—as is the usual case for earthquakes at sea or at great depth—the choice between the two possible solutions can be difficult. One solution may be more *likely*, based on the study of topography, other faults, or similar factors. But such a solution is by no means *proven* if the fault orientation is not known. As you read about first-motion studies, keep in mind that there are always two solutions to any pattern.

Earthquakes and Plate Tectonics

One of the great attractions of the concept of plate tectonics is its ability to explain the distribution of earthquakes and the rock motion associated with them.

As described briefly in chapter 1, the concept of plate tectonics is that the earth's surface is divided into a few giant *plates*. Plates are rigid slabs of rock, thousands of kilometers wide and 70 to 125 (or more) kilometers thick, that move across the earth's surface. Because the plates include continents

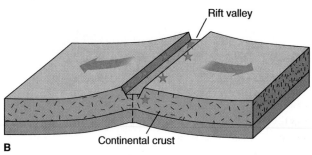

Figure 16.29

Divergent plate boundaries. (*A*) On the ocean floor. (*B*) On a continent. Each is marked by a rift valley and shallow-focus earthquakes (shown as stars). Depth of rift valleys is exaggerated.

← |‖ →   Rift valley on mid-oceanic ridge; a divergent boundary

Transform boundaries

Oceanic trench; a convergent boundary (triangles point down the subduction zone)

Figure 16.28

Relation of three types of plate boundaries to one another (shown in map view). Plate moves to right (as shown by the large arrows), away from rift valley at divergent boundary and toward oceanic trench at convergent boundary.

and sea floors on their upper surfaces, the plate-tectonics concept means that the continents and sea floors are moving. The plates change not only position but size and shape (as we discuss in more detail in chapter 19).

Earthquakes occur commonly at the edges of plates but only occasionally in the middle of a plate. The close correspondence between plate edges and earthquake belts can be seen by comparing the map of earthquake distribution in figure 16.22*A* with the plate map in figure 16.22*B*.

This correspondence is hardly surprising—plate boundaries are identified and *defined* by earthquakes. According to plate tectonics, earthquakes are caused by the interactions of plates along plate boundaries. Therefore, narrow bands of earthquakes are used to outline plates on plate maps. This can be clearly seen in the east Pacific Ocean off South America, where the Nazca plate (figure 16.22*B*) is almost completely outlined by earthquake epicenters (figures 16.22*A* and 16.23).

The earthquakes on the western border of the Nazca plate are shallow-focus quakes, and they occur in a narrow belt along the crest of the mid-oceanic ridge here, locally called the East Pacific Rise. The quakes along the eastern boundary occupy a broader belt that lies mostly within South America. This belt includes shallow-, intermediate-, and deep-focus earthquakes in a Benioff zone that begins at the Peru-Chile Trench just offshore and slopes steeply down under South America to the east

(figure 16.23). The Nazca plate moves eastward, away from the crest of the mid-oceanic ridge and toward the subduction zone at the trench, where the plate plunges down into the mantle. The plate's western boundary is located at the crest of the East Pacific Rise, and its eastern boundary is at the bottom of the Peru-Chile Trench.

Earthquakes at Plate Boundaries

As you have learned, there are three types of plate boundaries, *divergent boundaries* where plates move away from each other, *transform boundaries* where plates move horizontally past each other, and *convergent boundaries* where plates move toward each other. Each type of boundary has a characteristic pattern of earthquake distribution and rock motion.

One example of the relation of the three types of boundaries to each other is shown in figure 16.28. Note how the motion at each boundary (shown by the pairs of small arrows) agrees with the motion of the plate (shown by the large arrows).

Divergent Boundaries

At a divergent boundary, where plates move away from each other, earthquakes are shallow and restricted to a narrow band. A divergent boundary on the sea floor is marked by the crest of the mid-oceanic ridge and the *rift valley* that is often (but not always) found on the ridge crest (figure 16.29*A*). The earthquakes are located along the sides of the rift valley and beneath its floor. The rock motion that is deduced from first-motion studies shows that the faults here are normal faults, parallel to the rift valley. (This is the most likely solution of the first-

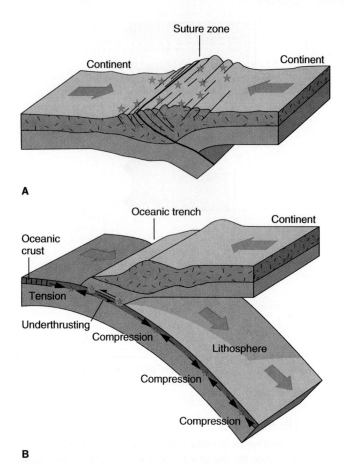

Figure 16.30

Transform boundaries. (*A*) Narrow band of shallow-focus earthquakes shown as stars along single fault. (*B*) Broad band of earthquakes along a system of parallel faults.

Figure 16.31

(*A*) A convergent boundary marked by the collision of two continents. A very broad zone of shallow-focus earthquakes occurs along a complex system of faults. (*B*) A convergent boundary with ocean floor subducting under a continent. Earthquakes occur near the top of the subducting plate due to tension, underthrusting, and compression.

motion studies. Keep in mind as you read this section that there are always two solutions for first motion; we give only one—the more likely solution.) If this interpretation of the first motions is correct, it implies that the ridge crest is undergoing horizontal extension, which is apparently tearing the sea floor open here, creating the rift valley and causing the earthquakes.

A divergent boundary within a *continent* is usually also marked by a rift valley, shallow-focus quakes, and normal faults (figure 16.29*B*). The African Rift Valleys in eastern Africa (figure 16.22*B*) seem to be such a boundary. Horizontal extension here may be tearing eastern Africa slowly apart, creating the rift valleys, some of which contain lakes.

Transform Boundaries

Where two plates move past each other along a transform boundary, the earthquakes are shallow. First-motion studies indicate strike-slip motion on faults parallel to the boundary. The earthquakes may be aligned in a narrow band along one fault (figure 16.30*A*), or they may form a broader zone if plate motion is taken up by movement along a system of parallel faults (figure 16.30*B*). The San Andreas fault in California (figure 16.22*B*) may be an example of a single fault forming a plate boundary, but many geologists believe that the broad zone of seismicity along the Basin and Range faults in the western United States (figure 16.12) shows that a system of faults both parallel and at an angle to the San Andreas forms the plate boundary here.

Convergent Boundaries

Convergent boundaries are of two general types, one marked by the *collision* of two continents (figure 16.31*A*), the other marked by *subduction* of the ocean floor under a continent (figure 16.31*B*) or another piece of sea floor. Each type has a characteristic pattern of earthquakes.

Collision boundaries are characterized by broad zones of shallow earthquakes on a complex system of faults (figure 16.31*A*). Some of the faults are parallel to the dip of the suture zone that marks the line of collision; some are not. One continent usually overrides the other slightly (continents are not dense enough to be subducted), creating thick crust and a mountain range. The Himalaya Mountains are thought to represent such a boundary (figure 16.22*B*). The seismic zone is so broad and complex at such boundaries that other criteria, such as detailed geologic maps, must be used to identify the position of the suture zone at the plate boundary.

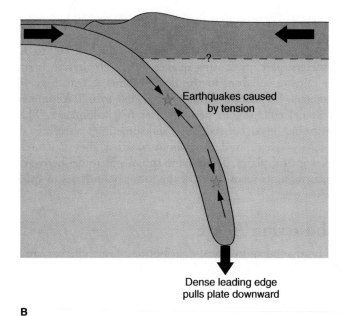

A　　　　　　　　　　　　　　　　　　　　**B**

670-km boundary acts as obstacle

Dense leading edge
pulls plate downward

Figure 16.32

(*A*) Many subducting plates have earthquakes caused by compression, perhaps from hitting an obstacle. (*B*) Other plates have tensional earthquakes, probably from the plate being pulled downward.

During *subduction,* earthquakes occur for several different reasons (figure 16.31*B*). As a dense oceanic plate bends to go down at a trench, it stretches slightly at the top of the bend, and normal faults occur as the rocks are subjected to *tension.* This gives a block-faulted character to the outer (seaward) wall of a trench. For some distance below the trench, the subducting plate is in contact with the overlying plate. First-motion studies of earthquakes at these shallow depths show that the quakes are caused by shallow-angle thrust-faulting. This is the motion expected as one plate slides beneath another, a process commonly called *underthrusting.*

At greater depths, where the descending plate is out of contact with the overlying plate, earthquakes are common, but the reasons for them are not obvious. The quakes are confined to a thin zone, only 20 to 30 kilometers thick, within the lithosphere of the descending plate, which is about 100 kilometers thick (figure 16.31*B*). This zone is thought to be near the top of the lithosphere, where the rock is colder and more brittle.

First motions of these quakes imply that along many subduction zones the quakes are caused by *compression* parallel to the dip of the plate (figure 16.31*B*). This could mean that the plates are running into some obstacle at depth (figure 16.32*A*).

Earthquake first motion on other subduction zones, however, implies *tension* parallel to the dip of the descending plates (figure 16.32*B*). This might mean that the plates are being pulled along by their dense leading edges. Cold, descending lithosphere is slightly denser than the surrounding hot mantle. The density of a subducting plate may also be increased by the loss of volatile materials or the pressure collapse of low-density minerals into denser forms.

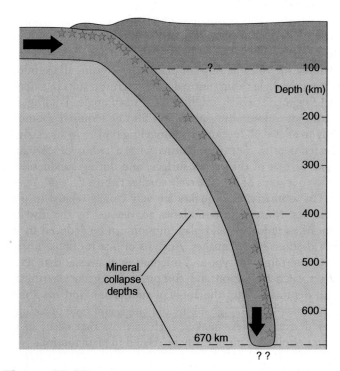

Figure 16.33

The distribution of earthquakes with depth.

The distribution of earthquakes with depth on a subducting plate suggests that mineral transformations may cause or control quakes (figure 16.33). Quakes are common near the surface and decrease exponentially to a depth of about 300 kilometers. There are few quakes from 300 to 450 kilometers, but many quakes from 450 to 670 kilometers. Serpentine on a subducting plate should lose its water at depths of 100 to 300 kilometers, perhaps causing quakes if the dehydration is sudden. Olivine collapses to the denser spinel structure at 400 kilometers. Serpentine forms glass at 450 to 670 kilometers, and spinel collapses to perovskite at 670 kilometers. These transformations occur suddenly, and sometimes noisily, in the laboratory. There is debate about whether they can create quakes in real plates. The olivine-spinel and spinel-perovskite transformations seem to mark the lower boundaries of quake depth zones.

Subduction Angle

The horizontal and vertical distribution of earthquakes can be used to determine the angle of subduction of a downgoing plate. Subduction angles vary considerably from trench to trench. Many plates start subducting at a gentle angle, which becomes much steeper with depth (figure 16.33). At a few trenches in the open Pacific, subduction begins (and continues) at almost a vertical angle. Subduction angle is probably controlled by plate density and by the rate of plate convergence.

Some plates crumple into folds as they descend, usually at depths below 200 kilometers. Other plates tear into segments during subduction (figure 16.34), perhaps as a result of density variations within a descending plate. The thick oceanic crust of a submarine volcanic ridge or plateau is buoyant, and subducts at a gentle angle. Thin, normal oceanic crust results in a denser plate segment, which subducts at a steep angle. Folding and tearing of a descending plate may also be required geometrically to allow subduction of a curved rigid plate to successively greater depths. The plate forms with a radius of curvature equal to that of the earth's surface, and during subduction it sinks to regions of progressively smaller radius.

In summary, earthquakes are very closely related to plate tectonics. Most plate boundaries are defined by the distribution of earthquakes, and plate motion can be deduced by the first motions of the quakes. Analysis of first motions can also help determine the type and orientation of stresses that act on plates, such as tension and compression. Quake distribution with depth indicates the angle of subduction and has shown that some plates change subduction angle and even break up as they descend. A few quakes, such as those that occur in the center of plates, cannot easily be related to plate motion.

Earthquake Prediction

People who live in earthquake-prone regions are plagued by unscientific predictions of impending earthquakes by popular writers and self-proclaimed prophets. Several techniques are being explored for *scientifically* forecasting a coming earthquake. One group of methods involves monitoring slight

Figure 16.34

A descending plate may tear into segments, some with steeper subduction angles than others.

changes that occur in rock next to a fault before the rock breaks and moves; these methods assume large amounts of strain are stored in rock before it breaks (figure 16.2).

Just as a bent stick may crackle and pop before it breaks with a loud snap, a rock may give warning signals that it is about to break. Before a large quake, small cracks may open within the rock, causing small tremors, or *microseisms,* to increase. The *properties of the rock* next to the fault may be changed by the opening of such cracks. Changes in the rock's magnetism, electrical resistivity, or seismic velocity may give some warning of an impending quake.

The opening of tiny cracks changes the rock's porosity, so *water levels in wells* often rise or fall before quakes. The cracks provide pathways for the release of radioactive radon gas from rocks (radon is a product of radioactive decay of uranium and other elements). An *increase in radon emission from wells* may be a prelude to an earthquake. A very local method of predicting quakes is to time the *interval between eruptions of Old Faithful geyser* in Yellowstone National Park. Long-term records of the time between eruptions have shown that this interval changes in a regular way before a large local earthquake, probably because of porosity changes within the surrounding rock.

In some areas the *surface of the earth tilts and changes elevation* slightly before an earthquake. Scientists use highly sensitive instruments to measure this increasing strain, in hopes of predicting quakes.

Chinese scientists claim successful predictions by watching *animal behavior*—horses become skittish and snakes leave their holes shortly before a quake. United States scientists are conducting a few pilot programs along these lines, but many are skeptical of the Chinese claims.

Japanese and Russian geologists were the first to predict earthquakes successfully, and Chinese geologists have made some very accurate predictions. In 1975 a 7.3-magnitude earthquake near Haicheng in northeastern China was pre-

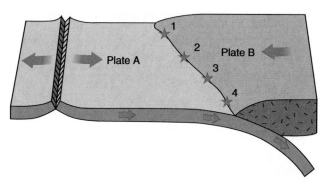

Figure 16.35

As plate *A* slides diagonally under plate *B*, earthquakes occur in a regular sequence from one to four. The entire sequence may repeat itself.

dicted five hours before it happened. Alerted by a series of *foreshocks,* authorities evacuated about a million people from their homes; many watched outdoor movies in the open town square. Half the buildings in Haicheng were destroyed, along with many entire villages, but only a few hundred lives were lost. In grim contrast, however, the Chinese program failed to predict the 1976 Tangshan earthquake (magnitude 7.6), which struck with no warning and killed an estimated 250,000 people.

But all of the above methods assume that great strain is being stored in rock before it breaks, and that that strain (or its effects) can be measured. As described in an earlier section, the new hypothesis that faults are weak (and break under small stress) contradicts this assumption. Most of these methods were considered very promising in the 1970s, but have proved to be of little real help in predicting quakes. A typical quake predictor, such as tilt of the land surface, may precede one quake, and then be absent for the next 10 quakes. In addition, each precursor can be caused by forces unrelated to earthquakes (land tilt is also caused by mountain building, mass wasting, and wetting and drying of the land). As a result, enthusiasm and funding for quake prediction are currently declining.

A fundamentally different method of quake risk analysis has been developed that does not depend upon whether faults are weak or strong. It involves determining the history of earthquakes along a fault, and using the history to calculate the probability of a quake at a given location. *Patterns of earthquakes in space and time* are being recognized, particularly on the faults along some plate boundaries. If one plate is sliding diagonally under another plate, a series of earthquakes may occur in a regular way along the plate boundary (figure 16.35). If such a pattern is known to repeat itself in the same area over several centuries, it becomes possible to forecast where the next large quake might occur. Along some long-active faults, there are short inactive segments called *seismic gaps* where earth-

quakes have not occurred for a long time. These gaps are widely thought to be the most likely sites for future earthquakes, so many of them are being carefully studied (see Box 16.2).

By studying the seismic history of faults, geologists in the United States are sometimes able to forecast earthquakes along some segments of some faults. In 1988 the U.S. Geological Survey estimated a 50 percent chance of a magnitude-7 quake along the segment of the San Andreas fault near Santa Cruz. In 1989 the magnitude-7 Loma Prieta quake occurred on this very section. Since the techniques are new and in some cases only partly understood, some errors will undoubtedly be made. Many faults are not monitored or studied historically because of lack of money and personnel, so we will never have a warning of impending quakes in some regions. For large urban areas near active faults such as the San Andreas, however, earthquake risk analysis may lessen the danger of an earthquake's occurring without warning.

Earthquake Control

Controlling the timing of earthquakes is an intriguing idea for the distant future. The potential for death and destruction in some cities near faults is so great that the planned triggering of earthquakes might be desirable. If people knew the exact time at which an earthquake was to occur, they could try to protect their belongings and temporarily evacuate the area, much as people do when a hurricane warning is given.

One method of earthquake control was discovered by accident when water under high pressure was pumped down into the ground. The water triggered earthquakes, released strain that had been accumulating in the rocks. The pressure of the water may reduce the stress holding the rock walls of a fault together. At the same time, the water can act as a lubricant, allowing the rocks to move, thus causing earthquakes. Such earthquakes have occurred around a disposal well near Denver and in an oil field in Colorado where water was forced underground. Without water, rock strain would build up until a large quake occurred. With the water reducing the friction, strain can be released in smaller, timed quakes. (Denver quakes reached Richter scale magnitude 4.3.)

Could such a method of quake release be used on the dangerous portions of the San Andreas fault? Perhaps, but only after years of testing, and then only as a trial in a relatively uninhabited section of the fault. The cost of such a program might prove to be prohibitive, and the liability problems would be enormous. Still, it is exciting to consider that in the future the danger of quakes along some faults might be greatly lessened by controlled release of strain.

It is important to realize that such experiments release strain that is already there. The advantage is in timing the release of the strain. The water does not *cause* the quake; it allows the timing of a quake that would occur sometime anyway to be *controlled.*

16.2

ENVIRONMENTAL GEOLOGY

Waiting for the Big One

The San Andreas fault, running north–south for 1,300 kilometers (800 miles) through California, is a right-lateral fault capable of generating great earthquakes of magnitude 8 or more. The 1906 earthquake (estimated Richter magnitude 8.25) near San Francisco caused a 450-kilometer (280-mile) scar in northern California (box figure 1). The portion of the fault nearest Los Angeles last broke in 1857 in a quake that was probably of comparable size. The ground has not broken in either of these regions since these quakes. Each old break is now a seismic gap, where rock strain is being stored prior to the next giant quake.

The recent California quakes were considerably smaller than the "Big One" long predicted by geologists to be in the magnitude-8 range. The 1906 quake in the north had an estimated Richter magnitude of 8.25, and a moment magnitude of 7.7. The Fort Tejon quake in the south was 7.6 Richter, 7.9 moment (see table 16.2). In contrast, the 1989 Loma Prieta quake on the San Andreas fault near San Francisco was 7.0 Richter, 7.2 moment. The 1994 Northridge quake (not on the San Andreas fault) was 6.4 Richter, 6.7 moment. So recent California quakes have been about magnitude 7 or less, and the Big One should be 8. A magnitude-8 quake has 10 times the ground shaking and 32 times the energy of a magnitude-7 quake. In other words it would take about 32 Loma Prieta quakes to equal the Big One. Comparing moment magnitudes for 1994 and 1857 in southern California, it would take nearly 64 Northridge quakes to equal the Fort Tejon quake.

A great earthquake of magnitude 8 could strike either the northern section or the southern section of the San Andreas fault tomorrow. Which section will break first? Because the southern section has been inactive longer, it may be the likelier candidate. A magnitude-8 quake here could cause hundreds of billions of dollars in damage and kill thousands of people if it struck during weekday business hours when Los Angeles area buildings and streets are crowded with people.

Spotty evidence suggests that great earthquakes have a recurrence interval of about 140 years on the southern portion of the San Andreas fault. Historic records in California do not go very far back in time, and much of the evidence involves radiometric dating of broken beds of carbon-rich lake muds. This number is only an average; 2 or 3 quakes spaced about 60 years apart occur in clusters, with the interval between clusters being 200 to 300 years. Prior to 1857 this portion last broke in about 1745. Adding this 112-year

Box 16.2 Figure 1

The two major breaks on the San Andreas fault in California. Each break occurred during a giant earthquake. Each old break is now a seismic gap (shown in red) where the fault is locked and may be the future site for another giant earthquake.

From U.S. Geological Survey.

difference to 1857 gives 1969, and adding the average of 140 years to 1857 gives 1997, so the present danger to Los Angeles is clear. If the 1857 quake ended a cluster, however, the next big quake may not occur till about 2050 or later.

The southern section of the fault is heavily instrumented to detect changes in ground tilt, water-well levels, and other earthquake precursors. In the mid-1970s most geologists thought that the Big One was imminent when it was discovered that most of southern California had risen as much as 0.4 meters (16 inches) from 1959 to 1974, with the greatest uplift occurring very close to the fault (the "Palmdale bulge"). Rapid subsidence from 1974 to 1976, however, eliminated at least half of this uplift, and the episode of rising and falling land passed without a large quake.

But the northern portion of the San Andreas fault is dangerous, too. Prior to 1906 this section of the fault broke in another giant quake in 1838. These quakes were only 68 years apart, and 1906 plus 68 equals 1974, so the northern section may actually be overdue for a big quake.

Another way of estimating the recurrence interval is by rock displacement. In 1906 rocks were displaced about 4.5 meters (15 feet) at the epicenter, and we know that plate motion across the San Andreas fault is about 5 centimeters (2 inches) per year. It should therefore take about 90 years to store enough strain to move rocks 5 meters, so the next quake should have occurred in 1996.

The U.S. Geological Survey recently estimated that the southern portion of the fault has about a 60 percent chance of an earthquake of magnitude 7.5 to 8.3 within the next 30 years. The 1992 Landers and the 1994 Northridge quakes have occurred since that prediction, and there is disagreement among geologists as to whether those quakes increase or decrease the likelihood of the Big One in southern California. Even more likely (86% chance) is a 7+-magnitude quake on any one of several faults that parallel the San Andreas fault and lie closer to (or even under) Los Angeles.

The probability of a repeat of the 1906 quake (8+) on the northern section of the San Andreas fault may be very low, perhaps only 10% for the next 30 years. Following the 1989 Loma Prieta quake, however, the chance of a 7-to-7.5 quake in the San Francisco Bay area in the next 30 years has been raised to 67%. A likely candidate for the quake is not the San Andreas but the Hayward fault across the Bay from San Francisco. Such a quake near or under Bay area cities such as Oakland and Berkeley would cause far greater death and destruction than the 1989 quake.

Although the chance of an 8+-magnitude Big One may be higher in southern California, the chance of a 7+-quake killing thousands of people is about equal in the north and the south.

Further Reading

U.S. Geological Survey, 1990. *The San Andreas fault system.* Professional Paper 1515.

Related Web Resources

For more information about the San Andreas Fault and the likelihood of it creating a large earthquake visit U.S. Geological Survey websites:

http://pubs.usgs.gov/gip/earthq3/safaultgip.html and
http://quake.wr.usgs.gov/hazprep/BayAreaInsert/

Summary

Earthquakes usually occur when rocks break and move along a fault to release strain that has gradually built up in the rock. Volcanic activity can also cause earthquakes. Deep quakes may be caused by mineral transformations.

Seismic waves move out from the earthquake's *focus. Body waves* (P waves and S waves) move through the earth's interior, and *surface waves (Love and Rayleigh waves)* move on the earth's surface.

Seismographs record seismic waves on *seismograms,* which can be used to determine an earthquake's strength, location, and depth of focus. Most earthquakes are shallow-focus quakes, but some occur as deep as 670 kilometers below the earth's surface.

The time interval between first arrivals of P and S waves is used to determine the distance between the seismograph and the *epicenter.* Three or more stations are needed to determine the location of earthquakes.

Earthquake *intensity* is determined by assessing damage and is measured on the *modified Mercalli scale.*

Earthquake *magnitude,* determined by the amplitude of seismic waves on a seismogram, is measured on the *Richter scale. Moment magnitudes,* determined by field work, are widely used today and often are larger than Richter magnitudes.

The most noticeable effects of earthquakes are ground motion and displacement (which destroy buildings and thereby injure or kill people), fire, landslides, and *tsunamis. Aftershocks* can continue to cause damage months after the main shock.

Earthquakes are generally distributed in belts. The *circum-Pacific belt* contains most of the world's earthquakes. Earthquakes also occur on the Mediterranean-Himalayan belt, the crest of the mid-oceanic ridge, and in association with basaltic volcanoes.

Benioff zones of shallow-, intermediate-, and deep-focus earthquakes are associated with andesitic volcanoes, oceanic trenches, and the edges of continents or island arcs.

The concept of plate tectonics explains most earthquakes as being caused by interactions between two plates at their boundaries. Plate boundaries are generally defined by bands of earthquakes.

Divergent plate boundaries are marked by a narrow zone of shallow earthquakes along normal faults, usually in a rift valley. Transform boundaries are marked by shallow quakes caused by strike-slip motion along one or more faults.

Convergent boundaries where continents collide are marked by a very broad zone of shallow quakes. Convergent boundaries involving deep subduction are marked by Benioff zones of quakes caused by tension, underthrusting, and compression.

The distribution of quakes indicates subduction angles and shows that some plates break up as they descend.

Although not yet too successful, earthquake prediction has been attempted through the measurement of rock properties near faults and the recognition of patterns of quakes at plate boundaries, including seismic gaps.

The timing of the release of accumulating rock strain may be controlled in the future by pumping water underground to trigger limited earthquakes.

Terms to Remember

aftershock 399

Benioff zone 403

body wave 386

circum-Pacific belt 401

depth of focus 390

earthquake 384

elastic rebound theory 385

epicenter 386

focus 386

intensity 390

island arc 404

Love wave 386

magnitude 390

Mediterranean-Himalayan belt 401

modified Mercalli scale 390

moment magnitude 392

P wave 386

Rayleigh wave 386

Richter scale 390

seismic sea wave 400

seismic wave 385

seismogram 387

seismograph 387

surface wave 386

S wave 386

travel-time curve 389

tsunami (seismic sea wave) 399

Testing Your Knowledge

Use the questions below to prepare for exams based on this chapter.

1. Describe in detail how earthquake epicenters are located by seismograph stations.

2. What causes earthquakes?

3. Compare and contrast the concepts of intensity and magnitude of earthquakes.

4. Name and describe the various types of seismic waves.

5. Discuss the distribution of earthquakes with regard to location and depth of focus.

6. Show with a sketch how the concept of plate tectonics can explain the distribution of earthquakes in a Benioff zone and on the crest of the mid-oceanic ridge.

7. Describe several techniques that may help scientists predict earthquakes.

8. How may the timing of earthquakes someday be controlled?

9. Describe several ways that earthquakes cause damage.

10. How do earthquakes cause tsunamis?

11. What are aftershocks?

12. The elastic rebound theory (a) explains folding of rocks (b) explains the behavior of seismic waves (c) involves the sudden release of progressively stored strain in rocks, causing movement along a fault (d) none of the above

13. The point within the earth where seismic waves originate is called the (a) focus (b) epicenter (c) fault scarp (d) fold

14. P waves are (a) compressional (b) transverse (c) tensional

15. What is the minimum number of seismic stations needed to determine the location of the epicenter of an earthquake? (a) 1 (b) 2 (c) 3 (d) 5 (e) 10

16. The Richter scale measures (a) intensity (b) magnitude (c) damage and destruction caused by the earthquake (d) the number of people killed by the earthquake

17. Benioff zones are found near (a) mid-ocean ridges (b) ancient mountain chains (c) interiors of continents (d) oceanic trenches

18. Most earthquakes at divergent plate boundaries are (a) shallow focus (b) intermediate focus (c) deep focus (d) all of the above

19. Most earthquakes at convergent plate boundaries are (a) shallow focus (b) intermediate focus (c) deep focus (d) all of the above

20. A zone of shallow earthquakes along normal faults is typical of (a) diverging boundaries (b) transform boundaries (c) subduction zones (d) collision boundaries

21. A seismic gap is (a) the time between large earthquakes (b) a segment of an active fault where earthquakes have not occurred for a long time (c) the center of a plate where earthquakes rarely happen

22. Which of the following is not true of tsunamis? (a) very long wavelength (b) high waveheight in deep water (c) very fast moving (d) continued flooding after wavecrest hits shore

Expanding Your Knowledge

1. What are some arguments in favor of and against predicting earthquakes? What would happen in your community if a prediction were made today that within a month a large earthquake would occur nearby?

2. Most earthquakes occur at plate boundaries where plates interact with each other. How might earthquakes be caused in the interior of a rigid plate?

3. How can you prepare for an earthquake in your own home?

4. Suppose you want to check for earthquake danger before buying a new home. How can you check the regional geology for earthquake dangers? The actual building site? The home itself?

Exploring Resources

Bolt, B. A. 1996. *Earthquakes.* New York: W. H. Freeman.

Canby, T. Y. 1990. California earthquake—prelude to the big one? *National Geographic* 177, no. 5:76–105.

Davidson, K. 1994. Predicting earthquakes: Can it be done? *Earth* 3, no. 3:56–63.

Frohlich, C. 1989. Deep earthquakes. *Scientific American* 260(1): 48–55.

Gere, J. M., and H. C. Shah. 1984. *Terra non firma—Understanding and preparing for earthquakes.* New York: W. H. Freeman.

Johnston, A. C., and L. R. Kanter. 1990. Earthquakes in stable continental crust. *Scientific American* 262(3): 68–75.

Keller, E. A., and N. Pinter. 1996. *Active tectonics: Earthquakes, uplift, and landscape.* Upper Saddle River, N.J.: Prentice Hall.

Kovach, R. L. 1995. *Earth's fury—An introduction to natural hazards and disasters.* Englewood Cliffs, N.J.: Prentice Hall.

Lomnitz, C. 1994. *Fundamentals of earthquake prediction.* New York: John Wiley & Sons.

Nuhfer, E. B., R. J. Proctor, and P. H. Moser. 1993. *The citizens' guide to geologic hazards.* Arvada, Colorado: American Institute of Professional Geologists.

Pakiser, L. C. 1991. *Earthquakes.* Washington, D.C.: U.S. Government Printing Office.

Penick, J. L., Jr. 1981. *The New Madrid earthquakes.* 2d ed. Columbia, Mo.: University of Missouri Press.

Stein, R. S., and R. S. Yeats. 1989. Hidden earthquakes. *Scientific American* 260(6): 48–57.

U.S. Geological Survey. *Earthquake information bulletin.*

U.S. Geological Survey, 1965–70. *The Alaska earthquake.* Professional Papers 541–546.

U.S. Geological Survey, 1971. *The San Fernando, California, earthquake of February 9, 1971.* Professional Paper 733.

U.S. Geological Survey, 1990. *The next big earthquake.* (This magazine-like pamphlet is available free from Earthquakes, U.S.G.S., 345 Middlefield Road, Menlo Park, CA 94025 or on the Internet at: http://quake.wr.usgs.gov/hazprep/BayAreaInsert/)

U.S. Geological Survey, 1990. *The San Andreas fault system.* Professional Paper 1515.

U.S. Geological Survey, 1993. *Seismicity of the United States, 1568–1989 (revised).* Professional Paper 1527.

Wesson, R. L., and R. E. Wallace. 1985. Predicting the next earthquake in California. *Scientific American* 252(2): 35–43.

http://www.geophys.washington.edu/seismosurfing.html
Exhaustive list of worldwide Internet sites for information about earthquakes.

http://quake.wr.usgs.gov/
U.S. Geological Survey Earthquake Information. Gives information on reducing earthquake hazards, earthquake preparedness, latest quake information, historical earthquakes, and how earthquakes are studied. Also a good starting place for links to other earthquake sites.

http://quake.wr.usgs.gov/recenteqs/faq.html
Frequently Asked Questions about recent earthquakes maintained by the U.S. Geological Survey.

http://www.seismo.unr.edu/
University of Nevada, Reno Seismological Laboratory site contains information about recent earthquakes, earthquake preparedness, and links to other earthquake sites.

http://www.seismo.berkeley.edu/seismo/Homepage.html
Seismographic information page maintained by U. C.–Berkeley that has many links to other earthquake sites (particularly in California), 3-D earthquake movie, Northridge earthquake rupture movies, and information on earthquake preparedness.

http://vquake.calstatela.edu
California State University, Los Angeles Virtual Earthquake. Create and analyze an earthquake.

http://pubs.usgs.gov/gip/earthq4/severitygip.html
General information about the size of an earthquake. Discussion of Richter and Mercalli scales.

http://pubs.usgs.gov/publications/text/dynamic.html
General information about plate tectonics.

Earthquake: Nova.
Day the Earth Shook: Nova.
Killer Quakes: The History Channel

Interacting with *Journey Through Geology* CD-ROM

Expand your knowledge of the concepts presented in this chapter by using the CD-ROM to answer the following questions.

1. Spreading Ridges module, Iceland: Observe the pattern of earthquakes since 1960. Why do these earthquakes form such a regular pattern? What kinds of faults did these earthquakes occur along?

2. Convergent Margins, Himalaya: Describe the geographic distribution of earthquakes in the Himalayan region and its relationship to plate tectonics. Why is the pattern of earthquakes narrowly defined bands in some areas (e. g. Southeast Asia) and broad diffuse areas elsewhere (e.g., north of the northern edge of the Indian Plate)?

3. Convergent Margins, Alaska: Carefully watch the footage of the 1964 Good Friday earthquake. Identify each of these effects of earthquakes in the footage: ground motion, landslides, fire, displacement, tsunamis.

Because diamonds form in the mantle and are brought to the surface in
kimberlite pipes, they give geologists a glimpse of the earth's interior.

Photo courtesy of the American Museum of Natural History

CHAPTER
17

The Earth's Interior

*T*he only rocks that geologists can study directly in place are those of the crust; and the earth's crust is but a thin skin of rock, making up less than 1% of the earth's total volume. Mantle rocks brought to the earth's surface in basalt flows, in diamond-bearing kimberlite pipes, and also the tectonic attachment of lower parts of the oceanic lithosphere to the continental crust, give geologists a glimpse of what the underlying mantle might look like. Meteorites also give clues about the possible composition of the core of the earth. However, to learn more about the deep interior of the earth, geologists must study it *indirectly*, largely by using the tools of geophysics—that is, seismic waves and the measurement of gravity, heat flow, and earth magnetism.

The evidence from geophysics suggests that the earth is divided into three major layers—the crust on the earth's surface, the rocky mantle beneath the crust, and the metallic core at the center of the earth. The study of plate tectonics has shown that the crust and uppermost mantle can be conveniently divided into the brittle lithosphere and the plastic asthenosphere.

You will learn in this chapter how gravity measurements can indicate where certain regions of the crust and upper mantle are being held up or held down out of their natural position of equilibrium. We will also discuss the earth's magnetic field and its history of reversals. We will show how magnetic anomalies can indicate hidden ore and geologic structures. We close with a discussion of the distribution and loss of the earth's heat.

Geologists who have considerable contact with the public (such as in government offices and universities) occasionally encounter people who are convinced that geologists are misled about the interior of the earth. These people hold strong convictions about what is deep inside the earth. Many believe, for example, that the earth is hollow and is a storage place for water or even flying saucers.

What *do* geologists know about the earth's interior? How do they obtain information about the parts of the earth beneath the surface? Is there even a slight chance that geologists are wrong, that the earth really is hollow?

Geologists, in fact, are not able to sample rocks very far below the earth's surface. Some deep mines penetrate 3 kilometers (2 miles) into the earth, and a deep oil well may go as far as 8 kilometers beneath the surface; the deepest scientific well has reached 12 kilometers in Russia. Rock samples can be brought up from a mine or a well for geologists to study.

A direct look at rocks from deeper levels can be gotten where mantle rocks have been brought up to the surface by basalt flows (see box 17.2), by the intrusion and erosion of diamond-bearing kimberlite pipes (see chapter 5), or where the lower part of the oceanic lithosphere (see chapter 18) has been tectonically attached to the continental crust at a convergent plate boundary. However, the earth has a radius of about 6,370 kilometers, so it is obvious that geologists can only scratch the surface when they try to study *directly* the rocks beneath their feet.

Deep parts of the earth are studied *indirectly,* however, largely through the branch of geology called **geophysics,** which is the application of physical laws and principles to a study of the earth. Geophysics includes the study of seismic waves and the earth's magnetic field, gravity, and heat. All these things tell us something about the nature of the deeper parts of the earth. Together they create a convincing picture of what makes up the earth's interior.

Evidence from Seismic Waves

Seismic waves from a large earthquake may pass through the entire earth. A nuclear bomb explosion also generates seismic waves. Geologists obtain new information about the earth's interior after every large earthquake and bomb test.

One important way of learning about the earth's interior is the study of **seismic reflection,** the return of some of the energy of seismic waves to the earth's surface after the waves bounce off a rock boundary. If two rock layers of differing densities are separated by a fairly sharp boundary, seismic waves reflect off that boundary just as light reflects off a mirror (figure 17.1). These reflected waves are recorded on a seismogram, which shows the amount of time the waves took to travel down to the boundary, reflect off it, and return to the surface. From the amount of time necessary for the round trip, geologists calculate the depth of the boundary.

Another method used to locate rock boundaries is the study of **seismic refraction,** the bending of seismic waves as they pass from one material to another, which is similar to the way that light waves bend when they pass through the lenses of eyeglasses. As a seismic wave strikes a rock boundary, much of

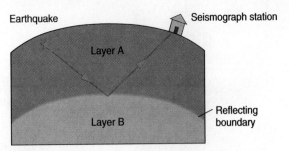

Figure 17.1

Seismic reflection. Seismic waves reflect from a rock boundary deep within the earth and return to a seismograph station on the surface.

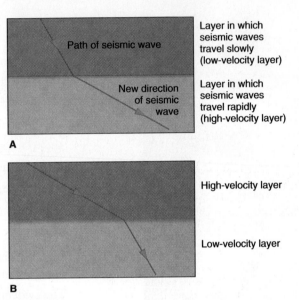

Figure 17.2

Seismic refraction occurs when seismic waves bend as they cross rock boundaries. (*A*) Low-velocity layer above high-velocity layer. (*B*) High-velocity layer above low-velocity layer.

the energy of the wave passes across the boundary. As the wave crosses from one rock layer to another, it changes direction (figure 17.2). This change of direction, or refraction, occurs only if the velocity of seismic waves is different in each layer (which is generally true if the rock layers differ in density or strength).

The boundaries between such rock layers are usually distinct enough to be located by seismic refraction techniques, as shown in figure 17.3. Seismograph station 1 is receiving seismic waves that pass directly through the upper layer A. Stations farther from the epicenter, such as station 2, receive seismic waves from two pathways: (1) a direct path straight through layer A and (2) a refracted path through layer A to a higher-velocity layer B and back to layer A. Station 2 therefore receives the same wave twice.

Seismograph stations close to station 1 receive only the direct wave or possibly two waves, the direct (upper) wave arriving before the refracted (lower) wave. Stations near station 2 receive both the direct and the refracted waves. At some point between station 1 and station 2 there is a transformation from receiving the direct wave first to receiving the *refracted* wave first. Even

Figure 17.3

Seismic refraction can be used to detect boundaries between rock layers. See text for explanation.

Figure 17.4

Curved paths of seismic waves caused by uniform rock with increasing seismic velocity with depth. (*A*) Path between earthquake and recording station. (*B*) Waves spreading out in all directions from earthquake focus.

though the refracted wave travels farther, it can arrive at a station first because most of its path is in the high-velocity layer *B*.

The distance between this point of transformation and the epicenter of the earthquake is a function of the depth to the rock boundary between layers *A* and *B*. A series of portable seismographs can be set up in a line away from an explosion (a *seismic shot*) to find this distance, and the depth to the boundary can then be calculated. The velocities of seismic waves within the layers can also be found.

Figure 17.2 shows how waves bend as they travel downward into higher velocity layers. But why do waves return to the surface, as shown in figure 17.3? The answer is that advancing waves give off energy in all directions. Much of this energy continues to travel horizontally within layer *B* (figure 17.3). This energy passes beneath station 2 and out of the figure toward the right. A small part of the energy "leaks" upward into layer *A,* and it is this pathway that is shown in the figure. There are many other pathways for this wave's energy that are not shown here.

A sharp rock boundary is not necessary for the refraction of seismic waves. Even in a thick layer of uniform rock, the increasing pressure with depth tends to increase the velocity of the waves. The waves follow curved paths through such a layer, as shown in figure 17.4. To understand the reason for the curving path, visualize the thick rock layer as a stack of very thin layers, each with a slightly higher velocity than the one above. The curved path results from many small changes in direction as the wave passes through the many layers.

The Earth's Internal Structure

It was the study of seismic refraction and seismic reflection that enabled scientists to plot the three main zones of the earth's interior (figure 17.5). The **crust** is the outer layer of rock, which forms a thin

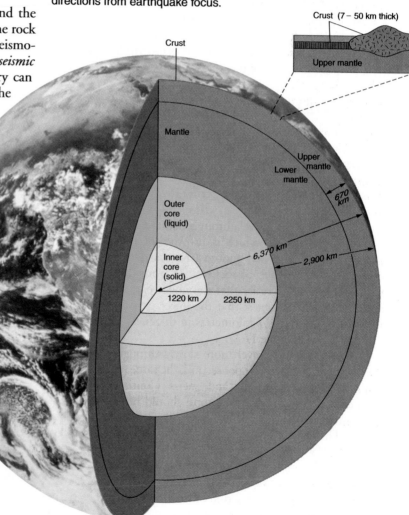

Figure 17.5

The earth's interior. Seismic waves show the three main divisions of the earth: the crust, the mantle, and the core.

Photo by NASA

Figure 17.6

Thin crust with a P-wave velocity of 7 kilometers per second underlies the ocean. Thick crust with a lower velocity makes up continents. Mantle velocities are about 8 kilometers per second.

Table 17.1	Characteristics of Oceanic Crust and Continental Crust	
	Oceanic Crust	Continental Crust
Average thickness	7 km	30 to 50 km (thickest under mountains)
Seismic P-wave velocity	7 km/second	6 km/second (higher in lower crust)
Density	3.0 gm/cm^3	2.7 gm/cm^3
Probable composition	Basalt underlain by gabbro	Granite, other plutonic rocks, schist, gneiss (with sedimentary rock cover)

skin on the earth's surface. Below the crust lies the **mantle,** a thick shell of rock that separates the crust above from the core below. The **core** is the central zone of the earth. It is probably metallic and the source of the earth's magnetic field.

The Earth's Crust

Studies of seismic waves have shown (1) that the earth's crust is thinner beneath the oceans than beneath the continents (figure 17.6) and (2) that seismic waves travel faster in oceanic crust than in continental crust. Because of this velocity difference, it is assumed that the two types of crust are made up of different kinds of rock.

Seismic P waves travel through oceanic crust at about 7 kilometers per second, which is also the speed at which they travel through basalt and gabbro (the coarse-grained equivalent of basalt). Samples of rocks taken from the sea floor by oceanographic ships verify that the upper part of the oceanic crust is basalt and suggest that the lower part is gabbro. The oceanic crust averages 7 kilometers in thickness, varying from 5 to 8 kilometers (table 17.1).

Seismic P waves travel more slowly through continental crust—about 6 kilometers per second, the same speed at which they travel through granite and gneiss. Continental crust is often called "granitic," but the term should be put in quotation marks because most of the rocks exposed on land are not granite. The continental crust is highly variable and complex, consisting of a crystalline basement composed of granite, other plutonic rocks, gneisses and schists, all capped by a layer of sedimentary rocks, like icing on a cake. Since a single rock term cannot accurately describe crust that varies so greatly in composition, some geologists use the term *sial* (rocks high in *si*licon and *al*uminum) for continental crust and *sima* (rocks high in *si*licon and *ma*gnesium) for oceanic crust.

Continental crust is much thicker than oceanic crust, averaging 30 to 50 kilometers in thickness, though it varies from 10 to 70 kilometers. Seismic waves show that the crust is thick-est under geologically young mountain ranges, such as the Andes and Himalayas, bulging downward as a *mountain root* into the mantle (figure 17.6). The continental crust is also less dense than oceanic crust, a fact that is important in plate tectonics (table 17.1).

The boundary that separates the crust from the mantle beneath it is called the **Mohorovičić discontinuity** (Moho for short). Note from figure 17.6 that the mantle lies closer to the earth's surface beneath the ocean than it does beneath continents. The idea behind an ambitious program called Project Mohole (begun during the early 1960s) was to use specially equipped ships to drill through the oceanic crust and obtain samples from the mantle. Although the project was abandoned because of high costs, ocean-floor drilling has become routine since then, but not to the great depth necessary to sample the mantle. Perhaps in the future the original concept of drilling to the mantle through oceanic crust will be revived. (Ocean drilling is discussed in more detail in the following two chapters.)

The Mantle

Because of the way seismic waves pass through the mantle, geologists believe that it, like the crust, is made of solid rock. Localized magma chambers of melted rock may occur as isolated pockets of liquid in both the crust and the upper mantle, but most of the mantle seems to be solid. Because P waves travel at about 8 kilometers per second in the upper mantle, it appears that the mantle is a different type of rock from either oceanic crust or continental crust. The best hypothesis that geologists can make about the composition of the upper mantle is that it consists of ultramafic rock such as peridotite. *Ultramafic rock* is dense igneous rock made up chiefly of ferromagnesian minerals such as olivine and pyroxene (see Box 17.2). Some ultramafic rocks contain garnet, and all of them lack feldspar.

The structure and composition of most of the continental crust is unknown. Surface mapping and seismic reflection and refraction suggest that the continents are largely igneous and metamorphic rock, such as granite and gneiss, overlain by a veneer of sedimentary rocks. This sedimentary cover is generally thin, like icing on a cake, but it may thicken to 10 kilometers or more in giant sedimentary basins where the underlying "basement rock" has subsided (see map in front inside cover). Although oil companies have drilled as deep as 8 kilometers on land, they drill in the sedimentary basins. The igneous and metamorphic basement, which averages 40 kilometers thick and makes up most of the continental crust, has rarely been sampled deeper than 2 or 3 kilometers (although uplift and erosion have exposed some rocks widely thought to have been formed much deeper in the crust).

Since 1970 Russia has been drilling a superdeep hole on the Kola Peninsula near Murmansk north of the Arctic Circle. The hole penetrates ancient Precambrian basement rocks, and is currently 12 kilometers deep. The Kola hole is the deepest of 11 holes currently being drilled in Russia to depths of 10 to 15 kilometers. Deep drilling is as technically complex as space exploration. High pressures and 300°C temperatures require special equipment and techniques. The Russians use a turbodrill that rotates under the pressure of circulating drilling mud. Unlike normal drilling operations, the lightweight aluminum drill pipe does not turn.

The drilling at Kola shows that seismic models for this area are wrong. The Russians expected 4.7 kilometers of metamorphosed sedimentary and volcanic rock, then a granitic layer to a depth of 7 kilometers, and a "basaltic" layer below that. The granite, however, appeared at 6.8 kilometers and extends to more than 12 kilometers; the "basalt" has not yet been found. These results (and other deep holes) show that seismic surveys of continental crust are being systematically misinterpreted.

The Russians unexpectedly found open fractures and circulating fluids throughout the borehole. The fluids include hydrogen, helium, and methane (natural gas), as well as mineralized waters forming ore bodies. Copper-nickel ore was found deeper than theory predicted, and gold mineralization was present from 9.5 to 11 kilometers down. These results will change geologists' models of ore formation and fluid circulation underground.

The earth's crust and uppermost mantle together form the **lithosphere,** the outer shell of the earth that is relatively strong and brittle. The lithosphere makes up the plates of plate-tectonic theory. The lithosphere averages about 70 kilometers thick beneath oceans and may be 125 to 250 kilometers thick beneath continents. Its lower boundary is marked by a curious mantle layer in which seismic waves slow down (figure 17.7).

Generally, seismic waves increase in velocity with depth as increasing pressure alters the properties of the rock. Beginning at a depth of 70 to 125 kilometers, however, seismic waves travel more slowly than they do in shallower layers, and so this zone has been called the *low-velocity zone* (figure 17.7). This zone, extending to a depth of perhaps 200 kilometers, is also called, in plate-tectonic theory, the **asthenosphere.** The rocks in this zone may be closer to their melting point than the rocks above or below the zone. (The rocks are probably not *hotter* than the rocks below—melting points are controlled by pressure as well as temperature.) Some geologists think that these rocks may actually be partially melted, forming a crystal-and-liquid slush; a very small percentage of liquid in the asthenosphere could help explain some of its physical properties.

If the rocks of the asthenosphere are close to their melting point, this zone may be important for two reasons: (1) it may represent a zone where magma is likely to be generated; and

(2) the rocks here may have relatively little strength and therefore are likely to flow. If mantle rocks in the asthenosphere are weaker than they are in the overlying lithosphere, then the asthenosphere can deform easily by plastic flow. Plates of brittle lithosphere probably move easily over the asthenosphere, which may act as a lubricating layer below.

There is widespread agreement on the existence and depth of the asthenosphere under oceanic crust, but considerable disagreement about asthenosphere under continental crust. Figure 17.7 shows asthenosphere at a depth of 125 kilometers below the continents. Some geologists think that the lithosphere is much thicker beneath continents than shown in the figure, and that the asthenosphere begins at a depth of 250 kilometers (or even more). A few geologists say that there is *no* asthenosphere beneath continents at all. The reasons for this disagreement are the results of the rapidly developing field of seismic tomography, which is described in a box in chapter 19. The figures throughout this book match the model in figure 17.7.

Data from seismic reflection and refraction indicate several concentric layers in the mantle (figure 17.7), with prominent boundaries at 400 and 670 kilometers (670 kilometers is also the depth of the deepest earthquakes). It is doubtful that the layering is due to the presence of several different kinds of rock. Most geologists think that the chemical composition of the mantle

IN GREATER DEPTH

Mantle Xenoliths—A Peek at the Deep

A *xenolith* is an inclusion of foreign rock within an igneous rock. Plutons sometimes contain xenoliths of nearby country rock, but xenoliths within some basalts may have come from deep in the earth's mantle.

Numerous basalts contain xenoliths of the relatively rare rocks *peridotite* and *dunite*. Peridotite is an igneous rock composed mostly of olivine and pyroxene; it lacks feldspar, which is common in most igneous rocks. Dunite is a related rock that is almost entirely olivine (box figure 1).

Because basalts appear to be generated in the mantle's asthenosphere, and because peridotite and dunite are uncommon at the earth's surface, most geologists suspect that these xenoliths may be solid pieces of the earth's mantle brought to

Box 17.2 Figure 1

Dunite xenoliths in basalt. The xenoliths are thought by many geologists to be pieces of mantle rock torn off and brought to the surface by ascending basaltic magma. Scale is 7 cm long.

the surface by the erupting basalt. The density, seismic velocity, and chemical composition of peridotite and dunite make them reasonable candidates for mantle rock. Peridotites (usually serpentinized by metamorphism) also occur at the base of *ophiolite* sequences on land (see next chapter), which are widely interpreted to be slivers of oceanic crust and mantle rock attached to continents at convergent plate boundaries.

Figure 17.7

The concentric shell structure of the upper mantle. The velocity of seismic P waves generally increases with depth except in the low-velocity zone. The lithosphere contains the crust and the uppermost mantle. The plastic asthenosphere slows down seismic waves. Velocity increases at 400 and 670 kilometers may be caused by mineral collapse.

rock is about the same throughout the mantle. Because pressure increases with depth into the earth, the boundaries between mantle layers possibly represent depths at which pressure collapses the internal structure of certain minerals into denser minerals (figure 17.7). For example, at a pressure equivalent to a depth of about 670 kilometers, the mineral *olivine* should collapse into the denser structure of the mineral *perovskite*. If the boundaries between mantle layers represent pressure-caused transformations of minerals, the entire mantle may have the same *chemical* composition throughout, although not the same

mineral composition. However, some geologists think that the 670-kilometer boundary represents a chemical change as well as a physical change and separates the *upper mantle* from the chemically different *lower mantle* below.

The Core

Seismic-wave data provide the primary evidence for the existence of the core of the earth. (See chapter 16 for a discussion of seismic P and S waves.) Seismic waves do not reach certain

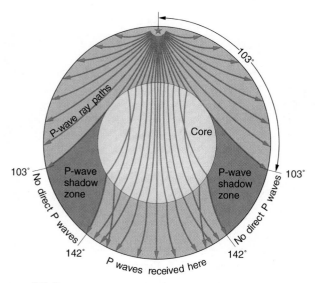

Figure 17.8

The P-wave shadow zone, caused by refraction of P waves within the earth's core.

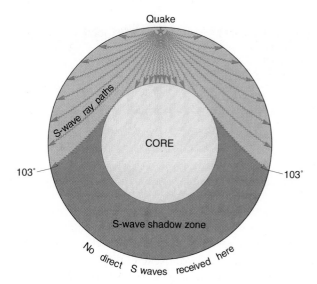

Figure 17.9

The S-wave shadow zone. Because no S waves pass through the core, the core is apparently a liquid (or acts like a liquid).

areas on the opposite side of the earth from a large earthquake. Figure 17.8 shows how seismic P waves spread out from a quake until, at 103° of arc (11,500 km) from the epicenter, they suddenly disappear from seismograms. At more than 142° (15,500 km) from the epicenter, P waves reappear on seismograms. The region between 103° and 142°, which lacks P waves, is called the **P-wave shadow zone.**

The P-wave shadow zone can be explained by the refraction of P waves when they encounter the core boundary deep within the earth's interior. Because the paths of P waves can be accurately calculated, the size and shape of the core can be determined also. In figure 17.8, notice that the earth's core deflects the P waves and, in effect, "casts a shadow" where their energy does not reach the surface. In other words, P waves are missing within the shadow zone because they have been bent (refracted) by the core.

The chapter on earthquakes explains that while P waves can travel through solids and fluids, S waves can travel only through solids. As figure 17.9 shows, an **S-wave shadow zone** also exists and is larger than the P-wave shadow zone. Direct S waves are not recorded in the entire region more than 103° away from the epicenter. The S-wave shadow zone seems to indicate that S waves do not travel through the core at all. If this is true, it implies that the core of the earth is a liquid, or at least acts like a liquid.

The way in which P waves are refracted within the earth's core (as shown by careful analysis of seismograms) suggests that the core has two parts, a *liquid outer core* and a *solid inner core* (figure 17.5).

Composition of the Core

When evidence from astronomy and seismic-wave studies is combined with what we know about the properties of materials, it appears that the earth's core is made of metal—not silicate rock—and that this metal is probably iron (along with a minor amount of oxygen, silicon, sulfur, or nickel). How did geologists arrive at this conclusion?

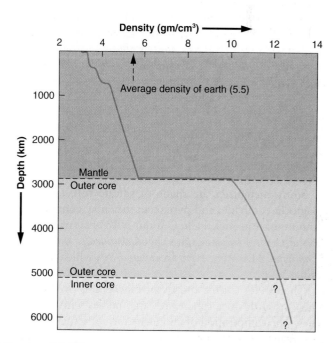

Figure 17.10

Density variations with depth into the earth. Note the great increase in density at the core-mantle boundary.

The overall density of the earth is 5.5 gm/cm³, according to astronomers who calculate the speed of the earth's revolution about the sun and the speed of its rotation on its own axis. The crustal rocks are relatively low density, from 2.7 gm/cm³ for granite to 3.0 gm/cm³ for basalt. The ultramafic rock thought to make up the mantle probably has a density of 3.3 gm/cm³ in the upper mantle, although rock pressure should raise this value to about 5.5 gm/cm³ at the base of the mantle (figure 17.10).

The Earth's Interior   **421**

17.1

Astrogeology

Meteorites

Small solid particles of rock, metal, and/or ice orbiting the Sun are called *meteoroids*. When these particles enter the Earth's atmosphere, they are heated to incandesence by friction; these glowing particles are called *meteors* (or "shooting stars" or "falling stars"). Most meteors are small and burn up while still in the atmosphere, but about 150 per year are large enough to strike the Earth's surface. Those that do are called *meteorites* (box figure 1). The largest fragment of a meteorite found (in South Africa) weighs 50 tons; much larger meteorites have hit the Earth in the past.

Three basic types of meteorite are iron, stony-iron, and stony meteorites. Stony meteorites are by far the most common, but they look like Earth rocks, so they are hard to find. Iron meteorites are rare, but look so unique that they are commonly found; most museum meteorites are of the iron type.

Iron meteorites are mostly iron alloyed (mixed) with a small percentage of nickel. Small amounts of other metals or minerals may be present.

Stony-iron meteorites are made of iron-nickel alloy and silicate minerals in about equal parts.

Stony meteorites are made of silicate minerals such as plagioclase, olivine, and pyroxene; they may contain a small amount of iron-nickel alloy. About 90% of stony meteorites contain round silicate grains called *chondrules* and are called *chondrites*. The other 10% are *achondrites*, which lack chondrules.

Chondrules consist mostly of olivine and pyroxene, and range from distinct spheres to large bodies with fuzzy outlines. The composition of chondrite meteorites resembles the ultramafic rock peridotite, but peridotite lacks the chondritic texture and iron-nickel content of the meteorites.

Astrogeology Box 17.1 Figure 1
A large meteorite from Meteor Crater, Arizona. Pocketknife for scale.
Photo by Frank M. Hanna

One kind of chondrite is composed mostly of serpentine or pyroxene and contains up to 5% organic materials, including carbon, hydrocarbon compounds, and amino acids. These meteorites are called *carbonaceous chondrites*. All available evidence indicates that the organic compounds were in fact produced by inorganic processes. Carbonaceous chondrites are of particular interest to scientists because they are believed to have the same composition as the original material from which the solar system was formed.

Achondrites are generally similar to terrestrial rocks in composition and texture. In composition they are most similar to basalts. Some have textures like ordinary igneous rocks, and others are breccias with fragments of different compositions and textures.

The origin of meteorites is controversial. Many meteorites have a coarse-grained texture, probably formed by slow cooling within a larger body, such as a planet. The similarity in iron-nickel composition among iron meteorites also suggests that they are fragments from a single, large body. The larger body may have differentiated into a heavy, iron-rich core and a lighter, rocky mantle before it fragmented into meteoroids. Isotopic dating shows that most meteorites have the same age, 4.6 billion years old.

If the crust and the mantle, which have approximately 85% of the earth's volume, are at or below the average density of the earth, then the core must be very heavy to bring the average up to 5.5 gm/cm^3.

Calculations show that the core has to have a density of about 10 gm/cm^3 at the core-mantle boundary, increasing to 12 or 13 gm/cm^3 at the center of the earth (figure 17.10). This great density would be enough to give the earth an average density of 5.5 gm/cm^3.

Under the great pressures existing in the core, iron would have a density slightly greater than that required in the core. Iron

mixed with a small amount of a lighter element, such as oxygen, sulfur or silicon, would have the required density. Therefore, many geologists think that such a mixture makes up the core.

But a study of density by itself is hardly convincing evidence that the core is mostly iron, for many other heavy substances could be there instead. The choice of iron as the major component of the core comes from looking at meteorites (see Astrogeology Box 17.1). Meteorites are thought by some scientists to be remnants of the basic material that created our own solar system. An estimated 10% of meteorites are composed of iron mixed with small amounts of nickel. Material similar to these meteorites may have helped create the earth, perhaps settling to the center of the earth because of metal's high density. The composition of these meteorites, then, may tell us what is in the earth's core. Nickel is denser than iron, however, so a mixture of just iron and nickel would have a density greater than that required in the core. (The other 90% of meteorites are mostly ultramafic rock and perhaps represent material that formed the earth's mantle.)

Seismic and density data, together with assumptions based on meteorite composition, point to a core that is largely iron, with at least the outer part being liquid. The existence of the earth's magnetic field, which is discussed later in this chapter, also suggests a metallic core. Of course, no geologist has seen the core, nor is anyone likely to in the foreseeable future. But since so many lines of indirect evidence point to a liquid metal outer core, most scientists accept this theory as the best conclusion that can be made about the core's composition. Geologists could be wrong about the earth's interior, but the current model of a solid rock mantle and a liquid metallic outer core and a solid inner core is widely accepted because it is consistent with all available knowledge. A hollow earth is not.

The Core-Mantle Boundary

The boundary between the core and mantle is marked by great changes in density (figure 17.10) and temperature, as we see later in the chapter. Here there is a transition zone up to 200 kilometers thick and an undulating border between mantle and core. Both the mantle and the core are undergoing **convection,** a circulation pattern in which low density material rises and high density material sinks. Heavy portions of the mantle (including subducting plates) might sink to its base, but are unable to penetrate the denser core. Light portions of the core may rise to its top, but not into the mantle above. Continent-sized blobs of liquid and liquid-crystal slush may accumulate here, perhaps interfering with (or even helping cause) heat loss from the core and causing changes in the earth's magnetic field. This boundary is an exciting frontier for geologic study, but data, of course, are sparse and hard to obtain.

Isostasy

Isostasy is a balance or *equilibrium* of adjacent blocks of brittle crust "floating" on the upper mantle. Since crustal rocks weigh less than mantle rocks, the crust can be thought of as floating on the denser mantle much as wood floats on water (figure 17.11).

Blocks of wood floating on water rise or sink until they displace an amount of water equal to their own weight. The weight of the displaced water buoys up the wood blocks, allowing them to float. The higher a wood block appears above

Figure 17.11

Isostatic balance. (*A*) Wood blocks float in water with most of their bulk submerged. (*B*) Crustal blocks "float" on mantle in approximately the same way. The thicker the block, the deeper it extends into the mantle.

the water surface, the deeper the block extends under water. Thus a tall block has a deep "root."

In a greatly simplified way, crustal rocks can be thought of as tending to rise or sink gradually until they are balanced by the weight of displaced mantle rocks. This concept of vertical movement to reach equilibrium is called **isostatic adjustment.** Just as with the blocks of wood, once crustal blocks have come into isostatic balance, a tall block (a mountain range) extends deep into the mantle (a *mountain root,* as shown in figure 17.11).

Figure 17.11 shows both the blocks of wood and the blocks of crustal rock in isostatic balance. The weight of the wood is equal to the weight of the displaced water. Similarly, the weight of the crustal blocks is equal to the weight of the displaced mantle. As a result, the rocks (and overlying sea water) in figure 17.11 can be thought of as separated into vertical columns, each with the same pressure at its base. At some *depth of equal pressure* each column is in balance with the other columns, for each column has the same weight. A column of thick continental crust (a mountain and its root) has the same weight as a column containing thin continental crust and some of the upper mantle. A column containing sea water, thin oceanic crust, and a thick section of heavy mantle weighs the same as the other two columns.

Figure 17.11 shows the crust as isolated blocks free to move past each other along vertical faults, but this is not really a good picture of crustal structure. It is more accurate to think of the crust as bending in broad uplifts and

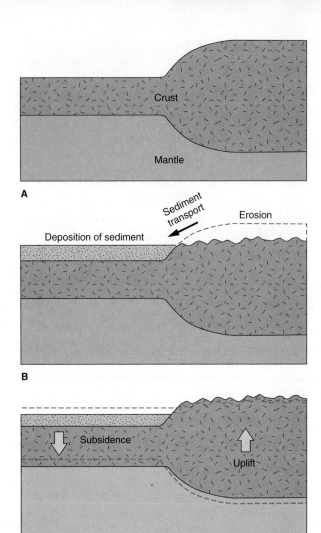

A

B

C

Figure 17.12

Isostatic adjustment due to erosion and deposition of sediment. Rock within the mantle must flow to accommodate vertical motion of crustal blocks. Mantle flow occurs in the asthenosphere, deeper than shown in *C*.

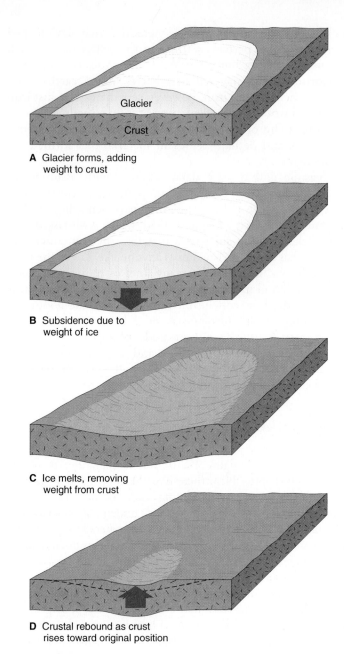

A Glacier forms, adding weight to crust

B Subsidence due to weight of ice

C Ice melts, removing weight from crust

D Crustal rebound as crust rises toward original position

Figure 17.13

The weight of glaciers depresses the crust, and the crust rebounds when the ice melts.

downwarps without vertical faults, as shown in figures 17.12 and 17.13.

Let us look at some examples of isostatic balance (equilibrium) in crustal rocks. Suppose that two sections of crust of unequal thickness are next to each other as in figure 17.12. Sediment from the higher part, which is subject to more rapid erosion, is deposited on the lower part. The decrease in weight from the high part causes it to rise, while the increase in weight on the low part causes it to sink. These vertical movements (isostatic adjustment) do in fact take place whenever large volumes of material are eroded from or deposited on parts of the crust.

Rising or sinking of the crust, of course, requires plastic flow of the mantle to accommodate the motion. By measuring

the rate of rising or sinking, the viscosity of the mantle can be calculated. The plastic flow of the mantle probably takes place within the asthenosphere in the upper mantle.

Another example of isostatic adjustment, caused by plastic mantle flow, is the upward movement of large areas of the crust since the glacial ages. The weight of the thick continental ice sheets during the Pleistocene Epoch depressed the crust underneath the ice (figure 17.13). After the melting of the ice, the crust rose back upward, a process that is still going on in some areas (figure 17.14). This rise of the crust after the removal of the ice is known as **crustal rebound.**

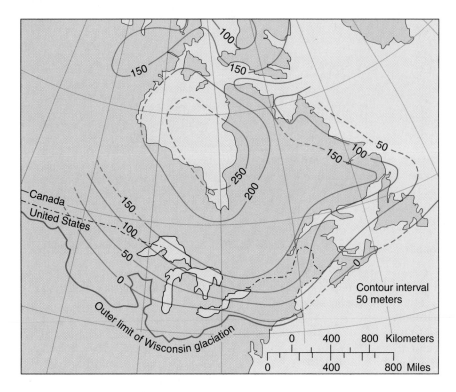

Figure 17.14

Uplift of land surface in Canada and the northern United States caused by crustal rebound after glaciers melted. Colored lines show the amount of uplift in meters since the ice disappeared.

From Phillip B. King, "Tectonics of Quaternary Time in Middle North America," in *The Quaternary of the United States*, H. C. Wright, Jr. and David G. Frey, eds,. fig. 4A, p. 836. Reprinted by permission of Princeton University Press.

Recent geophysical studies have shown that some mountains, such as the Rockies and southern Sierra Nevada, do not have thick roots and are instead buoyed by warm, less dense mantle. It appears that the upper mantle beneath some continents is not homogeneous, but has zones that are quite buoyant due to higher temperatures and less dense mineral phases.

Isostatic adjustment may occur at subduction zones. As explained in the chapter on igneous rocks, a subducting plate may generate molten magma, which sometimes rises all the way to the surface to erupt as lava. However, some geologists believe that in certain cases the magma stops at the base of an overlying continent (or perhaps within the continent near its base). This accumulation of magma can locally thicken the continent when the magma cools (figure 17.15A). The thickening of continents from below (a theory not accepted by all geologists) causes the crust to be out of isostatic equilibrium. So it rises, reestablishing equilibrium and forming a mountain range (figure 17.15B).

Gravity Measurements

The force of gravity between two objects varies with the masses of the objects and the distance between them (figure 17.16):

$$\text{Force of gravity between } A \text{ and } B = \text{constant}\left(\frac{\text{mass}_A \times \text{mass}_B}{\text{distance}^2}\right)$$

The force increases with an increase in either mass. The gravitational attraction between the earth and the moon, for example, is vastly greater than the extremely small attraction that exists between two bowling balls. The equation also shows that force decreases with the square of the distance between the two objects. The farther two objects are apart, the less gravitational attraction there is between them.

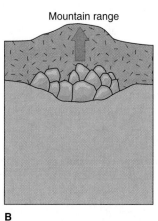

Figure 17.15

Isostatic uplift as a result of crustal thickening from below. (*A*) Rising blobs of magma accumulate at the base of a continent, thickening the crust (base of lithosphere below continent not shown). (*B*) Uplift produces a mountain range.

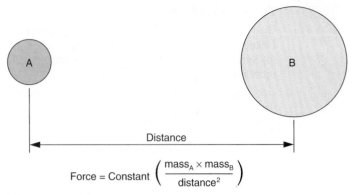

Figure 17.16

The force of gravitational attraction between two objects is a function of the masses of the objects and the distance between the centers of the objects.

$$Force = Constant \left(\frac{mass_A \times mass_B}{distance^2} \right)$$

A useful tool for studying the crust and upper mantle is the **gravity meter,** which measures the gravitational attraction between the earth and a mass within the instrument. One use of the gravity meter is to explore for local variations in rock density (mass = density × volume). Dense rock such as metal ores and ultramafic rock pulls strongly on the mass inside the meter (figure 17.17). The strong pull stretches a spring, and the amount of stretching can be very precisely determined. So a gravity meter can be used to explore for metallic ore deposits. A cavity or a body of low-density material such as sediment causes a much weaker pull on the meter's mass (figure 17.17). The use of a gravity meter to explore for salt domes and their associated traps for oil and gas is shown in Box 21.1.

Another important use of a gravity meter is to discover whether regions are in isostatic equilibrium. If a region is in isostatic balance, as in figure 17.18, each column of rock has the same mass. If a gravity meter were carried across the rock columns, it would register the same amount of gravitational attraction for

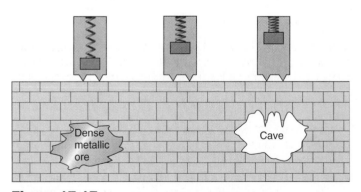

Figure 17.17

A gravity meter reading is affected by the density of the rocks beneath it. Dense rock pulls strongly on the mass within the meter, stretching a spring; a cavity exerts a weak pull on the mass. A gravity meter can be used to explore for hidden ore bodies, caves, and other features that have density contrasts with the surrounding rock.

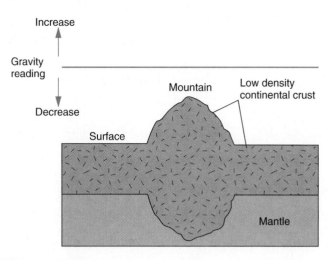

Figure 17.18

A region in isostatic balance gives a uniform gravity reading (no gravity anomalies).

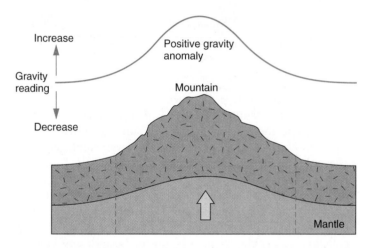

Figure 17.19

A region being held up out of isostatic equilibrium gives a positive gravity anomaly.

each column (after correcting for differences in elevation—gravitational attraction is less on a mountaintop than at sea level because the mountaintop is farther from the center of the earth).

Some regions, however, are held up out of isostatic equilibrium by deep tectonic forces. Figure 17.19 shows a region with uniformly thick crust. Tectonic forces are holding the center of the region up. This uplift creates a mountain range without a mountain root. There is a thicker section of heavy mantle rock under the mountain range than there is on either side of the mountain range. Therefore, the central "column" has more mass than the neighboring columns, and a gravity meter shows that the gravitational attraction is correspondingly greater over the central than over the side columns.

A gravity reading higher than the normal regional gravity is called a **positive gravity anomaly** (figure 17.19). It can indicate that tectonic forces are holding a region up out of isostatic equilibrium, as shown in figure 17.19. When the forces stop

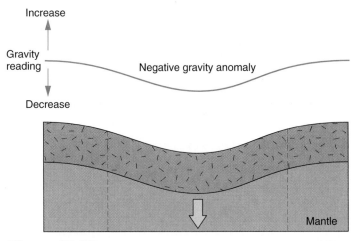

Figure 17.20

A region being held down out of isostatic equilibrium gives a negative gravity anomaly.

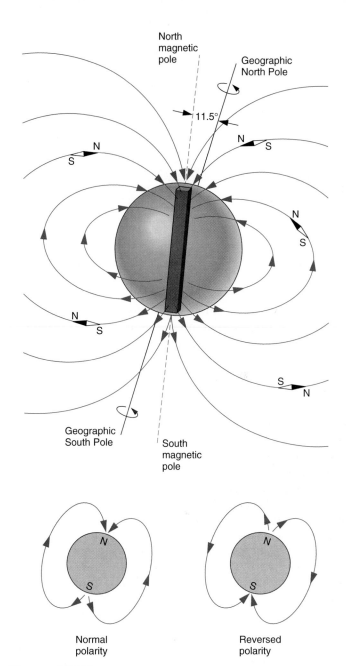

Figure 17.21

The earth's magnetic field. The depiction of the internal field as a large bar magnet is a simplification of the real field, which is more complex. N and S in the two small figures indicate the *geographic* poles.

acting, the land surface sinks until it reestablishes isostatic balance. The gravity anomaly then disappears. For the region shown in figure 17.19, equilibrium will be established when the land surface becomes level.

Positive gravity anomalies, particularly small ones, are also caused by local concentrations of dense rock such as metal ore. The gravity meter in figure 17.17 is registering a positive gravity anomaly over ore (the spring inside the meter is stretched). Since there can be more than one cause of a positive gravity anomaly, geologists may disagree about the interpretation of anomalies. Drilling into a region with a gravity anomaly usually discloses the reason for the anomaly.

A region can also be held down out of isostatic equilibrium, as shown in figure 17.20. The mass deficiency in such a region produces a **negative gravity anomaly**—a gravity reading lower than the normal regional gravity. Negative gravity anomalies indicate either that a region is being held down (figure 17.20) or that local mass deficiencies exist for other reasons (figure 17.17).

The greatest negative gravity anomalies in the world are found over oceanic trenches (see next two chapters). These negative anomalies are interpreted to mean that trenches are actively being held down and are out of isostatic balance.

The Earth's Magnetic Field

A region of magnetic force—a **magnetic field**—surrounds the earth. The invisible lines of magnetic force surrounding the earth deflect magnetized objects, such as compass needles, that are free to move. The field has north and south **magnetic poles,** one near the geographic North Pole, the other near the geographic South Pole. (Because it has two poles, the earth's field is called *dipolar.*) The strength of the magnetic field is greatest at the magnetic poles where magnetic lines of force appear to leave and enter the earth vertically (figure 17.21).

Because the compass is important in navigation, the earth's magnetism has been observed for centuries. It has long been known that the magnetic poles are displaced about $11\frac{1}{2}°$ from the geographic poles (about which the earth rotates). Furthermore, changes in the position of the magnetic poles have been well documented, especially since the time of the great explorations of the globe. The magnetic poles appear to be moving slowly around the geographic poles. The separation between the two types of poles has probably never been much greater than it is today.

More recently, geophysical studies have been directed toward the *source* of the earth's magnetism. The rate of the poles' changes in position, together with the strength of the magnetic field, strongly suggest that the magnetic field is

Figure 17.22

Some rocks preserve a record of the earth's magnetic field.

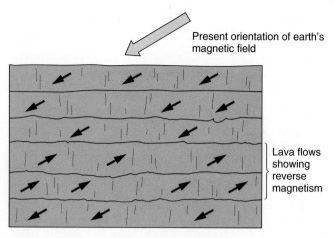

Figure 17.23

Cross section of stacked lava flows showing evidence of magnetic reversals.

generated within the liquid metal of the outer core rather than within the solid rock of the crust or the mantle.

How is the earth's magnetic field generated? A number of hypotheses have been put forth. One widely accepted hypothesis suggests that the magnetic field is created by electric currents within the liquid outer core. The outer core is extremely hot and flows at a rate of several kilometers per year in large convection currents, about one million times faster than mantle convection above it. Convecting metal creates electric currents, which in turn create a magnetic field. This hypothesis requires the core to be an electrical conductor. Metals are good conductors of electricity, whereas silicate rock is generally a poor electrical conductor. Indirectly, this is evidence that the core is metallic.

Magnetic Reversals

In the 1950s evidence began to accumulate that the earth's magnetic field has periodically reversed its polarity in the past. Such a change in the polarity of the earth's magnetic field is a **magnetic reversal.** During a time of *normal polarity,* magnetic lines of force leave the earth near the geographic South Pole and reenter the earth near the geographic North Pole (figure 17.21). This orientation is called "normal" polarity because it is the same as the present polarity. During a time of *reversed polarity,* the magnetic lines of force run the other way, leaving the earth near the North Pole and entering the earth near the South Pole (figure 17.21). In other words, during a magnetic reversal, the north magnetic pole and the south magnetic pole exchange positions.

Many rocks contain a record of the strength and direction of the earth's magnetic field *at the time the rocks formed.* When the mineral magnetite, for example, is crystallizing in a cooling lava flow, the atoms within the crystals respond to the earth's magnetic field and form magnetic alignments that "point" toward the north magnetic pole. As the lava cools slowly below the **Curie point** (580°C for magnetite), this magnetic record is permanently trapped in the rock (figure 17.22). Unless the rock is heated again above the Curie point or temperature this magnetic record is retained, and when studied reveals the direction of the earth's magnetic field at the time the lava cooled. Other rock types, including sedimentary rocks stained red by iron compounds, also record former magnetic field directions. The study of ancient magnetic fields is called **paleomagnetism.**

Most of the evidence for magnetic reversals comes from lava flows on the continents. Paleomagnetic studies of a series of stacked lava flows often show that some of the lava flows have a magnetic orientation directly opposite to the earth's present orientation (figure 17.23). That is, at the time these lava flows cooled, the magnetic poles had exchanged positions. During this time of magnetic reversal, a compass needle would have pointed south rather than north. Many periods of normal and reverse magnetization are recorded in continental lava flows. They are worldwide events. Since lava flows can be dated isotopically, the time of these reversals in the earth's past can be determined. The times of normal and reversed magnetization are shown for the past 4.5 million years in figure 17.24. Although reversals appear to occur randomly, records for tens of millions of years suggest that the earth's field reverses on average about once every 500,000 years. The present normal orientation has lasted for the past 700,000 years. It takes time for one magnetic orientation to die out and the reverse orientation to build up. Most geologists think that it takes 10,000 years for a reversal to develop, although new evidence suggests that a reversal can occur much faster than that.

What causes magnetic reversals? The question is difficult to answer because no one knows how the magnetic field is generated in the first place. Recent computer modeling and seismological research support the theory that the magnetic field is generated by convection currents in the liquid outer core (see Box 17.3). *If* the field is caused by convection currents within the liquid outer core, perhaps a reversal is caused when the currents change direction, or by a temporary current building up and then dying out. Some geologists think that reversals may be triggered by the impact of an asteroid or comet with the earth; other geologists disagree.

A magnetic reversal can have some profound effects on earth. The strength of the earth's magnetic field probably declines to near zero before the orientation reverses; then the field strength increases to its usual values, but in the opposite orientation. This collapse of the earth's magnetic field means that deadly cosmic radiation from the sun would be much

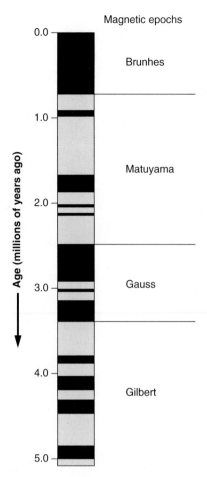

Figure 17.24

Magnetic reversals during the past 5 million years. Black represents normal magnetism; blue represents reverse magnetism.

From R. F. Butler, 1992, *Paleomagnetism,* Blackwell Scientific Publications, p. 212.

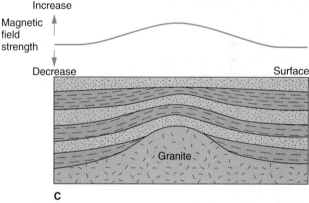

Figure 17.25

Positive magnetic anomalies can indicate hidden ore and geologic structures.

more intense at the earth's surface. When the magnetic field is at its usual strength, it shields the earth from these rays, but when the field collapses, this shielding is lost. Cosmic radiation affects organisms; the extinction of some species and the appearance of new species by mutation have been correlated with some magnetic reversals. However, there have been far more reversals than extinctions.

Magnetic Anomalies

A **magnetometer** is an instrument used to measure the strength of the earth's magnetic field. A magnetometer can be carried over the land surface or flown over land or sea. At sea, magnetometers can also be towed behind ships. They are also used as metal detectors in airports.

The strength of the earth's magnetic field varies from place to place. As with gravity, a deviation from average readings is called an *anomaly.* Very broad regional magnetic anomalies may be due to *circulation patterns in the liquid outer core* or to other deep-seated causes. Smaller anomalies generally reflect

variations in rock type, for the magnetism of near-surface rocks adds to the main magnetic field generated in the core. Rocks differ in their magnetism, depending upon their content of iron-containing mineral, particularly magnetite.

A **positive magnetic anomaly** is a reading of magnetic field strength that is higher than the regional average. Figure 17.25 shows three geologic situations that can cause positive magnetic anomalies. In figure 17.25*A,* a body of magnetite ore (a highly magnetic ore of the metal iron) has been emplaced in a bed of

17.3

IN GREATER DEPTH

The Earth's Spinning Inner Core

Recent studies have led to a new understanding of the dynamics of the earth's inner core and generation of the earth's magnetic field and periodic magnetic reversals. Gary A. Glatzmaier, of Los Alamos National Laboratory in New Mexico and Paul H. Roberts of the University of California, Los Angeles developed a very sophisticated computer model of convection in the earth's outer core that has been successful in simulating a magnetic field very similar to that measured on earth. The model utilizes circulating metallic fluids in the outer core, caused by cooling and heat loss, as the driving force of the earth's magnetic field. The circulation of metallic fluids in the outer core has been theorized for many years, and the computer model was successful in simulating and maintaining a magnetic field similar to that measured on the earth. The model also predicted that the earth's solid inner core spins faster than the rest of the earth, gaining a full lap on the rest of the earth every 150 years. Because the magnetic lines of force penetrate and connect both the inner and outer core, a faster rate of rotation of the inner core would play an important role in the generation of the earth's magnetic field and may also influence periodic magnetic reversals. Interestingly, Glatzmaier and Robert's model produced a magnetic reversal on its own without any additional input from the experimenters after about 35,000 years of simulated time (box figure 1).

The results of this computer model inspired seismologists Xiao Dong Song and Paul Richards from Columbia's Lamont-Doherty Earth Observatory to look for evidence that the earth's inner core actually spins at a more rapid rate than the rest of the planet. The seismologists knew that previous studies suggested seismic waves pass through the inner core faster along a nearly north-south route. This faster route, or high-velocity pathway, is similar to the grain in a piece of wood. This pathway is not aligned directly with the inner core's spin axis but is tilted about 10 degrees from it (box figure 2). Seismic waves tend to travel slower along other paths, such as in an east-west direction parallel to the inner core's equator.

The seismologists studied seismic wave records from 38 separate, closely spaced earthquakes from 1967 to 1995 near the Sandwich Islands, off Argentina to determine how long it took them to reach a monitoring station in College,

A

B

C

Box 17.3 Figure 1

Computer simulation of the earth's magnetic field and magnetic reversal. (*A*) Normal magnetic polarity with magnetic field lines leaving the north magnetic pole (orange) and reentering at the south pole. (*B*) Transitional magnetic field. (*C*) Reversed magnetic field.

Photos from the Geodynamo Computer Simulation, courtesy of G. A. Glatzmaier, Los Alamos National Laboratory, and P. H. Roberts, University of California, Los Angeles.

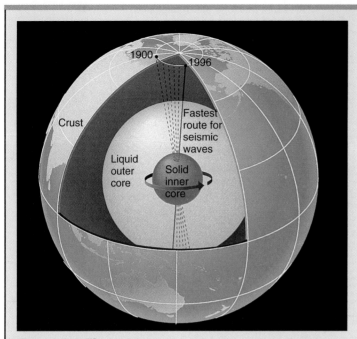

Box 17.3 Figure 2

Seismic waves indicate that the earth's core rotates faster than the rest of the planet by about a degree per year. The solid line indicates the 1996 position of a point in the core relative to the surface of the earth, and the dashed line indicates where the point was in 1900.

Courtesy of Lamont-Doherty Earth Observatory, Columbia University. Data from Michael Carlowicz, *Earth Magazine*, p. 21, 1996.

Alaska. The waves all took about the same amount of time to reach Alaska. However, the seismic waves in the 1990s arrived in Alaska about 0.3 seconds faster than the seismic waves in the 1960s. Since the seismic waves would have traveled through the inner core, the seismologists have explained the difference in travel time as indicating that the inner core had changed its position relative to the monitoring station in Alaska. That is, the inner core and the high velocity pathway had rotated slightly with respect to the rest of the earth.

Seismologists at Harvard looked at additional earthquake records and calculated that the inner core is rotating at approximately the same rate as Glatzmaier and Robert's model predicted. Future studies to examine earthquake records over a longer period of time are needed to confirm whether the inner core has been spinning faster than the rest of the earth. This is an exciting time for earth scientists since we may now have a better idea about the inner motion of the core and the generation of the earth's magnetic field.

Additional Resources:
Carlowicz, M. 1996. Spin control. *Earth* 12(21): 62–63.

Related Web Resources
Core convection and the Geodynamo web site discusses the recent model for reversals of the earth's magnetic field:
http://www.igpp.lanl/gov/Geodynamo.html

limestone by hot solutions rising along a fracture. The magnetism of the iron ore adds to the magnetic field of the earth, giving a stronger magnetic field measurement at the surface (a positive anomaly). In figure 17.25*B* a large dike of gabbro has intruded into granitic basement rock. Because gabbro contains more ferromagnesian minerals than granite, gabbro is more magnetic and causes a positive magnetic anomaly. Figure 17.25*C* shows a granitic basement high (perhaps originally a hill) that has influenced later sediment deposits, causing a draping of the layers as the sediments on the hilltop compacted less than the thicker sediments to the sides. Such a structure can form an *oil trap* (see chapter 21). The granite in the hill contains more iron in its ferromagnesian minerals than the surrounding sedimentary rocks, so a small positive magnetic anomaly occurs where the granite is closer to the surface. Note how each example shows horizontal sedimentary rocks at the surface, with no surface hint of the subsurface geology. The magnetometer helps find hidden ores and geologic structures.

A **negative magnetic anomaly** is a reading of magnetic field strength that is lower than the regional average. Figure 17.26 shows how a negative anomaly can be produced by a downdropped fault block (a *graben*) in igneous rock. The thick sedimentary fill above the graben is less magnetic than is the igneous rock, so a weaker field (a negative magnetic anomaly) develops over the thick sediment.

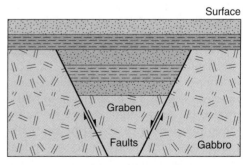

Figure 17.26

A graben filled with sediment can give a negative magnetic anomaly if the sediment contains fewer magnetic minerals than the rock beneath it.

Figure 17.27

Distribution of (*A*) temperature and (*B*) pressure with depth into the earth. Part (*A*) shows both the old and new estimates of the earth's internal temperature.

Not all local magnetic anomalies are caused by variations in rock type. The linear magnetic anomalies found at sea are apparently caused by a *variation in the direction of magnetism,* as you will see in chapter 19.

Heat Within the Earth

Geothermal Gradient

The temperature increase with depth into the earth is called the **geothermal gradient.** The geothermal gradient can be measured on land in abandoned wells or on the sea floor by dropping specially designed probes into the mud. The average temperature increase is 25°C per kilometer (about 75°F per mile) of depth. Some regions have a much higher gradient, indicating concentrations of heat at shallow depths. Such regions have a potential for generating *geothermal energy* (discussed in chapter 11).

The temperature increase with depth creates a problem in deep mines, such as in a 3-kilometer-deep gold mine in South Africa, where the temperature is close to the boiling point of water. Deep mines must be cooled by air-conditioning for the miners to survive. High temperatures at depth also complicate the drilling of deep oil wells. A well drilled to a depth of 7 or 8 kilometers must pass through rock with a temperature of 200°C. At such high temperatures, a tough steel drilling pipe will become soft and flexible unless it is cooled with a special mud solution pumped down the hole (see Box 17.1).

Geologists believe that the geothermal gradient must taper off sharply a short distance into the earth. The high values of 25°C/kilometer recorded near the earth's surface could not continue very far into the earth. If they did, the temperature would be 2500°C at the shallow depth of 100 kilometers. This temperature is above the melting point of all rocks at that depth—even though the increased pressure with depth into the earth increases the melting point of rocks. Seismic evidence seems to indicate a solid, not molten, mantle, so the geothermal gradient must drop to values as low as 1°C/kilometer within the mantle (figure 17.27*A*).

At the boundary between the inner core and the outer core, there would be some constraints on possible temperatures if the core is molten metal above the boundary and solid metal below. The weight of the thick rock layer of the mantle and the liquid metal of the outer core raises the pressure at this boundary (figure 17.27*B*) to about 3 million atmospheres. (An *atmosphere* of pressure is the force per unit area caused by the weight of the air in the atmosphere. It is about 1 kilogram per square centimeter, or 14.7 pounds per square inch.)

For many years geologists were reasonably certain that the temperature at the boundary between the inner core and the outer core was 3700°C ± 500°C. Calculations extrapolating from the surface downward showed that this temperature would melt the outer metal core at the 3-million-atmosphere pressure found there, while greater pressure below this boundary would keep the metal of the inner core solid. The temperature at the center of the earth was assumed to be about 4000°C.

Figure 17.28

Some regions have higher heat flow than others; the amount of heat flow is indicated by the length of the arrow. Regions of high heat flow may be underlain by cooling magma or uranium-rich igneous rock.

Figure 17.29

The average heat flow from oceans and continents is the same, but the origin of the heat differs from the ocean to continents.

Recent laboratory experiments with pressure anvils and giant guns have created (for a millionth of a second) the enormous pressures found at the center of the earth. The measured temperature at this pressure was far higher than expected. New estimates of the earth's internal temperatures have resulted: 4800°C at the core-mantle boundary, 6600°C at the inner-core/outer-core boundary, and 6900°C at the earth's center (1000° hotter than the surface of the sun). Both the old and new estimates of temperature are shown in figure 17.27.

Heat Flow

A small but measurable amount of heat from the earth's interior is being lost gradually through the earth's surface. This gradual loss of heat through the earth's surface is called **heat flow.** What is the origin of the heat? It could be "original" heat from the time that the earth formed, that is, *if* the earth formed as a hot mass that is now cooling down. Or the heat could be a by-product of the decay of radioactive isotopes inside the earth. Radioactive decay *may* actually be warming up the earth. Geologists are not sure whether the earth formed as a hot or cold mass, or whether the earth is now cooling off or warming up. Changes in the earth's internal temperature are extremely slow, and trying to work out its thermal history is a slow, often frustrating, job.

Some regions on earth have a high heat flow. More heat is being lost through the earth's surface in these regions than is normal. High heat flow is usually caused by the presence of a magma body or still-cooling pluton near the surface (figure 17.28). An old body of igneous rock that is rich in uranium and other radioactive isotopes can cause a high heat flow, too, because radioactive decay produces heat as it occurs. High heat flow over an extensive area may be due to the rise of warm mantle rock beneath abnormally thin crust.

The average heat flow from continents is the same as the average heat flow from the sea floor, a surprising fact if you consider the greater concentration of radioactive material in continental rock (figure 17.29). The unexpectedly high average heat flow under the ocean may be due to hot mantle rock rising slowly by convection under parts of the ocean (see chapter 1). Regional patterns of high heat flow and low heat flow on the sea floor (heat flow decreases away from the crest of the mid-oceanic ridge) may also be explained by convection of mantle rock, as we discuss in chapter 19.

Summary

The interior of the earth is studied indirectly by *geophysics*—a study of seismic waves, gravity, earth magnetism, and earth heat.

Seismic reflection and *seismic refraction* can indicate the presence of boundaries between rock layers.

The earth is divided into three major units—the *crust,* the *mantle,* and the *core.*

The earth's crust beneath oceans is 7 kilometers thick and made of basalt on top of gabbro. Continental crust is 30 to 50 kilometers thick and consists of a crystalline basement of granite and gneiss (and other rocks) capped by sedimentary rocks.

The *Mohorovičić discontinuity* separates the crust from the mantle.

The mantle is a layer of solid rock 2,900 kilometers thick and is probably composed of an ultramafic rock such as peridotite. Seismic waves show the mantle has a structure of concentric shells, perhaps caused by pressure transformations of minerals.

The *lithosphere,* which forms plates, is made up of brittle crust and upper mantle. It is 70 to 125 (or more) kilometers thick and moves over the plastic asthenosphere.

The *asthenosphere* lies below the lithosphere and may represent rock close to its melting point (seismic waves slow down here). It is probably the region of most magma generation and isostatic adjustment.

Seismic-wave shadow zones show the core has a radius of 3,450 kilometers and is divided into a liquid outer core and a solid inner core. A core composition of mostly iron is suggested by the earth's density, the composition of meteorites, and the existence of the earth's magnetic field.

Isostasy is the equilibrium of crustal columns floating on plastic mantle. *Isostatic adjustment* occurs when weight is added to or subtracted from a column of rock. *Crustal rebound* is isostatic adjustment that occurs after the melting of glacial ice.

A *gravity meter* can be used to study variations in rock density or to find regions that are out of isostatic equilibrium.

A *positive gravity anomaly* forms over dense rock or over regions being held up out of isostatic balance. A *negative gravity anomaly* indicates low-density rock or a region being held down.

The earth's *magnetic field* has two *magnetic poles*, probably generated by convection circulation and electric currents in the earth's outer core.

Some rocks record the earth's magnetism at the time they form. *Paleomagnetism* is the study of ancient magnetic fields.

Magnetic reversals of polarity occurred in the past, with the north magnetic pole and south magnetic pole exchanging positions. Isotopic dating of rocks shows the ages of the reversals.

A *magnetometer* measures the strength of the earth's magnetic field.

A *positive magnetic anomaly* develops over rock that is more magnetic than neighboring rock. A *negative magnetic anomaly* indicates rock with low magnetism.

Magnetic anomalies can also be caused by circulation patterns in the earth's core and by variations in the direction of rock magnetism.

The *geothermal gradient* is about 25°C/kilometer near the earth's surface but decreases rapidly at depth. The temperature at the center of the earth may be 6900°C. *Heat flow* measurements show that heat loss per unit area from continents and oceans is about the same, perhaps because of convection of hot mantle rock beneath the oceans.

Terms to Remember

asthenosphere 419
convection 423
core 418
crust 417
crustal rebound 424
Curie point 428
geophysics 416
geothermal gradient 432
gravity meter 426
heat flow 433

isostasy 423
isostatic adjustment 423
lithosphere 419
magnetic field 427
magnetic pole 427
magnetic reversal 428
magnetometer 429
mantle 418
Mohorovičić discontinuity 418
negative gravity anomaly 427

negative magnetic anomaly 431
paleomagnetism 428
positive gravity anomaly 426
positive magnetic anomaly 429
P-wave shadow zone 421
seismic reflection 416
seismic refraction 416
S-wave shadow zone 421

Testing Your Knowledge

Use the questions below to prepare for exams based on this chapter.

1. Describe how seismic reflection and seismic refraction show the presence of layers within the earth.

2. Sketch a cross section of the entire earth showing the main subdivisions of the earth's interior and giving the name, thickness, and probable composition of each.

3. What facts make it probable that the earth's core is composed of mostly iron?

4. Describe the differences between continental crust and oceanic crust.

5. What is a gravity anomaly, and what does it generally indicate about the rocks in the region where it is found?

6. Discuss seismic-wave shadow zones and what they indicate about the earth's interior.

7. Describe the earth's magnetic field. Where is it generated?

8. What is the temperature distribution with depth into the earth?

9. Heat flow has been found to be about equal through continents and the sea floor. Why was this unexpected? What might cause this equality?

10. What is the Mohorovičić discontinuity?

11. What is the asthenosphere? Why is it important?

12. How does the lithosphere differ from the asthenosphere?

13. What is a magnetic reversal? What is the evidence for magnetic reversals?

14. What is a magnetic anomaly? How are magnetic anomalies measured at sea?

15. Sial and sima are terms used by some geologists to describe (a) composition of continental and oceanic crust (b) behavior of earthquake waves (c) regions in the mantle

16. The boundary that separates the crust from the mantle is called the (a) lithosphere (b) asthenosphere (c) Mohorovičić discontinuity (d) none of the above

17. The core is probably composed mainly of (a) silicon (b) sulfur (c) oxygen (d) iron

18. The principle of continents being in a buoyant equilibrium is called (a) subsidence (b) isostasy (c) convection (d) rebound

19. A positive gravity anomaly indicates that (a) tectonic forces are holding a region up out of isostatic equilibrium (b) the land is sinking (c) local mass deficiencies exist in the crust (d) all of the above

20. A positive magnetic anomaly could indicate (a) a body of magnetic ore (b) the magnetic field strength is higher than the

regional average (c) an intrusion of gabbro (d) the presence of a granitic basement high (e) all of the above

21. Which of the following is not an example of the effects of isostasy? (a) deep mountain roots (b) magnetic reversals (c) the post-glacial rise of northeastern North America (d) mountain ranges at subduction zones

22. The S-wave shadow zone is evidence that (a) the core is made of iron and nickel (b) the inner core is solid (c) the outer core is fluid (d) the mantle is plastic

Expanding Your Knowledge

1. If the earth were hollow, what evidence would we have from seismic-wave shadow zones? Could a hollow earth have an average density of 5.5 gm/cm³? Could a hollow earth have a magnetic field?

2. Subsidence of the earth's surface sometimes occurs as reservoirs fill behind newly built dams. Why?

3. What geologic processes might cause the forces that can hold a region out of isostatic equilibrium?

4. Does the correlation of a species extinction with a magnetic reversal prove that the reversal caused the extinction?

5. If the upper mantle is chemically different from the lower mantle, is mantle convection possible?

Exploring Resources

 Bloxham, J., and D. Gubbins. 1989. The evolution of the earth's magnetic field. *Scientific American* 261(6): 68–75.

Bolt, B. A. 1982. *Inside the earth.* New York: W. H. Freeman.

Burchfiel, B. C. 1983. The continental crust. *Scientific American* 249(3): 86–98.

Carlowicz, M. 1996. Spin control. *Earth* 12(21): 62–63.

Carrigan, C. R., and D. Gubbins. 1979. The source of the earth's magnetic field. *Scientific American* 240(2): 118–30.

Fowler, C. M. R. 1990. *The solid earth.* New York: Cambridge Univ. Press.

Jacobs, J. A. 1992. *The deep interior of the earth.* New York: Chapman & Hall.

Jeanloz, R. 1990. The nature of the Earth's core. *Annual Review of Earth and Planetary Sciences* 18:357–86.

Jeanloz, R., and T. Lay. 1993. The core-mantle boundary. *Scientific American* 268(5): 48–55.

Kerr, R. A. 1991. Do plumes stir earth's entire mantle? *Science* vol. 252: 1068–1069.

Kerr, R. A. 1997. Deep-sinking slabs stir the mantle. *Science* vol. 275:613–15.

McKenzie, D. P. 1983. The earth's mantle. *Scientific American* 249(3): 50–62.

Wyession, M. E. 1996. Journey to the center of the earth. *Earth* 12:46–49.

 http://bang.lanl.gov/solarsys/edu/convect.htm
Jet Propulsion Laboratory's Guide to Convection

Interacting with Journey Through Geology CD-ROM

Expand your knowledge of the concepts presented in this chapter by using the CD-ROM to answer the following question.

1. Spreading Ridges, Iceland: How are changes in the earth's magnetic field recorded in the lavas of Iceland? How does this help geologists read the history of earth's plates?

Hydrothermal vent and deep-sea organisms from the Galápagos Rift in the Pacific Ocean. Photograph taken from the *Alvin* submersible

Photo © D. Foster/WHOI/VU

CHAPTER
18

The Sea Floor

*T*he rocks and topography of the sea floor are different from those on land. To understand the evidence for plate tectonics in the next chapter, you need to understand the nature of major sea-floor features such as mid-oceanic ridges, oceanic trenches, and fracture zones, as well as the surprisingly young age of the sea-floor rocks.

The material discussed in this chapter and the next are an excellent example of how the scientific method works. This chapter is concerned with the physical *description* of most sea-floor features—the data-gathering part of the scientific method. The next chapter shows how the theory of plate tectonics explains the *origin* of many of these features. Geologists generally agree upon the descriptions of features but often disagree on their interpretations. As you read, keep a clear distinction in your mind between *data* and the *hypotheses* used to explain the data.

Origin of the Ocean

As mentioned in the last chapter, geologists are unsure about the thermal history of the earth. The following scenario for the origin of the earth and the ocean is highly speculative. Many geologists would disagree with some of the statements; they are offered merely as an example of what could have happened in the earth's past.

According to a widely accepted theory, about 4.5 billion years ago the earth began to form by the accretion of small, cold chunks of rock and metal that surrounded the sun. As the earth grew it began to heat up because of the heat of collisional impact, gravita-

A

B

C

Figure 18.1

Bottom sampling devices. (*A*) Rock dredge for sampling hard rock. An unusually large rock is caught in the mouth of the dredge. Additional rocks are visible in the chain bag. (*B*) Corer for sampling sea-floor sediment. Sediment is caught inside the pipe when the corer is dropped to the sea floor. The bomb-shaped part of the corer adds weight to the core pipe (mostly under water). (*C*) A core of sea-floor sediment. The sediment has been forced out of the core barrel and sawed in half lengthwise, so that the sediment layers are visible.

Photo *A* from Scripps Institution of Oceanography, University of California, San Diego; photo *C* by D. Hopkins, U.S. Geological Survey

A   B

Figure 18.2

(*A*) The *Joides Resolution* is a ship specially built for sampling both sediment and rock from the deep ocean floor. (*B*) The small research submersible ALVIN of Woods Hole Oceanographic Institution in Massachusetts; it is capable of taking three oceanographers to a depth of 12,000 feet (about 4,000 meters).

Photo *A* by U.S. Geological Survey; photo *B* by Woods Hole Oceanographic Institution

tional compaction, and radioactive decay of elements such as uranium. The temperature of the earth rose until its iron melted and "fell" to the earth's center to form its core. Violent volcanic activity occurred at this time, releasing great quantities of water vapor and other gases from the earth's interior and perhaps even covering the earth's surface with a thick, red-hot sea of lava. The earth began to cool as its growth and internal reorganization slowed down and as the amount of radioactive material was reduced by decay. Eventually the earth's surface became solid rock, cool enough to permit the condensation of billowing clouds of volcanic water vapor to form liquid water, that fell as rain. Thus the modern oceans were born, perhaps 4 billion years ago. The oceans grew in size as volcanic *degassing* of the earth continued, and became salty as the water picked up chlorine from other volcanic gases, and sodium (and calcium and magnesium) from the chemical weathering of minerals on the earth's surface.

Geologists generally agree that the ocean is very old, and that it is salty for the reasons stated. Most would also agree that the oceans resulted from degassing of the earth's interior.

Methods of Studying the Sea Floor

Oceans cover more than 70% of the earth's surface. Even though the rocks of the sea floor are widespread, they are difficult to study. Geologists have to rely on small samples of rock taken from the sea floor and brought to the surface, or they must study the rocks indirectly by means of instruments on board ships. Although the sea floor is difficult to study, its overall structure is relatively simple (as we discuss later in the chapter), so the small number of samples is not nearly as much a problem as it would be in studying continental regions, where the structure is usually much more complex. The study of sea-floor rocks, sediment, and topography provided most of the information that led to the concept of plate tectonics.

Samples of rock and sediments can be taken from the sea floor in several ways. Rocks can be broken from the sea floor by a *rock dredge,* which is an open steel container dragged over the ocean bottom at the end of a cable. Sediments can be sampled with a *corer,* a weighted steel pipe dropped vertically into the mud and sand of the ocean floor (figure 18.1).

Both rocks and sediments can be sampled by means of *sea-floor drilling.* Offshore oil platforms drill holes in the relatively shallow sea floor near shore. A ship with a drilling derrick on its deck can drill a hole in the deep sea floor far from land (figure 18.2*A*). The drill cuts long, rodlike rock cores from the ocean floor. Thousands of such holes have been drilled in the sea floor, and the rock and sediment cores recovered from these holes have revolutionized the field of marine geology. In the 1950s more was known about the moon's surface than about the floor of the sea. Sea-floor drilling has been instrumental in expanding our knowledge of sea-floor features and history. Small research submarines, more correctly called *submersibles,* can take geologists to many parts of the sea floor to observe, photograph, and sample rock and sediment (figure 18.2*B*).

A basic tool for indirectly studying the sea floor is the *echo sounder,* which draws profiles of submarine topography (figure 18.3). A sound sent downward from a ship bounces off the sea floor and returns to the ship. The water depth is determined from the time it takes the sound to make the round trip. A *seismic profiler* works on essentially the same principles as the echo sounder but uses a louder noise at lower frequency. This sound penetrates the bottom of the sea and reflects from layers within the rock and sediment. The seismic profiler gives more information than the echo sounder. It records water depth and reveals the internal structure of the rocks and sediments of the sea floor, such as bedding planes, folds and faults, and unconformities (figure 18.4). *Magnetic, gravity,* and *seismic refraction* surveys (see chapter 17) also can be made at sea. *Deep-sea*

A

Figure 18.4

Seismic profiler record of an abyssal plain, showing sediment layers that have buried an irregular rock surface in the Atlantic Ocean.

From Vogt, et al, in Hart, *The Earth's Crust and Upper Mantle*, p. 574, 1969, copyrighted by the American Geophysical Union

B

Figure 18.3

Echo-sounding. (*A*) A sound bounces off the sea floor and returns to a ship. (*B*) An echo-sounder record of the continental shelf and continental slope off Virginia. The scattering layer is caused by a dense concentration of small marine organisms.

Echo-sounder record courtesy of Peter Rona

Figure 18.5

Photograph of the sea floor on the side of a South Pacific island at a depth of 1000 meters. The rocks are basalt. Rippled sediment between the rocks indicates that strong currents move sea-floor sediment here. Field of view is about 1.5 by 2 meters.

From Lamont-Doherty Geological Observatory

cameras can be lowered to the bottom to photograph the rock and sediment (figure 18.5).

Features of the Sea Floor

Figure 18.6, a simplified profile of the sea floor, shows that continents have two types of margins. A *passive continental margin,* as found on the east coast of North America, includes a continental shelf, continental slope, and continental rise. An *abyssal plain* usually forms a remarkably flat ocean floor beyond the continental rise. An *active continental margin,* mainly found around the Pacific rim, is associated with earthquakes and volcanoes and has a continental shelf and slope, but the slope extends much deeper to form one wall of an *oceanic trench.* Abyssal plains are seldom found off active margins. The deep ocean floor seaward of trenches is hilly and irregular, lacking the extreme flatness of abyssal plains. Encircling the globe is a *mid-oceanic ridge,* usually (but not always)

near the center of an ocean. Conical *seamounts* stick up from the sea floor in some regions. Some important submarine features do not show in this profile view—in particular, fracture zones, submarine canyons, and aseismic ridges.

Continental Shelves and Continental Slopes

Almost all continental edges are marked by a gently sloping continental shelf near shore, and a steeper continental slope that leads down to the deep ocean floor (figure 18.7). Figures

Figure 18.6

Profiles of sea-floor topography. The vertical scales differ from the horizontal scales, causing vertical exaggeration, which makes slopes appear steeper than they really are. The bars for the horizontal scale are 50 kilometers long, while the same distance vertically represents only 5 kilometers, so the drawings have a vertical exaggeration of 10.

Figure 18.7

Continental shelf and continental slope. (*A*) Vertical exaggeration 25 ×. The continental slope has an actual slope of only 4 or 5 degrees, but the great vertical exaggeration of this drawing makes it appear to be sloping about 60 degrees. (*B*) Same profile with no vertical exaggeration.

such as 18.6 and 18.7 are usually drawn with great *vertical exaggeration,* which makes submarine slopes appear much steeper than they really are.

A **continental shelf,** a shallow submarine platform at the edge of a continent, inclines very gently seaward, generally at an angle of 0.1°. Continental shelves vary in width. On the Pacific coast of North America the shelf is only a few kilome-

ters wide, but off Newfoundland in the Atlantic Ocean it is about 500 kilometers (300 miles) wide. Portions of the shelves in the Arctic Ocean off Siberia and northern Europe are even wider. Water depth over a continental shelf tends to increase regularly away from land, with the outer edge of the shelf being about 100 to 200 meters below sea level.

Continental shelves are *topographic features,* defined by their depth, flatness, and gentle seaward tilt. Their *geologic origin* varies from place to place and is related to plate tectonics, so it will be discussed in the next chapter. Some generalities about shelves, however, are worth noting here.

The continental shelves of the world are usually covered with relatively young sediment, in most cases derived from land. The sediment is usually sand near shore, where the bottom is shallow and influenced by wave action. Fine-grained mud is commonly deposited farther offshore in deeper, quieter water.

The outer part of a wide shelf is often covered with coarse sediment that was deposited near shore during a time of lower sea level. The advance and retreat of continental glaciers during the Pleistocene Epoch caused sea level to rise and fall many times by 100 to 200 meters. This resulted in a complex history of sedimentation for continental shelves as they were alternately covered with seawater and exposed as dry land.

Marine seismic surveys and drilling at sea have shown that the young sediments on many continental shelves are underlain by thick sequences of sandstone, shale, and less commonly limestone. These sedimentary rocks are of late Mesozoic and Cenozoic age, and appear to have been deposited in much the same way as the modern shelf sediments. Beneath these rocks is continental crust (figure 18.8); the continental shelves are truly part of the continents, even though today the shelves are covered by seawater.

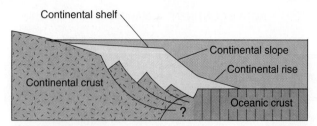

Figure 18.8

The continental shelf lies upon continental crust, and the continental rise lies upon oceanic crust. The complex transition from continental crust to oceanic crust lies under the continental slope.

A **continental slope** is a relatively steep slope that extends from a depth of 100 to 200 meters at the edge of the continental shelf down to oceanic depths. The average angle of slope for a continental slope is 4° to 5°, although locally some parts are much steeper.

Because the continental slopes are more difficult to study than the continental shelves, less is known about them. The greater depth of water and the locally steep inclines on the continental slopes hinder rock dredging and drilling and make the results of seismic refraction and reflection harder to interpret. This is unfortunate, for the rocks that underlie the slopes are of particular interest to marine geologists, who believe that in this area the thick continental crust (beneath the land and the continental shelves) grades into thin oceanic crust (underneath the deep ocean floor), as shown in figure 18.8.

Although relatively little is known about continental slopes, it is clear that their character and origin vary greatly from place to place. Because these variations are associated with the differences between divergent plate boundaries and convergent plate boundaries, we will discuss them in the next chapter.

Submarine Canyons

Submarine canyons are V-shaped valleys that run across continental shelves and down continental slopes (figure 18.9). On narrow continental shelves, such as those off the Pacific coast of the United States, the heads of submarine canyons may be so close to shore that they lie within the surf zone. On wide shelves, such as those off the Atlantic coast of the United States, canyon heads usually begin near the outer edge of the continental shelf tens of kilometers from shore. Great fan-shaped deposits of sediment called **abyssal fans** are found at the base of many submarine canyons (figure 18.9). Abyssal fans are made up of land-derived sediment that has moved down the submarine canyons. Along continental margins that are cut by submarine canyons, many coalescing abyssal fans may build up at the base of the continental slope.

Submarine canyons are erosional features, but how rock and sediment are removed from the steep-walled canyons is controversial. Erosional agents probably vary in relative importance from canyon to canyon. Divers have filmed *down-canyon movement of sand* in slow, glacierlike flow and in more rapid sand falls (figure 18.10). This sand movement, which has been observed to cause erosion of rock, is particularly common in

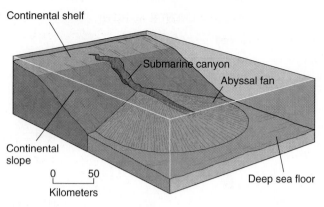

Figure 18.9

Submarine canyon and abyssal fan.

Figure 18.10

A 10-meter-high sand fall in a submarine canyon near the southern tip of Baja California, Mexico. The sand is beach sand, fed into the nearshore canyon head by longshore currents.

From Scripps Institution of Oceanography, University of California, San Diego

Pacific coast canyons, which collect great quantities of sand from longshore drift. *Bottom currents* have been measured moving up and down the canyons in a pattern of regularly alternating flow, in some cases apparently caused by ocean tides. The origin of these currents is not well understood, but they often move fast enough to erode and transport sediment. *River erosion* may have helped to cut the upper part of canyons when the drop in sea level during the Pleistocene glaciations left canyon heads above the water, such as the extension of the Hudson River into the Atlantic. Many (but not all) submarine canyons are found off land canyons or rivers, which tends to support the view that river erosion helped shape them. It is unlikely, however, that the deeper parts of submarine canyons were ever exposed as dry land.

Turbidity Currents

In addition to the canyon-cutting processes just described, turbidity currents probably play the major role in canyon erosion. **Turbidity currents** are great masses of sediment-laden water that are pulled downhill by gravity. The sediment-laden water is heavier than clear water, so the turbidity current flows down the continental slope until it comes to rest on the flat abyssal plain at the base of the slope (see figure 6.13). Turbidity currents are thought to be generated by underwater earthquakes and landslides, strong surface storms, and floods of sediment-laden rivers discharging directly into the sea on coasts with a narrow shelf. Although large turbidity currents have not been directly observed in the sea, small turbidity currents can be made and studied in the laboratory.

Indirect evidence also indicates that turbidity currents occur in the sea. The best evidence comes from the breaking of submarine cables that carry telephone and telegraph messages across the ocean floor. Figure 18.11 shows a downhill sequence of cable breaks that followed a 1929 earthquake in the Grand Banks region of the northwest Atlantic. This sequence of cable breaks has been interpreted to be the result of an earthquake-caused turbidity current flowing rapidly down the continental slope.

If cable breaks are caused by turbidity currents, they give good evidence of the currents' dramatic size, speed, and energy. Breaks in Grand Banks cables continued for more than thirteen hours after the 1929 earthquake, the last of the series occurring more than 700 kilometers from the epicenter. The velocity of the flow that caused the breaks has been calculated to be from 15 to 60 kilometers per hour. Sections of cable more than 100 kilometers long were broken off and carried away, both ends of a missing section being broken simultaneously. Attempts to find broken cable sections were fruitless, and it is assumed that they were buried by sediment.

The Grand Banks 1929 cable breaks are not unique. Cables crossing submarine canyons are broken frequently, particularly after river floods and earthquakes. In the submarine canyons off the Congo (Zaire) River of Africa and off the Magdalena River in Colombia, for example, cables break every few years.

Additional indirect evidence for the existence of turbidity currents comes from the graded bedding and shallow-water fossils in the sediments that make up the continental rise and the abyssal plains, as described in the next section.

Passive Continental Margins

A **passive continental margin** (figure 18.6) includes a continental shelf, continental slope, and continental rise and generally extends down to an abyssal plain at a depth of about 5 kilometers (15,000 feet). It is called a passive margin because it usually develops on geologically quiet coasts, that lack earthquakes, volcanoes, and young mountain belts.

Map

A

B

Figure 18.11

Submarine cable breaks following the Grand Banks earthquake of 1929. (*A*) Map view of the cable breaks. Black dots near the epicenter show locations of cable breaks that were simultaneous with the earthquake (cables not shown for these breaks). Colored dots show cable breaks that followed the earthquake, with the time of each break shown (on the 24-hour clock). Segments of cable more than 100 kilometers long were broken simultaneously at both ends and then carried away. Dashes show sea-floor channels that probably concentrated the flow of a turbidity current, increasing its velocity. (*B*) Profile showing the time elapsed between the quake and each cable break.

A From H. W. Menard, 1964, *Marine Geology of the Pacific,* copyright McGraw-Hill, Inc. *B* from B. C. Heezen and M. Ewing, 1952, *American Journal of Science.*

Passive margins are found on the edges of most land masses bordering the Atlantic Ocean. They also border most parts of the Arctic and Indian oceans and a few parts of the Pacific Ocean.

The Continental Rise

Along the base of many parts of the continental slope lies the **continental rise,** a wedge of sediment that extends from the

Figure 18.12

A contour current flowing along the continental margin shapes the continental rise by depositing fine sediment. Coarse layers within the continental rise were deposited by turbidity currents flowing down the continental slope.

lower part of the continental slope to the deep sea floor. The continental rise, which slopes at about 0.5°, more gently than the continental slope, typically ends in a flat abyssal plain at a depth of about 5 kilometers. The rise rests upon oceanic crust (figure 18.8).

Types of Deposition

Sediments appear to be deposited on the continental rise in two ways—by turbidity currents flowing *down* the continental slope and by *contour currents* flowing *along* the continental slope.

Cores of sediment recovered from most parts of the continental rise show layers of fine sand or coarse silt interbedded with layers of fine-grained mud. The mineral grains and fossils of the coarser layers indicate that the sand and silt came from the shallow continental shelf. Some transporting agent must have carried these sediments from shallow water to deep water. The coarse layers also exhibit graded bedding, which indicates that they settled out of suspension according to size and weight; therefore, the transporting agent for these sediments was most likely turbidity currents. The continental rise in these locations probably formed as turbidity currents deposited abyssal fans at the base of a continental slope.

Sediments in some other parts of the continental rise, however, are uniformly fine-grained and show no graded bedding. This sediment appears to have been deposited by the regular ocean currents that flow along the sea bottom rather than by the intermittent turbidity currents that occasionally flow downslope.

A **contour current** is a bottom current that flows parallel to the slopes of the continental margin—*along* the contour rather than *down* the slope (figure 18.12). Such a current runs south along the continental margin of North America in the Atlantic Ocean. Flowing at the relatively slow speed of a few centimeters per second, this current carries a small amount of fine sediment from north to south. The current is thickest along the continental slope and gets progressively thinner seaward. The thick landward part of the current carries and deposits the most sediment. The thinner seaward edge of the current deposits less sediment. As a result, the deposit of sediment beneath the current is wedge shaped, becoming thinner away from land. Similar contour currents apparently shape parts of the continental rise off other continents as well.

Abyssal Plains

Abyssal plains are very flat regions usually found at the base of the continental rise. Seismic profiling has shown that abyssal plains are formed of horizontal layers of sediment. The gradual deposition of sediment buried an older, more rugged topography that can be seen on seismic profiler records as a rock basement beneath the sediment layers (figure 18.4). Samples of abyssal plain sediment show that it is derived from land. Graded bedding within sediment layers suggests deposition by turbidity currents.

Abyssal plains are the flattest features on earth. They generally have slopes less than 1:1,000 (less than one meter of vertical drop for every 1,000 meters of horizontal distance) and some have slopes of only 1:10,000. Most abyssal plains are 5 kilometers deep.

Not all parts of the deep-ocean basin floor consist of abyssal plains. The deep floor is normally very rugged, broken by faults into hills and depressions and dotted with volcanic seamounts. Abyssal plains form only where turbidity currents can carry in enough sediment to bury and obscure this rugged relief. If the sediment is not available or if the bottom-hugging turbidity currents are stopped by a barrier such as an oceanic trench, then abyssal plains cannot develop.

Active Continental Margins

An **active continental margin,** characterized by earthquakes and by a young mountain belt and volcanoes on land, consists of a continental shelf, a continental slope, and an oceanic trench (figure 18.6). An active margin usually lacks a continental rise and an abyssal plain and is associated with convergent plate boundaries.

Active margins are found on the edges of most of the land masses bordering the Pacific Ocean and a few other places in the Atlantic and Indian oceans (figure 18.13). A notable exception in the Pacific Ocean is most of the coast of North America. The active margin's distinctive combination of oceanic trench and land volcanoes is found in three places along the Pacific coast of North America: southern Central America; Washington, Oregon, and northernmost California (the trench here is sediment-filled); and south-central Alaska. But passive margins with abyssal fans and abyssal plains occur along most of the coast of Mexico, California, British Colum-

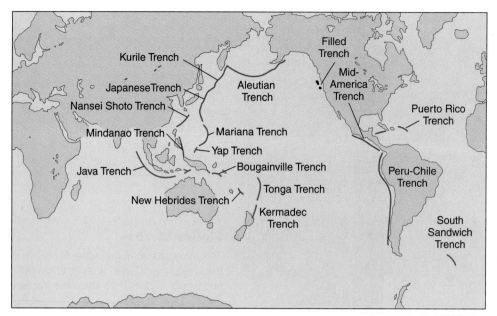

Figure 18.13

The distribution of oceanic trenches. Trenches next to continents mark active continental margins and convergent plate boundaries.

Oceanic Trenches

An **oceanic trench** is a narrow, deep trough parallel to the edge of a continent or an island arc (a curved line of islands like the Aleutians), as shown in figure 18.13. The continental slope on an active margin forms the landward wall of the trench, its steepness often increasing with depth. The slope is typically 4° to 5° on the upper part, steepening to 10° to 15° or even more near the bottom of the trench. The elongate oceanic trenches, often 8 to 10 kilometers deep, far exceed the average depth of abyssal plains on passive margins. The deepest spots on earth, more than 11 kilometers below sea level, are in oceanic trenches in the southwest Pacific Ocean.

Associated with oceanic trenches are the *earthquakes of the Benioff seismic zones* (see chapter 16), which begin at a trench and dip landward under continents or island arcs (figure 18.14). *Volcanoes* are found above the upper part of the Benioff zones and typically are arranged in long belts parallel to oceanic trenches. These belts of volcanoes form island arcs or erupt within young mountain belts on the edges of continents. The rock produced by these volcanoes is usually andesite, a type of extrusive rock intermediate in composition between basaltic oceanic crust and "granitic" continental crust.

Oceanic trenches are marked by abnormally *low heat flow* compared to normal ocean crust. This implies that the crust in trenches may be colder than normal crust.

As you learned in the previous chapter, oceanic trenches are also characterized by very large *negative gravity anomalies,* the largest in the world. This implies that trenches are being held down, out of isostatic equilibrium.

bia, and the Alaskan panhandle. These coasts *do* have earthquakes, mostly along strike-slip faults, and probably were full-fledged active margins in the geologic past, but are considered to be passive margins today.

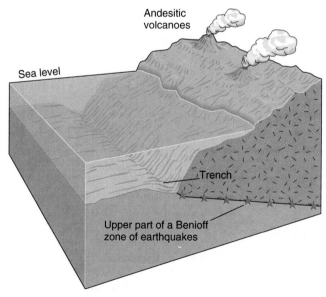

Figure 18.14

Active continental margin with an oceanic trench, a Benioff zone of earthquakes (only the upper part is shown), and a chain of andesitic volcanoes on land.

The Mid-Oceanic Ridge

The **mid-oceanic ridge** is a giant undersea mountain range that extends around the world like the seams on a baseball (figures 18.15 and 18.16). The ridge, which is made up mostly of basalt, is more than 80,000 kilometers long and 1,500 to 2,500 kilometers wide. It rises 2 to 3 kilometers above the ocean floor.

A **rift valley** of tensional origin runs down the crest of the ridge (figures 18.15 and 18.16). The rift valley is 1 to 2 kilometers deep and several kilometers wide—about the dimensions of the Grand Canyon in Arizona. The rift valley on the crest of the mid-oceanic ridge is a unique feature—no mountain range on land has such a valley running along its crest. The rift valley is present in the Atlantic and Indian oceans, but is generally absent in the Pacific Ocean. (This absence is related to the faster rate of plate motion in the Pacific, as we will discuss in chapter 19.)

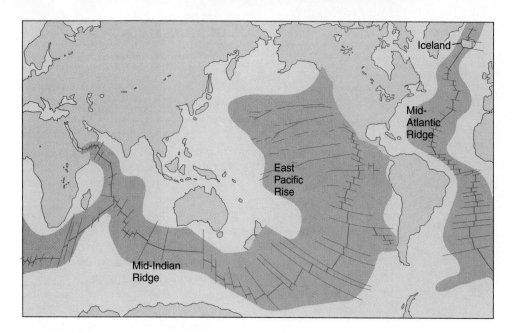

Figure 18.15

The mid-oceanic ridge, offset from fracture zones. Darker lines indicate the ridge crest; lighter lines show the fracture zones.

Geologic Activity on the Ridge

Associated with the rift valley at the crest of the mid-oceanic ridge and also with the riftless crest of the ridge in the Pacific Ocean, are *shallow-focus earthquakes* from 0 to 20 kilometers below the sea floor.

Careful measurements of the heat loss from the earth's interior through the crust have shown a very *high heat flow* on the crest of the mid-oceanic ridge. The heat loss at the ridge crest is many times the normal value found elsewhere in the ocean; it decreases away from the ridge crest.

Basalt eruptions occur in and near the rift valley on the ridge crest. Sometimes these eruptions build up volcanoes that protrude above sea level as oceanic islands. The large island of Iceland (figure 18.15), which is mostly basaltic, appears to be a section of the mid-oceanic ridge elevated above sea level. Many geologists have studied the active volcanoes, high heat flow, and central rift valley of Iceland to learn about the mid-oceanic ridge. As Iceland is above sea level, however, it may not be a typical portion of the ridge.

In the summer of 1974 geologists were able to get a first-hand view of part of the submerged ridge and rift valley. A series of more than forty dives by submersibles, including ALVIN, carried French and American marine geologists directly into the rift valley in the North Atlantic Ocean. The project (called FAMOUS for French-American Mid-Ocean Undersea Study) allowed the ridge rock to be seen, photographed, and sampled directly, rather than indirectly from surface ships (figure 18.17).

The geologists on the FAMOUS project saw clear evidence of extensional faults within the rift valley. These run parallel to the axis of the rift valley and range in width from hairline cracks to gaping fissures that ALVIN dived into. Fresh pillow basalts occur in a narrow band along the bottom of the rift valley, suggesting very recent volcanic activity there, although no active eruptions were observed. It appeared to the geologists that the rift was continuous, and that sporadic volcanic activity occurred as a result of the rifting.

Mid-oceanic ridges are often marked by lines of hot springs that carry and precipitate metals. Geologists in submersibles have observed hot springs in several localities along the rift valleys of the mid-oceanic ridges. The hot springs, caused by the high heat flow and shallow basaltic magma beneath the rift valley, range in temperature from about 20°C up to an estimated 350°C (660°F).

As the hot water rises in the rift valley, cold water is drawn in from the sides to take its place. This creates a circulation pattern in which cold seawater is actually drawn *downward* through the cracks in the basaltic crust of the ridge flanks and then moves horizontally toward the rift valley, where it reemerges on the sea floor after being heated (figure 18.18A). As the seawater circulates, it dissolves metals and sulfur from the crustal rocks. When the hot, metal-rich solutions contact the cold seawater, metal sulfide particles are discharged into the cold seawater at *black smokers,* which precipitate a chimney-like mound around the hot spring (figure 18.18B).

Biologic Activity on the Ridge

The occurrence of black smokers was a surprise to geologists exploring the mid-oceanic ridges, but an even bigger surprise was the presence of exotic, bottom-dwelling organisms surrounding the hot springs. The exotic organisms, including mussels, crabs, starfish, giant white clams and giant tube worms, are able to survive toxic chemicals, high temperatures, high pressures, and total darkness at depths of more than

Figure 18.16 (overleaf)

The floor of the ocean. Note continental margins, mid-oceanic ridge, fracture zones, and oceanic trenches (compare with figures 18.13 and 18.15). Conical mountains are volcanic seamounts; aligned seamounts are aseismic ridges.

From Floor of the Oceans, Bruce C. Heezen and Marie Tharp, 1975, © Marie Tharp 1980. By permission of Marie Tharp, 1 Washington Ave., South Nyack, NY 10960.

Figure 18.17

Underwater photograph of fresh pillow basalt on the floor of the rift valley of the mid-oceanic ridge in the North Atlantic Ocean. A white, tubular sponge grows in the foreground.

Photo by Woods Hole Oceanographic Institution

B

C

Cold sea water drawn into oceanic crust

RIFT VALLEY
Hot springs and black smokers

SEA FLOOR

Ore

Hydrothermally circulating sea water

1 Kilometer

Hot magma

A

Figure 18.18

(*A*) Hydrothermal circulation of sea water at ridge crest creates hot springs and metallic deposits in rift valley. Cold sea water is drawn into fractured crust on ridge flanks. (Size of ore deposit exaggerated.) (*B*) "Black smoker" or submarine hot spring on the crest of the mid-oceanic ridge in the Pacific Ocean near 21° North latitude. The "smoke" is a hot plume of metallic sulfide minerals being discharged into cold sea water from a chimney 0.5 meters high. The large mounds around the chimney are metallic deposits. The instruments in the foreground are attached to the small submarine from which the picture was taken. (*C*) Giant worms, crabs, and clams from Galápagos vent.

Photo (*B*) by U.S. Geological Survey. Photo (*C*) © WHOI/D. Foster/Visuals Unlimited

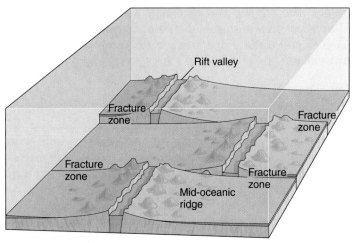

Figure 18.19
Fracture zones, which run perpendicular to the ridge crest and connect offset segments of the ridge are often marked by steep cliffs up to 3 or 4 kilometers high.

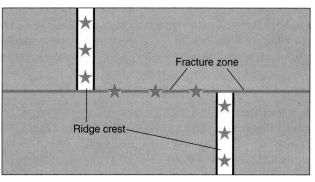

Figure 18.20
Shallow-focus earthquakes (shown by stars) occur on the ridge crest and on the portion of the fracture zone between the two segments of the ridge crest (map view).

2 kilometers (figure 18.18*C*). The organisms live on bacteria that thrive by oxidizing hydrogen sulfide from the hot springs. It is believed that the bacteria normally reside beneath the sea floor but are blown out of the hot spring when it erupts. Such sulfur-digesting bacteria have also been found in acidic water in mines containing sulfide minerals. Current research in the new field of *geomicrobiology* is examining the role such bacteria may have had in the precipitation of minerals and in the evolution of early life-forms on earth and possibly other places in the solar system.

Fracture Zones

Fracture zones are major lines of weakness in the earth's crust that cross the mid-oceanic ridge at approximately right angles (figures 18.15 and 18.16). The rift valley of the mid-oceanic ridge is offset in many places across fracture zones (figure 18.19), and the sea floor on one side of a fracture zone is often at a different elevation than the sea floor on the other side, producing steep cliffs. Shallow-focus earthquakes occur on fracture zones but are confined to those portions of the fracture zones between segments of the rift valley (figure 18.20). (The portion of the fracture zone that has earthquakes is known as a *transform fault;* the origin of these faults is discussed in the next chapter.)

Fracture zones extend for thousands of kilometers across the ocean floor, generally heading straight for continental margins. Although fracture zones are difficult to trace where they are buried by the sediments of the abyssal plain and the continental rise, some geologists think that they can trace the exten-

sions of fracture zones onto continents. Some major structural trends on continents appear to lie along the hypothetical extension of fracture zones onto the continents.

Seamounts, Guyots, and Aseismic Ridges

Conical undersea mountains that rise 1,000 meters or more above the sea floor are called **seamounts** (figures 18.16 and 18.21). Some rise above sea level to form islands. They are scattered on the flanks of the mid-oceanic ridge and on other parts of the sea floor, including abyssal plains. One area of the sea floor with a particularly high concentration of seamounts—one estimate is 10,000—is the western Pacific. Rocks dredged from seamounts are nearly always basalt, so it is thought that most seamounts are extinct volcanoes. Of the thousands of seamounts on the sea floor, only a few are active volcanoes. Most of these are on the crests of the mid-oceanic ridges. A few others, such as the two active volcanoes on the island of Hawaii, are found at locations not associated with a ridge.

Guyots are flat-topped seamounts (figure 18.21) found mostly in the western Pacific Ocean. Most geologists think that the flat summits of guyots were cut by wave action. These flat tops are now many hundreds of meters below sea level, well below the level of wave erosion. If the guyot tops were cut by waves, the guyots must have subsided after erosion took place. Evidence of such subsidence comes from the dredging of dead reef corals from guyot tops. Since such corals grow only in shallow water, they must have been carried to their present depths as the guyots sank.

Many of the guyots and seamounts on the sea floor are aligned in chains. Such volcanic chains, together with some other ridges on the sea floor, are given the name **aseismic ridges** (figures 18.16 and 18.22); that is, they are submarine ridges that are not associated with earthquakes. The name *aseismic* is used to distinguish these features from the much larger mid-oceanic ridge, where earthquakes occur along the rift valley.

Reefs

Reefs are wave-resistant ridges of coral, algae, and other calcareous organisms. They form in warm, shallow, sunlit water that is low in suspended sediment. Reefs stand above the surrounding sea floor, which is often covered with sediment derived from the reef (see figure 6.18). Three important types are *fringing reefs, barrier reefs,* and *atolls* (figure 18.23).

Fringing reefs are flat, tablelike reefs attached directly to shore. The seaward edge is marked by a steep slope leading down into deeper water. Many of the reefs bordering the Hawaiian Islands are of this type.

Barrier reefs parallel the shore but are separated from it by wide, deep lagoons. This type of reef is shown in figure 6.18. The lagoon has relatively quiet water because the reef shelters it by absorbing the energy of large, breaking waves. A barrier reef lies about 8 kilometers (5 miles) offshore of the Florida Keys, a string of islands south of Miami. On a much grander scale is the Great Barrier Reef off northeastern Australia. It extends for about 2,000 kilometers (1,200 miles) along the coast, and its seaward edge lies up to 250 kilometers (150 miles) from shore. Another long barrier reef lies along the eastern coast of the Yucatan Peninsula in Central America, and others surround many islands in the South Pacific.

Atolls are circular reefs that rim lagoons. They are surrounded by deep water. Small islands of calcareous sand may be built by waves at places along the reef ring. The diameter of atolls varies from 1 to more than 100 kilometers (60 miles). Numerous atolls dot the South Pacific. Bikini and Eniwetok atolls were used through 1958 for the testing of nuclear weapons by the United States.

Following the four-year cruise of the HMS *Beagle* in the 1830s, Charles Darwin proposed that these three types of reefs are related to one another by subsidence of a central volcanic island, as shown in figure 18.23. A fringing reef initially becomes established near the island's shore. As the volcano slowly subsides because of tectonic lowering of the sea floor, the reef becomes a barrier reef, because the corals and algae grow rapidly upward, maintaining the reef's position near sea level. Less and less of the island remains above sea level, but the reef grows upward into shallow, sunlit water, maintaining its original size and shape. Finally the volcano disappears completely below sea level, and the reef becomes a circular atoll.

Figure 18.21

(A) Seamounts are conical mountains on the sea floor, occasionally rising above sea level to form islands. (B) The flat summit of a guyot was probably eroded by waves when the top of a seamount was above sea level. (C) The present depth of a guyot is due to subsidence.

Figure 18.22

The distribution of the major aseismic ridges on the sea floor (compare with figure 18.16). The numbers on the Hawaiian-Emperor chain are ages in millions of years (see the end of chapter 19).

From W. Jason Morgan, 1972, *Geological Society of America Memoir 132,* and other sources.

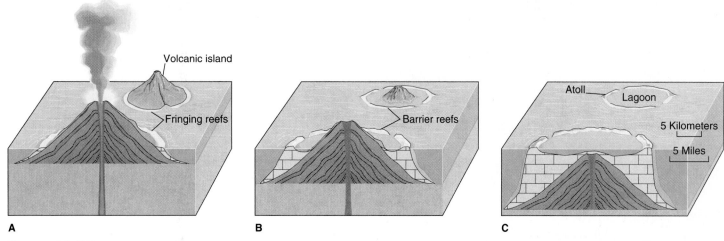

Figure 18.23

Types of coral-algal reefs. (*A*) Fringing reefs are attached directly to the island. (*B*) Barrier reefs are separated from the island by a lagoon. (*C*) Atolls are circular reefs with central lagoons. Charles Darwin proposed that the sequence of fringing, barrier, and atoll reefs form by the progressive subsidence of a central volcano, accompanied by the rapid upward growth of corals and algae.

Figure 18.23 continued

(*D*) Reefs on the island of Mooréa in the Society Islands, south-central Pacific (Tahiti in the background). Living corals appear brown. Fringing reef next to shoreline, barrier reef at breaker line; light blue lagoon between reefs is covered with carbonate sand. The island is an extinct volcano, heavily eroded by stream action.

Figure 18.24

Photograph (taken through a scanning-electron microscope) of pelagic sediment from the floor of the Pacific Ocean. The sediment is made up of microscopic skeletons of single-celled marine organisms (large objects are foraminifera; smaller, sievelike ones are radiolaria about 0.05 mm in diameter).

Drilling through atolls in the 1950s showed that these reefs were built on deeply buried volcanic cores, thus confirming Darwin's hypothesis of 120 years before.

Sediments of the Sea Floor

The basaltic crust of the sea floor is covered in many places with layers of sediment. This sediment is either *terrigenous,* derived from land, or *pelagic,* settling slowly through seawater.

Terrigenous sediment is land-derived sediment that has found its way to the sea floor. The sediment that makes up the continental rise and the abyssal plains is mostly terrigenous and apparently has been deposited by turbidity currents or similar processes. Once terrigenous sediment has found its way down the continental slope, contour currents may distribute it along the continental rise. On active continental margins, oceanic trenches may act as traps for terrigenous sediment and prevent it from spreading out onto the deep sea floor beyond the trenches.

Pelagic sediment is sediment that settles slowly through the ocean water. It is made up of fine-grained clay and the skeletons of microscopic organisms (figure 18.24). Fine-grained pelagic clay is found almost everywhere on the sea floor, although in some places it is masked by other types of sediments that accumulate rapidly. The clay is mostly derived from land; part of it may be volcanic ash. This sediment is carried out to sea primarily by wind, although rivers and ocean currents also help to distribute it.

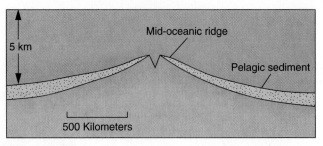

Figure 18.25

Pelagic sediment is thin or absent on the crest of the mid-oceanic ridge and becomes progressively thicker away from the ridge crest (distribution of sediment highly simplified).

Microscopic shells and skeletons of plants and animals also settle slowly to the sea floor when marine organisms of the surface waters die. In some parts of the sea, such as the polar and equatorial regions, great concentrations of these shells have built unusually thick pelagic deposits.

The constant slow rain of pelagic clay and shells occurs in all parts of the sea. Although the rate of accumulation varies from place to place, pelagic sediment should be expected on all parts of the sea floor.

Surprisingly, however, pelagic sediment is almost completely absent on the crest of the mid-oceanic ridge. Pelagic sediment is found on the flanks of the mid-oceanic ridge, often thickening away from the ridge crest (figure 18.25). But its absence on the ridge crest was an unexpected discovery about sea-floor sediment distribution.

Oceanic Crust and Ophiolites

As you have seen in the previous chapter, oceanic crust differs significantly from continental crust; it is both thinner and of a different composition than continental crust. Seismic reflection and seismic refraction surveys at sea have shown the oceanic crust to be about 7 kilometers thick and divided into three layers (figure 18.26).

The top layer (Layer 1), of variable thickness and character, is marine sediment. In an abyssal fan or on the continental rise, Layer 1 may consist of several kilometers of terrigenous sediment. On the upper flanks of the mid-oceanic ridge, there may be less than 100 meters of pelagic sediment. An average thickness for Layer 1 might be 0.5 kilometer.

Beneath the sediment is Layer 2, which is about 1.5 kilometers thick. It has been extensively sampled. This layer consists of pillow basalt overlying dikes of basalt. The basalt pillows, rounded masses that form when hot lava erupts into cold water (figure 18.26), are highly fractured at the top of Layer 2. The dikes in the lower part of Layer 2 are closely spaced, parallel, vertical dikes ("sheeted dikes"). Widely sampled by drilling and dredging, Layer 2 has also been observed directly by geologists in submersibles diving into the rift valley of the mid-oceanic ridge during the FAMOUS expedition.

The lowest layer in the crust is Layer 3. It is about 5 kilometers thick and thought to consist of sill-like gabbro bodies.

Layers

1 — Marine sediment

2 — Pillow basalt and sheeted dikes

Gabbro intrusions

3

Moho

Mantle — Ultramafic rock

7 kilometers

A

Thin-bedded chert

Pillow basalt and sheeted dikes

Gabbro intrusions

Serpentinized ultramafic rock

B

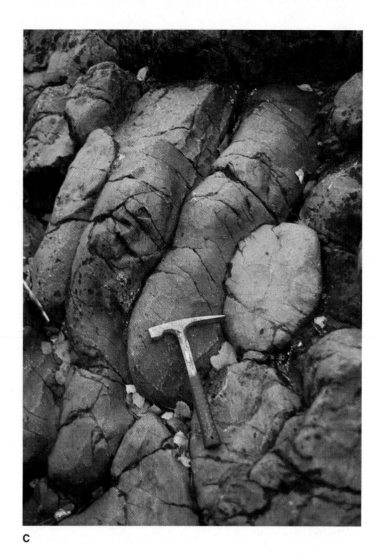

C

Figure 18.26

A comparison of oceanic crust and an ophiolite sequence. (*A*) Structure of oceanic crust, determined from seismic studies and drilling. Gabbro and ultramafic rock drilled and dredged from sea-floor fault blocks may be from Layer 3 and the mantle. (*B*) Typical ophiolite sequence found in mountain ranges on land. Thickness approximate—sequence is usually highly faulted. (*C*) Pillow basalt from the upper part of an ophiolite, northern California. These rocks formed as part of the sea floor, when hot lava cooled quickly in cold seawater.

The evidence for this interpretation is suggestive but not conclusive. Geologists have drilled through 2 kilometers of pillow basalt and basaltic dikes in oceanic crust, but have not yet been able to reach gabbro by drilling through basalt. Gabbro (and other rocks) are exposed on some fault blocks in rift valleys and on some steep submarine cliffs in fracture zones; 0.5 kilometer of gabbro has been drilled on one such cliff in the southern Indian Ocean. Some geologists presume that the gabbro represents Layer 3 exposed by faulting, but other geologists are not so sure. There is some evidence that rocks exposed in fracture zones may not be representative of the entire sea floor. Even if the drilled gabbro *does* represent the upper 1/10 of Layer 3, no one has yet sampled the lower 9/10 of the layer.

Seismic velocities of 7 kilometers per second are consistent with the choice of gabbro for Layer 3, but deep drilling on land (see box 17.1) has shown that seismic reflection and refraction records on land are routinely misinterpreted regarding rock type and depths to rock boundaries. These errors extend to oceanic crust as well. A drillhole in the eastern Pacific Ocean has been reoccupied four times in a 12-year span, and has now reached a total depth of 2,000 meters below the sea floor. Seismic evidence suggested that the Layer 2-Layer 3 boundary would be found at a depth of about 1,700 meters, but the drill went well past that depth without finding the con-

tact between the dikes of Layer 2 and the expected gabbro of Layer 3. Either the seismic interpretation or the model of Layer 3's composition must be wrong.

Geologists' ideas about the composition of oceanic crust are greatly influenced by a study of **ophiolites**, distinctive rock sequences found in many mountain chains on land (figure 18.26). The thin top layer of an ophiolite consists of marine sedimentary rock, often including thin-bedded chert. Below the sedimentary rock lies a zone of pillow basalt, which in turn is underlain by a sheeted-dike complex that probably served as feeder dikes for the pillowed lava flows above. The similarities between the upper part of an ophiolite and Layers 1 and 2 of oceanic crust are obvious.

Below the sheeted dikes of an ophiolite is a thick layer of pod-like gabbro intrusions, perhaps thick sills. Beneath the gabbro lies ultramafic rock such as peridotite, the top part of which has been converted to serpentine by metamorphism. Geologists have long thought that ophiolites represent slivers of oceanic crust somehow emplaced on land. If this is true, then the gabbro of ophiolites may represent oceanic Layer 3, and the peridotite may represent mantle rock below oceanic crust. The contact between the peridotite and the overlying gabbro would be the Mohorovičić discontinuity.

Geologic Riches in the Sea

Many resources are currently being extracted from the sea floor and from seawater, and in some instances there is a great potential for increased extraction.

Offshore oil and gas are the most valuable resources now being taken from the sea. More than one-sixth of the United States' oil production (and more than one-quarter of world production) comes from drilling platforms set up on the continental shelf (box figure 1). Oil and gas have been found within deeper parts of the sea floor, such as the continental slope and continental rise. Producing oil from these deeper regions is much more costly than present production; oil spills from wells in deep water would be especially hard to control.

Other important resources are dredged from the sea floor. *Phosphorite* can be recovered from shallow shelves and banks and used for fertilizers. *Gold, diamonds,* and heavy *black sands* (which are black because they contain metal-bearing minerals) are being separated from the surface sands and gravels of some continental shelves by specially designed ships.

Manganese nodules (box figure 2) cover many parts of the deep sea floor, notably in the central Pacific. These black, potato-sized lumps contain approximately 25% manganese, 15% iron, and up to 2% nickel and 2% copper, along with smaller amounts of cobalt. Although there are international legal problems concerning who owns

Box 18.1 Figure 1

Offshore oil drilling platform. As many as 50 different wells can be drilled from a single platform.

Photo by U.S. Geological Survey

Recent work has shown that many, if not most, ophiolites are *not* typical sea floor, but a special type of sea floor formed in marginal ocean basins next to continents by the process of backarc spreading (chapter 19). If ophiolites do not represent typical sea floor, then more extensive, deeper sea-floor drilling is needed before a clear picture of oceanic crust can be formed. This is an important goal, for oceanic crust is the most common surface rock, covering 60% of the earth's surface.

The Age of the Sea Floor

As marine geologists began to determine the age of sea-floor rocks (by isotopic dating) and sediments (by fossils), an astonishing fact was discovered. All the rocks and sediments of the sea floor proved to be younger than 200 million years old. This was true only for rocks and sediments from the *deep* sea floor, not those from the continental margins. The rocks and sediments presently found on the deep sea floor formed during the Mesozoic and Cenozoic eras, but not earlier.

By contrast, as discussed in chapter 8, the earth is estimated to be 4.5 billion years old. Every continent contains some rocks formed during the Paleozoic Era and the Precambrian. Some of the Precambrian rocks on continents are more than 3 billion years old, and a few are almost 4 billion years old. Continents, therefore, preserve rocks from most of the earth's history. In sharp contrast to the continents is the deep sea floor, which covthem, larger industrial countries such as the United States may mine them, particularly for copper, nickel, and cobalt. The United States is also interested in manganese; it imports 95% of its manganese, which is critical to producing some types of steel.

Metallic brines and sediments, first discovered in the Red Sea, are deposited by submarine hot springs active at the rift valley of the mid-oceanic ridge crest. The Red Sea sediments contain more than 1% copper and more than 3% zinc, together with impressive amounts of silver, gold, and lead (worth $25 billion according to a 1983 estimate). Because of their great value, the sediments will probably be mined even though they are at great depth. Deposits similar to those of the Red Sea—although not of such great economic potential—have been found on several other parts of the ridge.

A few substances can be extracted from the salts dissolved in seawater. Approximately two-thirds of the world's production of *magnesium* is obtained from seawater, and in many regions *sodium chloride* (table salt) is obtained by solar evaporation of seawater.

Box 18.1 Figure 2

Dense concentration of manganese nodules on an abyssal plain in the South Pacific, depth 5,300 meters. Field of view about 1.5 by 2 meters.

From Lamont-Doherty Geological Observatory

---

ers more than half of the earth's surface but preserves less than one-twentieth of earth's history in its rocks and sediment.

The Sea Floor and Plate Tectonics

As we mentioned at the beginning, this chapter is concerned with the *description* of the sea floor. The *origin* of most sea-floor features is related to plate tectonics. The next chapter shows you how the theory of plate tectonics explains the existence and character of continental shelves and slopes, trenches, the mid-oceanic ridge, and fracture zones, as well as the very young age of the sea floor itself. In the workings of the scientific method, this chapter largely concerns *data.* The next chapter shows how *hypotheses* and *theories* account for these data.

Summary

The *continental shelf* and the steeper *continental slope* lie under water along the edges of continents. They are separated by a change in slope angle at a depth of 100 to 200 meters.

Submarine canyons are cut into the continental slope and outer continental shelf by a combination of *turbidity currents,* sand flow and fall, bottom currents, and river erosion during times of lower sea level. Graded bedding and cable breaks suggest the existence of turbidity currents in the ocean.

Abyssal fans form as sediment collects at the base of submarine canyons.

A *passive continental margin* occurs off geologically quiet coasts and is marked by a continental rise and abyssal plains at the base of the continental slope.

The *continental rise* and *abyssal plains* may form from sediment deposited by turbidity currents.

The continental rise may also form from sediment deposited by *contour currents* at the base of the continental slope.

An *active continental margin* is marked by an *oceanic trench* at the base of the continental slope; the continental rise and abyssal plains are absent.

Oceanic trenches are twice as deep as abyssal plains, which generally lie at a depth of 5 kilometers. Associated with trenches are *Benioff zones* of earthquakes and *andesitic volcanism,* forming either an island arc or a chain of volcanoes in a young mountain belt near the edge of a continent. Trenches have low heat flow and negative gravity anomalies.

The *mid-oceanic ridge* is a globe-circling mountain range of basalt, located mainly in the middle of ocean basins. The crest of the ridge is marked by a *rift valley,* shallow-focus earthquakes, high heat flow, active *basaltic volcanism,* hydrothermal activity, and exotic organisms.

Fracture zones are lines of weakness that offset the mid-oceanic ridge. Shallow-focus quakes occur on the portion of the fracture zone between the offset ridge segments.

Seamounts are conical, submarine volcanoes, now mostly extinct. *Guyots* are flat-topped seamounts, probably leveled by wave erosion before subsiding.

Chains of seamounts and guyots form *aseismic ridges.*

Corals and algae living in warm, shallow water construct *fringing reefs, barrier reefs,* and *atolls.*

Terrigenous sediment is composed of land-derived sediment deposited near land by turbidity currents and other processes. *Pelagic sediment* is made up of wind-blown dust and microscopic skeletons that settle slowly to the sea floor.

The crest of the mid-oceanic ridge lacks pelagic sediment.

Oceanic crust consists of basalt pillows and dikes, probably overlying gabbro.

Ophiolites in continental mountain ranges probably represent slivers of somewhat atypical oceanic crust somehow emplaced on land.

The oldest rocks on the deep sea floor are 200 million years old. The continents, in contrast, contain some rock that is 3 to 4 billion years old.

Terms to Remember

abyssal fan 442

abyssal plain 444

active continental margin 444

aseismic ridge 450

atoll 450

barrier reef 450

continental rise 443

continental shelf 441

continental slope 442

contour current 444

fracture zone 449

fringing reef 450

guyot 449

mid-oceanic ridge 445

oceanic trench 445

ophiolite 453

passive continental margin 443

pelagic sediment 452

reef 450

rift valley 445

seamount 449

submarine canyon 442

terrigenous sediment 452

turbidity current 443

Testing Your Knowledge

Use the questions below to prepare for exams based on this chapter.

1. What is a submarine canyon? How do submarine canyons form?

2. Discuss the appearance, structure, and origin of abyssal plains.

3. Sketch a cross profile of the mid-oceanic ridge, showing the rift valley. Label your horizontal and vertical scales.

4. Sketch an active continental margin and a passive continental margin, labeling all their parts. Show approximate depths.

5. What is a fracture zone? Sketch the relation between fracture zones and the mid-oceanic ridge.

6. In a sketch, show the association between an oceanic trench, a Benioff zone of earthquakes, and volcanoes on the edge of a continent.

7. Describe two different origins for the continental rise.

8. What is a turbidity current? What is the evidence that turbidity currents occur on the sea floor?

9. Describe the appearance and origin of seamounts and guyots.

10. Describe the two main types of sea-floor sediment.

11. How does the age of sea-floor rocks compare with the age of continental rocks? Be specific.

12. Sketch a cross section of a fringing reef, a barrier reef, and an atoll.

13. What is the thickness and composition of oceanic crust?

14. Most ocean water probably came from (a) degassing of the earth's interior (b) outer space

15. Which is true of the continental shelf? (a) it is a shallow submarine platform at the edge of a continent (b) it inclines very gently seaward (c) it can vary in width (d) all of the above

16. The average angle of slope for a continental slope is (a) 1°–2° (b) 3°–4° (c) 4°–5° (d) greater than 10°

17. Great masses of sediment-laden water that are pulled downhill by gravity are called (a) contour currents (b) bottom currents (c) turbidity currents (d) traction currents

18. Oceanic trenches (a) are narrow deep troughs (b) run parallel to the edge of a continent or an island arc (c) are often 8–10 km deep (d) all of the above

19. Which is characteristic of mid-oceanic ridges? (a) shallow-focus earthquakes (b) high heat flow (c) basalt eruptions (d) all of the above

20. Reefs parallel to the shore but separated from it by wide, deep lagoons are called (a) fringing reefs (b) barrier reefs (c) atolls (d) lagoonal reefs

21. Pelagic sediment could be composed of (a) fine-grained clay (b) skeletons of microscopic organisms (c) volcanic ash (d) all of the above

22. What part of the continental margin marks the true edge of the continent (a) continental shelf (b) continental slope (c) continental rise (d) abyssal plain

23. Distinctive rock sequences of basalt and marine sedimentary rock that may be slices of the ocean floor are (a) guyots (b) ophiolites (c) seamounts (d) fracture zones

Expanding Your Knowledge

1. How many possible origins can you think of for the rift valley on the mid-oceanic ridge?

2. What factors could cause sea level to rise? To fall?

3. Why is the rock of the deep sea floor (60% of the earth's surface) basalt? Where did the basalt come from?

4. How could fracture zones have formed?

5. How many hypotheses can you think of to explain the relatively young age of the sea floor?

Exploring Resources

 Anderson, R. N. 1986. *Marine geology—A planet earth perspective.* New York: John Wiley & Sons.

Dietz, R. S. 1964. Origin of continental slopes. *American Scientist* 52:50–69.

Heezen, B. C., and C. D. Hollister. 1971. *The face of the deep.* New York: Oxford University Press.

Lutz, R. A., and R. M. Haymon. 1994. Rebirth of a deep-sea vent. *National Geographic* 186(5): 114–26.

Macdonald, J. B., and P. J. Fox. 1990. The mid-ocean ridge. *Scientific American* 262(6): 72–79.

Menon, S. 1997. Ocean Watch: Deep sea rebirth. *Discover* no. 7:34.

Nicolas, A. 1995. *The mid-ocean ridges: Mountains below sea level.* New York: Springer-Verlag.

Pinet, P. 1996. *Invitation to Oceanography.* St. Paul, Minn.: West Publishing.

Rona, P. A. 1986. Mineral deposits from sea-floor hot springs. *Scientific American* 254(1): 84–93.

———. 1992. Deep-sea geysers of the Atlantic. *National Geographic* 182(4): 105–109.

Scientific American Editors. 1983. *The ocean.* New York: W. H. Freeman.

Scrutton, R. A., and M. Talwani, eds. 1982. *The ocean floor.* New York: Wiley-Interscience.

Seibold, E., and W. H. Berger. 1982. *The sea floor.* New York: Springer-Verlag.

Waters, T. 1995. The other Grand Canyon. *Earth* 4(6): 44–51.

Yulsman, T. 1996. The seafloor laid bare. *Earth* 5(3): 42–51.

(See also the list of readings for the next chapter.)

 http://seawifs.gsfc.nasa.gov:80/
OCEAN_PLANET/HTML/
ocean_planet_ocean_science.html

Smithsonian Ocean Planet Exhibit containing many links to sea floor topics such as hydrothermal vents and microbiology of deep-sea vents.

http://www.esdim.noaa.gov/ocean_page.html
National Oceanographic and Atmospheric Administration (NOAA) has many links to numerous sites relating to the oceans.

http://www.ngdc.noaa.gov/mgg/mggd.html
NOAA Marine Geology and Geophysics Division website contains world sea floor maps and information on ocean-drilling data and samples.

http://www.whoi.edu/VideoGallery/
Woods Hole Oceanographic Institution contains video clips of black smokers, exotic organisms at mid-oceanic ridges, oceanographic research using vessels and submersibles.

Interacting with Journey Through Geology CD-ROM

Expand your knowledge of the concepts presented in this chapter by using the CD-ROM to answer the following questions.

1. Spreading Ridges module, Iceland: What kind of volcano is associated with this spreading center? What kind of lava (basalt, andesite, rhyolite) would you expect to find at these volcanoes?

2. Changing Earth: Continental shelf area changes through time as sea level and the plate setting of continents changes. Use this module to answer these questions:

   —What continent has the most continental shelf area today? The least?

   —Watch North America's history carefully and find a time of maximum and minimum continental shelf area.

Space shuttle photo looking east toward the Red Sea. Plate motion has torn the Arabian Peninsula (top of photo) away from Africa (shown in foreground) to form the Red Sea.

CHAPTER
19

Plate Tectonics

*A*s you studied volcanoes, igneous, metamorphic and sedimentary rocks, and earthquakes, you learned how these topics are related to plate tectonics. In this chapter we take a closer look at plates and plate motion. We will pay particular attention to plate boundaries and the possible driving mechanisms for plate motion.

The history of the concept of plate tectonics is a good example of how scientists think and work and how a hypothesis can be proposed, discarded, modified, and then reborn. In the first part of this chapter we trace the evolution of an idea—how the earlier hypotheses of moving continents (continental drift) and a moving sea floor (sea-floor spreading) were combined to form the theory of plate tectonics.

Figure 19.1

The major plates of the world. The western edge of the map repeats the eastern edge so that all plates can be shown unbroken. Double lines indicate spreading axes on divergent plate boundaries. Single lines show transform boundaries. Heavy lines with triangles show convergent boundaries, with triangles pointing down subduction zones.

Modified from W. Hamilton, U.S. Geological Survey.

Tectonics is the study of the origin and arrangement of the broad structural features of the earth's surface, including not only folds and faults, but also mountain belts, continents, and earthquake belts. Tectonic models such as an expanding earth or a contracting earth have been used in the past to explain *some* of the surface features of earth. Plate tectonics has come to dominate geologic thought today because it can explain so *many* features. The basic idea of **plate tectonics** is that the earth's surface is divided into a few large, thick plates that move slowly and change in size. Intense geologic activity occurs at *plate boundaries* where plates move away from one another, past one another, or toward one another. The eight large plates shown in figure 19.1, plus a few dozen smaller plates, make up the outer shell of the earth (the crust and upper part of the mantle).

The concept of plate tectonics was born in the late 1960s by combining two preexisting ideas—continental drift and sea-floor spreading. **Continental drift** is the idea that continents move freely over the earth's surface, changing their positions relative to one another. **Sea-floor spreading** is a hypothesis that the sea floor forms at the crest of the mid-oceanic ridge, then moves horizontally away from the ridge crest toward an oceanic trench. The two sides of the ridge are moving in opposite directions like slow conveyor belts.

Before we take a close look at plates, we will examine the earlier ideas of moving continents and a moving sea floor because these two ideas embody the theory of plate tectonics.

The Early Case for Continental Drift

Continents can be made to fit together like pieces of a picture puzzle. The similarity of the Atlantic coastlines of Africa and South America has long been recognized. The idea that continents were once joined together, and have split and moved apart from one another, has been around for more than 130 years (figure 19.2).

In the early 1900s Alfred Wegener, a German meteorologist, made a strong case for continental drift. He noted that South America, Africa, India, and Australia had almost identical late Paleozoic rocks and fossils (including the plant *Glossopteris*). He reassembled the continents to form a giant supercontinent *Pangaea* (also spelled Pangea today). Wegener thought that the similar rocks and fossils were easier to explain if the continents were joined together, rather than in their present, widely scattered positions.

Pangaea initially separated into two parts. *Laurasia* was the northern supercontinent, containing what is now North America and Eurasia (excluding India). *Gondwanaland* was the southern supercontinent, composed of all the present-day southern-hemisphere continents and India (which has drifted north).

The distribution of Late Paleozoic glaciation strongly supports the idea of Pangaea (figure 19.3). The Gondwanaland

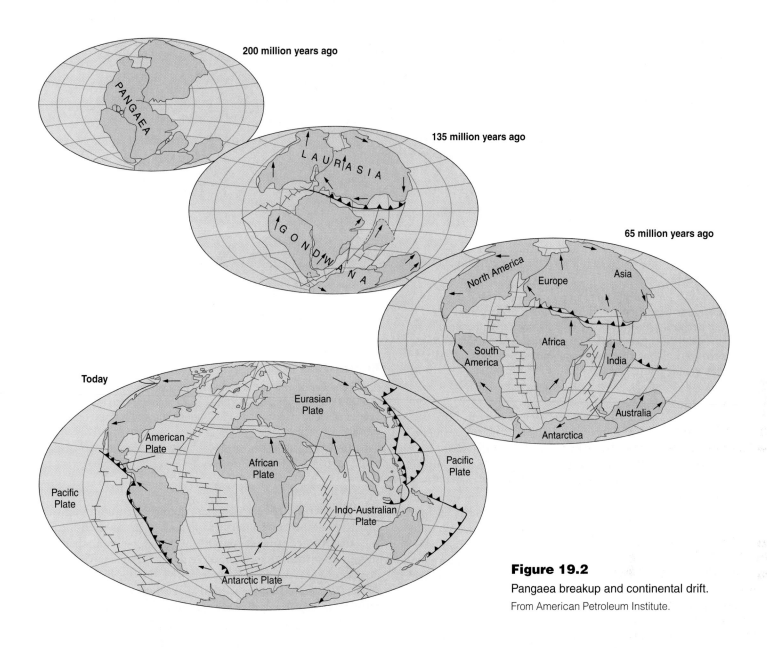

200 million years ago

PANGAEA

135 million years ago

LAURASIA

GONDWANA

65 million years ago

North America Europe Asia

South America Africa India

Australia

Antarctica

Today

Eurasian Plate

American Plate

African Plate

Pacific Plate

Pacific Plate

Indo-Australian Plate

Antarctic Plate

Figure 19.2

Pangaea breakup and continental drift.

From American Petroleum Institute.

A

Africa

South America

India

Antarctica Australia

B

Figure 19.3

Distribution of late Paleozoic glaciations; arrows show direction of ice flow. (*A*) Continents in present positions show wide distribution of glaciation (white land areas with flow arrows). (*B*) Continents reassembled into Pangaea. Glaciated region becomes much smaller.

A from Arthur Holmes, 1965, *Principles of Physical Geology,* 2d ed., Ronald Press.

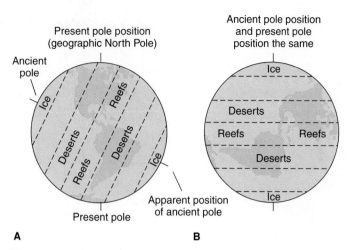

Figure 19.4

Two ways of interpreting the distribution of ancient climate belts. (*A*) Continents fixed, poles wander. (*B*) Poles fixed, continents drift. For simplicity, the continents in *B* are shown as having moved as a unit, without changing positions relative to one another. If continents move, they should change relative positions, complicating the pattern shown.

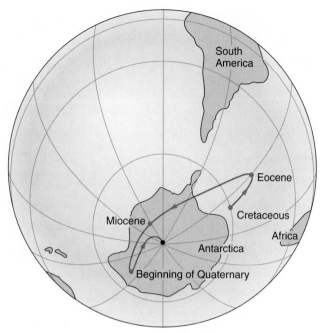

Figure 19.5

Apparent wandering of the South Pole since the Cretaceous Period as determined by Wegener from paleoclimate evidence. Wegener, of course, believed that *continents* rather than poles moved.

From A. Wegener, 1928, *The Origins of Continents and Oceans*, reprinted and copyrighted. 1968, Dover Publications.

continents (the southern-hemisphere continents and India) all have glacial deposits of Late Paleozoic age. If these continents were spread over the earth in Paleozoic time as they are today, a climate cold enough to produce extensive glaciation would have had to prevail over almost the whole world. Yet no evidence has been found of widespread Paleozoic glaciation in the northern hemisphere. In fact, the late Paleozoic coal beds of North America and Europe were being laid down at that time in swampy, probably warm environments. If the continents are arranged according to Wegener's Pangaea reconstruction, then glaciation in the southern hemisphere is confined to a much smaller area (figure 19.3), and the absence of widespread glaciation in the northern hemisphere becomes easier to explain.

Wegener also reconstructed old climate zones (the study of ancient climates is called *paleoclimatology*). Glacial till and striations indicate a cold climate near the North or South Pole. Coral reefs indicate warm water near the Equator. Cross-bedded sandstones can indicate ancient deserts near 30° North and 30° South latitude. If ancient climates had the same distribution on earth that modern climates have, then sedimentary rocks can show where the ancient poles and Equator were located.

Wegener determined the positions of the North and South Poles for each geologic period. He found that ancient poles were in different positions than the present poles (figure 19.4*A*). This apparent movement of the poles he called **polar wandering.** Polar wandering, however, is a deceptive term. The evidence can actually be explained in two different ways:

1. The continents remained motionless and the poles actually *did* move—polar wandering (figure 19.4*A*).
2. The poles stood still and the continents moved—continental drift (figure 19.4*B*).

Wegener plotted curves of apparent polar wandering (figure 19.5). Since one interpretation of polar wandering data was that the continents moved, Wegener believed that this supported his concept of continental drift. (Notice that in only one interpretation of polar wandering do the poles actually move. You should keep in mind that when geologists use the term *polar wandering* they are referring to an *apparent* motion of the poles, which may or may not have actually occurred.)

Skepticism about Continental Drift

Although Wegener presented the best case possible in the early 1900s for continental drift, much of his evidence was not clear-cut. Fossil plants, for example, could have been spread from one continent to another by winds or ocean currents. Their distribution over more than one continent does not *require* that the continents were all joined in the supercontinent, Pangaea. In addition, polar wandering might have been caused by moving poles rather than by moving continents. Because his evidence was not conclusive, Wegener's ideas were not widely accepted. This was particularly true in the United States, largely because of the mechanism Wegener proposed for continental drift.

Wegener proposed that continents plowed through the oceanic crust (figure 19.6), perhaps crumpling up mountain ranges on the leading edges of the continents where they pushed against the sea floor. Most geologists in the United States thought that this idea violated what was known about the strength of rocks at the time. The driving mechanism

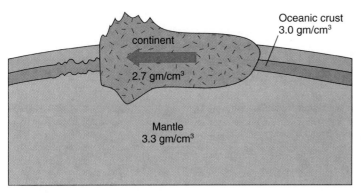

Figure 19.6

Wegener's concept of continental drift implied that the less dense continents drifted *through* oceanic crust, crumpling up mountain ranges on their leading edges as they pushed against oceanic crust.

Figure 19.7

Some rocks preserve a record of the earth's magnetic field.

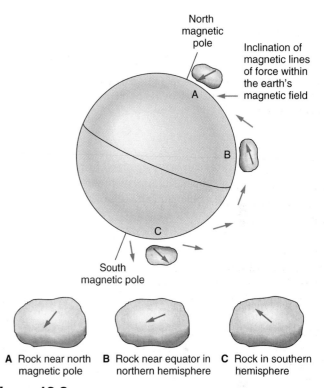

Figure 19.8

Magnetic dip (inclination) increases toward the north magnetic pole. Rocks in bottom part of figure are small samples viewed horizontally at locations A, B, and C on the globe. The magnetic dip can therefore be used to determine the distance from a rock to the north magnetic pole.

proposed by Wegener for continental drift was a combination of centrifugal force from the earth's rotation and the gravitational forces that cause tides. Careful calculations of these forces showed them to be too small to move continents. Because of these objections, Wegener's ideas received little support in the United States or much of the northern hemisphere (where the great majority of geologists live) in the first half of the twentieth century. The few geologists in the southern hemisphere, however, where Wegener's matches of fossils and rocks between continents were more evident, were more impressed with the concept of continental drift.

Paleomagnetism and the Revival of Continental Drift

Much work in the 1940s and 1950s set the stage for the revival of the idea of continental drift and its later incorporation, along with sea-floor spreading, into the new concept of plate tectonics. The new investigations were in two areas: (1) study of the sea floor and (2) geophysical research, especially in relation to rock magnetism.

Convincing new evidence about polar wandering came from the study of rock magnetism. Wegener's work dealt with the wandering of the earth's *geographic* poles of rotation. The earth's *magnetic* poles are located close to the geographic poles, as you saw in the chapter on the earth's interior. Historical measurements show that the position of the magnetic poles moves from year to year, but that the magnetic poles stay close to the geographic poles as they move. As we discuss magnetic evidence for polar wandering, we are referring to an apparent motion of the magnetic poles. Because the magnetic and geographic poles are close together, our discussion will refer to apparent motion of the geographic poles as well.

As we discussed in chapter 17, many rocks record the strength and direction of the earth's magnetic field at the time the rocks formed. Magnetite in a cooling basaltic lava flow acts like a tiny compass needle, preserving a record of the earth's

magnetic field when the lava cools below the *Curie point* (figure 19.7). Iron-stained sedimentary rocks such as red sandstone can also record earth magnetism. The magnetism of old rocks can be measured to determine the direction and strength of the earth's magnetic field in the past. The study of ancient magnetic fields is called *paleomagnetism.*

Because magnetic lines of force dip more steeply as the north magnetic pole is approached, the inclination (dip) of the magnetic alignment preserved in the magnetite minerals in the lava flows can be used to determine the distance from a flow to the pole at the time that the flow formed (figure 19.8).

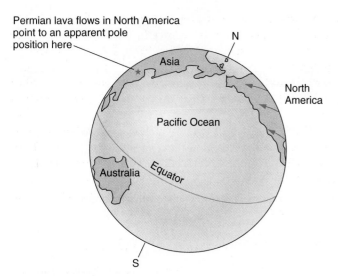

Figure 19.9

Paleomagnetic studies of Permian lava flows on North America indicate an apparent position for the north magnetic pole in eastern Asia.

Figure 19.10

Polar wandering of the north magnetic pole as determined from measurements of rocks from North America and Europe.

From A. Cox and R. R. Doell, 1960, *Geological Society of America Bulletin.*

Old pole positions can be determined from the magnetism of old rocks. The magnetic alignment preserved in magnetite minerals points to the pole, and the dip of the alignment tells how far away the pole was. Figure 19.9 shows how Permian lava flows in North America indicate a Permian pole position in eastern Asia.

For each geologic period, North American rocks reveal a different magnetic pole position; this path of the *apparent* motion of the north magnetic pole through time is shown in figure 19.10. Paleomagnetic evidence thus verifies Wegener's idea of polar wandering (which he based on paleoclimatic evidence).

Like Wegener's paleoclimatic evidence, the paleomagnetic evidence from a *single* continent can be interpreted in two ways: either the continent stood still and the magnetic pole moved or the pole stood still and the continent moved. At first glance, paleomagnetic evidence does not seem to be a significant advance over paleoclimatic evidence. But when paleomagnetic evidence from *different* continents was compared, an important discovery was made.

Although Permian rocks in North America point to a pole position in eastern Asia, Permian rocks in *Europe* point to a different position (closer to Japan), as shown in figure 19.10. Does this mean there were *two* north magnetic poles in the Permian Period? In fact, every continent shows a different position for the Permian pole. A different magnetic pole for

each continent seems highly unlikely. A better explanation is that a single pole stood still while continents split apart and rotated as they diverged.

Note the polar wandering paths for North America and Europe in figure 19.10. The paths are of similar shape, but the path for European poles is to the east of the North American path. If we mentally push North America back toward Europe, closing the Atlantic Ocean, and then consider the paths of polar wandering, we find that the path for North America lies exactly on the path for Europe. This strongly suggests that there was one north magnetic pole and that the continents were joined together. There appear to be two north magnetic poles because the rocks of North America moved west; their magnetic minerals now point to a different position than they did when the minerals first formed.

Recent Evidence for Continental Drift

As paleomagnetic evidence revived interest in continental drift, new work was done on fitting continents together. By defining

Figure 19.11

Jigsaw puzzle fit and matching rock types between South America and Africa. Light-blue areas around continents are continental shelves (part of continents). Colored areas within continents are broad belts of rock that correlate in type and age from one continent to another. Arrows show direction of glacier movement as determined from striations.

the edge of a continent as the middle of the continental slope, rather than the present (constantly changing) shoreline, a much more precise fit has been found between continents (figure 19.11).

The most convincing evidence for continental drift came from greatly refined rock matches between now-separated continents. If continents are fitted together like pieces of a jigsaw puzzle, the "picture" should match from piece to piece.

The matches between South America and Africa are particularly striking. Some distinctive rock contacts extend out to sea along the shore of Africa. If the two continents are fitted together, the identical contacts are found in precisely the right position on the shore of South America (figure 19.11). Isotopic ages of rocks also match between these continents.

Glacial striations show that during the late Paleozoic Era continental glaciers moved from Africa toward the present Atlantic Ocean, while similar glaciers seemingly moved *from* the Atlantic Ocean *onto* South America (figure 19.11).

Continental glaciers, however, cannot move from sea onto land. If the two continents had been joined together, the ice that moved off Africa could have been the ice that moved onto South America. This hypothesis has now been confirmed; from their lithology, many of the boulders in South American tills have been traced to a source that is now in Africa.

Some of the most detailed matches have been made between rocks in Brazil and rocks in the African country of Gabon. These rocks are similar in type, structure, sequence, fossils, ages, and degree of metamorphism. Such detailed matches are convincing evidence that continental drift did, in fact, take place.

History of Continental Positions

Rock matches show when continents were together; once the continents split, the new rocks formed are dissimilar. Paleomagnetic evidence indicates the direction and rate of drift, allowing maps of old continental positions, such as figure 19.2, to be drawn.

Although Pangaea split up 200 million years ago to form our present continents, the continents were moving much earlier. Pangaea was formed by the collision of many small continents long before it split up. Recent work shows that continents have been in motion for the past 2 billion years (some geologists say 4 billion years), well back into Precambrian time. For half or more of earth's history, the continents appear to have collided, welded together, then split and drifted apart, only to collide again, over and over, in an endless, slow dance.

Sea-Floor Spreading

At the same time that many geologists were becoming interested again in the idea of moving *continents*, Harry Hess, a geologist at Princeton University, proposed that the *sea floor* might be moving, too. This proposal contrasted sharply with the earlier ideas of Wegener, who thought that the ocean floor remained stationary as the continents plowed through it (figure 19.6). Hess's 1962 proposal was quickly named sea-floor spreading, for it suggests that the sea floor moves away from the mid-oceanic ridge as a result of mantle convection (figure 19.12).

According to the concept of sea-floor spreading, the sea floor is moving like a conveyor belt away from the crest of the mid-oceanic ridge, down the flanks of the ridge, and across the deep-ocean basin, to disappear finally by plunging beneath a

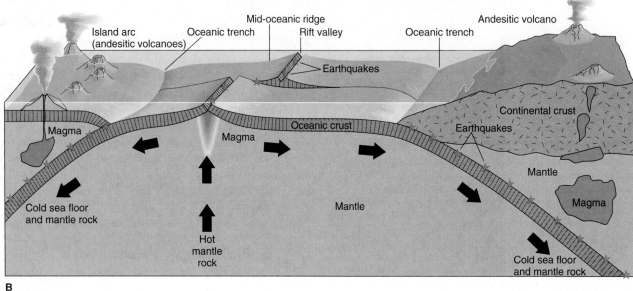

Figure 19.12

Sea-floor spreading. (*A*) Hess proposed that convection extended throughout the mantle. (Scale of ridge and trenches is exaggerated.) (*B*) Hot mantle rock rising beneath the mid-oceanic ridge (a spreading axis) causes basaltic volcanism and high heat flow. Divergence of sea floor splits open the rift valley and causes shallow-focus earthquakes (stars on ridge). Sinking of cold rock causes subduction of older sea floor at trenches, producing Benioff zones of earthquakes and andesitic magma.

continent or island arc (figure 19.12). The ridge crest, with sea floor moving away from it on either side, has been called a *spreading axis* (or *spreading center*). The sliding of the sea floor beneath a continent or island arc is termed **subduction.** The sea floor moves at a rate of 1 to 6 centimeters per year (your fingernail grows at about 1 cm/year). Although this may seem to be quite slow, it is rapid compared to most geologic processes.

Hess's Driving Force

Why does the sea floor move? Hess's original hypothesis was that sea-floor spreading is driven by deep mantle convection. **Convection** is a circulation pattern driven by the rising of hot material and/or the sinking of cold material. Hot material has a low density, so it rises; cold material has a high density and sinks. The circulation of water heating in a pan on a stove is an example of convection. Convection in the mantle was a controversial

idea in 1962; for although convection can be easily demonstrated in a pan of water, it was hard to visualize the solid rock of the earth's mantle behaving as a liquid. Over very long periods of time, however, it is possible for the hot mantle rock to flow plastically. A slow convective circulation is set up by temperature differences in the rock, and convection can explain many sea-floor features as well as the young age of the sea-floor rocks. (The heat that flows outward through the earth to drive convection is both original heat from the earth's formation and heat from the decay of radioactive isotopes, as discussed in chapter 17.)

Explanations

The Mid-Oceanic Ridge

If convection drives sea-floor spreading, then hot mantle rock must be rising under the mid-oceanic ridge. Hess showed how the *existence of the ridge* and its *high heat flow* are caused by the rise of this hot mantle rock. The *basalt eruptions* on the ridge crest are also related to this rising rock, for here the mantle rock is hotter than normal and therefore close to its melting point.

As hot rock continues to rise beneath the ridge crest, the circulation pattern splits and diverges near the surface. Mantle rock moves horizontally away from the ridge crest on each side of the ridge. This movement creates tension at the ridge crest, cracking open the oceanic crust to form the *rift valley* and its associated *shallow-focus earthquakes*.

Oceanic Trenches

As the mantle rock moves horizontally away from the ridge crest, it carries the sea floor (the basaltic oceanic crust) piggyback along with it. As the hot rock moves sideways, it cools and becomes denser, sinking deeper beneath the ocean surface. Hess thought it would become cold and dense enough to sink back into the mantle. This downward plunge of cold rock accounts for the *existence of the oceanic trenches* as well as their *low heat flow* values. It also explains the large *negative gravity anomalies* associated with trenches, for the sinking of the cold rock provides a force that holds trenches out of isostatic equilibrium (see chapter 17).

As the sea floor moves downward into the mantle along a subduction zone, it interacts with the stationary rock above it. This interaction between the moving sea-floor rock and the stationary rock can cause the *Benioff zones of earthquakes* associated with trenches. It can also produce *andesitic volcanism*, which forms volcanoes either on the edge of a continent or in an island arc (figure 19.12).

Hess's ideas have stood up remarkably well over more than 30 years. We now think of plates moving instead of sea floor riding piggyback on convecting mantle, and we think that several mechanisms cause plate motion, but Hess's explanation of sea-floor topography, earthquakes, and age remain valid today.

Age of the Sea Floor

The *young age of sea-floor rocks* (see the previous chapter) is neatly explained by Hess's sea-floor spreading. New, young sea floor is continually being formed by basalt eruptions at the ridge crest. This basalt is then carried sideways by convection and is subducted into the mantle at an oceanic trench. Thus old sea floor is continually being destroyed at trenches, while new sea floor is being formed at the ridge crest. (This is also the reason for the puzzling lack of pelagic sediment at the ridge crest. Young sea floor at the ridge crest has little sediment because the basalt is newly formed. Older sea floor farther from the ridge crest has been moving under a constant rain of pelagic sediment, building up a progressively thicker layer as it goes.)

Note that sea-floor spreading implies that the youngest sea floor should be at the ridge crest, with the age of the sea floor becoming progressively older toward a trench. This increase in

Figure 19.13

The rigid lithosphere includes the crust and uppermost mantle; it forms the plates. The plastic asthenosphere acts as a lubricating layer beneath the lithosphere. Oceanic lithosphere averages 70 kilometers thick; continental lithosphere varies from 125 to 250 kilometers thick. Asthenosphere may not be present under continents.

age away from the ridge crest was not known to exist at the time of Hess's proposal but was an important prediction of his hypothesis. This prediction has been successfully tested, as you shall see later in this chapter when we discuss marine magnetic anomalies.

Plates and Plate Motion

By the mid-1960s the twin ideas of moving continents and a moving sea floor were causing great excitement and emotional debate among geologists. By the late 1960s, these ideas had been combined into a single theory that revolutionized geology by providing a unifying framework for earth science—the theory of plate tectonics.

As described earlier, a **plate** is a large, mobile slab of rock that is part of the earth's surface (figure 19.1). The surface of a plate may be made up entirely of sea floor (as is the Nazca plate), or it may be made up of both continental and oceanic rock (as is the North American plate). Some of the smaller plates are entirely continental, but all the large plates contain some sea floor.

Plate tectonics has added some new terms, based on rock behavior, to the zones of the earth's interior, as we have discussed in some previous chapters. The plates are part of a relatively rigid outer shell of the earth called the **lithosphere.** The lithosphere includes the rocks of the earth's crust and uppermost mantle (figure 19.13).

The lithosphere beneath oceans increases in both age and thickness with distance from the crest of the mid-oceanic ridge. Young lithosphere near the ridge crest may be only 10 kilometers thick, while very old lithosphere far from the ridge crest may be as much as 100 kilometers thick. An average thickness for oceanic lithosphere might be 70 kilometers, as shown in figure 19.13.

Continental lithosphere is thicker, varying from perhaps 125 kilometers thick to as much as 200 to 250 kilometers thick beneath the oldest, coldest, and most inactive parts of the continents.

Below the rigid lithosphere is the **asthenosphere,** a zone of low seismic-wave velocity that behaves plastically because of increased temperature and pressure. Some geologists think that the asthenosphere is partially molten; the melting of just a few percent of the asthenosphere's volume could account for its properties and behavior. The plastic asthenosphere acts as a lubricating layer under the lithosphere, allowing the plates to move. The asthenosphere, made up of upper mantle rock, is the low-velocity zone described in chapter 17. It may extend from a depth of 70 to 200 kilometers beneath oceans; its thickness, depth, and even existence under continents is vigorously debated. Below the asthenosphere is more rigid mantle rock.

The idea that plates move is widely accepted by geologists, although the reasons for this movement are debated. Plates move away from the mid-oceanic ridge crest or other spreading axes. Some plates move toward oceanic trenches. If the plate is made up mostly of sea floor (as are the Nazca and Pacific plates), the plate can be subducted down into the mantle, forming an oceanic trench and its associated features. If the leading edge of the plate is made up of continental rock (as is the South American plate), that plate will not subduct. Continental rock, being less dense (specific gravity 2.7) than oceanic rock (specific gravity 3.0), is too light to be subducted.

A plate is a rigid slab of rock that moves as a unit. As a result, the interior of a plate is relatively inactive tectonically. Plate interiors generally lack earthquakes, volcanoes, young mountain belts, and other signs of geologic activity. According to plate-tectonic theory, these features are caused by plate interactions at plate boundaries.

Earthquakes, volcanoes, and young mountains are distributed in narrow belts separated by broad regions of inactivity, as you have seen in previous chapters. This distribution puzzled geologists for a long time, and many hypotheses were advanced to explain it. The plate-tectonic concept is the latest explanation. Plate boundaries are defined and located by mapping narrow belts of earthquakes, volcanoes, and young mountains. The plates themselves are the broad, inactive regions outlined by these belts of geologic activity.

Plate tectonics has become a unifying theory of geology because it can explain so many diverse features of the earth. Earthquake distribution, the origin of mountain belts, the origin of sea-floor topography, the distribution and composition of volcanoes, and many other features can all be related to plate tectonics. It is a convenient framework that unifies geologic thought, associating features that were once studied separately and relating them to a single cause: plate interactions at plate boundaries.

Plate boundaries are of three general types, based on whether the plates move away from each other, move toward each other, or move past each other. A **divergent plate boundary** is a boundary between plates that are moving apart. A **convergent plate boundary** lies between plates that are moving toward each other. A **transform plate boundary** is one at which two plates move horizontally past each other.

Figure 19.14

Marine magnetic anomalies. (*A*) The red line shows positive and negative magnetic anomalies as recorded by a magnetometer towed behind a ship. In the cross section of oceanic crust, positive anomalies are drawn as black bars and negative anomalies are drawn as blue bars. (*B*) Perspective view of magnetic anomalies shows that they are parallel to the rift valley and symmetric about the ridge crest.

How Do We Know That Plates Move?

The proposal that the earth's surface is divided into moving plates was an exciting, revolutionary hypothesis, but it required testing to win acceptance among geologists. You have seen how the study of paleomagnetism supports the idea of moving continents. In the 1960s two critical tests were made of the idea of a moving sea floor. These tests involved marine magnetic anomalies and the seismicity of fracture zones. These two, successful tests convinced most geologists that plates do indeed move.

Marine Magnetic Anomalies

In the mid-1960s, magnetometer surveys at sea disclosed some intriguing characteristics of marine magnetic anomalies. Most magnetic anomalies at sea are arranged in bands that lie parallel to the rift valley of the mid-oceanic ridge. Alternating positive and negative anomalies (chapter 17) form a stripelike pattern parallel to the ridge crest (figure 19.14).

The Vine-Matthews Hypothesis

Two British geologists, Fred Vine and Drummond Matthews, made several important observations about these anomalies.

A Time of normal magnetism

B Time of reverse magnetism

C Time of normal magnetism

Figure 19.15

Magnetic reversals during the past 4.5 million years. Black represents normal magnetism; blue represents reverse magnetism.

Modified from A. Cox, 1969, *Science,* vol. 163, p. 240, copyright 1969 by the American Association for the Advancement of Science and Robert Butler, 1992, *Paleomagnetism,* Blackwell Scientific Publications, p. 212.

Figure 19.16

The origin of magnetic anomalies. (*A*) During a time of normal magnetism, a series of basaltic dikes intrudes the ridge crest, becoming normally magnetized. (*B*) The dike zone is torn in half and moved sideways, as a new group of reversely magnetized dikes forms at the ridge crest. (*C*) A new series of normally-magnetized dikes forms at the ridge crest. The dike pattern becomes symmetric about the ridge crest.

They recognized that the pattern of magnetic anomalies was symmetrical about the ridge crest. That is, the pattern of magnetic anomalies on one side of the mid-oceanic ridge was a mirror image of the pattern on the other side (figure 19.14). Vine and Matthews also noticed that the same pattern of magnetic anomalies exists over different parts of the mid-oceanic ridge. The pattern of anomalies over the ridge in the northern Atlantic Ocean is the same as the pattern over the ridge in the southern Pacific Ocean.

The most important observation that Vine and Matthews made was that the pattern of magnetic *anomalies* at sea matches the pattern of magnetic *reversals* already known from studies of lava flows on the continents (figure 19.15 and chapter 17). This correlation can be seen by comparing the pattern of colored bands in figure 19.15 (reversals) with the pattern in figure 19.14 (anomalies).

Putting these observations together with Hess's concept of sea-floor spreading, which had just been published, Vine and Matthews proposed an explanation for magnetic anomalies.

They suggested that there is continual opening of tensional cracks within the rift valley on the mid-oceanic ridge crest. These cracks on the ridge crest are filled by basaltic magma from below, which cools to form dikes. Cooling magma in the dikes records the earth's magnetism at the time the magnetic minerals crystallize. The process is shown in figure 19.16.

When the earth's magnetic field has a *normal polarity* (the present orientation), cooling dikes are normally magnetized. Dikes that cool when the earth's field is reversed (figure 19.16) are reversely magnetized. So each dike preserves a record of the polarity that prevailed during the time the magma cooled. Extension produced by the moving sea floor then cracks a dike in two, and

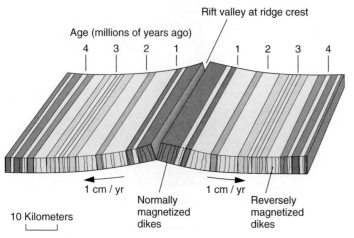

Figure 19.17

Correlation of magnetic anomalies with magnetic reversals allows anomalies to be dated. Magnetic anomalies can therefore be used to predict the age of the sea floor and to measure the rate of sea-floor spreading (plate motion).

the two halves are carried away in opposite directions down the flanks of the ridge. New magma eventually intrudes the newly opened fracture. It cools, is magnetized, and forms a new dike, which in turn is split by continued extension. In this way a system of reversely magnetized and normally magnetized dikes forms parallel to the rift valley. These dikes, in the Vine-Matthews hypothesis, are the cause of the anomalies.

The magnetism of normally magnetized dikes adds to the earth's magnetism, and so a magnetometer carried over such dikes registers a stronger magnetism than average—a *positive* magnetic anomaly. Dikes that are reversely magnetized subtract from the present magnetic field of the earth, and so a magnetometer towed over such dikes measures a weaker magnetic field—a *negative* magnetic anomaly. Since sea-floor motion separates these dikes into halves, the patterns on either side of the ridge are mirror images.

Measuring the Rate of Plate Motion

There are two important points about the Vine-Matthews hypothesis of magnetic anomaly origin. The first is that it allows us to measure the *rate of sea-floor motion* (which is the same as plate motion, since continents and the sea floor move together as plates).

Because magnetic reversals have already been dated from lava flows on land (figure 19.15), the anomalies caused by these reversals are also dated and can be used to discover how fast the sea floor has moved (figure 19.17). For instance, a piece of the sea floor representing the reversal that occurred 4.5 million years ago may be found 45 kilometers away from the rift valley of the ridge crest. The piece of sea floor, then, has traveled 45 kilometers since it formed 4.5 million years ago. Dividing the distance the sea floor has moved by its age gives 10 km/million years, or 1 cm/year for the rate of sea-floor motion here. In other words, on each side of the ridge, the sea floor is moving away from the ridge crest at a rate of 1 cen-

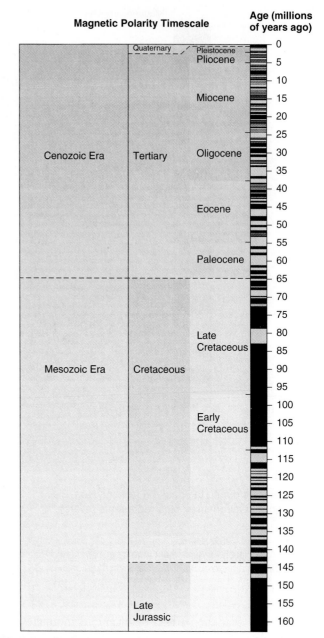

Figure 19.18

Marine magnetic anomaly pattern from the Cenozoic and Mesozoic Eras. Black indicates positive anomalies (and therefore normal polarity). Blue indicates negative anomalies (reverse polarity).

Modified from R. L. Larson and W. C. Pitman, III, 1972, *Geological Society of America Bulletin.*

timeter per year. Such measured rates generally range from 1 to 6 centimeters per year.

Predicting Sea-Floor Age

The other important point of the Vine-Matthews hypothesis is that it *predicts the age of the sea floor* (figure 19.17). Magnetic reversals are now known to have occurred back into Precambrian time. Sea floor of *all* ages is therefore characterized by parallel bands of magnetic anomalies. Figure 19.18 shows the

Figure 19.19

The age of the sea floor as determined from magnetic anomalies.

After *The Bedrock Geology of the World* by R. L. Larson, W. C. Pitman, III, et al. W. H. Freeman.

pattern of marine magnetic anomalies (and the reversals that caused them) during the past 160 million years. The distinctive pattern of these anomalies through time allows them to be identified by age, a process similar to dating by tree rings.

Now, even before they sample the sea floor, marine geologists can predict the age of the igneous rock of the sea floor by measuring the magnetic anomalies at the sea surface. Most sections of the sea floor have magnetic anomalies. By matching the measured anomaly pattern with the known pattern that is shown in figure 19.18, the age of the sea floor in the region can be predicted, as shown on the map in figure 19.19.

This is a very powerful test of the hypothesis that the sea floor moves. Suppose, for example, that the sea floor in a particular spot is predicted to be 70 million years old from a study of its magnetic anomalies. If the hypothesis of sea-floor motion and the Vine-Matthews hypothesis of magnetic anomaly origin are correct, a sample of igneous rock from that spot *must* be 70 million years old. If the rock proves to be 10 million years old or 200 million years old or 1.2 billion years old, or any other age except 70 million years, then both these hypotheses are wrong. But if the rock proves to be 70 million years old, as predicted, then both hypotheses have been successfully tested.

Hundreds of rock and sediment cores recovered from holes drilled in the sea floor were used to test these hypotheses. Close correspondence has generally been found between the predicted age and the measured age of the sea floor. (The sea-floor age is usually measured by fossil dating of sediment in the cores rather than by isotopic dating of igneous rock.) This evidence from deep-sea drilling has been widely accepted by geologists as verification of the hypotheses of plate motion and magnetic anomaly origin. Most geologists now think that these concepts are no longer hypotheses but can now be called theories. (A *theory*, as discussed in Box 1.3 in connection with the scientific method, is a concept with a much higher degree of certainty than a hypothesis.)

Another Test: Fracture Zones and Transform Faults

Deep-sea drilling tested plate motion by allowing us to compare the actual age of the sea floor with the age predicted from magnetic anomalies. Another rigorous test of plate motion has been made by studying the seismicity of fracture zones.

The mid-oceanic ridge is offset along fracture zones (see figure 18.18). Conceivably, the mid-oceanic ridge was once continuous across a fracture zone but has been offset by strike-slip

Figure 19.20

Two possible explanations for the relationship between fracture zones and the mid-oceanic ridge. (*A*) The expected rock motions and earthquake distribution assuming that the ridge was once continuous across the fracture zone. (*B*) The expected rock motions and earthquake distribution assuming that the two ridge segments were never joined together and that the sea floor moves away from the rift valley segments. Only explanation *B* fits the data. The portion of the fracture zone between the ridge segments is a transform fault.

motion along the fracture zone (figure 19.20*A*). If such motion is occurring along a fracture zone, we would expect to find two things: (1) earthquakes should be distributed along the entire length of the fracture zone, and (2) the motion of the rocks on either side of the fracture zone should be in the direction shown by the arrows in figure 19.20*A*.

In fact, these things are not true about fracture zones. Earthquakes do occur along fracture zones, but only in those segments between offset sections of ridge crest. In addition, first-motion studies of earthquakes (see chapter 16) along fracture zones show that the motion of the rocks on either side of the fracture zone during an earthquake is exactly opposite to the motion shown in figure 19.20*A*. The actual motion of the rocks as determined from first-motion studies is shown in figure 19.20*B*. The portion of a fracture zone between two offset portions of ridge crest is called a **transform fault.**

The motion of rocks on either side of a transform fault was predicted by the hypothesis of a moving sea floor. Note that sea floor moves away from the two segments of ridge crest (figure 19.20*B*). Looking along the length of the fracture zone, you can see that blocks of rock move in opposite directions only on that section of the fracture zone between the two segments of ridge crest. Earthquakes, therefore, occur only on this section of the fracture zone, the transform fault. The direction of motion of rock on either side of the transform fault is exactly predicted by the assumption that rock is moving away from the ridge crests. Verification by first-motion studies of this predicted motion along fracture zones was another successful test of plate motion.

Measuring Plate Motion Directly

In recent years the motion of plates has been directly measured using satellites, radar, lasers, and the Global Positioning System. These techniques can measure the distance between two widely separated points to within one centimeter. If two plates move toward each other at individual rates of 2 cm/year and 6 cm/year, the combined rate of convergence is 8 cm/year. The measurement techniques are sensitive enough to easily measure

such a rate if measurements are repeated each year. Such measured rates match closely the predicted rates from magnetic anomalies.

Divergent Plate Boundaries

Divergent plate boundaries, where plates move away from each other, can occur in the middle of the ocean or in the middle of a continent. The result of divergent plate boundaries is to create, or open, new ocean basins. This dynamic process has occurred throughout the geologic past.

When a supercontinent such as Pangaea breaks up, a divergent boundary can be found in the middle of a continent. The divergent boundary is marked by rifting, basaltic volcanism, and uplift. During rifting, the continental crust is stretched and thinned. This extension produces shallow-focus earthquakes on normal faults, and a rift valley forms as a central *graben* (a down-dropped fault block). The faults act as pathways for basaltic magma, which rises from the mantle to erupt on the surface as cinder cones and basalt flows. Uplift at a divergent boundary is usually caused by the upwelling of hot mantle beneath the crust; the surface is elevated by the thermal expansion of the hot, rising rock and of the surface rock as it is warmed from below.

There is current debate on the sequence of events as continents split. Some geologists think that rifting comes first (figure 19.21). The thinning of cold crust would reduce pressure

Figure 19.21

Two models of the beginning of continental divergence. (*A*) Rifting comes first; uplift may follow. (*B*) Uplift comes first; rifting may follow.

A Continent undergoes extension. The crust is thinned and a rift valley forms (East African Rift Valleys).

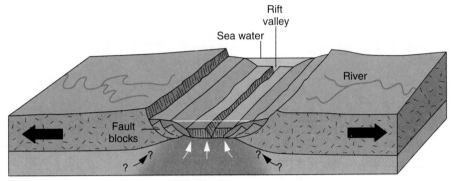

B Continent tears in two. Continent edges are faulted and uplifted. Basalt eruptions form oceanic crust (Red Sea).

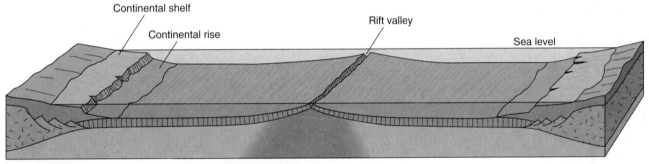

C Continental sediments blanket the subsiding margins to form continental shelves and rises. The ocean widens and a mid-oceanic ridge develops (Atlantic Ocean).

Figure 19.22

A divergent plate boundary forming in the middle of a continent will eventually create a new ocean.

on the underlying rock and allow hot mantle rock to rise passively, elevating the land by thermal expansion; in this case rifting *causes* uplift. Other geologists believe that thermal uplift comes first as hot mantle rock rises actively under uniformly thick crust. As the uplifted crust is stretched, rifting occurs; in this case rifting is the *result* of uplift.

Figure 19.22 shows how a continent might rift to form an ocean. The figure shows rifting before uplift, because recent work indicates that this was the sequence for the opening of the Red Sea. The crust is initially stretched and thinned. Numerous normal faults break the crust, and the surface subsides into a central graben (figure 19.22*A*). Shallow earthquakes and basalt eruptions occur in this rift valley, which also

has high heat flow. An example of a boundary at this stage is the African Rift Valleys in eastern Africa (figure 19.23). The valleys are grabens that may mark the site of the future breakup of Africa.

As divergence continues, the continental crust on the upper part of the plate clearly separates, and seawater floods into the linear basin between the two divergent continents (figure 19.22*B*). A series of fault blocks have rotated along curved fault planes at the edges of the continents, thinning the continental crust. The rise of hot mantle rock beneath the thinned crust causes continued basalt eruptions that create true oceanic crust between the two continents. The center of the narrow ocean is marked by a rift valley with its typical high heat flow

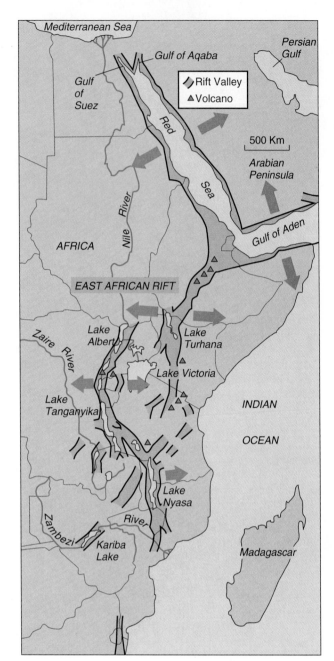

Figure 19.23

The East African Rift Valleys and the Red Sea.

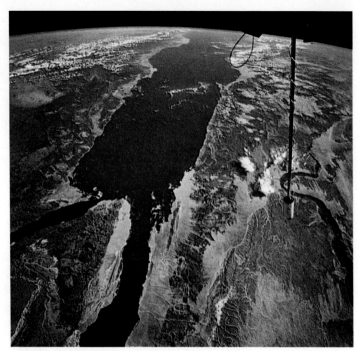

Figure 19.24

Spacecraft photograph looking south along the Red Sea. Gulf of Suez at bottom, Gulf of Aqaba at lower left. Note similarities in the shorelines of the Arabian peninsula (left) and Africa (right), suggesting that the Red Sea was formed by splitting of the continent.

Photo by NASA

Figure 19.25

Water rises upward to fill the gap when (*A*) two floating blocks are (*B*) moved apart. The water moves as a result of the motion of the blocks.

and shallow earthquakes. The Red Sea is an example of a diverging margin at this stage (figures 19.23 and 19.24).

The upward rise of basaltic magma from the mantle to form oceanic crust between two diverging continents is analogous to the rise of water between two floating blocks of wood that are moved apart (figure 19.25).

After modest widening of the new ocean, uplift of the continental edges may occur. As continental crust thins by stretching and faulting, the surface originally subsides. At the same time, hot mantle rock wells up beneath the stretched

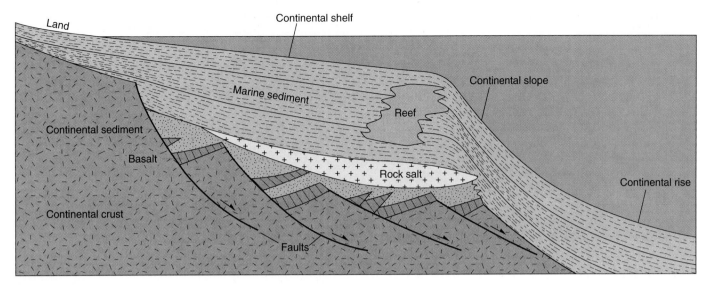

Figure 19.26

A passive continental margin formed by continental breakup and divergence. Downfaulted continental crust forms basins, which fill with basalt and sediment. A layer of rock salt may form if a narrow ocean evaporates. A thick sequence of marine sediments covers these rocks and forms the continental shelf, slope, and rise. A reef may form at the shelf edge if the water is warm; buried reefs occur on many parts of the Atlantic shelf of North America.

crust (figure 19.22*B*). The rising diapir of hot mantle rock would cause uplift by thermal expansion.

The new ocean is narrow, and the tilt of the adjacent land is away from the new sea, so rivers flow away from the sea (figure 19.22*B*). At this stage the seawater that has flooded into the rift may evaporate, leaving behind a thick layer of rock salt overlying the continental sediments. The likelihood of salt precipitation increases if the continent is in one of the desert belts or if one or both ends of the new ocean should become temporarily blocked, perhaps by volcanism. Not all divergent boundaries contain rock salt, however.

The plates continue to diverge, widening the sea. Thermal uplift creates a mid-oceanic ridge in the center of the sea (figure 19.22*C*). The flanks of the ridge subside as the sea-floor rock cools as it moves.

The trailing edges of the continents also subside as they are lowered by erosion and as the hot rock beneath them cools. Subsidence continues until the edges of the continents are under water. A thick sequence of marine sediment blankets the thinned continental rock, forming a *passive continental margin* (figures 19.22*C* and 19.26; see also previous chapter). The sediment forms a shallow continental shelf, which may contain a deeply buried salt layer. The deep continental rise is formed as sediment is carried down the continental slope by turbidity currents and other mechanisms. The Atlantic Ocean is currently at this stage of divergence (figure 18.16).

A divergent boundary on the sea floor is located on the crest of the mid-oceanic ridge. If the spreading rate is slow, as it is in the Atlantic Ocean (1 cm/year), the crest has a rift valley.

Fast spreading, as in the Pacific Ocean (6 cm/year), prevents a rift from forming. A divergent boundary at sea is marked by the same features as a divergent boundary on land—tensional cracks, normal faults, shallow earthquakes, high heat flow, and basaltic eruptions. The basalt forms dikes within the cracks and pillow lavas on the sea floor, creating new oceanic crust on the trailing edge of plates.

Transform Boundaries

At transform boundaries, where one plate slides horizontally past another plate, the plate motion can be taken up on a single fault or on a group of parallel faults. Transform boundaries are marked by shallow-focus earthquakes in a narrow zone for a single fault, or in a broad zone for a group of parallel faults (see figure 16.30). First-motion studies of the quakes indicate strike-slip movement parallel to the faults.

The name *transform fault* comes from the fact that the displacement along the fault abruptly ends or transforms into another kind of displacement. The most common type of transform fault occurs along fracture zones and connects two divergent plate boundaries at the crest of the mid-oceanic ridge (figures 19.27 and 19.20*B*). The spreading motion at one ridge segment is transformed into the spreading motion at the other ridge segment by strike-slip movement along the transform fault.

Not all transform faults connect two ridge segments. As you can see in figures 16.28 and 19.27, a transform fault can connect a ridge to a trench (a divergent boundary to a convergent

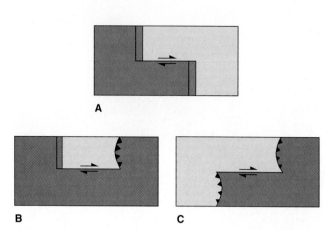

Figure 19.27

Transform boundaries (*A*) between two ridges; (*B*) between a ridge and a trench; and (*C*) between two trenches. Triangles on trenches point down subduction zones. Color tones show two plates in each case. (*D*) The San Andreas fault is a transform plate boundary between the North American plate and the Pacific plate. The south end of the San Andreas fault is a ridge segment (shown in red) near the U.S.-Mexico border. The north end of the fault is a "triple junction" where three plates meet at a point. The relative motion along the San Andreas fault is shown by the large black arrows, as the Pacific plate slides horizontally past the North American plate.

Modified from U.S. Geological Survey.

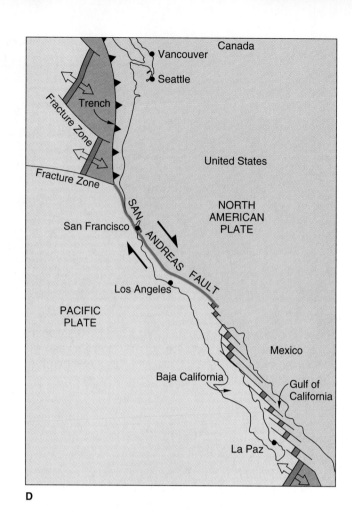

boundary), or can connect two trenches (two convergent boundaries). The San Andreas fault in California is a transform fault with a complex history (figure 19.27*D*).

What is the origin of the offset in a ridge-ridge transform fault? The offsets appear to be the result of irregularly shaped divergent boundaries (figure 19.28). When two oceanic plates begin to diverge, the boundary may be curved on a sphere. Mechanical constraints prevent divergence along a curved boundary, so the original curves readjust into a series of right-angle bends. The ridge crests align perpendicular to the spreading direction, and the transform faults align parallel to the spreading direction. An old line of weakness in a continent may cause the initial divergent boundary to be oblique to the spreading direction when the continent splits. The boundary will then readjust into a series of transform faults parallel to the spreading direction.

Convergent Plate Boundaries

At convergent plate boundaries two plates move toward each other (often obliquely). The character of the boundary depends partly on the type of plates that converge. A plate

Figure 19.28

Divergent boundaries form ridge crests perpendicular to the spreading direction and transform faults parallel to the spreading direction. (*A*) Oceanic plates. (*B*) Continental plates.

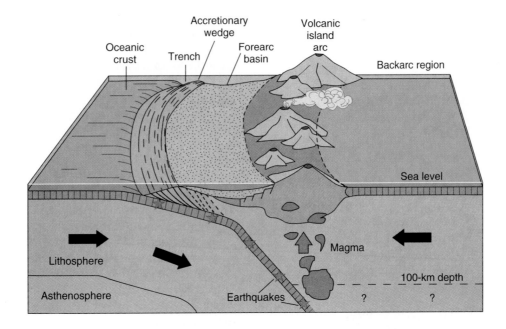

Figure 19.29

Ocean-ocean convergence forms a trench, a volcanic island arc, and a Benioff zone of earthquakes.

Modified from W. R. Dickinson, 1977, in *Island Arcs, Deep Sea Trenches and Back-Arc Basins* (pp. 33–40), copyrighted by American Geophysical Union.

capped by oceanic crust can move toward another plate capped by oceanic crust, in which case one plate dives (subducts) under the other. If an oceanic plate converges with a plate capped by a continent, the dense oceanic plate subducts under the continental plate. If the two approaching plates are both carrying continents, the continents collide and crumple but neither is subducted.

Ocean-Ocean Convergence

Where two plates capped by sea floor converge, one plate subducts under the other (the Pacific plate sliding under the western Aleutian Islands is an example). The subducting plate bends downward, forming the outer wall of an oceanic trench, which usually forms a broad curve convex to the subducting plate (figures 19.29 and 19.30).

As one plate subducts under another, a Benioff zone of shallow-, intermediate-, and deep-focus earthquakes is created within the upper portion of the downgoing lithosphere (figure 19.29). The reasons for these quakes are discussed in chapter 16. The existence of deep-focus earthquakes to a depth of 670 kilometers tells us that brittle plates continue to (at least) that depth. The pattern of quakes shows that the angle of subduction changes with depth, usually becoming steeper (figure 19.29). Some plates crumple or break into segments as they descend.

As the descending plate reaches depths of at least 100 kilometers, magma is generated in the overlying asthenosphere (figure 19.29). The magma probably forms by partial melting of the asthenosphere, perhaps triggered by dewatering of the downgoing oceanic crust as it is subducted, as described in

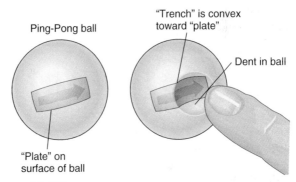

Figure 19.30

A dented Ping-Pong ball can show why trenches are curved on a sphere.

chapter 3. Differentiation and assimilation may also play an important role in the generation of the magma, which is typically andesitic to basaltic in composition.

The magma works its way upward to erupt as an **island arc,** a curved line of volcanoes that form a string of islands parallel to the oceanic trench (figure 19.29). Beneath the volcanoes are large plutons in the thickened arc crust.

The distance between the island arc and the trench can vary, depending upon where the subducting plate reaches the 100-kilometer depth. If the subduction angle is steep, the plate reaches this magma-generating depth at a location close to the trench, so the horizontal distance between the arc and trench is short (figure 19.31). If the subduction angle is gentle, the arc-trench distance is greater. A thick, buoyant plate (such as a subducting aseismic ridge) may subduct at such a gentle angle that

Figure 19.31

Andesitic magma is generated where the top of the subducting lithosphere reaches a depth of 100 kilometers, so the subduction angle determines the arc-trench spacing.

it merely slides horizontally along under another plate. Because the top of the subducting plate never reaches the 100-kilometer depth, such very shallow subduction zones lack volcanism.

When a plate subducts far from a mid-oceanic ridge, the plate is cold, with a low heat flow. Oceanic plates form at ridge crests, then cool and sink as they spread toward trenches. Eventually they become cold and dense enough to sink back into the mantle. Oceanic trenches are marked by strong negative gravity anomalies. These show that trenches are not currently in isostatic equilibrium, but are being actively pulled down. Hess thought that this pulling was caused by a down-turning convection current in the mantle. Today most geologists think that the pulling is caused by the sinking of cold, dense lithosphere.

The inner wall of a trench (toward the arc) consists of an *accretionary wedge* (or *subduction complex*) of thrust-faulted and folded marine sediment (figure 19.29). The sediment is snow-plowed off the subducting plate by the overlying plate. New slices of sediment are continually added to the bottom of the accretionary wedge, pushing it upward to form a ridge on the sea floor. A relatively undeformed *forearc basin* lies between the accretionary wedge and the volcanic arc. (The trench side of an arc is the forearc; the other side of the arc is the backarc.)

Trench positions change with time (figure 19.32). As one plate subducts, the overlying plate may be moving toward it. The motion of the leading edge of the overlying plate will force the trench to migrate horizontally over the subducting plate (figure 19.32*A*). The Peru-Chile trench is moving over the Nazca plate in this manner as South America moves westward (figure 19.1). There is another reason that trenches move. It is now widely believed that a subducting plate does not sink in a direction parallel to the length of the plate, but falls through the mantle at an angle that is *steeper* than the dip of the down-going plate (figure 19.32*B*). This steep sinking pulls the subducting plate progressively away from the overlying plate, and causes the hinge line of bending and the oceanic trench to migrate seaward onto the subducting plate. The location at which the subducting plate contacts the 100-kilometer depth to generate andesite also migrates seaward toward the subducting plate, and may cause the position of the island arc to migrate toward the subducting plate as well.

Ocean-Continent Convergence

When a plate capped by oceanic crust is subducted under the *continental* lithosphere, an accretionary wedge and forearc basin form an *active continental margin* between the trench and the continent (figure 19.33). A Benioff zone of earthquakes dips under the edge of the continent, which is marked by andesitic volcanism and a young mountain belt. An example of this type of boundary is the subduction of the Nazca plate under western South America.

The magma that is created by ocean-continent convergence forms a **magmatic arc,** a broad term used both for island arcs at sea and for belts of igneous activity on the edges of continents. The surface expression of a magmatic arc is either a line of andesitic islands (such as the Aleutian Islands) or a line of andesitic continental volcanoes (such as the Cascade volcanoes of the Pacific Northwest). Beneath the volcanoes are large plutons in thickened crust. We see these plutons as batholiths on land when they are exposed by deep erosion. The igneous processes that form the granitic and intermediate magmas of batholiths are described in chapter 3.

The hot magma rising from the subduction zone thickens the continental crust and makes it weaker and more mobile than cold crust. Regional metamorphism takes place within this hot, mobile zone. Crustal thickening causes uplift, so a young mountain belt forms here as the thickened crust rises isostatically.

Another reason for the growth of the mountain belt is the stacking up of thrust sheets on the continental (backarc) side of the magmatic arc (figure 19.33). The thrust faults, associated with folds, move slivers of mountain-belt rocks landward over the continental interior (the *craton*). Underthrusting of the rigid craton beneath the hot, mobile core of the mountain belt may help form the fold-thrust belt.

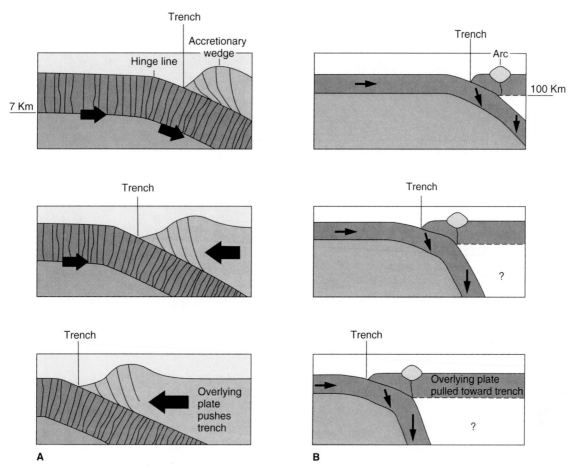

Figure 19.32

Migration of trench, hinge line, accretionary wedge, and volcanic arc. (*A*) The motion of the overlying plate can force this migration. (*B*) The cooling, subducting plate sinks at a steeper angle than its dip, pulling the overlying plate toward the subducting plate.

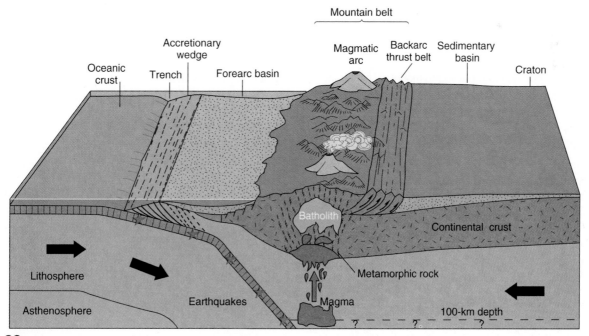

Figure 19.33

Ocean-continent convergence forms an active continental margin with a trench, a Benioff zone, a magmatic arc, and a young mountain belt on the edge of the continent.

A Ocean-continent convergence

B Ocean-continent convergence

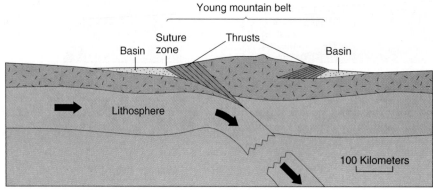

C Continent-continent collision

Figure 19.34

Continent-continent collision forms a young mountain belt in the interior of a new, larger continent.

Modified from W. R. Dickinson, 1977, in *Island Arcs, Deep Sea Trenches and Back-Arc Basins* (pp. 33–40), copyrighted by American Geophysical Union.

Inland of the backarc fold-thrust belt, the craton subsides to form a sedimentary basin (sometimes called a *foreland basin*). The weight of the stacked thrust sheets depresses the craton isostatically. The basin receives sediment, some of which may be marine if the craton is forced below sea level. This basin extends the effect of subduction far inland. Subduction of the sea floor off California during the Mesozoic Era produced basin sedimentation as far east as the central Great Plains.

Continent-Continent Convergence

Two continents may approach each other and collide. They must be separated by an ocean floor that is being subducted under one continent and that lacks a spreading axis to create

new oceanic crust (figure 19.34). The edge of one continent will initially have a magmatic arc and all the other features of ocean-continent convergence.

As the sea floor is subducted, the ocean becomes narrower and narrower until the continents eventually collide and destroy or close the ocean basin. Oceanic lithosphere is heavy and can sink into the mantle, but continental lithosphere is less dense and cannot sink. One continent may slide a short distance under another, but it will not go down a subduction zone. After collision the heavy oceanic lithosphere breaks off the continental lithosphere and continues to sink, leaving the continent behind (figure 19.34C).

The two continents are welded together along a dipping *suture zone* that marks the old site of subduction. Thrust belts

Figure 19.35

Backarc spreading. Regional extension in the overlying plate of a subduction zone can split an arc, move an arc offshore, or split a continent.

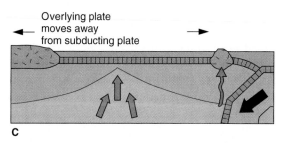

Figure 19.36

Causes of backarc spreading. Extension may be caused by a rising mantle diapir, by secondary convection, or by relative plate motions.

and subsiding basins occur on both sides of the original magmatic arc, which is now inactive. The presence of the original arc thickens the crust in the region of impact. The crust is thickened further by the shallow underthrusting of one continent beneath the other and also by the stacking of thrust sheets in the two thrust belts. The result is a mountain belt in the interior of a continent (a new large continent formed by the collision of the two smaller continents). The entire region of impact is marked by a broad belt of shallow-focus earthquakes along the numerous faults, as shown in figure 16.31A. A few deeper quakes may occur within the sinking oceanic lithosphere beneath the mountain range.

The Himalaya Mountains in central Asia are thought to have formed in this way, as India collided with and underthrust Asia to produce exceptionally thick crust and high elevations. Paleomagnetic studies show that India was once in the southern hemisphere and drifted north to its present position. The collision with Asia occurred after an intervening ocean was destroyed by subduction (figure 19.2).

Backarc Spreading

Regional extension occurs within or behind many arcs. This extension can tear an arc in two, moving the two halves in opposite directions (figure 19.35). If it occurs behind an arc, it can move the arc away from a continent. It can split the edge of a continent, moving a narrow strip of the continent seaward (this is apparently how Japan formed). In each case the spread-

ing creates new oceanic crust that is similar, but not identical, to the oceanic crust formed at the crest of mid-oceanic ridges. This backarc oceanic crust is apparently the type of crust found in most ophiolites (chapter 18).

The reason for backarc extension is energetically debated. One suggestion is that extension is caused by a rising and spreading mantle diapir of hot rock or magma somehow generated by the downgoing plate (figure 19.36). The spreading diapir tears open the backarc basin, and the rising magma forms new oceanic crust. Another suggestion is that the subducting plate drags on the overlying asthenosphere, causing it to move in secondary convection cells that stretch and fracture the overlying oceanic crust. A third suggestion, which seems the best explanation for the most rapidly spreading backarc basins in the Pacific, is that the overlying plate is retreating away from the subducting plate. If the arc on the overlying plate stays fixed near the subducting plate, the retreat of the overlying plate will tear open the backarc basin.

The Motion of Plate Boundaries

Almost nothing is fixed in plate tectonics. Not only do plates move, but plate boundaries move as well. Plates may move away from each other at a divergent boundary on a ridge crest for tens of millions of years, but the ridge crest can be migrating across the earth's surface as this occurs. Ridge crests can also jump to new positions. The original ridge crest may suddenly become inactive; the divergence will jump quickly to a new position and create a new ridge crest (the evidence lies in the sea-floor magnetic anomaly pattern).

Convergent boundaries migrate, too, as shown in figure 19.32. As they migrate, trenches and magmatic arcs migrate along with the boundaries. Convergent boundaries can also jump; subduction can stop in one place and begin suddenly in a new place.

Transform boundaries change position, too. California's San Andreas fault has been in its present position about 5 million years. Prior to that, the plate motion was taken up on sea-floor faults parallel to the San Andreas (figure 19.37). In the future, the San Andreas may shift eastward again. The 1992 Landers earthquake, on a new fault in the Mojave Desert, and its pattern of aftershocks extending an astonishing 500 miles northward, suggest that the San Andreas may be trying to jump inland again. If it eventually does, most of California will be newly attached to the Pacific plate instead of the North American plate, and California will slide northwestward relative to the rest of North America.

Plate Size

Plates can change in size. For example, new sea floor is being added on the trailing edge of the North American plate at the spreading axis in the central Atlantic Ocean. Most of the North American plate is not being subducted along its leading edge because this edge is made up of lightweight continental rock. Thus the North American plate is growing in size as it moves slowly westward.

The Nazca plate probably is getting smaller. The spreading axis is adding new rock along the trailing edge of the Nazca plate, but the leading edge is being subducted down the Peru-Chile Trench. If South America were stationary, the Nazca plate might remain the same size, because the rate of subduction and the rate of spreading are equal. But South America is slowly moving westward, pushing the Peru-Chile Trench in front of it. This means that the site of subduction of the Nazca plate is gradually coming closer to its spreading axis to the west, and so the Nazca plate is getting smaller. The same thing is probably happening to the Pacific plate as the Eurasian plate moves eastward into the Pacific Ocean.

The Attractiveness of Plate Tectonics

Most geologists accept the general concept of plate tectonics because it can explain in a general way the distribution and ori-

Figure 19.37

The San Andreas fault (a transform boundary) has changed position through time. Prior to 5 million years ago, the fault was offshore (blue line). In the future it may jump inland again (green zone).

gin of many earth features. These features are discussed throughout this book, and we summarize them here.

The distribution and composition of the world's *volcanoes* can be explained by plate tectonics. *Basaltic* volcanoes and lava flows form at divergent plate boundaries when hot mantle rock rises at a spreading axis. *Andesitic* volcanoes, particularly those in the circum-Pacific belt, result from subduction of an oceanic

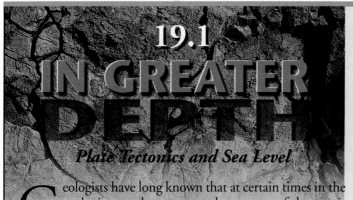
plate beneath either a continental plate or another oceanic plate. Although most of the world's volcanoes occur at plate margins, some do not (Hawaii being an example). We will discuss some of these isolated volcanoes later in the chapter when we describe mantle plumes.

Earthquake distribution and first motion can largely be explained by plate tectonics. Shallow-focus earthquakes along normal faults are caused by extension at divergent plate boundaries. Shallow-focus earthquakes occur on transform faults when plates slide past one another. Broad zones of shallow-focus earthquakes are located where two continents collide. Dipping Benioff zones of shallow-, intermediate-, and deep-focus quakes are found along the giant thrust faults formed when an oceanic plate is subducted beneath another plate. Most of the world's earthquakes (like most volcanoes) occur along plate boundaries, although a few take place within plates and are difficult to explain in terms of plate tectonics.

Young mountain belts—with their associated igneous intrusions, metamorphism, and fold-thrust belts—form at convergent boundaries. "Subduction mountains" form at the edges of continents where sea floor is sliding under continents. "Continental-collision" mountains form in continental interiors when two continents collide to form a larger continent. Old mountain belts mark the position of old, now inactive, plate boundaries.

The major features of the sea floor can also be explained by plate tectonics. The *mid-oceanic ridge* with its rift valley forms at divergent boundaries. *Oceanic trenches* are found where oceanic plates are subducted at convergent boundaries. *Fracture zones* are created at transform boundaries.

Other hypotheses can explain some of these features, but not all of them. Belts of folded mountains have been attributed to compression caused by a contracting earth. Rift valleys, on the other hand, have been explained by extension caused by an *expanding* earth. The hypotheses are incompatible with each other and do not give a unifying view of the earth. Plate tectonics explains more features than any other hypothesis or theory, and it provides a unifying framework for the study of the earth. That is why so many geologists support the concept, at least as a working model of how the earth works.

What Causes Plate Motions?

There is currently a great deal of speculation about why plates move. There may be several reasons for plate motion. Any mechanism for plate motion has to explain why:

1. mid-oceanic ridge crests are hot and elevated, while trenches are cold and deep;
2. ridge crests have tensional cracks;

3. the leading edges of some plates are subducting sea floor, while the leading edges of other plates are continents (which cannot subduct).

Convection in the mantle, proposed as a mechanism for sea-floor spreading (figure 19.12), can account for these facts, as we have shown earlier in this chapter. Mantle convection is quite likely because heat loss from the earth's core should heat the overlying mantle, causing it to overturn. Debate about the size and shape of mantle convection cells led to several models (figure 19.38). The old seafloor-spreading model assumed mantle-deep convection. Proponents of a layered mantle think that the 670-kilometer boundary may be a compositional boundary that prevents mantle-deep convection, so two-layer convection models have been proposed. Rising hot rock in the lower mantle may heat the base of the upper mantle, causing the upper mantle to rise directly over the rising lower mantle (figure 19.38B). Alternatively, the horizontal motion of the lower mantle just below the boundary may drag the upper mantle rock along in the same direction (figure 19.38C). Note that this creates *sinking* upper mantle directly over rising lower mantle.

Many geologists now think that mantle convection is a *result* of plate motion rather than a cause of it. The sinking of a cold, subducting plate can create mantle convection (convection can be driven by either hot, rising material or by cold, sinking material). Hot mantle rock rises at divergent boundaries to take the place of the diverging plates (figure 19.25). Such plate-caused convection could be shallow, rather than mantle-deep.

The basic question in plate motion is why do plates diverge and sink? Two or three different mechanisms may be at work here.

One proposal is called "*ridge-push.*" As a plate moves away from a divergent boundary, it cools and thickens. Cooling sea floor subsides as it moves, and this subsidence forms the broad side slopes of the mid-oceanic ridge. An even more important slope forms on the base of the lithosphere as new igneous rock is added to the base of the lithosphere by progressive cooling of the asthenosphere below (figure 19.39). The oceanic plate is thought to slide down this slope at the base of the lithosphere, which may have a relief of 80 to 100 kilometers.

Another mechanism is called "*slab-pull*" (figure 19.40). Cold lithosphere sinking at a steep angle through hot mantle should pull the surface part of the plate away from the ridge crest and then down into mantle as it cools. A subducting plate sinks because it is denser than the surrounding mantle. This density contrast is partly due to the fact that the sinking lithosphere is cold. The subducting plate may also increase its density while it sinks, as low-density materials such as water are lost, and as plate minerals collapse into denser forms during subduction. Slab-pull is thought to be at least twice as important as ridge-push in moving an oceanic plate away from a ridge crest. Slab-pull causes rapid plate motion.

If subducting plates fall into the mantle at angles steeper than their dip (figure 19.32), then trenches and the overlying

A

B

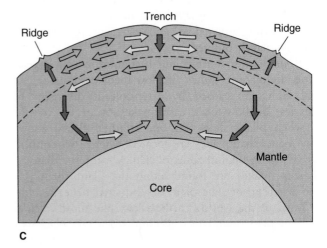

C

Figure 19.38

Models of mantle convection. (*A*) Mantle-deep convection. (*B*) Two-tiered convection separated at 670-kilometer boundary. The lower mantle delivers heat to the upper mantle. (*C*) Two-tiered convection. The lower mantle drags on the upper mantle.

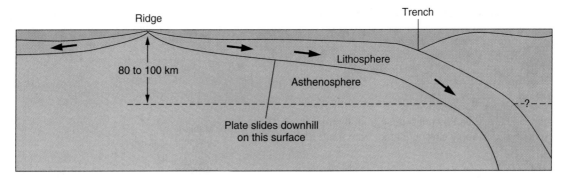

Figure 19.39

Ridge-push. A plate may slide downhill on the sloping boundary between the lithosphere and the asthenosphere at the base of the plate.

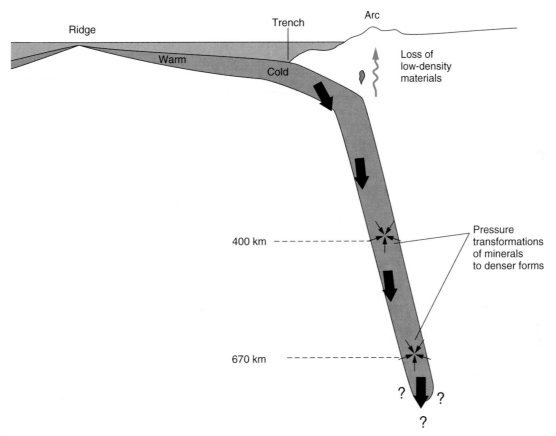

Figure 19.40

Slab-pull. The dense, leading edge of a subducting plate pulls the rest of the plate along. Plate density increases due to cooling, loss of low-density material, and pressure transformation of minerals to denser forms.

plates are pulled horizontally seaward toward the subducting plates. This mechanism has been termed "*trench-suction*." It is probably a minor force, but may be important in moving continents apart. Divergent continents at the leading edges of plates cannot be moved by slab-pull, because they are not on subducting plates. They might be moved by ridge-push from the rear, and/or trench-suction from the front (figure 19.41). They move much more slowly than subducting plates.

All three of these mechanisms (ridge-push, slab-pull, and trench-suction), particularly in combination, are compatible with high, hot ridges; cold, deep trenches; and tensional cracks at the ridge crest. They can account for the motion of both oceanic and

Figure 19.41

Divergent continents may be moved by ridge-push from the rear, or trench-suction (caused by steeply sinking plates) from the front.

Figure 19.42

Mantle plumes rise upward through the mantle. When the large head contacts a continent, it causes uplift and the eruption of flood basalts.

continental plates. In this scheme, plate motions are controlled by variations in lithosphere density and thickness, which, in turn, are controlled largely by cooling. In other words, the reasons for plate motions are the properties of the plates themselves and the pull of gravity. This idea is in sharp contrast to most convection models, which assume that plates are dragged along by the movement of mantle rock beneath the plates.

Mantle Plumes and Hot Spots

A modification of the convection process has been suggested by W. Jason Morgan of Princeton University. Morgan believes that convection occurs in the form of **mantle plumes,** narrow columns of hot mantle rock that rise through the mantle,

much like smoke rising from a chimney (figure 19.42). Mantle plumes are now thought to have large spherical or mushroom-shaped heads above a narrow rising tail. They are essentially stationary with respect to moving plates and to each other.

Plumes form "hot spots" of active volcanism at the earth's surface. Note in figure 19.43 that many plumes are located in

Figure 19.43

Distribution of hypothesized mantle plumes, identified by volcanic activity and structural uplift within the past few million years. The hot spots near the poles are not shown.

Compiled by W. S. F. Kidd and K. Burke, SUNY at Albany.

Figure 19.44

Continental breakup caused by a mantle plume. (*A*) A dome forms over a mantle plume rising beneath a continent. (*B*) Three radial rifts develop due to outward radial flow from the top of the mantle plume. (*C*) Continent separates into two pieces along two of the three rifts, with new ocean floor forming between the diverging continents. The third rift becomes an inactive "failed rift" (or aulacogen) filled with continental sediment.

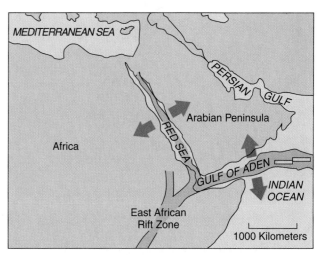

Figure 19.45

An example of radial rifts. The Red Sea and the Gulf of Aden are the active rifts, as the Arabian peninsula drifts away from Africa. The Gulf of Aden contains a mid-oceanic ridge and central rift valley. The inactive failed rift is the rift valley shown in Africa.

volcanic regions such as Iceland, Yellowstone, and Hawaii. When the large head of the plume nears the surface, it causes uplift and the eruption of vast fields of flood basalts. As the head widens beneath the crust the flood-basalt area widens and the crust is stretched. The tail that follows the head produces a narrow spot of volcanic activity, much smaller than the head.

The outward, radial flow of the expanding head may be strong enough to break the lithosphere and start plates moving. In Morgan's view a few plumes, such as those on the mid-oceanic ridge in the Atlantic Ocean in figure 19.43, are enough to drive plates apart (in this case, to push the American plates westward).

A mantle plume rising beneath a continent should heat the land and bulge it upward to form a dome marked by volcanic eruptions. As the dome forms, the stretched crust typically fractures in a three-pronged pattern (figure 19.44). Continued radial flow outward from the rising plume eventually separates the crust along two of the three fractures but leaves the third fracture inactive. In this model of continental breakup, the two active fractures become continental edges as new sea floor forms between the divergent continents. The third fracture is a *failed rift* (or *aulacogen*), an inactive rift that becomes filled with sediment.

An example of this type of fracturing may exist in the vicinity of the Red Sea (figure 19.45). The Red Sea and the Gulf of Aden are active diverging boundaries along which the Arabian Peninsula is being separated from northeastern Africa. The third, inactive, rift is the northernmost African Rift Valley, lying at an angle of about 120° to each of the narrow seaways.

Figure 19.46 shows how two plumes might split a continent and begin plate divergence. Local uplift *causes* rifting over each plume. The rifts lengthen with time until the land is torn in two. The two halves begin to diverge either from sliding down the uplifts or from being dragged along from below by the outward radial flow of the plume. Along the long rift segments between plumes, rifting occurs *before* uplift.

A place where a mantle plume might now be rising beneath a continent is in Yellowstone National Park in northwestern Wyoming. The area's volcanism, high elevation, high heat flow, and hot spring and geyser activity all may be due to this plume. Radial flow of mantle rock beneath the western United States may be tearing the continent apart and causing the earthquakes in the region, including the 1959 earthquake near Madison Canyon, Montana. Eventually an ocean may form here as North America is split apart by the plume.

A

B

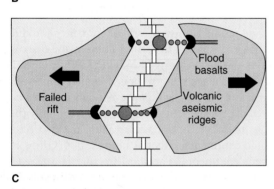

C

Figure 19.46

(*A*) Two mantle plumes beneath a continent. (*B*) The rifts lengthen and flood basalts erupt over the plumes. (*C*) The continent splits and failed rifts form. The new ocean is marked by ridge crests, fracture zones, and aseismic ridges (chains of volcanoes).

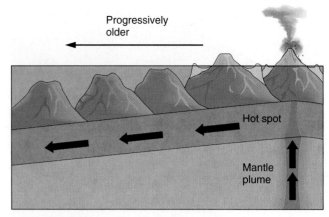

Figure 19.47

Sea floor moving over a hot spot forms an aseismic ridge as a chain of volcanoes and guyots.

Figure 19.48

Ages of volcanic rock of the Hawaiian island group. Ages increase to northwest. Two active volcanoes on Hawaii shown by red dots. See also figure 18.22.

Some plumes rise beneath the centers of oceanic plates. A plume under Hawaii rises in the center of the Pacific plate. As the plate moves over the plume, a line of volcanoes forms, creating an aseismic ridge (figure 19.47 and chapter 18). The volcanoes are gradually carried away from the eruptive center, sinking as they go because of cooling. The result is a line of extinct volcanoes (seamounts and guyots) increasing in age away from an active volcano directly above the plume.

In the Hawaiian island group, the only two active volcanoes are in the extreme southeastern corner (figure 19.48). The isotopic ages of the Hawaiian basalts increase regularly to the northwest, and a long line of submerged volcanoes forms an aseismic ridge to the northwest of Kauai (figure 18.22). Most aseismic ridges on the sea floor appear to have active volcanoes at one end, with ages increasing away from the eruptive cen-

ters. Deep-sea drilling has shown, however, that not all aseismic ridges increase in age along their lengths. This evidence has led to alternate hypotheses for the origin of aseismic ridges. It may pose difficulties for the plume hypothesis itself.

Note in figure 18.22 that the three large aseismic ridges in the Pacific Ocean change direction abruptly. If these ridges formed from three separate stationary plumes, they suggest that the Pacific plate has moved in two directions in the past. Early movement approximately northward, followed by more westerly movement, could have produced the ridge patterns.

The Relationship Between Plate Tectonics and Ore Deposits

The plate tectonic theory provides an overall model for the origin of metallic ore deposits that has been used to explain the occurrence of known deposits and in exploration for new

Figure 19.49

Divergent oceanic plates carry metallic ores away from rift valley. (Size of ore deposits exaggerated.)

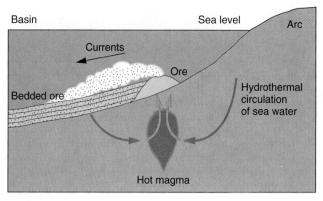

Figure 19.50

On island arcs metallic ores can form over hot springs and be redistributed into layers by currents in shallow basins.

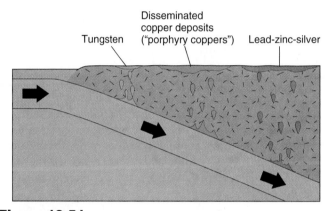

Figure 19.51

Possible relation of ore belts in the western United States to depth along the subduction zone. Different metallic ores (and different igneous rocks) are generated at different depths along a subducted plate.

deposits. Because many ore deposits are associated with igneous activity, there is a close relationship between plate boundaries and metallic ore deposits.

As discussed in chapter 18, *divergent plate boundaries* are often marked by lines of active hot springs in rift valleys that carry and precipitate metallic minerals in mounds around the hot springs. The metals in rift-valley hot springs are predominantly iron, copper, and zinc, with smaller amounts of manganese, gold, and silver. Although the mounds are nearly solid metal sulfide, they are small and widely scattered on the sea floor, so commercial mining of them may not be practical. Occasionally, the ore minerals may be concentrated in richer deposits. On the floor of the Red Sea metallic sediments have precipitated in basins filled with hot-spring solutions. Although the solutions are hot (up to 60°C or 140°F), they are very dense because of their high salt content (they are seven times saltier than sea water), so they collect in sea-floor depressions instead of mixing with the overlying sea water. Although not currently mined, the metallic sediments were estimated in 1983 to be worth $25 billion.

Hot metallic solutions are also found along some divergent continental boundaries. Near the Salton Sea in southern California, which lies along the extension of the mid-oceanic ridge inland, hot water very similar to the Red Sea brines has been discovered underground. The hot water is currently being used to run a geothermal power plant. The high salt and metal content is corrosive to equipment, but metals such as copper and silver may one day be recovered as valuable by-products.

Sea-floor spreading carries the metallic ores away from the ridge crest (figure 19.49), perhaps to be subducted beneath island arcs or continents at *convergent plate boundaries.* Slivers of *ophiolite* on land may contain these rich ore minerals in relatively intact form. A notable example of such ores occurs on the island of Cyprus in the Mediterranean Sea. Banded chromite ores may also be contained in the serpentinized ultramafic rock at the bottom of ophiolites.

Volcanism at *island arcs* can also produce hot-spring deposits on the flanks of the andesitic volcanoes. Pods of very rich ore collect above local bodies of magma, and the ore is sometimes distributed as sedimentary layers in shallow basins (figure 19.50). The circulation pattern and the ore-forming

processes are quite similar to those of spreading centers, but the island arc ores usually contain more lead. Rich *massive sulfide deposits* overlying fractured volcanic rock in the Precambrian shield area of Canada may have formed in this way on ancient island arcs.

Subduction of the sea floor beneath a *continent* produces broad belts of metallic ore deposits on the edge of the continent. Figure 19.51 shows how the distribution of some metals in the western United States might be related to depth along a subduction zone (the figure shows only one of several competing models relating continental ore deposits to plate tectonics). The pattern of ore belts in the United States has probably been disturbed by changing subduction angles, strike-slip faulting, and backarc spreading. Similar patterns of ore belts occur in other subduction mountain ranges, notably the Andes.

The origin of the continental ores above a subduction zone is not clear. The hot-spring deposits from the ridge crest

19.2

IN GREATER DEPTH

A CAT Scan of the Mantle

A new technique for looking at the mantle is similar to the medical technique of CAT scanning (CAT stands for computed axial tomography), which builds up a three-dimensional picture of soft body tissues such as the brain by taking a series of X-ray pictures along successive planes in the body.

Seismic tomography uses earthquake waves and powerful computers to study planar cross sections of the mantle following large earthquakes. Slight variations from expected arrival times at distant seismograph stations can be used to find temperature variations in the mantle. Hot rock slows down seismic waves, so a late arrival of a seismic wave shows that the wave went through hot rock. Cold rock is dense and strong, so it speeds up seismic waves, resulting in early arrivals. Sophisticated computer analysis of hundreds of sections through the mantle allows maps of seismic-wave velocity (and therefore mantle rock temperature) to be drawn for various depths.

Box figure 1, top shows mantle velocities at a depth of 100 kilometers. Red areas show low velocities (probably caused by hot rock) in generally expected positions—along the crest of the mid-oceanic ridge and beneath hot spots. Blue areas show high-velocity (probably cold) rock under continents and old sea floor such as the western Pacific. Box figure 1, bottom shows that these patterns are dramatically different at a depth of 300 kilometers. High-velocity rock extends to this depth below most continents, implying that continents have very deep roots. Some areas that appear hot at 100 kilometers are cold at 300 kilometers, such as the ridge crest just south of Australia. Areas such as the central Pacific and the Red Sea region appear cold at 100 kilometers and hot at 300 kilometers.

In box figure 2, vertical cross sections of seismic velocity are shown to a depth of 670 kilometers for two regions. Note in box figure 2A that high-velocity (cold) roots beneath North America, Asia, and Antarctica extend 400 to 600 kilometers downward. This finding casts doubt on our simple

Box 19.2 Figure 1

Map views of seismic-wave velocities in the mantle at depths of 100 and 300 kilometers, as determined by seismic tomography. Blue indicates high velocity (cold rock), red indicates low velocity (hot rock). White lines outline plates; white circles are major hot spots.

From Dziewonski and Anderson, *American Scientist,* 1984, 72:483–94.

lithosphere-asthenosphere model of plate behavior—continental plates here seem to be hundreds of kilometers thick. Notice, too, how some low-velocity hot spots near Greenland (box figure 2, top) and in the south Atlantic and south Pacific (box figure 2, bottom) are underlain by apparently cold rock. This pattern suggests to some geologists that mantle plumes may be quite shallow and may not extend vertically throughout the mantle. On the other hand, plume tails may be too narrow to be detected by this technique.

More recent, deeper CAT scans of the mantle (box figure 3) seem to indicate that some mantle plumes emanate from the core-mantle boundary and are fed by heat loss from the core. It is likely that the hot plumes originate from various depths in the mantle.

The new tomographic images reveal high-velocity areas, which are interpreted as cold sinking slabs of subducted

are subducted with oceanic crust and could become remobilized to rise into the continent above. The ores may also "distill" off other parts of the descending oceanic crust or upper mantle. The metals may also derive from the continental crust itself or the mantle below it. The metals may be concentrated

somehow by the heat of a rising blob of magma or the hydrothermal circulation associated with it.

The connection between some hydrothermal ore deposits and plate tectonics is tenuous at best. The "Mississippi Valley-type" lead-zinc deposits of the continental interior are very puzlat.: 6.0 lon.: 125.0   lat.: −2.5 lon.: 35.0

Box 19.2 Figure 3

Cross section of seismic-wave velocities from the earth's surface (upper curve) to core. Blue indicates fast seismic velocities (cold rock), and red indicates low velocities (hot rock). There is a presumed cold slab of rock, shown on the left side of the cross section, that is sinking into the lower mantle into other slabs that rest on the core-mantle boundary. Hot rocks, believed to represent mantle plumes, also emanate from the core-mantle boundary, on the right side of the cross section.

Photo Courtesy of Stephen Grand, University of Texas at Austin

Box 19.2 Figure 2

Vertical cross sections of seismic-wave velocities to a depth of 670 kilometers in the mantle. The orange lines show the locations of the cross sections.

From Dziewonski and Anderson, *American Scientist*, 1984, 72:483–94.

plates, also extend all the way to the core-mantle boundary (box figures 3 and 4).

Other plates stop descending at the 670-kilometer boundary within the mantle. Perhaps the depth of sinking is controlled by plate density. The older the subducting rock is, the colder and denser it is. Old, dense plates may sink to the base of the mantle, while younger plates, being less dense, stop at a depth of 670 kilometers (box figure 4).

It is becoming increasingly apparent that the core-mantle boundary may play an important role in the overall mechanism of plate movement.

Additional Reading
Kerr, R. A. 1991. Do plumes stir earth's entire mantle? *Science* vol. 252: 1068–1069.

———. 1997. Deep-sinking slabs stir the mantle. *Science* vol. 275:613–15.

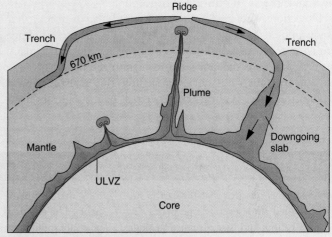

Box 19.2 Figure 4

Seismic data suggest some plates sink to the base of the mantle, whereas other plates are impeded by the increase in density of the mantle at 670 km. Deep mantle plumes emanating from the core-mantle boundary are thought to be underlain by an ultralow-velocity zone (ULVZ).

zling features. Over broad areas metal ore has been emplaced in limestone and dolomite, both by cavity filling and replacement (both are usually considered to be hydrothermal processes). There is no obvious connection between the ores and any igneous rocks, which may be absent in the ore regions. The presence of the ores in the thin sedimentary cover of a supposedly inactive interior of a plate is difficult to explain. The ores do group roughly around the New Madrid earthquakes of southeastern Missouri (figure 16.12); deep faults, perhaps on a failed rift, may have provided pathways for ore-bearing fluids.

It is tempting to think that *mantle plumes* might cause ore deposition, for plumes provide a source of both magma and hydrothermal solutions. The locations of supposed plumes, however, such as Yellowstone and Hawaii, are notable for their *lack* of ore deposits.

A Final Note

Geologists, like other people, are susceptible to fads. Although most geologists believe plate tectonics is an exciting theory and accept it as a working model of the earth, the theory may or may not be correct. Most geologists today believe that plates exist and move. But widespread belief in a theory does not make it true. Two hundred years ago geologists "knew" that basalt crystallized out of seawater. In the 1800s glacial deposits were thought to be deposited by Noah's flood. Both of these incorrect ideas were finally disproved by decades of exacting field work and often bitter debate.

Forty years ago continental drift rated only a footnote in most introductory textbooks. Now there are many believers in continental motion, and textbooks use it as a framework for the entire field of physical geology. Although the idea of continental stability provided the framework for many past textbooks, today the idea that continents are fixed in position rates only a footnote as an outmoded concept.

Objections were raised to the concept of plate tectonics after it was proposed in the late 1960s. Some sea-floor features did not seem compatible with a moving sea floor. The geology of many continental regions did not seem to fit into the theory of plate tectonics, in some cases not even slightly. But a revolutionary, new idea in science is always controversial. As it progresses from an "outrageous hypothesis" to a more widely accepted theory, after much discussion and testing, a new idea evolves and changes. The newness of the idea wears off, and successful tests and predictions convert skeptics to supporters (sometimes grudgingly). Perhaps equally importantly, dissenters die off.

As refinements were made to plate tectonics, and as more was learned about the puzzling sea-floor features and continental regions, they began to seem more compatible with plate tectonics. Objections died out, and plate tectonics became widely accepted.

It is wise to remember that at the time of Wegener most geologists vehemently disagreed with continental drift. Because Wegener proposed that continents plow through sea-floor rock, and because his proposed forces for moving continents proved inadequate, most geologists thought that continental drift was wrong. Although these geologists had sound reasons for their dissent, we now think that due to the mounting evidence, continental drift is more acceptable and that the early *geologists* were wrong.

Science should not depend on majority vote. The arguments of a dissenting minority, such as opponents of plate tectonics, should be carefully studied and rationally and scientifically answered, if possible. This is the very heart of science, the careful consideration of *all* possible explanations of natural phenomena.

The evidence for plate tectonics is very convincing. The theory has been rightly called a revolution in earth science, comparable to the development of the theory of evolution in the biological sciences. It is an exciting time to be a geologist. Our whole concept of earth dynamics has changed in the last forty years.

Summary

Plate tectonics is the idea that the earth's surface is divided into several large plates that change position and size. Intense geologic activity occurs at plate boundaries.

Plate tectonics combines the concepts of *sea-floor spreading* and *continental drift*.

Alfred Wegener proposed continental drift in the early 1900s. His evidence included coastline fit, similar fossils and rocks in now-separated continents, and paleoclimatic evidence for *apparent polar wandering*. Wegener proposed that all continents were once joined together in the supercontinent *Pangaea*.

Wegener's ideas were not widely accepted until the 1950s, when work in paleomagnetism revived interest in polar wandering.

Evidence for continental drift includes careful fits of continental edges and detailed rock matches between now-separated continents. The positions of continents during the past 200 million years have been mapped.

Hess's hypothesis of *sea-floor spreading* suggests that the sea floor moves away from the ridge crest and toward trenches as a result of mantle convection.

According to the concept of sea-floor spreading, the high heat flow and volcanism of the ridge crest are caused by hot mantle rock rising beneath the ridge. Divergent *convection* currents in the mantle cause the rift valley and earthquakes on the ridge crest, which is a *spreading axis* (or *center*). New sea floor near the rift valley has not yet accumulated pelagic sediment.

Sea-floor spreading explains trenches as sites of sea-floor *subduction*, which causes low heat flow and negative gravity anomalies. Benioff zones and andesitic volcanism are caused by interaction between the subducting sea floor and the rocks above.

Sea-floor spreading also explains the young age of the rock of the sea floor as caused by the loss of old sea floor through subduction into the mantle.

Plates are composed of blocks of *lithosphere* riding on a plastic *asthenosphere*. Plates move away from spreading axes, which add new sea floor to the trailing edges of the plates.

An apparent confirmation of plate motion came in the 1960s with the correlation of marine *magnetic anomalies* to *magnetic reversals* by Vine and Matthews. The origin of magnetic anomalies at sea apparently is due to the recording of normal and reverse magnetization by dikes that intrude the crest of the mid-oceanic ridge, then split and move sideways to give anomaly patterns a mirror symmetry.

The Vine-Matthews hypothesis gives the rate of plate motion (generally 1 to 6 cm/year) and can predict the age of the sea floor before it is sampled.

Deep-sea drilling has apparently verified plate motions and the age predictions made from magnetic anomalies.

Earthquake distribution and first-motion studies on *transform faults* on fracture zones also verify plate motions.

Divergent plate boundaries are marked by rift valleys, shallow-focus earthquakes, high heat flow, and basaltic volcanism.

Transform boundaries between plates sliding past one another are marked by strike-slip (transform) faults and shallow-focus earthquakes.

Convergent plate boundaries can cause *subduction* or *continental collision*. Subducting plate boundaries are marked by trenches, low heat flow, Benioff zones, andesitic volcanism, and young mountain belts or island arcs. Continental-collision boundaries have shallow-focus earthquakes and form young mountain belts in continental interiors.

The distribution and origin of most volcanoes, earthquakes, young mountain belts, and major sea-floor features can be explained by plate tectonics.

Plate motion was once thought to be caused by *mantle convection* (either shallow or deep), but is now attributed to the cold, dense, leading edge of a subducting plate pulling the rest of the plate along with it (*slab-pull*). Plates near mid-oceanic ridges also slide down the sloping lithosphere-asthenosphere boundary at the ridge (*ridge-push*). *Trench-suction* may help continents diverge.

Mantle plumes are narrow columns of hot, rising mantle rock. They cause flood basalts and may split continents, causing plate divergence.

An aseismic ridge may form as an oceanic plate moves over a mantle plume acting as an eruptive center (hot spot).

Terms to Remember

asthenosphere 468

continental drift 460

convection 466

convergent plate boundary 468

divergent plate boundary 468

island arc 477

lithosphere 467

magmatic arc 478

mantle plume 486

plate 467

plate tectonics 460

polar wandering 462

sea-floor spreading 460

subduction 466

transform fault 472

transform plate boundary 468

Testing Your Knowledge

Use the questions below to prepare for exams based on this chapter.

1. What was Wegener's evidence for continental drift?

2. What is polar wandering? What is the paleoclimatic evidence for polar wandering? What is the magnetic evidence for polar wandering? Does polar wandering require the poles to move?

3. What is the evidence that South America and Africa were once joined?

4. In a series of sketches show how the South Atlantic Ocean might have formed by the movement of South America and Africa.

5. What is Pangaea?

6. In a single cross-sectional sketch, show the concept of sea-floor spreading and how it relates to the mid-oceanic ridge and oceanic trenches.

7. How does sea-floor spreading account for the age of the sea floor?

8. What is a plate in the concept of plate tectonics?

9. Define *lithosphere* and *asthenosphere*.

10. What is the origin of marine magnetic anomalies according to Vine and Matthews?

11. Why does the pattern of magnetic anomalies at sea match the pattern of magnetic reversals (recorded in lava flows on land)?

12. How has deep-sea drilling tested the concept of plate motion?

13. How has the study of fracture zones tested the concept of plate motion?

14. Explain how plate tectonics can account for the existence of the mid-oceanic ridge and its associated rift valley, earthquakes, high heat flow, and basaltic volcanism.

15. Explain how plate tectonics can account for the existence of oceanic trenches as well as their low heat flow, their negative gravity anomalies, the associated Benioff zones of earthquakes, and andesitic volcanism.

16. What is a transform fault?

17. Discuss possible driving mechanisms for plate tectonics.

18. Describe the various types of plate boundaries and the geologic features associated with them.

19. What is a mantle plume? What is the geologic significance of mantle plumes?

20. The southern supercontinent is called (a) Gondwanaland (b) Pangaea (c) Laurasia (d) Glossopteris

21. The sliding of the sea floor beneath a continent or island arc is called (a) rotation (b) tension (c) subduction (d) polar wandering

22. In cross section, the plates are part of a rigid outer shell of the earth called the (a) lithosphere (b) asthenosphere (c) crust (d) mantle

23. The Vine-Matthews hypothesis explains the origin of (a) polar wandering (b) sea floor magnetic anomalies (c) continental drift (d) mid-ocean ridges

24. The San Andreas fault in California is a (a) normal fault

(b) reverse fault (c) transform fault (d) thrust fault

25. What would you most expect to find at ocean-ocean convergence? (a) suture zone (b) island arc (c) mid-ocean ridge

26. What would you most expect to find at ocean-continent convergence? (a) magmatic arc (b) suture zone (c) island arc (d) mid-ocean ridge

27. What would you most expect to find at continent-continent convergence? (a) magmatic arc (b) suture zone (c) island arc (d) mid-ocean ridge

28. Passive continental margins are created at (a) divergent plate boundaries (b) transform faults (c) convergent plate boundaries

29. The Hawaiian islands are thought to be the result of (a) subduction (b) mid-ocean ridge volcanics (c) mantle plumes (d) ocean-ocean convergence

30. Metallic ores are created at diverging plate boundaries (a) through hydrothermal processes (b) in lava flows (c) in sedimentary deposits (d) through metamorphism

Expanding Your Knowledge

1. Plate tectonics helps cool the earth as hot mantle rock rises near the surface at ridge crests and mantle plumes. What can we assume about the internal temperature of other planets that do not seem to have plate tectonics? What would happen to earth's internal temperature if the plates stopped moving?

2. Are ridge offsets along fracture zones easier to explain with mantle-deep convection *causing* plate motion or with shallow convection occurring as a *result* of plate motion?

3. Why are mantle plumes narrow? What conditions at the core-mantle boundary could cause the formation and rise of a mushroom-shaped plume?

4. The slab-pull and ridge-push mechanisms of plate motion may operate only after a plate starts to move. What starts plate motion?

5. If subducting plates can penetrate the 670-kilometer mantle boundary, and perhaps sink all the way to the base of the mantle, why are there no earthquakes deeper than 670 kilometers?

Exploring Resources

Beloussov, V. V. 1979. Why I do not accept plate tectonics. *EOS,* American Geophysical Union 60:207–10.

Bird, J. M. ed. 1980. *Plate tectonics.* 2d ed. Washington, D.C.: American Geophysical Union.

Bonatti, E. 1987. The rifting of continents. *Scientific American* 256 (3): 96–103.

———. 1994. The Earth's mantle below the oceans. *Scientific American* 270 (3): 44–51.

Bonatti, E., and K. Crane. 1984. Ocean fracture zones. *Scientific American* 250 (5): 40–51.

Butler, R. F. 1992. *Paleomagnetism.* Palo Alto, Calif: Blackwell Scientific Publications.

Condie, K. 1997. *Plate tectonics and crustal evolution.* 4th ed. New York: Butterworth/Heinemann.

Courtillot, V., and G. E. Vink. 1983. How continents break up. *Scientific American* 249 (1): 42–47.

Cox, A., ed. 1973. *Plate tectonics and geomagnetic reversals.* New York: W. H. Freeman.

Cox, A., and R. B. Hart. 1986. *Plate tectonics: How it works.* Palo Alto, Calif: Blackwell Scientific Publications.

Dalziel, I. W. D. 1995. Earth before Pangea. *Scientific American* 272 (1): 58–63.

Francheteau, J. 1983. The oceanic crust. *Scientific American* 249 (3): 68–84.

Hallam, A. 1973. *A revolution in the earth sciences: From continental drift to plate tectonics.* London: Oxford University Press.

Kearey, P., and F. J. Vine. 1996. *Global tectonics.* 2d ed. Palo Alto, Calif: Blackwell Scientific Publications.

Klein, G. D. ed. 1994. Pangaea: Paleoclimate, tectonics, and sedimentation during accretion, zenith, and breakup of a supercontinent. *Geological Society of America Special Paper* 288.

MacDonald, K. C., and P. J. Fox. 1990. The mid-ocean ridge. *Scientific American* 262 (6), 72–79.

Moores, E. M., and R. J. Twiss. 1995. *Tectonics.* New York: W. H. Freeman.

Morgan, W. J. 1968. Rises, trenches, great faults, and crustal blocks. *Journal of Geophysical Research* 73: 1959–1982.

Press, F., and R. Siever. 1974. *Planet earth— Readings from Scientific American.* New York: W. H. Freeman.

Uyeda, S. 1978. *The new view of the earth— Moving continents and moving oceans.* New York: W. H. Freeman.

Van Andel, T. H. 1985. *New views of an old planet.* New York: Cambridge University Press.

Vine, F. J., and D. H. Matthews. 1963. Magnetic anomalies over ocean ridges. *Nature* 199: 947–49.

Vink, G. E., W. J. Morgan, and P. R. Vogt. 1985. The earth's hot spots. *Scientific American* 252 (4): 50–56.

Wegener, A. 1924. *The origin of continents and oceans.* London: Methuen.

Wilson, J. T., ed. 1976. *Continents adrift and continents aground.* New York: W. H. Freeman.

Windley, B. F. 1984. *The evolving continents.* New York: Wiley.

 http://pubs.usgs.gov/publications/ text/dynamic.html
This Dynamic Earth: The Story of Plate Tectonics (general information)

http://cddis.gsfc.nasa.gov/926/slrtecto.html
Tectonic Plate Motion (explains how plate motion is calculated)

Continental Drift and Plate Tectonics (UC Santa Barbara)

The Birth of a Theory (Annenburg/CPB Project)

Plate Dynamics (Annenburg/CPB Project)

Interacting with Journey Through Geology CD-ROM

Expand your knowledge of the concepts presented in this chapter by using the CD-ROM to answer the following questions.

1. Hot Spots module, Hawaii, Watch Hot Spots Animation: Why do the Hawaiian islands become older to the northwest? Why is the Hawaiian Island chain bent?

2. Hot Spots, Hawaii, Next Volcano: Predict where the next Hawaiian island will emerge. Justify your answer.

3. Convergent Margins, Alaska, Earthquakes Through Time: Describe the pattern of earthquakes near Cook Inlet in terms of their depth of focus and geographic distribution. What plate tectonic process creates this pattern? Draw a cross-section through this plate boundary, showing where earthquakes of different depths occur.

4. Transform Faults, Watch Transform Faults Animation: Transform faults usually run between offset segments of a mid-ocean ridge, yet motion along the fault does not further offset the ridge segments. Why not?

5. Transform Faults, Watch the Fault Evolve: How was the San Andreas Fault created?

Climber in the Karakoram Himalaya, Pakistan.

Photo © Galen Rowell/Mountain Light Photography

CHAPTER
20

Mountain Belts and the Continental Crust

Mountain belts can evolve from marine-deposited rocks to towering peaks during periods of tens of millions of years. Ultimately the peaks are eroded to plains and become part of the stable interior of a continent. To appreciate the long and complex process of mountain building, you need to know much of the material covered in previous chapters. For instance, you must understand structural geology to appreciate what a particular pattern of folds and faults can tell us about the history of mountain building in a particular region. To understand how the rocks formed during the various stages of a mountain belt's history, you must know about volcanism, plutonism, sedimentation, and metamorphism. Your earlier study of weathering and erosion will help you understand how mountains are worn away. Plate-tectonic theory has been strikingly effective in helping geologists make sense of all the complex aspects of mountain belts and the continental crust. For this reason, you need to thoroughly understand the material in chapter 19 before you can appreciate how continents evolve.

In this chapter, we first point out what geologists have observed of mountain belts. Next, we describe how these observations are interpreted, particularly in light of plate tectonic theory. Finally, we discuss current perceptions of how continents change and grow.

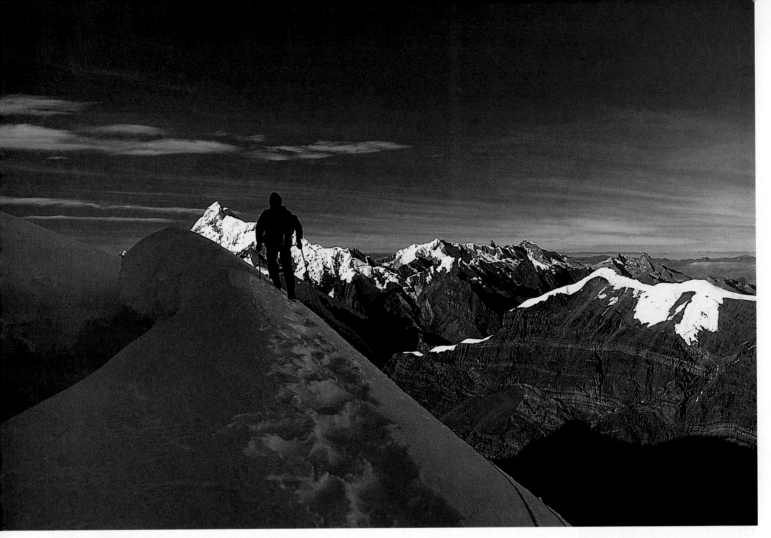

Figure 20.1

View of glaciated peaks in one of the mountain ranges in the Andes mountain belt. A parallel but much lower range is visible at the extreme right skyline of the picture.

Photo by C. C. Plummer

A mountain, as you know, is a large terrain feature that rises more or less abruptly from surrounding levels. Volcanoes are mountains, so are erosional remnants of plateaus (mesas). In this chapter, we will not focus on individual mountains; rather, we are concerned here with the earth's **major mountain belts,** chains thousands of kilometers long composed of numerous mountain ranges. A **mountain range** is a group of closely spaced mountains or parallel ridges (figure 20.1). A mountain range is likely to be composed of tectonically deformed sedimentary, volcanic or metamorphic rocks. It may also show a history of intrusive igneous activity.

The map in figure 20.2 shows that most of the world's mountains are in long chains that extend for thousands of kilometers. The Himalaya, the Andes, the Alps, and the Appalachians are examples of major mountain belts, each comprising numerous mountain ranges.

Geologists find working in mountain ranges to be physically and intellectually challenging. High mountains have steep faces and broad exposures of bedrock. This is good because they allow a geologist to decipher complex interrela-

tionships between rock units. But the geologist may have to become a proficient mountain climber to access the good exposures. (Conversely, mountain climbers who develop an interest in the rocks they climb sometimes become geologists.) On the other hand, exposures of bedrock critical to interpreting the local geology may be buried beneath glaciers or talus from rockfall. Furthermore, even in the highest and best exposed mountains, we never see bedrock representative of all of a mountain range. Significant amounts of rock (usually thousands of meters) once overlying the rocks we now see have been eroded away. Moreover, the present exposures are like the proverbial tips of icebergs—there is much more rock below the exposed mountain range that we cannot observe. For instance, the Himalaya, earth's highest mountain range, rise to 8,000 meters above sea level; yet, their roots (the earth's crust beneath the mountains) extend downward 65,000 meters. In other words, at best, we have exposed to us less than 1/8 of the thickness of a mountain range.

Our models of how major mountain belts evolve use data from over a century of studying the geologic structures and

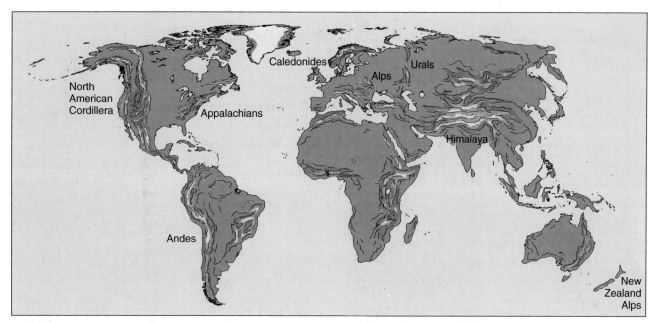

Figure 20.2

Map of the world showing major mountain belts.

Characteristics of Major Mountain Belts

rocks exposed in the world's many mountain ranges. Often, a particular study aims to piece together the geologic history of a single mountain range or part of a range. In other field studies, a geologist focuses on a particular type of rock exposed in a mountain range. For instance, a geologist may study the variations in metamorphic rocks in a mountainous area with the intent of gaining more insight into metamorphic processes. Geologists working on the "big picture," developing hypotheses of how major mountain belts evolve, might use the published results of hundreds or thousands of local studies, using them as pieces of a puzzle. (Science works largely because scientists build on the work of others.) Models that currently are widely accepted regarding the evolution of mountain belts are cast within the broader framework of plate-tectonic theory and will be described later in this chapter.

Despite the complexities of individual ranges, there are some characteristics that are shared by most or all major mountain belts. These are described below. As you read through each of the characteristics, think of what you have learned from previous chapters (especially the chapter on plate tectonics) that might help explain an observed characteristic of mountain belts.

Characteristics of Major Mountain Belts

Size and Alignment

Major mountain belts are very long compared to their width. For instance, the mountain belt that forms the western part of North America (the *North American Cordillera*) starts as the Aleutian Island arc in southwestern Alaska. The trend of the

Aleutians is continuous with the trend of the mountain ranges that make up much of the Alaskan mainland. (See figure 20.3. You will also want to refer to figure 1.5 and the geologic map on the inside cover as you read this chapter.) Through Alaska the ranges trend more or less east-west, but in northern Canada they curve into a more north-south trend. The north-south ranges include the many individual ranges of western Canada and western United States. The belt is widest in the United States where it extends from the Coast Ranges of California and Oregon eastward to the Rocky Mountains of Montana, Wyoming, and Colorado. The belt narrows as the north-south trend continues through Mexico.

Ages of Mountain Belts and Continents

Major mountain belts with higher mountain ranges tend to be geologically younger than those where mountains are lower. Mountain building for the Himalaya, our highest mountain belt, began only about 45 million years ago and is still taking place. Mountain building in the much lower Appalachians ceased around 250 million years ago. Individual ranges within a mountain belt, however, may vary considerably in height even though they are about the same age.

Mountain regions commonly show evidence that they were once high above sea level, were eroded to hills or low plains, and then rose again in a later episode of uplift. Such episodes of uplift and erosion may occur a number of times during the long history of a mountain range. Ultimately, mountain ranges stabilize and are eroded to plains.

On the North American continent, the Appalachian Mountains extend from eastern Canada southward through the eastern United States into Alabama (figure 20.3). In the Appalachians, fossils and isotopic ages of rocks indicate that

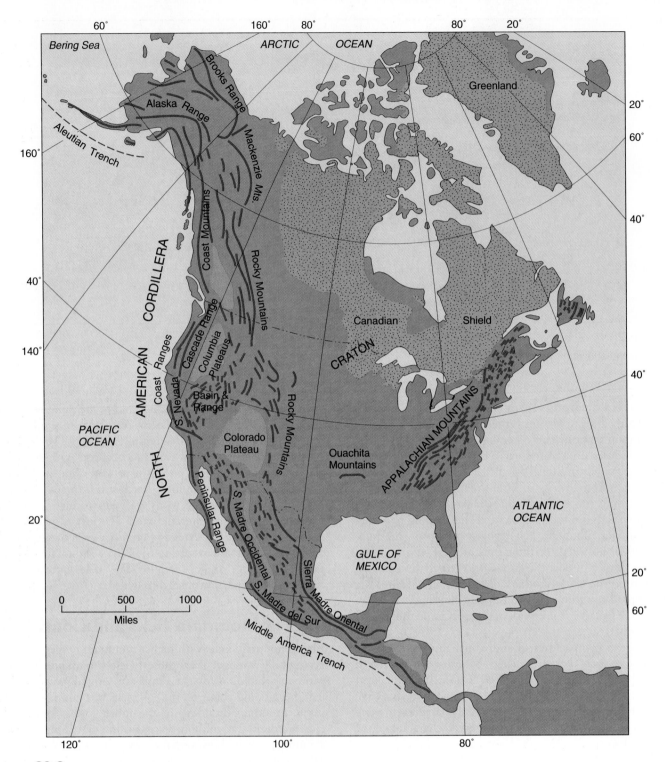

Figure 20.3

The mountain belts of North America. Some of the major ranges in the Cordillera are shown.

these mountains began to evolve earlier than the mountain belt along the western coast of North America. The interior plains between the Appalachians and the Rockies are considered to have evolved from mountain belts in the very distant geologic past (during the Precambrian). The once deep-seated roots of the former Precambrian mountain belts are the *basement* rock for the now stable central part of the continent. Layers of Pa-

leozoic and younger sedimentary rock cover most of that basement. The great age of the mountain-building episodes that preceded the Paleozoic sedimentation is confirmed by isotopically determined dates of over one billion years obtained from rocks in the few scattered locations where the basement is exposed. (These are the Grand Canyon in Arizona, the Ozark dome in Missouri, the Black Hills of South Dakota, several

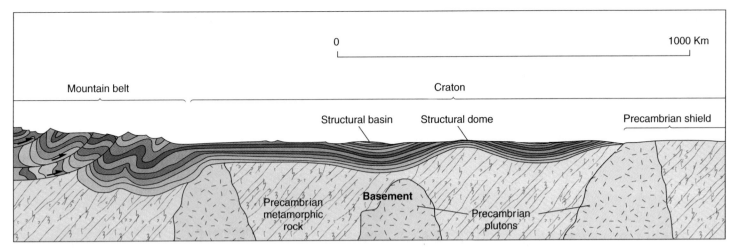

Figure 20.4

Schematic cross section through part of a mountain belt (left) and part of a continental interior (craton). Vertical scale is exaggerated.

ranges in the Rocky Mountains, and the Adirondacks in New York.) The region of a continent that has been structurally stable for a prolonged period of time is called a **craton** (figures 20.3, 20.4 and 20.5). The central part of the United States and Canada is all part of a craton. Other continents are similarly cored by cratons.

Most of the craton in the central United States has a very thin blanket—only 1,000 to 2,000 meters—of sedimentary rock layers overlying its basement. Sediment was mostly deposited in shallow inland seas during Paleozoic time. However, for the craton in much of eastern and northern Canada, as well as Greenland, no sedimentary rocks cover the eroded remnants of old mountain ranges. This region is a **Precambrian shield**—that is, a complex of Precambrian metamorphic and plutonic rocks exposed over a large area. Such shields and basement complexes of cratons represent the roots of mountain ranges that completed the deformation process more than a billion years ago.

Thickness and Characteristics of Rock Layers

The relatively thin cover of sedimentary rock overlying the basement in the craton contrasts sharply with the thick sedimentary sequence typical of a mountain belt. In mountain belts, layered sedimentary rock commonly is more than 10 kilometers thick. The sedimentary rock in cratons may show no deformation, or it may have been gently warped into basins and domes above the basement (figure 20.4). By contrast, mountain belts are characterized by a variety of folds and faults that indicate moderate to very intense deformation.

Most of the sedimentary rock in mountains is of marine origin, indicating that today's highlands were once below sea level. In part of a mountain belt, the rocks are similar to the sedimentary rocks found on the craton—limestones, shales, and sandstones. By contrast, another wide zone in a mountain belt contains great thicknesses of volcanically derived rock—

Figure 20.5

Satellite image of part of a craton in Western Australia. Metamorphic rock (dark gray) that is 3.5 to 3 billion years old surrounds oval-shaped domes of granite and gneiss (white) that are 2.8 to 3.3 billion years old. Gently dipping sedimentary and volcanic rocks (tan and reddish) unconformably overlie the granite-metamorphic basement complex. The area is 400 km across.

Landsat mosaic produced by the Remote Sensing Applications Centre Department of Land Administration, Western Australia volcanic ash, sediments from eroded volcanic rock, as well as lava flows.

Patterns of Folding and Faulting

Reconstructing the original position and determining the thickness of layers of sedimentary and volcanic rock in mountain belts is complicated because in most instances the layered rocks have been folded and faulted at some time after they were deposited. (Refer to figure 20.6 as you read through the following paragraphs.) Folds will be open in those parts of a mountain belt where deformation is not very intense. Tighter folds (figure 20.7) indicate greater stresses. Large overturned and recumbent folds (figure 20.8) may be

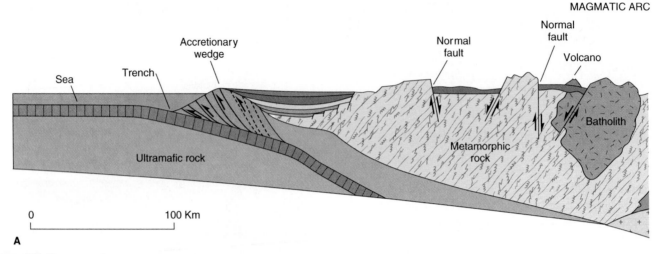

Figure 20.6

Cross section of a "typical" mountain belt. For simplicity, only a few of the many layers of sedimentary rock are shown. (*B*) is the continuation of the mountain belt to the right of (*A*).

Figure 20.7

False color satellite image of part of the Valley and Ridge province of the Appalachian mountain belt, near Harrisburg, Pennsylvania. The ridges are sedimentary beds resistant to erosion. The pattern indicates tight and open folding occurred prior to erosion.

U.S. Geological Survey.

exposed in more intensely deformed portions of mountain belts. Reverse faults are common, particularly in the intensely folded regions. Especially noteworthy are the **fold and thrust belts** found in many mountainous regions. These are characterized by large thrust faults (reverse faults at a low angle to horizontal), stacked one upon another; the intervening rock usually was folded while it was being transported during faulting. Beneath the lowermost fault, called a *detachment fault,* the rock has remained in place.

Overall, the folds and thrust faults in a mountain belt suggest tremendous squeezing or *crustal shortening* and *crustal thickening.* The sedimentary rocks of the Alps, for instance, are estimated to have covered an area of ocean floor about 500 kilometers wide when they were deposited. They were later

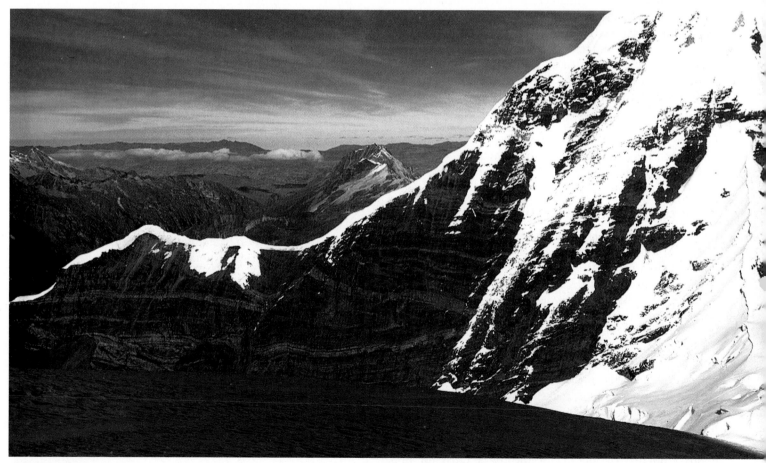

Figure 20.8

Recumbent folds exposed on a mountainside in the Andes.

Photo by C. C. Plummer

Ultramafic rocks (described in chapter 3 on intrusive rocks) occur commonly in the portions of mountain belts occupied by metamorphic and plutonic rocks. Ultramafic rocks tend to crop out in long, narrow zones that parallel the trend of a mountain belt. Most geologists believe the bodies of ultramafic rocks represent mantle material that was faulted into the crust during the mountain-building process. Some of the ultramafic bodies are found associated with marine-deposited volcanic and sedimentary rocks in an *ophiolite sequence* (described in chapter 18 about the sea floor). This may represent a segment of a former oceanic crust together with its underlying mantle.

Ultramafic rocks in mountain belts commonly show the effects of the metamorphism that has altered adjacent rock units. Two of the foliated metamorphic products of ultramafic rocks are of special interest. One is *serpentinite*, a rock composed of the mineral serpentine. Another is a rock composed mostly of the mineral talc, commonly known as *soapstone*.

Serpentinite is a shiny, mottled, dark green and black rock that looks rather like a snake's skin. It splits apart easily along irregular, slippery, foliation surfaces. Hillsides or slopes with serpentinite as bedrock are sparsely vegetated because constant sliding prevents soil and vegetation from building up. Houses built on serpentinite hillsides (by people without a knowledge of geology) also slide downslope. Serpentinite is the official state rock of California—a state in which a large number of homes have been destroyed because they were built on sliding hillsides. (Serpentinite, however, is seldom to blame.)

Soapstone, which is less common than serpentinite, is valuable mainly because of talc's softness (number 1 on Mohs scale). Many sculptures (most notably Eskimo carvings) are made from soapstone because of the ease with which it can be cut. The best-known product of talc, however, is talcum powder.

Talc's perfect cleavage has led to its unique application in reflecting highway signs and center road stripes. Finely crushed talc is mixed with paint. The many fine, parallel cleavage flakes reflect light in a specific direction depending on how the mixture is brushed on the sign. For the signs showing different daytime and nighttime speed limits, the daytime speed limit is painted with ordinary black paint, and the nighttime limit is painted with talc mixed into the white paint. At night the light from a car's headlights reflects back to the car from the cleavage surfaces of many tiny flakes of talc.

compressed into the present width of the Alps, which is less than 200 kilometers.

Metamorphism and Plutonism

A complex of regional metamorphic and plutonic rock is generally found in the mountain ranges of the most intensely deformed portions of major mountain belts. Most of the metamorphic rocks were originally sedimentary and volcanic rocks that had been deeply buried and subjected to intense stress and high temperature. Where found, *migmatites* (interlayered granitic and metamorphic rocks) may represent those parts of the mountain belts that were once at even deeper levels in the crust, where higher temperatures caused partial melting of the rocks (as described in chapter 3). Where found, rocks such as these that formed in the lower crust must have been transported into much higher levels of the crust during a mountain-building episode. Batholiths, mostly of granite, attest to magma generated from partial melting of the lower crust (or upper mantle). Magma diapirs traveled upward to collect and solidify at a higher level in the crust.

Normal Faulting

Older portions of some major mountain belts have undergone normal faulting (figure 20.9). Cross-cutting relationships show that the normal faulting occurs after the intense deformation that resulted in tight folding, thrust faulting, and metamorphism and after most batholiths had formed. As shown in figure 20.10, this late stage of normal faulting is a result of *vertical uplift* or *extension* (regional tensional stress). Either of these contrasts with the overall compression that prevailed during folding, thrust faulting, and metamorphism of the original marine rocks.

In addition to the clearly post-compressional normal faulting described above, normal faulting takes place in the high, central part of a major mountain belt *simultaneous* with the folding and thrust faulting of the outer part of the belt (figure 20.6). This implies that *extension* occurs in the central portion at the same time that *compression* takes place in the outer portion of an active major mountain belt.

Thickness and Density of Rocks

Geophysical investigations yield additional information about mountains and the continental crust. As discussed in chapter 17

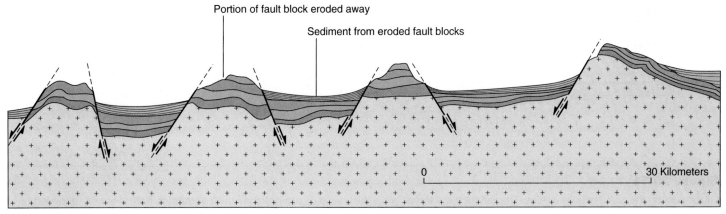

Portion of fault block eroded away

Sediment from eroded fault blocks

0   30 Kilometers

Figure 20.9

Fault-block mountains with movement along normal faults.

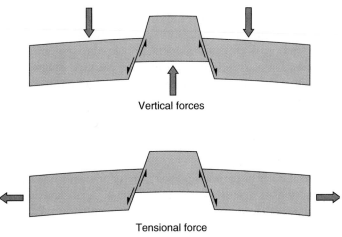

Vertical forces

Tensional force

Figure 20.10

Normal faults bounding fault-block mountain ranges. Two ways in which forces might be distributed.

on the earth's interior, gravity measurements indicate that the rocks of the continental crust (including mountain belts) are lighter (less dense) than those of the oceanic crust. Seismic velocities indicate a composition approximating that of granite for continental crust. Furthermore, evidence from seismic studies supports the view that this lighter crust is much thicker beneath mountain belts than under the craton and that the crust is thicker under younger mountain belts than under older ones.

Features of Active Mountain Ranges

Frequent earthquakes are characteristic of portions of mountain belts that are geologically young and considered still active. Also, deep ocean trenches are found parallel to many young mountain belts (the Andes, for example). Trenches lie off the coasts of island arcs, which can be regarded as very young mountain ranges. Isolated active volcanoes perched on top of older rock in a mountain range suggest that melting is still taking place at depth.

The Evolution of a Mountain Belt

Although each mountain belt differs in details, geologists have developed (and continue to modify) models that provide logical explanations of how mountain belts evolve. Plate-tectonic theory provides the framework for most of our models that explain various aspects of mountain belts.

The evolution of a mountain belt begins when sediment is deposited; it ends hundreds of millions of years later when the former mountain belt has become part of a craton. For instructional purposes we describe the history of a mountain belt in three stages: (1) the accumulation, (2) the orogenic, and (3) the uplift and block-faulting stages. However, there are no sharp time boundaries between the stages. Processes associated with more than one stage may be occurring at the same time in different parts of a mountain belt.

The Accumulation Stage

As we noted earlier, mountain belts typically contain thick sequences of sedimentary and volcanic rocks. The accumulation of these great thicknesses (several kilometers) of sedimentary or volcanic rocks takes place in the **accumulation stage** of mountain building. Most of the sedimentary rocks and much of the volcanic material accumulate in a marine environment. The source for the sediment deposited in the water must be a nearby landmass, belonging to either an adjoining continent or part of a volcanic island arc.

Accumulation in an Opening Ocean Basin

The east coast of North America is a *passive continental margin* (described in the previous chapter; see figures 19.22 and 19.26). Europe and North America have been moving away from each other for well over 100 million years. During this time sediment eroded from the continents has washed into the adjoining seas. The resulting very thick sequences of sedimentary rock make up continental shelves and slopes. The sedimentary rocks that form at passive margins are predominantly shales, limestones, and

Figure 20.11

Cross section through part of the Alps. Thicker lines are thrust faults. Lesser folds are not shown. Movement is from the right to the left of the diagram (southeast to northwest). Only a few arrows are shown to indicate movement of the overriding thrust block.

From S. E. Boyer and D. Elliot, 1982. *AAPG Bulletin.* Reprinted by permission of American Association of Petroleum Geologists.

sandstones (mostly quartz sandstone). Volcanic rocks are rare or absent. As long as the Atlantic Ocean basin continues to open, sedimentation will continue and more sedimentary layers will accumulate on both sides of the Atlantic.

Where we find a thick sequence of shale, sandstone, and limestone exposed in a mountain range, it is reasonable to assume these rocks were originally deposited at a similar passive continental margin.

Accumulation Along a Convergent Boundary

A large variety of rock types are created near a convergent plate boundary. Volcanic rocks—most characteristically andesites—accumulate near the boundary either as pyroclastic layers or as flows. Sedimentary rocks, moreover, may accumulate in equal or greater amounts than volcanic rocks. Limestone, however, is usually absent or present only in minor proportions. Shales and sandstones are the predominant sedimentary rock types. The sandstones are mostly *graywackes* ("dirty" looking sandstones with sand-sized grains of volcanic or other rocks in a fine-grained matrix) rather than quartz sandstone.

The source of this sedimentary as well as volcanic material is a *magmatic arc,* which, as described in chapter 19 (see figure 19.33), is a chain of volcanoes or plutons along a line (usually curved as seen from above).

Sediment eroded from the magmatic arc, as well as newly erupted pyroclastic debris, is transported to and deposited in the basins on either side of the arc. Usually the basin to the seaward side will be underwater, at least initially, and in time will fill with sediment. Once the basin has filled, sediment may be transported beyond it and accumulate outward onto the deep ocean floor.

The Orogenic Stage

Intense deformation follows or is contemporaneous with the accumulation stage. An **orogeny** is an episode of intense deformation of the rocks in a region; the deformation is usually accompanied by metamorphism and igneous activity. Layered rocks are compressed into folds. Reverse faulting (especially thrust faulting) is widespread during an orogeny. Normal faulting may also occur but is not as widespread.

The more deeply buried rocks, subjected to regional metamorphism, are converted to schists and gneisses. Magma generated in the deep crust or the upper mantle may work its way upward to erupt in volcanoes or form batholiths.

Orogenies and Ocean-Continent Convergence

The relationships among igneous activity, metamorphism, and subduction are described in chapters 3 and 7. Plate convergence also accounts for the folded and reverse-faulted layered rocks found in mountain belts. An *accretionary wedge* develops where newly-formed layers of marine sediment are folded and faulted as they are snowplowed off the subducting oceanic plate (see figure 19.32 and explanation in the previous chapter).

Rock caught in and pulled down the subduction zone is subjected to intense shearing. If rock is carried further down the subduction zone it becomes metamorphosed (as described in chapter 7).

Fold and thrust belts may develop on the craton (backarc) side of the mountain belt (figure 20.11). Thrusting is away from the magmatic arc toward the craton. The magmatic arc is at a high elevation, because the crust is thicker and composed largely of hot igneous and metamorphic rocks. The large thrust sheets move toward and sometimes over the craton. (In the Rocky Mountains, thrust faulting of the craton itself has taken place.) The thrusting probably is largely due to the crustal shortening caused by convergence. There is, however, some controversy among geologists over additional processes that may take place. Some geologists regard gravity flow (from the high and mobile magmatic arc outward over the low and rigid craton) to contribute significantly to the process. Others think that the expanding magmatic arc pushes the sedimentary (and sometimes metamorphic and igneous) rocks outward to become the fold and thrust belt. (The magmatic arc is likened to a bulldozer pushing a wedge of loose material outward.)

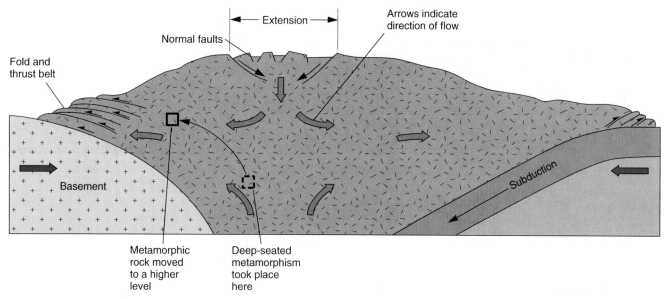

Figure 20.12

Schematic cross section of a mountain belt in which gravitational collapse and spreading are taking place during plate convergence. Red arrows indicate flowage of rock. Faulting occurs in brittle rock near the surface. Rock that was metamorphosed at depth flows to a higher level in the mountain belt.

Normal faulting may take place due to tensional forces in the high, central part.

In the late 1980s and early 1990s geologists developed a model that explains (1) fold and thrust belts, (2) simultaneous normal faulting, and (3) how once deep-seated metamorphic rocks rise to an upper level in a mountain belt. What is believed to occur is that the central part of the mountain belt becomes too high and gravitationally unstable resulting in **gravitational collapse and spreading.** The mobile central portion becomes increasingly elevated during plate convergence. This is due to compression of sedimentary and metamorphic rocks, as well as to volcanic eruptions and emplacement of plutons. After some time, the welt in the mountain belt becomes too high to be supported by the underlying rocks and collapse begins. (Oxford University geologist John Dewey suggests that collapse begins when the welt exceeds 3 kilometers above sea level.) As shown in figure 20.12 the gravitational collapse forces rock outward as well as downward. At deeper levels in the mountain belt, the rock is *plastic* (or *ductile*) and flows; nearer the surface, rock fractures, so movement is through faulting. The rock is pushed outward and helps create, along with crustal shortening, the fold and thrust belt.

In the high, central part, the outward flowing rock results in extension (figure 20.12), therefore the brittle, near-surface rocks fracture and, because forces there are tensional, normal faulting takes place.

The flowage pattern (as shown in figure 20.12) can also explain how once deep-seated metamorphic rocks (migmatites, for example) are found in upper levels of a mountain belt. Lower crustal rocks are squeezed, forcing them to flow upward and outward, bringing them closer to the surface.

It is important to note that where there are convergent boundaries, accumulation and deformation are occurring simultaneously. In other words, the accumulation stage and the orogenic stage are taking place at the same time. This was not evident to early geologists, who did not have the perspective of plate-tectonic theory for understanding mountain building. They generally assumed that the accumulation stage ended before the orogenic stage began.

Arc-Continent Convergence

Sometimes an island arc collides with a continent (figure 20.13). If an intervening ocean is destroyed by subduction, (the subduction also causes the arc) the arc will approach the continent. When collision occurs, the arc, like a continent, is too buoyant to be subducted. Continued convergence of the two plates may cause the remaining sea floor to break away from the arc and create a new site of subduction and a new trench seaward of the arc (figure 20.13C). Note that the direction of the new subduction is opposite to the direction of the original subduction (these are sometimes called flipping subduction zones), but it still may supply the arc with magma. However, the arc has now become welded to the continent, increasing the size of the continent.

This type of collision apparently occurred in northern New Guinea (north of Australia). A similar collision may have added an island arc to the Sierra Nevada complex in California during Mesozoic time, when a subduction zone may have existed in what now is central California. Many geologists think that much of westernmost North America has formed from a series of arcs colliding with North America (discussed later in this chapter under "terranes").

Mountain Belts and the Continental Crust   **507**

C

Figure 20.13

Arc-continent convergence can weld an island arc onto a continent. The direction of subduction changes after impact.

Orogenies and Continent-Continent Convergence

As described in the previous chapter (see figure 19.34), some mountain belts form when an ocean basin closes and continents collide. Mountain belts that we find within continents (with cratons on either side) are believed to be products of continent-continent convergence. The Ural Mountains resulted from the collision of Asia and Europe. Convergence of the African and European plates created the Alps. Our highest mountains are in the Himalayan belt. The Himalayan orogeny started around 45 million years ago as India began colliding with Asia (India was originally in the southern hemisphere). The thick sequences of sedimentary rocks that had built up on both continental margins were intensely faulted and folded. Fold and thrust belts developed and were carved by erosion into the mountain ranges that make up the Himalaya. The mountains are still rising and frequent earthquakes attest to continuing tectonic activity. North of the Himalaya, Tibet rose to become what is now the highest plateau in the world. Normal faults in the Tibetan plateau indicate that gravitational collapse is taking place.

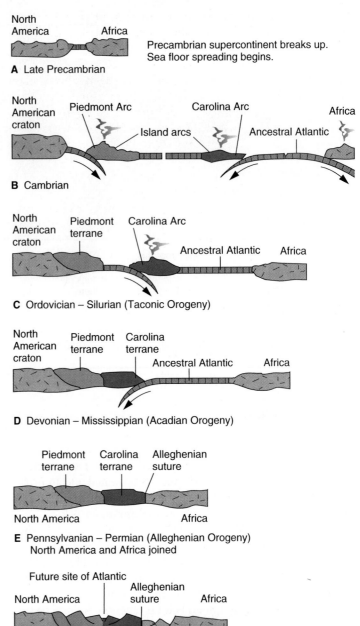

Figure 20.14

The geologic evolution of the southern Appalachians. Modified from R. D. Hatcher, Jr., 1989. Tectonic synthesis of the U.S. Appalachians, figure 9, volume F-2, *Geology of North America.* By permission of Geological Society of America.

The Appalachian Mountains are an example of continent-continent convergence, but with a more complicated history. Arc-continent convergence was also involved and the mountain belt was later split apart by plate divergence.

A condensed version of orogeny in the Appalachians is as follows (figure 20.14): During late Precambrian and earliest Paleozoic time the ancestral Atlantic Ocean developed as sea-floor spreading forced the passive margins of North America,

Europe, and Africa away from one another. During the Paleozoic, plate motion shifted, subduction began, and the ocean basin began closing. Island arcs developed between the continents. These became plastered onto the North American craton as the ancestral Atlantic basin closed. A couple hundred million years after subduction began, the ocean basin closed completely, first with Europe and later with Africa crashing into North America. By the end of the Paleozoic, the three continents were sutured together. The Appalachians and what is now the Caledonide mountain belt of Great Britain and Norway were part of a single mountain belt within the supercontinent, *Pangaea*. The mountain belt was comparable to the present-day Himalaya.

Early in the Mesozoic Era, the supercontinent split, roughly parallel to the old suture zone. The present continents moved (and continue to move) farther and farther away from their present diverging boundary, the mid-Atlantic ridge.

What happened to the Appalachians would seem too implausible even for a science fiction plot. Yet, if one accepts the principles of plate-tectonic theory and examines the rocks and structures in the Appalachians (and their counterparts in Europe and Africa), the argument for this sequence of events is not only plausible but convincing.

The cycle of splitting of a supercontinent, opening of an ocean basin, followed by closing of the basin and collision of continents is known as the *Wilson Cycle*. Canadian geologist J. Tuzo Wilson proposed the cycle in the 1960s for the tectonic history of the Appalachians.

The Wilson Cycle apparently has occurred before (see Box 20.3). A question raised is why would a continent split apart more or less along a suture zone where one would expect the crust to be thickest? One recently proposed hypothesis is that this is the zone that is weakened and thinned somewhat by outward flow of rock during gravitational collapse and spreading. (Another hypothesis involves *delamination*, detachment and sinking downward of the underlying lithospheric mantle, as described on p. 513.)

The Uplift and Block-faulting Stage

After plate convergence stops and the compressive force of the orogeny is relaxed, there is a long period of uplift accompanied by erosion. During this stage, which lasts many millions of years, large regions in the mountain belt move vertically upward. Erosion may keep pace with uplift and the area remain low. Alternatively, uplift may temporarily outpace erosion, resulting in plateaus or mountain ranges. The present Appalachian Mountains are the result of uplift and erosion that have taken place since the orogenic stage ended more than 250 million years ago. However, the coastal plains east of the present Appalachians were also a part of the original mountain belt, yet they were eroded down and remained as lowlands for over 70 million years. Eventually, the entire Appalachian mountain belt will be eroded to a plain and become part of the North American craton.

Isostasy

For the Appalachians and most other older mountain belts, we can attribute the geologically most recent uplift to *isostatic*

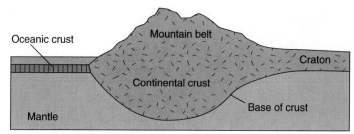

A At end of orogenic stage

B Mountain belt moves upward

C Erosion and renewed uplift continue until crust beneath mountain belt is the same thickness as that of the craton

Figure 20.15

Isostasy in a mountain belt. The thickness of the continental crust is exaggerated.

adjustment of the crust that was thickened during the orogenic stage (as explained in chapter 17 and as shown in figure 20.15). According to the concept of isostasy, lighter, less dense continental crust "floats" higher on the mantle than the denser oceanic crust. The craton has achieved an equilibrium and is floating at the proper level for its thickness. Mountains, being thicker continental crust, "float" higher than the stable continent. As material is removed from mountains by erosion, the range floats upward to regain its isostatic balance. Isostatic adjustment does not take place instantaneously. Usually there will be a considerable time lag between erosion and isostatic adjustment.

At most places on continents, the altitude above sea level is related to local crustal thickness. Beneath the 5-kilometer-high Tibetan Plateau, the crust is 75 kilometers thick. Under Kansas, the crust is 44 kilometers thick and beneath Denver, the "mile high city" the crust is 50 kilometers thick. (If the United States ever joins the rest of the world and goes metric, Denver will be known as the "1.6 kilometer high city.") Just west of Denver, the altitude of the Rocky Mountains jumps to 2 kilometers higher than that at Denver. Scientists expected to find a corresponding thickening of the crust beneath these

20.2 ENVIRONMENTAL GEOLOGY

A System Approach to Understanding Mountains

During recent years, geologists have used a system approach to gain insight into the growth and wearing away of mountains. This approach regards mountains as products of three closely interdependent components. The components of the system are: (1) tectonics (plate tectonics and isostasy), (2) climate, and (3) erosion.

The tendency in the past has been to concentrate on tectonics to explain the growth of a mountain belt and to relegate climate and erosion to relatively minor roles. Through mountain system analyses, we gain insight into the extent to which each of the three components interacts with and changes the other two components. Climate influences erosion in obvious ways. For instance, if there is a wet climate there will be erosion due to abundant running water at lower elevation and heavy glaciation at higher elevation. If the climate is arid, erosion will be much slower. Tectonics affects climate because if a region is uplifted to a high elevation, the climate there will be cold and glaciers can develop. With less uplift and lower mountains, erosion will be mainly due to running water. A moist climate can also result in heavy vegetation at lower elevations, which would tend to retard erosion.

Erosion and climate can influence tectonics as well. For example, the extent and type of erosion can help determine whether a highland grows higher or lower with time. If a high plateau, dissected by only a few valleys, undergoes erosion, the plateau is eroded downward uniformly (box figure 1). Following erosion, isostasy results in the plateau floating upward, but not up to its original level. Its average surface, which essentially is its actual surface, is at a lower elevation than before erosion took place. If erosion carves many deep valleys and leaves relatively few mountains between the valleys, the entire regional block will have less mass and will float isostatically upward. As in the case of the plateau, its average surface would rise to a level lower than before erosion. However, its average surface is somewhere between the peaks and bottoms of valleys. Although the average height of the block rises to a level below its previous average height, the mountains rise to heights greater than before.

Climate enters the picture, because, interacting with tectonics, it helps control the type and extent of weathering that takes place. For instance, heavy precipitation takes place in the Himalaya because of the flow of very humid air from the south during the summer monsoon. At the higher elevations, the precipitation in the form of heavy snowfall contributes to extensive and very active glaciation. As described in the chapter on glaciation, glaciers are extremely effective agents of erosion. Deep, glacially carved valleys separate high mountains. The mountains will grow higher during isostatic adjustment at the same time the region as a whole is lowered by erosion. The Tibetan Plateau is north of the Himalaya. It is the highest, largest plateau in the world, with an average elevation of around 5 kilometers—higher than any mountain in the United States except for Alaska. The plateau has not been carved into mountains, because the climate is quite different from that of the Himalaya. The moist air from the south is blocked by the Himalaya and the Tibetan Plateau is in its rain shadow (see chapter 13). Without water, there are no glaciers or large rivers to carve the plateau into mountains and valleys. So this region is slowly being eroded downward, getting progressively lower as erosion and isostasy balance each other out.

Additional Reading

Pinter, N., and M. T. Brandon. How erosion builds mountains. *Scientific American* (April, 1997):pp. 74–79.

mountains. They were surprised by 1995 seismic studies that indicated that the crust is no thicker beneath the Rockies than at Denver. (Similar discrepancies between crustal thickness and mountain elevations have been reported for the southern Sierra Nevada.) To explain the higher elevations, geologists regard the mantle as hotter and therefore less dense beneath that part of the Rockies. The crust plus less dense mantle are floating on deeper denser mantle. Seismic wave studies verify that the mantle here is hot and appears to be asthenosphere that is at a shallower level in the earth than usual.

It is important to mention that isostatic adjustment takes place during the orogenic stage as well, but the isostatic forces during an orogeny are overshadowed by forces due to plate convergence.

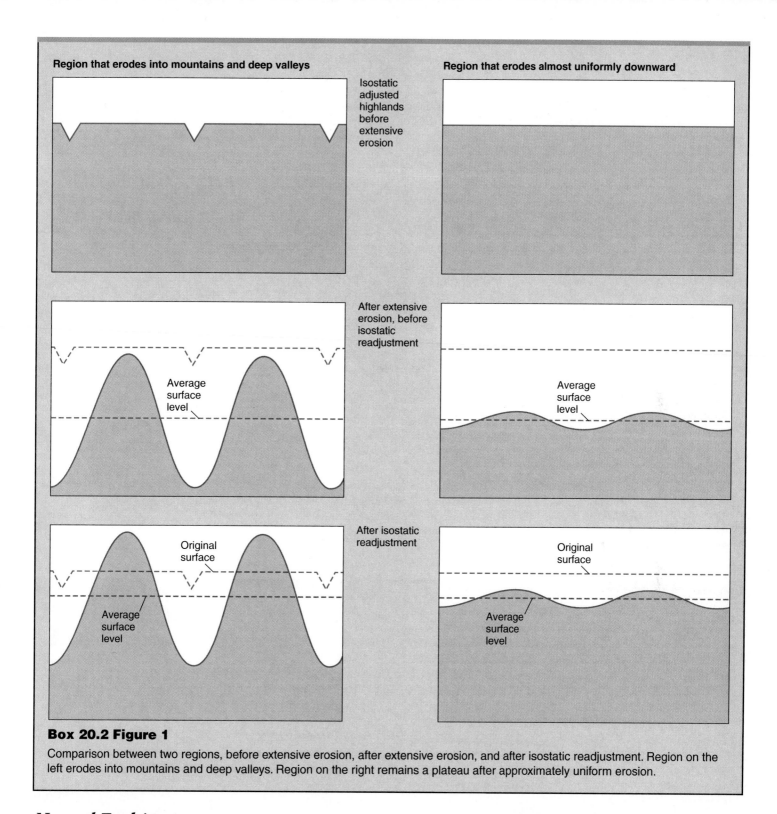

Box 20.2 Figure 1

Comparison between two regions, before extensive erosion, after extensive erosion, and after isostatic readjustment. Region on the left erodes into mountains and deep valleys. Region on the right remains a plateau after approximately uniform erosion.

Labels in figure:
- Region that erodes into mountains and deep valleys
- Region that erodes almost uniformly downward
- Isostatic adjusted highlands before extensive erosion
- After extensive erosion, before isostatic readjustment
- Average surface level
- After isostatic readjustment
- Original surface
- Average surface level

Normal Faulting

Normal faults are characteristic of this stage. The crust breaks into fault-bounded blocks. If an upthrown block is large enough, it becomes a **fault-block mountain range.** The normal faulting implies horizontal tensional force, therefore, fault-block mountain ranges are caused by *extension,* the regional pulling apart of the crust. Isostatic vertical adjustment of a fault block probably occurs at the same time.

Although most fault-block mountain ranges are bounded by normal faults on either side of the range, some are tilted fault blocks in which the uplift has been great along one side of the range while the other side of the range has pivoted as if hinged (figure 20.16). The Sierra Nevada (California) and Teton (Wyoming) Range are tilted fault-block mountains (figure 20.17).

Erosion surface

Trace of fold moved by erosion

Granite

A Before block faulting. Folding and intrusion of a pluton during an orogeny has been followed by a period of erosion.

0   10 Kilometers

Mountain peak

Portion of uplifted block
removed by erosion

Sediment deposited

B The same area after block-faulting. Tilted fault-block mountain range on left. Range to right is bounded by normal faults.

Figure 20.16

Development of fault-block mountain ranges.

Figure 20.17

The Teton Range, Wyoming, a fault-block range. The rocks exposed are Precambrian metamorphic and igneous rocks that were faulted upward. Extensive past glaciation is largely responsible for their rugged nature.

Photo by C. C. Plummer

512   *Chapter 20*

Figure 20.18

The Basin and Range and adjoining geological provinces.

Satellite and computer generated image prepared for *Scientific American* by Thomas H. Jordan and J. Bernard Minster

Isolated volcanic activity may be associated with this stage of a mountain belt's evolution. Eruptions occur along faults extending deep into the crust or the upper mantle.

Uplift is neither rapid nor continuous. Part of a mountain range may suddenly move upward a few centimeters (or, more rarely, a few meters) and then not move again for hundreds of years. Erosion works relentlessly on newly uplifted mountains, carving the block into peaks during the long, spasmodic rise. Over the long time period, the later episodes of renewed faulting and uplift involve successively less and less vertical movement.

Block-faulting is taking place in much of the western United States—the Basin and Range province (also called the Great Basin) of Nevada and parts of Utah, Arizona, New Mexico, Idaho, and California (figure 20.18). Hundreds of small, block-faulted mountain ranges are in evidence. They are separated by valleys that are filling with debris eroded from the mountains. Extension in the Basin and Range is probably due to hotter mantle beneath the crust as shown in figure 20.19.

Figure 20.19

Hot, buoyant mantle causes extension, thinning, and block-faulting of the overlying crust.

Delamination

Researchers in the 1990s were paying considerable attention to the hypothesis of lithospheric delamination and the role it may play in the evolution of mountain belts. For instance, delamination is used to explain the block-faulting, thin crust, and geologically young volcanic activity of the Basin and Range.

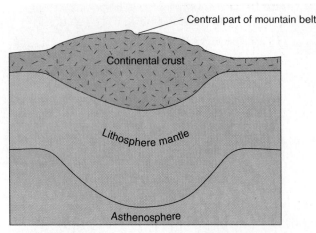

A Thick continental crust of a mountain belt produced during orogeny.

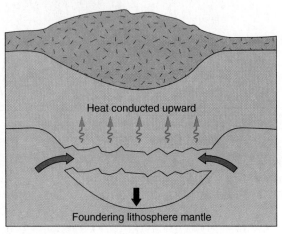

B Delamination of gravitationally unstable lithosphere mantle. Hot asthenosphere flows into place and heats overlying lithosphere.

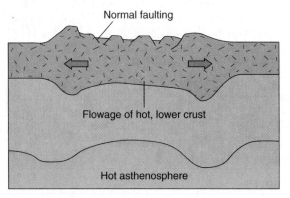

C Extension with hot lower crust flowing outward.

Figure 20.20

Delamination and thinning of continental crust following orogeny. Not drawn to scale.

Based on J. F. Dewey, 1988, Extensional collapse of orogens. *Tectonics*, v. 7, pp. 1123–1139, and K. D. Nelson, 1992, Are crustal thickness variations in old mountain belts like the Appalachians a consequence of lithospheric delamination? *Geology*, v. 20. pp. 498–502.

Lithospheric delamination (or simply **delamination**) is the detachment of part of the mantle portion of the lithosphere beneath a mountain belt (figure 20.20). As you know, the lithosphere consists of the crust and the underlying, rigid mantle. Beneath the lithosphere is the hotter, plastic mantle of the asthenosphere. During an orogeny the crust as well as the underlying lithosphere mantle thickens. The lithosphere mantle is cooler and denser than the asthenosphere mantle. As indicated in figure 20.20, the thickened portion of the lithosphere mantle is gravitationally unstable so it breaks off and sinks through the asthenosphere to a lower level in the mantle. Hot asthenosphere mantle flows in to replace the foundered, colder mantle. Heating of the crust follows, allowing the lower crust to flow. The once thick crust becomes thinner than that of adjoining regions of the mountain belt. Extension results in block-faulting in the upper part of the crust (as in figure 20.19).

Delamination beneath the Basin and Range helps explain the extensive rhyolitic and basaltic eruptions that occurred tens of million years after the end of the orogenic stage. Heating in the lower part of the crust to 700°C would have generated silicic magma that erupted as rhyolite. Basaltic magma would have formed from partial melting of the asthenosphere when it moved upward (replacing the foundered lithosphere mantle) and pressure was reduced (as explained in chapter 3). That the crust was once thicker in the Basin and Range is supported by recent studies of fossil plants indicating that the Basin and Range was 3 kilometers higher than at present.

Delamination is also being invoked to help explain why, when Pangaea broke up, North America split from Europe and Africa more or less along the old suture zone. *Gravitational collapse* could have contributed to the weakening and thinning of this once thick part of the mountain belt during the orogenic stage. The breakup of the supercontinent began around 30 million years after the orogenic stage ended. Delamination of the underlying lithosphere mantle would have resulted in heating and thinning of the overlying, remaining lithosphere. Rifting of the supercontinent began with normal faulting (see previous chapter, figure 19.22) and was accompanied by basaltic eruptions and intrusions. The Appalachians split from the European Caledonides. Europe, Africa, and North America went their separate ways as the Atlantic opened and widened.

Delamination (like gravity collapse) is an example of an hypothesis that builds on plate-tectonic theory. It was thought up and proposed because it explains data better than other concepts do. It still needs further testing to become widely accepted as a theory.

The Growth of Continents

Continents grow bigger as mountain belts evolve along their margins. Accumulation and igneous activity add new continental crust beyond former coastlines. In the Paleozoic Era the Appalachians were added to eastern North America, and during the Mesozoic and Cenozoic eras the continent grew westward because of accumulation and orogenic processes in many

20.3 IN GREATER DEPTH

Dance of the Continents (with SWEAT)

During recent years, geologists have proposed that the Wilson Cycle has taken place earlier in the earth's history and that Pangaea is only the youngest supercontinent. Canadian geologist, Paul Hoffman, has described an earlier cycle of formation and breakup of supercontinents, which he likens to a dance. The dancers are continental fragments and newly created crust that come together and later separate. Each dance cycle takes about a billion years and involves the building and dispersal of a supercontinent. Based on his studies of the Canadian Shield, Hoffman has described how the North American craton evolved through one "dance" cycle. The cycle is set to a symphony in four movements. It began two billion years ago (b.y.) and ended a billion years ago.

In the first movement (2.0–1.8 b.y.), seven microcontinents came together to form the beginnings of the North American (and part of what is now Europe's) craton. (Hoffman calls this the theory of the united plates of America.) New crust formed along some of the margins of the new continent through plate tectonic accretionary processes. Hoffman speculates that Precambrian shields now in distant continents (e.g., Australia) were part of this supercontinent.

The second movement took place between 1.8 and 1.6 b.y. This was characterized by tectonic accretion of new crust along the edge of the supercontinent. The third movement (1.6–1.3 b.y.) featured silicic magmatism in the interior of the continent, away from converging boundaries. In

the fourth movement, basalt floods and rifting heralded the breakup of the supercontinent. The symphony, or cycle, ended when the supercontinent fragmented—the "dancers" broke away.

The breaking away of the "dancers" may have been vigorous enough to cause SWEAT. SWEAT stands for Southwest United States–East Antarctic connection. According to the SWEAT hypothesis, proposed in 1991 by geologist Eldridge Moores, the North American and East Antarctic cratons were adjoining as part of the Precambrian supercontinent. The North American craton fled to the northern hemisphere, while the East Antarctic craton, along with the other Gondwanaland continents, remained in the southern hemisphere.

The cycle would be repeated as the fragments came together during the Paleozoic to form the supercontinent Pangaea. Pangaea, of course, split apart into the present continents.

Hoffman also cites evidence that a similar cycle took place in the earlier Precambrian, prior to 2 b.y.

How is the cycle explained? The gathering of continental fragments is attributed to downwelling of cold mantle. The continental fragments are carried to the area of sinking mantle where they collect, much like leaves washed along a gutter collect over a drain. Once the supercontinent has formed, downwelling mantle (and subduction) moves to the edges of the continent. The mantle beneath the interior of the supercontinent now heats up, due to the insulation provided by the continental crust and the fact that the downwelling cold mantle is now at the periphery of the large continent. Once the mantle beneath the interior of the continent heats sufficiently, hot spots develop resulting in magmatism and eventual rifting and breakup of the supercontinent.

Further Reading

Hoffman, P. F. Speculations on Laurentia's first gigayear (2.0 to 1.0 Ga). *Geology* Vol. 17 (1989): pp. 135–38.

parts of what is now the Cordillera. Therefore, if we isotopically dated rocks that had been through an orogeny, starting in the Canadian Shield and working toward the east and west coasts, we should find the rocks to be progressively younger. In a very general way, this seems to be the case. However, there are some rather glaring exceptions.

Suspect and Exotic Terranes

In many parts of mountain belts are regions where the age and characteristics of the bedrock appear unrelated to that of adjacent regions. To help understand the geology of moun-

tain belts, geologists have in recent years begun dividing major mountain belts into **tectonostratigraphic terranes** (or, more simply, **terranes**), regions within which there is geologic continuity. The geology in one terrane is markedly different from a neighboring terrane. Terrane boundaries are usually faults. Typically, a terrane covers thousands of square kilometers, but some terranes are considerably smaller. Alaska and western Canada have been subdivided by some geologists into over fifty terranes (figure 20.21). Terranes are named after major geographic features; for instance, Wrangellia, parts of which are now in Alaska and Canada (and with fragments in Washington and Idaho, according to

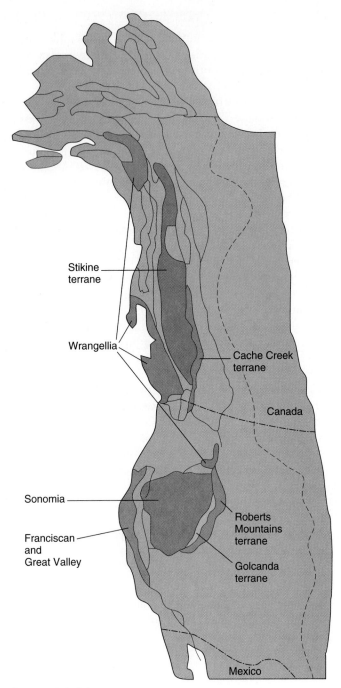

Figure 20.21

Some terranes in western North America. Note Wrangellia is in Alaska, British Columbia and Idaho.

After U.S. Geological Survey Open File Map 83–716.

Labels on figure: Stikine terrane, Wrangellia, Cache Creek terrane, Canada, Sonomia, Roberts Mountains terrane, Franciscan and Great Valley, Golconda terrane, Mexico

and have been called **suspect terranes,** that is, terranes that may not have formed at their present site. If evidence indicates that a terrane did not form at its present site on a continent, it is regarded as an **accreted terrane.** Accreted terranes that can be shown to have traveled great distances are known as *exotic terranes.*

A suspect terrane will have rock types and ages different from adjoining terranes, but to prove that it came from elsewhere in the world (and therefore is an accreted terrane), geologists may compare fossil assemblages or determine the paleomagnetic poles (see chapter 19) of the terrane's rocks. If the terrane is exotic, its fossil assemblage should indicate a very different climatic or environmental setting compared to that of the adjoining terrane. For an exotic terrane, the paleomagnetic poles for the rocks in the terrane will plot at some part of the world very distant from poles of adjoining terranes that formed in place. This indicates that a particular terrane formed in a different part of the earth and drifted into the continent of which it is now a part. Some accreted terranes were island arcs and some might have been *microcontinents* (such as present-day New Zealand) that moved considerable distances before crashing into other landmasses. Others may have been fragments of distant continents that split off and moved a long distance because of transform faulting. Imagine what might happen if the San Andreas fault remains active for another 100 million years or so. Not only would Los Angeles continue northward toward San Francisco and bypass it in about 25 million years (see Box 15.3), but the block of coastal California west of the fault would continue moving out to sea, becoming a large island with continental crust that drifts northward across the Pacific. Ultimately it would crash into and suture onto Alaska.

Figure 20.22 shows a tentative reconstruction of how parts of Alaska might have migrated in time. This is based on paleomagnetic data that indicate that parts of Alaska originally formed south of the equator and moved many thousands of kilometers to become part of the Cordillera. Note from the diagram that the path of migration was not simple. Plates split, plates joined, and the direction of movement changed from time to time.

The Appalachians as well as mountain belts in other continents have also been divided into terranes. Even the Canadian Shield has been subdivided into terranes. Some geologists think they can determine, despite the great age and complexity of the shield's rocks, the extent to which some terranes traveled before crashing together.

We should caution the reader that geologists do not always agree on the nature and boundaries of terranes. While most would probably agree that some terranes are exotic, many geologists think the subdividing of Alaska and western Canada into fifty terranes is overdoing it and not supported by sufficient evidence. Only time and more painstaking gathering of evidence will allow geologists to determine the history of each alleged terrane.

some geologists) was named after the Wrangell Mountains of Alaska.

Many terranes appear to have formed essentially in place as a result of accumulation and orogeny along the continent's margin. Other terranes have rock types and ages that do not seem related to the rest of the geology of the mountain belt

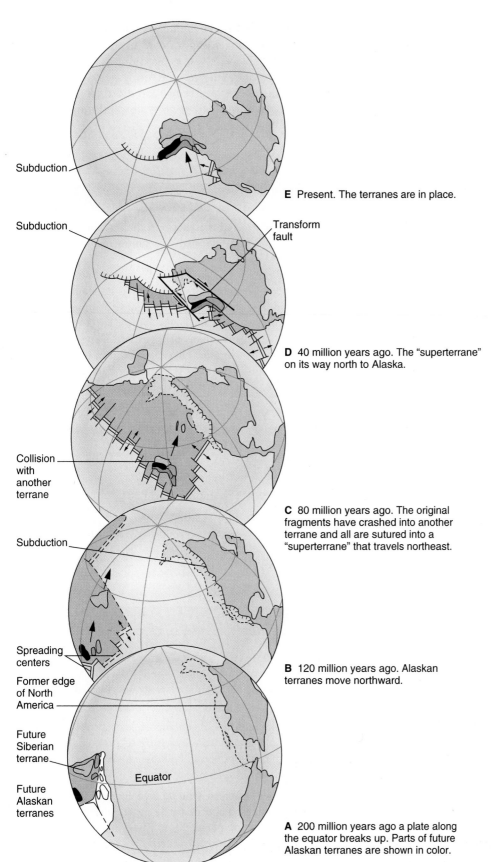

E Present. The terranes are in place.

Subduction

Transform fault

D 40 million years ago. The "superterrane" on its way north to Alaska.

Subduction

Collision with another terrane

C 80 million years ago. The original fragments have crashed into another terrane and all are sutured into a "superterrane" that travels northeast.

Subduction

Spreading centers

Former edge of North America

Future Siberian terrane

Equator

Future Alaskan terranes

B 120 million years ago. Alaskan terranes move northward.

A 200 million years ago a plate along the equator breaks up. Parts of future Alaskan terranes are shown in color.

Figure 20.22

How fragments of the Southern Pacific Crust may have become part of Alaska.

Modified from D. B. Stone, B. C. Panuska, and D. R. Packer, 1982, "Paleolatitudes versus time for southern Alaska," *Journal of Geophysical Research,* vol. 87 (pp. 3697–3707), copyrighted by American Geophysical Union.

Concluding Comment

Only a couple decades ago, many geologists thought that through the application of plate tectonic theory, we could easily determine the processes at work in each mountain belt and work back in time to understand the history of each of the earth's continents. Some suggested that there would hardly be major problems for earth scientists to solve in the future. Plate tectonics was a breakthrough, and a great many problems were solved; but with this great leap forward in the science we have identified new problems. New generations of geologists will have no shortage of challenges and no less excitement from solving newly discovered problems than did their predecessors who saw the dawn of the plate tectonics breakthrough. Science present builds on science past.

Summary

Major mountain belts are made up of a number of *mountain ranges.* Mountain belts are generally several thousand kilometers long but only a few hundred kilometers wide.

Mountain belts generally evolve as follows. First, a thick sequence of sedimentary and volcanic rock accumulates (the *accumulation stage*). Second, the accumulation stage is either accompanied or followed by an *orogenic stage,* which involves intense deformation of the layered rocks into folds and reverse (including thrust) faults, along with metamorphism and igneous activity. Third, the area is then subjected to a long period of uplift, often with block-faulting, and erosion. Eventually the mountain belt is eroded down to a plain and incorporated into the *craton,* or stable interior of the continent.

According to the theory of plate tectonics, mountains on the edge of continents are formed by continent-oceanic convergence, and mountains in the interior of continents are formed by continent-continent collisions.

The uplift of a region following termination of an orogeny is generally attributed to isostatic adjustment of continental crust.

Continents grow larger when new mountain belts evolve along continental margins. They may also grow by the addition of terranes that may have traveled great distances before colliding with a continent.

Terms to Remember

accreted terrane 516
accumulation stage 505
craton 501
fault-block mountain range 511
fold and thrust belts 503

gravitational collapse and spreading 507
lithospheric delamination
    (or delamination) 514
major mountain belt 498
mountain range 498

orogeny 506
Precambrian shield 501
suspect terrane 516
terrane (tectonostratigraphic terrane) 515

Testing Your Knowledge

Use the questions below to prepare for exams based on this chapter.

1. What does a fold and thrust belt tell us about what occurred during an orogeny?

2. What is the difference between the forces that could explain fault-block mountains and the forces that could account for an orogenic stage?

3. Explain how erosion and isostasy eventually produce stable, relatively thin, continental crust.

4. How do the sequences of sedimentary rocks in cratons differ from those in mountain belts?

5. What sequence of events accounts for a mountain belt that is bounded on either side by cratons?

6. The mountain belt that forms the western part of North America is called the (a) Appalachians (b) North American Cordillera (c) Himalaya (d) Andes (e) Rockies

7. The craton (a) covers the central part of the United States and Canada (b) has only 1,000–2,000 m of sedimentary rock overlying basement rock (c) is relatively undeformed (d) all of the above

8. The Precambrian shield (a) contains geologically young rocks (b) occurs only in mountainous regions (c) is a complex of Precambrian metamorphic and plutonic rocks exposed over a large area (d) all of the above

9. Folds and reverse faults in a mountain belt suggest (a) crustal shortening (b) tensional stress (c) deep water deposition of the sediment (d) all of the above

10. Which is not a stage in the history of a mountain belt?
(a) subsidence (b) accumulation (c) orogenic (d) uplift and block-faulting

11. To explain fold and thrust belts, simultaneous normal faulting, and how once deep-seated metamorphic rocks rise to an upper

level in a mountain belt, geologists use a model called (a) tectonism (b) gravitational collapse and spreading (c) orogeny (d) faulting

12. The Wilson Cycle describes (a) the cycle of uplift and erosion of mountains (b) the movement of asthenosphere (c) the block-faulting that occurs at mountains (d) the cycle of splitting of a supercontinent, opening of an ocean basin, followed by closing of the basin and collision of continents

13. The detachment of part of the mantle portion of the lithosphere beneath a mountain belt is called (a) gravitational collapse (b) rifting (c) lithospheric delamination (d) none of the above

14. Which is not a type of terrane? (a) accumulated (b) exotic (c) suspect (d) tectonostratigraphic (e) accreted

15. Which is a source for terranes? (a) microcontinents (b) fragments of distant continents (c) ocean crustal fragments (d) all of the above

16. Block-faulting may be due to (choose all that apply) (a) isostatic adjustment (b) subduction (c) gravitational collapse (d) lithospheric delamination

17. A mountain belt formed through ocean-continent convergence may contain (choose all that apply) (a) fold and thrust belts (b) thick accumulations of marine sediment (c) normal faults (d) metamorphism

18. Place these stages of development of a mountain belt in order (a) uplift and block faulting (b) accumulation (c) orogeny

Expanding Your Knowledge

1. How are unconformities used to determine when orogenies occurred?

2. How has seismic tomography contributed to our understanding of mountain belts?

3. How do basalt and ultramafic rocks from the oceanic lithosphere become part of mountain belts?

4. Why is a craton locally warped into basins and domes?

5. How could fossils in a terrane's rocks be used to indicate that it is an exotic terrane?

Exploring Resources

Condie, K. C. 1997. *Plate tectonics and crustal evolution.* 4th ed. Stoneham, MA: Butterworth/Heinemann.

Maher, H. D., Jr. 1994. The role of extension in mountain-belt life cycles. *Journal of Geological Education,* v. 42, pp. 212–19.

McPhee, J. A. 1981. *Basin and range.* New York: Farrar, Straus & Giroux.

———. 1983. *In suspect terrain.* New York: Farrar, Straus & Giroux.

———. 1986. *Rising from the plains.* New York: Farrar, Straus & Giroux.

———. 1993. *Assembling California.* New York: Farrar, Straus & Giroux.

Moores, E. M. ed. 1990, *Shaping the Earth—Tectonics of Continents and Oceans.* New York: W. H. Freeman.

Geological Society of America 1989–94. *The geology of North America.* For information on a particular part of North America refer to the appropriate volume in 19 vol. Boulder, Colorado: Geological Society of America.

The Appalachian story. Atlantic Geoscience Society, Atlantic Independent Media: Halifax, Nova Scotia.

Continental crust: Ancient and modern. The Media Guild: San Diego, California.

 http://www.hartwick.edu/geology/work/VFT-so-far/VFT.html Hartwick College Virtual Field Trip through part of the Appalachians in upstate New York. Includes a geologic history of the Appalachians.

http://geology.usgs.gov/geography.html U.S. Geological Survey site to get information on geologic research activities by geographic location in the United States.

Interacting with Journey Through Geology CD-ROM

Expand your knowledge of the concepts presented in this chapter by using the CD-ROM to answer the following questions.

1. Changing Earth module, Indian Ocean: This sequence shows the birth and growth of the Himalaya. Watch carefully as India moves northward to identify first the accumulation and then the orogenic phases in the growth of this mountain chain.

2. Changing Earth: Find two examples of continents growing through accretion.

Copper mine at San Zavier, Arizona.

CHAPTER
21

Geologic Resources

*T*hroughout this book, we have mentioned human use of earth materials, most of which are nonrenewable. Our purpose in this chapter is to survey briefly some important geologic resources of economic value.

We first look at energy resources to see which ones might help replace our disappearing supplies of oil. Then we discuss metals and their relation to igneous rocks and plate tectonics and conclude with nonmetallic resources such as sand and gravel.

Nearly every manufactured object requires geologic resources. An automobile contains substantial amounts of iron, chromium, manganese, nickel, platinum, tin, copper, lead, and aluminum in its body and engine and quartz sand in its window glass. It consumes petroleum in several forms—as fuel and lubricants, as synthetic rubber for tires, and as plastic for electrical parts, upholstery, and steering wheels. Dozens of other geologic resources go into automobiles, from tungsten in light bulb filaments to sulfur in battery acid.

People are beginning to realize how heavily dependent on geologic resources they are. Some have tried to limit their consumption of resources, or at least the rate at which their consumption *increases*. But it is impossible to stop consuming geologic resources. Think about some of the objects near you as you are reading this chapter. Your shirt may be partly polyester, which is made from petroleum, as are the plastic buttons. If you're wearing jeans, they may be made of fabric that is 100% cotton, but many jeans are made of shrinkproof fabrics that blend cotton with petroleum-based synthetic fibers. The brass zipper is made of copper and zinc. Some brands of jeans have pocket rivets made of copper. To make the leather tags on the back of some jeans, either aluminum or chromium was used during tanning. The fabric dye almost certainly came from petroleum. A pencil uses many geologic resources (figure 21.1) and a transistor radio many more. Eyeglasses are made of quartz sand and petroleum. Dental fillings are made of mercury, silver, and other metals. Cassettes and videotapes are made of vinyl, a petroleum derivative, and iron or chromium particles from metal mines. Every one of us uses geologic resources, and therefore is directly responsible for the existence of mines and oil wells.

Just how much of our resources do we use? According to the Mineral Information Institute, for every person living in the United States we annually mine 18,000 kilograms (40,000 pounds) of resources, not including energy resources. The amount of each commodity mined is 4,400 kg stone, 3,500 kg sand and gravel, 325 kg limestone for cement, 160 kg clays, 165 kg salt, 760 kg other nonmetals, 545 kg iron, 19 kg aluminum, 9 kg copper, 5 kg each for lead and zinc, 3 kg manganese, and 11 kg other metals. Our yearly per capita consumption of energy resources is over 8,000 kg (17,000 pounds); of this, 3,500 kg is petroleum, 2,300 kg coal, 2,250 natural gas, and .02 kg uranium.

Types of Resources

Geologic resources are valuable materials of geologic origin that can be extracted from the earth. There are three main categories of geologic resources:

1. Energy resources—petroleum (oil and natural gas), coal, uranium, and a few others, such as geothermal resources.
2. Metals—iron, copper, aluminum, lead, zinc, gold, silver, and many more.
3. Nonmetallic resources—sand and gravel, building stone, limestone (for cement), sulfur, gems, gypsum, fertilizers, and many more. Ground water (chapter 11) should also be regarded as an important geologic resource.

Figure 21.1

Mineral resources necessary to make a wooden pencil.

Geologic resources are sometimes called *mineral resources*, but the term, though widely used, is not accurate. Oil, natural gas, and coal, for example, are not formed of solid, crystalline minerals.

A very important point about geologic resources is that they are **nonrenewable resources**, except for ground water. They form so slowly that, at the present rapid rates of consumption, they can easily become exhausted. Some earth resources, such as food or timber, usually can be produced as fast as they are consumed. These are *renewable resources*. But petroleum, iron, lead, uranium, sulfur, sand and gravel, and

Figure 21.2

The difference between reserves and resources. Reserves (discovered deposits that are economically and legally extractable) form a small part of total resources (discovered and undiscovered deposits that are economical to extract now or may be economical in the future).

essentially all other geologic resources are being used at rates far greater than the rates at which new deposits form. In some areas even ground water is being consumed faster than it is being replenished. This means that eventually these resources are going to either run out or be priced so high because of their scarcity that their use will drop to insignificant levels. Recycling, new discoveries, and substitutes can help prolong the life of some resources, but it is likely that within your lifetime some of the resources in common use today will be scarce. This need not mean that civilization will come to a standstill, but it does mean that we must find inexpensive substitutes for these resources. Otherwise, skyrocketing costs will cause a drop in nearly everyone's material standard of living.

Resources and Reserves

Two different terms are used to describe the amount of geologic material not yet extracted from the earth. **Resources** is a broad term used for the total amount of a geologic material in all deposits, discovered and undiscovered. It includes both the deposits that can be economically extracted under present conditions and those that may be extracted economically in the future (figure 21.2). It is very difficult to estimate resources because educated guesses must be made about the existence and sizes of deposits yet undiscovered as well as about what type of deposit might someday be economical to extract. **Reserves** are a small part of resources. They are the *discovered* deposits that can be extracted economically and legally under present conditions. That is, they are the short-term supply of a geologic material.

Once *resources* have been carefully estimated, the amount should not change from year to year, for an estimate of resources is basically an estimate of the total future supply. Estimates of *reserves,* however, change all the time. The extraction and use of a substance lowers reserves. New discoveries add to reserves, for part of the definition of reserves is that we have to know the deposit is there. (*Reserves* are like your bank balance, rising and falling with time; *resources* are like your future, lifetime income.)

A deposit also has to be profitable to extract, and many factors determine whether a profit can be made. The costs of extraction, including workers' salaries and the fuel used to run equipment, are important. New inventions that make extraction cheaper can increase reserves. The final price a company can receive for its product is also important. When prices are kept artificially low by the government, as was the case in the 1980s for natural gas and oil, reserves are naturally low because few deposits can be extracted profitably at that price. When prices rise, more deposits become economical, and reserves increase, even without new discoveries. Reducing the amount of taxes that extraction companies pay can also increase reserves.

Changes in laws can affect reserves. Large areas of government-owned land are off-limits to mining and drilling, so any geologic materials under these areas are not legally extractable and cannot be included in reserves. Opening more land to extraction can therefore increase reserves.

Energy Use

The United States consumes huge amounts of energy. By 1990 the United States was consuming ten times the amount of energy that it used in 1900. In 1973 and 1978 came steep price rises for petroleum. The need for conservation and fuel-efficient buildings and transportation became more and more apparent. From 1979 to 1985 total United States energy use declined and so did its dependence on petroleum (particularly imported oil). Since a precipitous drop in oil prices at the beginning of 1986, however, there has been a sharp increase in both United States energy use and its reliance on imported oil.

The sources of United States energy have been changing in recent years in response to changing prices.

	1975	1985	1993	1997
Oil	46%	42%	40%	36%
Natural gas	28	24	25	31
Coal	18	24	23	21
Nuclear	3	5.5	8	7
Hydroelectric	5	4.5	4	4

Note that oil and natural gas account for two-thirds of the United States' energy supply; the fossil fuels (coal, oil, and gas) provide almost 90% of our energy.

Oil and Natural Gas

Within the petroleum industry, **petroleum** is a broad term that includes both crude oil and natural gas. That will be our usage. (In common usage *petroleum* is synonymous with crude oil; many geologists also use it this way.)

A Anticline

B Normal fault

C Thrust fault

D Sandstone lenses

E Sandstone pinchout

F Unconformity

G Reef (a small "patch" reef)

Figure 21.3

Traps for oil and gas. *A–C* are structural traps. *D–G* are stratigraphic traps. Oil floats on water, and is often capped by natural gas.

Crude oil is a liquid mixture of naturally occurring hydrocarbons (compounds containing hydrogen and carbon), which can be distilled to yield a great variety of products.

Natural gas is a *gaseous* mixture of naturally occurring hydrocarbons. Its origin and occurrence closely parallel that of oil. Many wells that recover oil also recover natural gas, although either may exist alone.

The Occurrence of Oil and Gas

The origin of oil and gas in sediment and sedimentary rocks was described in chapter 6.

Oil pools are valuable underground accumulations of oil. They are found only where four specific conditions occur together: (1) a **source rock** (such as shale) containing organic matter that is converted to petroleum by burial and other postdepositional changes; (2) a **reservoir rock** (usually sandstone or limestone) that is sufficiently porous and permeable to store and transmit the petroleum; (3) an **oil trap,** a set of conditions to hold petroleum in a reservoir rock and prevent its escape by migration; and (4) deep enough burial (or *thermal maturity*) to "cook" the oil and gas out of the organic matter. All four conditions must be met; if one of the four is missing, the rock will hold no oil or gas.

Natural gas requires the same conditions as oil for accumulation. Gas can exist at greater depth than oil, however, and variations in source rock, depth of burial, and thermal history of the organic matter probably control whether gas, oil, or both accumulate in a region.

Figure 21.3 depicts several types of *structural traps* for oil and gas. *Anticlines* (described in chapter 15) are the most common oil traps. Where oil and water occur together in folded sandstone beds, the oil droplets, being less dense than water, rise within the permeable sandstones toward the top of the fold. There the oil may be trapped by impermeable shale overlying the sandstone reservoir rocks. Because natural gas is less dense than oil, the gas collects in a pocket, under fairly high pressure, on the top of the oil.

Faults may create oil traps when permeable reservoir rocks break and slide next to impermeable rocks. Thrust faults are often associated with folds because both are caused by compression. The backarc thrust belts on the landward side of both the Cordilleran and Appalachian mountain belts are currently being intensively explored for oil and gas.

A *stratigraphic trap* is a result of natural sedimentation rather than folding or faulting. It may be a lens of sandstone within a larger bed of shale. Another such trap is the narrow edge of a gently dipping sandstone layer where it pinches out

Box 21.1 Figure 1

A bed of rock salt flows because it is loaded unevenly. Salt domes form as parts of the original layer flow upward due to rock salt being less dense than the overlying layers of sedimentary rock.

S alt domes are vertical columns of rock salt rising upward from a deeply buried bed of salt. Subsidence of divergent continental margins allows rock salt to be covered with a thick sequence of sedimentary rocks (see figures 19.22 and 19.26). The great weight of the dense overburden allows the low-density salt to flow plastically upward (box figure 1). Rising columns (which may break off from the salt bed) deform the surrounding sedimentary rock into folds and faults that create oil traps (box figure 2).

Circulating ground water dissolves the salt, preventing most salt domes from reaching the surface. An insoluble *cap rock* may develop if the rock salt contains gypsum. This calcium sulfate can convert to yellow native sulfur if it contacts petroleum and bacteria.

In addition to oil and gas, salt domes can be mined for salt (underground) and sulfur (which is melted by superheated water and pumped to the surface). In the United States a broad belt of salt domes extends both onshore and offshore along the Texas and Louisiana coasts.

There are many ways to find these valuable salt domes. Some push up the surface, forming low, round islands in the coastal swamps. Others dissolve so rapidly that they form surface depressions. Most are found by geophysical surveys.

The density of rock salt is 2.2 grams per cubic centimeter, even when it is compressed. The surrounding sedimentary rock has a density of 2.0 at the surface, increasing to 2.5 or more at depth due to compaction. The salt dome rises because it is out of isostatic equilibrium with the heavier rock around it. This density contrast produces a *negative*

Box 21.1 Figure 2

Cross-sectional view of a salt dome. The rising salt dome caused folding and faulting of the adjacent sedimentary strata resulting in oil and gas traps. Sulfur and gypsum are concentrated in cap rock.

within a shale sequence. Oil can collect under *unconformities* if shale seals off a reservoir rock. Limestone *reefs* can form a variety of traps. The core of a reef is usually full of large openings formed by the irregular growth of coral and algae. Oil can collect both in a reef core and in the dipping beds of wave-broken debris on the reef flanks. *Salt domes* create a variety of traps described in Box 21.1.

Oil fields are regions underlain by one or more oil pools. Figure 21.4 shows the location of most of the major North American oil fields. The two largest oil fields within the United States are in eastern Texas and on Alaska's North Slope. Most of the world's remaining oil lies in giant fields in the Middle East (especially in Saudi Arabia and Kuwait), Russia and Azerbaijan, Venezuela, and Mexico.

(Continued from page 525)

gravity anomaly that can be quickly detected by an inexpensive gravity survey (box figure 3). *Seismic reflection* can detect the curved surface of the dome or the deformed rocks above the dome. *Seismic refraction* has been used in the past to find domes. Seismic waves speed up in salt; therefore, the arrival of waves at a recording station before they are expected indicates that a high-speed material, such as salt, occurs below the surface.

Drilling is much more expensive than geophysical surveys, so several types of surveys are usually run first, to see if drilling is a good gamble (many domes have no oil, gas, or sulfur).

Box 21.1 Figure 3

A salt dome can be detected by a gravity survey, by seismic reflection, and by seismic refraction.

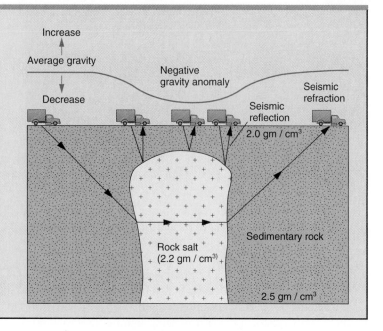

Recovering the Oil

When an oil pool or field has been discovered, wells are drilled into the ground. Permanent derricks used to be built to handle the long sections of drilling pipe. Now portable drilling rigs are set up to drill (figure 21.5), and are then dismantled and moved. When the well reaches a pool, oil usually rises up the well because of its density difference with water or because of the pressure of the expanding gas cap above the oil. Although this rise of oil is almost always carefully controlled today, spouts of oil, or *gushers,* were common in the past. Gas pressure gradually dies out, and oil is then pumped from the wells. Water or steam may be pumped down adjacent wells to help push the oil out. At a refinery the crude oil from underground is separated into natural gas, gasoline, kerosene, lubricating oil, fuel oil, grease, asphalt, and paraffin. *Petrochemicals,* manufactured from petroleum, include dyes, fertilizers, medicines, synthetic rubber, explosives, perfumes, paints, saccharin, solvents, synthetic fibers, and plastics used for such varied products as compact discs, cassettes, floor tile, and garbage bags.

As oil becomes increasingly difficult to find, the search for it is extended into more hostile environments. The development of the oil field on the North Slope of Alaska and the construction of the Alaska pipeline are examples of the great expense and difficulty involved in new oil discoveries. Offshore platforms extend the search for oil to the ocean's continental shelves and even beyond. More than one-quarter of the world's oil and almost one-fifth of the world's natural gas comes from offshore, even though offshore drilling is six to seven times as expensive as drilling on land.

Environmental Effects

Getting petroleum out of the ground and to the consumer can create environmental problems anywhere along the line.

Figure 21.4

Major oil fields in North America. The amount of oil in a field is not necessarily related to its areal extent on a map. It is also governed by the vertical "thickness" of the oil pools in a field. The fields with the most oil are in Alaska and east Texas.

From U.S. Geological Survey and other sources.

Salty *brine,* a by-product of most oil wells, can pollute surface water. Pipelines carrying oil can be broken by faulting, landsliding, or acts of war, causing an *oil spill.* Tanker spillage from collisions or groundings (such as the *Exxon Valdez* off Alaska in 1989) can create oil slicks at sea. Offshore platforms may also lose oil (such as the Santa Barbara blowout off southern California in 1969). Oil slicks can drift ashore, fouling the beaches. Natural oil seeps can also cause pollution. Tar-like lumps collect on some California beaches as a result of natural seepage of petroleum from the ground. At one place in Alaska, oil seeps build a crude oil lake behind ice during the winter; in the summer, when the ice melts, the oil flows out to a nearby bay.

Subsidence of the ground can occur as oil is removed. The Wilmington field near Long Beach, California, has subsided 9 meters in fifty years; dikes have had to be built to prevent sea water from flooding the area. The refining and burning of petroleum and its products can cause *air pollution.* Advancing technology and strict laws are helping control some of these adverse environmental effects.

How Much Oil Do We Have Left?

The World Situation

The world's oil *reserves* are estimated to be 1,000 to 1,100 billion barrels (a barrel contains 42 gallons). Total *resources,* of course, are much larger than reserves. One estimate is 3,000 billion barrels (including reserves).

Since the world's present consumption of oil is 24 billion barrels per year, known *reserves* will last about 40 to 45 years. Total *resources* may last well over 100 years. The world is not running rapidly out of oil; the United States *is,* however, as we will see in the next section.

There is far more oil underground than is listed in the resources estimate; some oil cannot be recovered. The oil may be in a pool too small, or lie too far from market, to justify the expense of drilling. Oil lies under regions where drilling is forbidden, such as national parks or other public lands. Even using the best extraction techniques, only about 30% to 40% of the oil in a given pool can be brought to the surface. The rest is far too difficult to extract, and has to remain underground.

The Outlook in the United States

United States consumption of oil is 6.3 billion barrels of oil per year (with 6% of the world's population, the United States uses 29% of the world's oil production each year). Sixty-five percent of U.S. consumption is used for transportation. United States *reserves* are currently about 25 billion barrels. The United States imports 51% of the oil it uses; domestic production is 3.1 billion barrels per year. At current production rates, we have an 8-year supply in reserves (less if the production rate goes up).

Because the reserve estimate changes each year, an estimate of *resources* is more important for planning purposes. Such esti-

Figure 21.5

Drilling rig on Alaska's North Slope.

Photo by BP America, Inc.

mates have fluctuated wildly in the past 25 years. In recent years the estimate of resources has been going down rapidly as promising drill sites have proved to be dry. The current resource estimate is 74 billion barrels (including reserves). This means that at the current production rate we have a 24-year supply of oil left, counting both reserves and resources. The United States *is* truly running out of oil, and we will need alternate fuels and/or more imported oil in the future (some energy experts expect the United States to be importing 65% to 70% of its oil by the year 2000).

The situation becomes more serious if the temporary ban on most new offshore drilling, enacted after the *Exxon Valdez* spill, becomes permanent. About 16 billion barrels of the 74-billion total are in undiscovered offshore oil. If this oil is placed off-limits, our oil resources will run out in 19 years at the current production rate.

Past estimates of *natural gas* have been similarly gloomy, largely because estimators assumed that gas discoveries are linked to oil discoveries. In the past, gas has been found in a fixed ratio to oil, so some estimators assume that because we are running out of oil, we are also running out of gas.

The United States produces more than 18 trillion cubic feet (TCF) of natural gas per year. Reserve estimates are 173 TCF, with total resources perhaps 570 TCF. By these estimates, natural gas will be gone in the United States in 30 years.

The recent decontrol of natural gas prices, however, has led to renewed interest in gas and revised estimates of resources. The new estimates are based on unconventional gas sources, not linked to oil. There may be 300 to 500 TCF of gas within coal beds, for example (the gas is a hazard to coal miners because it is toxic and tends to explode). Since 1995, two areas have begun to be important producers of coal bed gas, the San Juan Basin of northwestern New Mexico and the Warrior Basin of Alabama. Tightly compacted sandstones ("tight sands") in the western United States may hold 200 to 800 TCF. Devonian shales rich in organic matter in the eastern United States may hold 500 to 600 TCF. Gas can exist at greater depth than oil, so it is likely that deep drilling (beyond 5 kilometers) in sedimentary basins can produce substantial amounts of gas. Gas is also dissolved in hot salty water within highly fractured rock along the Gulf Coast ("geopressured zones") that may contain as much as 3,000 TCF of gas. Truly staggering amounts of gas may be tied up as gas hydrates, a peculiar solid that forms from natural gas and water in two unusual environments: (1) on the deep sea floor and (2) under permafrost. Recent estimates of worldwide gas in hydrate form range from 100,000s to 1,000,000s of TCF. All these estimates are for total gas in place, not for recoverable reserves, which would be far lower. Some geologists discount all these estimates as wildly optimistic and misleading. All these unconventional gas sources are getting serious attention today, however, and although some difficult extraction problems remain unsolved, there is at least cautious optimism that the United States may not be as short of gas as we used to think.

Heavy Crude and Oil Sands

Heavy crude is dense, viscous petroleum. It may flow into a well, but its rate of flow is too slow to be economical. As a result, heavy crude is left out of reserve and resource estimates of less viscous "light oil," or regular oil. Heavy crude can be made to flow faster by injecting steam or solvents down wells, and if it can be recovered, it can be refined into gasoline and many other products just as light oil is. Most California oil is heavy crude.

Oil sands (or *tar sands*) are asphalt-cemented sand or sandstone deposits. The asphalt is solid, so oil sands are often mined rather than drilled into, although the techniques for reducing the viscosity of heavy crude often work on oil sands as well.

The origin of heavy crude and oil sands is uncertain. They may form from regular oil if the lighter components are lost by evaporation or other processes. Oil sands and asphalt seeps at the earth's surface (such as the Rancho La Brea Tar Pits in Los Angeles) probably formed from evaporating oil. But some heavy crudes and oil sands are found as far as 4,000 meters underground. Most of them have much higher concentrations of sulfur and metals, such as nickel and vanadium, than does regular oil. These facts suggest that heavy crude and oil sands may have a somewhat different origin than light oil.

The best-known oil sand deposit is the Athabasca Oil Sand in northern Alberta, Canada (figure 21.6). The deposit, currently being strip-mined, contains 900 billion barrels of oil, of which at least 80 billion barrels are recoverable. About 10% of Canada's oil production comes from these oil sands.

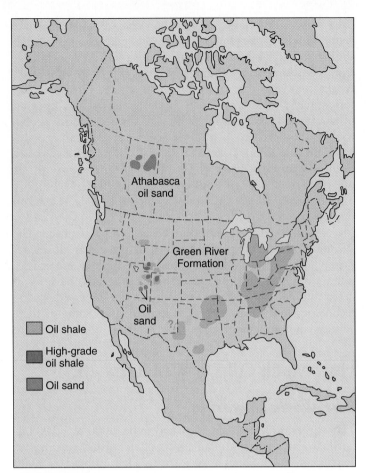

Figure 21.6

Distribution of major deposits of oil sand and oil shale in the United States and Canada.

From U.S. Geological Survey and other sources

Venezuela has even more oil sand than Canada. The United States has more than 100 billion barrels of heavy crude and oil in oil sands, including about 30 billion in the form of oil sands in Utah. Half of it may be ultimately recoverable. This means that heavy crude and oil sands may supply as much oil in the future as our light oil reserves.

Oil Shale

Oil shale is a black or brown shale with a high content of solid organic matter from which oil may be extracted by distillation. The best-known oil shale in the United States is the Green River Formation, which covers more than 40,000 square kilometers in Colorado, Wyoming, and Utah, with deposits up to 650 meters thick (figures 21.6 and 21.7). The oil shale, which includes numerous fossils of fish skeletons, formed from mud deposited on the bottom of large, shallow Eocene lakes. The organic matter came from algae and other organisms that lived in the lakes.

The Green River Formation includes more than 400 billion barrels of oil in rich beds that yield over 25 gallons of oil

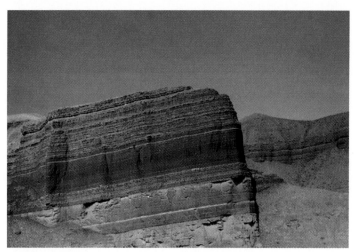

Figure 21.7

Cliffs of oil shale near Green River, Wyoming

Photo by Frank M. Hanna

per ton of rock. Another 1,400 billion barrels of oil occur in lower-grade beds yielding 10 to 25 gallons per ton. An estimated 300 to 600 billion barrels may be recoverable.

Relatively low-grade oil shales in Montana contain another 180 billion barrels of recoverable oil in shale that should be economical to mine because of its high content of vanadium, nickel, and zinc. Therefore, oil shale can supply potentially vast amounts of oil in the future as our liquid petroleum runs out.

A few distillation plants extract shale oil, but the current low price for oil makes shale oil uneconomical. A price increase for oil may make large-scale production of shale oil feasible in the future.

The mining of oil shale can create environmental problems. There is a space problem, for during distillation the shale expands. Spent shale could be piled in valleys and compacted, but land reclamation would be troublesome. A great amount

of water is required, both for distillation and for reclamation, and water supply is always a problem in the arid West. New processing techniques that extract the oil in place without bringing the shale to the surface may eventually help solve some of the problems and lower the water requirements. It is possible to burn fractured oil shale in large underground excavations. The heat separates most of the oil from the rock; the oil can be collected as a liquid. (The fires, however, would be hard to control and would affect ground-water levels.) Another proposal involves heating the shale with radio waves or microwaves to separate the liquid oil from the rock.

Coal

Coal, as described in chapter 6, is a sedimentary rock that forms from the compaction of plant material that has not completely decayed. After oil and natural gas, coal is our third major energy resource. It provided 90% of the energy used in the United States in 1900 but provides only 21% today. Coal use has increased as petroleum becomes scarcer and more expensive.

About 88% of the present use of coal in the United States is for generating electricity. Coal is also used to make *coke*, which is used in steel making. In the future, coal may be used instead of petroleum in the manufacture of some chemicals. *Coal gas* and *coal oil* made from coal are reasonable approximations of natural gas and oil and can be used for some of the same purposes, although they are still very expensive to produce. Coal can also be powdered and mixed with water to form a *slurry*, which can be transported through pipelines and burned as a liquid fuel.

Varieties of Coal

Table 21.1 shows the common varieties (ranks) of coal. *Peat*, a mat of unconsolidated plant material, is not coal but probably represents the initial stage of coal development (figure 21.8*A*). When dry, it can be burned as a fuel (peat fires used to dry

Table 21.1	**Varieties (Ranks) of Coal**				
	Color	Water Content (%)	Other Volatiles (%)	Fixed Carbon[2] (%)	Approximate Heat Value (BTUs of heat per pound of dry coal)
Peat[1]	Brown	75	10	15	Varies
Lignite	Brown to brownish-black	45	25	30	7,000
Subbituminous coal	Black	25	35	40	10,000
Bituminous coal (soft coal)	Black	5 to 15	20 to 30	50 to 75	12,000 to 15,000
Anthracite (hard coal)	Black	5	5	90	14,000

1. Peat is not a coal.

2. "Fixed carbon" means solid combustible material left after water, volatiles, and ash (noncombustible solids) are removed.

A

B

Figure 21.8

(*A*) A layer of peat being cut and dried for fuel, island of Mull, Scotland. Coal often forms from peat. (*B*) A bed of subbituminous coal near Gillette, Wyoming.

malted barley give Scotch whisky its smoky flavor). With compaction, peat can become *lignite* (*brown coal*), which may still contain visible pieces of wood. Lignite is soft and often crumbles as it dries in air. It is subject to spontaneous combustion as it oxidizes in air, and this somewhat limits its use as a fuel. *Subbituminous coal* and *bituminous coal* (*soft coal*) are black and often banded with layers of different plant material (figure 21.8*B*). They are dusty to handle, ignite readily, and burn with a smoky flame. *Anthracite* (*hard coal*) is actually a metamorphic rock, generally formed only under the regional compression associated with folding. It is hard to ignite but is dust-free and smokeless.

Occurrence of Coal

Coal occurs in beds that range in thickness from a few cm to 30 m or more. If the beds are deeply buried, underground mines are dug to extract the coal (figure 21.9). If the coal beds are close to the land surface, the coal is mined in a **strip mine,** in which the overburden is removed to expose coal at the surface (figure 21.10). When a strip of coal has been uncovered and removed, the resulting trench is filled in with the overburden from the adjoining strip.

Figure 21.11 shows the coalfields of the "lower 48" states. We will discuss three major regions, the Appalachian fields, the interior fields, and the far western fields.

The major coal-producing fields in the United States are the *Appalachian fields,* which stretch from Pennsylvania to Alabama and contain extensive beds of bituminous coal. The

Figure 21.9

An underground coal mine.

Photo by Larry Lee/West Light

coals are mostly of Pennsylvanian age; a few are Mississippian and Permian. The coal beds, which thin westward, were included in the late Paleozoic folding and faulting of the Appalachian orogeny, so they are strongly deformed in the eastern part of the belt. Steeply dipping coal beds here are mostly mined underground. The folds are gentler to the west, where the coals can be extracted by both underground and strip mining. Northeastern Pennsylvania has some anthracite that resulted from intense folding.

Figure 21.10

Coal strip mine near Gillette, Wyoming. The upper half of a 30-meter-thick bed of subbituminous coal is shown here, overlain by 3 to 10 meters of overburden.

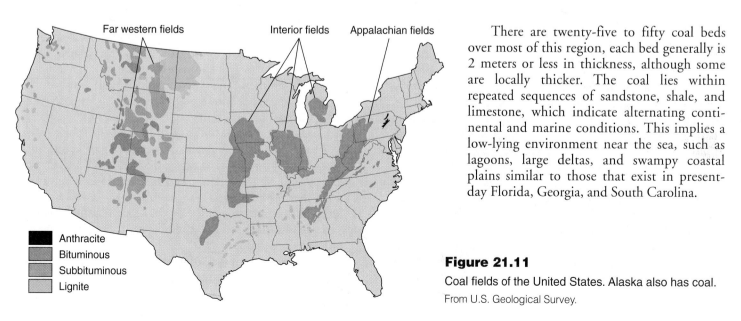

Far western fields   Interior fields   Appalachian fields

■ Anthracite
■ Bituminous
■ Subbituminous
□ Lignite

There are twenty-five to fifty coal beds over most of this region, each bed generally is 2 meters or less in thickness, although some are locally thicker. The coal lies within repeated sequences of sandstone, shale, and limestone, which indicate alternating continental and marine conditions. This implies a low-lying environment near the sea, such as lagoons, large deltas, and swampy coastal plains similar to those that exist in present-day Florida, Georgia, and South Carolina.

Figure 21.11

Coal fields of the United States. Alaska also has coal.

From U.S. Geological Survey.

The *interior fields* extend from Michigan through Illinois to Texas and are extensions of the Appalachian rocks westward onto the continental interior. In Michigan and Illinois the rocks are preserved in large basins; the fields from Iowa to Texas are mostly horizontal. The coals are usually strip-mined, particularly near major industrial centers around the basin's edge in Illinois, Indiana, and Kentucky.

The *far western fields* extend from New Mexico northward through the Rocky Mountains to Montana and the Great Plains of North Dakota (and also into Canada). These coal beds are thicker and younger than eastern coals. They are generally of Cretaceous or Eocene age and are up to 30 meters thick. The coals occur in large basins and are generally of low rank, being either lignite or subbituminous coal, although some good-quality bituminous coals occur in Colorado and Utah. Many thick beds are very close to the surface and are strip-mined, although underground mining is common in some states. Western coals are attractive fuels, presently in very high demand, because they typically have less sulfur than eastern coals (sulfur compounds can pollute the air when coal burns).

Environmental Effects

Extracting and using coal creates environmental problems. The presence of a mine usually lowers the local water table as ground water is pumped out of the mine. The drainage out of the mines tends to be highly acidic, polluting surface streams and water supplies. In the past, strip mines have been refilled as barren, unsightly piles, but new techniques of recontouring the overburden and restoring the topsoil and vegetation help reclaim mined-out areas for other uses. When coal is burned, ash and sulfur gases can pollute air, but most of the harmful components can be removed with existing technology. Solving environmental problems associated with coal, of course, raises the immediate cost of extraction and the price to the consumer. Long-term costs may be *decreased*, however; preventing environmental damage is usually less expensive than cleaning it up.

Reserves and Resources

Coal production in the United States is about 1 billion tons per year (the United States uses 0.9 billion tons and exports the rest). Sixty percent of the production is from surface strip mines. Although Kentucky, West Virginia, and Pennsylvania still produce almost half the coal, the rapid development of the huge beds of low-sulfur western coal may soon cause the Northern Rockies states to surpass the Appalachian states.

Recoverable *reserves* in the United States are about 265 billion tons (40% of this is in Montana and Wyoming). As you can see, there are centuries of coal left at the present rate of production.

Coal *resources* within 1,000 meters of the surface are 1,700 billion tons. A greater amount is estimated to lie deeper, so the total United States resources are an impressive 3,900 billion

tons (much of which, of course, is not presently usable). Even as we step up extraction for the export market, it seems obvious that our coal supplies will last a very long time. The United States has been called the Saudi Arabia of coal (there are also huge deposits in China, Australia, Ukraine, and Russia).

Uranium

The metal *uranium,* which powers nuclear reactors, occurs as *pitchblende,* a black uranium oxide found in hydrothermal veins, or, much more commonly in the United States, as yellow *carnotite,* a complex hydrated oxide found as incrustations in sedimentary rocks. Ground water easily transports oxidized uranium, which is highly soluble. Organic matter reduces uranium, making it relatively insoluble, so uranium precipitates in association with organic matter.

Most of the easily recoverable uranium in the United States is found in sandstone in New Mexico, Utah, Colorado, and Wyoming, some of it in and near petrified wood. In the 1950s uranium boom, western prospectors looked for petrified logs and checked them with Geiger counters. Some individual logs contained tens of thousands of dollars worth of uranium. Some petrified logs have so much uranium that it would be dangerous to keep them as souvenirs. Most of the uranium, however, is in sandstone channels that contain plant fragments.

Organic phosphorite deposits of marine origin in Idaho and Florida also contain uranium. The uranium is not very concentrated, but the deposits are so large that overall they contain a substantial amount of uranium. The black Devonian shales of the eastern United States also contain uranium. These shales are really low-grade oil shales (figure 21.6), and they contain large amounts of natural gas, as you have seen. Uranium may be recovered from phosphorites or shales as a byproduct of another mining operation.

Uranium is used in nuclear reactors to produce electricity and in nuclear weapons and some naval craft. At present, nuclear reactors produce 7% of the energy needs of the United States. In France, over half of its electricity is produced by nuclear power. Use of nuclear power may not rise appreciably in the future, depending on public acceptance. Nuclear plants produce long-lived waste products that remain dangerous for centuries, and no suitable storage site yet exists, although Yucca Mountain in Nevada is being intensively studied as a possible nuclear waste repository (see chapter 11). Accidents at the Three Mile Island reactor in Pennsylvania in 1979 and the Chernobyl reactor near Kiev in Ukraine in 1986 caused a major rethinking of the desirability of nuclear power. In 1993 the United States had 109 operational reactors and 7 other permits granted. This total of 116 is down from 119 in 1991 and compares with 236 built or planned before the Three Mile Island accident. There have been no new plants ordered since 1978.

Recoverable reserves of nearly 500,000 tons of uranium oxide in western sandstones seem adequate to power the exist-

Table 21.2	Common Ore Minerals	
Metal	**Ore Mineral**	**Composition**
Aluminum	Bauxite (a mineral mixture)	$Al_2O_3 \cdot nH_2O$
Chromium	Chromite	$FeCr_2O_4$
Copper	Native copper	Cu
	Chalcocite	Cu_2S
	Chalcopyrite	$CuFeS_2$
Gold	Native gold	Au
Iron	Hematite	Fe_2O_3
	Magnetite	Fe_3O_4
Lead	Galena	PbS
Manganese	Pyrolusite	MnO_2
Mercury	Cinnabar	HgS
Nickel	Pentlandite	(Fe, Ni)S
Silver	Native silver	Ag
	Argentite	Ag_2S
Tin	Cassiterite	SnO_2
Uranium	Pitchblende	U_3O_8
	Carnotite	$K(UO_2)_2(VO_4)_2 \cdot 3H_2O$
Zinc	Sphalerite	ZnS

Table 21.3	Some Ways Ore Deposits Form
Type of Ore Deposit	**Some Metals Found in This Type of Ore Deposit**
Crystal settling within cooling magma	Chromium, platinum, iron
Hydrothermal deposits (contact metamorphism, hydrothermal veins, disseminated deposits, hot-spring deposits)	Copper, lead, zinc, gold, silver, iron, molybdenum, tungsten, tin, mercury, cobalt
Pegmatites	Lithium, rare metals
Chemical precipitation in layers	Iron, manganese, copper
Placer deposits	Gold, tin, platinum, titanium
Concentration by weathering and ground water	Aluminum, nickel, copper, silver, uranium, iron, manganese, lead, tin, mercury

ing reactors in the United States well into the next century. Phosphorites and black shales contain millions of tons of low-grade uranium-oxide deposits as resources.

Alternative Sources of Energy

Several other sources may contribute enough energy in the future to help reduce the expected demand for oil, natural gas, coal, and uranium. *Hydroelectric power* contributes about 4% of the energy needs of the United States. Electricity is generated by turbines turned by water falling from dammed reservoirs. Hydroelectric power will probably not increase because most suitable rivers in the United States have already been dammed. Public pressure is growing to preserve most of the remaining undammed rivers in their wild state, despite the danger of floods. *Geothermal power* (chapter 11) may contribute substantially to our power needs, particularly if successful techniques are developed for tapping the heat of areas not marked by surface hot springs. Currently, geothermal power contributes only 0.2% of our energy use. *Solar power* and *wind power* may contribute to our needs in the future, particularly if improved methods of storing the energy are devised. At the present, a great effort is being made to improve the technology for collecting solar energy. More exotic forms of generating energy include harnessing *tidal power, wave power, ocean current power,* and the energy represented by *vertical temperature differences in the sea.* The

burning of *hydrogen from the dissociation of water* also may be developed. However, major technological problems are slowing the adoption of this method.

One great attraction of many of these sources of energy is that they are *renewable.* We do not use up sunlight, wind, or tides by harnessing their power. Widespread use of renewable resources in the next century may lessen the world's demand for fossil fuels.

Metals and Ores

The successful search for metals depends on finding **ores,** which are naturally occurring materials that can be profitably mined. Whether or not a mineral (or rock) is considered a metal ore depends on its chemical composition, the percentage of extractable metal, and the market value of the metal. The mineral hematite (Fe_2O_3), for example, is usually a good *iron ore* because it contains 70% iron by weight; this high percentage is profitable to extract at current prices for iron. Limonite ($Fe_2O_3 \cdot nH_2O$) contains less iron than hematite and hence is not as extensively mined. Even a mineral containing a high percentage of metal is not described as an *ore* if the metal is too difficult to extract or the site is too far from a market; profit is part of what defines an ore. As the prices of metals and the energy used to extract them fluctuate, so do the potential profits from minerals. Some of the common ore minerals are listed in table 21.2.

Origin of Metallic Ore Deposits

Table 21.3 summarizes most of the ways ore deposits form. Several of the processes have been discussed in earlier chapters.

Figure 21.12
Early-forming minerals such as chromite may settle through magma to collect in layers near the bottom of a cooling sill.

 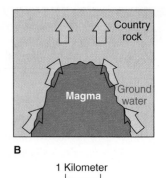

Figure 21.13
Two possible origins of hydrothermal fluids. (*A*) Residually concentrated magmatic water moves into country rock when magma is nearly all crystallized. (*B*) Ground water becomes heated by magma (or by a cooling solid pluton), and a convective circulation is set up.

Ores Associated with Igneous Rocks

Crystal Settling

Crystal settling occurs as early-forming minerals crystallize and settle to the bottom of a cooling body of magma (figure 21.12). This process was described under differentiation in chapter 3. The metal chromium comes from chromite ore bodies near the base of sills and other intrusions. Most of the world's chromium and platinum come from a single intrusion, the huge Bushveldt Complex in South Africa. In Montana, another huge Precambrian sill called the Stillwater Complex contains similar but lower grade deposits of these two metals.

Hydrothermal Fluids

Hydrothermal fluids, discussed in chapter 7, are the most important source of metallic ore deposits. The hot water and other fluids might be part of the magma itself, injected into the surrounding country rock during the last stages of magma crystallization (figure 21.13*A*; see also figure 7.21). Atoms of metals such as copper and gold, which do not fit into the growing crystals of feldspar and other minerals in the cooling pluton, would be concentrated residually in the remaining water-rich magma. Eventually a hot solution, rich in metals and silica (quartz is the lowest-temperature mineral on Bowen's reaction series), could move into the country rock to create ore deposits. Most hydrothermal ores are metallic sulfides, often mixed with milky quartz. The origin of the sulfur is widely debated.

Some hydrothermal fluids may be regular ground water, circulating in a convection pattern caused by the heat of the pluton. As water becomes heated by the pluton, it rises above it, drawing in new water from the sides to take its place. The new water then becomes heated and rises, and a continual circulation of water past the cooling pluton is set up (figure 21.13*B*). The hot water moving near and through the pluton could leach metals from the pluton and carry them upward to

be deposited elsewhere as ore. However the hydrothermal solutions form, they tend to create four general types of hydrothermal ore deposits: (1) contact metamorphic deposits, (2) hydrothermal veins, (3) disseminated deposits, and (4) hot-spring deposits.

Contact metamorphism can create ores of tungsten, copper, lead, zinc, silver, and other metals in country rock. The country rock may be completely or partially removed and replaced by ore (figure 21.14*A*). This is particularly true of limestone beds, which react readily with hydrothermal solutions. (The metasomatic addition of ions to country rock is described in chapter 7.) The ore bodies can be quite large and very rich.

Hydrothermal veins are narrow ore bodies formed along joints and faults (figure 21.14*B*). They can extend great distances from their apparent plutonic sources. Some extend so far that it is questionable whether they are even associated with plutons. The fluids can precipitate ore (and quartz) within cavities along the fractures and may also replace the wall rock of the fractures with ore. Hydrothermal veins form most of the world's great deposits of lead, zinc, silver, gold, tungsten, tin, mercury, and, to some extent, copper.

Hot solutions can also form *disseminated deposits* in which metallic sulfide ore minerals are distributed in very low concentration through large volumes of rock, both above and within a pluton (figure 21.14*C* and box 7.4). Most of the world's copper comes from disseminated deposits (also called *porphyry copper deposits* because the associated pluton is usually porphyritic). Along with the copper are deposited many other metals, such as lead, zinc, molybdenum, silver, and gold (and iron, though not in commercial quantities).

Where hot solutions rise to the earth's surface, *hot springs* form. Hot springs on land may contain large amounts of dissolved metals. Some California hot springs contain so much mercury that the water is unfit to drink. More impressive are hot springs on the sea floor (figure 21.14*D*), which can precipitate large mounds of metallic sulfides, sometimes in commercial quantities. We will look at submarine hot springs later in the chapter in connection with plate tectonics.

Figure 21.14

Hydrothermal ore deposits. (*A*) Contact metamorphism in which ore replaces limestone. (*B*) Ore emplaced in hydrothermal veins. (*C*) Disseminated ore within and above a pluton. (*D*) Ore precipitated around a submarine hot spring (size of ore deposit exaggerated). (*E*) Hydrothermal veins of milky quartz in granite, northern California.

Pegmatites (box 3.1) are another type of ore deposit associated with igneous rocks. They may contain important concentrations of minerals containing lithium, beryllium, and other rare metals, as well as gemstones such as emeralds and sapphires.

Ores Formed by Surface Processes

Chemical precipitation in layers is the most common origin for ores of iron and manganese. A few copper ores form in this way too. Banded iron ores, usually composed of alternating layers of iron minerals and chert, formed as sedimentary rocks in many parts of the world during the Precambrian, apparently in shallow, water-filled basins (figure 21.15). Later folding, faulting, metamorphism, and solution have destroyed many of the original features of the ore, so the origin of the ore is difficult to interpret. The water may have been fresh or marine, and the iron may have come from volcanic activity or deep weathering of the surrounding continents. The alternating bands may have been created by some rhythmic variation in volcanic activity, river runoff, basin water circulation, growth of organisms, or some other factor. Since banded iron ores are all Precambrian, their origin might be connected to an ancient atmosphere or ocean chemically different from today's.

Placer deposits in which streams have concentrated heavy sediment grains in a river bar are described in chapter 10. Wave action can also form placers at beaches. Placers include gold nuggets and dust, native platinum, diamonds and other gemstones, and worn pebbles or sand grains composed of the heavy oxides of titanium and tin.

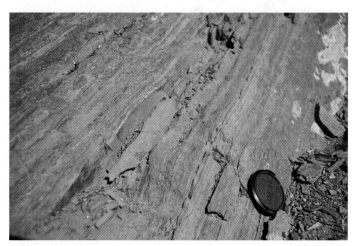

Figure 21.15

Banded iron ores of Precambrian age probably accumulated in shallow basins.

Ore deposits due to *concentration by weathering* were described in chapter 5. Aluminum (in bauxite) forms through weathering in tropical climates.

Another type of concentration by weathering is the *supergene enrichment* of disseminated ore deposits. The major ore

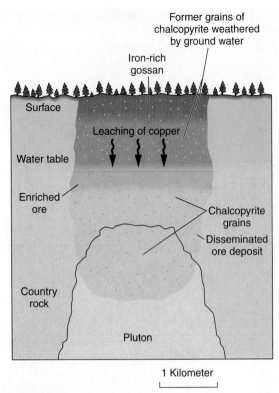

Figure 21.16

Supergene enrichment. Ground water leaches copper from upper part of disseminated deposit and precipitates it at or below the water table, forming rich ore.

mineral in a disseminated copper deposit is chalcopyrite, a copper-iron sulfide containing about 35% copper. Near the earth's surface, downward-moving ground water can leach copper and sulfur from the ore, leaving the iron behind (figure 21.16). At or below the water table the dissolved copper can react with chalcopyrite in the lower part of the disseminated deposit, forming a richer ore mineral such as chalcocite, which is about 80% copper:

$$3\,Cu^{++} \quad + \quad CuFeS_2 \quad \rightarrow \quad 2\,Cu_2S \quad + \quad Fe^{++}$$

| Copper dissolved in ground water | Chalcopyrite | Chalcocite | Iron in solution |

In this way copper is removed from the top of the deposit and added to the lower part. The ore below the water table may be several times richer than in the rest of the deposit (silver can move with the copper). The iron left behind at the earth's surface forms a rusty cap called a *gossan,* or "iron hat," which is a visible clue to the ore below.

Metal Ores and Plate Tectonics

Plate tectonics has provided a model that is important to geologists searching for new metallic ore deposits. The relationships between plate tectonics and ore deposits were described in chapter 19 and are summarized below.

Figure 21.17

Metallic ores form at hot springs in a rift valley. Previously formed ores are carried away on moving plates. (Ore deposit sizes are exaggerated.)

At *divergent* plate boundaries, metals form in rift valley hot springs due to the circulation of ground water, driven by magma. Nearly solid metal sulfide mounds precipitate from the hot water as it is cooled by sea water. The metals that accumulate in these mounds are predominantly iron, copper, and zinc, with smaller amounts of manganese, gold, and silver. Mining them is presently impractical because they are small and scattered on the sea floor.

Metallic ores that form at a divergent boundary are carried away from the ridge crest (figure 21.17) toward another plate boundary. Some *ophiolites* contain rich ore deposits that may have originally been metallic mounds. Ophiolites represent former crust and mantle created at spreading axes that has accreted onto continents (chapter 18). Some ophiolites also contain chromite ores within serpentinized ultramafic rock.

At *convergent* plate boundaries metallic ore deposits form in a variety of settings. Hot-spring deposits form on the flanks of andesitic volcanoes at *island arcs.* The ore sometimes collects as sedimentary layers in shallow basins above local magma bodies (figure 19.51). These usually contain more lead than the mounds that form along spreading axes. Rich *massive sulfide* ore deposits may have originally formed in this manner.

In western North America, metallic ore deposits tend to be distributed in broad belts that parallel past or present convergent plate boundaries. The type of ore deposit that forms seems to be related to the depth of subduction as described in chapter 19 (see figure 19.51). One model postulates that metallic minerals originally formed in a divergent plate boundary setting are carried down a subduction zone. Rock becomes hotter as it progresses down a subduction zone. The metals are remobilized and rise into the overlying continental crust. Metals that mobilized at relatively low temperatures rise closer to the plate boundary than those that mobilize ar higher temperatures. Other metals may also be derived directly from the continental crust or underlying mantle. Rising magma and associated hot water circulation redistribute and concentrate the ores into veins and other hydrothermal deposits.

Mining

Mining can be carried out on the earth's surface or underground (figure 21.19). Two forms of surface mines are *strip mines,* used for mining some beds of coal, and **open-pit**

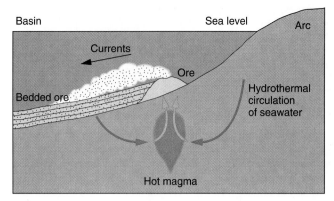

Figure 21.18

Metallic ores that form over hot springs on the flanks of island arcs are redistributed into layers in shallow basins.

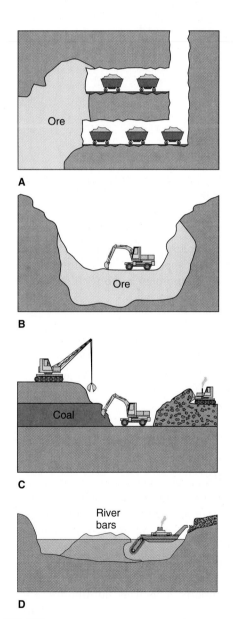

Figure 21.19

Types of mines. (*A*) Underground. (*B*) Open pit. (*C*) Strip. (*D*) Placer (being mined by a floating dredge).

mines, in which ore is exposed in a large excavation. Some geologists use **strip mine** and *open-pit mine* interchangeably. However, strip mines generally expose coal or another resource in a long band that is later filled in, while open-pit mines are roughly circular and are excavated to extract huge amounts of material. They are not usually refilled. **Placer mines** are surface mines in which valuable sediment grains are extracted from stream bar or beach deposits.

Environmental Effects

Some of the environmental problems associated with mining can be partially solved if care is taken. For example, in the past *waste rock* (or *tailings*) was routinely left in unsightly heaps and dumps (figure 21.20*A*). The excavations for strip mines and placer mines can be filled in with waste rock, leveled or graded,

A

B

Figure 21.20

(*A*) Waste rock piles from coal mining in the 1950s, Sweetwater County, Wyoming. (*B*) Subsidence caused by the collapse of an underground coal mine, Acme, Wyoming. Mining continued from 1900 to 1943 (1976 photo).

Photo *A* by H. E. Maide, U.S. Geological Survey; photo *B* by C. R. Dunrud, U.S. Geological Survey

Geologic Resources **537**

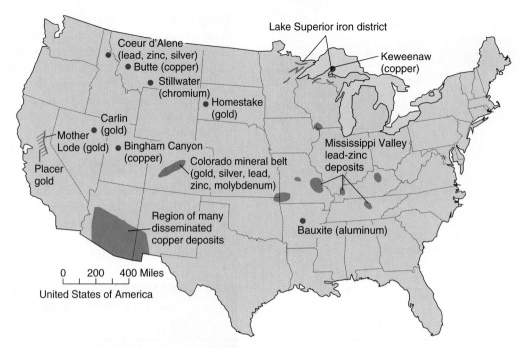

Figure 21.21

Some metallic ore localities mentioned in the text.

and then covered over with topsoil to restore the land to usable condition. In some cases crops can be grown on reclaimed land within one year after mining operations are completed. Open pits, being larger, are rarely filled in, for the filling cannot be done gradually while mining is in progress. Underground mines are sometimes back-filled with waste rock to prevent land *subsidence* after ore is removed. Figure 21.20*B* shows extensive subsidence caused by mine collapse.

One of the more difficult problems to deal with is *acid drainage* from mines, caused by ground water running or being pumped out of a mine. Sulfide ore minerals (table 21.2) are most often the source of the trouble. Ground water dissolves some of the sulfide, which oxidizes in air to form sulfuric acid. In some mines, expensive programs of holding and neutralizing drainage water in ponds or artificial wetlands prevent pollution of surface streams and harm to forests and wildlife. The worst problem is with long-abandoned mines that are still draining acid waters. Many of these may never be neutralized.

Some Important Metals

Iron

The basis for any modern, industrialized economy is *iron,* which is used to make steel. Iron and steel are used in a huge variety of products, from cast-iron frying pans to locomotives.

The major iron ore minerals are hematite and magnetite. Most of the iron in the United States comes from the region around Lake Superior (figure 21.21), most notably, Minnesota. The ores are banded iron ores of Precambrian age, typical of most iron ores in the world. Mining is done mostly by open-pit methods.

Copper

Less abundant is *copper,* another important metal for industry. More than half the copper used in the United States goes into electrical wire and equipment and one-third into the manufacture of brass, a copper-zinc alloy.

Most copper ores are sulfides. Chalcopyrite is the most important copper ore mineral. Some vein deposits of copper exist (as at Butte, Montana), but most major deposits are disseminated through large volumes of rock; so most copper mines are open pits (figure 21.22). Arizona, Utah, and a few other western states are the major producers in the United States (figure 21.21). The concentration of copper averages about 0.5% in most currently worked deposits; that is, 1 kg of copper is recovered for every 200 kg of rock processed.

Aluminum

Widely used in the United States, *aluminum* is consumed in the manufacture of beer and soft drink cans, airplanes, electrical cable, and many other products. The use of aluminum is increasing rapidly.

The ore of aluminum is bauxite, which forms under tropical weathering conditions. The United States has very little bauxite, so it imports 90% of its aluminum ore from tropical countries. The largest aluminum mine in the United States is in Arkansas (figure 21.21). Open-pit mining is the usual technique for extracting bauxite.

Figure 21.22

Open-pit copper mine, Morenci, Arizona.

Figure 21.23

Sluice box used to separate gold from gravel, Alaska.

Photo by D. J. Miller, U.S. Geological Survey

Lead

The main use of *lead* (79%) is in batteries. Substantial amounts are recycled, largely from automobile batteries.

The most important ore of lead is galena. Major deposits occur in Missouri, Idaho, Utah, and Colorado (figure 21.21). The Missouri deposits are mostly found in limestone beds; their precise origin is a matter of some controversy. The ore is mined both underground and from open pits. Deposits in Idaho occur mostly as veins and are usually mined underground.

Zinc

Widely used in industry, *zinc* is necessary for galvanizing and the manufacture of brass and other alloys.

The major zinc ore is sphalerite. As sphalerite usually is found closely associated with galena, most lead mines also extract zinc. Zinc occurs without lead in some areas, however.

Silver

Coins, tableware, jewelry, photographic film, and many other products are made of *silver*.

Silver, found as a native metal and in sulfide ores, is a common by-product of lead and copper mining. The lead-zinc mines of Idaho (the Coeur d'Alene district) are the largest silver producers in the United States (figure 21.21).

Gold

The rare and valuable metal *gold* is used in coins, jewelry, decoration, dentistry, electronics, and the space program. Gold bars are stored to back national currency, although this use is rapidly disappearing.

Gold is found as a native element in the form of nuggets and grains. In some parts of the world these are concentrated

Figure 21.24

Hydraulic mining for gold, Alaska.

Photo by T. L. Péwé, U.S. Geological Survey

in placer deposits (California's Gold Rush of 1849 was triggered by discoveries of placer gold). Gold nuggets, flakes, and dust can be separated from the other sediments by (1) *panning;* (2) *sluice boxes* (figure 21.23), which catch the heavy gold on the bottom of a box as gravel is washed through it; (3) *hydraulic mining* (figure 21.24), which washes gold-bearing gravel from a hillside into a sluice box; or (4) floating *dredges* (figure 21.25), which separate gold from gravel aboard a large barge, piling the spent gravel behind. When gold is found in hydrothermal veins associated with milky quartz, as it is in parts of Colorado and in California's Mother Lode (figure 21.21), it is mined underground. The largest underground gold mine in the United States is the Homestake Mine in South Dakota, where finely disseminated gold is extracted

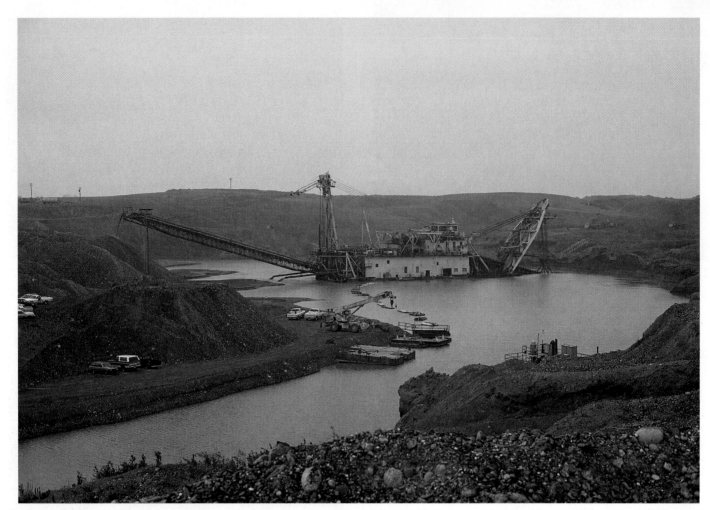

Figure 21.25

A gold dredge separates gold from gravel.

from folded metamorphic rock. A large amount of disseminated gold is mined in open pits near Carlin, Nevada, and at several other localities in Nevada and California.

The world has large reserves of iron and aluminum, moderate reserves of copper, lead, and zinc, and scanty reserves of gold and silver.

Other Metals

Many other metals are vital to our economy. *Chromium, nickel, cobalt, manganese, molybdenum, tungsten,* and *vanadium* are all important in the steel industry, particularly in the manufacture of specialty products such as stainless steel and armor plate. Most of these metals have other uses as well. *Tin* is used in solder and for plating steel in "tin cans." *Mercury* is used in thermometers, silent electrical switches, medical compounds, and batteries. *Magnesium* is used in aircraft and flashbulbs. *Titanium,* as strong as steel but weighing half as much, is used in aircraft. *Platinum* is used in catalytic converters to clean automobile exhaust.

Nonmetallic Resources

Nonmetallic resources are earth resources that are not mined to extract a metal or as a source of energy. Most rocks and minerals contain metals, but when nonmetallic resources are mined it is usually to use the rock (or mineral) as is (for example, using gravel and sand for construction projects); whereas, metallic ores are processed to extract metal. With the exception of the gemstones such as diamonds and rubies, nonmetallic resources do not have the glamour of many metals or energy resources. Nonmetallic resources are generally inexpensive and are needed in large quantities (again, except for gemstones). The large demand and low unit price means that these resources are best taken from local sources. Transportation over long distances would add significantly to the cost.

Construction Materials

Sand and *gravel* are both needed for the manufacture of concrete for building and highway construction. Sand is also used

Figure 21.26

Sand and gravel pit in a glacial esker, near Saranac Lake, New York.

Courtesy Ward's Natural Science Est., Inc., Rochester, NY

in mortar, which holds bricks and cement blocks together. The demand for sand and gravel in the United States has more than doubled in the last twenty-five years. Sand dunes, river channel and bar deposits, glacial outwash, and beach deposits are common sources for sand and gravel. Cinder cones are mined for "gravel" in some areas. Sand and gravel are ordinarily mined in open pits (figure 21.26).

Stone refers to rock used in blocks to construct buildings or crushed to form roadbed. Most stone in buildings is limestone or granite, and most crushed stone is limestone. Huge quantities of stone are used each year in the United States. Stone is removed from open pits called *quarries* (figure 21.27).

Limestone has many uses other than building stone or crushed roadbed. Cement, used in concrete and mortar, is made from limestone. Pulverized limestone is in demand as a soil conditioner and is the principal ingredient of many chemical products.

Fertilizers and Evaporites

Fertilizers (phosphate, nitrate, and potassium compounds) are extremely important to agriculture today, so much so that they are one of the few nonmetallic resources transported across the sea. *Phosphate* is produced from phosphorite, a sedimentary rock formed by the accumulation and alteration of the remains of marine organisms. Major phosphate deposits in the United States are in Idaho, Wyoming, and Florida. *Nitrate* can form directly as an evaporite deposit but today is usually made from atmospheric nitrogen. *Potassium* compounds are often found as evaporites. Figure 21.28 shows an underground mine for potassium compounds.

Rock salt is coarsely crystalline halite formed as an evaporite. Salt beds are mined underground in New York, Ohio, and Michigan; underground salt domes are mined in Texas and

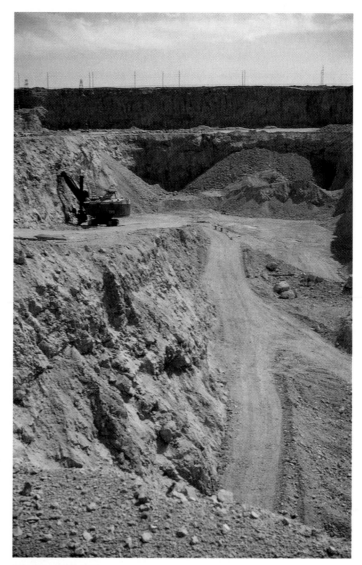

Figure 21.27

A limestone quarry in northern Illinois. The horizon marks the original land surface before the rock was removed.

Louisiana. (Some salt is also extracted from sea water by evaporation.) Rock salt is used in many ways—de-icing roads in winter, preserving food, as table salt, and in manufacturing hydrochloric acid and sodium compounds for baking soda, soap, and other products. Rock salt is heavily used by industry.

Gypsum forms as an evaporite. Beds of gypsum are mined in many states, notably California, Michigan, Iowa, and Texas. Gypsum, the essential ingredient of plaster and wallboard (Sheetrock), is used mainly by the construction industry, although there are other uses.

Sulfur occurs in bright-yellow deposits of elemental sulfur. Most of its commercial production comes from the cap rock of salt domes (see Box 21.1). Sulfur is widely used in agriculture as a fungicide and fertilizer and by industry to manufacture sulfuric acid, matches, and many other products.

Figure 21.28

Loading potassium minerals in an underground mine, Eddy County, New Mexico.

Photo by H. I. Smith, U.S. Geological Survey

Other Nonmetallics

Gemstones (called *gems* when cut and polished) include precious stones such as diamonds, rubies, emeralds, and sapphires and semiprecious stones such as beryl, garnet, jade, spinel, topaz, turquoise, and zircon. Gems are used for jewelry, bearings and abrasives (most are above 7 on Mohs' scale of hardness). Diamond drills and diamond saws are used to drill and cut rock. Old watches and other instruments often have hard gems at bearing points of friction ("17-jewel watches"). Gemstones are often found in pegmatites or in close association with other igneous intrusives. Some are recovered from placer deposits.

Asbestos is a fibrous variety of serpentine or chain silicate minerals. The fibers can be separated and woven into fireproof fabric used for firefighters' clothes and theater curtains. It is also used in manufacturing ceiling and sound insulation, shingles, and brake linings, although the use of asbestos is being rapidly curtailed because of concern about its connection with lung cancer (Box 2.3). The United States produces little asbestos, mostly from belts of serpentinized ultramafic rocks in the Appalachians and the Pacific Coast states. Large amounts are mined in Canada, chiefly in Quebec. *Talc,* used in talcum powder and other products, is often found associated with asbestos (see Box 20.1).

Other nonmetallic resources are important. *Mica* is used in electrical insulators. *Barite* ($BaSO_4$), because of its high specific gravity, is used to make heavy drilling mud to prevent oil gushers. *Borates* are boron-containing evaporites used in fiberglass, cleaning compounds, and ceramics. *Fluorite* (CaF_2) is used in toothpaste, Teflon finishes, and steel-smelting. *Clays* are used in ceramics and as filters and absorbents. *Diatomite* is used in swimming pool filters and to filter out yeast in beer and wine. *Glass sand,* which is over 95% quartz, is the main component of glass. *Graphite* is used in foundries, lubricants, steel-making, batteries, and pencil "lead."

Some Future Trends

Ocean mining, now rather uncommon, will increase in the near future. Mining of *manganese nodules* from the deep sea floor (see chapter 18) could provide the United States with substantial amounts of copper and nickel, together with far more manganese than United States industry can consume. The copper content of many nodules is four to five times higher than most deposits on land. *Metallic brines and deposits* of the Red Sea type may be a source of several metals in the future.

Several tools are of great help in mineral exploration on land. Highly sophisticated *geochemical tests* of soils and soil gases point to ore bodies underground. *Geophysical techniques* continue to be refined for resource exploration. Satellites and space-shuttle crews photograph the earth's surface in many different wavelengths of energy, and careful analysis of satellite imagery is proving to be of great help in prospecting. The economic returns from the satellite photography program will far outweigh its costs.

As the relation of plate tectonics to the distribution of resources such as metals and petroleum becomes clearer, selecting areas for exploring for these materials should become more sophisticated.

The Human Perspective

There is a tendency for humans to take a one-sided view toward geologic resources and the problems created by the extraction and transportation of those resources. The conflicts are between (1) maintaining or increasing our standard of living and raising the quality of life not only for ourselves, but for poverty-stricken people throughout the world; (2) maintaining the environment; and (3) making sure that we do not deprive future generations of the resources that sustain us.

The extreme position for each of the above could be stated as follows: (1) Extreme exploiter: "Let's mine what we can and make ourselves as rich as possible now. Technological breakthroughs will provide for future generations. Damage to the environment due to extraction is insignificant (or it's where it doesn't bother me)." (2) Extreme environmentalist: "Any mine or oil well does environmental damage and therefore should not be permitted. We can maintain our lifestyles by recycling or by leading less technologically dependent lifestyles." (3) Extreme conservationist: "Let's not mine anything now because there are many future generations that need to rely on these resources."

You probably agree that none of the three extreme positions is reasonable. The middle ground among the three is where almost everyone thinks we should be. The challenge is deciding where in the middle ground. Should we lean toward more exploitation and away from environmental concerns in order that underdeveloped countries can raise their standards of living? Should we minimize mining and energy consumption so as to reduce any impact on air pollution or wildlife? Our hope is that we can at least understand the perspective of those who may disagree to strike a balance and try to deal with each case with enlightenment. Your understanding of geology is an important step in your being able to help resolve moral dilemmas that we face in which there is no ideal solution.

21.2 IN GREATER DEPTH

Substitutes, Recycling, and Conservation

Substitutes for many geologic resources now exist, and others will be found. Aluminum is replacing the more expensive copper in many electrical uses, particularly in transmission lines. Glass fibers also are replacing copper for telephone lines. Aluminum has largely replaced tin-coated steel for beverage containers. Cotton and wool use could increase, replacing polyester and other petroleum-based synthetic fibers in clothes.

Suitable substitutes, however, seem unlikely for some resources. Nothing yet developed can take the place of steel in bridges or mercury in thermometers. Cobalt is vital for strong, permanent magnets. Although substitutes may help prolong the life of the supplies of some resources, they are not available for many others.

Recycling helps augment the supply of some resources such as aluminum, gold, silver, lead, plastics, and glass. No resource, however, receives even half its supply from recycling. Increased volunteer recycling on the part of the public and waste reclaiming of urban trash could increase these percentages. New ore will always be needed, however, because some uses of products prevent the material from being recyclable. A steel can that rusts away beside a road will never be recycled. The iron oxides are scattered in low percentage in the soil and can never be recovered. Many resources, such as petroleum and coal, are consumed during use and cannot be recycled.

Conservation of scarce resources is extremely important. The United States' use of oil from 1979 to 1985 declined as a result of conservation efforts such as the 55 mile-per-hour speed limit, more fuel-efficient automobiles, the upgrading of insulation in buildings, and the elimination of unneeded heating and lighting. But in 1986 oil use rose sharply as its price declined. Politically caused shortages and gluts of oil can obscure the fact that the supply of oil in the United States is severely limited. In the difficult times ahead as the United States faces up to this limit, the need for conserving fuel should be repeatedly stressed. The price of gasoline will rise substantially as the United States becomes increasingly dependent on foreign oil. Conservation of metals, particularly those imported in large quantities, will become increasingly important. Smaller automobiles and more durable appliances can help conserve metals.

Summary

Geologic resources include energy resources, metals, and nonmetallic resources. All are nonrenewable, except ground water.

Reserves are known deposits that can be legally and economically recovered now—the short-term supply. *Resources* include reserves as well as other known and undiscovered deposits that may be economically extracted in the future.

Petroleum (*oil* and *natural gas*) supplies almost two-thirds of the energy used by the United States.

The occurrence of oil and natural gas is limited to regions having these conditions together: *source rocks, reservoir rocks, traps,* such as anticlines, faults, stratigraphic traps, and salt domes, and *thermal maturity* from burial.

Natural gas, *heavy crude, oil sand,* and *oil shale* may all help replace liquid petroleum in the future. Most of these resources are in the western states.

The United States has huge *coal* reserves, enough for centuries of use at the present rate. Coal, now used mostly for generating electricity, will probably be used more widely in the future as oil runs out.

The United States has ample *uranium* for its reactor program, mostly in sandstones in western states.

Metallic ores, which can be profitably mined, are often associated with igneous rocks, particularly their hydrothermal fluids, which can form *contact metamorphic deposits, hydrothermal veins, disseminated deposits,* and submarine *hot-spring deposits.* Iron occurs in sedimentary layers, and aluminum ores form from weathering.

Metallic ores form from hot springs at divergent plate boundaries, on the flanks of island arcs, and in belts on the edges of continents above subduction zones.

Ores are mined *underground* and also at the earth's surface in *strip mines, open-pit mines,* and *placer mines.*

Metals are vital to an industrial economy, particularly *iron* for steel production and *copper* for electrical equipment.

Nonmetallic resources such as *sand and gravel* and *limestone* for crushed rock and cement are used in huge quantities. *Fertilizers, rock salt, gypsum, sulfur,* and *clays* are also widely used.

Substitutes, recycling, and conservation can help cut consumption of some resources but will not eliminate the need for finding new deposits.

Deep-ocean mining and increasingly sophisticated exploration techniques will help supply some of our future resource needs.

crude oil 524

geologic resources 522

heavy crude 528

natural gas 524

nonrenewable resource 522

oil field 525

oil pool 524

oil sand 528

oil shale 528

oil trap 524

open-pit mine 536

ore 533

petroleum 523

placer mine 537

reserves 523

reservoir rock 524

resources 523

source rock 524

strip mine 530, 537

Testing Your Knowledge

Use the questions below to prepare for exams based on this chapter.

1. Name the three major classes of geologic resources. Give four examples of each class.

2. Discuss the United States' supplies and potential use of natural gas, heavy crude, oil sand, oil shale, and uranium.

3. List in decreasing order of use the energy resources used in the United States. Discuss possible future trends in this ranking.

4. What geologic conditions are necessary for the accumulation of oil and natural gas?

5. Differentiate between *reserves* and *resources*. Can reserves be increased? Can resources be increased?

6. Compare oil reserves with coal reserves. What might this indicate for the future use of each?

7. Describe several ways in which ore deposits form. Which are the most important?

8. Describe four ways in which resources are mined.

9. Discuss the environmental effects of oil extraction and coal mining.

10. Discuss common uses for iron, copper, lead, zinc, and aluminum.

11. Describe the potential of substitutes, conservation, recycling, and deep-ocean mining for meeting an increasing need for geologic resources.

12. All of the following are nonrenewable resources except (a) ground water (b) oil (c) coal (d) iron

13. The major source of energy of the United States is (a) natural gas (b) coal (c) oil (d) nuclear power

14. Which is not a type of coal? (a) peat (b) lignite (c) bituminous (d) anthracite

15. Which metal would most likely be found in an ore deposit formed by crystal settling? (a) copper (b) gold (c) silver (d) chromium

16. Which metal would not be found in hydrothermal veins? (a) aluminum (b) lead (c) zinc (d) silver (e) gold

17. Metal ore deposits have been found at all these tectonic settings except (a) mid-oceanic ridges (b) island arcs (c) subduction zones (d) mantle plumes

18. The main use of lead is in (a) coins (b) gasoline (c) batteries (d) pencils

19. What factors can increase reserves of earth resources (choose all that apply) (a) extraction of the resource (b) new discoveries (c) price controls (d) new mining technology

20. The largest use of sand and gravel is (a) glassmaking (b) extraction of quartz (c) construction (d) ceramics

21. Oil accumulates when the following conditions are met (choose all that apply) (a) source rock where oil forms (b) permeable reservoir rock (c) impermeable oil trap (d) shallow burial

22. Coal forms (a) by crystal settling (b) through hydrothermal processes (c) by compaction of plant material (d) on the ocean floor

Expanding Your Knowledge

1. Many underdeveloped countries would like to have the standard of living enjoyed by the United States, which has 6% of the world population and uses 15% to 40% of the world production of many resources. As these countries become industrialized, what happens to the world demand for geologic resources? Where will these needed resources come from?

2. If driven 12,000 miles per year, how many more gallons of gasoline per year does a sports car or pickup truck rated at 12 miles per gallon use than a minicompact car rated at 52 mpg? Over 5 years how much more does it cost to buy gasoline at $1.20 per gallon for the low-mileage car? At $4 per gallon (the price in many European countries)?

Brobst, D. A., and W. P. Pratt. 1973. *United States mineral resources.* U.S. Geological Survey Professional Paper 820.

Cameron, E. N. 1986. *At the crossroads—the mineral problems of the United States.* New York: John Wiley & Sons.

Craig, J. R., D. J. Vaughan, and B. J. Skinner. 1988. *Resources of the Earth.* Englewood Cliffs, N.J.: Prentice-Hall.

Dixon, C. J. 1979. *Atlas of economic mineral deposits.* Ithaca, N.Y.: Cornell University Press.

Guilbert, J. M., and C. F. Park. 1986. *The geology of ore deposits.* New York: W. H. Freeman.

Hodges, C. A. 1995. Mineral resources, environmental issues, and land use. *Science* 268: 1302–1312.

Jensen, M. L., and A. M. Bateman. 1981. *Economic mineral deposits.* 3d ed. New York: John Wiley & Sons.

Keary, P., and F. J. Vine. 1990. Plate tectonics and economic geology. In *Global Tectonics.* London: Blackwell Scientific Publications.

McKelvey, V. E. 1986. *Subsea mineral resources.* U.S. Geological Survey Professional Paper 1689-A.

North, F. K. 1985. *Petroleum geology.* Boston: Allen & Unwin.

Park, C. F., Jr. 1968. *Affluence in jeopardy: Minerals and the political economy.* New York: Freeman, Cooper.

———. 1975. *Earthbound: Minerals, energy, and man's future.* New York: Freeman, Cooper.

Rona, P. A. 1986. Mineral deposits from sea-floor hot springs. *Scientific American* 254(1), 84–93.

Sawkins, F. J. 1984. *Metal deposits in relation to plate tectonics.* New York: Springer-Verlag.

Selley, R. C. 1985. *Elements of petroleum geology.* New York: W. H. Freeman.

U.S. Bureau of Mines (now part of U.S. Geological Survey). *Minerals yearbook* and *Mineral commodity summary* (published annually). Washington, D.C.: U.S. Government Printing Office.

Minerals and Materials Information. U.S. Geologic Survey. This is the electronic version of the Minerals Yearbook and other U.S. Government publications. For information check:

http://minerals.er.usgs.gov/minerals/
pubs/mmi_cd.html

http://www.microserve.net/~doug/
Mine Net. Gateway to World Wide Mining Information.

http://www.klws.com/gold/gold.html
Gold prospecting

http://www.NRCan.gc.ca
Natural Resources Canada site includes mineral and energy resource information.

http://minerals.er.usgs.gov
The U.S. Geological Survey maintains a site for mineral resources with links to domestic and international sites.

http://www.eia.doe.gov/
The U.S. Energy Information Administration site.

http://cg.ensmp.fr/pointeurs/petroleum.
html
Petroleum Internet resources. A website with a huge number of links related to petroleum.

Interacting with Journey Through Geology CD-ROM

Expand your knowledge of the concepts presented in this chapter by using the CD-ROM to answer the following questions.

1. Pegmatite Pocket module: What resource do miners seek in pegmatites?

   What kind of plate boundary would be likely to create pegmatites?

2. Spreading Ridges, Action at Spreading Ridges, Visit a Black Smoker: What causes minerals to precipitate at black smokers? What kind of plate boundary produces black smokers? What kinds of ores might someday be mined here?

Appendix A

Identification of Minerals

Each mineral is identified by a unique set of physical or chemical properties. To determine some of these properties requires specialized equipment and techniques. Most common minerals, however, can be distinguished from one another by tests involving simple observations. Cleavage is an especially useful property. If cleavage is present, you should determine the number of cleavage directions, estimate the angles between cleavage directions, and note the quality of each direction of cleavage. Other easily performed tests and observations check for hardness (abbreviated H), luster, and color, and determining crystal form (if present). A simple chemical test can be made using dilute hydrochloric acid to see if the mineral effervesces.

The identification tables included here can be used to identify the most common minerals (the rock-forming minerals) and some of the most common ore minerals. For identifying less common minerals, refer to one of the books or websites on mineralogy listed at the end of chapter 2. Mineral identification takes practice, and you will probably want to verify your mineral identifications with a geology instructor.

Because the common rock-forming minerals are the ones you are most likely to encounter, we have included a simple key for identifying them. The key is based on first determining whether or not the mineral is harder than glass and then checking other properties that should lead to identification of the mineral. You should verify your identification by seeing whether other properties of your sample correspond to those listed for the mineral in table A.1.

Ore minerals are usually distinctive enough that a key is unnecessary. To identify an ore mineral, read through table A.2 and determine which set of properties best fits the unknown mineral.

Key for Identifying Common Rock-Forming Minerals

Determine whether a fresh surface of the mineral is harder or softer than glass. If you can scratch the mineral with a knife blade, the mineral is softer than glass.

I. Harder than glass—knife will not scratch mineral. (If softer than glass, go to II.)

  A. Determine if cleavage is present or absent (this may require careful examination). If cleavage is absent proceed below; if cleavage is present, proceed to B.
    1. Vitreous luster
      a. Olive green or brown—*olivine*
      b. Reddish brown or in equidimensional crystals with twelve or more faces—*garnet*
      c. Usually light-colored or clear—*quartz*
    2. Metallic luster
      a. Bright yellow—*pyrite*
    3. Greasy or waxy luster
      a. Mottled green and black—*serpentine*
  B. Cleavage present. Determine the number of directions of cleavage in an individual grain or crystal.
    1. Two directions, good, at or near 90°—*feldspar*
      a. If striations are visible on cleavage surfaces—*plagioclase*
      b. If pink or salmon-colored—*potassium feldspar* (or *orthoclase*)
      c. If white or light gray without striations, it could be either type of feldspar
    2. Two directions, fair, at 90°
      a. Dark green to black—*pyroxene* (usually augite)
    3. Two directions, excellent, not near 90°
      a. Dark green to black—*amphibole* (usually hornblende)

II. Softer than glass—knife scratches mineral
  A. No cleavage detectable
    1. Earthy luster, in masses too fine to distinguish individual grains—*clay group* (for instance, *kaolinite*)
  B. Cleavage present
    1. One direction
      a. Perfect cleavage in flexible sheets—*mica:*
      Clear or white—*muscovite mica*
      Black or dark brown—*biotite mica*
    2. Three directions
      a. All three perfect and at 90° to each other (cubic cleavage)—*halite*
      b. All three perfect and not near 90° to each other:
      If effervesces in dilute acid—*calcite*
      If effervesces in dilute acid only after being pulverized—*dolomite*

Name (mineral groups shown in capitals)	Chemical Composition	Chemical Group	Diagnostic Properties	Other Properties
AMPHIBOLE (A mineral group in which *hornblende* is the most common member.)	$XSi_8O_{22}(OH)_2$ (*X* is a combination of Ca, Na, Fe, Mg, Al)	Chain silicate	2 good cleavage directions at 60° (120°) to each other.	H = 5–6 (barely scratches glass). Hornblende is dark green to black; tends to form in needle-like or elongate crystals; vitreous luster.
Augite (see Pyroxene)				
Biotite (see Mica)				
Calcite	$CaCO_3$	Carbonate	3 excellent cleavage directions, *not* at right angles (they define a rhombohedron). H = 3. Effervesces vigorously in weak acid.	Usually white, gray, or colorless; vitreous luster. Clear crystals show double refraction.
CLAY MINERALS (*Kaolinite* is a common example of this large mineral group.	Compositions include $XSi_4O_{10}(OH)_8$ (*X* is Al, Mg, Fe, Ca, Na, K)	Sheet silicate	Generally microscopic crystals. Masses of clay minerals are softer than fingernail. Earthy luster. Claylike smell when damp.	Seen as a chemical weathering product of feldspars and most other silicate minerals. A constituent of most soils.
Dolomite	$CaMg(CO_3)_2$	Carbonate	Identical to calcite (rhombohedral cleavage, H = 3) except effervesces in weak acid only when pulverized.	Usually white, gray, or colorless. Vitreous luster.
FELDSPAR (Most common group of minerals.) The group includes:	Framework	Framework silicates	H = 6 (scratch glass). 2 good cleavage directions at about 90° to each other	Vitreous luster but surface may be weathered to clay, giving an earthy luster. Perfect crystal, shaped like an elongated box.
Potassium feldspar (orthoclase)	$KAlSi_3O_8$		White, pink, or salmon-colored.	Never has striations on cleavage surfaces.
Plagioclase (sodium and calcium feldspar)	Mixture of: $CaAl_2Si_2O_8$ and $NaAlSi_3O_8$		White, light to dark gray, rarely other colors. *May* have striations on cleavage surfaces.	Calcium-rich varieties generally a darker gray and may show an iridescent play of colors.
GARNET	$XSiO_4$ (*X* is a combination of Ca, Mg, Fe, Al, Mn)	Isolated silicate	No cleavage. Usually reddish brown. Tends to occur in perfect equidimensional crystals, usually 12 sided. H = 7.	Rarely yellow, green, or black. Usually found in metamorphic rocks. Vitreous luster.
Gypsum	$CaSo_4 \cdot 2H_2O$	Sulfate	H = 2. 1 good and 2 perfect cleavage directions. Vitreous or silky luster.	Clear, white, or pastel colors. Flexible cleavage fragments.

Table A.1 *Continued*

Name (mineral groups shown in capitals)	Chemical Composition	Chemical Group	Diagnostic Properties	Other Properties
Halite	NaCl	Halide	3 excellent cleavage directions at 90° to each other (cubic). H = 2 1/2. Salty taste. Soluble in water.	Usually clear or white.
Hematite (*see* Ore mineral table)				
Hornblende (*see* Amphibole)				
Kaolinite (*see* Clay)				
MICA The group includes:	$K(X)(AlSi_3O_{10})(OH)_2$	Sheet silicate	1 perfect cleavage direction (splits easily into flexible sheets).	H = 2–3. Vitreous luster.
Biotite	(*X* is Mg, Fe, and Al)		Black or dark brown.	
Muscovite	(*X* is Al)		White or transparent.	
Olivine	X_2SiO_4 (*X* is Fe, Mg)	Isolated silicate	No cleavage. Generally olive green or brown. H = 6–7 (scratches glass). Vitreous luster.	Usually as small grains in mafic or ultramafic igneous rocks.
Orthoclase (*see* Feldspar)				
Plagioclase (*see* Feldspar)				
Pyrite ("fool's gold")	FeS_2	Sulfide	H = 6 (scratches glass). Bright, yellow, metallic luster. Black streak.	Commonly occurs as perfect crystals: cubes or crystals with five-sided faces. Weathers to brown.
PYROXENE (A mineral group; *Augite* is most common member.)	$XSiO_3$ (*X* is Fe, Mg, Al, Ca)	Chain silicate	2 fair cleavage directions at 90° to each other.	H = 6. Augite is dark green to black. Vitreous luster; usually stubby crystals.
Quartz	SiO_2	Framework silicate	H = 7. No cleavage. Vitreous luster. Does not weather to clay.	Almost any color but commonly white or clear. Good crystals have six-sided "column" with complex "pyramid" on top.
Serpentine	$Mg_6Si_4O_{10}(OH)_8$	Sheet silicate	Hardness variable but softer than glass. Mottled green and black. Greasy luster. Fractures along smooth curved surfaces.	Sometimes fibrous (asbestos).

	Table A.2	Diagnostic Properties of the Most Common Ore Minerals	
Name	**Chemical Composition**	**Diagnostic Properties**	**Other Properties**
Azurite	$Cu_3(CO_3)_2(OH)_2$	Azure blue; effervesces in weak acid.	H = 3-4.
Bauxite	$Al_2O_3 \cdot nH_2O$	Earthy luster. A variety of clay. Generally pea-sized spheres included in a fine-grained mass.	
Bornite	Cu_3FeS_4	Metallic luster, tarnishes to iridescent purple color.	Gray streak; H = 3 (softer than glass).
Chalcopyrite	$CuFeS_2$	Metallic luster, brass-yellow. Softer than glass.	Black streak.
Cinnabar	HgS	Scarlet red, bright red streak.	Softer than glass. Generally an earthy luster.
Galena	PbS	Metallic luster, gray; 3 directions of cleavage at 90° (cubic). High specific gravity.	Softer than glass; gray streak.
Gold	Au	Metallic luster, yellow. H = 3 (softer than glass, can be pounded into thin sheets, easily deformed).	Yellow streak; high specific gravity.
Halite	$NaCl$	Salty taste; 3 cleavage directions at 90° (cubic).	Clear or white; easily soluble in water.
Hematite	Fe_2O_3	Red-brown streak.	Either in earthy reddish masses or in metallic, silver-colored flakes or crystals.
Limonite	$Fe_2O_3 \cdot nH_2O$	Earthy luster; yellow-brown streak.	Yellow to brown color; softer than glass.
Magnetite	Fe_3O_4	Metallic luster, black; magnetic.	Harder than glass; black streak.
Malachite	$Cu_2(CO_3)(OH)_2$	Bright-green color and streak.	Softer than glass; effervesces in weak acid.
Sphalerite	ZnS	Brown to yellow color; 6 directions of cleavage.	Lusterlike resin; yellow or cream-colored streak; softer than glass.
Talc	$Mg_3Si_4O_{10}(OH)_2$	White, gray, or green; softer than fingernail (H=1).	Greasy feel.

Appendix B
Identification of Rocks

Igneous Rocks

Igneous rocks are classified on the basis of texture and composition. For some rocks, texture alone suffices for naming the rock. For most igneous rocks, composition as well as texture must be taken into account. Ideally, the mineral content of the rock should be used to determine composition; but for fine-grained igneous rocks, accurate identification of minerals may require a polarizing microscope or other special equipment. In the absence of such equipment, we rely on the color of fine-grained rocks and assume the color is indicative of the minerals present.

To identify a common igneous rock, use either table 3.1 or follow the key given below.

Key for Identifying Common Igneous Rocks

I. What is the texture of the rock?
  A. Is it glassy (a very vitreous luster)? If so, it is *obsidian,* regardless of its chemical composition. Obsidian exhibits a pronounced conchoidal fracture.
  B. Does it have a frothy appearance? If so, it is *pumice.* Pumice is light in weight and feels abrasive (it probably will float on water).
  C. Does it have angular fragments of rock embedded in a volcanic-derived matrix? If so, it is a *volcanic breccia.* If the precise nature of the rock fragments and matrix can be identified, modifiers may be used; for instance, the rock may be an *andesite* breccia or a *rhyolite* breccia.
  D. Is the rock composed of interlocking, very coarse-grained minerals? (The minerals should be more than 1 centimeter across.) If so, the rock is a *pegmatite.* Most pegmatites are mineralogically equivalent to granite, with feldspars and quartz being the predominant minerals.
  E. Is the rock entirely coarse-grained? (That is, does it have an interlocking crystalline texture in which nearly all grains are more than 1 mm across?) If so, go to part II of this key.
  F. Is the rock *entirely* fine-grained? (Are grains less than 1 mm across or too fine to distinguish with the naked eye?) If so, go to part III of this key.
  G. Is the matrix fine-grained with some coarse-grained minerals visible in the rock? If so, go to part III and add the adjective *porphyritic* to the name of the rock.
II. Igneous rocks composed of interlocking coarse-grained minerals.
  A. Is quartz present? If so, the rock is a *granite.*
    Confirmation: Granite should be composed predominantly of feldspar—generally white, light gray, or pink (indicating high amounts of potassium or sodium in the feldspar). Rarely are there more than 20% ferromagnesian minerals in a granite.
  B. Are quartz and feldspar absent? If so, the rock should be composed entirely of ferromagnesian minerals and is *ultramafic.*

Confirmation: Identify the minerals as being olivine or pyroxene (or less commonly, amphibole or biotite).
  C. Does the rock have less than 50% feldspar and no quartz? If so, the rock should be a *gabbro.*
    Confirmation: Most of the rock should be ferromagnesian minerals. Plagioclase can be medium or dark gray. There would be no pink feldspars.
  D. Is the rock composed of 30% to 60% feldspar (and no quartz)? If so, the rock is a *diorite.*
    Confirmation: Feldspar (plagioclase) is usually white to medium gray but never pink.
III. Igneous rocks that are fine-grained.
  A. Can quartz be identified in the rock? If so, the rock is a *rhyolite.*
  B. If the rock is too fine-grained for you to determine whether quartz is present but is white, light gray, pink, or pale green, the rock is most likely a *rhyolite.*
  C. Is the rock composed predominantly of ferromagnesian minerals? If so, the rock is *basalt.*
  D. If the rock is too fine-grained to identify ferromagnesian minerals but is black or dark gray, the rock is probably a *basalt.*
    1. Does the rock have rounded holes in it? If so, it is a *vesicular basalt* or *scoria.*
  E. Is the rock composed of roughly equal amounts of white or gray feldspar and ferromagnesian minerals (but no quartz)? If so, the rock is an *andesite.*
    Confirmation: Most andesite is porphyritic, with numerous identifiable crystals of white or light gray feldspar and lesser amounts of hornblende crystals within the darker, fine-grained matrix. Andesite is usually medium to dark gray or green.

Sedimentary Rocks

The following key shows how sedimentary rocks are classified on the basis of texture and composition. The descriptions of the rocks in the main body of the text provide additional information, such as common rock colors. *Equipment* needed for identification of sedimentary rocks includes a bottle of dilute hydrochloric acid, a hand lens or magnifying glass, a millimeter scale, a glass plate for hardness tests, and a pocketknife or rock hammer.

Begin by testing the rock for carbonate minerals by applying a small amount of dilute hydrochloric acid (0.1 molar HCl) to the surface of the rock (CAUTION—dilute HCl can burn eyes and clothing; use only with supervision).

1. The rock does not effervesce (fizz) in acid, or effervesces weakly, but when powdered by a knife or hammer, the powder effervesces strongly. If so, the rock is *dolomite.*

2. The rock does not effervesce at all, even when powdered, or effervesces only in some places, such as the cement between grains. Go to part I of this key.

3. The rock effervesces strongly. The rock is *limestone.* Go to part II of this key to determine limestone type.

I. With a hand lens or magnifying glass, determine if the rock has a clastic texture (grains cemented together) or a crystalline texture (visible, interlocking crystals).

A. If clastic:
1. Most grains are more than 2 mm in diameter.
   a. Angular grains *sedimentary breccia.*
   b. Rounded grains—*conglomerate.*
2. Most grains are between 1/16 and 2 mm in diameter. Rock feels gritty to the fingers. *Sandstone.*
   a. More than 90% of the grains are quartz—*quartz sandstone.*
   b. More than 25% of the grains are feldspar—*arkose.*
   c. More than 25% of the grains are fine-grained rock fragments, such as shale, slate, and basalt—*lithic sandstone.*
   d. More than 15% of the rock is fine-grained matrix—*graywacke.*
3. Rock is fine-grained (grains less than 1/16 mm in diameter). Feels smooth to fingers.
   a. Grains visible with a hand lens—*siltstone.*
   b. Grains too small to see, even with a hand lens.
      1. Rock is laminated, fissile—*shale.*
      2. Rock is unlayered, blocky—*mudstone.*

B. If crystalline:
1. Crystals fine to coarse, hardness of 2—*rock gypsum.*
2. Coarse crystals that dissolve in water—*rock salt.*

C. Hard to determine if clastic or crystalline:
1. Very fine-grained, smooth to touch, conchoidal fracture, hardness of 6 (scratches glass), nonporous—*chert* (*flint* if dark)
2. Very fine-grained, smooth to touch, breaks into flat chips—*shale.*
3. Black or dark brown, readily broken, soils fingers—*coal.*

II. *Limestone* may be clastic or crystalline, fine- or coarse-grained, and may or may not contain visible fossils. Usually gray, tan, buff, or white. Some distinctive varieties are:

A. *Bioclastic limestone*—clastic texture, grains are whole or broken fossils. Two relatively rare varieties are:
1. *Coquina*—very coarse, recognizable shells, much open pore space.
2. *Chalk*—very fine-grained, white or tan, soft and powdery.

B. *Oolitic limestone*—grains are small spheres (less than 2 mm in diameter), all about the same size.

C. *Travertine*—coarsely crystalline, no pore space, often contains different-colored layers (bands).

Metamorphic Rocks

The characteristics of a metamorphic rock are largely governed by (1) the composition of the parent rock and (2) the particular combination of temperature, confining pressure, and directed pressure. These factors cause different textures in rocks formed under different sets of conditions. For this reason, texture is usually the main basis for naming a metamorphic rock. Determining the composition (e.g., mineral content) is necessary for naming some rocks (e.g., *quartzite*), but for others, the minerals present are used as adjectives to describe the rock completely (e.g., *biotite* schist).

Metamorphic rocks are identified by determining first whether the rock has a *foliated* or *nonfoliated* texture.

Key for Identifying Metamorphic Rocks

I. If the rock is *nonfoliated,* then it is identified on the basis of its mineral content.

A. Does the rock consist of mostly quartz? If so, the rock is a *quartzite.* A quartzite has a mosaic texture of interlocking grains of quartz and will easily scratch glass.

B. Is the rock composed of interlocking coarse grains of calcite or dolomite? If so, it is *marble.* (The individual grains should exhibit rhombohedral cleavage; the rock is softer than glass.)

C. Is the rock a dense, dark mass of grains mostly too fine to identify with the naked eye? If so, it probably is a *hornfels.* A hornfels may have a few larger crystals of uncommon minerals enclosed in the fine-grained mass.

II. If the rock is *foliated,* determine the type of foliation and then, if possible, identify the minerals present.

A. Is the rock very fine-grained and does it split into sheetlike slabs? If so, it is *slate.* Most slate is composed of extremely fine-grained clay minerals, and the rock has an earthy luster.

B. Does the rock have a silky sheen but otherwise appear similar to slate? If so, it is a *phyllite.*

C. Is the rock composed mostly of visible grains of platy or needle-like minerals that are approximately parallel to one another? If so, the rock is a *schist.* If the rock is composed mainly of mica, it is a *mica schist.* If it also contains garnet, it is called a *garnet mica schist.* If hornblende is the predominant mineral, the rock is a *hornblende schist.* If talc prevails, it is a *talc schist* (sometimes called soapstone). A schistose rock composed of serpentine is called a *serpentinite.*

D. Are dark and light minerals found in separate lenses or layers? If so, the rock is a *gneiss.* The light layers are composed of feldspars and perhaps quartz, whereas the darker layers commonly are formed of biotite, amphibole, or pyroxene. A gneiss may appear similar to granite or diorite but can be distinguished from these igneous rocks by the foliation.

Appendix C

The Elements Most Significant to Geology

Table C.1

Atomic Number	Name	Symbol	Atomic Weight	Some Usual Charge of Ions	Atomic Number	Name	Symbol	Atomic Weight	Some Usual Charges of Ions
1	Hydrogen	H	1.0	+1	29	Copper	Cu	63.5	+2
2	Helium	He	4.0	0 inert	30	Zinc	Zn	65.4	+2
3	Lithium	Li	6.9	+1	33	Arsenic	As	74.9	+3
4	Beryllium	Be	9.0	+2	35	Bromine	Br	79.9	—
5	Boron	B	10.8	+3	37	Rubidium	Rb	85.5	+1
6	Carbon	C	12.0	+4	38	Strontium	Sr	87.3	+2
7	Nitrogen	N	14.0	+5	40	Zirconium	Zr	91.2	—
8	Oxygen	O	16.0	-2	42	Molybdenum	Mo	95.9	+4
9	Fluorine	F	19.0	-1	47	Silver	Ag	107.9	+1
10	Neon	Ne	20.2	0 inert	48	Cadmium	Cd	112.4	—
11	Sodium	Na	23.0	+1	50	Tin	Sn	118.7	+4
12	Magnesium	Mg	24.3	+2	51	Antimony	Sb	121.8	+3
13	Aluminum	Al	27.0	+3	52	Tellurium	Te	127.6	—
14	Silicon	Si	28.1	+4	55	Cesium	Cs	132.9	—
15	Phosphorus	P	31.0	+5	56	Barium	Ba	137.4	+2
16	Sulfur	S	32.1	-2	60	Neodymium	Nd	144	
17	Chlorine	Cl	35.5	-1	62	Samarium	Sm	150	
18	Argon	Ar	39.9	0 inert	74	Tungsten	W	183.9	—
19	Potassium	K	39.1	+1	78	Platinum	Pt	195.2	—
20	Calcium	Ca	40.1	+2	79	Gold	Au	197.0	—
22	Titanium	Ti	47.9	+4	80	Mercury	Hg	200.6	+2
23	Vanadium	V	50.9		82	Lead	Pb	207.2	+2
24	Chromium	Cr	52.0		83	Bismuth	Bi	209.0	—
25	Manganese	Mn	54.9	+4, +3	86	Radon	Rn	222	0 inert
26	Iron	Fe	55.8	+2, +3	88	Radium	Ra	226.1	
27	Cobalt	Co	58.9		90	Thorium	Th	232.1	—
28	Nickel	Ni	58.7	+2	92	Uranium	U	238.1	—
					94	Plutonium	Pu	239.0	—

Appendix D

Periodic Table of Elements

1 Group IA																	18 VIIIA
1 **H** 1.008	2 IIA											13 IIIA	14 IVA	15 VA	16 VIA	17 VIIA	**2** **He** 4.00

6 C 12.01 — Atomic number, Symbol, Atomic mass

Legend: Representative elements · Noble gases · Transition metals · Metalloids · Lanthanides · Actinides

Transition metals

		3 IIIB	4 IVB	5 VB	6 VIB	7 VIIB	8	9 VIIIB	10	11 IB	12 IIB						

3 **Li** 6.94	**4** **Be** 9.01											**5** **B** 10.81	**6** **C** 12.01	**7** **N** 14.01	**8** **O** 16.00	**9** **F** 19.00	**10** **Ne** 20.18
11 **Na** 22.99	**12** **Mg** 24.31											**13** **Al** 26.98	**14** **Si** 28.09	**15** **P** 30.97	**16** **S** 32.06	**17** **Cl** 35.45	**18** **Ar** 39.95
19 **K** 39.10	**20** **Ca** 40.08	**21** **Sc** 44.96	**22** **Ti** 47.90	**23** **V** 50.94	**24** **Cr** 52.00	**25** **Mn** 54.94	**26** **Fe** 55.85	**27** **Co** 58.93	**28** **Ni** 58.69	**29** **Cu** 63.54	**30** **Zn** 65.37	**31** **Ga** 69.72	**32** **Ge** 72.61	**33** **As** 74.92	**34** **Se** 78.96	**35** **Br** 79.91	**36** **Kr** 83.80
37 **Rb** 85.47	**38** **Sr** 87.62	**39** **Y** 88.91	**40** **Zr** 91.22	**41** **Nb** 92.91	**42** **Mo** 95.94	**43** **Tc** (98)	**44** **Ru** 101.07	**45** **Rh** 102.90	**46** **Pd** 106.42	**47** **Ag** 107.87	**48** **Cd** 112.41	**49** **In** 114.82	**50** **Sn** 118.69	**51** **Sb** 121.75	**52** **Te** 127.60	**53** **I** 126.90	**54** **Xe** 131.29
55 **Cs** 132.91	**56** **Ba** 137.34	**57** **La*** 138.91	**72** **Hf** 178.49	**73** **Ta** 180.95	**74** **W** 183.85	**75** **Re** 186.21	**76** **Os** 190.2	**77** **Ir** 192.22	**78** **Pt** 195.08	**79** **Au** 196.97	**80** **Hg** 200.59	**81** **Tl** 204.37	**82** **Pb** 207.19	**83** **Bi** 208.98	**84** **Po** (209)	**85** **At** (210)	**86** **Rn** (222)
87 **Fr** (223)	**88** **Ra** 226.03	**89** **Ac†** 227.03	**104** **Rf** (261)	**105** **Ha** (262)	**106** **Sg** (263)	**107** **Ns** (262)	**108** **Hs** (265)	**109** **Mt** (266)									

*Lanthanides

58 **Ce** 140.12	**59** **Pr** 140.91	**60** **Nd** 144.24	**61** **Pm** (145)	**62** **Sm** 150.36	**63** **Eu** 151.96	**64** **Gd** 157.25	**65** **Tb** 158.92	**66** **Dy** 162.50	**67** **Ho** 164.93	**68** **Er** 167.26	**69** **Tm** 168.93	**70** **Yb** 173.04	**71** **Lu** 174.97

†Actinides

90 **Th** 232.04	**91** **Pa** 231.04	**92** **U** 238.03	**93** **Np** (237)	**94** **Pu** (244)	**95** **Am** (243)	**96** **Cm** (247)	**97** **Bk** (247)	**98** **Cf** (251)	**99** **Es** (252)	**100** **Fm** (257)	**101** **Md** (258)	**102** **No** (259)	**103** **Lr** (260)

Appendix E

Selected Conversion Factors

Table E.1

	English Unit	Conversion Factor	Metric Unit	Conversion Factor	English Unit
Length and Distance	inch (in)	2.54	centimeters (cm)	0.4	inch (in)
	foot (ft)	0.3048	meter (m)	3.28	feet (ft)
	inch (in)	0.026	meter (m)	39.4	inches (in)
	mile, statute (mi)	1.61	kilometers (km)	0.62	mile (mi)
Area	square inch (in^2)	6.45	square centimeters (cm^2)	0.16	square inch (in^2)
	square foot (ft^2)	0.093	square meter (m^2)	10.8	square feet (ft^2)
	square mile (mi^2)	2.59	square kilometers (km^2)	0.39	square mile (mi^2)
	acre	0.4	hectare	2.47	acres
Volume	cubic inch (in^3)	16.4	cubic centimeters (cm^3)	0.06	cubic inch (in^3)
	cubic yard (yd^3)	0.76	cubic meter (m^3)	1.3	cubic yards (yd^3)
	cubic foot (ft^3)	0.0283	cubic meter (m^3)	35.3	cubic feet (ft^3)
	quart (qt)	0.95	liter	1.06	quarts (qt)
Weight	ounce (oz)	28.3	grams (g)	0.04	ounce (oz)
	pound (lb)	0.45	kilogram (kg)	2.2	pounds (lb)
	ton, short (2,000 lb)	907	kilograms (kg)	0.001	ton, short
	ton, short	0.91	ton, metric	1.1	ton, short
Temp.	degrees Fahrenheit (°F)	$-32° \times 5/9$	degrees Celsius (°C) (centigrade)	$\times 1.8 + 32°$	degrees Fahrenheit (°F)

Appendix F

Rock Symbols

Shown below are the rock symbols used in the text. In general, these symbols are used by all geologists, although they sometimes are modified slightly.

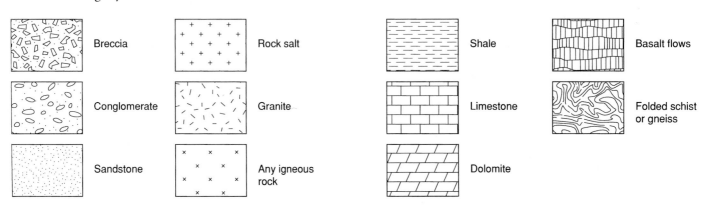

Breccia

Conglomerate

Sandstone

Rock salt

Granite

Any igneous rock

Shale

Limestone

Dolomite

Basalt flows

Folded schist or gneiss

Appendix G
Commonly Used Prefixes, Suffixes, and Roots

abyss   deep (Greek)

alluvium   deposited by flowing water (Latin)

anti-   opposite (Greek)

archea (archaeo)-   ancient (Greek)

astheno-   weak, lack of strength (Greek)

ceno   recent (Greek)

circum-   about, around, round about (Latin)

clast   broken (Greek)

-cline   tilted, gradient (Greek)

de-   lower, reduce, take away (Latin)

dis-   separation, opposite of (Latin)

ex-   out of, away from (Greek)

feld   field (Swedish, German)

folium   leaf (Latin)

geo-   earth (Greek)

glomero-   cluster (Latin)

hydro-   water (Greek)

iso-   equal (Greek)

-lith   stone or rock (Greek)

meso-   middle (Greek)

meta-   change (Greek)

-morph   form, shape (Greek)

paleo-   ancient (Greek)

ped-   foot (Latin)

pelagic   pertaining to the ocean (Greek)

petro-   stone or rock (Greek)

phanero-   visible, evident (Greek)

pheno-   large, conspicuous ("to show" in Greek)

pluto-   deep-seated (from Roman god of the underworld or infernal regions)

pre-   before, in front (Latin)

proto-   first, primary, primative (Greek)

pyro-   fire (Greek)

spar   crystalline material (German)

-sphere   ball (Greek)

stria   small groove, streak, band (Latin)

sub-   under, less than (Latin)

super-   above, more than, in addition to (Latin)

syn-   together, at the same time (Latin, Greek)

tecto-   means building or constructing in Greek and Latin; in geology it means movement or structures caused by internal forces.

terra, terre   pertaining to the earth (Latin)

thermal   pertaining to heat (Greek)

trans-   over, beyond, through, across (Latin)

xeno-   strange, foreign (Greek)

zoo, zoic-   animal (Greek)

Glossary

A

aa   A lava flow that solidifies with a spiny, rubbly surface.

abrasion   The grinding away of rock by friction and impact during transportation.

absolute age   Age given in years or some other unit of time.

abyssal fan   Great fan-shaped deposit of sediment on the deep-sea floor at the base of many submarine canyons.

abyssal plain   Very flat sediment-covered region of the deep-sea floor, usually at the base of the continental rise.

accreted terrane   Terrane that did not form at its present site on a continent.

accretionary wedge (subduction complex)   A wedge of thrust-faulted and folded sediment scraped off a subducting plate by the overlying plate.

accumulation stage   Stage in the evolution of major mountain belts characterized by the accumulation of great thicknesses (several kilometers) of sedimentary or volcanic rocks.

active continental margin   A margin consisting of a continental shelf, a continental slope, and an oceanic trench.

actualism   The principle that the same processes and natural laws that operated in the past are those we can actually observe or infer from observations as operating at present. Under present usage, *uniformitarianism* has the same meaning as actualism for most geologists.

advancing glacier   Glacier with a positive budget, so that accumulation results in the lower edges being pushed outward and downward.

aftershock   Small earthquake that follows a main shock.

A horizon   The top layer of soil, characterized by the downward movement of water; also called *zone of leaching*.

alkali soil   Soil containing such a great quantity of sodium salts precipitated by evaporating ground water that it is generally unfit for plant growth.

alluvial fan   Large fan-shaped pile of sediment that usually forms where a stream's velocity decreases as it emerges from a narrow canyon onto a flat plain at the foot of a mountain range.

alpine glaciation   Glaciation of a mountainous area.

amphibole group   Mineral group in which all members are double-chain silicates.

amphibolite   Amphibole (hornblende), plagioclase schist.

andesite   Fine-grained igneous rock of intermediate composition. Up to half of the rock is plagioclase feldspar with the rest being ferromagnesian minerals.

angle of dip   A vertical angle measured downward from the horizontal plane to an inclined plane.

angular   Sharp-edged; lacking rounded edges or corners.

angular unconformity   An unconformity in which younger strata overlie an erosion surface on tilted or folded layered rock.

anorthosite   A crystalline rock composed almost entirely of calcium-rich plagioclase feldspar.

antecedent stream   A stream that maintains its original course despite later deformation of the land.

anticline   An arched fold in which the rock layers usually dip away from the axis of the fold.

aquifer   A body of saturated rock or sediment through which water can move readily.

arch (sea arch)   Bridge of rock left above an opening eroded in a headland by waves.

Archean Eon   The oldest eon of the earth's history.

arête   A sharp ridge that separates adjacent glacially carved valleys.

arid region   An area with less than 25 cm of rain per year.

arkose   A sandstone in which more than 25% of the grains are feldspar.

artesian aquifer   See confined aquifer.

artesian well   A well in which water rises above the aquifer.

artificial recharge   Ground-water recharge increased by engineering techniques.

aseismic ridge   Submarine ridge with which no earthquakes are associated.

ash (volcanic)   Fine pyroclasts (less than 4 mm).

assimilation   The process in which very hot magma melts country rock and assimilates the newly molten material.

asthenosphere   A region of the earth's outer shell beneath the lithosphere. The asthenosphere is of indeterminate thickness and behaves plastically.

atoll   A circular reef surrounding a deeper lagoon.

atom   Smallest possible particle of an element that retains the properties of that element.

atomic mass number   The total number of neutrons and protons in an atom.

atomic number   The total number of protons in an atom.

atomic weight   The sum of the weight of the subatomic particles in an average atom of an element, given in atomic mass units.

augite   Mineral of the pyroxene group found in mafic igneous rocks.

aulacogen   See failed rift.

aureole   Zone of contact metamorphism adjacent to a pluton.

axial plane   A plane containing all of the hinge lines of a fold.

axis   See hinge line.

B

backarc spreading   A type of sea-floor spreading that moves an island arc away from a continent, or tears an island arc in two, or splits the edge of a continent, in each case forming new sea floor.

backshore   Upper part of the beach, landward of the high-water line.

bajada   A broad, gently sloping, depositional surface formed at the base of a mountain range in a dry region by the coalescing of individual alluvial fans.

bar   A ridge of sediment, usually sand or gravel, that has been deposited in the middle or along the banks of a stream by a decrease in stream velocity.

barchan   A crescent-shaped dune with the horns of the crescent pointing downwind.

barrier island   Ridge of sand paralleling the shoreline and extending above sea level.

barrier reef   A reef separated from the shoreline by the deeper water of a lagoon.

basal sliding Movement in which the entire glacier slides along as a single body on its base over the underlying rock.

basalt A fine-grained, mafic, igneous rock composed predominantly of ferromagnesian minerals and with lesser amounts of calcium-rich plagioclase feldspar.

base level A theoretical downward limit for stream erosion of the earth's surface.

batholith A large discordant pluton with an outcropping area greater than 100 square kilometers.

bauxite The principal ore of aluminum; $Al_2O_3 \cdot nH_2O$.

baymouth bar A ridge of sediment that cuts a bay off from the ocean.

beach Strip of sediment, usually sand but sometimes pebbles, boulders, or mud, that extends from the low-water line inland to a cliff or zone of permanent vegetation.

beach face The section of the beach exposed to wave action.

bedding An arrangement of layers or beds of rock.

bedding plane A nearly flat surface separating two beds of sedimentary rock.

bed load Heavy or large sediment particles in a stream that travel near or on the stream bed.

bedrock Solid rock that underlies soil.

Benioff zone Distinct earthquake zone that begins at an oceanic trench and slopes landward and downward into the earth at an angle of about 30° to 60°

bergschrund The crevasse that develops where a glacier is pulling away from a cirque wall.

berm Platform of wave-deposited sediment that is flat or slopes slightly landward.

B horizon A soil layer characterized by the accumulation of material leached downward from the A horizon above; also called *zone of accumulation*

biochemical Precipitated by the action of organisms.

bioclastic limestone A limestone consisting of fragments of shells, corals, and algae.

biotite Iron/magnesium-bearing mica.

block Large angular pyroclast.

blowout A depression on the land surface caused by wind erosion.

body wave Seismic wave that travels through the earth's interior.

bomb Large spindle- or lens-shaped pyroclast.

bonding Attachment of an atom to one or more adjacent atoms.

bottomset bed A delta deposit formed from the finest silt and clay, which are carried far out to sea by river flow or by sediments sliding downhill on the sea floor.

boulder A sediment particle with a diameter greater than 256 mm.

Bowen's reaction series The sequence in which minerals crystallize from a cooling basaltic magma.

braided stream A stream that flows in a network of many interconnected rivulets around numerous bars.

breaker A wave that has become so steep that the crest of the wave topples forward, moving faster than the main body of the wave.

breakwater An offshore structure built to absorb the force of large breaking waves and provide quiet water near shore.

brittle strain Cracking or rupturing of a body under stress.

butte A narrow pinnacle of resistant rock with a flat top and very steep sides.

C

calcareous Containing calcium carbonate.

calcite Mineral with the formula $CaCO_3$.

caldera A volcanic depression much larger than the original crater.

capacity (of stream) The total load that a stream can carry.

capillary action The drawing of water upward into small openings as a result of surface tension.

capillary fringe A thin zone near the water table in which capillary action causes water to rise above the zone of saturation.

carbonaceous chondrite Stony meteorite containing chondrules and composed mostly of serpentine and large quantities of organic materials.

carbonic acid H_2CO_3, a weak acid common in rain and surface waters.

cave (cavern) Naturally formed underground chamber.

cement The solid material that precipitates in the pore space of sediments, binding the grains together to form solid rock.

cementation The chemical precipitation of material in the spaces between sediment grains, binding the grains together into a hard rock.

Cenozoic Era The most recent of the eras; followed the Mesozoic Era.

chain silicate structure Silicate structure in which two of each tetrahedron's oxygen ions are shared with adjacent tetrahedrons, resulting in a chain of tetrahedrons.

chalk A very fine-grained bioclastic limestone.

channel (*Mars*) Feature on the surface of the planet Mars that resembles very closely certain types of stream channels on the earth.

chaotic terrain (*Mars*) Patch of jumbled and broken angular slabs and blocks on the surface of Mars.

chemical sedimentary rock A rock composed of material precipitated directly from solution.

chemical weathering The decomposition of rock resulting from exposure to water and atmospheric gases.

chert A hard, compact, fine-grained sedimentary rock formed almost entirely of silica.

chill zone In an intrusion, the finer-grained rock adjacent to a contact with country rock.

chondrule Round silicate grain within some stony meteorites.

C horizon A soil layer composed of incompletely weathered parent material.

cinder (volcanic) Pyroclast approximately the size of a sand grain. Sometimes defined as between 4 and 32 millimeters in diameter.

cinder cone A volcano constructed of loose rock fragments ejected from a central vent.

circum-Pacific belt Major belt around the edge of the Pacific Ocean on which most composite volcanoes are located and where many earthquakes occur.

cirque A steep-sided, amphitheater-like hollow carved into a mountain at the head of a glacial valley.

clastic sedimentary rock A sedimentary rock composed of fragments of preexisting rock.

clastic texture An arrangement of rock fragments bound into a rigid network by cement.

clay Sediment composed of particles with diameter less than 1/256 mm.

clay mineral A hydrous aluminum-silicate that occurs as a platy grain of microscopic size with a sheet silicate structure.

clay mineral group Collective term for several clay minerals.

cleavage The ability of a mineral to break along preferred planes.

coal A sedimentary rock formed from the consolidation of plant material. It is rich in carbon, usually black, and burns readily.

coarse-grained rock Rock in which most of the grains are larger than 1 millimeter (igneous) or 2 millimeters (sedimentary).

coast The land near the sea, including the beach and a strip of land inland from the beach.

coastal straightening The gradual straightening of an irregular shoreline by wave erosion of headlands and wave deposition in bays.

cobble A sediment particle with a diameter of 64 to 256 mm.

column A dripstone feature formed when a stalactite growing downward and a stalagmite growing upward meet and join.

columnar structure Volcanic rock in parallel, usually vertical columns, mostly six-sided; also called *columnar jointing.*

comet   Small object in space, no more than a few kilometers in diameter, composed of frozen methane, frozen ammonia, and water ice, with small solid particles and dust imbedded in the ices.

compaction   A loss in overall volume and pore space of a rock as the particles are packed closer together by the weight of overlying material.

competence   The largest particle that a stream can carry.

composite volcano (stratovolcano)   A volcano constructed of alternating layers of pyroclastics and rock solidified from lava flows.

compressive directed pressure   Directed pressure that tends to compress some portions of a body more than other portions.

compressive stress   A stress due to a force pushing together on a body.

conchoidal fracture   Curved fracture surfaces.

concordant   Parallel to layering or earlier developed planar structures.

concretion   Hard, rounded mass that develops when a considerable amount of cementing material precipitates locally in a rock, often around an organic nucleus.

cone of depression   A depression of the water table formed around a well when water is pumped out; it is shaped like an inverted cone.

confined aquifer (artesian aquifer)   An aquifer completely filled with pressurized water and separated from the land surface by a relatively impermeable confining bed, such as shale.

confining pressure   Pressure applied equally on all surfaces of a body; also called *geostatic* or *lithostatic pressure*

conglomerate   A coarse-grained sedimentary rock (grains coarser than 2 mm) formed by the cementation of rounded gravel.

consolidation   Any process that forms firm, coherent rock from sediment or from liquid.

contact   Boundary surface between two different rock types or ages of rocks.

contact (thermal) metamorphism   Metamorphism under conditions in which high temperature is the dominant factor.

continental crust   The thick, granitic crust under continents.

continental drift   A concept suggesting that continents move over the earth's surface.

continental glaciation   The covering of a large region of a continent by a sheet of glacial ice.

continental rise   A wedge of sediment that extends from the lower part of the continental slope to the deep sea floor.

continental shelf   A submarine platform at the edge of a continent, inclined very gently seaward generally at an angle of less than 1°.

continental slope   A relatively steep slope extending from a depth of 100 to 200 meters at the edge of the continental shelf down to oceanic depths.

contour current   A bottom current that flows parallel to the slopes of the continental margin (along the contour rather than down the slope).

contour line   A line on a topographic map connecting points of equal elevation.

convection (convection current)   A very slow circulation of a substance driven by differences in temperature and density within that substance.

convergent plate boundary   Boundary between two plates that are moving toward each other.

coquina   A limestone consisting of coarse shells.

core   The central zone of the earth.

correlation   In geology, correlation usually means determining time equivalency of rock units. Rock units may be correlated within a region, a continent, and even between continents.

country rock   Any rock that was older than and intruded by an igneous body.

covalent bonding   Bonding due to the sharing of electrons by adjacent atoms.

crater (of a volcano)   A basinlike depression over a vent at the summit of a volcanic cone.

craton   Portion of a continent that has been structurally stable for a prolonged period of time.

creep   Very slow, continuous downslope movement of soil or debris.

crest (of wave)   The high point of a wave.

crevasse   Open fissure in a glacier.

cross-bedding   An arrangement of relatively thin layers of rock inclined at an angle to the more nearly horizontal bedding planes of the larger rock unit.

cross-cutting relationship   A principle or law stating that a disrupted pattern is older than the cause of disruption.

cross section   *See* geologic cross section.

crude oil   A liquid mixture of naturally occurring hydrocarbons.

crust   The outer layer of rock, forming a thin skin over the earth's surface.

crustal rebound   The rise of the earth's crust after the removal of glacial ice.

crystal   A homogeneous solid with an orderly internal atomic arrangement.

crystal form   Arrangement of various faces on a crystal in a definite geometric relationship to one another.

crystalline   Describing a substance in which the atoms are arranged in a regular, repeating, orderly pattern.

crystalline texture   An arrangement of interlocking crystals.

crystallization   Crystal development and growth.

crystal settling   The process whereby the minerals that crystallize at a high temperature in a cooling magma move downward in the magma chamber because they are denser than the magma.

cuesta   A ridge with a steep slope on one side and a gentle slope on the other side.

Curie point   The temperature below which a material becomes magnetized.

D

daughter product   The isotope produced by radioactive decay.

debris   Any unconsolidated material at the earth's surface.

debris avalanche   Very rapid and turbulent mass wasting of debris, air, and water.

debris fall   A free-falling mass of debris.

debris flow   Mass wasting in which motion is taking place throughout the moving mass (flow). The common varieties are earthflow, mudflow, and debris avalanche.

debris slide   Rapid movement of debris as a coherent mass.

deflation   The removal of clay, silt, and sand particles from the land surface by wind.

delamination   *See* lithospheric delamination.

delta   A body of sediment deposited at the mouth of a river when the river velocity decreases as it flows into a standing body of water.

dendritic pattern   Drainage pattern of a river and its tributaries, which resembles the branches of a tree or veins in a leaf.

density   Weight per given volume of a substance.

deposition   The settling or coming to rest of transported material.

depth of focus   Distance between the focus and the epicenter of an earthquake.

desert   A region with low precipitation (usually defined as less than 25 cm per year).

desertification   The expansion of barren deserts into once-populated regions.

desert pavement   A thin layer of closely packed gravel that protects the underlying sediment from deflation; also called *pebble armor*

detachment fault   Major fault in a mountain belt above which rocks have been intensely folded and faulted.

diapir   Bodies of rock (e.g., rock salt) or magma that ascend within the earth's interior because they are less dense than the surrounding rock.

differential stress   When pressures on a body are not of equal strength in all directions.

differential weathering Varying rates of weathering resulting from some rocks in an area being more resistant to weathering than others.

differentiation Separation of different ingredients from an originally homogeneous mixture.

dike A tabular, discordant intrusive structure.

diorite Coarse-grained igneous rock of intermediate composition. Up to half of the rock is plagioclase feldspar and the rest is ferromagnesian minerals.

dip *See* angle of dip, direction of dip.

dip-slip fault A fault in which movement is parallel to the dip of the fault surface.

directed pressure *See* differential stress.

direction of dip The compass direction in which the angle of dip is measured.

discharge In a stream, the volume of water that flows past a given point in a unit of time.

disconformity A surface that represents missing rock strata but beds above and below that surface are parallel to one another.

discordant Not parallel to any layering or parallel planes.

dissolved load The portion of the total sediment load in a stream that is carried in solution.

distributary Small shifting river channel that carries water away from the main river channel and distributes it over a delta's surface.

divergent plate boundary Boundary separating two plates moving away from each other.

divide Line dividing one drainage basin from another.

dolomite A sedimentary rock composed mostly of the mineral dolomite.

dolomitic marble Marble in which dolomite, rather than calcite, is the prevalent mineral.

dome *See* structural dome.

double refraction The splitting of light into two components when it passes through certain crystalline substances.

downcutting A valley-deepening process caused by erosion of a stream bed.

drainage basin Total area drained by a stream and its tributaries.

drainage pattern The arrangement in map view of a river and its tributaries.

drawdown The lowering of the water table near a pumped well.

dripstone Deposits of calcite (and, rarely, other minerals) built up by dripping water in caves.

drumlin A long, streamlined hill made of till.

dust (volcanic) Finest-sized pyroclasts.

dynamic pressure *See* directed pressure.

E

earthflow Slow-to-rapid mass wasting in which debris moves downslope as a very viscous fluid.

earthquake A trembling or shaking of the ground caused by the sudden release of energy stored in the rocks beneath the surface.

earthy luster A luster giving a substance the appearance of unglazed pottery.

echo sounder An instrument used to measure and record the depth to the sea floor.

elastic limit The maximum amount of stress that can be applied to a body before it deforms in a permanent way by bending or breaking.

elastic rebound theory The sudden release of progressively stored strain in rocks results in movement along a fault.

elastic strain Strain in which a deformed body recovers its original shape after the stress is released.

electron A single, negative electric charge that contributes virtually no mass to an atom.

element A substance that cannot be broken down to other substances by ordinary chemical methods. Each atom of an element possesses the same number of protons.

emergent coast A coast in which land formerly under water has recently been placed above sea level, either by uplift of the land or by a drop in sea level.

end moraine A ridge of till piled up along the front edge of a glacier.

environment of deposition The location in which deposition occurs, usually marked by characteristic physical, chemical, or biological conditions.

eon The largest unit of geological time.

epicenter The point on the earth's surface directly above the focus of an earthquake.

epoch Each period of the standard geologic time scale is divided into epochs (e.g., Pleistocene Epoch of the Quaternary Period).

equilibrium Material is in equilibrium if it is adjusted to the physical and chemical conditions of its environment so that it does not change or alter with time.

era Major subdivision of the standard geologic time scale (e.g., Mesozoic Era).

erosion The physical removal of rock by an agent such as running water, glacial ice, or wind.

erratic An ice-transported boulder that does not derive from bedrock near its present site.

esker A long, sinuous ridge of sediment deposited by glacial meltwater.

estuary Drowned river mouth.

etch-pitted terrain (*Mars*) A terrain on the surface of Mars characterized by small pits.

eugeosyncline *See* volcanic portion of a geosyncline.

evaporite Rock that forms from crystals precipitating during evaporation of water.

exfoliation The stripping of concentric rock slabs from the outer surface of a rock mass.

exfoliation dome A large, rounded landform developed in a massive rock, such as granite, by the process of exfoliation.

exotic terrane Terrane that did not form at its present site on a continent and traveled a great distance to get to its present site.

expansive clay Clay that increases in volume when water is added to it.

extension Strain involving an increase in length. Extension can cause crustal thinning and faulting.

extrusive rock Any igneous rock that forms at the earth's surface, whether it solidifies directly from a lava flow or is pyroclastic.

F

faceted A rock fragment with one or more flat surfaces caused by erosive action.

failed rift (aulacogen) An inactive, sediment-filled rift that forms above a mantle plume. The rift becomes inactive as two other rifts widen to form an ocean.

fall The situation in mass wasting that occurs when material free-falls or bounces down a cliff.

fault A fracture in bedrock along which movement has taken place.

fault-block mountain range A range created by uplift along normal or vertical faults.

faunal succession A principle or law stating that fossil species succeed one another in a definite and recognizable order; in general, fossils in progressively older rock show increasingly greater differences from species living at present.

feldspars Group of most common minerals of the earth's crust. All feldspars contain silicon, aluminum, and oxygen and may contain potassium, calcium, and sodium.

felsic rock Silica-rich igneous rock with a relatively high content of potassium and sodium.

ferromagnesian mineral Iron/magnesium-bearing mineral, such as augite, hornblende, olivine, or biotite.

fine-grained rock A rock in which most of the mineral grains are less than one millimeter across (igneous) or less than 1/16 mm (sedimentary).

fiord A coastal inlet that is a glacially carved valley, the base of which is submerged.

firn A compacted mass of granular snow, transitional between snow and glacier ice.

firn limit *See* snow line.

fissility   The ability of a rock to split into thin layers.

flank eruption   An eruption in which lava erupts out of a vent on the side of a volcano.

flash flood   Flood of very high discharge and short duration; sudden and local in extent.

flood plain   A broad strip of land built up by sedimentation on either side of a stream channel.

flow   A type of movement that implies that a descending mass is moving downslope as a viscous fluid.

flowstone   Calcite precipitated by flowing water on cave walls and floors.

focus   The point within the earth from which seismic waves originate in an earthquake.

fold   Bend in layered bedrock.

fold and thrust belt   A portion of a major mountain belt characterized by large thrust faults, stacked one upon another. Layered rock between the faults was folded when faulting was taking place.

fold axis   *See* hinge line.

foliation   Parallel alignment of textural and structural features of a rock.

footwall   The underlying surface of an inclined fault plane.

foreland basin   A sediment-filled basin on a continent, landward of a magmatic arc, and caused indirectly by ocean-continent convergence.

foreset bed   A sediment layer in the main part of a delta, deposited at an angle to the horizontal.

foreshock   Small earthquake that precedes a main shock.

foreshore   The zone that is regularly covered and uncovered by the rise and fall of tides.

formation   A body of rock of considerable thickness that has a recognizable unity or similarity making it distinguishable from adjacent rock units. Usually composed of one bed or several beds of sedimentary rock, although the term is also applied to units of metamorphic and igneous rock. A convenient unit for mapping, describing, or interpreting the geology of a region.

fossil   Traces of plants or animals preserved in rock.

fossil assemblage   Various different species of fossils in a rock.

fracture   The way a substance breaks where not controlled by cleavage.

fracture zone   Major line of weakness in the earth's crust that crosses the mid-oceanic ridge at approximately right angles.

fracturing   Cracking or rupturing of a body under stress.

framework silicate structure   Crystal structure in which all four oxygen ions of a silica tetrahedron are shared by adjacent ions.

fretted terrain (*Mars*)   Flat lowland with some scattered high plateaus on the surface of Mars.

fringing reef   A reef attached directly to shore. (*See* barrier reef.)

frost action   Mechanical weathering of rock by freezing water.

frost heaving   The lifting of rock or soil by the expansion of freezing water.

frost wedging   A type of frost action in which the expansion of freezing water pries a rock apart.

G

gabbro   A mafic, coarse-grained igneous rock composed predominantly of ferromagnesian minerals and with lesser amounts of calcium-rich plagioclase feldspar.

gaining stream   A stream that receives water from the zone of saturation.

geode   Partly hollow, globelike body found in limestone or other cavernous rock.

geologic cross section   A representation of a portion of the earth in a vertical plane.

geologic map   A map representing the geology of a given area.

geologic resources   Valuable materials of geologic origin that can be extracted from the earth.

geology   The scientific study of the earth.

geophysics   The application of physical laws and principles to a study of the earth.

geosyncline   A very long (thousands of kilometers) and relatively narrow (a few hundred kilometers) basin that acts as a collecting area for marine sedimentary and volcanic rocks.

geothermal energy   Energy produced by harnessing naturally occurring steam and hot water.

geothermal gradient   Rate of temperature increase associated with increasing depth beneath the surface of the earth (normally about 25°C/km).

geyser   A type of hot spring that periodically erupts hot water and steam.

geyserite   A deposit of silica that forms around many geysers and hot springs.

glacier   A large, long-lasting mass of ice, formed on land by the compaction and recrystallization of snow, which moves because of its own weight.

glacier ice   The mosaic of interlocking ice crystals that form a glacier.

glassy (or vitreous) luster   A luster that gives a substance a glazed, porcelainlike appearance.

gneiss   A metamorphic rock composed of light and dark layers or lenses.

gneissic   The texture of a metamorphic rock in which minerals are separated into light and dark layers or lenses.

goethite   The commonest mineral in the limonite group; $Fe_2O_3 \cdot nH_2O$.

Gondwanaland   The southern part of *Pangaea* that formed South America, Africa, India, Australia, and Antarctica.

graben   A down-dropped block bounded by normal fault.

graded bed   A single bed with coarse grains at the bottom of the bed and progressively finer grains toward the top of the bed.

graded stream   A stream that exhibits a delicate balance between its transporting capacity and the sediment load available to it.

granite   A felsic, coarse-grained, intrusive igneous rock containing quartz and composed mostly of potassium- and sodium-rich feldspars.

granitization   The process by which granite is created from other rock without a melt being involved.

gravel   Rounded particles coarser than 2 mm in diameter.

gravity   The force of attraction that two bodies exert on each other that is proportional to the product of their masses and inversely proportional to the square of the distance from the centers of the two bodies.

gravity anomaly   A deviation from the average gravitational attraction between the earth and an object. *See* negative gravity anomaly, positive gravity anomaly.

gravity meter   An instrument that measures the gravitational attraction between the earth and a mass within the instrument.

graywacke   A sandstone with more than 15% fine-grained matrix between the sand grains.

groin   Short wall built perpendicular to shore to trap moving sand and widen a beach.

ground moraine   A blanket of till deposited by a glacier or released as glacier ice melted.

groundwater   The water that lies beneath the ground surface, filling the cracks, crevices, and pore space of rocks.

guyot   Flat-topped seamount.

H

half-life   The time it takes for a given amount of a radioactive isotope to be reduced by one-half.

hanging valley   A smaller valley that terminates abruptly high above a main valley.

hanging wall   The overlying surface of an inclined fault plane.

hardness   The relative ease or difficulty with which a smooth surface of a mineral can be scratched; commonly measured by Mohs' scale.

headland   Point of land along a coast.

headward erosion   The lengthening of a valley in an uphill direction above its original source by gullying, mass wasting, and sheet erosion.

heat engine   A device that converts heat energy into mechanical energy.

heat flow   Gradual loss of heat (per unit of surface area) from the earth's interior out into space.

heavy crude   Dense, viscous petroleum that flows slowly or not at all.

hematite   A type of iron oxide that has a brick-red color when powdered; Fe_2O_3.

highland (*Moon*)   A rugged region of the lunar surface representing an early period in lunar history when intense meteorite bombardment formed craters.

hinge line   Line about which a fold appears to be hinged. Line of maximum curvature of a folded surface.

hinge plane   *See* axial plane.

hogback   A sharp-topped ridge formed by the erosion of steeply dipping beds.

horn   A sharp peak formed where cirques cut back into a mountain on several sides.

hornblende   Common amphibole frequently found in igneous and metamorphic rocks.

hornfels   A fine-grained, unfoliated metamorphic rock.

horst   An up-raised block bounded by normal faults.

hot spot   An area of volcanic eruptions and high heat flow above a rising mantle plume.

hot spring   Spring with a water temperature warmer than human body temperature.

hydraulic action   The ability of water to pick up and move rock and sediment.

hydrologic cycle   The movement of water and water vapor from the sea to the atmosphere, to the land, and back to the sea and atmosphere again.

hydrology   The study of water's properties, circulation, and distribution.

hydrothermal metamorphism   Alteration of a rock by hot water passing through it.

hydrothermal rock   Rock deposited by precipitation of ions from solution in hot water.

hypocenter   Synonym for the focus of an earthquake.

hypothesis   A tentative theory.

I

iceberg   Block of glacier-derived ice floating in water.

ice cap   A glacier covering a relatively small area of land but not restricted to a valley.

icefall   A chaotic jumble of crevasses that split glacier ice into pinnacles and blocks.

ice sheet   A glacier covering a large area (more than 50,000 square kilometers) of land.

igneous rock   A rock formed or apparently formed from solidification of magma.

incised meander   A meander that retains its sinuous curves as it cuts vertically downward below the level at which it originally formed.

inclusion   A fragment of rock that is distinct from the body of igneous rock in which it is enclosed.

inclusions, principle of   Fragments included in a host rock are older than the host rock.

index fossil   A fossil from a very short-lived species known to have existed during a specific period of geologic time.

intensity   A measure of an earthquake's size by its effect on people and buildings.

intermediate rock   Rock with a chemical content between felsic and mafic compositions

intrusion (intrusive structure)   A body of intrusive rock classified on the basis of size, shape, and relationship to surrounding rocks.

intrusive rock   Rock that appears to have crystallized from magma emplaced in surrounding rock.

ion   An electrically charged atom or group of atoms.

ionic bonding   Bonding due to the attraction between positively charged ions and negatively charged ions.

iron meteorite   A meteorite composed principally of iron-nickel alloy.

island arc   A curved line of islands.

isoclinal fold   A fold in which the limbs are parallel to one another.

isolated silicate structure   Silicate minerals that are structured so that none of the oxygen atoms are shared by silica tetrahedrons.

isostasy   The balance or equilibrium between adjacent blocks of crust resting on a plastic mantle.

isostatic adjustment   Concept of vertical movement of sections of the earth's crust to achieve balance or equilibrium.

isotherm   A line along which the temperature of rock (or other material) is the same.

isotopes   Atoms (of the same element) that have different numbers of neutrons but the same number of protons.

isotopic dating   Determining the age of a rock or mineral through its radioactive elements and decay products (previously and somewhat inaccurately called *radiometric or radioactive dating*).

J

jetty   Rock wall protruding above sea level, designed to protect the entrance of a harbor from sediment deposition and storm waves; usually built in pairs.

joint   A fracture or crack in bedrock along which essentially no displacement has occurred.

joint set   Joints oriented in one direction approximately parallel to one another.

K

karst topography   An area with many sinkholes and a cave system beneath the land surface and usually lacking a surface stream.

kettle   A depression caused by the melting of a stagnant block of ice that was surrounded by sediment.

KREEP (*Moon*)   A lunar basalt enriched in potassium (K), the rare earth elements (REE), and phosphorus (P).

L

laccolith   A concordant intrusive structure, similar to a sill, with the central portion thicker and domed upward. Laccoliths are not common and are not discussed in this textbook.

laminar flow   Slow, smooth flow, with each drop of water traveling a smooth path parallel to its neighboring drops.

laminated terrain (*Mars*)   Area where series of alternating light and dark layers can be seen on the surface of Mars.

lamination   A thin layer in sedimentary rock (less than one centimeter thick).

landform   A characteristically shaped feature of the earth's surface, such as a hill or a valley.

landslide   The general term for a slowly to very rapidly descending mass of rock or debris.

lateral continuity   Principle that states that an original sedimentary layer extends laterally until it tapers or thins at its edges.

lateral erosion   Erosion and undercutting of stream banks caused by a stream swinging from side to side across its valley floor.

lateral moraine   A low ridgelike pile of till along the side of a glacier.

laterite   Highly leached soil that forms in regions of tropical climate with high temperatures and very abundant rainfall.

lava   Magma on the earth's surface.

left-lateral fault   A strike-slip fault in which the block seen across the fault appears displaced to the left.

limb   Portion of a fold shared by an anticline and a syncline.

limestone   A sedimentary rock composed mostly of calcite.

limonite   A type of iron oxide that is yellowish-brown when powdered; $Fe_2O_3 \cdot nH_2O$.

liquefaction   A type of ground failure in which water-saturated sediment turns from a solid to a liquid as a result of shaking, often caused by an earthquake.

lithification   The consolidation of sediment into sedimentary rock.

lithosphere   The rigid outer shell of the earth, 70 to 125 or more kilometers thick.

lithospheric delamination   The detachment of part of the mantle portion of the lithosphere beneath a mountain belt.

lithostatic pressure   Confining pressure due to the weight of overlying rock.

loam   Soil containing approximately equal amounts of sand, silt, and clay.

loess   A fine-grained deposit of wind-blown dust.

longitudinal dune (seif)   Large, symmetrical ridge of sand parallel to the wind direction.

longitudinal profile   A line showing a stream's slope, drawn along the length of the stream as if it were viewed from the side.

longshore current   A moving mass of water that develops parallel to a shoreline.

longshore drift   Movement of sediment parallel to shore when waves strike a shoreline at an angle.

losing stream   Stream that loses water to the zone of saturation.

Love waves   A type of surface seismic wave that causes the ground to move side to side in a horizontal plane perpendicular to the direction the wave is traveling.

low-velocity zone   Mantle zone at a depth of about 100 kilometers where seismic waves travel more slowly than in shallower layers of rock.

luster   The quality and intensity of light reflected from the surface of a mineral.

M

mafic rock   Silica-deficient igneous rock with a relatively high content of magnesium, iron, and calcium.

magma   Molten rock, usually mostly silica. The liquid may contain dissolved gases as well as some solid minerals.

magmatic arc   A line of batholiths or volcanoes. Generally the line, as seen from above, is curved.

magmatic underplating   *See* underplating.

magnetic anomaly   A deviation from the average strength of the earth's magnetic field. *See* negative magnetic anomaly, positive magnetic anomaly.

magnetic field   Region of magnetic force that surrounds the earth.

magnetic pole   An area where the strength of the magnetic field is greatest and where the magnetic lines of force appear to leave or enter the earth.

magnetic reversal   A change in the earth's magnetic field between normal polarity and reversed polarity. In normal polarity the north magnetic pole, where magnetic lines of force enter the earth, lies near the geographic North Pole. In reversed polarity the south magnetic pole, where lines of force leave the earth, lies near the geographic North Pole (the magnetic poles have exchanged positions).

magnetite   An iron oxide that is attracted to a magnet.

magnetometer   An instrument that measures the strength of the earth's magnetic field.

magnitude   A measure of the energy released during an earthquake.

major mountain belt   A long chain (thousands of kilometers) of mountain ranges.

mantle   A thick shell of rock that separates the earth's crust above from the core below.

mantle diapir   A body of mantle rock, hotter than its surroundings, that ascends because it is less dense than the surrounding rock.

mantle plume   Narrow column of hot mantle rock that rises and spreads radially outward.

marble   A coarse-grained rock composed of interlocking calcite (or dolomite) crystals.

maria (Moon)   Lava plains on Moon's surface (singular, *mare*).

marine terrace   A broad, gently sloping platform that may be exposed at low tide.

mass wasting (or mass movement)   Movement, caused by gravity, in which bedrock, rock debris, or soil moves downslope in bulk.

matrix   Fine-grained material found in the pore space between larger sediment grains.

meander   A pronounced sinuous curve along a stream's course.

meander cutoff   A new, shorter channel across the narrow neck of a meander.

meander scar   An abandoned meander filled with sediment and vegetation.

mechanical weathering   The physical disintegration of rock into smaller pieces.

medial moraine   A single long ridge of till on a glacier, formed by adjacent lateral moraines joining and being carried downglacier.

Mediterranean-Himalayan belt (Mediterranean belt)   A major concentration of earthquakes and composite volcanoes that

runs through the Mediterranean Sea, crosses the Mideast and the Himalaya, and passes through the East Indies.

melt   Liquid rock resulting from melting in a laboratory.

Mercalli scale   *See* modified Mercalli scale.

mesa   A broad, flat-topped hill bounded by cliffs and capped with a resistant rock layer.

Mesozoic Era   The era that followed the Paleozoic Era and preceded the Cenozoic Era.

metallic bonding   Bonding, as in metals, whereby atoms are closely packed together and electrons move freely among atoms.

metallic luster   Luster giving a substance the appearance of being made of metal.

metamorphic facies   Metamorphic rocks that contain the same set of pressure or temperature sensitive minerals are regarded as belonging to the same facies, implying that they formed under broadly similar pressure and temperature conditions.

metamorphic rock   A rock produced by metamorphism.

metamorphism   The transformation of preexisting rock into texturally or mineralogically distinct new rock as a result of high temperature, high pressure, or both, but without the rock melting in the process.

metasomatism   Metamorphism coupled with the introduction of ions from an external source.

meteor   Fragment that passes through the earth's atmosphere, heated to incandescence by friction; sometimes incorrectly called "shooting" or "falling" stars.

meteorite   Meteor that strikes the earth's surface.

meteoroid   Small solid particles of stone and/or metal orbiting the sun.

mica group   Group of minerals with a sheet silicate structure.

microcline (potassium) feldspar   A feldspar with the formula $KAlSi_3O_8$.

mid-oceanic ridge   A giant mountain range that lies under the ocean and extends around the world.

migmatite   Mixed igneous and metamorphic rock.

mineral   A naturally occurring, inorganic, crystalline solid that has a definite chemical composition.

mineraloid   A substance that is not crystalline but otherwise would be considered a mineral.

miogeosyncline   *See* nonvolcanic portion of a geosyncline.

model   In science, a model is an image—graphic, mathematical, or verbal—that is consistent with the known data.

modified Mercalli scale   Scale expressing intensities of earthquakes (judged on amount of damage done) in Roman numerals ranging from I to XII.

Mohorovičić discontinuity   The boundary separating the crust from the mantle beneath it (also called *Moho*).

Mohs' hardness scale   Scale on which ten minerals are designated as standards of hardness.

molecule   The smallest possible unit of a substance that has the properties of that substance.

moment magnitude   An earthquake magnitude calculated from the strength of the rock, surface area of the fault rupture, and the amount of rock displacement along the fault.

monocline   A local steeping in a gentle regional dip; a steplike fold in rock.

moraine   A body of till either being carried on a glacier or left behind after a glacier has receded.

mountain range   A group of closely spaced mountains or parallel ridges.

mud   Term loosely used for silt and clay, usually wet.

mud crack   Polygonal crack formed in very fine-grained sediment as it dries.

mudflow   A flowing mixture of debris and water, usually moving down a channel.

mudstone   A fine-grained sedimentary rock that lacks shale's laminations and fissility.

multi-ringed basin (*Moon*)   Large lunar crater surrounded by a series of concentric rings with intervening lowlands.

muscovite   Transparent or white mica that lacks iron and magnesium.

N

natural gas   A gaseous mixture of naturally occurring hydrocarbons.

natural levee   Low ridges of flood-deposited sediment formed on either side of a stream channel, which thin away from the channel.

nebula   A large volume of interstellar gas and dust.

negative gravity anomaly   Less than normal gravitational attraction.

negative magnetic anomaly   Less than average strength of the earth's magnetic field.

neutron   A subatomic particle that contributes mass to an atom and is electrically neutral.

nonconformity   An unconformity in which an erosion surface on plutonic or metamorphic rock has been covered by younger sedimentary or volcanic rock.

nonmetallic luster   Luster that gives a substance the appearance of being made of something other than metal (e.g., glassy).

nonrenewable resource   A resource that forms at extremely slow rates compared to its rate of consumption.

nonvolcanic portion of a geosyncline   The part of a geosyncline formed of a thick sequence of limestone, sandstone, and shale, with little or no volcanic rock; called *miogeosyncline*.

normal fault   A fault in which the hanging-wall block moved down relative to the footwall block.

nucleus   Protons and neutrons form the nucleus of an atom. Although the nucleus occupies an extremely tiny fraction of the volume of the entire atom, practically all the mass of the atom is concentrated in the nucleus.

O

oblique-slip fault   A fault with both strike-slip and dip-slip components.

obsidian   Volcanic glass.

oceanic crust   The thin, basaltic crust under oceans.

oceanic trench   A narrow, deep trough parallel to the edge of a continent or an island arc.

O horizon   Dark-colored soil layer that is rich in organic material and forms just below surface vegetation.

oil   *See* crude oil.

oil field   An area underlain by one or more oil pools.

oil pool   Underground accumulation of oil.

oil sand   Asphalt-cemented sand deposit.

oil shale   Shale with a high content of organic matter from which oil may be extracted by distillation.

oil trap   A set of conditions that hold petroleum in a reservoir rock and prevent its escape by migration.

olivine   A ferromagnesian mineral with the formula $(Fe, Mg)_2SiO_4$.

oolite (ooid)   A small sphere of calcite precipitated from seawater.

oolitic limestone   A limestone formed from oolites.

opal   A mineraloid composed of silica and water.

open fold   A fold with gently dipping limbs.

open-pit mine   Mine in which ore is exposed at the surface in a large excavation.

ophiolite   A distinctive rock sequence found in many mountain ranges on continents.

ore   Naturally occurring material that can be profitably mined.

ore mineral   A mineral of commercial value.

organic sedimentary rock   Rock composed mostly of the remains of plants and animals.

original horizontality   The deposition of most water-laid sediment in horizontal or near-horizontal layers that are essentially parallel to the earth's surface.

orogeny   An episode of intense deformation of the rocks in a region, generally accompanied by metamorphism and plutonic activity.

orthoclase (potassium) feldspar   A feldspar with the formula $KAlSi_3O_8$.

outcrop   A surface exposure of bare rock, not covered by soil or vegetation.

outwash   Material deposited by debris-laden meltwater from a glacier.

overburden   The upper part of a sedimentary deposit. Its weight causes compaction of the lower part.

overturned fold   A fold in which both limbs dip in the same direction.

oxbow lake   A crescent-shaped lake occupying the abandoned channel of a stream meander that is isolated from the present channel by a meander cutoff and sedimentation.

P

pahoehoe   A lava flow characterized by a ropy or billowy surface.

paired terraces   *Stream terraces (see* definition) that occur at the same elevation on each side of a river.

paleomagnetism   A study of ancient magnetic fields.

Paleozoic Era   The era that followed the Precambrian and began with the appearance of complex life, as indicated by fossils.

Pangaea   A supercontinent that broke apart 200 million years ago to form the present continents.

parabolic dune   A deeply curved dune in a region of abundant sand. The horns point upwind and are often anchored by vegetation.

parent rock   Original rock before being metamorphosed.

partial melting   Melting of the components of a rock with the lowest melting temperatures.

passive continental margin   A margin that includes a continental shelf, continental slope, and continental rise that generally extends down to an abyssal plain at a depth of about 5 kilometers.

paternoster lakes   A series of rock-basin lakes carved by glacial erosion.

peat   A brown, lightweight, unconsolidated or semi-consolidated deposit of plant remains.

pebble   A sediment particle with a diameter of 2 to 64 mm.

pedalfer   A soil characterized by the downward movement of water through it, downward leaching, and abundant humus. Found in humid climates.

pediment   A gently sloping erosional surface cut into the solid rock of a mountain range in a dry region; usually covered with a thin veneer of gravel.

pedocal   A soil characterized by little leaching, scant humus, the upward movement of water through it, and the precipitation of salts. Found in dry climates.

pegmatite   Extremely coarse-grained igneous rock.

pelagic sediment   Sediment made up of fine-grained clay and the skeletons of microscopic organisms that settle slowly down through the ocean water.

peneplain   A nearly flat erosional surface presumably produced as mass wasting, sheet erosion, and stream erosion reduce a region almost to base level.

perched water table   A water table separated from the main water table beneath it by a zone that is not saturated.

period   Each era of the standard geologic time scale is subdivided into periods (e.g., the Cretaceous Period).

permafrost   Ground that remains permanently frozen for many years.

permeability   The capacity of a rock to transmit a fluid such as water or petroleum.

petrified wood   A material that forms as the organic matter of buried wood is either filled in or replaced by inorganic silica carried in by ground water.

petroleum   Crude oil and natural gas. (Some geologists use petroleum as a synonym for oil.)

Phanerozoic   Eon of geologic time. Includes all time following the Precambrian.

phenocryst   Any of the large crystals in porphyritic igneous rock.

phyllite   A metamorphic rock in which clay minerals have recrystallized into microscopic micas, giving the rock a silky sheen.

physical continuity   Being able to physically follow a rock unit between two places.

physical geology   A large division of geology concerned with earth materials, changes of the surface and interior of the earth, and the forces that cause those changes.

piezoelectricity   Electrical current generated when pressure is applied to certain minerals.

pillow structure   Rocks, generally basalt, formed in pillow-shaped masses fitting closely together; caused by underwater lava flows.

placer mine   Surface mines in which valuable mineral grains are extracted from stream bar or beach deposits.

plagioclase feldspar   A feldspar containing sodium and/or calcium in addition to aluminum, silicon, and oxygen.

plastic   Capable of being molded and bent under stress.

plastic flow   Movement within a glacier in which the ice is not fractured.

plastic strain   Strain in which a body is molded or bent under stress and does not return to its original shape after the stress is released.

plate   A large, mobile slab of rock making up part of the earth's surface.

plateau   Broad, flat-topped area elevated above the surrounding land and bounded, at least in part, by cliffs.

plateau basalts   Layers of basalt flows that have built up to great thicknesses.

plate tectonics   A theory that the earth's surface is divided into a few large, thick plates that are slowly moving and changing in size. Intense geologic activity occurs at the plate boundaries.

playa   A very flat surface underlain by hard, mud-cracked clay.

playa lake   A shallow temporary lake (following a rainstorm) on a flat valley floor in a dry region.

Pleistocene Epoch   An epoch of the Quaternary Period characterized by several glacial ages.

plunging fold   A fold in which the hinge line (or axis) is not horizontal.

pluton   An igneous body that crystallized deep underground.

plutonic rock   Igneous rock formed at great depth.

pluvial lake   A lake formed during an earlier time of abundant rainfall.

point bar   A stream *bar* (*see* definition) deposited on the inside of a curve in the stream, where the water velocity is low.

polarity   *See* magnetic reversal.

polar wandering   An apparent movement of the earth's poles.

pore space   The total amount of space taken up by openings between sediment grains.

porosity   The percentage of a rock's volume that is taken up by openings.

porphyritic rock   An igneous rock in which large crystals are enclosed in a matrix (or ground mass) of much finer-grained minerals or obsidian.

positive gravity anomaly   Greater than normal gravitational attraction.

positive magnetic anomaly   Greater than average strength of the earth's magnetic field.

potassium (orthoclase) feldspar   A feldspar with the formula $KAlSi_3O_8$.

pothole   Depression eroded into the hard rock of a stream bed by the abrasive action of the stream's sediment load.

Precambrian   The vast amount of time that preceded the Paleozoic Era.

Precambrian shield   A complex of old Precambrian metamorphic and plutonic rocks exposed over a large area.

pressure release   A significant type of mechanical weathering that causes rocks to crack when overburden is removed.

progressive metamorphism   Metamorphism in which progressively greater pressure and temperature act on a rock type with increasing depth in the earth's crust.

Proterozoic   Eon of Precambrian time.

proton   A subatomic particle that contributes mass and a single positive electrical charge to an atom.

pumice   A frothy volcanic glass.

P wave   A compressional wave (seismic wave) in which rock vibrates parallel to the direction of wave propagation.

P-wave shadow zone   The region on the earth's surface, 103° to 142° away from an earthquake epicenter, in which P waves from the earthquake are absent.

pyroclast   Fragment of rock formed by volcanic explosion.

pyroclastic debris   Rock fragments produced by volcanic explosion.

pyroclastic flow   Turbulent mixture of pyroclastics and gases flowing down the flank of a volcano.

pyroxene group   Mineral group, all members of which are single-chain silicates.

Q

quartz   Mineral with the formula SiO_2.

quartzite   A rock composed of sand-sized grains of quartz that have been welded together during metamorphism.

quartz sandstone   A sandstone in which more than 90% of the grains are quartz.

Quaternary Period   The youngest geologic period; includes the present time.

R

radial pattern   A drainage pattern in which streams diverge outward like spokes of a wheel.

radioactive decay   The spontaneous nuclear disintegration of certain isotopes.

radioactivity   The spontaneous nuclear disintegration of atoms of certain isotopes.

radon   A radioactive gas produced by the radioactive decay of uranium.

rain shadow   A region on the downwind side of mountains that has little or no rain because of the loss of moisture on the upwind side of the mountains.

rampart crater (*Mars*)   Meteorite crater that is surrounded by material that appears to have flowed from the point of impact.

Rayleigh waves   A type of surface seismic wave that behaves like a rolling ocean wave and causes the ground to move in an elliptical path.

rayed crater (*Moon*)   Crater with bright streaks radiating from it on the moon's surface.

receding glacier   A glacier with a negative budget, which causes the glacier to grow smaller as its edges melt back.

Recent (Holocene) Epoch   The present epoch of the Quaternary Period.

recessional moraine   An end moraine built during the retreat of a glacier.

recharge   The addition of new water to an aquifer or to the zone of saturation.

reclamation   Restoration of the land to usable condition after mining has ceased.

recrystallization   The development of new crystals in a rock, often of the same composition as the original grains.

rectangular pattern   A drainage pattern in which tributaries of a river change direction and join one another at right angles.

recumbent fold   A fold overturned to such an extent that the limbs are essentially horizontal.

reef   A resistant ridge of calcium carbonate formed on the sea floor by corals and coralline algae.

regional (dynamothermal) metamorphism   Metamorphism that takes place at considerable depth underground.

regolith   Loose, unconsolidated rock material resting on bedrock.

relative time   The sequence in which events took place (not measured in time units).

relief   The vertical distance between points on the earth's surface.

reserves   The discovered deposits of a geologic material that are economically and legally feasible to recover under present circumstances.

reservoir rock   A rock that is sufficiently porous and permeable to store and transmit petroleum.

residual clay   Fine-grained particles left behind as insoluble residue when a limestone containing clay dissolves.

residual soil   Soil that develops directly from weathering of the rock below.

resources   The total amount of a geologic material in all its deposits, discovered and undiscovered (*see* reserves).

reverse fault   A fault in which the hanging-wall block moved up relative to the footwall block.

rhyolite   A fine-grained, felsic, igneous rock made up mostly of feldspar and quartz.

Richter scale   A numerical scale of earthquake magnitudes.

ridge push   The concept that oceanic plates diverge as a result of sliding down the sloping lithosphere-asthenosphere boundary.

rift valley   A tensional valley bounded by normal faults. Rift valleys are found at diverging plate boundaries on continents and along the crest of the mid-oceanic ridge.

right-lateral fault   A strike-slip fault in which the block seen across the fault appears displaced to the right.

rigid zone   Upper part of a glacier in which there is no plastic flow.

rille (*Moon*)   Elongate trenched or cracklike valley on the lunar surface.

rip current   Narrow currents that flow straight out to sea in the surf zone, returning water seaward that has been pushed ashore by breaking waves.

ripple mark   Any of the small ridges formed on sediment surfaces exposed to moving wind or water. The ridges form perpendicularly to the motion.

rock   Naturally formed, consolidated material composed of grains of one or more minerals. (There are a few exceptions to this definition.)

rock avalanche   A very rapidly moving, turbulent mass of broken-up bedrock.

rock-basin lake   A lake occupying a depression caused by glacial erosion of bedrock.

rock cycle   A theoretical concept relating tectonism, erosion, and various rock-forming processes to the common rock types.

rockfall   Rock falling freely or bouncing down a cliff.

rock flour   A powder of fine fragments of rock produced by glacial abrasion.

rock gypsum   An evaporite composed of gypsum.

rock salt   An evaporite composed of halite.

rockslide   Rapid sliding of a mass of bedrock along an inclined surface of weakness.

rotational slide   In mass wasting, movement along a curved surface in which the upper part moves vertically downward while the lower part moves outward. Also called a *slump*.

rounded knobs (*glacial*)   Bedrock that is more resistant to glacial erosion stands out as rounded knobs, usually elongated parallel to the direction of glacier flow. These are also known as *roche moutonnées* (French for "rock sheep").

rounding   The grinding away of sharp edges and corners of rock fragments during transportation.

rubble   Angular sedimentary particles coarser than 2 mm in diameter.

S

saltation   A mode of transport that carries sediment downcurrent in a series of short leaps or bounces.

sand   Sediment composed of particles with a diameter between 1/16 mm and 2 mm.

sand dune   A mound of loose sand grains heaped up by the wind.

sandstone   A medium-grained sedimentary rock (grains between 1/16 mm and 2 mm) formed by the cementation of sand grains.

saturated zone   A subsurface zone in which all rock openings are filled with water.

scale   The relationship between distance on a map and the distance on the terrain being represented by that map.

schist   A metamorphic rock characterized by coarse-grained minerals oriented approximately parallel.

schistose   The texture of a rock in which visible platy or needle-shaped minerals have grown essentially parallel to each other under the influence of directed pressure.

scientific method   A means of gaining knowledge through objective procedures.

scoria   A basalt that is highly vesicular.

sea cave   A cavity eroded by wave action at the base of a sea cliff.

sea cliff   Steep slope that retreats inland by mass wasting as wave erosion undercuts it.

sea-floor spreading   The concept that the ocean floor is moving away from the mid-oceanic ridge and across the deep ocean basin, to disappear beneath continents and island arcs.

seamount   Conical mountain rising 1,000 meters or more above the sea floor.

seawall   A wall constructed along the base of retreating cliffs to prevent wave erosion.

sediment   Loose, solid particles that can originate by (1) weathering and erosion of preexisting rocks, (2) chemical precipitation from solution, usually in water, and (3) secretion by organisms.

sedimentary breccia   A coarse-grained sedimentary rock (grains coarser than 2 mm) formed by the cementation of angular rubble.

sedimentary facies Significantly different rock types occupying laterally distinct parts of the same layered rock unit.

sedimentary rock Rock that has formed from (1) lithification of any type of sediment, (2) precipitation from solution, or (3) consolidation of the remains of plants or animals.

sedimentary structure A feature found within sedimentary rocks, usually formed during or shortly after deposition of the sediment and before lithification.

seismic gap A segment of a fault that has not experienced earthquakes for a long time; such gaps may be the site of large future quakes.

seismic profiler An instrument that measures and records the subbottom structure of the sea floor.

seismic reflection The return of part of the energy of seismic waves to the earth's surface after the waves bounce off a rock boundary.

seismic refraction The bending of seismic waves as they pass from one material to another.

seismic sea wave *See* tsunami.

seismic wave A wave of energy produced by an earthquake.

seismogram Paper record of earth vibration.

seismograph A seismometer with a recording device that produces a permanent record of earth motion.

seismometer An instrument designed to detect seismic waves or earth motion.

serpentine A magnesium silicate mineral. Most asbestos is a variety of serpentine.

shale A fine-grained sedimentary rock (grains finer than 1/16 mm in diameter) formed by the cementation of silt and clay (mud). Shale has thin layers (laminations) and an ability to split (fissility) into small chips.

shear force In mass wasting, the component of gravitational force that is parallel to an inclined surface.

shearing Movement in which parts of a body slide relative to one another and parallel to the forces being exerted.

shear strength In mass wasting, the resistance to movement or deformation of material.

shear stress Stress due to forces that tend to cause movement or strain parallel to the direction of the forces.

sheet erosion The removal of a thin layer of surface material, usually topsoil, by a flowing sheet of water.

sheet joints Cracks that develop parallel to the outer surface of a large mass of expanding rock, as pressure is released during unloading.

sheet silicate structure Crystal structure in which each silica tetrahedron shares three oxygen ions.

sheetwash Water flowing down a slope in a layer.

shield volcano Broad, gently sloping cone constructed of solidified lava flows.

sial Rock rich in silicon and aluminum; characteristic of continental crust.

silica A term used for oxygen plus silicon.

silicate A substance that contains silica as part of its chemical formula.

silica tetrahedron *See* silicon-oxygen tetrahedron.

silicic rock or magma Silica-rich igneous rock or magma with a relatively high content of potassium and sodium.

silicon-oxygen tetrahedron Four-sided, pyramidal object that visually represents the four oxygen atoms surrounding a silicon atom; the basic building block of silicate minerals. Also called a silica tetrahedron or a silicon tetrahedron.

sill A tabular intrusive structure concordant with the country rock.

silt Sediment composed of particles with a diameter of 1/256 to 1/16 mm.

siltstone A sedimentary rock consisting mostly of silt grains.

sima Rock rich in silicon and magnesium; characteristic of oceanic crust.

sinkhole A closed depression found on land surfaces underlain by limestone.

sinter A deposit of silica that forms around some hot springs and geysers.

slab pull The concept that subducting plates are pulled along by their dense leading edges.

slate A fine-grained rock that splits easily along flat, parallel planes.

slaty Describing a rock that splits easily along nearly flat and parallel planes.

slaty cleavage The ability of a rock to break along closely spaced parallel planes.

slide In mass wasting, movement of a relatively coherent descending mass along one or more well-defined surfaces.

slip face The steep, downwind slope of a dune; formed from loose, cascading sand that generally keeps the slope at the angle of repose (about 34°).

slump In mass wasting, movement along a curved surface in which the upper part moves vertically downward while the lower part moves outward. Also called a *rotational slide*.

snow line An irregular line marking the highest level to which the winter snow cover on a glacier is lost during a melt season.

soil A layer of weathered, unconsolidated material on top of bedrock; often also defined as containing organic matter and being capable of supporting plant growth.

soil horizon Any of the layers of soil that are distinguishable by characteristic physical or chemical properties.

solifluction Flow of water-saturated debris over impermeable material.

solution Usually slow but effective process of weathering and erosion in which rocks are dissolved by water.

sorting Process of selection and separation of sediment grains according to their grain size (or grain shape or specific gravity).

source area The locality that eroded to provide sediment to form a sedimentary rock.

source rock A rock containing organic matter that is converted to petroleum by burial and other postdepositional changes.

spatter cone A small, steep-sided cone built from lava spattering out of a vent.

specific gravity The ratio of the mass of a substance to the mass of an equal volume of water, determined at a specified temperature.

spheroidally weathered boulder Boulder that has been rounded by weathering from an initial blocky shape.

spit A fingerlike ridge of sediment attached to land but extending out into open water.

spreading axis (or spreading center) The crest of the mid-oceanic ridge, where sea floor is moving away in opposite directions on either side.

spring A place where water flows naturally out of rock onto the land surface.

stable Describing a mineral that will not react with or convert to a new mineral or substance, given enough time.

stack A small rock island that is an erosional remnant of a headland left behind as a wave-eroded coast retreats inland.

stalactite Iciclelike pendant of dripstone formed on cave ceilings.

stalagmite Cone-shaped mass of dripstone formed on cave floors, generally directly below a stalactite.

standard geologic time scale A worldwide relative scale of geologic time divisions.

static pressure *See* confining pressure.

stock A small discordant pluton with an outcropping area of less than 100 square kilometers.

stony-iron meteorite A meteorite composed of silicate minerals and iron-nickel alloy in approximately equal amounts.

stony meteorite A meteorite made up mostly of plagioclase and iron-magnesium silicates.

stoping Upward movement of a body of magma by fracturing of overlying country rock. Magma engulfs the blocks of fractured country rock as it moves upward.

storm surge High sea level caused by the low pressure and high winds of hurricanes.

strain Change in size (volume) or shape of a body (or rock unit) in response to stress.

stratovolcano *See* composite volcano.

streak Color of a pulverized substance; a useful property for mineral identification.

stream A moving body of water, confined in a channel and running downhill under the influence of gravity.

stream capture *See* stream piracy.

stream channel A long, narrow depression, shaped and more or less filled by a stream.

stream discharge Volume of water that flows past a given point in a unit of time.

stream-dominated delta A delta with finger-like distributaries formed by the dominance of stream sedimentation; also called a birdfoot delta.

stream gradient Downhill slope of a stream's bed or the water surface, if the stream is very large.

stream headwaters The upper part of a stream near the source.

stream mouth The place where the stream enters the sea, a large lake, or a larger stream.

stream piracy The natural diversion of the headwaters of one stream into the channel of another.

stream terrace Steplike landform found above a stream and its flood plain.

stream velocity The speed at which water in a stream travels.

stress A force acting on a body, or rock unit, that tends to change the size or shape of that body, or rock unit. Force per unit area within a body.

striations (1) On minerals, extremely straight, parallel lines; (2) Glacial—straight scratches in rock caused by abrasion by a moving glacier.

strike The compass direction of a line formed by the intersection of an inclined plane (such as a bedding plane) with a horizontal plane.

strike-slip fault A fault in which movement is parallel to the strike of the fault surface.

strip mine A mine in which the valuable material is exposed at the surface by removing a strip of overburden.

structural basin A structure in which the beds dip toward a central point.

structural dome A structure in which beds dip away from a central point.

structural geology The branch of geology concerned with the internal structure of bedrock and the shapes, arrangement, and interrelationships of rock units.

subduction The sliding of the sea floor beneath a continent or island arc.

subduction complex *See* accretionary wedge.

subduction zone Elongate region in which subduction takes place.

submarine canyon V-shaped valleys that run across the continental shelf and down the continental slope.

submergent coast A coast in which formerly dry land has been recently drowned, either by land subsidence or a rise in sea level.

subsidence Sinking or downwarping of a part of the earth's surface.

superposed stream A river let down onto a buried geologic structure by erosion of overlying layers.

superposition A principle or law stating that within a sequence of undisturbed sedimentary rocks, the oldest layers are on the bottom, the youngest on the top.

surf Breaking waves.

surface wave A seismic wave that travels on the earth's surface.

suspect terrane A terrane that may not have formed at its present site.

suspended load Sediment in a stream that is light enough in weight to remain lifted indefinitely above the bottom by water turbulence.

S wave A seismic wave propagated by a shearing motion, which causes rock to vibrate perpendicular to the direction of wave propagation.

S-wave shadow zone The region on the earth's surface (at any distance more than 103° from an earthquake epicenter) in which S waves from the earthquake are absent.

swelling clay *See* expansive clay.

syncline A fold in which the layered rock usually dips toward an axis.

T

talus An accumulation of broken rock at the base of a cliff.

tarn *See* rock-basin lake.

tectite Small, rounded bits of glass formed from rock melting and being thrown into the air due to a meteorite impact.

tectonic forces Forces generated from within the earth that result in uplift, movement, or deformation of part of the earth's crust.

tectonostratigraphic terrane *See* terrane.

tensional stress A stress due to a force pulling away on a body.

tephra *See* pyroclastic debris.

terminal moraine An end moraine marking the farthest advance of a glacier.

terminus The lower edge of a glacier.

terrane (tectonostratigraphic terrane) A region in which the geology is markedly different from that in adjoining regions.

terrigenous sediment Land-derived sediment that has found its way to the sea floor.

theory An explanation for observed phenomena that has a high possibility of being true.

theory of glacial ages At times in the past, colder climates prevailed during which significantly more of the land surface of the earth was glaciated than at present.

thermal metamorphism *See* contact (thermal) metamorphism.

thrust fault A reverse fault in which the dip of the fault plane is at a low angle to horizontal.

tidal delta A submerged body of sediment formed by tidal currents passing through gaps in barrier islands.

"tidal wave" An incorrect name for a tsunami.

tide-dominated delta A delta formed by the reworking of sand by strong tides.

till Unsorted and unlayered rock debris carried by a glacier.

tillite Lithified till.

time-transgressive rock unit An apparently continuous rock layer in which different portions formed at different times.

tombolo A bar of marine sediment connecting a former island or stack to the mainland.

topographic map A map on which elevations are shown by means of contour lines.

topset bed In a delta, a nearly horizontal sediment bed of varying grain size formed by distributaries shifting across the delta surface.

traction Movement by rolling, sliding, or dragging of sediment fragments along a stream bottom.

transform fault The portion of a fracture zone between two offset segments of a mid-oceanic ridge crest.

transform plate boundary Boundary between two plates that are sliding past each other.

translational slide In mass wasting, movement of a descending mass along a plane approximately parallel to the slope of the surface.

transportation The movement of eroded particles by agents such as rivers, waves, glaciers, or wind.

transported soil Soil not formed from the local rock but from parent material brought in from some other region and deposited, usually by running water, wind, or glacial ice.

transverse dune A relatively straight, elongate dune oriented perpendicular to the wind.

travel-time curve A plot of seismic-wave arrival times against distance.

travertine A porous deposit of calcite that often forms around hot springs.

trellis pattern   A drainage pattern consisting of parallel main streams with short tributaries meeting them at right angles.

trench   *See* oceanic trench.

trench suction   The concept that overlying plates move horizontally toward oceanic trenches as subducting plates sink at an angle steeper than their dip.

tributary   Small stream flowing into a large stream, adding water to the large stream.

trough (of wave)   The low point of a wave.

truncated spur   Triangular facet where the lower end of a ridge has been eroded by glacial ice.

tsunami   Huge ocean wave produced by displacement of the sea floor; also called seismic sea wave.

tufa   A deposition of calcite that forms around a spring, lake, or percolating ground water.

tuff   A rock formed from fine-grained pyroclastic particles (ash and dust).

turbidity current   A flowing mass of sediment-laden water that is heavier than clear water and therefore flows downslope along the bottom of the sea or a lake.

turbulent flow   Eddying, swirling flow in which water drops travel along erratically curved paths that cross the paths of neighboring drops.

U

ultramafic rock   Rock composed entirely or almost entirely of ferromagnesian minerals.

unconfined aquifer   A partially filled aquifer exposed to the land surface and marked by a rising and falling water table.

unconformity   A surface that represents a break in the geologic record, with the rock unit immediately above it being considerably younger than the rock beneath.

unconsolidated   In referring to sediment grains, loose, separate, or unattached to one another.

underplating   The pooling of magmas at the base of the continental crust.

uniformitarianism   Principle that geologic processes operating at present are the same processes that operated in the past. The principle is stated more succinctly as "The present is the key to the past." *Also, see* actualism.

unloading   The removal of a great weight of rock.

unpaired terraces   *Stream terraces (see definition)* that do not have the same elevation on opposite sides of a river.

unsaturated zone   *See* vadose zone.

U-shaped valley   Characteristic cross-profile of a valley carved by glacial erosion.

V

vadose zone   A subsurface zone in which rock openings are generally unsaturated and filled partly with air and partly with water; above the saturated zone.

valley glacier   A glacier confined to a valley. The ice flows from a higher to a lower elevation.

Van der Waal's bonds or forces   Weak bonds in crystals, such as those that hold adjacent sheets of mica together.

varve   Two thin layers of sediment, one dark and the other light in color, representing one year's deposition in a lake.

vent   The opening in the earth's surface through which a volcanic eruption takes place.

ventifact   Boulder, cobble, or pebble with flat surfaces caused by the abrasion of wind-blown sand.

vertical exaggeration   An artificial steepening of slope angles on a topographic profile caused by using a vertical scale that differs from the horizontal scale.

vesicle   A cavity in volcanic rock caused by gas in a lava.

viscosity   Resistance to flow.

vitreous luster   *See* glassy luster.

volcanic breccia   Rock formed from large pieces of volcanic rock (cinders, blocks, bombs).

volcanic dome   A steep-sided, dome- or spine-shaped mass of volcanic rock formed from viscous lava that solidifies in or immediately above a volcanic vent.

volcanic neck   An intrusive structure that apparently represents magma that solidified within the throat of a volcano.

volcanic portion of a geosyncline   A thick sequence of submarine lava flows and marine sedimentary rock, much of which derives from volcanic ash and from sediment eroded from volcanic rocks; called *eugeosyncline*.

volcanism   Volcanic activity, including the eruption of lava and rock fragments and gas explosions.

volcano   A hill or mountain constructed by the extrusion of lava or rock fragments from a vent.

W

wastage   Glacial ice or snow that is lost by melting, evaporation, or breaking off into icebergs.

water table   The upper surface of the zone of saturation.

wave crest   *See* crest.

wave-cut platform   A horizontal bench of rock formed beneath the surf zone as a coast retreats because of wave erosion.

wave-dominated delta   A delta formed by the reworking of sand by wave action.

wave height   The vertical distance between the crest (the high point of a wave) and the trough (the low point).

wavelength   The horizontal distance between two wave crests (or two troughs).

wave refraction   Change in direction of waves due to slowing as they enter shallow water.

wave trough   *See* trough.

weathering   The group of processes that change rock at or near the earth's surface.

welded tuff A rock composed of pyroclasts welded together.

well A hole, generally cylindrical and usually walled or lined with pipe, that is dug or drilled into the ground to penetrate an aquifer below the zone of saturation.

Wilson cycle The cycle of splitting of a continent, opening of an ocean basin, followed by closing of the basin and collision of the continents.

wind ripple Small, low ridge of sand produced by the saltation of windblown sand.

wrinkle ridge (*Moon*) Wrinkle on lunar maria surface.

X

xenolith Fragment of rock distinct from the igneous rock in which it is enclosed.

Z

zone of accumulation (1) That portion of a glacier with a perennial snow cover; (2) *See* B horizon (a soil layer).

zone of leaching *See* A horizon (a soil layer).

zone of saturation *See* saturated zone.

zone of wastage That portion of a glacier in which ice is lost or wasted.

zoning Orderly variation in the chemical composition within a single crystal.

Index

A

A horizon, 114
Aa, 88
Abrasion, 106, 228
Absolute age, 174, 186, 190
Abyssal fan, 442
Abyssal plain, 444
Accreted terrane, 516
Accretionary wedge, 478, 506
Accumulation stage, 505
Acid mine drainage, 268, 538
Acid rain, 112
Aconcagua, 90
Active continental margin, 444, 478
Actualism, 174
Adirondacks, 501
Aftershock, 399
Agassiz, Louis, 284
Ages of continents, 499
Ages of mountain belts, 499
Alaska pipeline, 9, 210
Alaska, terranes, 515–516
Aleutian Island Arc, 499
Alkali soil, 116, 276
Alluvial fan, 142, 241, 322
Alpine glaciation, 254, 295
Alps, 498, 503
Aluminum, 33, 116, 538
Alvarez, Luis and Walter, 184
Amphibole, 35–36, 43, 67
Amphibolite, 159
Andalusite, 161
Andes, 201, 498, 505
Andesite, 55–57, 59, 63, 69,
        84–85, 90, 94, 445 467,
        478, 482
    origin of, 71–72
Angle of repose, 329
Angular (pyroclast), 87
Angular unconformity, 185
Animal behavior, 408
Antarctic ice sheet,
        292–294, 298
Antarctica, 285–286, 292–298,
        303–304
Anthracite, 530

Anticline, 363, 524
Apatite, 40
Appalachian Mountains, 62,
        498, 499, 508–509, 514
    glaciation, 310
Aquifer, 263
    confined, 264
    unconfined, 264
Aquitard, 264
Arc-continent convergence, 507
Arch, 348
Archean, 192
Arctic Ocean, 307
Arete, 297–298
Arid climate, 316
Arkose, 128
Arroyo, 317
Artesian aquifer, 265
Artesian well, 265
Artificial recharge, 273
Asbestos, 35–36, 504
Aseismic ridge, 450, 488
Ash, volcanic, 79–80, 87, 98
Ashfall, 84
Assimilation, 69, 72
Asteroid, 184, 194
Asthenosphere, 15, 70–72, 419,
        468, 514
Atlantis, 92
Atoll, 450
Atom, 28–46
Atomic mass number, 28
Atomic number, 29
Atomic weight, 30
Augite, 38–39
Aulacogen, 487
Aureole, 157
Avalanche, 201, 297
Axial plane, 363
Axis, fold, 363

B

B horizon, 114
Backarc spreading, 481
Backwash (waves), 341
Badland topography, 318
Bajada, 322

Baker, Mt., 78, 81
Bar, 230
Barchan, 331
Barrier island, 144, 348
Barrier reef, 450
Basal sliding, 290
Basalt, 55–58, 63, 70–73,
        84–85, 95–98, 446, 463,
        482, 514
    metamorphism of, 156
Base level, 249, 317
Basement, 63, 500
Basin and Range province, 321,
        513, 514
Basin, structural, 365
Batholith, 61–63, 72–73, 504
Bauxite, 117, 535
Baymouth bar, 343
Beach, 144, 341
    seasonal cycle, 342
Beach face, 341
Bed load, 229
Bedding, 135
Bedding plane, 136
Benioff zone, 403,
        445, 467
Berm, 342
Beryl, 59
Big Thompson Canyon, 245
Bingham Canyon Mine,
        Utah, 168
Biochemical, 126
Biochemical limestones, 130
Biotite mica, 38–39
Birdfoot delta, 239
Bituminous coal, 530
Black Hills, South Dakota,
        63, 500
Black sand, 342, 454
Black smokers, 446
Block, volcanic, 87
Block-faulting, 513
Blowout, 327
Body waves, 386
Bomb, volcanic, 87
Bonding
    chemical, 31–32
    covalent, 32

    ionic, 32
    metallic, 32
Bottomset beds, 240
Boulder, 122
Bowen, N.L., 66–67
Bowen's reaction theory, 66
Braided stream, 230
Breaker, 339
Breakwater, 344
Brittle, 359
    behavior, 385
    strain, 367
Brunton pocket transit, 362
Buchwald, Art, 209
Buried erosion surface, 183
Butte, 320

C

C horizon, 114
Calcite, 38–40, 45–46, 110,
        125, 247, 274
Calcium, 32
Caldera, 87
Caledonide mountains,
        509, 514
Caliche, 116
Cambrian, 182, 183, 192
Canadian Rocky Mountains, 11
Canadian Shield, 515, 516
Capillary fringe, 261
Carbon, 32, 189
Carbon dioxide, 274
    metamorphism, 153
Carbonate, 38–39, 46
Carbonate rock, 130
Carbonic acid, 110
Carnotite, 532
Cascade Mountains, 79–81, 82
Cascade volcanoes, 91
Cave, 274
Cavern, 274
Cement, 125, 276
Cementation, 125
Cenozoic Era, 182–183, 192
Chain silicates, 35
Chalcopyrite, 38
Chalk, 132

Emerald
Aquamarine
Calcite (2)

Channel shape, 226
Channeled scablands, 309
Chaotic terrain, 210
Chemical activity, 30
Chemical sedimentary rock, 126, 130
Chemical weathering, 104, 108
Chert, 133
Chicxulub, 184
Chief Mountain, Montana, 375
Chill zone, 54
Chlorine, 28–31
Chromium, 68
Chrysotile, 36
Cinder cone, 88, 98
Circum-Pacific belt, 91, 401
Cirque, 297
Clastic sedimentary rock, 126
Clastic texture, 125
Clay, 122
Clay minerals, 37, 39, 111, 113, 122
Cleavage, 43
Climate, arid, 316
Climate change, metamorphism, 153
Climate changes, record in glaciers, 298
Climate, semiarid, 319
Coal, 135, 529
Coal fields, 530
Coal reserves, 533
Coal resources, 532
Coarse-grained texture, 55–56
Coast, 344
    depositional, 348
    drowned, 349
    erosion rate, 347
    erosional, 345
    uplifted, 352
Coastal straightening, 346
Cobble, 122
Collision boundary, 406
Colorado Plateau, 320
Columbia Plateau, 70, 95, 309
Column, 272
Columnar jointing, 96, 368
Columnar structure, 96
Comet, 289
Compaction, 125
Composite volcano, 90
Compression, 407

Compressive differential stress, 154
Compressive stress, 154, 163, 358, 368
Conchoidal fracture, 45
Concordant, 61
Concretion, 276
Cone of depression, 265
Confined aquifer, 264
Confining pressure, 153, 163
Conglomerate, 126
Constancy of interfacial angles, 42
Contact, 140
    intrusive, 54
Contact metamorphism, 156, 179, 534
Contacts, 174
Continental collision, 480
Continental crust, 15, 417, 497
Continental divide, 223
Continental drift, 20, 460, 463, 464
Continental glaciation, 284, 298
Continental Growth, 514
Continental margin, 402, 440, 443
Continental rise, 443, 475
Continental shelf, 441, 475
Continental slope, 442, 475
Continent-continent convergence, 508
Contour current, 444
Convection, 423, 466, 484
Convergence
    continent-continent, 480
    ocean-continent, 478
    ocean-ocean, 477
Convergent boundary, 18, 162, 506
Convergent boundary, igneous processes, 71
Convergent plate boundaries, 145
Convergent plate boundary, 406, 476, 489, 536
Copper, 32, 59, 164–165, 167, 538
Coquina, 130
Cordillera, 516
Core, 15, 417, 420
Core, inner, 421, 430
Core, outer, 421
Core-mantle boundary, 423
Corer, 439

Correlation, 179
    by fossils, 180
    similarity of rock type, 180
Corundum, 40
Country rock, 54, 60
Covalent bonding, 32
Crater, 87
Crater Lake, 81–92
Craton, 478, 501, 515
Creep, 200, 205
Crest, of wave, 338
Cretaceous, 182–183, 192
Crevasse, 290, 297
Cross-bedding, 136
Cross-cutting relationships, 176–177, 185, 458
Crude oil, 524
Crust, 15, 417
    chemical composition, 31
    continental, 497
    oceanic, 20
Crustal rebound, 310, 424
Crustal shortening, 503
Crystal, 33, 41, 46–47, 58
Crystal face, 41
Crystal form, 41
Crystal settling, 67–68, 534
Crystal symmetry, 43
Crystal system, 43
Crystalline, 33, 37, 41, 46–47
Crystalline texture, 125
Crystallization, 125
Cuesta, 321

D

D horizon, 114
Dacite, 84
Dance of the continents, 515
Darcy's Law, 262
Darwin, 450
Daughter product, 188
Debris, 200, 201
Debris avalanche, 200, 212
Debris fall, 200, 214
Debris flow, 206
Debris, mass wasting of, 214
Debris slide, 200, 214–215
Deflation, 327
Degassing, 439
Delamination, 509, 513
Delta, 144, 239, 349
    stream-dominated, 239
    tide-dominated, 239
    wave-dominated, 239

Dendritic drainage pattern, 224
Density, 45
Deposition, 123
Depositional coast, 348
Depth of equal pressure, 423
Depth of focus, 390
Desert, 316
Desert distribution, 316
Desert, expanding, 318
Desert pavement, 329
Desert, polar, 316
Desert, Southwestern United States, 319
Desert varnish, 329
Desertification, 318
Detachment fault, 503
Devil's Postpile, California, 97
Devonian, 182–183, 192
Dewey, John 507
Diamond, 32, 38–39, 45
Diamond pipes, 113
Diapir, 62, 70, 73, 369
Differential stress, 154, 162
Differential weathering, 104
Differentiation, 66
Dike, 61
Dinosaur, 22, 184
Diorite, 55–58, 69
Dip
    angle of, 362
    direction of, 362
Dip-slip fault, 370
Discharge, 227
Disconformity, 183
Discordant, 61
Disseminated deposits, 534
Disseminated ore deposit, 167
Dissolved load, 230
Distributary, 239
Divergent boundary, 17
    igneous processes, 70
Divergent plate boundaries, 146
Divergent plate boundary, 405, 472, 489, 536
Diverging boundary, 97
    hydrothermal activity, 163
Dolomite, 38, 133
Dolomitic marble, 157–158
Dolostone, 133
Dome, 365
Double refraction, 46
Downcutting, 248
Dowsing, 270

Drainage
  basin, 224
  divide, 224
Drainage pattern, 224
  dendritic, 224
  radial, 225
  rectangular, 225
  trellis, 225
Drawdown, 265
Dredge, 539
Drilling, 417, 419, 439
Drilling, for petroleum, 526
Drilling through ice, 292, 298
Dripstone, 274
Drowned Coast, 349
Drumlin, 302, 307
Dry wash, 317
Ductile, 359
Ductile behavior, 385
Dune, 144
Dune types, 331
Dunes, 327
Dunite, 420
Dust Bowl, 324
Dust storm, 324
Dust, volcanic, 87
Dynamothermal
      metamorphism, 158

E

Earth, 194
  age of, 186, 191
  internal temperature, 432
Earth's interior, 15
Earthflow, 200, 207
Earthquake, 384
  Alaska, 384
  control of, 409
  distribution, United
      States, 393
  distribution, world,
      401, 483
  effects, 390
  first motion, 404
  intensity, 390
  Kobe, Japan, 383
  location, 389
  Loma Prieta, 384
  magnitude, 390
  Mexico City, 398
  Northridge, 384
  patterns, 409
  prediction of, 408, 409
  San Francisco, 384
  "The Big One", 410
Earthquake size, 390

Earthquake waves, 385
Echo sounder, 439
Elastic, 359
Elastic rebound theory, 385
Electron, 29–31
Element, 28–32
Emerald, 59
End moraine, 300
Energy resources, 523
Energy sources, 7
Energy use, 523
Environment of deposition,
      124, 142
Eocene, 192
Eon, 191–192
Epicenter, 386
Epoch, 182
Equilibrium, 21, 48, 108,
      152, 423
Era, 182
Erebus, Mt. (Antarctica), 91
Erosion, 21, 48, 104
Erosional coast, 345
Erratic, glacial, 299
Eruptions, Submarine, 96
Escher, M. C., 33
Esker, 302
Estuary, 349
Etch-pitted terrain, 289
Etna, Mt., 92
Europa, 289
Evaporite, 369
Evaporites, 134, 541
Everest, Mount, 16
Evolution, 183
Exfoliation, 107
Exfoliation domes, 107
Exotic terrane, 516
Extension, 504, 511, 514
Extrusive rock, 18, 44, 54,
      78–99
Extrusive rock, identification
      of, 84

F

Faceted rock, 294
Facies, metamorphic, 164
Failed rift, 393, 487
Fall, 200, 202
FAMOUS, 446
Fault, 322, 360, 370, 385, 524
  active, 370
  block mountain range, 511
  detachment, 503
  dip-slip, 370
  left lateral, 375

normal, 371, 506,
      507, 511
  oblique-slip, 370
  reverse, 374, 503
  right lateral, 375
  strike-slip, 375, 370
  strong, 385
  thrust, 374, 503
  weak, 385
Faunal succession, 181, 185
Feeder current, 341
Feldspar, 37, 38, 41, 43,
      84–85, 111
Felsic, 56, 59
Ferromagnesian minerals, 40
Fine-grained rock, 55–56, 85
Fiord, 310, 352
Firn, 286
Fissility, 128
Flank eruption, 87
Flash flood, 245, 317
Flipping subduction
      zone, 507
Flood basalt, 487
Flood deposits, 237, 242
Flood erosion, 242
Flood frequency curve, 244
Flood plain, 143, 222, 237
Flood recurrence interval,
      242, 243
Flooding, 242
Flow, 200, 202
Flowstone, 274
Fluid potential, 263
Fluorite, 40
Focus, 386
  depth of, 390
Fold
  axial plane, 363
  hinge line, 363
  isoclinal, 366
  limb, 363
  open, 366
  overturned, 367, 502
  plunging, 364
  recumbent, 367, 502
Fold and thrust belt, 503,
      506
Foliation, 154, 156, 162–163
Footwall, 370
Forearc basin, 478
Foreland basin, 480
Foreset beds, 239
Foreshock, 399, 409
Formation, 139, 174, 362
Fossil, 22, 180–183, 276
  assemblage, 181

correlation by, 180
  index, 181
Fossil fuels, 139, 523
Fossils, 139
Fracture, 45
Fracture zone, 449, 471, 483
Fractured rock, 367
Fragmental texture, 87
Framework silicates, 37
Fringing reef, 450
Frost action, 104
Frost heaving, 106
Frost wedging, 106, 294, 297
Fujiyama (Japan), 92

G

Gabbro, 55–58, 63, 70
Galena, 38, 40, 45
Galloping glacier, 292
Galveston, 351
Garibaldi, Mt., 81
Garnet, 38, 41, 45
Gas, in magma, 54, 64
Gas, volcanic, 78–80, 84,
      86, 88
Gasoline, 269
Geode, 276
Geologic
  cross section, 363
  map, 361
  resources, 522
  structures, 357
  time, 21
  time scale, 182, 192
Geology, defined, 6
Geophysics, 416
Geothermal energy, 65, 78,
      278, 533
Geothermal gradient, 64, 163,
      277, 432
Geyser, 277, 408
Geyserite, 278
Glacial ages, 303
Glacial budget, 286
Glacial deposition, 299
Glacial erosion, 294
Glacial lakes, 303
Glacial valley, 295
Glaciation, 283
Glacier, 283
  advancing, 286
  formation and growth, 285
  lake beneath, 292
  outburst flood, 292
  receding, 286
  rigid zone, 290

surge, 292
terminus, 288
wastage, 286
water resource, 287
zone of accumulation, 288, 293
zone of wastage, 288
Global warming, 350
Gneiss, 157, 160
Gneissic texture, 155, 157
Gold, 38–39, 45, 59, 164, 167, 368, 539
Gondwanaland, 460
Graben, 371, 431, 430
Graded bed, 136
Grand Canyon, Arizona, 63, 174, 180–183, 186, 500
Granite, 54–59, 61, 63, 65–66, 68, 70–73
origin of, 71
Granodiorite, 58
Gravel, 122
Gravitaional collapse, 162, 514
and spreading, 507
Gravity, 14, 18, 425
Gravity anomaly, 426, 445, 467
Gravity meter, 426
Graywacke, 128
Great Basin, 513
Great Flood of 1993, 247
Great Lakes, 308
Great Plains, glaciation of, 305
Great Salt Lake, Utah, 310
Greenhouse effect, 153, 306
Greenland ice sheet, 298
Greenschist, 159
Greywacke, 506
Groin, 344
Gros Ventre Mountains, Wyoming, 214
Ground displacement, 397
Ground mass, 86
Ground motion, 395
Ground water, 167, 260
pollution, 266
Ground-water movement, 262
Guyot, 449
Gypsum, 38, 40

H

Hadean, 191
Half-life, 188–189
Halite, 28, 32–33, 38, 42, 45
Hanging valley, 296
Hanging wall, 370

Hard water, 271
Hardness, mineral, 40
Hardness scale, Mohs, 40
Hardpan, 116
Hawaii, 64, 70
Hawaii, volcanism, 78, 88
Headland, 346
Headward erosion, 251
Heat, 432
Heat engine, 11, 14, 18, 44
Heat flow, 406, 423, 433, 445, 467
Heavy crude, 528
Heavy metals, 267
Helium, 29–30
Hematite, 38, 40, 109
Herbicides, 266
Himalaya, 153, 498, 508, 510
Hinge line, 363
Historical geology, 181
Hoffman, Paul, 515
Hogback, 321
Holocene, 182, 192
Hood, Mt., 78, 81, 82
Horn, 297
Hornblende, 38–39
Hornfels, 157
Horst, 374
Hot spot, 486
Hot spring, 277, 455, 534, 536
Huascaran, 201
Hudson Bay, 305
Hutton, James, 174
Hydraulic action, 228
Hydraulic gradient, 262
Hydraulic head, 262
Hydraulic mining, 539
Hydroelectric power, 533
Hydrogen, 29–30, 32
Hydrologic cycle, 222
Hydrothermal fluids, 534
Hydrothermal mineral, 163
Hydrothermal rocks, 163
Hydrothermal veins, 59, 534
Hypocenter, 386
Hypothesis, 19–20, 184, 471, 492, 515

I

Ice, 29
Ice cap, 285, 293
Ice sheet, 285, 293
movement of, 293
Iceberg, 286
Icefall, 290

Iceland, volcanism, 98
Igneous, 53–99
Igneous rock, 44, 54
grain size, 85
identification, 55
texture, 85
Impact crater, 158
Incised meanders, 253
Index fossil, 181
Index mineral, 161
Interglacial periods, 303
Intermediate rock or magma, 56, 59, 69, 71, 85
Internal drainage, 317
Internal processes, 14
Intraplate igneous activity, 70
Intrusion, 59
Intrusive body, 59
Intrusive contact, 54
Intrusive rock, 18, 44, 53–73
Intrusive structure, 59–61
Io, 93
Ion, 31–32
Ionic bond, 32
Iridium, 184
Iron, 33, 164–165, 538
Island arc, 404, 477, 489, 507
Isoclinal fold, 366
Isolated silicate structure, 35
Isostasy, 423, 509
Isostatic adjustment, 423, 509
Isotherm, 163
Isotope, 30, 187
Isotopic dating, 186
reliability, 190

J

Jet stream, 324
Jetty, 344
Joint, 367
Joint set, 368
Jupiter, 93, 289
Jurassic, 182, 192

K

Kaolinite, 39
Karst topography, 275
Kettle, 302
Kilauea Volcano, Hawaii, 78
Kobe, Japan, earthquake, 5
Krakatoa, 81, 84, 307
K-T boundary, 184
Kyanite, 161

L

Lagoon, 144
Lake, 143
Lake Bonneville, 310
Lake Missoula (glacial), 309
Laminated terrain, 289
Laminations, 128
Landslide, 7, 80, 200–202, 207, 396
preventing, 215
Lassen Peak, 79
Lateral continuity, 175–176
Lateral erosion, 251
Laterite, 116
Laurasia, 460
Lava, 54, 78, 92
flood, 95
flow, 78, 88
viscosity of, 83, 85
Lead, 40, 59, 165, 167, 539
Left-lateral fault, 375
Lepidolite, 59
Lignite, 530
Limestone, 130
Limonite, 109
Liquefaction, 396
Lithification, 125
Lithium, 59
Lithosphere, 15–18, 70, 72, 419, 467, 514
Lithospheric delamination, 514
Lithostatic pressure, 154
Loam, 114
Loess, 327
Loma Prieta, earthquake, 5, 384
Longitudinal dune, 332
Longshore current, 340
Longshore drift, 342
Los Angeles, landslides, 209
Low-velocity zone, 419
Lunar crater, 158
Luster, 40

M

Mafic, 56, 58
Mafic rock or magma, 66–67, 70–73, 84–85, 95
Magma, 17–18, 44, 54, 78–80
evolution of, 66
gas, 64
how it forms, 63
mixing of, 69, 72
Magmatic arc, 478, 506
Magmatic energy, 65

Magmatic underplating, 71
Magnesium, 33
Magnetic anomaly, 429, 469
Magnetic field, 427
Magnetic polarity, 427
Magnetic pole, 427
Magnetic reversal, 428, 469
Magnetism, 427
   normal, 428, 469
   reversed, 428, 469
Magnetite, 38, 40, 46
Magnetometer, 429
Mammoths, 310
Manganese, 164
Manganese nodules, 454
Mangrove, 353
Mantle, 15, 417, 514
   diapir, 481
   lower, 420
   plume, 64, 90, 486, 492
   upper, 420
Map, geologic, 361
Marble, 152, 157
Mare, 92
Marine terrace, 341, 347
Marine terrace, uplifted, 352
Mars, 93, 194, 254, 289, 304
Mars, craters, 158
Mars, life on, 304
Mars, permafrost, 210
Mass wasting, 179, 294
   classification of, 200
   controls of, 203
   types, 205
Massive sulfide deposits,
   489, 536
Matrix, 86, 128
Matterhorn, 297
Mauna Loa, Hawaii, 78
Mayon volcano (Phillipines),
   91
Mazama, Mt., 81
Meander cutoff, 235
Meanders, 235
Mechanical weathering, 104
Mediterranean belt, 92
Mediterranean-Himalayan
   belt, 401
Melting temperature, 64–65
Member, 142
Mercalli scale, 390
Mercury, 194
   craters, 158
Mesa, 320
Mesozoic, 182–183, 192
Metallic bonding, 32

Metallic brines, 455
Metals, 533
Metamorphic facies, 164
Metamorphic rock, 18, 48,
   151–171, 186, 506
   classification of, 156
   factors controlling, 152
Metamorphism, 151–171
   contact, 156
   dynamothermal, 158
   effects of fluids, 155
   in mountain belts, 504
   plate tectonics, 162
   progressive, 159
   regional, 158
   shale, 157, 159–161
   temperature, 153
   thermal, 156
   types of, 156
Metasomatism, 165
Meteor Crater, Arizona, 158
Meteorite, 158, 191, 304, 422
Meteorite crater, 184
Meteorite, from Mars, 304
Mexico, earthquake, 5
Mica, 35, 38–39, 43–44
Microcontinent, 516
Microseisms, 408
Mid-Atlantic ridge, 18, 509
Mid-oceanic ridge, 17, 20,
   96–97, 401, 445,
   466, 483
Migmatite, 160, 504
Milankovitch, Milutin, 306
Mineral, 28–46
   ore, 40, 164–165, 167
   physical properties, 40
   rock-forming, 38
Mineral collapse, 365, 407
Mineral, definition, 37
Mineraloid, 37
Mining, 536
Miocene, 182, 192
Mississippian, 182–183
Model, scientific, 28
Modified Mercalli scale, 390
Mohorovicic discontinuity, 417
Mohs harness scale, 40
Molecule, 28–29
Moment magnitude, 391, 392
Monocline, 320
Montmorillonite, 39
Montserat, 94
Moon, 194
Moores, Eldridge, 515
Moraine, 300, 306–308

end, 300
ground, 300
lateral, 300
medial, 300
recessional, 300
terminal, 300
Mountain, 498
Mountain belts, 438, 440, 478,
   497, 498
   evolution of, 506
Mountain range, 498
   fault-block, 511
   metamorphism, 158
Mountain root, 417, 423
Mountains, systems approach,
   510
Mt. Logan, Yukon Territory,
   Canada, 285, 291, 299
Mud, 122
Mud cracks, 136
Mudflow, 6, 79, 200,
   210, 317
Mudpot, 278
Mudstone, 129
Muscovite, 59, 67
Muscovite mica, 38–39

N

Native element, 39
Natural gas, 523, 527
Natural levee, 237
Nebula, 194
Neptune, 93
Neutron, 28
Nevado del Ruiz, Colombia, 6,
   91, 212
Nodules, 133
Nonconformity, 186
Nonmetallic resources, 540
Nonrenewable resource, 522
Normal fault, 371, 511
Normal faulting, 506, 507
Normal force, 204
North American Cordillera, 499
Northridge, earthquake,
   4–5, 384
Nuclear power, 532
Nucleus, 29–30

O

Obsidian, 85, 93
Ocean, origin of, 438
Ocean-continent convergence,
   506

Oceanic crust, 15, 20, 417, 452
Oceanic trench, 403, 445, 467,
   478, 483
Offshore oil, 454
Oil, 523
Oil, Alaskan, 9
Oil field, 525
Oil pool, 372, 524
Oil resources, 527
Oil reverves, 527
Oil sand, 528
Oil shale, 528
Oil spills, 527
Oil trap, 372, 431, 524
Oligocene, 182, 192
Olivine, 35, 38, 67–68
Olympus Mons, 93, 194
Ooolites, 132
Opal, 37
Open fold, 366
Open-pit mine, 536–537
Ophiolite, 420, 452, 453, 489,
   504, 536
Orbital variation theory, 306
Ordovician, 182–183, 192
Ore deposit, 164, 488
   crystal settling, 68
   disseminated, 167
Ore minerals, 40, 164–165,
   167
Ores, 533
   precipitated, 535
   weathered, 535
Organic sedimentary rock, 126
Original horizontality, 135,
   175–176, 361
Orizaba, volcano, 91
Orogenic stage, 505, 506
Orogeny, 506, 516
Orthoclase feldspar, 38
Outcrop, 361
Outwash, 302
Overburden, 125
Overturned fold, 367, 502
Oxbow lake, 235
Oxygen, 28–37
Ozark dome, 500

P

P wave, 386
   shadow zone, 421
Pahoehoe, 88
Paleocene, 182, 192
Paleoclimates, 462
Paleocurrents, 142

Paleomagnetic pole, 516
Paleomagnetism, 428, 463
Paleontologist, 181
Paleosol, 117
Paleozoic, 182–183, 192
Palmdale bulge, 411
Pangaea, 460, 465, 509, 515
Parabolic dune, 332
Parallel retreat, 322
Parent rock, 152, 157
Partial melting, 68, 70–73, 160
Passive continental margin, 443, 475, 505
Paternoster lake, 297
Patterned ground, 210
Peat, 135, 529
Pebble, 122
Pedalfer, 116
Pediment, 322
Pedocal, 116
Pegmatite, 58–59, 535
Pelagic sediment, 452, 467
Pelee, Mt. (Martinique), 84, 94
Pennsylvanian, 182–183, 192
Perched water table, 261
Peridotite, 420
Period, 182
Permafrost, 208
Permeability, 260, 263
Permian, 182–183, 192
Pesticides, 266
Petrified wood, 276
Petrochemicals, 526
Petroleum, 523
Petroleum reserves, 8
Phanerozoic, 191–192
Phenocryst, 86
Phyllite, 157, 160
Physical continuity, 180
Physical geology, 5
    overview, 11
Physical properties of
    minerals, 40
Piezoelectricity, 47
Pillow basalt, 97
Pillow structure, 97
Pinatubo, Mount, 7, 78, 91, 211, 298
Pitchblende, 532
Placer deposits, 231, 535
Placer mine, 537
Plagioclase feldspar, 38–39, 46, 67–68
Planet, 194
Plastic, 152, 359

Plastic flow of ice, 290
Plastic strain, 363
Plastically, 14
Plate, 404, 467
Plate, boundaries, 17–18, 460, 468
    convergent, 406, 476, 489, 536
    divergent, 405, 475, 489, 536
    motion of, 482
    transform, 405, 475
Plate motion
    cause, 483
    rate, 470, 472
Plate, size, 482
Plate tectonics, 16–18, 19–21, 48, 97, 404, 460
    igneous activity, 70–73
    metamorphism, 162
    mountain belts, 505
Plateau, 320
Plateau basalt, 95
Platinum, 68
Playa, 322
Playa lake, 322
Pleistocene Epoch, 182, 192, 303
Pliocene, 182, 192
Pluton, 61, 73
Plutonic rock, 55, 63, 186
Plutonism, in mountain
    belts, 504
Pluvial lake, 310
Point bars, 235
Polar wandering, 462
Pollution, 266
Pompeii, 81
Popocatepetl, 91
Pore pressure, 204
Pore space, 125
Pores, 125
Porosity, 258
Porphyritic, 86
Porphyry copper deposits, 534
Potassium, 33
Potassium feldspar, 38
Potassium-argon, 190
Pothole, 228
Prearchean, 191–192
Precambrian, 182–183, 191–192, 515
Precambrian shield, 501
Pressure
    confining, 153
    release, 107

Progressive metamorphism, 159
Proterozoic, 191–192
Proton, 29–31
Pumice, 87
Pyroclast, 78, 84, 87, 90, 211
Pyroclastic flow, 7, 84, 94
Pyroxene, 35–39 43, 67

Q

Quartz, 31, 37–38, 40, 41, 45, 47, 58–59, 67, 112, 166–167
Quartz sandstone, 127
Quartz watch, 47
Quartzite, 157–158
Quaternary, 182, 192

R

Radial drainage pattern, 225
Radioactive decay, 187
Radioactive waste, 268
Radioactivity, 190–191
Radiocarbon dating, 181
radiometric dating, 174, 187
Radon, 408
Rain shadow, 280, 316
Rainier, Mt., 78, 81, 82, 90
Rampart crater, 210
Recent Epoch, 192
Recharge, 263, 270
Recrystallization, 126, 132
Rectangular drainage pattern, 225
Recumbent fold, 367, 502
Reef, 130, 144, 352, 450, 525
Regional metamorphism, 158
Regolith, 113
Regression, 144
Relative age, 190
Relative time, 174
Relief, 203
Renewable resources, 522, 533
Reserves, 523
Reservoir rock, 372, 525
Residual clay, 274
Residual soil, 114
Resources, 523
Reverse fault, 374
Rhyolite, 55–59, 63, 70, 84–85, 93
Richter scale, 390
Ridge-push, 484
Rift, 371
Rift valley, 405, 445, 467, 472

Right-lateral fault, 375
Rigid zone, of a glacier, 290
Rille, 92
Rills, 224
Rip current, 341
Ripple marks, 136
River channel, 143
Rock avalanche, 200, 213
Rock basin lake, 296
Rock cycle, 46–48
Rock, definition of, 28
Rock flour, 294
Rock gypsum, 134
Rock salt, 28, 134, 369
Rock steps, 296
Rock symbols, 126
Rockfall, 200, 212
    prevention of, 216
Rockslide, 200, 213
    prevention of, 216
Rocky Mountains, 308, 499–501, 506
Rotational debris slide, 215
Rotational slide, 202–203, 207
Rounding, 122
Rubble, 122
Ryolite, 514

S

S wave, 386
    shadow zone, 421
Sag pond, 376
Salt dome, 369, 525
Saltation, 229, 326
Saltwater intrusion, 272
San Andreas fault, 18, 360, 376, 384, 406, 410, 476
San Francisco, earthquake, 5
Sand, 122
Sand and gravel, 540
Sand dune, 327
Sand Hills, 320
Sand sea, 328
Sandstone, 127
Sandstorm, 326
Saturated zone, 261
Scarp, 207, 397
Schist, 156, 157, 160
Schistose texture, 155, 157
Scientific method, 19
Scoria, 86
Sea cave, 347
Sea cliff, 347
Sea-floor age, 454, 467, 470
Sea-floor drilling, 439

Sea-floor spreading, 460, 465
Sea level, 350
    lowering by glaciation, 310
Seamount, 449
Seas, 338
Seawall, 347
Sediment, 21, 48, 122
Sedimentary breccia, 126
Sedimentary rock, 21, 48, 126
    melting of, 72
Sedimentary structures, 135
Seif, 332
Seismic
    gap, 409
    moment, 391
    profiler, 439
    reflection, 416
    refraction, 416
    sea wave, 399
    tomography, 490
    wave, 345, 386
Seismogram, 387
Seismograph, 387
Seismometer, 387
Serpentine, 36, 504
Shadow zone, 421
Shale, 128
    metamorphism of,
      159–161
Shasta, Mt., 82
Shear
    force, 204
    strength, 204
    stress, 359
Shearing, 149, 154
Sheet
    erosion, 223
    joints, 107, 368
    silicate, 35
Sheetwash, 223
Shield, Precambrian, 501
Shield volcano, 88
Shingle, 342
Ship Rock, New Mexico, 60
Shock metamorphism, 158
Sial, 417
Sierra Nevada, California, 62,
    507, 511
Silica, 31
Silica tetrahedron, 33
Silicate, 31–37
Silicic, 56, 59
Silicic rock, 84–85
Silicic rock or magma, 66–67,
    70, 466
Silicon, 33–37

Silicon-oxygen tetrahedron, 33
Sill, 61
Sillimanite, 161
Silt, 122
Siltstone, 129
Silurian, 182–183, 192
Silver, 59, 164–165, 167, 539
Sima, 417
Sinkhole, 274
Sinter, 278
Slab-pull, 484
Slate, 157, 159–160
Slaty, 155, 157
Slaty cleavage, 155
Slide, 202
Slip face, 329
Slot canyon, 248
Sluice-box, 539
Slump, 203
Snow line, 288
Soapstone, 504
Sodium, 28–31
Soft water, 271
Soil, 113
Soil horizons, 114
Soil moisture, 261
Solar energy, 306
Solar power, 18, 533
Solar system, origin of, 194
Solifluction, 208
Solution, 228
Sorting, 123
Soufriere Hills volcano, 94
Source area, 140
Source rock, 372, 524
South Pole, 294
Spatter cone, 88
Specific gravity, 45
Sphalerite, 38, 40, 45
Spheroidally weathered
    boulder, 104
Spit, 343
Spreading axis (center), 466
Spring, 266
Spring, hot, 277
St. Francis Dam, 215
St. Helens, Mt., 78–81, 87, 91,
    93, 212
St. Pierre, Martinique, 84, 94
Stability field, 164
Stack, 347
Stalactite, 274
Stalagmite, 274
Steno, N., 41–42
Stock, 61
Storm surge, 350

Strain, 358, 359, 385
Stratigraphic trap, 524
Stratovolcano, 90
Streak, 40
Stream, 222
    banks, 222
    bed, 222
    channel, 222
    channel roughness, 226
    channel shape, 226
    deposition, 230, 244
    discharge, 227
    erosion, 228
    gaining, 266
    graded, 250
    gradient, 226
    losing, 266
    piracy, 251
    profile, 222
    terraces, 252
    transportation, 229
    underground, 275
    valleys, 248
    velocity, 225
Stress, 154, 358, 359, 385
Striations, glacial, 294, 311
Striations, mineral, 46
Strike, 361
Strike-slip fault, 375
Strip mine, 530, 536
Structural basin, 365
Structural dome, 365
Structural Geology, 358
Structural trap, 524
Subbituminous coal, 530
Subduction, 406, 466, 467,
    478, 491, 507, 536
    angle, 408
    complex, 478
    zone, 18, 506–507
Submarine canyon, 442
Submarine eruptions, 96
Submersibles, 439
Subsidence, 272, 527, 538
Subsoil, 114
Sulfate, 39
Sulfide, 39
Supergene enrichment, 535
Superposed stream, 255
Superposition, 175–176, 180
Surf, 338, 339
Surface tension, 204
Surface waves, 386
Surficial processes, 14, 18
Surtsey, Iceland, 98
Suspect terrane, 515

Suspended load, 229
Suture, 509
Suture zone, 480
Swash (waves), 341
SWEAT, 515
Swell, 338
Swelling clay, 39
Syncline, 363

T

Tailings, 537
Talc, 40, 504
Talcum powder, 504
Talus, 212
Tambora (Indonesia), 81
Tar sand, 528
Tarn, 296
Tectonic forces, 16, 48,
    358, 385
Tectonics, 460
Tectonostratigraphic
    terrane, 515
Tektite, 158
Temperature, metamorphism,
    153
Tension, 407
Tensional stress, 358
Tephra, 78
Terminus, of a glacier, 288
Terrace, marine, 341, 347
Terrane, 515
Terrigenous sediment, 452
Tertiary period, 182, 192
Teton Range, Wyoming, 511
Theory, 19–22, 471, 492
Thera, Mt., 92
Thermal maturity, 524
Thermal metamorphism, 156
Thrust fault, 374
Tidal delta, 350
Tidal flat, 144
"Tidal wave," 399
Till, 142, 299, 303,
    305–306, 311
Tillite, 311
Time, geologic, 21, 173
Time, metamorphism, 156
Time scale, 182
Tombolo, 343
Topaz, 40
Topset beds, 240
Topsoil, 114
Traction, 229
Transform fault, 449, 471,
    472, 475

Transform plate boundary, 18, 145, 405, 475
Transgression, 144
Translational slide, 202
Transportation, 104, 122
Transported soil, 114
Transverse dune, 331
Travel-time curve, 389
Travertine, 278
Trellis drainage pattern, 225
Trench, oceanic, 505
Trench-suction, 485
Triassic, 182, 192
Tributary, 224
Trough of wave, 338
Truncated spur, 296
Tsunami, 338, 399
Tuff, 87
Tungsten, 165, 167
Turbidity current, 128, 136, 443
Turbulence, 229
Twain, Mark, 22

U

U-shaped valley, 295
Ultramafic, 56–59, 68, 70–71
Ultramafic rock, 417
  in mountain belts, 504
Unconfined aquifer, 264
Unconformity, 177, 183, 525
Underground stream, 275
Underplating, magmatic, 71

Underthrusting, 407
Uniformitarianism, 174
Unit cell, 43
Unloading, 107
Unzen, Mt. (Japan), 84
Uplifted coast, 352
Uplifted marine terrace, 352
Ural Mountains, 508
Uranium, 59, 187–188, 532

V

V-shaped valley, 295
Vadose zone, 261
Vaiont rockslide (Italy), 213
Valley glacier, 285, 296
  movement of, 288
Varve, 186, 303
Vein, 163, 166–167
Vent, 87
Ventifact, 326
Venus, 93, 194
  craters, 158
Vertical exaggeration, 441
Vesicle, 86
Vesicular basalt, 86
Vesuvius, Mt., 81, 92
Vine-Matthews hypothesis, 468
Viscosity of lava, 85
Viscosity of magma, 93
Volcanic ash, 324
Volcanic breccia, 87
Volcanic dome, 81, 93
Volcanic neck, 60

Volcanic record in glaciers, 298
Volcanism, 77–99
  extraterrestial, 92
Volcano, 77–99
  effect on climate, 78
  type of, 87
Vostok, 292, 298

W

Wastage, 286
Waste rock, 537
Water, 29
Water table, 261, 408
  perched, 261
Water waves, 338
Water witching, 270
Wave crest, 338
Wave height, 338
Wave orbit, 339
Wave refraction, 340, 346
Wave trough, 338
Wave-cut platform, 347, 352
Wavelength, 338
Waves, 338
  earthquake, 386
Weathering, 48
  chemical, 104, 106
  mechanical, 104, 106
Wegener, Alfred, 20, 460
Well, 264
White Mountains, New Hampshire, 310
Wilson cycle, 509

Wilson, J. Tuzo, 509
Wind action, 322
Wind deposition, 327
Wind erosion, 324
Wind power, 533
Wind ripples, 330
Wollastonite, 152

X

X ray, 46
Xenolith, 54, 420

Y

Yellowstone National Park, 64, 70, 284, 296, 308
Yosemite, rockfall, 213
Yucca Mountain, Nevada, 268, 532
Yungay, Peru, 201

Z

Zinc, 40, 59, 164–165, 167, 539
Zion National Park, 180
Zone of accumulation
  soil, 114
  glacier, 288
Zone of leaching (soil), 114
Zone of wastage, 288
zoning (mineral), 37